Marmes Rockshelter

A Final Report on 11,000 Years of Cultural Use

Edited by
Brent A. Hicks

Washington State University Press
Pullman, Washington

Washington State University Press
PO Box 645910
Pullman, Washington 99164-5910
Phone: 800-354-7360
Fax: 509-335-8568
E-mail: wsupress@wsu.edu
Web site: wsupress.wsu.edu

© 2004 by the Board of Regents of Washington State University
All rights reserved
First printing 2004

Printed and bound in the United States of America on pH-neutral, acid-free paper. Reproduction or transmission of material contained in this publication in excess of that permitted by copyright law is prohibited without permission in writing from the publisher.

Library of Congress Cataloging-in-Publication Data

Marmes Rockshelter : a final report on 11,000 years of cultural use / edited by Brent A. Hicks.
   p. cm.
 Includes bibliographical references and index.
 ISBN 0-87422-275-3 (alk. paper)
   1. Paleo-Indians—Washington (State)—Franklin County. 2. Archaeological geology—Washington (State)—Franklin County. 3. Geology, Stratigraphic—Holocene. 4. Franklin County (Wash.)—Antiquities. 5. Palouse River Valley (Idaho and Wash.)—Antiquities. I. Hicks, Brent A., 1963-

E78.W3M35 2004
979.7'33—dc22
                       2004004112

Marmes Rockshelter

A Final Report on 11,000 Years of Cultural Use

Edited by
Brent A. Hicks

Confederated Tribes of the Colville Reservation
History/Archaeology Department

With Contributions by
Virginia L. Butler, John L. Fagan, Pamela J. Ford,
Shawn Gibson, Carl E. Gustafson, Brent A. Hicks,
Gary Huckleberry, Joy Mastrogiuseppe, Guy Moura, Terry L. Ozbun,
Matthew J. Root, Daniel O. Stueber, Robert M. Wegener,
Peter E. Wigand, and Maureen Zehendner

**Acknowledgement of Federal Assistance
and Publication Disclaimer**

This study of the Marmes Site was performed for the Confederated Tribes of the Colville Reservation. The Confederated Tribes of the Colville Reservation conducted and funded the study through a contract with the United States Army Corps of Engineers. Additionally, publication was assisted by a grant from the Department of the Interior, National Park Service.

Any opinions, findings, and conclusions or recommendations expressed in this material are those of the author/editor and do not necessarily reflect the views of the Department of the Interior, the Army Corps of Engineers, or the Confederated Tribes of the Colville Reservation.

MARMES ROCKSHELTER APPENDICES AVAILABLE ON-LINE AT WSU MUSEUM OF ANTHROPOLOGY WEB SITE:

WWW.ARCHAEOLOGY.WSU.EDU

The appendices listed in the Table of Contents, pages vi-vii, are not included in this volume. They are available for viewing on the Web site, Museum of Anthropology, Washington State University: www.archaeology.wsu.edu.

ABSTRACT

This report, presenting the results and interpretation of the 1960s excavations of the Marmes Rockshelter, represents the culmination of five years work and the contributions of dozens of people. It attempts to be a thorough presentation of the data that resulted from those excavations. The work was conducted by the Confederated Tribes of the Colville Reservation's History/Archaeology Department under a contract with the U.S. Army Corps of Engineers, Walla Walla District (DACW68-98-C-0004). Two preliminary reports were prepared in the course of this study (Hicks 1998, 2000); those reports are supplanted by this final report.

This study involved analysis and interpretation of a curated archaeological collection and its associated records. The Marmes site is currently underwater and not available for examination by this study. The site contains two distinct landforms, a modest-sized rockshelter and an area of riverine floodplain in front of the rockshelter. A previous preliminary descriptive report (Rice 1969) had been prepared for the rockshelter materials. This study describes and interprets the cultural uses of both areas of the site and the relationship between them. Research questions were organized under the following topics: Cultural Chronology, Past Environment and Climate Change, Site Function, Trade, and Human Remains/Burials.

The majority of the research was conducted in two phases. Radiocarbon dating of selected samples was conducted in each phase and during follow-up research as well. In Phase One, background research into the site records and the collection was conducted. Research into the site records found problems that affected all of the project research to varying degrees. In particular, the field catalogs are incomplete, many items are missing from the collection, and provenience is lacking for many other items, especially the formed tools. In addition, variation in the sampling methods used in the field from year to year has had some effect on representation of certain material types. Background research located records relevant to the material analyses, and for describing the site's features and human remains. Analysis of human remains or grave goods was not conducted for this study; only previous research has been summarized. The nature of the site's stratigraphy and sediments was investigated and all of the stratigraphic sections prepared in the field in the 1960s were located and compiled, and ultimately were converted to digital formats. Faunal analysis included revisiting a previous analysis of the rockshelter fauna and analysis of faunal remains collected from floodplain bulk samples. Bone tool analysis also was conducted during Phase One.

Phase Two included analysis of the stone tools and debitage, fish bone, shellfish, and botanical materials in the collection. The few historic-era artifacts in the collection were found to correlate closely with burial features and were not analyzed. Phase Three involved conducting follow-up lab analyses and preparing the final report.

The 1960s investigations at the Marmes site resulted in the recording of eight distinct geological strata interpreted to extend back ca. 10,000 years in time. This study confirms the stratigraphic assignments and extends the earliest date of use of the site back to 11,230 B.P. Stone tool analysis confirms the general outline of Leonhardy and Rice's (1970) model of a culture-historical sequence for the lower Snake River region. Stone tool and debitage analysis interpreted the site's assemblage, from the earliest defined cultural phase (Windust) through to the late prehistoric period, as reflecting a habitation site where people performed a wide range of domestic tasks. The lithic artifacts represent tool manufacturing and maintenance (of stone tools and tools of hard organic materials such as bone, antler, and wood), food procurement (projectile points, bola stones, possible net weights), and processing of food and other materials, probably including hides and fiber. These combinations of activities characterize a residential camp. The evolution of technologies at the Marmes site appears to be deliberate and conservative. While there are differences between successive occupations, they are interpreted as gradual and incremental; what distinguishes one period from another is simply the relative frequency of certain artifact types and traits. A previous interpretation of Stratigraphic Unit V representing use only for burials is not supported.

Faunal remains analysis found significant differences between the rockshelter and the floodplain, both in their quantitative taxonomic characteristics and in the condition of the remains. Exotic taxa (i.e. Arctic Fox, pine marten) were present in the floodplain deposits but not in the rockshelter. Large mammal bones, with evidence of butchery, were common in the rockshelter deposits, while small animals are better represented in the floodplain deposits. Fish bone analysis found relatively few salmon bones but a high representation of small fish species. This may be a result of field sampling methods or a reflection of the particular activities for which the site was used at different times correlated with fluctuations in river environments near the site. Shellfish trends at the Marmes site largely match those described for the region by other researchers. Few botanical items are present in the collection, hampering use of these materials for interpreting site use through time.

A model for the changes in the use of the Marmes site through time was presented positing it as a "residential base" for a tethered mobile foraging settlement and subsistence strategy (after Binford 1980) during the early periods of site use. The model proposed that this changed to use as a "field camp" and then a "cache" by logistical collectors as the region experienced a similar change in strategy. Assessment of the data resulting from the different material analyses supports the early use of the site by tethered foragers. But the site appears to continue to be used as a foragers' base or collectors' field camp well beyond the time expected, although it is certainly also used as a cache. It is hypothesized that the location of the site in the vicinity of several kinds of higher-volume food resources attracted the initial forager use of the site and also contributed to its persistent use well into the period hypothesized for a logistical collector strategy. Although the cultural assemblage in the site changed through time, this is viewed as in situ response to environmental and technological change and not population replacement.

ACKNOWLEDGEMENTS

This project would not have been possible without the involvement of many people and the agencies or institutions they represent. The Confederated Tribes of the Colville Reservation (CCT) provided essential support. Adeline Fredin, the CCT's Tribal Historic Preservation Officer and Manager of the History/Archaeology Department, was the focus of that support, providing guidance throughout as well as the initial stimulus for this project by seeing and convincing others of its need. Upon Adeline's retirement in April of 2003, Camille Pleasants took over as Program Manager and Tribal Historic Preservation Officer and the related responsibilities in regard to this report. Guy Moura, who is only listed once as a contributor, provided far more than he gets credit for. Judy Zunie helped with the graphics and production of several of the project reports; V.J. Seymour also assisted with graphics. Vera Morgan helped format the final report to the Corps.

The U.S. Army Corps of Engineers' Walla Walla District Archaeologist, John Leier, served as the Contracting Officers Technical Representative and provided documents, reports, editorial assistance, contractual interpretations, and forbearance as needed. Considerable patience also is acknowledged from the members of Payos Kuus C'uukwe', the Federal Columbia River Power System cultural resources working group for the Walla Walla District.

Much assistance was given by Mary Collins, the Director of the Museum of Anthropology at Washington State University, and her staff who located and retrieved site materials and prepared them for loan to the CCT. Mary's attention to detail and her willingness to oblige me was much appreciated. WSU Department of Anthropology staff and students gave much of their time to this project in the initial year. Carl Gustafson not only helped author several chapters in this report, but took the time to sit and converse about Marmes fieldwork and events over the intervening years. It was that willingness to give his time that has made Gus of such value to his students at WSU. Bill Lipe and Matt Root also stepped up when it was most needed.

The numerous people who conducted the material analyses are credited for their patience, especially John Fagan and Terry Ozbun for providing follow-up information when requested. I hope I have presented your results in a manner that reflects the extent of your efforts and professionalism.

Paul Solimano provided graphics assistance, cleaning up the digitized stratigraphic profile scans and revising or creating a number of the digital drawings in the text. Sarah Moore photographed many of the artifacts, and worked with old prints and negatives from the Museum of Anthropology archives to create others.

Ken Ames and Ken Reid provided peer reviews of a draft of this report. John Leier and Mona Wright at the Corps, and Guy Moura of the CCT, reviewed and commented on the draft as well. This report was greatly improved by their assistance; Ken Ames contributions are gratefully acknowledged in particular.

The CCT provided funds toward publication, as did the Corps of Engineers and the National Park Service through Heritage Preservation Grant No. 53-02-NA-115D.

The support of Jill, and our children Alexander and Salal, was essential to completion of this project. Their encouragement cannot be measured.

I apologize to the many others not mentioned in this statement. I am grateful for the efforts of all who participated. Any errors are entirely my own.

Brent Hicks
December 2003

CONTENTS

Abstract	i
Acknowledgements	iii
List of Figures	viii
List of Tables	xi
Preface	xiii

1. Project Background   1

Site Description	1
History of Investigations	4

2. Research Design   25

Research Themes and Questions	25
Cultural Chronology	25
Past Environment and Climate Change	28
Site Function	28
Trade	31
Human Remains/Burials	31
The Study Plan	32
Anticipated Problems	33
Background Research and Project Constraints	34
Sampling	38

3. Environmental Overview   43

Geological Setting	44
Glacial-era Floods and Tephra Chronology	46
Climatic and Vegetation History	49
The Nature of the Data	49
Late Pleistocene	52
Latest Pleistocene/Early Holocene (12,500-8000 B.P.)	54
Middle Holocene (8000-5500 B.P.)	57
Early Late Holocene (5500 to 2000 B.P.)	59
Late Holocene (2,000 to Present)	61
Study Area Vegetation	63
Study Area Faunal Resources	64

4. Cultural Context   65

Prehistory	65
Ethnographic Period	70
Historic Period	74

5. Stratigraphy and Site Formation Processes   77

Geomorphic Context	77
Methods	78
Rockshelter Stratigraphy	82
The Data (1968 and 1998)	83
Interpretations	89
Colluvial Slope Stratigraphy	95
The Data	95
Interpretations	98
Floodplain Stratigraphy	103
The Data	104

Interpretations	113
Site Formation Processes at the Marmes Site	118
Summary and Conclusions	120

6. Features — 123

Rockshelter Features	124
Hearths/Fire Pits	124
Storage Pit Features	128
Floodplain Features	134

7. Human Remains — 139

Methods	139
Rockshelter	140
Floodplain	143
Results of Analysis of Rockshelter Remains	144
Results of Analysis of Floodplain Remains	156
Conclusions	157

8. Lithic Debitage and Formed Tools — 159

Methods of Analyses	162
Raw Materials	163
Reduction Technology	164
Use-wear	164
Obsidian X-ray Fluorescence	165
Blood Residue Analysis	166
Results of Analyses	167
Lithic Technology	167
Obsidian Sourcing	203
Blood Residue Analysis	209
Summary of Analyses	209
Conclusions	214
Culture-History	214
Index Fossils and Tool Functions	217
The Cascade Technique	223
Gravels and Travels	224
Site Function	225
Comparisons With Other Sites	226

9. Modified Bone and Antler — 229

Methods and Techniques	229
Manufacturing Technology and Function	230
Use-Life Classes	230
Description of the Modified Bone Collection	231
Changes Through Time	249
Comparison of the Windust Phase Floodplain and Rockshelter Deposits	252

10. Faunal Remains — 253

Current Research	253
The Marmes Fauna	260
Summary of Taxa Identified	280
Intersite Analyses	298
Summary and Conclusions	309

11. Fish Remains — 319

Methods and Materials — 319
Descriptive Summary of Fish Remains — 320
Results — 322
Comparison of Rockshelter and Floodplain Fish Fauna — 336

12. Invertebrate Fauna (Shellfish) — 339

Materials and Methods — 339
Results — 340
Discussion of Results by Stratum — 342
Discussion — 344

13. Botanical Materials — 347

Methods of Analysis — 347
Results and Discussion — 352
Temporal Considerations — 368
Contaminants — 371

14. Summary of Results — 373

Background Research — 373
Site Setting — 373
Feature Review Results — 375
Human Remains — 377
Material Analysis Results — 380
Conclusions — 386

15. Interpretation — 389

Cultural Chronology — 389
Trade — 400
Past Environment and Climate Change — 403
Site Function — 408
Subsistence — 412
Marmes Rockshelter in a Regional Perspective — 414
Recommendations — 420

References — 423

Appendices (available online at www.archaeology.wsu.edu)

Appendix A	Marmes Collection Inventory Database (provided on disk)
Appendix B	Stratigraphic Profile Drawings
Appendix C	Particle-Size Data
Appendix D	Chemical Data
Appendix E	Physical Descriptions
Appendix F	Feature Data Tables
Appendix G	List of Abbreviations and Glossary of Terms – Lithic Analysis
Appendix H	Lithic Technological Analysis Data
Appendix I	Obsidian Source Analysis
Appendix J	List of Samples Analyzed in This Study from the Marmes Floodplain
Appendix K	List of Unanalyzed Presorted Mammal Remains from the Marmes Floodplain
Appendix L	List of Unanalyzed Bird Remains from the Marmes Floodplain
Appendix M	List of Unanalyzed Samples Containing Reptile and Amphibian Bones
Appendix N	Fish Remains Analysis Data
Appendix O	All Identified Shellfish

Appendix P	Marmes Horizon Swan Bone Identification
Appendix Q	Radiocarbon Dating Reports
Appendix R	Pollen Analysis Report

LIST OF FIGURES

Figure 1.1 Map showing the location of Marmes Rockshelter — 2
Figure 1.2 View of the lower Palouse River Canyon — 3
Figure 1.3 Photograph of Marmes Rockshelter in 1962 — 3
Figure 1.4 Photograph showing topography of the lower Palouse River canyon — 4
Figure 1.5 View across the berm of Marmes Rockshelter — 5
Figure 1.6 Block diagram illustrating the eight major stratigraphic units — 6
Figure 1.7 Photograph of location depicted in block diagram in Figure 1.6 — 7
Figure 1.8 Plan drawings of rockshelter with excavation grid overlain — 8
Figure 1.9 North face of Marmes Rockshelter control block (85N/30W) — 9
Figure 1.10 View to the north showing bulldozer trench — 12
Figure 1.11 Block diagram showing relationships among rockshelter deposits and those on the adjacent floodplain — 13
Figure 1.12 View of the floodplain area from above the rockshelter — 14
Figure 1.13 View of floodplain stratigraphy in the trench wall — 16
Figure 1.14 Photograph of the water screening apparatus — 16
Figure 1.15 View of the east end of the floodplain excavation area — 17
Figure 1.16 View of the west end of the floodplain excavation area — 17
Figure 1.17 View of the floodplain area in the winter — 18
Figure 1.18 Diagram of the site grid and the units excavated — 19
Figure 1.19 Plan drawing of the main floodplain excavation area — 20
Figure 1.20 Photograph showing an example of the care given to documenting materials as features during the floodplain excavations — 21
Figure 1.21 Photograph of excavators and tractor removing floodplain overburden in the summer of 1968 — 21
Figure 1.22 Photograph showing plastic sheeting and sand used to protect the site — 22
Figure 3.1 Map of the Columbia Plateau showing locations of paleo-environmental data — 45
Figure 3.2 Road-cut exposing soils spanning the late Pliocene through Pleistocene — 45
Figure 3.3 Map showing the area of the Pacific Northwest affected by the Spokane/Missoula flooding — 47
Figure 3.4 Depiction of vegetation changes through time for a cross-section of the northwest — 51
Figure 3.5 Pollen diagram of Carp Lake, Washington near Mount Adams — 54
Figure 3.6 Pine pollen percentage between 11,250 B.P. (the age of Glacier Peak volcanic ash) and 10,200 years ago — 56
Figure 3.7 Photograph of remnant frost cracks in the Marmes floodplain area. — 56
Figure 3.8 Hansen's record from the northern Great Basin — 58
Figure 3.9 Drought index generated from pollen at Diamond Pond — 62
Figure 4.1 Comparison of Cultural Chronologies for the Southern Plateau — 68
Figure 4.2 Palouse territory during the 19th century — 72
Figure 5.1 Schematic cross-section of the Marmes site — 79
Figure 5.2 Particle-size histograms from rockshelter control block — 85
Figure 5.3 Vertical particle-size distribution from rockshelter control block — 87
Figure 5.4 Vertical distribution of mean particle size (mm) at rockshelter control block — 88
Figure 5.5 pH, electrical conductivity, cation exchange capacity, organic matter, nitrogen, phosphorus, and calcium carbonate data — 90
Figure 5.6 Typical disturbance in the sediments near the back of the rockshelter — 93
Figure 5.7 Cumulative rockshelter sedimentation rate — 94
Figure 5.8 Stratigraphic profile of Monoliths A and B localities — 97
Figure 5.9 Photograph of one of the soil monolith being prepared for collection — 98
Figure 5.10 Vertical particle-size distribution from Monoliths A and B — 99
Figure 5.11 pH, electrical conductivity, exchangeable sodium percentage, cation exchange capacity, organic matter, nitrogen, and calcium carbonate measurements with depth for Monolith A — 100
Figure 5.12 pH, electrical conductivity, exchangeable sodium percentage, ation exchange

capacity, organic matter, nitrogen, gypsum, and calcium carbonate data for Monolith B	101
Figure 5.13 Stratigraphic diagram of stream overbank deposits	105
Figure 5.14 Nearly identical topography of the Marmes and Harrison horizons	106
Figure 5.15 Excavation grid showing relationship of stream channel to excavated area	107
Figure 5.16 Stratigraphic profile at Monolith G	108
Figure 5.17 Stratigraphic profile of Monolith K	109
Figure 5.18 Stratigraphic profile of Monolith M, N, and O localities	110
Figure 5.19 Vertical particle-size distribution from Monolith G	111
Figure 5.20 pH, electrical conductivity, exchangeable sodium percentage, cation exchange capacity, organic matter, nitrogen, gypsum, and calcium carbonate data for Monolith G	112
Figure 5.21 Vertical particle-size distribution from Monoloth K	114
Figure 5.22 Vertical particle-size distribution from Monolith N	115
Figure 5.23 Cross section of stream channel illustrating cut-and-fill relationship to Marmes and Harrison horizons	117
Figure 6.1 Plan and oblique views of the features by stratigraphic unit	125
Figure 6.2 Stratigraphic profile showing the complexity of feature relationships near the back of the rockshelter	129
Figure 6.3 Drawing of typical storage pit feature in cross-section	131
Figure 6.4 Stick "platform" in a storage pit at McGregor Cave	131
Figure 6.5 Diagram of hind limb bones of a large elk from the northern portion of the Marmes floodplain excavation	137
Figure 6.6 Diagram of vertebrae and ribs of the large elk from the south-central portion of the Marmes floodplain excavation	138
Figure 7.1 Photograph of feature being encased in plaster	141
Figure 7.2 Photograph of completed plaster cast	141
Figure 7.3 Photograph of Grover Krantz excavating the Marmes I cranium	145
Figure 7.4 Location of burials in Marmes Rockshelter by stratigraphic unit	146
Figure 8.1 Digital enhancement of photographic prints of projectile points	161
Figure 8.2 Frequency distribution of stage-diagnostic debitage from the Floodplain Harrison Horizon assemblage	170
Figure 8.3 Spatial distribution of high-density lithic artifact deposits and stone tools associated with the Floodplain Harrison Horizon assemblage	172
Figure 8.4 Frequency distribution of stage-diagnostic debitage from the Floodplain Marmes Horizon assemblage	174
Figure 8.5 Spatial association of high-density lithic artifact deposits and stone tools with faunal remains from the Floodplain Marmes Horizon	175
Figure 8.6 Frequency distribution of stage-diagnostic debitage from the Rockshelter Stratum I assemblage	178
Figure 8.7 Spatial association of high-density artifact deposits and features from the Rockshelter Stratum I assemblage	179
Figure 8.8 Frequency distribution of stage-diagnostic debitage from the Rockshelter Stratum II assemblage	182
Figure 8.9 Spatial association of high-density artifact deposits and features from the Rockshelter Stratum II assemblage	184
Figure 8.10 Frequency distribution of stage-diagnostic debitage from the Rockshelter Stratum III assemblage	187
Figure 8.11 Spatial association of high-density artifact deposits and features from the Rockshelter Stratum III assemblage	188
Figure 8.12 Frequency distribution of stage-diagnostic debitage from the Rockshelter Stratum IV assemblage	190
Figure 8.13 Frequency distribution of stage-diagnostic debitage from the Rockshelter Stratum V assemblage	194
Figure 8.14 Example of a Cold Springs side-notched point	194
Figure 8.15 Spatial association of high-density artifact deposits and features from the	

Rockshelter Stratum V assemblage	195
Figure 8.16 Frequency distribution of stage-diagnostic debitage from the Rockshelter Stratum VI assemblage	198
Figure 8.17 Frequency distribution of stage-diagnostic debitage from the Rockshelter Stratum VII assemblage	202
Figure 8.18 Spatial association of high-density artifact deposits and features from the Rockshelter Stratum VII assemblage	203
Figure 8.19 Frequency distribution of stage-diagnostic debitage from the Rockshelter Stratum VIII assemblage	205
Figure 8.20 Location of obsidian sources identified in XRF analysis of obsidian artifacts from the Marmes site	207
Figure 8.21 Proportional representation of the major lithic material composition types in the rockshelter strata	217
Figure 8.22 Frequency distribution of projectile point types	218
Figure 8.23 Reduction sequence for Cascade projectile points	220
Figure 8.24 Windust, Cascade, and Cold Springs points illustrating impact fracture patterns	221
Figure 8.25 Lithic tools from the Marmes site	222
Figure 8.26 Ratio of flakes to tools for rockshelter strata	226
Figure 8.27 Percentage of lithic tools representing flaked stone manufacturing byproducts for rockshelter strata	227
Figure 9.1 Photograph of bone and antler artifacts	234
Figure 9.2 Photograph of bone and antler artifacts	235
Figure 9.3 Photograph of bone artifacts and unmodified bone items	236
Figure 9.4 Eyed needles, a needle preform, and a distal needle tip	238
Figure 9.5 Photograph of bone needles	238
Figure 9.6 Photograph of bone artifacts and unmodified bone item	241
Figure 9.7 Computer scanned image of fully articulated owl foot	251
Figure 10.1 Plan map showing Marmes Floodplain excavation units and strata that contained bulk samples analyzed in this study	255
Figure 10.2 Plan map showing locations of Marmes Floodplain excavation units sampled by this study and by Caulk	259
Figure 10.3 Graph showing percentage of ascribed shapes by vertical stratum	297
Figure 10.4 Graph showing percentage of each ascribed condition by vertical stratum	299
Figure 10.5 Computer scanned images of modern Arctic fox skull with blackened fragment of Marmes floodplain specimen	310
Figure 10.6 Computer scanned images of typical bones from the Marmes Rockshelter compared with a sample of floodplain faunal remains	311
Figure 10.7 Photograph of the elk bone feature	314
Figure 13.1 Botanical artifacts from the Marmes site	366
Figure 15.1 Olivella shell beads	402
Figure 15.2 Regional environmental trends	405
Figure 15.3 Some Snake River region economic resources trends	415

LIST OF TABLES

Table 2.1 Marmes site radiocarbon dating results (through Sheppard et al. 1987)	27
Table 5.1 Inventory of sediment data for the Marmes site	80
Table 5.2 Particle-size statistics for rockshelter control block	84
Table 5.3 Correlation of stratigraphic units I-VIII with the rockshelter control block based on particle-size data	92
Table 5.4 Estimates of rockshelter sedimentation rates	96
Table 7.1 Human remains feature summary	147
Table 8.1 Excavation contexts for analyzed lithic assemblage	160
Table 8.2 List of species that react to the antisera used in the Marmes site blood residue analysis	168
Table 8.3 Summary of debitage analysis data for the floodplain Harrison Horizon assemblage	169
Table 8.4 Summary of tool analysis data for the floodplain Harrison Horizon assemblage	170
Table 8.5 Summary of debitage analysis data for the floodplain Marmes Horizon assemblage	173
Table 8.6 Summary of tool analysis data for the floodplain Marmes Horizon assemblage	174
Table 8.7 Summary of debitage analysis data for the rockshelter Stratum I assemblage	176
Table 8.8 Summary of tool analysis data for the rockshelter Stratum I assemblage	177
Table 8.9 Summary of debitage analysis data for the rockshelter Stratum II assemblage	180
Table 8.10 Summary of tool analysis data for the rockshelter Stratum II assemblage	181
Table 8.11 Summary of debitage analysis data for the rockshelter Stratum III assemblage	185
Table 8.12 Summary of tool analysis data for the rockshelter Stratum III assemblage	186
Table 8.13 Summary of debitage analysis data for the rockshelter Stratum IV assemblage	189
Table 8.14 Summary of tool analysis data for the rockshelter Stratum IV assemblage	190
Table 8.15 Summary of debitage analysis data for the rockshelter Stratum V assemblage	192
Table 8.16 Summary of tool analysis data for the rockshelter Stratum V assemblage	193
Table 8.17 Summary of debitage analysis data for the rockshelter Stratum VI assemblage	196
Table 8.18 Summary of tool analysis data for the rockshelter Stratum VI assemblage	197
Table 8.19 Summary of debitage analysis data for the rockshelter Stratum VII assemblage	200
Table 8.20 Summary of tool analysis data for the rockshelter Stratum VII assemblage	201
Table 8.21 Summary of debitage analysis data for the rockshelter Stratum VIII assemblage	204
Table 8.22 Summary of tool analysis data for the rockshelter Stratum VIII assemblage	205
Table 8.23 List of obsidian sourcing samples from the Marmes site	206
Table 8.24 Distribution of obsidian samples from the Marmes site	208
Table 8.25 Blood residue analysis comparative results	210
Table 8.26 Seriation of Marmes projectile point styles	216
Table 9.1 Definitions of wear types recorded in the modified bone collection	231
Table 9.2 Modified bone and antler by morpho-functional and use-life classes	232
Table 9.3 Measurements of split metapodial and splinter awls	233
Table 9.4 Measurements and proveniences of needles from the Marmes site	239
Table 9.5 Measurements and geologic contexts of spatulate bone tools	242
Table 9.6 Measurements and geologic contexts of narrow-diameter bone pins	242
Table 9.7 Measurements and associations of antler or bone rods	244
Table 9.8 Measurements and geologic context of pointed bone tools	246
Table 9.9 Measurements and geologic contexts of ornaments and polished bone	247
Table 9.10 Measurements and geologic contexts of bone manufacturing debris	249
Table 9.11 Measurements and geologic context of unmodified bone	250
Table 9.12 Bone tool morpho-functional classes by geologic unit	250
Table 10.1 Summary of Marmes (45FR50) floodplain units sampled	256
Table 10.2 Number and percentage of sampled floodplain excavation units and levels by strata from this study and Caulk	258
Table 10.3 Scientific and common names for all taxa from the Marmes site	261
Table 10.4 Number of identified specimens…from the Marmes floodplain	262
Table 10.5 Number of identified specimens…from Marmes rockshelter	264
Table 10.6 Number of identified terrestrial taxa reported by Caulk from the Marmes and	

Harrison Horizons of the Marmes floodplain	265
Table 10.7 Number of identified specimens summary…	266
Table 10.8 Number of identified specimens…from the Marmes floodplain	272
Table 10.9 Number of identified specimens…from each sub-area analyzed	283
Table 10.10 Number of identified specimens…from Caulk for each sub-area	285
Table 10.11 Number of identified specimens…from the Marmes floodplain sub-areas analyzed	287
Table 10.12 Number and percentage of identified economic mammals per-geologic unit and taxon from Marmes Rockshelter	291
Table 10.13 Number of identified specimens…from the Marmes floodplain compared to the rockshelter	292
Table 10.14 Number and percentage of bone fragments by shape and stratum	296
Table 10.15 Number and percentage of bone fragments by condition and stratum	298
Table 10.16 Number of identified specimens…per-cultural unit and taxon from 45WT2	301
Table 10.17 Number of identified specimens…per-cultural unit and taxon from Granite Point	302
Table 10.18 Number of identified specimens…for the Lind Coulee site	303
Table 10.19 Total number and percentage of identified specimens…from pre-Mazama contexts at Marmes Rockshelter and floodplain…	304
Table 10.20 Total number of identified specimens…from post-Mazama contexts at Marmes Rockshelter…	307
Table 10.21 Spatial distribution of bone elements from large elk	313
Table 10.22 Habitat preferences of mammals identified from Marmes	315
Table 10.23 Pre-Mazama and post-Mazama ratios of pronghorn to elk	317
Table 11.1 Frequency of fish remains by taxon, rockshelter	323
Table 11.2 Frequency of fish taxa by strata/unit, rockshelter	326
Table 11.3 Frequency of fish taxa across three strata/units, rockshelter	326
Table 11.4 Mean width of *Oncorhynchus* vertebrae across strata/unit	326
Table 11.5 Standard length and vertebra widths of selected species	327
Table 11.6 Mean width of Cyprinidae/Catostomidae vertebrae across strata/unit, rockshelter	327
Table 11.7 Standard length and vertebra widths of selected *Cypriniformes*	328
Table 11.8 Frequency of fish taxa, floodplain	329
Table 11.9 Mean width of *Cyprinidae/Catostomidae* vertebrae across horizon	331
Table 11.10 Frequency of burned specimens across the floodplain horizons	331
Table 11.11 Frequency of skeletal elements with landmarks from *Cyprinidae* and *Catostomidae*, floodplain	332
Table 11.12 Frequency of fish taxa by horizon, Marmes floodplain	334
Table 11.13 Frequency of fish remains by horizon, floodplain	335
Table 11.14 Frequency of fish taxa in the floodplain and contemporary units of the rockshelter	336
Table 12.1 Summary of invertebrate remains from 45FR50	343
Table 12.2 Columbia Basin bivalve sequence	345
Table 13.1 Provenience of botanical samples, Marmes archaeological site	348
Table 13.2 Botanical samples from the Palouse River floodplain Marmes site	351
Table 13.3 Common names of plants	352
Table 13.4 Botanical samples containing seeds	353
Table 13.5 Botanical samples containing cordage and matting	355
Table 13.6 Botanical samples, wood and other materials	356
Table 13.7 Plant materials recovered from sediment samples	363
Table 13.8 Plant materials in sediment samples from the Marmes site	364
Table 13.9 Plant materials by age/depositional unit, Marmes Rockshelter	369
Table 13.10 Plant materials by depositional unit from the Marmes floodplain	372
Table 15.1 All radiocarbon dates from Marmes Rockshelter	390
Table 15.2 Correlation of radiocarbon dates of geological strata…with Leonhardy and Rice's cultural chronology phases	398

PREFACE

Through a combination of circumstances, the discoveries at Marmes Rockshelter had an impact on archaeology and historic preservation far beyond the recovery of early human and cultural remains. Valuable as those discoveries are, even more valuable is the site's influence on legislation for the preservation of our cultural past and on sources of funding for research.

The Marmes Rockshelter story really begins in 1952 when rancher John McGregor of Hooper, Washington took me to the lower Palouse River area to see a number of caves and rockshelters in the cliffs above the river. They were impressive and the next year—with support from the McGregor Family—I directed the excavation of a large site we named McGregor Cave. A productive site, the cave held several storage pits including some with the remains of dried eels. We excavated only that one season; the cave is still an excellent possibility for yielding very early cultural remains, and perhaps even human remains, if excavation were continued deeper into the floor deposits.

Ten years passed before I returned to the Palouse River canyon. Ongoing construction of Lower Monumental Dam on the Snake River would soon cause a reservoir to flood into the canyon and that meant the chance to investigate part of a large Indian village site at the mouth of the Palouse River that would be lost. We developed an excavation plan and while most of the crew was erecting tents and establishing a summer field camp, two students went with me to check nearby Marmes Rockshelter. It was immediately apparent that this was a rich site. That fact, together with finding that the river-mouth area where we'd intended to excavate had been badly disturbed years ago by railroad construction, prompted abandoning the original plan and switching our efforts to the rockshelter.

The first season proved the richness of the deposits in Marmes Rockshelter and therefore excavation continued through the next two summers, 1963 and 1964. Without reaching the bottom of the deposits, we recovered a number of burials including some over 7,000 years old. But funding for archaeological research behind the Snake River dams was extremely limited during those years, so by the end of the '64 field season we ceased excavating at Marmes. Eighty other sites were known within the Lower Monument Dam reservoir basin, and I couldn't justify spending all of the time remaining before the flooding at the one site.

As a last effort, our project geologist Roald Fryxell wanted a bulldozer trench dug from inside the rockshelter down through the talus slope and into the floodplain below. Roland Marmes did the job for us. At the deepest point cut into the floodplain—twelve feet below the surface—the crew monitoring the dozer work found two dozen or so pieces of human bone. The question was: had the blade of the bulldozer brought them down from inside the rockshelter, or did they belong to the floodplain deposits? If the latter, they were very old. A test pit provided no answer.

Fryxell had wanted the trench so that he could geologically tie the deposits within the rockshelter to those of the floodplain, thus exposing a complete profile from late Pleistocene Missoula Flood events to the present. That profile made the trench a favorite place for Fryx to take students. The puzzle of the human bones' point of origin remained, however. Then, in the early spring of 1968, three years after the bones' discovery, students dug a 2-x-2-meter test pit in the bottom of the trench closer to the rockshelter than the previous test pit had been. This time, in undisturbed deposits, they found four pieces of a bone tool, bits of animal bone cracked open for marrow, *and* human bone. Dating gave an age of 10,000 RCYBP, an antiquity confirmed by geologic methods based on six years of previous study.

While all this was taking place, work at Lower Monumental Dam was continuing and the flooding of the reservoir was scheduled for year-end. That made it urgent to find funds for continuing the archaeological research.

Since 1962 I had been a member of the national Committee for the Recovery of Archaeological Remains (CRAR), a four-member group, each person representing a scientific society and each well-connected politically. Dr. J.O. Brew from Harvard, Dr. Emil Haury from the University of Arizona, Henry Hamilton, a banker from Missouri, and I were the members. CRAR assisted the National Park Service archaeology program in several ways, but our primary duty was to lobby Congress in support of the NPS archaeology budget. I was also a member of the Advisory Board of the National Trust for Historic Preservation and had just been appointed by

President Lyndon Johnson to the newly created National Advisory Council on Historic Preservation, the membership of which consisted of the Secretary of the Interior, the Secretary of Agriculture, the Secretary of Housing and Urban Development, the U.S. Attorney General, the head of the General Services Committee, and ten citizen members, of which I was one.

The significance of this for Marmes Rockshelter is that for several years I had been spending a considerable amount of time in Washington, D.C. I was well acquainted with members of our Congressional delegation and their staffs, and had made a number of lasting friendships. Consequently, as soon as we knew the early date of the Marmes human remains, I called Carl Downing, Press Secretary for Senator Warren G. Magnuson. He was a bright, thinking person and a good friend. His response to my call was: "How soon can you be in Washington, D.C.?"

Fryxell could get away one day before I could, so he flew east with our discovery carefully wrapped in a foam plastic box buckled into the seat next to him. I arrived the next day to find that Carl and Senator Magnuson had scheduled a meeting in the Senator's office. Among those in attendance were the Director of the National Science Foundation, the Chief of the Corps of Engineers, and the Director of the National Park Service and key members of his staff. *Time* magazine, in a brief article about the Marmes discovery and the meeting, made a wry comment that this was an interesting way to announce an important scientific discovery. Magnuson chaired the Senate Appropriations Committee. The upshot was that the Corps of Engineers transferred funds to the National Park Service, and major excavation at the Marmes site got underway immediately.

Those of us in charge of the project had no idea at the time of the impact the discoveries would have on the press, and through the press on citizens of the state and the nation. The 10,000-year-old date—the oldest then known for human remains in the new world—generated immense interest. We had carloads of interested people arriving at the site for a look and busloads of students brought from as far away as Seattle. We gave them all tours and explanations, and that seemed to spread the word and bring more people. This level of interest didn't escape our senators' attention. Magnuson commented later that he'd never carried a Whitman County election until Marmes cranked up.

The search for human and cultural remains centered below the rockshelter on the floodplain, where the ancient human bone had been found. But we also continued work in the rockshelter itself and there we made another discovery: a cremation hearth 10,000 years old, the oldest yet known in the new world. The date for flooding—and losing the site—was getting closer and closer but the search was far from concluded. That led me to again approach our senators, asking about the possibility of building a coffer dam in front of the rock shelter so that work could continue there and on the floodplain despite the rise of the reservoir. More than a million dollars were appropriated and construction of a coffer dam began right away with Senator Magnuson coming to dedicate the project. Senator Henry M. Jackson had earlier led dedication of the rockshelter as a National Historic Landmark.

The coffer dam was completed as the reservoir behind Lower Monumental Dam began to rise, but unfortunately the water rose as fast inside of it as outside. The problem was geologic. Pleistocene floods caused a gigantic waterfall to pour over the cliff above the rock shelter, plucking basalt, enlarging the rock shelter, creating a plunge pool, and depositing gravel. Roland Marmes had once drilled for a well there and found he had to go through 130 feet of gravel. Unfortunately, this was all within the area of the coffer dam. We knew we were losing the site. In late February 1969 we built heavy wooden cribs to protect two unexcavated squares that seemed particularly promising (they were next to where the Marmes III skull cap had lain; we lined the trenches with plastic and for five around-the-clock days dumped truckloads of sand to backfill our excavations. If it ever becomes possible for work at Marmes to resume, our excavations will be found clearly marked.

While we had been excavating, we thought of the site—and the work, which was interdisciplinary—in terms of archaeological results, the human and environmental record of the last 10,000 years. In retrospect, the site's political impact is also clear. Science, public interest, and politics converged at Marmes Rockshelter. Cultural preservation is an issue everyone can rally behind: senators and congressman on both sides of the aisle and from all states can favor discoveries about our human selves and can ease passage of cultural resource bills through committees. Our state delegation had already been supportive of such legislation and Marmes added vindication. Public appreciation of Senator Magnuson's and Senator Jackson's support for Marmes strengthened support of archaeology and historic preservation in general, and that favorably affected legislation.

An editorial in the *Seattle PI* dated February 26, 1969 stated:

> Certified archaeological sites supposedly are protected by the National Park Service, but federal funding of archaeological preservation and exploration is a pittance. Also, there is no excuse for lack of long-range planning in such projects as Lower Monumental Dam to insure that terrain as valuable to science as the Marmes site is not buried under tons of water. . . The immediate lesson to be learned . . . is that there is a need for much improved analysis at the federal level of the negative side effects of such public works as dams, and a need for early planning to avoid those effects.

Magnuson called Marmes "a landmark precedent in our nation's responsibility to its own heritage, which will be felt for decades to come."

Richard D. Daugherty, Ph.D.
Lacey, Washington
January 21, 2003

Preface figure: Roald Fryxell (at left), Carl Downing (3rd from left), Senator Warren Magnuson (grey trenchcoat), Richard Daugherty (foreground at right) at dedication of the coffer dam at Marmes Rockshelter, 1968.

1
PROJECT BACKGROUND

Brent A. Hicks

Marmes Rockshelter is one of the more important archaeological sites in Washington State and the Pacific Northwest, not only because of the archaeological information it was found to possess, but also because of the attention it generated towards American archaeology throughout the Northwest, the nation, and the world. The political and administrative story behind the excavations that followed the discovery of late-Pleistocene/early-Holocene human skeletal remains in the Marmes site encapsulates, perhaps even incited, the manner of the transition in archaeological studies that played out in the 1970s following passage of the National Historic Preservation Act of 1966. That transition led to the current era of archaeological studies epitomized by Cultural Resources Management approaches driven by federal government mandates to preserve archaeological materials and information, and to seek greater public involvement in the management process. It is these mandates that have led full-circle to the current study and completion of a final report some thirty-five years after the Marmes site was excavated.

Site Description

The Marmes Rockshelter site (45FR50) is located in the southern Columbia Basin within the lower Palouse River canyon about 1 1/2 miles north of its confluence with the Snake River (Figures 1.1 and 1.2). The site is in the SE ¼, NW ¼, Sec. 17, T13N, R37E, W.M., Franklin County, in southeastern Washington. The site is currently inundated by Lake Herbert G. West, the reservoir behind the U.S. Army Corps of Engineers' Lower Monumental Lock and Dam. Efforts to protect the site with a levee prior to inundation were unsuccessful. Normal pool level of the reservoir is 540 feet above sea level, approximately 10 m (ca. 35 feet) above the Palouse River at its nearest point to the rockshelter, and fills the rockshelter to the approximate level of the original surface when excavations first began (Rice 1969:2).

Although dubbed "Marmes Rockshelter" on the site form, the site actually encompasses the rockshelter itself, the slope in front of the rockshelter, and the adjacent floodplain of the Palouse River lying immediately in front of the rockshelter mouth (Figure 1.3). The floodplain portion of the site is on a remnant of the earliest floodplain terrace present in that part of the Palouse River Canyon (Rice 1969:1). The terms *Marmes Rockshelter* and the *Marmes site* will be used interchangeably with specific distinctions made to the different portions of the site when necessary.

Marmes Rockshelter is located within the lower Palouse River canyon, at the north end of what now appears as a wide bay or inlet of the reservoir. The rockshelter's aspect is generally in a southerly direction. From the Palouse River mouth at the Snake River to a point ca. 10 km (6 miles) north marked by Palouse Falls, the canyon is a striking reminder of the impact of the glacial-era catastrophic flooding that created the scablands of the Columbia Basin (Figure 1.4). Like the Grand Coulee at the north end of the basin, the walls of the canyon are generally exposed basalt bedrock rising up to 700 feet above the Palouse River floodplain. Prior to Euro-American settlement of the area, the floodplain was covered by willow thickets and brush, which was stripped away to permit agricultural use of the valley floor during the historic period. Remnants of the original vegetation community can be observed upstream from the Marmes site where the Palouse River flows in its natural channel for several kilometers through the canyon before entering the reservoir. Above the floodplain, the native vegetation, consisting of sagebrush-bunchgrass communities, grows in stony, carbonate-rich, shallow soils. The native vegetation has been thoroughly invaded by non-native plant species such as *cheatgrass (Bromus tectorum)*, probably due to the use of the canyon for grazing cattle and sheep.

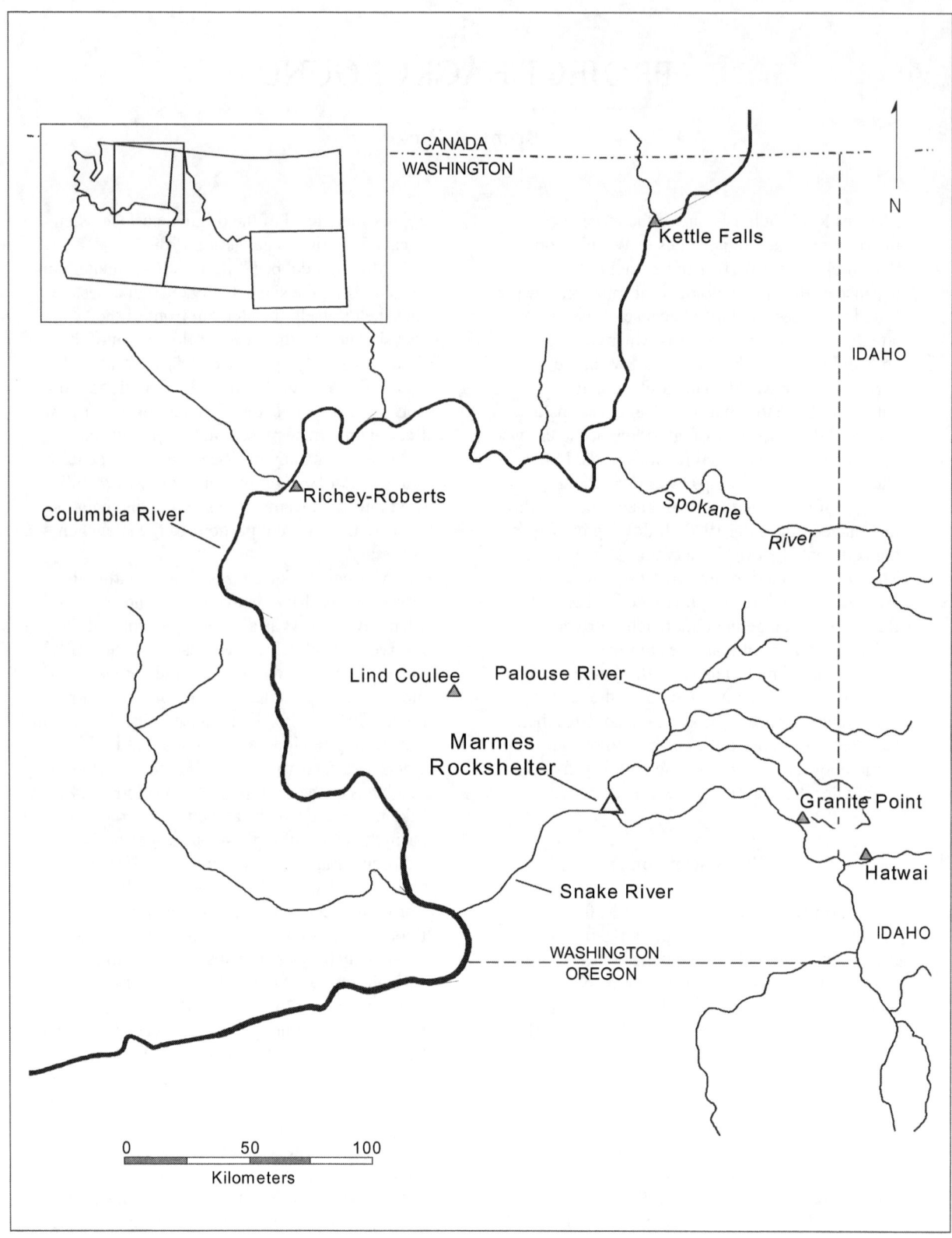

Figure 1.1 Map showing the location of Marmes Rockshelter and other prominent Columbia Plateau archaeological sites.

Figure 1.2 View of the lower Palouse River Canyon, facing northeast. Arrow points to location of Marmes Rockshelter.

Figure 1.3 Photograph of Marmes Rockshelter in 1962. View is to the northwest.

Figure 1.4 Photograph showing topography of the lower Palouse River canyon with Marmes Rockshelter in foreground. View is to the northwest.

History of Investigations

The Marmes Rockshelter site was excavated between 1962 and 1964 by Washington State University (WSU) under contract with the National Park Service, and in 1968 with funds from the U.S. Army Corps of Engineers. The investigations were initiated as the first part of an archaeological survey intended to reconstruct the record of human occupation of the area prior to its inundation by the backwaters of Lower Monumental Dam. Construction of the dam, already underway when the archaeological project began, was expected to be completed within five years. The 1962 investigations began as a testing program at the Palus Village site (45FR36A) on the west side of the Palouse River at its mouth. However, initial testing produced a conclusion that the deposits were disturbed and excavation efforts were moved to a rockshelter that a nearby resident John McGregor had shown to WSU's archaeologist Dr. Richard Daugherty (personal communication 1998) a number of years before.

Marmes Rockshelter, named for the landowner at the time of the archaeological investigations, was approximately 12 m (40 ft) wide by 8 m (25 ft) deep when excavations began (Figure 1.5). Daugherty directed excavations in the rockshelter during the summers of 1962 through 1964 while Roald Fryxell conducted the geological and stratigraphic studies. Fryxell was keenly interested in the details of the Marmes site's stratigraphy not only for providing a basic relative cultural chronology but also for assessing landscape evolution, paleoclimate, and

Figure 1.5. View across the berm of Marmes Rockshelter, facing west.

human adaptation on the Columbia Plateau (Gustafson and Gibson 1998:2).

All site measurements were made in units of feet and tenths of feet. An arbitrary datum of 100 feet vertical elevation was established at the highest point in the rockshelter, located at the N80/W0 grid point of the site grid. Excavation units consisted of squares five feet (ca. 1.5 m) on a side dug in 0.5-foot (ca. 15 cm) level intervals. To ensure uniformity, this system was continued throughout the entire period of excavation (Gustafson and Gibson 1998:6). During the 1962 to 1964 field seasons, excavations were confined to the rockshelter and sediment was dry-screened through ¼- inch (6-mm) wire mesh. In 1962, seventeen units were excavated to a depth of about five feet (ca. 1.5 m) over the course of ten weeks. The excavators recovered between 700 and 800 artifacts and eleven burials. During these excavations, materials found in the rockshelter were keyed to eight stratigraphic layers designated with Roman numerals I to VIII from bottom to top that Fryxell believed indicated a time depth of perhaps 10,000 years at the site (Fryxell and Daugherty 1962) (Figures 1.6, 1.7 and 1.8).

Human remains became a focus of the Marmes Rockshelter investigations within two weeks of the initiation of excavations at the site in 1962. Eleven individuals were located that year, three stratigraphically below pumicite believed to be from Mount Mazama, the eruption of which at that time was thought to date to ca. 6,500 uncalibrated years before present (B.P.) (Fryxell and Daugherty 1962:16). Cultural pit features dominated the uppermost layers and were recognized as intrusive into earlier deposits resulting in varying degrees of mixing of sediment and cultural materials. These pit features likely represented use of the rockshelter predominantly for storage, a pattern identified in other rockshelters throughout the Palouse River canyon (Draper and Morgenstein 1993; Endacott 1992; Hicks and Morgenstein 1994; Hicks 1995, 1996; Mallory 1966), but a few were burial features (Rice 1969:Figure 58). Due to the mixing of sediment and cultural materials, interpretation of these features and much of the upper sediments was considered untenable.

A local media sensation was created by the recovery of the deepest human skeletons from Marmes Rockshelter that in 1962 were thought

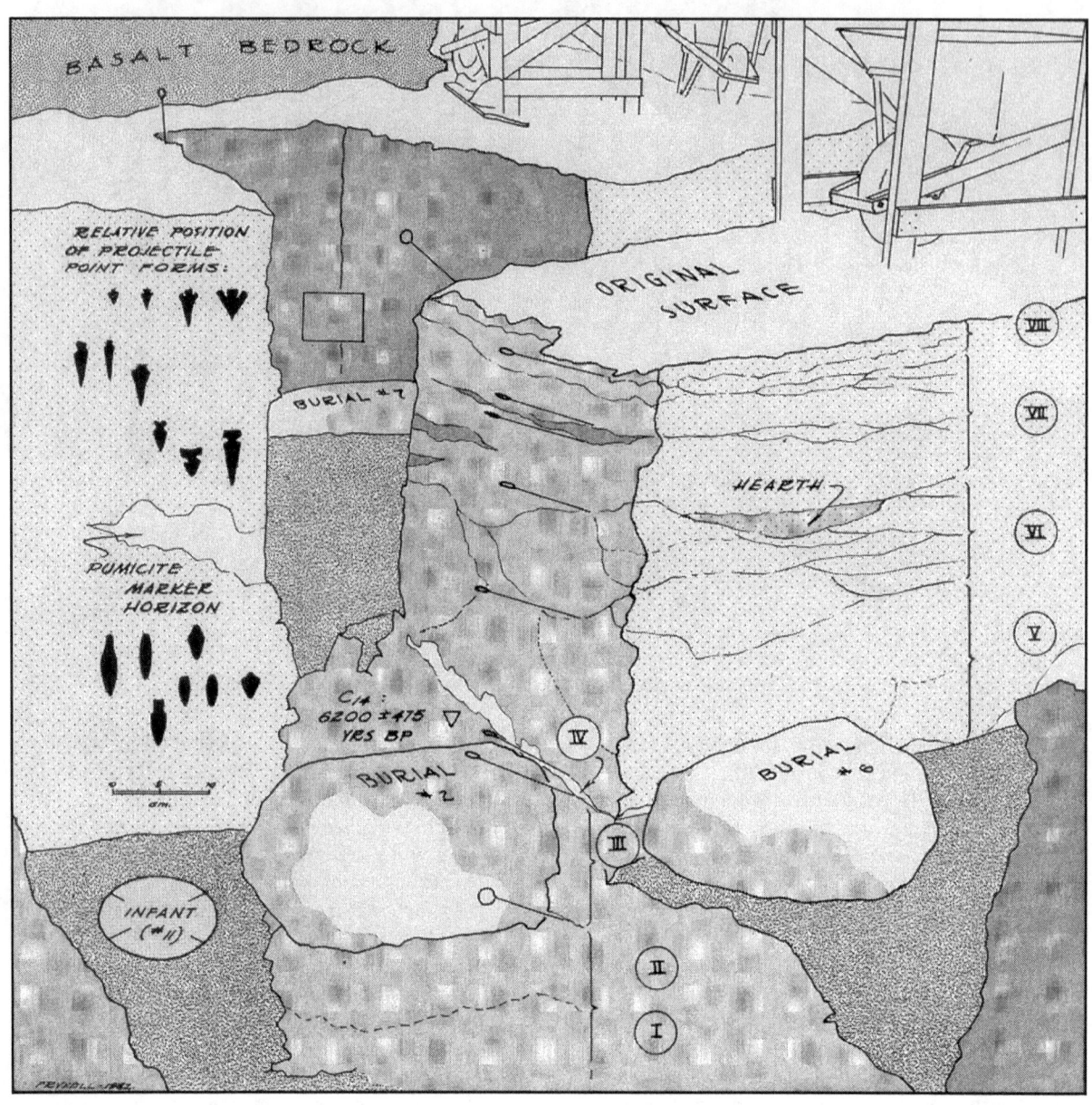

Figure 1.6 Block diagram illustrating the eight major stratigraphic units within Marmes Rockshelter (drawing by Fryxell from Fryxell and Daugherty 1962:19).

Figure 1.7 Photograph of location depicted in block diagram in Figure 1.6 (from Fryxell and Daugherty 1962:18).

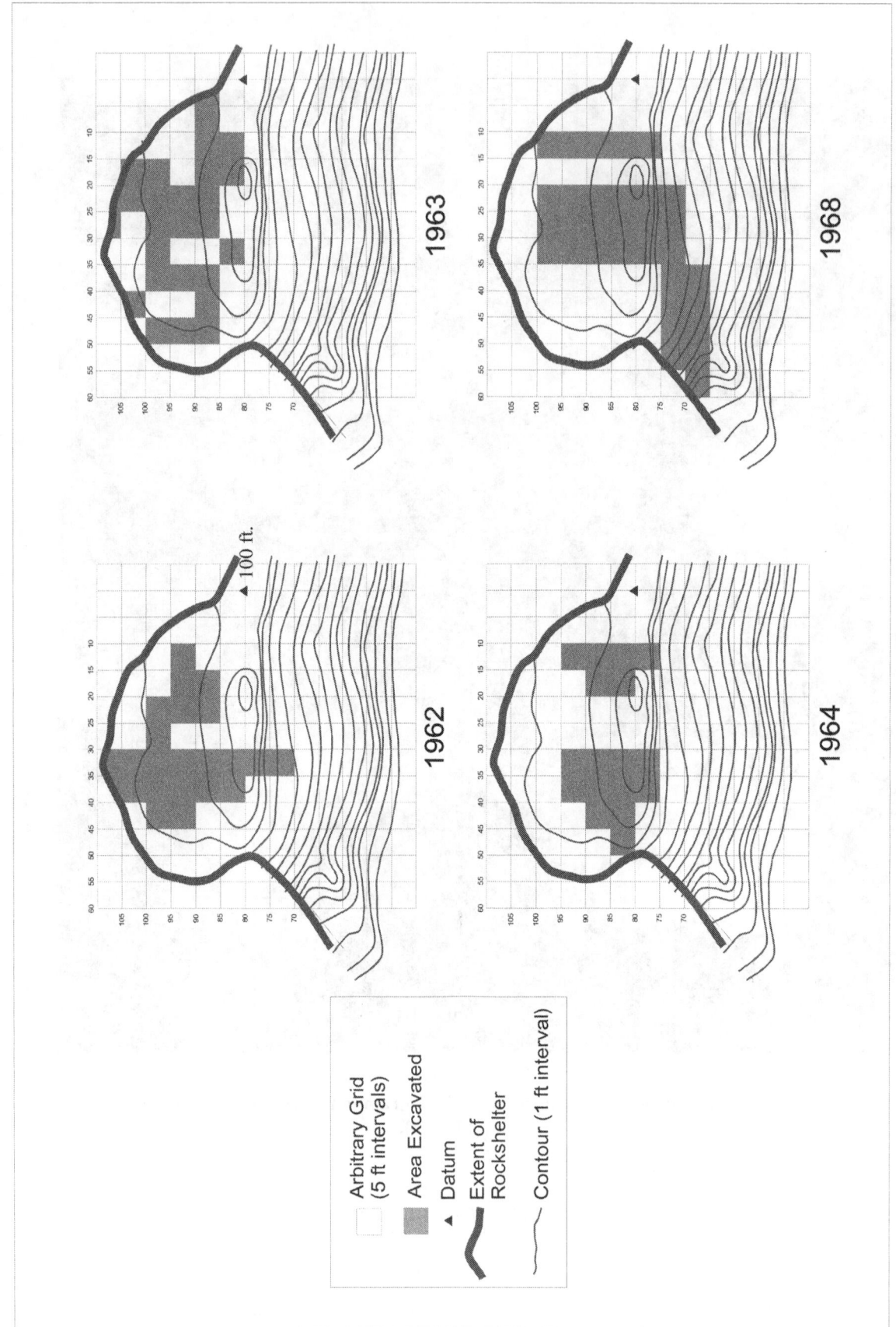

Figure 1.8 Plan drawings of rockshelter with excavation grid overlain, showing units excavated in 1962-1964 and 1968.

to be ca. 7,000 to 8,000 years old and therefore quite rare. But national attention wasn't brought to the project until an article in *Natural History* (Grosso 1967:38-43) described the results of the first three years work at the rockshelter and a number of other sites nearby. By that time, three seasons of excavation in the rockshelter had been conducted and cultural deposits spanning at least 10,000 years were recognized (Figure 1.9). At the time, investigation of the site was considered completed and the extensive excavations were thought to have recovered all of the human remains and enough archaeological materials and other data to fully analyze and interpret the rockshelter's

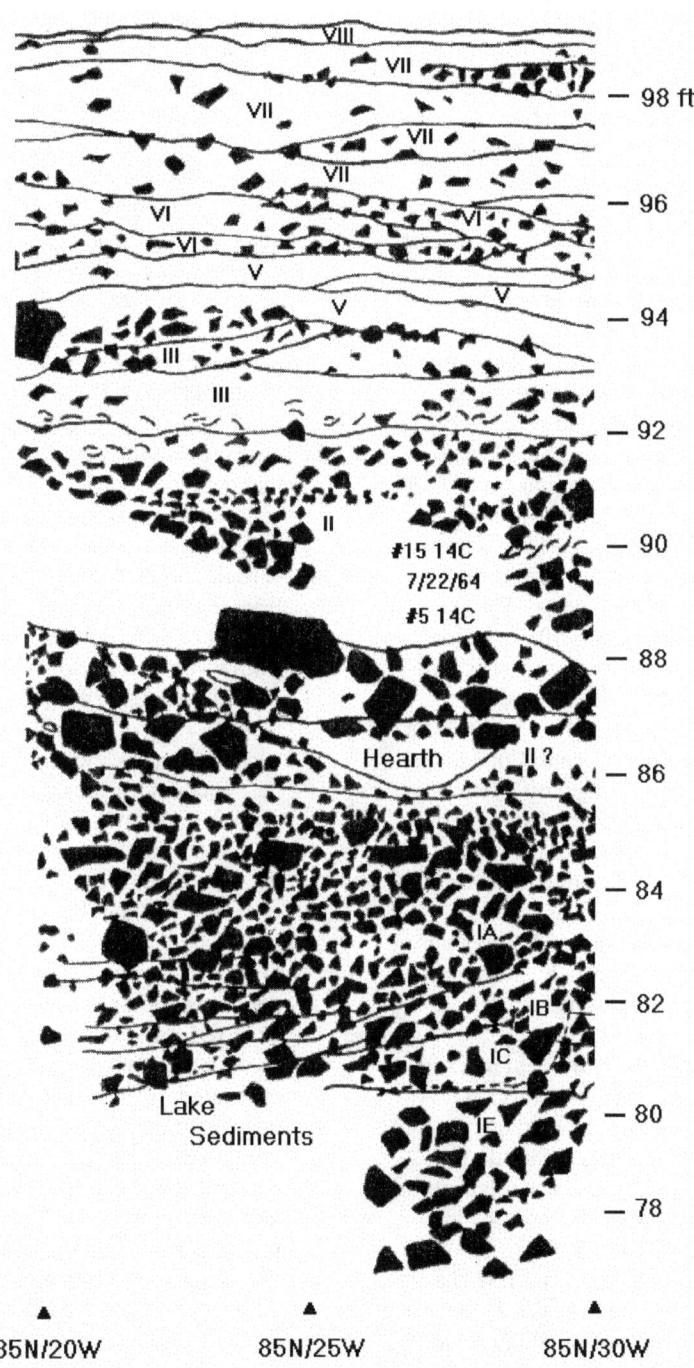

Figure 1.9 North face of Marmes Rockshelter control block (85N/30W).

cultural use. In addition, WSU needed to move on to investigating other sites that also would be inundated by the planned reservoir (Richard Daugherty, personal communication 1998).

Unfortunately, other than the 1962 progress report (Fryxell and Daugherty 1962), the only reporting of the early years' investigations is by David Rice (1969) and little information on the 1963 and 1964 excavations is available. WSU, like most university archaeology programs at the time, had minimal funding and facilities. Contracts such as the one with the National Park Service were seen as ways to collect data for advanced degree students (Richard Daugherty, personal communication 1998). Reporting of archaeological projects often took the form of students' theses or dissertations.

For the current study, determining what methods were applied in the early years of the project largely has come from discussions with Daugherty and Carl Gustafson (who joined the project in 1964 as a zoologist), and review of the minimal field records that remain from those two years. Daugherty (personal communications 1999, 2000) recounted that daily field notes were kept, that all features were recorded on feature forms that included space for detailed notes, and artifacts were recorded on individual artifact forms. Those records were kept in Daugherty's office at least until his retirement from the university (Richard Daugherty, personal communication 1998). Marmes project records and many of the artifacts were maintained in a corner office in College Hall for a number of years after Daugherty's retirement before the room was cleared out and all materials catalogued into the WSU Museum of Anthropology's archives. The volume of records on file in the archives do not amount to the magnitude described by Daugherty despite many searches conducted over the years by WSU Anthropology Department and Museum of Anthropology staff.

The lack of supervisors' daily field notes in particular has had a great impact on this study's ability to determine the particular methods applied in the field in the different seasons the site was investigated. As examples: there is no indication in any records why there were so few botanical materials collected in 1963 and 1964; or, why the presence of shell varies greatly in unit level bags where the field notes indicate little variance of shell in the sediments; and, the decision not to excavate, but to shovel out the upper strata in parts of the rockshelter in 1964 is only documented in two excavators' field notes by single sentences.

Regarding the small number of botanical materials, Daugherty (personal communication 2002) recounted that the upper strata at Marmes Rockshelter were much more disturbed than other nearby rockshelter sites where botanical materials had been observed to be an important element in cultural features (e.g., storage pit features in McGregor Cave). The botanical materials observed at Marmes Rockshelter were thought to be associated with the disturbance of the upper strata since they came from the pit features that were intrusive into lower strata. This is related to the decision in 1964 to shovel out areas of the upper strata; the principals were certain 1964 would be the last season for excavation in Marmes, and investigating the lower strata were of greater interest to the culture-historically focused study. It was thought that the 1964 investigations might encounter additional older burials, but what was particularly desired was the continued refinement of their understanding of rockshelter stratigraphy that could be applied at other sites that were to be excavated in the next few years ahead of reservoir inundation.

Fewer excavators worked at the site in 1964 and most were new to the site. There also is great variation in the level of detail in the field notes from that year. Some of the variation appears to be a result of different supervisors keeping the records from week to week and their varying knowledge about the items being found (especially identification of specific human skeletal elements) and the features encountered. Shifting supervision may account for the apparent variable collection of shell noted above. The use of excavators who were new to the site had a cumulative effect as units adjacent to units excavated in previous seasons were dug without the benefit of 'institutional knowledge' of what had been found there in the previous year(s). For example, a small concentration of human remains found in 1964 occurred at the same level and directly adjacent to the unit that contained Burial 13, which was excavated in 1963. But the probable relationship between the two concentrations was not identified until research for the current study.

In 1965, Fryxell returned to the site with an idea to trace the stratigraphy that he had meticulously recorded for the rockshelter out onto the adjacent floodplain. He thought this would allow more reliable correlation with open sites, some of which were being excavated at different parts of the lower Snake River drainage. In part, because the previous years' excavations had used the area in front of the rockshelter as the backdirt area, a bulldozer was required to provide the exposure Fryxell envisioned. Roland Marmes, the landowner, provided the bulldozer and carefully bulldozed a trench through the berm at the mouth of the rockshelter, down the talus slope in front, and onto the floodplain. Fryxell followed the blade as the bulldozer removed about 10 cm (four inches) at a time. About 12 m (40 ft) in front of the shelter mouth, at a depth of about four meters (14 ft) below the surface of the floodplain, they encountered a dark, oval concentration containing a few bone fragments which Gustafson later identified as elk (*Cervus* sp.) and human (Fryxell et al. 1968a,b; Gustafson and Gibson 1998:3).

Although Fryxell had watched the bulldozer blade expose the bones, he could not prove beyond doubt that they had not fallen out of the trench wall from above the base of the trench. Fryxell and others revisited the site often over the next two years attempting to find more remains in situ in the trench wall (Figure 1.10). Finally, in April 1968, they discovered more fragments of both elk and human bone in situ in the trench wall. Figure 1.11 shows the general stratigraphic relationships from the rockshelter to the adjacent floodplain (grid line 20 west). The square pit near the base of the talus slope approximates the location of the elk and human bone fragments discovered in 1965. A carbon-14 date of $10,750 \pm 100$ B.P.[1] (WSU 211) had been obtained from shell recovered from the earliest cultural stratum in the rockshelter (Fryxell 1968a; Sheppard et al. 1987). Lake sediments - indicated by the horizontal lines at the base of the rockshelter deposits in Figure 1.11 - contained volcanic ash attributed to Glacier Peak then believed to have erupted between 12,000 and 13,000 B.P. Tracing the deposits outward onto the floodplain, Fryxell demonstrated that the human remains lay between these deposits making them among the oldest yet discovered in the North America. Subsequent radiocarbon dating placed the age of these remains at about 10,000 B.P. (Fryxell et al. 1968a,b; Gustafson and Gibson 1998:4; Sheppard et al. 1987).

The nearly completed Lower Monumental Dam (the dam was finished in February 1969) threatened inundation of the site so Fryxell and Daugherty obtained emergency salvage funds to recover data from deep and early deposits on the floodplain and remaining early deposits in the rockshelter. Excavations began in May 1968 and continued through February despite one of the coldest winters on record (Fryxell and Keel 1969; Keel and Fryxell 1969). Fryxell and Henry Irwin directed the excavations. The rockshelter grid was extended to the floodplain and artifacts found in situ were plotted to their exact location using the same arbitrary datum established in 1962. Gridline 60 north, located about halfway down the slope in front of the rockshelter, was arbitrarily selected as the division between the rockshelter and floodplain areas of the site.

The floodplain sediments consisted of several feet of aeolian silts and sands that overlay alluvial deposits representing stream overbank deposits. Fryxell presumed these deposits were from an early postglacial channel of the Palouse River. The aeolian sediments were found to be largely sterile, but the earlier alluvial deposits where the elk and human bones had been found were relatively rich in cultural materials. Fryxell and the other supervisory personnel, faced with the knowledge that in less than eight months the site would be flooded by water impounded behind the Lower Monumental Dam, elected to sacrifice the upper deposits, and most were removed by bulldozer (Figure 1.12). Removal of the overburden revealed portions of relict Palouse River channels that confirmed Fryxell's earlier interpretations that only an area of floodplain about 300 x 400 feet remained undisturbed by post-glacial erosion that could retain culture-bearing deposits (Fryxell and Keel 1969:40). The bulldozer work also encountered a shell deposit below Mazama ash west of the main floodplain excavation that attracted a brief exploration and represents the only excavation of Cascade Phase materials from the floodplain. In the rockshelter, sediments above the Mazama ash layer (Unit IV) were removed by a bulldozer or were shoveled out.

The deep cultural deposits in the floodplain were darker in color when first exposed or when dampened. Fryxell concluded that the dark-stained overbank deposits had remained exposed long enough - perhaps a few tens of years - for soil development to begin with an accumulation of organic matter. It was on these longer-lived surfaces that people and animals had left their remains. Fryxell called these darker-colored flood deposits incipient soil "A" horizons. He named the two uppermost (A_1 and A_2) the Marmes horizon. These were separated from the next lower three A horizons by sterile alluvial silts. Layers A_3, A_4 and A_5 were named the Harrison horizon after the

---

[1] All radiocarbon results are presented as uncalibrated radiocarbon years unless otherwise noted.

Figure 1.10 View to the north showing bulldozer trench. Fryxell is standing in the trench near the location of the initial floodplain find. Photograph courtesy of Ruth Kirk.

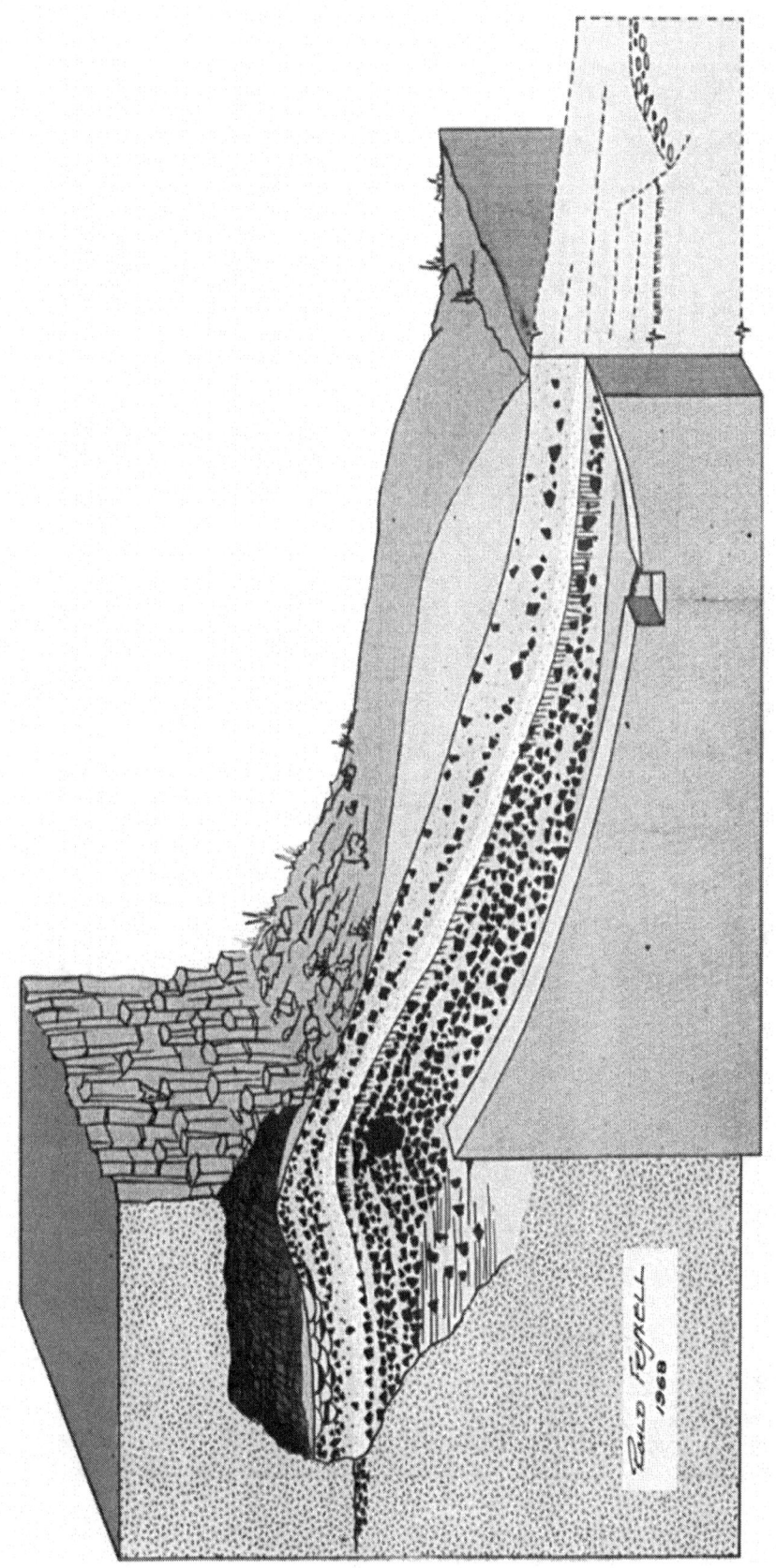

Figure 1.11 Block Diagram showing relationships among rockshelter deposits and those on the adjacent floodplain. The square pit near the base of the talus slope indicates the position of the elk and human bone fragments (after Fryxell, from Gustafson 1972).

Figure 1.12 View of the floodplain area from above the rockshelter looking south. Excavators and tractor are removing overburden above the Marmes horizon sediment layer. Note the small excavated area in the west trench. Photograph courtesy of Ruth Kirk.

owners of the adjacent land (Figure 1.13). Artifacts and charred animal bones were found in both horizons, but only the Marmes horizon contained human bone fragments (Gustafson and Gibson 1998:8). All hand-excavated sediment from the floodplain area was passed through 1-mm plastic netting in a water screen to collect cultural as well as microfloral and faunal material (Figure 1.14). The use of this fine mesh for screening resulted in the recovery of much smaller cultural items by comparison with the rockshelter excavations. Over 10,000 cubic feet of sediment was excavated at the Marmes site in 1968 (Fryxell and Keel 1969:28, 30, 37) (Figure 1.15 and 1.16) with the work continuing into the winter months (Figure 1.17). Figure 1.18 shows the imprint of the excavation areas for the rockshelter and the floodplain below.

A great deal of attention was dedicated to identifying features in the floodplain area of the site in 1968 (Figures 1.19 and 1.20). This was certainly in response to the desire to associate the human remains found in the floodplain with cultural deposits of equal antiquity. At the time of the 1969 progress report, which does not reflect all of the results of the post-field laboratory processing, Fryxell and Keel (1969:38) report that 52 "Special Features" were identified in the floodplain while only 96 artifacts had been recovered. The field records indicate that individual artifacts often were "featured" by the floodplain excavators.

It is clear from the field notes that the floodplain excavations began with great expectations and excitement at the prospect of finding very old cultural material, probably fueled in part by the media attention directed at the project. In addition, very different site preparation approaches were applied in 1968 than had been applied at the site in prior years and probably than at any site the excavators had worked at in their (mostly short) careers. The project principal investigators were mostly on-site in the weeks leading up to the beginning of hand excavation in the floodplain, supervising the careful extension of the site grid from the rockshelter datum, the very detailed identification of the sediments above the initial find of human remains in the bulldozer trench, and the establishment of a field processing laboratory on site. Then, following the principals' decision to sacrifice any archaeological materials that might exist in sediments above the stratigraphic layer in which those early human remains were found, the excavators observed and participated in the careful bulldozing and backhoe removal of sediments to just above that layer and the hand-shoveling of the remaining overburden (Figure 1.21). All of this, together with the continuous stream of visitors to the site throughout the summer, lent an air of excited purposefulness to the work at the site that is evident in the excavators' notes; a feeling that cutting-edge science was occurring. This sense of purposefulness carried over to the fieldwork techniques, including the emphasis on defining features based on individual items before their context could be revealed. In a way, such an emphasis on featuring items was not unjustified, and if the same level of emphasis had been continued in the post-field analysis, would have resembled more contemporary archaeological research methodologies. However, this did not happen.

As the construction of Lower Monumental Dam neared completion, word of the discovery and impending loss of the important archaeological resource traveled through political circles, aided in particular by Daugherty, and public pressure to protect the site from inundation grew. Additional funds were granted to construct a protective levee around the site, but water passed through a gravel layer beneath the levee and the site was flooded in early January 1969 by waters of the Lower Monumental Reservoir (Gustafson and Gibson 1998:4-5). Despite the failure of the levee, it did provide some protection of the site's remaining deposits since the levee prevents destruction from the high, wind-driven waves that are common in the broad bay south of the site.

Daugherty and Fryxell had agreed at the beginning of the 1968 season to be co-directors of the effort and both participated in the field until Daugherty left the project in late 1968 to dedicate his attention to the Ozette Village site excavations. At that time Daugherty (personal communication 2000) relinquished the project lead role to Fryxell. But Fryxell, a geologist, was not an archaeologist and had little experience with the post-field analysis and reporting of archaeological investigations. At the end of the 1968 fieldwork, Fryxell became immersed in the effort to protect the site through construction of the coffer dam; and when he concluded that the Corps of Engineers was not accounting for deeply buried gravel layers that would allow water to seep under the coffer dam, he lead the effort to cover the site with plastic sheeting and sand (Figure 1.22). As such, the post-field laboratory investigations consisted largely of the completion of processing of the 1968 excavation

Figures 1.13 View of floodplain stratigraphy in the trench wall. The Marmes (M) and Harrison (H) horizons are marked at the center of the photograph.

Figure 1.14 Photograph of the water screening apparatus used during much of the floodplain excavations.

Figure 1.15 View of the east end of the floodplain excavation area, showing the boulders exposed at lower left. Photograph courtesy of Ruth Kirk.

Figure 1.16 View of the west end of the floodplain excavation area, showing the boulders exposed at lower left. Photograph courtesy of Ruth Kirk.

Figure 1.17 View of the floodplain area in the winter as the fieldwork neared the end. Note the coffer dam in the background. View is to the east.

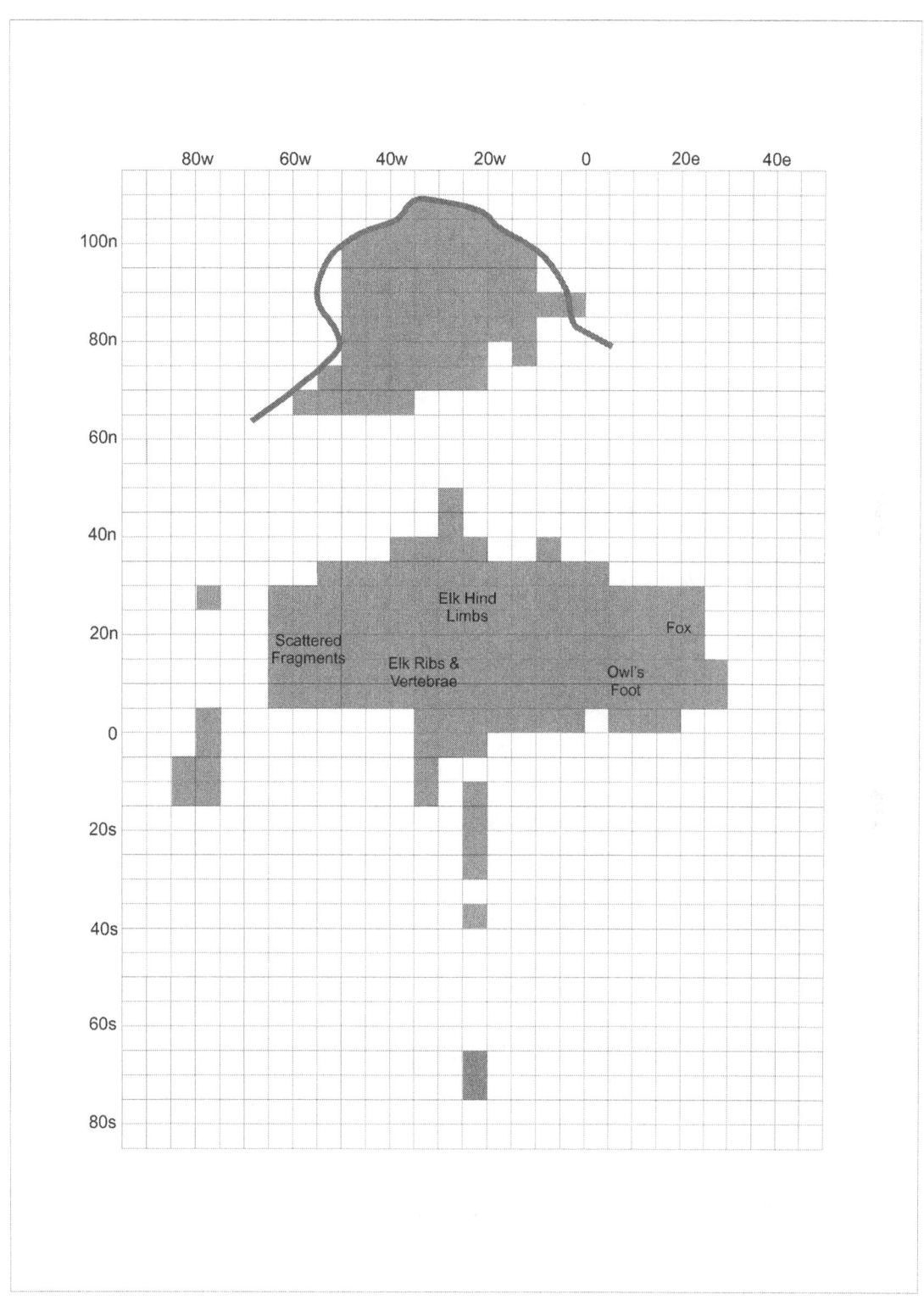

Figure 1.18 Diagram of the site grid and the units excavated (shaded).

Figure 1.19 Plan drawing of the main floodplain excavation area showing the principal concentrations of cultural material.

Figure 1.20 Photograph showing an example of the care given to documenting materials as features during the floodplain excavations.

Figure 1.21 Photograph of excavators and tractor removing floodplain overburden in the summer of 1968. View is to the north up the trench into the rockshelter.

Figure 1.22 Photograph showing plastic sheeting and sand used to protect the site. View is to the north.

level bags (but not the water screened bulk samples) and geological analysis. Archaeological materials analysis was limited except where such materials contributed to Fryxell's geological interests (e.g., sediment studies, radiocarbon dating, volcanic ash identification). Gustafson continued his involvement with the site's materials by analyzing a selection of the faunal bone, principally that from within the rockshelter, for his dissertation (1972). Other than a few select pieces of fauna addressed by Gustafson, the floodplain faunal materials were not analyzed until Caulk conducted a limited study for his Masters thesis (1988).

Under pressure from the National Park Service to provide reporting on the Marmes site to justify their funding of the excavations, Daugherty convinced Rice to prepare a final report for WSU that would elaborate on analysis that Rice had begun of the 1962-64 materials from the site and including artifacts from the 1968 excavations in the rockshelter portion of the site. It was expected that analysis of the floodplain materials would be an extended, expansive study largely directed by Fryxell (David Rice, personal communication, 1998). The result was a descriptive report (Rice 1969) with limited site interpretation as his analysis addressed only the formed tools, burials and archaeological features. Rice conducted analysis of the formed tools, developing a seriation of the projectile points that was a major contribution to the typology he and Leonhardy developed for the region (Leonhardy and Rice 1970), and that ultimately provided the foundation for Rice's dissertation on the Windust Phase in the lower Snake River region (1972).

Rice's analysis of the formed tools focused on research questions typical of a cultural chronological focus. The difficulty of the task Rice took on in conducting the analysis of the Marmes materials cannot be overstated. The combined stone, bone, and botanical artifact counts from all of the years of excavations was in the thousands. While Rice had visited the site during the investigations, he did not work at the site, and so had to recreate everything from the field records. In addition, because the National Park Service was pressuring for a report, his work had to be conducted quickly. Many kinds of analyses typical of a large site investigation were not done, and recognizing this and in anticipation of a report on the floodplain investigations, Rice termed his report "preliminary." Rice (1969) tentatively tabulated the features encountered during the four seasons of excavation in the rockshelter as 56 hearths, 38 fire pits, at least 20 storage pits, 22 burials, and a cremation hearth. The floodplain finds included, in addition to stone and bone artifacts, four concentrations of human remains and an elk-butchering feature (Fryxell and Keel 1969:33). Rice concluded that the rockshelter had been "occupied intermittently, but continuously, over a period of 10,000 to 11,000 years" during which it "was used as a shelter for living, as a burial depository, and as a storage shelter" (1969:88).

Regarding WSU's plans for a final report that would include interpretation of the floodplain materials, it appears clear from a detailed "Tentative Dissertation Outline" prepared by Fryxell in June 1969 (found in Fryxell archives, Box 6, University of Wyoming Archive of Contemporary History) that much of that reporting would have formed his dissertation through the Department of Geology at the University of Idaho. However, the high profile of the Marmes site discoveries soon brought Fryxell numerous other opportunities to be involved in exciting projects, including analysis of stratigraphy at "early man" sites in Mexico and Spain, and, in particular, analysis of rocks and cores collected from the moon by the Apollo missions. These projects diverted his attention from following-up on the Marmes site. The impetus to produce a final report that presented all of the results and conclusions of the investigations at the Marmes site dissipated with Fryxell's tragic death at age 40 in a 1974 car accident.

As noted above, several interim reports were submitted for the excavations at the Marmes site (Fryxell and Daugherty 1962; Fryxell and Keel 1969; Keel and Fryxell 1969), and Rice (1969) prepared a preliminary summary report for the rockshelter portion of the site only. While Rice's report was limited to presenting information related to research questions that dealt with the artifacts (formed tools), burials and other archaeological features, Rice presented the information in the context of the geological sequence of the site's deposits as determined by Fryxell (Fryxell and Daugherty 1962:14). However, until the current study, a final report on all site materials and incorporating the results of all analyses into an interpretive framework had not been prepared. In particular, the artifacts from the floodplain and the lithic debitage from the whole site were not included in Rice's report; presentation of botanical material data was limited to description of a few wood and textile artifacts; presentation of feature data is limited to descriptions of their locations and contents; and, there is no correlation of artifact and feature data by stratum. Despite these shortcomings (and Rice not having been a participant in the excavations at the site), the report bears witness to a great effort to develop the data that is presented and represents the most comprehensive report on the site's investigations.

Studies incorporating information obtained from the Marmes site investigations were published by Fryxell and Daugherty during the years the site was being excavated (Daugherty 1962), most of which focused on burial topics (Fryxell et al. 1968a, 1968b) and geology topics (Fryxell 1963a, 1963b, 1964, 1965, 1971; Fryxell and Cook 1964; Mullineaux et al. 1978). A number of additional studies have been conducted that addressed specific assemblages or material types from the site (Bense 1972; Breschini 1975, 1979; Caulk 1988; Gustafson 1972; Hess 1997; Krantz 1979; Rice 1972; Sheppard et al. 1984; Sheppard et al. 1987) or that incorporate other aspects of the site (e.g., Hammett 1977; Marshall 1971), often for thesis or dissertation projects. Importantly, Fryxell's conclusions regarding the rockshelter's stratigraphic layers were correlated with radiocarbon ages and the content of each layer was used (along with data from other sites) to define a material culture-historical sequence for the lower Snake River region of southeastern Washington (Leonhardy and Rice 1970).

2
RESEARCH DESIGN

Brent A. Hicks

It is important for archaeologists to address curated assemblages from previously excavated sites that lack reporting. This follows the conservation ethic in attempting to gain new insight on past human activities without having to excavate additional archaeological sites, and thereby damage cultural properties. Many of the unreported sites in the Columbia Basin are currently inaccessible (e.g., inundated by reservoirs) or were destroyed in the course of project development. It is only in the curated assemblages that we can hope to learn more about the human activities that created such sites. The Marmes site analysis and reporting project illustrates this kind of research endeavor.

Research Themes and Questions

Cultural Chronology

Research orientation early in the history of the field of Archaeology centered on culture history. Archaeological research in the Interior Plateau was slow to develop a prehistoric chronology despite systematic work occurring prior to the turn of the twentieth century (Campbell 1989; also see Lohse and Sprague 1998 for a detailed account of the history of Plateau anthropological research). By the time of the initial Marmes Rockshelter investigations in 1962, Daugherty (1962) had presented his theoretical framework for the Northwest Riverine Areal Tradition that built on his own fieldwork in the Columbia Plateau and that responded to the work of others throughout the Pacific Northwest (e.g., Borden 1952, 1954; Butler 1958, 1961; Cressman 1960; Osborne 1957) (see Chapter Four). In turn, Leonhardy and Rice (1970) developed a culture typology for the Lower Snake River drainage that built on Daugherty's periodization. Their creation has survived with minimal alteration and continues to be the dominant typology referenced in work from throughout the southern Columbia Plateau.

Marmes Rockshelter can uniquely gauge the adequacy of Leonhardy and Rice's typology because the rockshelter deposits represent a long sequence of cultural use dating back to the transition between PaleoIndian and Archaic periods. While other sites in the Columbia Plateau have been found that represent cultural use prior to the earliest published dates for the Marmes site (e.g., Richey-Roberts Clovis site), none of those sites possess the long record of cultural use that Marmes does. Chronology also must be an important focus of this study's site interpretation because of the culture-historical focus of the original excavations, which emphasized vertical relationships rather than horizontal ones.

Applicable questions in the chronology research theme include:

- What is the archaeologically demonstrable first use of the site?
- Does the initial assemblage correlate with other early site assemblages in the region?
- What is the temporal relationship between the initial cultural materials in the rockshelter and on the floodplain?
- What is the continuing relationship between the two areas of the site through time?
- Was the site utilized continuously throughout the whole of its dated record? If so, was it utilized by the same cultural group? If not, what explanations for a break in site use are suggested by the site data?

Cultural continuity in the Plateau can be a contentious topic. Different kinds of information have been used to make arguments on both sides of the debate; a debate that has engaged archaeologists since the region's initial period of research (Lohse and Sprague 1998). Ames (2000) notes that inferences of archaeological continuity or discontinuity can be made based on analogies (similarities due to common function) or homologies (similarities due to common history/ancestry – a historical tie between cultural groups), but the archaeological record generally provides poor evidence when the factors of change occur over just a few centuries. Because of its long sequence, the Marmes site deposits have a contribution to make to questions of archaeological continuity or discontinuity in the region. In particular, it may clarify Leonhardy and Rice's (1970) perception that the Tucannon Phase did not evolve out of the preceding Cascade Phase.

Fryxell and Daugherty obtained a good sequence of radiocarbon dates from the site that were solidified by subsequent dating in the 1980s (Sheppard et al. 1987) (Table 2.1). Together with other chronological indicators, especially a layer of primary Mazama ash in the middle of the stratigraphic sequence, the site's chronological boundaries are fairly well established. The question persists, however, whether the site's initial use has been firmly established. This is largely due to the presence of a small number of cultural items in undated stratigraphic layers below those demonstrated to represent the earliest (dated) period of site use, both in the floodplain and in the rockshelter. In 1968, a few units in the rockshelter and the floodplain were excavated much deeper than the rest, ostensibly to address geological more than archaeological curiosity. The cultural items recovered were catalogued appropriately, but received little additional attention. In the rockshelter, it was assumed that these items had originated in stratigraphic Unit I, but had sifted down through the coarse rooffall rock and settled on top of the firm layer of glacial lake silts that form the lowest sediment (Carl Gustafson, personal communication 1998). No similar explanation has been found for the items found deep in the floodplain. This study has attempted to determine the age of these remains or, where no dateable materials could be located in the collection, the sediments from which the items were collected.

Related to this, Gustafson (personal communication 1998) recalled Fryxell being intrigued by a volcanic ash he located deep in nearby off-site soil profiles that might be present in the floodplain stratigraphic layers. Fryxell believed the ash was older than the Glacier Peak ash (11,200 B.P.; Foit et al. 1992) that underlay the oldest cultural sediments in the site. Sheppard et al. (1987) considered identification of this tephra to be unresolved at the time of their article, speculating that the off-site ash could be St. Helens S series (ca. 13,000 B.P.; Mullineaux 1996), which, if found in the floodplain cultural deposits, would set the deeper floodplain stratigraphy back another ca. 1,800 years. However, letters in the site records at the WSU archives indicate that Fryxell processed a number of ash samples from the site that were identified as Glacier Peak, but he did not seem content with the identification, and submitted follow-up samples several times. Unfortunately, few of Fryxell's field notes for his Marmes site work were located and they make no reference to this topic. In particular, it is not clear whether any of the ash samples he submitted that were identified as Glacier Peak came from the site's deposits or were only from off-site profiles. Detailed study of the site's unit grid and areas where Fryxell seemed to concentrate deep stratigraphic studies has offered hints of from where potential ash samples may have been collected. While no tephra samples from these locations were found in the collection, several soil samples from these locations were submitted to WSU's Geochronology Laboratory to determine if they contained volcanic ash and its identity. None of the submitted samples were found to contain volcanic ash.

Radiocarbon Dating. A total of seven new dates was obtained in the course of this study to supplement those previously obtained (see Table 2.1). Dating for this study sought to address several unresolved chronology issues at the site:

1) the desire for a clarifying date for Stratum V,
2) whether the human remains found on the surface of the floodplain Marmes Horizon derived from the cremation feature in the rockshelter,
3) when did use of the site by people begin?

No appropriate (non-burial related), dateable sample of material was located in the site collection from which to obtain a date of Stratum V deposits. Samples were found and processed that addressed the other two issues.

More refined dating than had been conducted previously of the locations of the Marmes Horizon human remains and the rockshelter cremation feature was necessary to determine whether there was a direct association between them. Two samples of wood charcoal picked from feature sediments from two different areas of the cremation hearth and two faunal bone samples from the Marmes A1 horizon were processed using the AMS method of dating. Correlation of the calibrated dates indicates that the floodplain human remains did not derive from the cremation feature. The results are discussed in detail in Chapter Seven.

Three samples were submitted to investigate the earliest use of the site, one from the deepest

Table 2.1 Marmes Site Radiocarbon Dating Results (through Sheppard et al. 1987[a])

Sample #	Sample Material[b]	Age Years B.P.	Depth (feet)	North	West	Stratum
Rockshelter Dates						
WSU-362	charcoal	Modern	Surface			
WSU-206	charcoal	1,100 +/- 300	97.7	82.5	44	VII
WSU-212	charcoal	1,300 +/- 300	97.7	82.5	44	VII
WSU-3034	charcoal	660 +/- 75	96.5	97.5	12.5	VII[c]
WSU-3033	shell	1,600 +/- 100	96.5	97.5	12.5	VII[c]
WSU-205	charcoal	1,300 +/- 300	95.9	92.5	25.8	VI
WSU-3035	bone	1,940 +/- 70	94.3	102.5	42.5	VI[c]
WSU-207	shell	4,250 +/- 300	94	82.5	44	V[d]
----------Mazama Tephra 6700 B.P. (Stratum IV)----------						
I-638	shell	6,200 +/- 475	92.7	97.5	27.5	III[e]
WSU-3036	shell	7,070 +/- 110	93.9	83.5	25	III[c,e]
WSU-209	shell	7,400 +/- 300	92.4	85	45.5	III
WSU-3037	shell	7,805 +/- 130	92.1	84	25	III[c]
WSU-210	shell	7,870 +/- 300	89.7	87.5	40	III
WSU-3038	shell	8,525 +/- 100	89.8	82.5	27.5	II[c]
W-2208	charcoal	8,700 +/- 300	85.6	75	50	II / I
W-2207	shell	9,010 +/- 300	85.6	75	50	II / I
Y-2482	charcoal	9,200 +/- 110	?	78	20	II / I
W-2210	shell	9,540 +/- 300	?	78	20	II / I
WSU-120	shell	7,550 +/- 300	89	90	25	I[f]
WSU-363	shell	10,810 +/- 300	89	74.5	20	I
WSU-366	shell	10,475 +/- 300	89	74.5	20	I
WSU-211	shell	10,750 +/- 300	88.3	87.5	24	I
----------Unidentified Tephra Layer----------						
Floodplain Dates						
W-2213	shell	7,980 +/- 300	69	0	71	=III[e]
W-2209	shell	9,820 +/- 300	?	-20	24.7	=I[g]
Y-2481	shell	9,970 +/- 110	?	-20	24.7	=I[g]
W-2212	charcoal	9,840 +/- 300	?	-10	-30	=I[h]
W-2218	charcoal	10,130 +/- 300	?	-10	-30	=I[h]

[a] C^{13} isotopic ratios were not determined for the radiocarbon dates obtained in the 1960s or by Sheppard et al. (1987).

[b] All shell dates are on western-river pearl mussel (*Margaritifera falcata*) from which the outer 20% was removed with hydrochloric acid before generation of the counting gas. Bone collagen, obtained by a modified Longin procedure (1971), was dated.

[c] Dates obtained by Sheppard et al. (1987).

[d] Stratigraphically anomalous, originally assigned to Stratum VII, its age and depth argue for its placement in Stratum V (Strata V through VII deposits were disturbed by numerous storage pits).

[e] Immediately beneath Mazama tephra.

[f] Stratigraphically anomalous, probably from Stratum III.

[g] Marmes horizon.

[h] Harrison horizon.

recorded levels in the floodplain, and two from the deepest potentially-cultural levels excavated in the rockshelter. The floodplain sample produced a date of 10,570 ± 70 B.P. (Beta-156697; wood charcoal; $\delta^{13}C/^{12}C$ = -24.9%). A date of 11,230 ± 50 B.P. (Beta-156698; faunal bone; $\delta^{13}C/^{12}C$ = -17.9%) was obtained from a swan bone recovered from deep in the rockshelter atop glacial lake sediments impregnated with volcanic ash that Fryxell believed to be from the Glacier Peak eruption ca. 11,250 B.P. (Johnson et al. 1994). A confirming date was attempted, but another dateable sample from directly atop the glacial lake sediments was not available. Three samples, one shell and two bone, were submitted from nearby proveniences, but only one returned an applicable date. The shell sample, recovered from within the sediments of the same unit and level as the swan bone, returned a conventional radiocarbon date of 9610 ± 40 B.P. (Beta-168491; shell; $\delta^{13}C/^{12}C$ = -10.5%). This date is significantly later than the date on the swan bone, and this difference is maintained when calibrated to 2 sigma (9210-8800 BC), despite being from close to the same location. The differing result may be due to the shell sifting down from above, perhaps during the excavations, the possibility of which had been a concern of Fryxell and Daugherty's (Carl Gustafson, personal communication 1998). This shell date is out of sequence with several previously obtained, older dates from material recovered well above the location where the shell was recovered. Interpretation of the new dates obtained by this study is presented in Chapter Fifteen in the context of the relevant research questions.

Past Environment and Climate Change

Most interpretations of archaeological data include some reference to climate and vegetation as causative factors for changes in assemblages over time. Given the long cultural record available at Marmes Rockshelter, questions about past environment are particularly relevant.

- What is the sequence of climatic change in the Columbia Basin over the course of aboriginal use of Marmes Rockshelter?
- What effects did the Mazama ashfall have on vegetation and soil development in the Marmes site area?

- What is the geological and sedimentary history of the Marmes site area and how might that have dictated/impacted its cultural use?

Site Function

A basic question:

- What is the nature of the Marmes site's cultural materials, how does it change through time, and what does it indicate about how the site was used?

Basic, but it invokes one of the principal debates about the Plateau cultural record, that of the transition of the subsistence and settlement pattern from one of mobile foraging to a semi-sedentary lifeway and the subsistence changes that caused the change or accompanied it. The places where people lived have the greatest potential for yielding reliable cultural and environmental evidence. At the most basic explanatory level, prehistoric people made decisions about how best to position themselves to obtain critical resources such as food, water, and technological raw materials. Archaeological sites are a tangible outcome of those decisions. Since Archaeology became a field of study, numerous constructs have been devised for explaining these material outcomes of people's interaction with the environment. Citing Binford's (1980) hunter-gatherer subsistence-settlement strategies intersite variability model, foragers use two kinds of sites: residential bases and locations. Simply put, locations are the source locations of resources; extractive and some processing activities occur at these sites. Residential bases are the "hub of subsistence activities…where most processing, manufacturing, and maintenance activities…" (Binford 1980:9) and food consumption takes place. Because the timing of their visit to the site is expected to reflect to some degree the status of target resources in the vicinity (e.g., roots in season), the breadth of resources found in foragers' residential bases should be relatively narrow. The range of resources addressed by collectors would vary depending on whether the site functioned as a location, a field camp (where people stay while resources from a location are processed in bulk for storage and/or transportation), or a residential base (a long-term domestic site where consumption of stored resources occurs along with domestic and economic manufacturing and maintenance activities).

The forager-collector strategies occur along a continuum, with some aspects of a foraging strategy employable within a collector strategy. Likewise, within the collector strategy, each type of site noted above can reflect evidence of activities associated with the other types. Plateau researchers have hypothesized different time periods for the shift from mobile foraging to semi-sedentary foraging and/or logistical collecting land-use strategies, as well as different causative factors (e.g., environmental change and/or change in subsistence resources, social factors, population replacement, technological advances, etc.), and differing degrees of success at achieving the transition, citing different data sets (e.g., Ames 1985; Ames and Marshall 1980; Campbell 1985; Chatters 1989, 1995; Lohse and Sammons-Lohse 1986; Reid 1991; Schalk 1983).

Initially, the emergence of the semi-sedentary settlement pattern in combination with resource intensification and storage as an adaptive strategy was viewed as a pattern that, having developed, remained the strategy until disruption by historic era forces. In addition, the presence of pithouses was viewed as the calling card for this new strategy throughout the region (Ames et al. 1998). However, Chatters (1989) has shown that a more sedentary pattern was adopted twice by Columbia Plateau peoples, ca. 4500 B.P. and again after 3500 B.P. Ames et al. (1998) refer to the initial pithouses as representative of a "preliminary shift to semisedentism" and note that there are few dated dwellings in the region from 4,000 to 3,800 B.P.

For his sample, Chatters (1989) states that different adaptive strategies based on resource intensification were used during the two periods of semi-sedentism. Pithouse I utilized a semi-sedentary foraging adaptation that included housepits while Pithouse II represents the familiar semi-sedentary collector pattern that persisted to contact. As such, it may be that the Plateau peoples' subsistence strategy varied considerably within the data parameters that we recognize archaeologically as semi-sedentary settlement coupled with resources intensification and storage, revealing it as a truly adaptive strategy that compensated for localized environmental and resource fluctuations on an ongoing basis.

Any one archaeological site can offer only limited conclusions about regional settlement and subsistence strategies. A single assemblage can uphold hypotheses about what foragers' residential bases or collectors' residential bases and field camps should contain (after Binford's [1980] hunter-gatherer subsistence-settlement strategies intersite variability model). The Marmes site's long record has the potential to exhibit multiple assemblages that represent both forager and collector use of the site through time. Marmes assemblages also may reflect what the transitional period(s) between foraging and collecting entailed. If the transition period involved both strategies, only a regional perspective will reveal this.

Because the Marmes site saw some level of consistent use throughout the time period hypothesized to have represented a foraging lifeway on the Plateau, it may represent a "tethered" site (Binford 1980:9) in the seasonal round; a residential base that was returned to regularly. If "regular occupation" entailed staying just once a year, then the visit was probably tied to a particular resource or set of resources that was at its prime at that time. If so, the resulting assemblage should be relatively limited and point to certain resources and activities related to those resources (in addition to some generalized activities expected at every site people occupied). If "regular occupation" entailed visiting the site several times a year, then the resources addressed probably were more numerous and the activities represented by the assemblage less tightly focused. In this case, the assemblage may appear more like that expected for a field camp within a collector strategy. Differentiation of strategies would then rest on whether resources were brought to the site for processing and consumption (as expected of forager residential bases), or resources were processed elsewhere (i.e. at a field camp) and activities related to consumption of food resources predominate (as expected of collector residential bases). This is the general model examined for this study, with the period when the site was a collector field camp expected to be brief (or ephemeral) due to the proximity of recorded habitation sites nearby. Depending on the intensity of activities at forager residential bases and collector field camps, it may not be possible to differentiate between them; particularly where assemblages represent palimpsests of multiple periods of occupation, such as at Marmes Rockshelter.

There are problems with the Marmes site collection that may have impacted the ability of the different analyses to collect the information necessary to pursue description of the model described above. In particular, materials were found to be missing from the collection or otherwise lacking probably due to differences in field collection methods. The degree to which these problems may have impacted site interpretation are discussed in the *Background Research and Project Constraints* and *Sampling* sections below). As a result, the best source of data relevant to these questions may be in the site's features.

- What does the site's feature data offer toward discerning changes in predation focus or the intensity of use of different resources through time? How might this reflect changes in subsistence and settlement patterns?

It has been suggested (Hicks and Morgenstein 1994) that such strategic lifeway compensations, and the fluctuations in resources that may have spawned them, would not be detectable by the culture-historically focused excavation strategies applied during the early period of archaeological investigations on the Plateau (including those applied at Marmes Rockshelter). These investigations emphasized vertical excavation and the collection of dateable sediments and diagnostic artifacts resulting in a narrow picture of the activities that contributed to the lifeway at any given time and limiting the detection of subtle, short-term changes in that lifeway. More recent studies have sought the collection of spatially oriented data that are more responsive to research themes addressing site function and subsistence and settlement strategies. However, these more recent studies typically have been limited in scope due to funding limitations or the specific requirements of project-driven studies aimed at regulatory compliance rather than research.

Chatters (1989:3) suggests that description of the ethnographic subsistence and settlement pattern have had a great influence on archaeologists' interpretations of the adaptive strategy hypothesized for late prehistoric period Plateau sites. This influence may have produced inappropriate conclusions regarding the earlier sites; apparent similarities in sites from both of these time periods should not be considered evidence that the sites were equivalents in the same adaptive strategy. Hicks and Morgenstein (1994) note that a number of major events occurred during the ethnographic and protohistoric periods that certainly altered the lifeway of Plateau peoples, in particular depopulation due to epidemic diseases and the introduction of the horse. Given when the information was collected from living native peoples, published ethnographies must reflect the effects of these events on subsistence round activities, activities that must have changed from the pre-ethnographic adaptive strategy to some extent. The resulting Plateau ethnographic pattern may have been less complex in terms of resource acquisition than that which preceded it, requiring less intensive collection techniques or fewer resource locations, among other things.

Perusal of a number of ethnographic accounts from the southern Plateau suggests that the pattern of seasonally divergent occupation sites (logistical, collector strategy [Binford 1980]) may be oversimplified for the region's prehistoric inhabitants in general and the Palouse groups in particular.

> All accounts agree that Plateau groups spent the winter months in large residential sites, typically at the confluences of the Columbia and Snake Rivers and their major tributaries. In early spring most of the people moved to upland root collection camps, then returned to the rivers in late spring for the salmon runs. Also at this time and into summer, the berries ripened, followed by other root crops. In the fall, time was split between root and hunting grounds and the smaller of the annual salmon runs. Most of these accounts describe wholesale abandonment of the winter villages from spring to fall, as the occupants moved to seasonal camps at these other resources areas (e.g., Harbinger 1964; Schwede 1970; Spinden 1908; Trafzer and Scheuerman 1986). [Hicks and Morgenstein 1994:9]

This brief description illustrates the potential logistical difficulties that such a seasonal round would present in the spring and fall (and perhaps in the summer months as well). Multiple task groups would have to be in the different resource areas at least part of the time. This would have presented even greater difficulties for the larger populations that existed prior to the introduction of the horse and western diseases. Bartholomew (1982:69) estimates the Palouse Indians population at roughly 5,400 individuals around A.D. 1780, and Swanton (1952) estimates their number at approximately 1,600 in 1805 declining to about 500 by 1854. No estimates have been advanced prior to 1780, but extensive population loss in a series of recurring epidemics from ca. 1770 on the Plateau has been documented (Boyd 1985).

A large population occupying a given geographic area, however organized, would have needed to collect and store foodstuffs on an intensive basis to avoid widespread deprivation (and sometimes starvation) in the winter months along with feeding themselves throughout the year. Whether Plateau peoples could have accomplished this within the ethnographically described subsistence and settlement pattern is uncertain. Without horses (obtained ca. 1730 [Haines 1938]), trips to higher elevation root camps and hunting grounds, such as those in the Blue Mountains and in Idaho, which were both resource areas used by Snake River groups prior to contact, would have been long journeys. More importantly, they would not have had the horses to transport the collected resources back to the winter village location for storage. While resource areas closer to the winter food storage locations likely were used prior to the acquisition of the horse, the intensive use of a small number of widely dispersed root locations (the ethnographic pattern) may not have occurred until such time as smaller populations could be sustained by intensively collecting from only a few of the most prolific fields and horses were available to haul the resources.

At a time of higher population density, it may be that many more task groups spread out of the main village, collecting resources to be returned to the village, while others stayed in these settlements year-round, not only for the purpose of exploiting *all* of the salmon runs (it makes sense that a predictable resource that could be collected near the location of the need, the winter village, would be gathered and stored whenever it was available) but to fish and hunt species available locally as well. People remaining in the village also could collect and process the many plants used for basketry and other textiles that a larger population would require. In this regard, Plateau culture would share a trait with nearly every other culture around the world where collection of storable surpluses of food led to the development of task specialists within the population. Further, the ethnographic record notes that a few of the older people would stay in the villages year-round. It is certainly true that without the horse the elderly would have been less able to make the long trips to the dispersed resource areas and could have stayed in the villages instead. Thus, the pattern of the bulk of the populations moving from one task area to another may certainly be an ethnographic period pattern that was different from the initial semi-sedentary pattern(s) that preceded it (Hicks and Morgenstein 1994).

While the vertical emphasis of the Marmes site excavations hinders analysis of data types sensitive to the discernment of hunter-gatherer subsistence and land use patterns, analysis is more seriously impacted by the poor condition of the site's records and the relative lack of some of the material types that would contribute most to these questions. Still, this study will attempt to address the following questions specific to subsistence, but that offer information to the questions regarding the transition from mobile foraging to a more sedentary lifestyle:

- What indications of seasonality of use are present in the data?
- What is the relative dietary importance of different terrestrial faunal species through time?
- What is the contribution of fish to the diet and how does species utilization change through time, in particular, the changing role of salmon and the cause?
- What is the dietary contribution of plant foods, shellfish, etc. at different time periods?

Trade

Several related questions address trade:

- What materials/items were obtained outside the Lower Snake River vicinity?
- Where were these items/materials obtained from and what does this indicate about social contacts/networks through time?
- Can the site's data inform on whether the trade was motivated by technological or subsistence needs, or are other cultural factors indicated?

Human Remains/Burials

The human remains from the Marmes site are the best published portion of the collection. However, the published record of the remains and the accounts of the analyses that have been conducted is limited.

- Do the human remains found in the rockshelter and on the floodplain represent intentional burials?
- In what period(s) were human remains buried in the rockshelter? Does the mortuary pattern(s) change through time?
- What do the previous osteological analyses of the remains indicate about population group, pathologies, etc. through time?

Finally, a cautionary statement: this study attempts to address questions and themes with regional implications. But when addressed from the perspective of a single site, even one with a long, rich record such as Marmes Rockshelter, it is still only a single perspective. Correlation of the Marmes Rockshelter results with other site information from the Palouse Canyon and Snake River areas can assist in establishing the veracity of observations made from a single site perspective, but this is still a "big-river" perspective and does not incorporate and benefit from the important data being collected in upland settings. Much more focused study using the data from this site and others is needed to adequately address the regional application of the broader questions presented above.

The Study Plan

The current study, by the Confederated Tribes of the Colville Reservation's (Colville Tribe) History/Archaeology Department under contract with the Corps of Engineers Walla Walla District (Corps), seeks to produce a comprehensive report of the findings of the investigations at the Marmes Rockshelter site. An equally important goal of this study is to make the data from the site available for research. Refinement of our understanding of the prehistoric Plateau lifeway must come from the sum of the similarities and differences among Plateau archaeological sites. The data from the Marmes site can offer much to such research efforts. The Corps recognized the importance of the Marmes site's information and the need to provide appropriate reporting of the results of the work conducted at the site to legitimate researchers and other interested parties. The preliminary reports by WSU (Fryxell and Keel 1969; Rice 1969) do not meet the level of reporting expected for salvage or data recovery projects and, as such, have limited the use of the site's data in other studies.

Two preliminary reports were prepared by the Colville Tribe in the course of this study (Hicks 1998, 2000), each reporting on the first two phases of materials analysis and other research into the collection. Phase One focused on collecting, organizing, and reviewing all records of the field investigations, and conducting geoarchaeological and faunal analysis. Phase Two addressed the bulk of the materials analyses. The work was conducted on this schedule because it was expected that the first year's research would reveal if the data generated from Marmes site investigations would address the initial study plan needs and would guide the sampling strategy for the material analyses. Some aspects of the study plan changed as the project moved forward and background research clarified the limitations presented by data gaps in the collection and records. Phase Three involved conducting follow-up lab analyses and preparing the final report.

The scope of work for the project specifically excluded analysis of human remains. During the study, handling or otherwise disturbing such remains was minimized. In addition, efforts were made to identify and exclude from study items that may have been associated with Marmes site burials. However, a number of tools that probably were associated with burials were inadvertently included in the bone and antler tool sample and a draft report of the analysis performed was received before the error was noted. The results of the analysis of these items are included in this report because the handling of the items had already occurred and omitting reference to specific items in the analyst's report seemed a pointless solution; these items have been excluded from photographs. A historic artifact analysis included in the original study plan proposal was not conducted because the records research and observation of a number of these artifacts indicated that most were probably associated with protohistoric or early historic-era burial features. Other more recent items (e.g., cloth fibers, match sticks, cigarette butts, etc.) are probably from the excavators or other visitors to the site. This report includes a summary of all previous studies of the Marmes site human remains and burial features. While a portion of the collection is certainly subject to the requirements of the Native American Graves Protection and Repatriation Act (NAGPRA), those materials must be addressed separately through the appropriate NAGPRA process.

Phase One activities included research into the site collection and records, compilation and development of digital versions of the site's stratigraphic profiles, site formation/ sedimentological analysis, faunal analysis (fish bone was analyzed separately in Phase Two), and bone artifact analysis (Hicks 1998). Two samples also were submitted for radiocarbon dating analysis. For the faunal analysis, Gustafson was asked to revisit the data he compiled for his dissertation project (1972) for the rockshelter fauna and his conclusions in light of his nearly three decades of subsequent research on Plateau fauna. In addition, Gustafson directed analysis of a sample of the floodplain faunal (non-fish) material, contributed to the analysis of bone implements from the site, and oversaw the compilation of the site profile drawings. The latter were essential for the sedimentological analysis and all subsequent site research and material analyses. Huckleberry conducted the sedimentological analysis including soil chemistry studies and dissection of mounted soil columns collected from the site, as well as description of all of Fryxell's analyses related to the geology of the Marmes site.

Phase Two activities included all other material analyses, particularly analysis of the stone artifacts from the site, both the formed tools and debitage. The chipped and ground stone formed tools had been inventoried by artifact type for the 1969 preliminary report (Rice 1969), but no additional interpretive work was done for these artifacts. The debitage from the site, over 9,000 items, had not been examined for the 1969 report and needed a complete analysis. A number of these artifacts were selected for immunological residue analysis and obsidian sourcing; hydration studies were not conducted due to a predicted lack of enough items from a single source (see Hess 1997). Immunological analysis can provide information regarding the use of faunal and some plant species through time. Immunological analysis of selected soil samples was planned, but adequate provenience information was not available for both target soil samples and non-target control samples.

Because of the very large number of shell items only a sample of the shell materials was analyzed (the sampling plan is discussed below). Formed shell artifacts and shell from features was emphasized in the sample provided to the analyst. It was anticipated that the shellfish analysis would buttress other sources of information on past environment and land-use. Palynological analysis was attempted but adequate pollen counts were not available despite the availability of an extensive series of sediment samples from the central control block in the rockshelter that provides a near continuous vertical sequence to well below the lowest cultural stratum.

Analysis of botanical items was conducted to obtain additional environmental information and examine organic materials technology and specific feature functions. This analysis also included flotation analysis of a small number of sediment bulk samples and feature fill samples. Following the sorting of selected bulk samples from the floodplain in Phase One, the fish bone collection was analyzed but the results were hampered by the small sample of fish bone available from the rockshelter (probably due to the use of larger screen sizes during the field investigations).

Anticipated Problems

A number of problems were anticipated when developing the initial study plan proposal. Some of these problems were encountered during a recent inventory of the Marmes site collection, which is housed in the collections storage facility at WSU, in partial fulfillment of the Native American Graves Protection and Repatriation Act (Collins and Andrefsky 1995). Many of these problems are attributable to the extent of time that has passed since the site was excavated, and the different, often less precise, data collection and recording methods commonly used at that time. In the case of the Marmes site excavations, the data is now 30 or more years old, two of the site's investigators have passed away, and most of the site supervisors and excavators who worked at the site are too far distanced - both in time and space - from their efforts to consult with reliably.

The site collection and associated excavation records have been moved a number of times over the years. Collins and Andrefsky (1995) found many items were missing (n=695) or had been separated from their provenience information rendering them useless for any chronological interpretation (n=1,527). It was determined prior to inception of the current study that some analyses conducted during the initial investigations should be repeated using contemporary analysis techniques and equipment, both to verify and perhaps refine the previous results (e.g., dating). Other previous analyses could benefit from repeated analysis where an additional 30 years of data is available (e.g., tephra analysis). In addition, analysis methods that were not available to the previous analysts also could be applied to the Marmes site materials in this study (e.g., protein-residue analysis, lithic material sourcing). But it was expected that the number of samples available for these tests would be limited and the lack of provenience would pare the lists down further, perhaps preventing some analyses from being conducted.

Background Research and Project Constraints

Project research included background studies to become familiar with the fieldwork and the research questions/directions that were being pursued by the previous investigators of the Marmes site. Background research also allowed early assessment of the adequacy to which the collection could address this study's proposed research questions. This involved research into the artifacts and other materials gathered during the excavations and the records produced by the project.

Efforts were made to find and review all reports, theses, and articles that had been produced from and about the 1960s Marmes Rockshelter excavations (see History of Investigations). Most of these were available at the Holland Library at WSU, in the WSU Museum of Anthropology archives in College Hall, or were made available by Dr. Carl Gustafson. In addition to professional or academic documents, there are a number of magazine articles, newspaper accounts, and even educational films that brought information about the project to the public. The site's records include the original field logs, field and lab notes, copies of project reports and articles (including drafts), print photographs and slides, a few feature forms, maps, stratigraphic profiles and other field drawings, data sheets and other analysis forms, computer punch cards (faunal analysis), cultural material lists, student papers, and many files of correspondence and administrative paperwork (e.g., supplies requisitions, student work applications, grant applications, etc.).

Dr. Richard Daugherty, the directing archaeologist of the project in the 1960s was interviewed, as was Dr. David Rice who had prepared the preliminary report at the end of the field effort (Rice 1969). Dr. Daugherty compiled his remaining files on the Marmes project for entry into the permanent collection at WSU. In 1998, Dr. Rice reported locating several boxes of material related to the Marmes project in his personal stores and was in the process of reviewing their contents. Dr. Gustafson also related a considerable amount of the project's history during many meetings throughout the Phase One work conducted at WSU (i.e. faunal analysis and compilation of the stratigraphic profiles). In addition, a visit was made to the American Heritage Center at the University of Wyoming where the Roald Fryxell Collection (ACC. #6108-85-07-25) is archived. After reviewing the American Heritage Center's box inventories, the contents of more than 60 of the 100+ boxes of material in the Fryxell Collection were examined and copies were made of materials in support of the sedimentological analysis and stratigraphic profile compilation tasks. The rest of the boxes contained materials unrelated to Marmes Rockshelter or the part of Fryxell's life at WSU.

Project records also were examined for information related to items that either are missing from the collection or for which their provenience information was never recorded or has been lost. The records were examined to determine the methods used in fieldwork and to establish the extent of post-field analysis conducted. Field notes and logs were examined for detail on features and other areas of the site's deposits of particular interest to the current study. Unfortunately, the 1962, 1963, and 1964 field catalogs are incomplete (although a student tried to recreate them in 1984) and only include formed tools and shell artifacts; useful detail that would allow matching items to their collected locations is generally lacking (e.g., provenience, artifact descriptions) (Collins and Andrefsky 1995:50). The 1964 catalog is marginally better at the beginning, with reference to stratigraphic units for the collected artifacts and a few words of detail for each item, but after a while also reverts to entries of only catalog number and artifact type. Level forms were not used from 1962-1964 at the site, although a form used by the supervisors in 1964 for their field notes may have been intended to be used as level forms. A version of this same form was used in 1968 as the level record form but generally includes only text statements that resemble field excavator's notes. There is considerable variability in the kinds of information and the amount of detail presented in these level forms/notes. The lack of uniformity of information limits fruitful comparisons between units and levels throughout the site. Feature forms were only used for burial features or suspected burial features. Only stratigraphic profiles were consistently prepared and by only a small pool of the more experienced students, a result of Fryxell's influence on the project.

The detail in the excavators' field notes varies considerably and rarely provides enough information to correlate with the catalogs. The 1962 and 1963 field notes consist of about a dozen pages for each field season with long gaps between the dates of the entries after about the first week of the season. The 1964 field notes are better, but only provide useful details on the burial features encountered. In 1968, only the supervisors kept notes consistently (although useful information is available in some of the level records kept by the excavators), and they were responsible for giving site tours to the constant stream of visitors (as many as 500 visitors on a weekend was not uncommon during the summer and fall). In summary, much that happened archaeologically is not reported and this hampered attempts to assign, define, or refine the status of spatially-related accumulations of cultural material in the collection. The lack of provenience information for many formed tools and other material in level bags had a profound effect on site interpretation for this study.

Finally, archaeological, ethnographic, and historic-era literature on the southern Plateau and lower Snake River area was reviewed and collected, mostly from the Holland Library, but also from the Suzallo Library at the University of Washington. This material contributed most to the interpretive discussion. Research materials relevant to more general topics, such as applicable theoretical and methodological treatises, as well as articles/manuscripts on types of analyses and data presentations also were collected; some of these were located in the Owen Library, also at WSU.

Research into the Marmes site collections continued throughout the four years of this study and involved efforts to review both the material collected and the records developed during the original project and by subsequent researchers, as well as locating and retrieving materials to be used in each year's analyses. In the first phase of the project the status of the existing field and laboratory records was assessed and the collections inventory database reported by Collins and Andrefsky (1995) was studied in detail. WSU researchers largely worked independently of the Colville Tribe's researchers. WSU personnel located and retrieved the stratigraphic profile drawings that had been drawn in the field and developed a list of the bulk soil samples collected from the floodplain for sorting. They also located the sediment monoliths collected by Fryxell, interpreted their designations, and determined from where in the site they had been collected by consulting the profile drawings. This separation of purpose created a problem later. Museum of Anthropology policy dictated that no researchers have direct access to original collection inventory databases, a prudent policy. Therefore, copies of the Marmes site database were provided to both Colville Tribe and WSU personnel. Colville Tribe personnel found it necessary to revise the database to improve the ability to search and sort records (see discussion below). However, WSU researchers changed the field types and sizes and added a field in the database copy they used. This complicated the later united efforts because the materials identified from the bulk samples sorted in 1998 were entered into the WSU database format and could not be imported directly into the Colville Tribe's revised database.

One of the greatest concerns for this study was the observation reported in Collins and Andrefsky (1995) that a large number of items/materials were missing (n=695) or without provenience (n=1,527). The latter number is taken from Collins and Andrefsky's (1995) Appendix G listing, but their Table 14 gives the number as 1,587 (see discussion below). A few of the missing items had been found in other collections' boxes since the 1995 inventory was completed. Many of the missing and non-provenienced items are human remains and apparent grave goods (e.g., beads). Although this is not a Native American Graves Protection and Repatriation Act (NAGPRA) study, making an effort to systematically locate these items was a priority of the Colville Tribe. Other (non-burial related) missing or non-provenience items (if found) are relevant to the interpretation of the site and were desired for inclusion in the materials analyses in Phase Two of the project. Of particular concern was that over half of all the formed stone tools were missing or were without provenience information. This had the potential to impact a number of the proposed research questions, so a considerable amount of time was spent attempting to locate the missing items and the provenience information.

The fact that items are missing may in part be attributed to the many moves the collection has been subject to since it was brought to WSU. These moves were necessitated by the changing storage needs of the Anthropology Department. A number of buildings around campus served as temporary storage facilities for archaeological collections until the current collections facility was put into service in the 1980s. In addition, parts of the collection have been used for teaching purposes and were kept in several different labs used by Anthropology Department faculty over the past 30 years. Also, portions of the collection were used for material studies (e.g., graduate theses and other student and researcher projects) and access to and the use of those materials were not governed by standardized loan procedures as are common for collections storage facilities today.

Research into the items in the collection that are listed as 'without provenience' was more complex. First, there was a discrepancy in the numbers of items without provenience in Collins and Andrefsky (1995). In trying to determine which was accurate, a number of things were discovered. When the database was sorted for all non-provenienced items, with "0" in the UNIT and LEVEL fields (and excluding Project Records items), 1,761 records resulted. A number of procedures identified additional records/items that should not be considered non-provenienced for our purposes and analyses, including: fully processed level bags that had no provenience at the time of the 1995 inventory (n=4) as well as the bags that resulted from sorting those non-provenienced level bags during the 1995 inventory (n≈37); tracking down the grid units in which a number of "test squares" designated as only "TSQs" in the database had been excavated (n=8); using Gustafson's faunal analysis catalog of 1962 materials allowed non-faunal bags to be matched to provenience (n=15); matching additional CATALOG numbers from 1962 with obvious similar associations (n=8); dropping all bags derived from "backdirt" as listed in the COMMENTS field (n=65); and, reinstating the provenience data for 44 records of materials recovered from the area of the Marmes I feature previously known only as having come from the bulldozer trench (Gustafson provided the provenience of these items). Finally, there are 83 records of materials without provenience that were received in the early 1990s from a researcher; it has not been verified that these are from the Marmes site. If these also are excluded, this brings the total count of non-provenienced bags or items to 1,497. These efforts also confirmed that no additional records from the 1960s investigations were available at WSU that could resolve the provenience of the remaining items. Unfortunately, nearly half of the formed stone tools recovered from the site remained without provenience.

Determining how so many items became separated from their provenience information also was complex, and for many of the items we are likely never to know. But in terms of the formed tools, after piecing together information gathered from discussions with Daugherty, Rice, and staff of the Museum of Anthropology and Center for Northwest Anthropology (CNA), the following appears to have occurred. Rice conducted his descriptive analysis in 1968-1969 when the Anthropology Department was located in Pine Manor at WSU. Sometime after Rice's analysis and before the Department moved out of Pine Manor, someone put many of the formed tools from the Marmes site onto trays that were placed in drawers in several wooden cabinets, presumably for teaching purposes. The position of each item on the trays was noted alphabetically and the drawers were numbered. When the Anthropology Department moved into College Hall, the cabinets were moved into a third floor room reserved as a Marmes laboratory. Later, that room became CNA's laboratory and, in need of storage space, Dr. William Andrefsky had the Marmes tools removed from the cabinets and stored in the collections facility with the balance of the Marmes collections. CNA staff carefully recorded all of the information present on the trays, including artifact descriptions, in hopes that a key would someday be found. But no key or listing has been found that correlates those letters and drawer numbers to bag/catalog numbers, units and levels, or any other location or provenience data. And even were a key to the cabinet drawer contents to be found, if it only keyed back to the field catalogs we would be no better off since the lack of detail in the field catalogs (especially the lack of artifact descriptions) would prevent matching up artifacts to catalog numbers. In 1984, Heidi Hill, under the direction of Daugherty, conducted an inventory while attempting to recreate the missing catalog information but was not able to locate any sources of provenience data beyond those records present today (personal communication with Guy Moura 1998).

The original bags that these items had been in also have not been found, although many empty bags were located in the collection during the 1995 inventory and bagged together. Even if these empty bags could be determined to have housed the non-provenienced items, the same problem would exist as with a key to the cabinet artifacts - the lack of information in the field notes and dearth of forms from the excavations means that the tools likely could not be correlated with catalog numbers by descriptions.

Rice included a descriptive analysis of the formed tools from the rockshelter in the preliminary report of the Marmes excavations (1969). He presented the data as tool classes and many of the tools are shown as outline drawings. But the artifacts are demarcated only by their tool class number and a consecutive alphabetical designation that ties the object back to the tool class description; catalog numbers of the items in the classes are not given. However, Rice must have had the formed tool provenience information for the analysis since he shows the distribution of the items in each class by the stratigraphic units defined by Fryxell and Daugherty (1962). In a few cases, the number of items within a tool class are low enough that all of the members are illustrated and, therefore, could be identified to Rice's tool classes by viewing all of the tools of that type in the collection that are without provenience data. However, because the catalog numbers are not used in Rice's reports (and provenience often is not included in the catalogs) the provenience information for those artifacts could only be determined if all items in the tool class were in the same stratigraphic unit, or less precisely, if only one item were from each of several stratigraphic units. In the latter case, this would only give the vertical provenience of the *class* of tool, not of individual items. Horizontal provenience would not be determined in either case unless both the catalog number could be determined and the catalogs included the provenience of that item, a rare case. There is no record in the collection at WSU that correlates Rice's analysis results with artifact catalog numbers or any other artifact designation or site record that would allow corroboration of artifact provenience.

It is easy to see from this discussion, that the greatest potential for resolving the provenience of these many items would be to find a key that correlates catalog numbers to drawer numbers and the alphabetical designators for the location of each item on the trays as recorded by Andrefsky. All records associated with the site have been searched for such a key and it may be that there was no key; that the letters were placed only to guide students in identifying the tools in a classroom exercise. The last place where provenience information may exist rests in Rice's analysis working notes and data tables, documents that existed when that analysis was conducted (see Rice 1969:10) but that are not in the Marmes site collections at WSU. It is unknown if records that may contribute to identifying the missing provenience are available in those materials.

In the case of the missing items, verification of their absence was the obvious first step. This would have required a considerable amount of searching through boxes of the Marmes site materials and other sites as well. Unfortunately, the WSU collections facility's procedures for gaining access to collection materials, while necessary for maintaining the security of the collections and the integrity of the facility's inventory database, largely prevented examination of specific materials in this manner. Those procedures dictate that collection materials cannot be viewed in the storage facility. Instead, a loan application must be submitted that lists each of the inventory (bag) numbers for the items desired. The items then must be retrieved from the facility by museum staff and delivered to the applicant. Further, the collection is not boxed by either material type or provenience, so items from a single excavation unit can be located in multiple (sealed) storage boxes. This made collection searches impractical for more than a small number of items.

One early complicating factor of the efforts to resolve the missing and non-provenienced items was that the collection inventory database coding for the UNIT field did not differentiate between grid numbers with one or two numbers. The entries lacked blank spaces or "0"'s before grid numbers which prevented them from sorting in proper numerical sequence. Instead, the grid numbers sorted by the first numbers encountered in the column. Thus the N1's were followed by the N10's, N100's, N15's, N2's, N20's, N25's, N3's, and so on. This was further compounded by the fact that this system extended to the entire entry. An example is N95-100/W5-10: in this instance "N95" sorted first on the "9" placing it after all the "N100" units, but also within the N95's, the "-100" would sort before anything except the 1's and 10's, and the same was true of all West codes.

In addition, a small number of entries used different symbols, such as: N10/W10; N10,W10; or N10-W10. While the entries for this example would have been grouped consecutively, it disrupted the level sequence. Also, with compound entries, the data was entered with the southern most grid reading preceding the northern most grid reading for a unit, and the eastern most grid reading preceding the western most grid reading. This created a difficulty because the unit datum and designation used throughout the excavations was the northwest corner. For example, an entry of N95-100/W5-10 is actually unit N100/W10. However, any "query" of the original database seeking units greater than or equal to the North 100 grid line would not include the entry used in this example.

For these reasons, the collections inventory database was re-coded for the use of Colville Tribe researchers. Re-entering data also was considered necessary for the development of the Phase Two material analyses samples, where knowing the number of samples available from a given provenience unit, and the number of samples available in associated provenience units was needed. This was particularly true since representative samples, rather than a 100 percent sample, was planned for most of the material analyses. The reordered database was provided to the WSU Museum of Anthropology and, together with additional revisions that resulted from the material analyses, is included in Appendix A. To assist the reader in understanding the sample designations used throughout this report, the following conventions, used in re-coding the database, are illustrated:

1) The northwest corner designates all regular 5-x-5-foot units or entries where provenience was given in 5-x-5-foot units. If only a single Northing and Westing was provided, it was retained. There were many entries that originally only had a single Northing and Westing, suggesting a 5-x-5-foot unit, but it is clear from information provided in the COMMENTS field that these units were not 5-x-5-foot squares.
2) The Northing always has three digits to the left of the decimal (i.e., 003., 030., or 300.), because unit grid numbers extend into the hundred's range. The Westing only has two digits to the left of the decimal because no grid number reached 100 east or west of the zero line. "0" was used instead of blank spaces to make it clear when a record had been re-coded. The same procedure was used in re-coding many of the level designations.
3) All compound entries have the northern most and western most points as the leading number.
4) Point provenience has been left as entered, but modified to adhere to number 2, above.
5) Any unit which does not conform to the standard 5-x-5-foot units contains all points originally entered, but has been modified to adhere to number 2, above.
6) All "E0" unit designations have been changed to "W0".
7) Several entries were discovered which had only Northings (e.g., N95-100/N10-15). The COMMENTS made it clear that these were entry or field errors with the Westing portion of the unit designation being incorrect. Checking unit maps proved that some should definitely be "W" and all were uniformly changed to "W". Using the example above, and following previous conventions listed, the new designation would be N100/W15. While there may still be errors associated with the arbitrary assigning of "W", at least half will be correct, whereas previously all were wrong.

Once the database had been modified, a considerable amount of time was spent conducting database operations preparing the Phase Two analyses sample lists.

Sampling

The sampling strategy for the materials analyses conducted in this study was driven by a desire to:

- Incorporate the results of previous reported analyses of Marmes site materials (i.e. Gustafson's (1972) dissertation project for the rockshelter fauna and Caulk's (1988) thesis project for a limited sample of the floodplain fauna), while avoiding redundant analysis of these materials where possible.
- Include as many items as possible within the restrictions imposed by the vertical (culture-historical) emphasis of the excavation strategy, the lack of records from the site excavations, and the available funds.
- Assess data patterns demonstrative of both culture-historical and functional research themes, also within the restrictions cited above.

For the current study, Gustafson revisited his rockshelter fauna data and conclusions and directed the analysis of a sample of the floodplain faunal (non-fish) material. Radiocarbon dating conducted in the 1960s indicated that both the Marmes and Harrison horizons in the floodplain dated to the approximate time period of stratigraphic Unit I in the rockshelter. As such, for all material analyses, the Marmes floodplain materials were considered a single component representative of the earliest period at the site.

The floodplain faunal analysis focused primarily on the water-screened bulk samples described earlier in this chapter. During the 1968 excavations on the floodplain, recognizable artifacts and larger cultural items such as pieces of bone were mapped *in situ*, and bagged by the excavators. In addition, the crewmembers overseeing the water screen apparently plucked and bagged occasional items before they were bagged with the screen residue. It is unclear if there was a consistent collection policy regarding the floodplain items that were bagged separately. It is apparent that much of the lithic debitage found in the floodplain units were bagged in this fashion, but other materials of equivalent size (e.g., salmonid vertebrae) are rare in the collection and few were found during this study's sorting of water-screened bulk samples. Unfortunately, many of the items bagged separately from the floodplain, including many fragments of human remains, have gone missing over the years and most were never analyzed. It is clear that some macro-sized faunal items visible in photographs taken during the excavations are not included in the analysis results. Some of these items may be in the collection but could be items that lack provenience designations. But whether missing or lacking provenience information, the opportunity to correlate these larger items with the small fraction materials sorted out of the screen residue samples has been lost. This had the potential to impact the results of several analyses (i.e. faunal, fish bones, lithics) with implications for several research questions.

Initially, we intended to analyze about 300 samples from excavation units that represented the greater aerial material concentrations from throughout the floodplain exposure. These four material concentrations are dubbed areas of 'specific cultural context' because they include most of the features found in the floodplain (see Figure 1.19):

1. The northeast quadrant of the excavation where Arctic fox remains and the owl's foot artifact were found and lithic flakes were scattered about among some large boulders;
2. The north-central area near where the original floodplain discovery of human and elk bone was made in 1965;
3. The central area where semiarticulated remains of a large elk were found; and,
4. The southwest quadrant exhibiting a concentration of smaller bone fragments.

Sampling the floodplain materials in this way offered the best opportunity for interpretation of the cultural use of the floodplain given the overall low density of material available from this part of the site.

A full accounting of how WSU created the final sample is presented in Chapter Ten; ultimately, the floodplain faunal sample deviated from that proposed by excluding much of the north-central concentration. The site inventory indicated that 164 of the 1,198 floodplain bulk samples lacked provenience information reducing the number available for analysis to 1,034, still a huge number given the time required to pick through the several liters of carbonate rootlet casts and small stone clasts in each sample. The final sample analyzed by WSU included 221 bulk samples sorted for this study and 18 samples previously sorted in the field.

While the lithics were not sampled, the shell, botanical, and fishbone materials from the rockshelter and floodplain were sampled differently. For the Phase Two floodplain analyses samples, proveniences were selected that would allow correlation of results with the Phase One faunal analysis.

The Phase Two rockshelter analyses samples targeted grid units from areas with limited disturbance and that had been excavated through all or the majority of the eight stratigraphic units identified by Fryxell. In this way, we attempted to overcome the effect on materials representation of the decision in 1968 (and variously in the previous years' excavations) to not excavate stratigraphic Units IV – VIII. This sampling strategy also had the effect of emphasizing vertical representation over horizontal representation, a factor that greatly limits some intra-site interpretation, but a factor we accepted for three reasons: 1) the great antiquity of the site means that it possesses a greater vertical span than is usually available for modern studies and it is what makes the Marmes site unique and worthy of this final reporting; 2) the culture-historical focus of the 1960s field investigations resulted in a collection that offers more to eliciting vertical rather than horizontal commonalities and contrasts. The collection lacks records that would allow detailed analysis of spatial data to elicit patterns of interest to contemporary function focused interpretation; and, 3) economics. The collection from the site is large and could not be fully analyzed with the funds available.

One of the principle goals that will be realized by this study is to make the collection's data available to researchers who may be able to more fully investigate specific research questions about the site and the region. Discussion of horizontal material representation is addressed in the form of features and materials patterning identified within the eight stratigraphic units. For all of the materials analyses samples, lots identified as disturbed (e.g., "from slump") were excluded and units of non-standard size (i.e. units dug as 'trenches' greater than 5-x-5 feet [1.5 m], partial units along the bedrock) were generally avoided.

Overall, this sampling approach meant that most of the same grid units were included in the three Phase Two materials analyses and all of the shell, botanical, and fishbone from the sample units was analyzed. Exceptions to this approach included:

Shellfish Analysis – a reduction in redundant bags from the same levels in the same units. An initial examination of the shell collection by Dr. Ford indicated that the collection was quite large (many levels produced multiple two gallon-size bags of shell), but there were few species represented. Therefore, for units with a single bag of shell, all of the shell was analyzed, while only one full bag of shell from units with multiple bags was analyzed. As a check, excluded bags were occasionally examined to verify that species representation was consistent. Because Stratum III represented the only shell midden identified in the field as part of a 'living floor', the initial selection of grid units emphasized those thicker Stratum III deposits.

Botanical Analysis – two grid units from the rockshelter were included in the botanical analysis sample that were not in the shell sample because "organics" lot bags existed in the collection indicating that they had been excavated. As noted above, the uppermost strata (with 'cache pits') were not always excavated and these grid units produced substantial bags of "organics", a relative rarity in the overall site collection. In addition, several bulk sediment samples from the deepest levels in the rockshelter were analyzed to gather botanical information that might contribute to discerning the natural environment and local vegetation during the earliest uses of the site. Also, several bulk collection bags marked "feature fill" from different stratigraphic units were analyzed in hopes of finding material relevant to interpretation of subsistence.

The lithics were not sampled; rather, the entire lithic collection was given to the analysts. Because the excavation strategy was culture-historical in focus, which favored the collection of lithics over many other material types, and because so many formed tools are missing or lack provenience, the CCT did not want to further complicate the representation of lithic functions at the site by sampling. It is acknowledged that since we do not know what kind of representation is missing in the form of the non-provenienced items, we cannot know how lithic representation is altered. The non-provenienced items were included in the analysis to make the data available to future researchers, but the analytical discussions in the lithics chapter of this report emphasize materials securely identified with the major stratigraphic units defined for the rockshelter and the floodplain.

Ames (2002) noted in a review of the draft report for this study the potential impact on site interpretation of the missing and non-provenienced tools, and of missing materials representing the resources used at the site. He cites in particular the relationship between taxonomic richness and sample size. Ames conducted several regression analyses to analyze the potential impact of the missing formed lithic tools and found that some of the variation of several tool types among the site strata could be explained by variation in the total number of artifacts in each strata (2002:5). This can have a particular effect on the apparent representation of rare artifacts that may apply to the faunal analyses and lithic material (e.g., obsidian) representation as well; where assemblages are small, rare types are less likely to be included in the analysis sample and vice versa. Since assemblage size is often the result of the volume excavated, Ames conducted a regression analysis to test (roughly) whether large formed tool assemblages correlated with thick strata. The result was negative; thickness of strata does not explain the variation in assemblage size (2002:5). This study did not employ any post-analysis measurement or controls of this potential interpretive distortion. The reader is cautioned to pay careful note of the differing assemblage sizes by strata cited by the analysts.

While developing the samples for the different material analyses to be conducted in Phase Two, the author made additional attempts (see above) to establish provenience for items of these material types that lacked provenience data. The material analysts made similar attempts to establish provenience for the specific materials they analyzed. Some of the consultants' efforts were redundant to those attempted by the Colville Tribe for the whole collection. But the additional efforts were considered worthy, and all shared concern that the many non-provenienced items could impact the functional and chronological interpretation of the site.

Specific efforts at determining provenience for items lacking that information during Phase Two included matching grid unit and depth information for artifacts with identified strata to those with the same grid unit and depth but that lacked the strata designations. This was generally unsuccessful because some of the stratum-provenienced artifacts that had the same grid unit and depth had different designated strata as observed in the field. A second method involved using the final stratigraphic profile drawings developed in Phase One and attempting to correlate Fryxell's strata with excavation unit and level depth information for items in the database that had at least excavation level provenience. Some items could be confidently assigned using this method, but only when stratigraphic unit boundaries were horizontally matched across specific grid units from where the item was recovered and the entire level fell within a single stratigraphic unit. Other items could be assigned provenience within a stratigraphic unit with less confidence; usually this was where the upper and lower extent of a stratigraphic unit varied somewhat calling into question whether an entire level fell within a single stratigraphic unit. For these items, provenience assignments were made conservatively and because it was not known where within the grid unit the artifacts were recovered they often were assigned to multiple stratigraphic units (e.g., IV/V).

Items from the upper stratigraphic units in the rockshelter failed to be confidently assigned more often than items from the deeper stratigraphic units. The upper strata boundaries are irregular and discontinuous due to the presence of intrusive cultural pit features in these levels preventing tracking of the stratum depths across the 5-foot square grid units. Many of the items collected from upper strata were considered to have come from 'disturbed' sediments, a conclusion based on the presence of the many intrusive pit features (Fryxell and Daugherty 1962). It was, in part, based on this conclusion that the upper stratigraphic units were bulldozed off to increase access to the older cultural materials for the 1968 investigations. Materials from levels or areas within levels described as "disturbed" or in "slump" were not included in the current analyses. Although provenience cannot be assigned for a greater proportion of the upper strata items in comparison to the lower strata items, and disparity is likely to have some effect on site interpretation, this follows from the overall excavation strategy implemented at the Marmes site. As noted above, that strategy reflected the cultural-historical focus of Archaeology in the early 1960s. The kinds of artifacts associated with storage (i.e. botanical materials), rather than occupation (e.g., lithics and faunal materials), did not garner equal attention from researchers and these items were often not collected. Indeed, despite descriptions of considerable amounts of 'grasses' and 'reeds' in the upper levels by people involved in the early excavation work, there is little of such materials in the Marmes collection.

For this study, many unprovenienced items did not meet the conservative level of confidence for provenience assignment and are noted as such (e.g., "unassigned") or were not included in the sample to be analyzed. This approach was not applied to the lithic analysis where all lithic items were analyzed. All items assigned provenience through records research are considered provisional and have been noted as such (e.g., 'probably Unit #'). The intent of making these provenience assignments was to increase the number of items included in the material analyses and assist interpretation of the site. The author is confident that these provenience assignments are as accurate as can be determined without field designations, and that any discrepancies will have little effect on the overall interpretation of the site's use through time. Still, any discrepancies are the sole responsibility of the author.

3
ENVIRONMENTAL OVERVIEW

Peter E. Wigand with Brent A. Hicks

The geological investigations conducted by Fryxell during the Marmes site excavations, and subsequent research (e.g., Gustafson 1972 [see Chapter Ten]; Hammatt 1977; Marshall 1971) that references data from the Marmes site, contributed much to the current understanding of the environmental history of the Columbia Basin. Fryxell was particularly interested in landscape evolution and paleoclimate in the Columbia Basin. Based on evidence from Marmes Rockshelter and other rockshelters in the Lower Snake River area Fryxell and Daugherty (1962) concluded that the period before ca. 8000 B.P. was characterized by cooler and moister climatic conditions, indicated by higher frequencies of roof-fall rock as a result of increased freeze-thaw cycles. After 8000 B.P., aeolian deposits were interpreted to indicate more arid conditions until ca. 4000 B.P. (Fryxell and Daugherty 1962). Increased roof-fall again from ca. 4000 to 2000 B.P. argues for cooler climates (Fryxell and Daugherty 1962).

This chronology agreed in broad terms with Antevs' (1948) climatic model developed for the Great Basin. But after examining the alluvial deposits in the Palouse River canyon, Marshall (1971) proposed a slightly different sequence based on discharge rates of the Snake and Palouse Rivers as indicators of rainfall amounts during the Holocene. Marshall attributed the amount of runoff solely to the amount of rainfall (1971:27), with the different periods of Palouse River terrace-building a result of periods of increased rainfall. Marshall (1971:36) proposed that cool, moist conditions initially prevailed, followed by a trend toward drier conditions that stabilized around 6700 B.P. with minimum precipitation and runoff. Marshall (1971:37) proposed that this period of aridity then reversed quickly after which moist but fluctuating conditions prevailed, with a possible dry trend, until present. The historic period is characterized as a time of increased runoff and erosion.

Hammatt (1977) extended Marshall's (1971) work to the Lower Snake River and identified two distinct alluvial cycles during the Holocene. He noted that aggregation occurred between 10,000 and 8000 B.P. and again between 4000 and 2500 B.P. Hammatt links these aggregation phases to the Pinedale glacial advance (Rocky Mountains; 10,000-8000 B.P.) and the Neoglacial Temple Lake (4000-2500 B.P.) advance. Each phase of aggregation was followed by lower seasonal runoff and/or downcutting of the rivers, correlated to retreating glaciers.

Both Marshall (1971) and Hammatt (1977) inferred broad climatic periods based on their data sets. More recent paleoenvironmental research has refined the understanding of the Interior Plateau's environmental history. For example, refinement of the dating of glacial activity has found that the Pinedale glaciation occurred earlier than Hammatt (1977) knew at the time, so it now seems that the earlier phase of aggregation he observed may reflect cooler, moister conditions than during the middle Holocene and may also reflect the post-flood readjustment of the entire fluvial system within the Columbia Basin. In addition, recent research may indicate that the greater alluvial activity in mid-Holocene Palouse River terraces reflects a period of protracted dry conditions rather than increased rainfall as inferred by Marshall (1971). Drier conditions can decrease vegetation cover leaving the ground surface exposed to greater runoff and more intensive erosion. Davis (1983) documented stacks of alluvium that accumulated in Gatecliff Rockshelter in Nevada during the dry middle Holocene that he interpreted were a result of a torrential rainfall pattern. Analysis of pollen influx in records from the region indicates a much drier climate at this time (see discussion below).

Much of the recent research that contributes to our understanding of the Columbia Basin's environmental history has been conducted in adjacent regions, especially the Great Basin. Research has indicated that even when broad climatic periods are demonstrated, there are sometimes drastic variations in climate that affect human adaptations to their environment (Mehringer 1985:174). This chapter presents a summary of the history of the natural setting in the vicinity of the Marmes site and surrounding area using applicable paleoenvironmental studies.

Geological Setting

Marmes Rockshelter is located within the lower Palouse Canyon, a deep, glacial flood-carved canyon that opens to the Snake River Canyon approximately 96 km above the confluence of the Snake and Columbia Rivers. The area is within the Columbia Basin physiographic province (Freeman et al. 1945; Franklin and Dyrness 1973), a semi-arid, upland dominated by grassy, shrub steppe which is bounded by forested mountain ranges; on the east by the mountains of Idaho, on the south by the Blue Mountains, on the west by the Cascade Range, and on the north by the Okanogan Highlands (Baker 1978a:18).

At its confluence with the Palouse River, the Snake River Canyon lies both within the Cheney-Palouse Tract of the Channeled Scablands and within the main channel of the pluvial Lake Bonneville flood system, both of which formed during a series of late Pleistocene floods (see discussion below). The steep basalt cliffs forming the canyon rise over 1,000 feet above the canyon floor, which is incised into the gently rolling Palouse Hills (McKee 1972).

The basalt bedrock in the area was deposited during the Miocene as successive flows of lava that covered over 32,000 square kilometers (20,000 square miles) of area in Washington, Oregon, and Idaho. The lava flows, referred to as Columbia River basalts, vary from 600 to over 1,500 m in thickness and were erupted from fissures that are presently visible as dikes (Franklin and Dyrness 1973:29; Swanson and Wright 1978).

Deformation of the basalt occurring during the Pliocene and early Pleistocene may have been related to the uplift of the Cascades, resulting in the present ridges and hills (Franklin and Dyrness 1973; Grolier 1965; Hammatt 1977:7-8; Mackin 1961). The basalt bedrock dips to the southwest with a low point near the Tri-Cities area of about 100 m (328 ft amsl) and a high point of about 800 m (2,625 ft amsl) near Pullman (Figure 3.1). The basalt bedrock has been incised by the Snake and Columbia Rivers and their tributaries along natural faults and weak zones. The effects of weathering on individual basalt flows within the overall exposed basaltic sequence vary. Highly vesicular portions of the basalt flows weather faster than the balance of the flow, leading to the formation of overhangs and rockshelters. Colluvial and alluvial processes, including considerable plucking and scouring action during Missoula Flood episodes, are responsible for the full development of rockshelters.

The basalts of the area are capped by Plio-Pleistocene deposits that cover over 8,000 km^2 and is over 75 m thick near the center (Treasher 1925; Ringe 1970). The hills that characterize much of the Palouse began forming during the late Pliocene through the deflation of fluvial/lacustrine sediments of the ancestral Columbia River from the Pasco Basin, the deepest part of the basin near the confluence of the Snake and Columbia rivers (Bryan 1927). From there they were transported by the prevailing winds and deposited downwind in a broad arc ranging from the northeast to southeast. Because these sediments were rich in sands and coarse silts they were deposited into dune-like hills whose texture fines with distance from the sediment source area. Busacca (1991) indicates that although the source area for these sediments remained the same through time their changing mineralogy reflects the changing nature of sediments within the Pasco and Quincy basins. The lower and middle sediments in the Palouse Hills reflect an origin within the Pliocene alluvial deposits of the Mio/Pliocene age Ringold Formation. Sediments in the upper section of the Hills reflect an origin within the catastrophic flood sediments that were deposited in the southwestern Columbia Basin that may be mixed with glacial outwash sediments.

Layered within these hills are numerous ancient soils (paleosols), each representing periods of moister climate when vegetation stabilized the soil surface, and weathering processes were able to out pace that accumulation rate of new sediments (Busacca 1991). As a result, layer upon layer of soil reflects the wet climatic cycles that have acted upon the region since at least the middle Pliocene (Figure 3.2). Although great differences are seen between the potential climates that formed the soils at single points within the Columbia Basin, the aerial distribution of soil types found within paleosols of the Palouse Hills reflect a climatic gradient

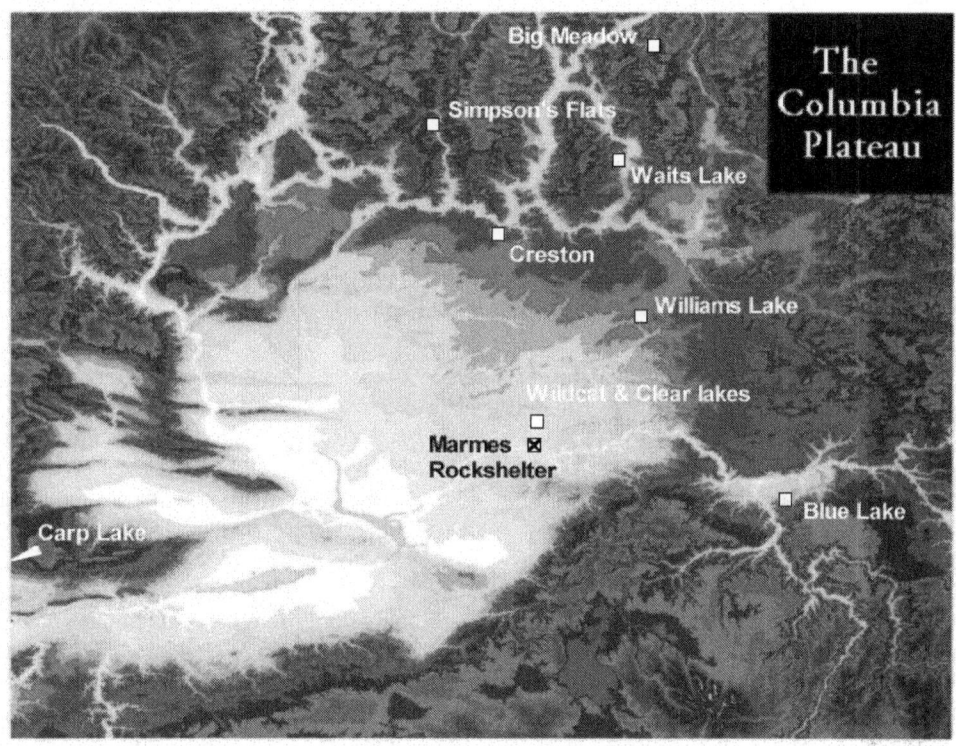

Figure 3.1 Map of the Columbia Plateau showing locations of paleoenvironmental data described herein, with elevation progressing from white (lowest) to dark (highest).

Figure 3.2 Road-cut exposing soils spanning the late Pliocene through Pleistocene in the central Palouse Hills. These reflect clear cycles of arid period sediment erosion and deposition alternating with episodes of wet climate sediment stability (P. Wigand photo).

across the Columbia Basin similar to that of today (Busacca 1991). That is, the driest climate soil types are found near the Pasco and Quincy basins and the wettest near the northern and eastern margins. Today rainfall on the Columbia Plateau rises with elevation eastward ranging from 15.2 - 22.9 cm (6 - 9 inches) per year in the immediate rain-shadow of the Cascade Mountains to 45.7 – 58.4 cm (18 – 23 inches) per year along its eastern margins. Thus although rainfall values varied dramatically since the Pliocene the distribution was very similar.

Soils in the vicinity of Marmes Rockshelter are derived primarily from loess and only rarely from basalt parent materials occurring on benches and foothills. South-facing slopes not subject to erosion generally contain deeper soil deposits with clay enriched subsoils. In contrast, north-facing slopes are characterized by shallow, rocky soils.

Glacial-era Floods and Tephra Chronology

The glacial activity that led to the formation of the Channeled Scablands began roughly 100,000 years ago (U.S. Geological Survey 1982:9). At that time, large ice lobes began to move southward from southern British Columbia into the Okanogan, Columbia, Sanpoil, Colville, Priest, and Pend Orielle River valleys, as well as the Purcell Trench, containing the Kootenai River valley. The Okanogan ice lobe advanced far enough south to block the Columbia River, redirecting its flow south along the Coulee Mountains. The Spokane River was dammed by the Colville lobe, resulting in the formation of Glacial Lake Spokane, which had a surface elevation of ca. 2,300 feet (ca. 700 meters) above mean sea level (amsl) (U.S. Geological Survey 1982:9-10). The Pend Orielle basin, presently containing Lake Pend Orielle, was formed when the Purcell lobe moved southwest near to the eastern city limits of Spokane (U.S. Geological Survey 1982:10). Hayden, Liberty, Newman, Twin, Spirit, and Coeur d'Alene Lakes, were formed when stream valleys were blocked or partially blocked by glacial debris. Glacial activity in the Columbia Basin itself is recorded in its northwestern most corner just west of Moses Coulee on the Waterville Plateau. Large erratics and morainal deposits document the presence of Pleistocene ice.

Within the Purcell Trench, the Purcell lobe blocked the Clark Fork valley, damming the Clark Fork River near its entry point into Pend Oreille Lake. Glacial Lake Missoula, the source of the Missoula Flood waters, formed behind the ice dam. This lake, estimated to have covered nearly 3,000 square miles, or 500 cubic miles of water, had a surface elevation of slightly more than 4,100 feet amsl. The lake was fed by melting ice lobes and alpine glaciers in the Bitterroot Range, melting snow, and rainfall (U.S. Geological Survey 1982:11). The formation of the glacial lake occurred on several occasions, resulting in a series of catastrophic scabland flood events between 22,000 to 12,700 B.P. (Baker 1978b). The Purcell lobe may have blocked the Clark Fork River as many as seven times, corresponding with the end of each of the Pleistocene glacial cycles, before the largest lake was formed ca. 14,000 years ago. Each flood seems to have scoured out much of the sediment remaining from previous floods and deepened existing channels (Baker and Nummedal 1978). Although Waitt (1985) suggests that 40 or more floods originated from glacial Lake Missoula between 15,300 and 12,700 B.P. others suggest that fewer floods are represented, in part, because several rhythmites might form during a single flood events (Lovett 1984; Kietzman 1985).

When the lake level reached its maximum, the ice dam was breached by the water, which flowed south and southwest down the Spokane Valley, and spread out over the Columbia Plateau (Figure 3.3). The floodwaters crossed the relatively tilted, loess-covered lava flows of eastern Washington in three main flows between 22,000 and 18,000 B.P., stripping the loess cover and exposing the underlying basalt deposits. Large blocks of basalt were plucked loose and carried downstream, resulting in the formation of coulees. Fryxell believed that it was this hydraulic plucking that created Marmes Rockshelter (Fryxell and Daugherty 1962:13), probably during one of the last flood events given its location low on the cliffs in the canyon. The easternmost flood flow carved the Cheney-Palouse Tract, the widest tract in the scablands, measuring up to 40 km (25 miles) wide and ca. 180 m (600 feet) deep in places. The floodwaters entered the Snake River Canyon via the Palouse River, Washtucna Coulee, and Devils Canyon. The steep terrain, multiple coulees, mesas, and rockshelters in the Palouse Canyon are products of the Missoula Floods.

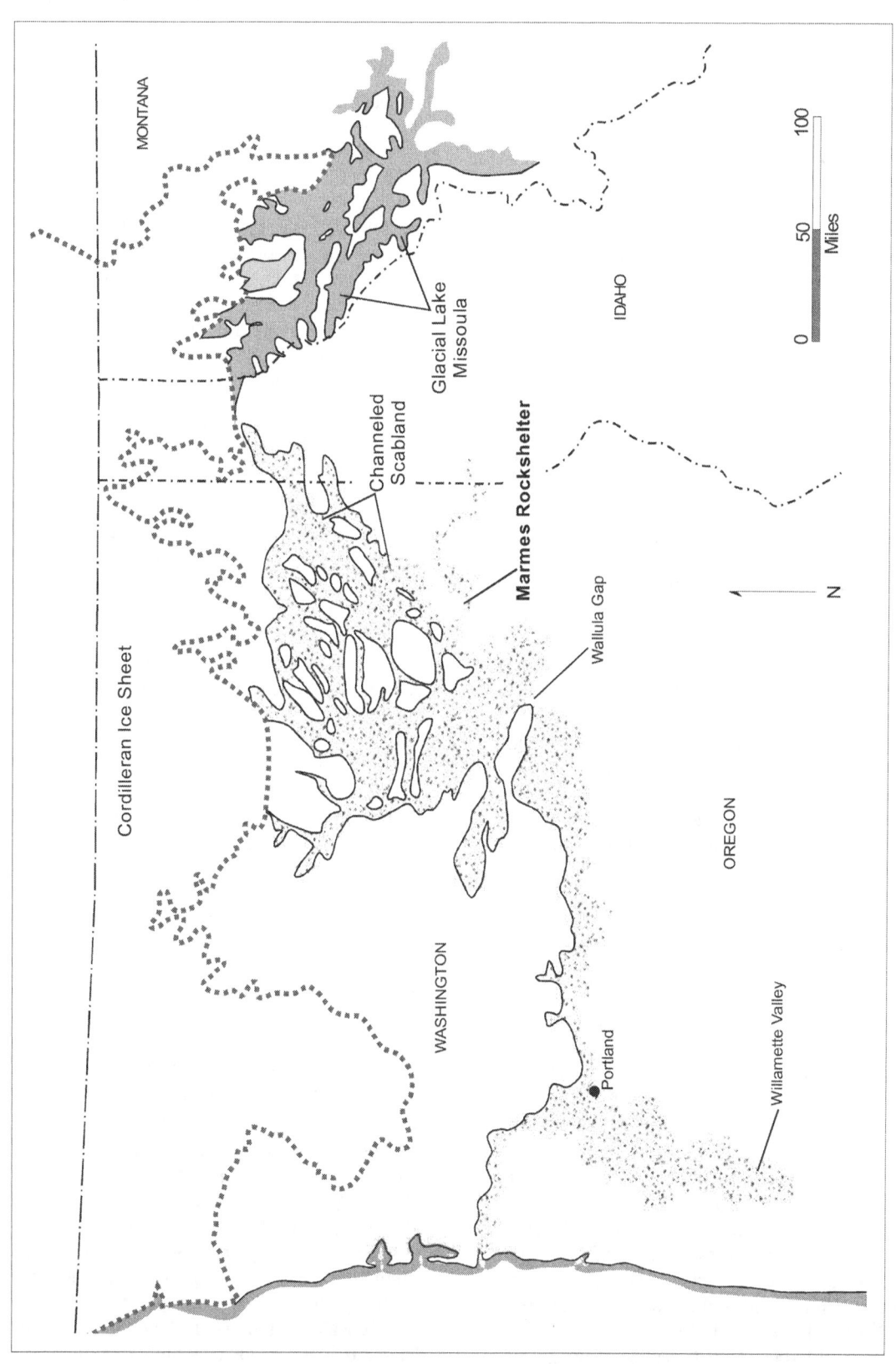

Figure 3.3 Map showing the area of the Pacific Northwest affected by the Spokane/Missoula flooding (based on Waitt 1985, from Reid and Gallison 1995:Figure 2.5).

Other flood signatures include flood bar deposits, ice-rafted erratics, eroded channel margins, and divide crossings (Baker 1978b:60-62). The floods deposited gravels and cobbles on the slope below the mouth of Marmes Rockshelter and at depth beneath Palouse River alluvium.

All of the scabland floodwaters flowed into the Pasco basin toward the Horse Heaven Hills. The Horse Heaven Hills deflected the water toward Wallula Gap, a narrow opening along the Columbia River Valley. Owing to the small size of the opening at Wallula Gap, a large lake formed in the Pasco Basin that extended up the Snake River as far as Lewiston, Idaho. Ponding action from this lake may be responsible for the deposits of fine textured sediments exposed in the deepest excavations of the Marmes rockshelter and elsewhere over a large part of central and eastern Washington (Baker et al. 1991; Marshall 1971:19). The lake subsided as the floodwaters flowed west down the Columbia Gorge toward the present site of Portland, Oregon.

Leonhardy (1970) placed the floodwater lake finally draining between 11,000 and 10,000 B.P., allowing human occupation of the Snake and Palouse River canyons by ca. 10,000 B.P. (1970:71-73); this would preclude the occupation of the study area by Clovis peoples. Hammatt (1977) set the date of the last glacial flood back to between 14,000 and 13,000 B.P. with the backwater lake draining sometime before $10,600 \pm 200$ B.P. Hammatt argued that Clovis people may have been able to move into the valley, but that any traces of their occupation would have been removed by later flooding of the landforms available at that time (1977:173). However, radiocarbon dates from the floodplain deposits at Marmes Rockshelter indicate the use of the highest terrace surfaces beginning about 11,000 B.P. (Sheppard et al. 1987) (see Chapter Two). In addition, Reid and Gallison (1995:2.19) note that a tephra layer was observed below the dated floodplain deposits at the Marmes site that is likely to be either Glacier Peak (ca. 11,250 B.P. [Johnson et al. 1994]) or St. Helens S ash (ca. 13,000 B.P. [Mullineaux 1986]). This indicates that potential occupation surfaces in Palouse Canyon were not subject to catastrophic backwater flooding by 11,250 B.P., and perhaps not since 13,000 B.P. However, discrepancies occur in recent data relevant to timing of the final cross-Scabland floods. A radiocarbon date from the base of a peat bog overlying Missoula Flood deposits in the Willamette Valley in Portland, Oregon, appears to indicate that no catastrophic flooding occurred after $13,080 \pm 300$ B.P. (Mullineaux et al. 1978). But a date of $12,800 \pm 60$ B.P. on bone from a ruminant killed by one of the flood events has been reported along with a date of 12,100 B.P. on a large mammal immediately overlying rhythmically bedded flood sediments within the same depositional sequence (Lenz, Gentry, and Clingman 2002). The latter dates would support exposure of the Marmes floodplain prior to the Glacier Peak eruption.

The distinct physiographic features of the Palouse River canyon and surrounding area all are a product of floods, both the latest Pleistocene floods and much earlier catastrophic flooding documented within the Columbia Basin Cheney-Palouse tract (Kiver et al. 1991). In addition, the Snake River channel and the mouth of the Palouse River around Marmes Rockshelter was impacted by the Bonneville Flood. Dating to ~15,000 years ago, the gravels deposited by this flood are overlain by those of the Missoula flood within the Snake River channel (Webster et al. 1982).

After the last flood event occurred, the Snake River began down-cutting and reworking the flood deposits. In the Snake River canyon, meander bars formed as the river continued to down-cut, and floodplain terraces were formed, providing the earliest landforms in the area suitable for human habitation (Fryxell and Daugherty 1962; Fryxell et al. 1968a). These terraces were eroded and covered by Mazama ash emanating from Mt. Mazama (Crater Lake, Oregon) around 6730 B.P. (Hallett et al. 1997). Floodplain deposition postdating the eruption of Mt. Mazama and subsequent accumulations of aeolian silt and sand and loess complete the depositional history of the area.

Several volcanic tephras deposited from the eruption of volcanoes in the Cascade Range provide time-stratigraphic markers that can be used to date both cultural and depositional sequences in the study area. In the Channeled Scablands, Glacier Peak and Mt. Mazama set O tephras are the two most common ashes (Baker 1978a). Radiocarbon dates place the eruption of Glacier Peak at 11,250 years B.P. (Johnson et al. 1994). The Mt. Mazama Set O represents two ashfalls, one at 7015 B.P. and a much larger eruption at 6730 B.P. (Hallett et al. 1997). While the Glacier Peak tephra is not verified as the lower tephra observed in the deep alluvial deposits at the Marmes site, the Mazama ash has been confirmed as the tephra that makes up the stratigraphic Unit IV defined by Fryxell and Daugherty (1962:16). This Mazama ash is observable in stratigraphic profiles throughout the southern Plateau and as a primary deposit can be up to 80 cm thick.

Repeated eruptions of Mount St. Helens also have provided a sequence of tephra deposits in the region (Mullineaux 1986:Figure 2). St. Helens set K ashes are the first of the last glacial cycle volcanic ashes deposited in the Columbia Basin. Crandell et al. (1981) suggest a date between 20,350 ± 350 B.P. and 18,560 ± 180 B.P. as the age of this set. Moody (1978) suggested that Mount St. Helens set S was a result of a series of eruptions between 18,000 and 12,000 B.P. that often occurs as couplets and triplets in slackwater sediments. Mullineaux et al. (1975) indicate that the Sg and So layers of the set S eruption stem from an eruption around 13,000 years ago. Crandell et al. (1981) bracket set S between 13,650 ±120 B.P. and 12,120 ± 100 B.P. The Mount St. Helens tephra set J was deposited from eruptions between 12,000 and 8000 B.P. (Crandell et al. 1981; Mullineaux et al. 1975; Mullineaux 1986). Tephra set Ye has been identified in the Imnaha basin just west of Hells Canyon and is dated at about 4000 B.P. (Reid and Gallison 1995:2.21). Crandell et al (1981) bracket set Y between 3,900 ± 50 B.P. and 3,350 ± 50 B.P. A set P is dated at 2,670 ± 70 B.P. (Crandell et al. 1981) and has been traced as far as Canada (Westgate 1977). A deposit of set W ashfall, dated at 450 B.P., was identified in an alluvial terrace in the Palouse Canyon (Marshall 1971). Tree-ring data indicate that layers Wn and We probably were erupted in A.D. 1480 and 1482, respectively (Yamaguchi 1983). The 1980 ashfall is observed consistently as a thin layer in the bottom of the duff throughout the region.

Climatic and Vegetation History

Changing weather from year to year affects the health of plant communities. It affects annual biomass production and is even reflected in small-scale changes in species abundance. Over the long-term, changing climates effect major changes in the plant distribution and plant community composition of the Columbia Plateau forest, woodland, and shrub- and grassland ecosystems. Analysis of paleobotanical data, primarily from pollen records and potentially from woodrat middens, allows the reconstruction of major changes in plant community composition and in vegetation distribution patterns for the region.

Plant community composition, structure, and patterns in ecosystems today reflect not only the underlying climates, soils and geology, but also the impacts of disturbances both natural (fire, insect infestation, and disease) and anthropogenic (human-caused such as grazing). The long-term dynamics of vegetation communities reflect the successional processes resulting from these inputs. These disturbances are part of the ongoing processes that have altered and are altering plant communities and the overall ecosystem, and that have shaped and are shaping the lives of peoples throughout the world.

Although Euro-American settlement altered natural processes through intensified land management practices and fire suppression, periodic vegetation burning by Native American peoples already constituted land management on a limited scale during prehistoric times. Euro-American land management practices have interacted with natural ecosystem process to accelerate the rate and direction of modern successional processes and resulting plant community composition, structure and pattern. As such, modern (during the last 100 years) plant community succession processes contrast dramatically with those of the prehistoric past.

The Nature of the Data

This is an extrapolated reconstruction of the vegetation and climate of the Plateau region from sources within the region and from further south in the Intermountain West. Extrapolation of data from other parts of the Intermountain West is appropriate as paleoenvironmental research during the last decade has revealed very strong correspondences in the timing of climatic events throughout the region (e.g., the Middle Holocene drought or the Late Holocene cool/wet episode). In addition, in contrast to the Great Basin, there is limited applicable data published for the Columbia Plateau. Although there are several records published for the Interior Plateau region (additional ones are in press as of this writing) some of the more crucial records still remain unpublished or are works in progress. Chatters (1998) provided the Environment section to the Smithsonian Institution's Plateau Handbook and in this chapter his information is correlated with the author's where appropriate. It is an affirmation of the state of paleoenvironmental research in the intermountain region that these two chronologies generally agree; however, magnitudes, timing, and sometimes directions of change vary somewhat.

In part, this is due to the fact that the Intermountain West can be viewed as a region within which climates receive influences from two major sources: 1) the northern Pacific, and 2) the central and southern Pacific via the Gulf of California or from the Gulf of Mexico via the Southwest. These differences become especially clear during periods of El Niño/La Niña patterns when the northern and southern regions of the Intermountain West manifest exactly opposite climates. During El Niño cycles the southern part of the Intermountain West is affected by much wetter climate originating from the central Pacific, while the northern portion remains relatively unaffected by these storms and has much drier climate. During La Niña episodes the reverse is true. When these cycles are weak the central portion of the region (the Great Basin) tends to be minimally affected by what occurs in either area. As a result it has generally drier climate. When El Niño/La Niña cycles are stronger storms will spill over into the central Great Basin from either the south during El Niño cycles and from the north during La Niña cycles. This pattern seems to hold even for general climatic patterns as well (e.g., the "neoglacial/neopluvial" period as discussed below).

This discussion makes use of what is currently available from the Columbia Plateau, the Great Basin and the northern Mojave Desert. It includes published pollen diagrams, modifications of published diagrams and summary diagrams that will hopefully clarify some of the relationships between the regions. In addition, this presentation makes use of the pollen diagrams of Henry Hansen, a pioneer palynologist in the northern portion of the Intermountain West, who conducted research before the advent of radiocarbon dating. However, using the one clear time-stratigraphic marker that occurs in most of his records, Mazama volcanic ash, relatively good age approximations can be made for many of the vegetation shifts in his pollen sequences. When compared with published radiocarbon dated records they provide excellent paleoenvironmental coverage for the Columbia Plateau/Columbia Basin. The issue of the identification of pine to species level aside, the directions and relative timings of changes in climate reflected in his records are worth investigation. Finally, woodrat midden data from the central and southern Intermountain West are coupled with the pollen data from those regions to clarify the similarities and contrasts between the Plateau and the rest of the Intermountain West. These relationships are explored primarily because it is clear that at times movement of peoples between these regions may have been highly fluid. At times these movements may have occurred as a result of dramatic changes in local environments. For example, Baumhoff and Heizer (1965) suggested that much of the Great Basin may have been abandoned during the middle Holocene. Did these peoples retreat to adjacent regions or to higher elevations? Knowledge of the environmental dynamics of adjacent regions might explain some of the cultural changes or possible external intrusions that we see at times in the Plateau.

The region's vegetation is very sensitive to changing climate. This arises primarily from the physiological characteristics of its species. As the result of millions of years of increasingly semi-arid and arid climates, many of the plants growing here became increasingly opportunistic. They evolved the capability to take advantage of brief episodes of increased precipitation. Many species respond to wetter conditions through increased pollen and seed production, and rapid vegetative growth. Because the annual production of biomass is linearly related to effective precipitation up to 600 cm (Walter 1954), its production in arid and semiarid environments is a sensitive indicator of changing climate. This is reflected not only in annual growth-ring size, the abundance and size of foliage, but also in pollen production. These changes are all reflected in some manner in the paleobotanical record.

Vegetation changes observed in the Intermountain West include large-scale latitudinal shifts of plant taxa that have characterized the Mojave Desert and southern Great Basin regions, or altitudinal shifts that typified the northern Great Basin and Columbia Plateau areas (Figure 3.4). Smaller scale changes in the abundance of major (and minor) taxa within plant associations have characterized the entire region during both low and high magnitude climate changes. The degree of vegetation response to variation in climate is due to several factors. These include: 1) the magnitude and duration of the new climatic regime, 2) the seasonality of the climatic regime, 3) the variable effects of topographic relief, and 4) the distance to source areas from which plant species would have to expand during climatic transitions. Apparent differences between the Great Basin and the Plateau regions' response to climate trends may represent variation in these factors.

The preservation, and therefore the distribution of paleobotanical data sets within the Intermountain West is constrained primarily by moisture, that is, either very dry or very wet environments are the most favorable for the preservation of paleobotanical data. Plant macrofossils (*e.g.*, seeds, twigs, leaves, and bark) preserved in dry cave deposits and well-protected woodrat middens provide intermittent records of local vegetation spanning tens of thousands of years in the southern Intermountain West, and only a few thousand years in the north. Wood from living or long-dead trees, preserved by the cold- dry conditions that often characterize upper tree lines in the higher

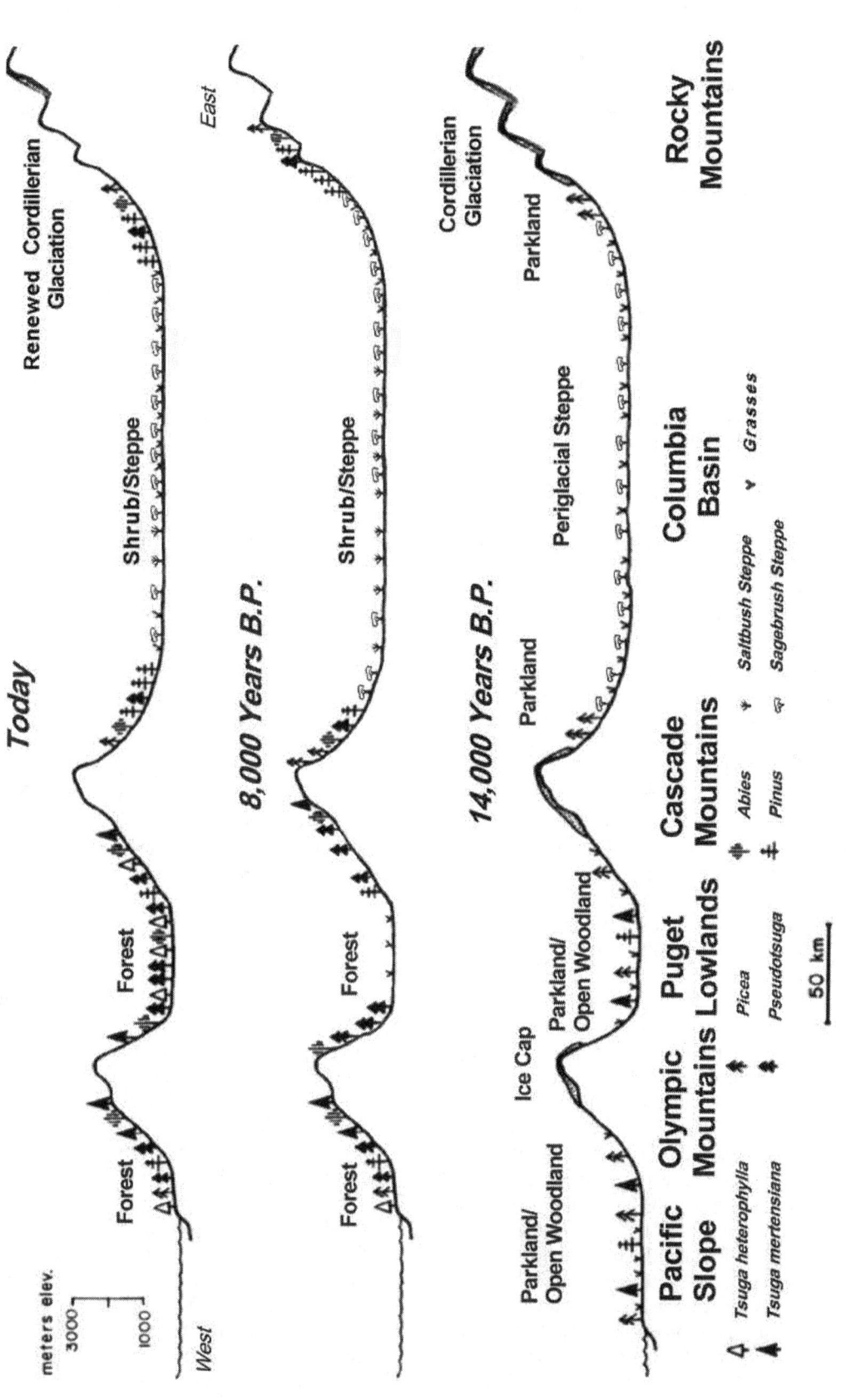

Figure 3.4 Depiction of vegetation changes through time for a cross-section of the northwest (provided by Wigand).

mountains of the margins of the Intermountain West, provides continuous tree-ring width data mirroring climatic variation. In the south, tree-ring series span much of the Holocene, but only the last thousand years or so in the mountains surrounding the Columbia Plateau.

A few lakes and, occasionally, springs provide continuous though sometimes complex sedimentary records that can span thousands to hundreds of thousands of years. High organic production in some wet environments leads to rapid deposition rates with records that are particularly amenable for the examination of high frequency pollen and local aquatic-plant macrofossil records. These can be compared with tree-ring data to generate long, detailed records not only of regional climate but also of the resulting vegetation response (e.g., Wigand 1997). In general, the number of well-preserved palynological records decreases southward (reflecting the greater rarity of lakes and marshes in the regions of increasingly arid climate) and the number of woodrat midden localities decreases northward (reflecting increasingly wetter conditions that favor dissolution of the protective urine coating of the nests).

Currently available paleobotanical proxy data reveals regional vegetation dynamics in great detail for the last 4,000 years, and more coarsely for the last 35,000 years. Beyond 40,000 years only a few pollen or midden records are available (e.g., from the Summer/Chewaucan Lake Basin in southern Oregon, Owens Lake in the western Great Basin, Searles Lake in the northern Mojave Desert, and the Bonneville Basin in the eastern Great Basin). For the Columbia Plateau a roughly 70,000-year pollen record is available from Carp Lake near Mount Adams. Several high-resolution pollen records in the northwestern Intermountain West provide excellent coverage from 6,000 years ago to the present of both terrestrial and aquatic vegetation history. In the Great Basin and southward, vegetation history since 40,000 years ago, reconstructed from pollen records, is supplemented by evidence obtained from ancient woodrat middens, and in the Lahontan and Bonneville basins it can be correlated with well-dated late Quaternary lake histories. These data all record changes in the degree, directions and rates of vegetation change during the late Quaternary.

In general, evidence of the expansion and contraction of semi-arid and sub-alpine woodlands is most obvious for the late Pleistocene/early Holocene transition, especially as it is revealed in the woodrat midden record of the Great Basin and northern Mojave Desert mountain ranges (Van Devender et al. 1987). Changes in the distribution of individual shrub species, and changes in the composition of shrub communities have also been very dramatic during the late Holocene. These shifts are most obvious in lowland and intermediate elevation woodrat midden records, but some less obvious changes in shrub communities are also apparent in the pollen records of the region. These data are enhanced, both qualitatively and quantitatively, by other data sources, including geomorphology, paleosols, stable isotope records, rock spall deposition in caves, surface water hydrology and ground water dating in describing past variation in potential available moisture (Chatters 1998; Wharton et al. 1990)

This discussion includes a brief presentation on late Pleistocene environments. Although they may not directly have impacted early peoples within the region, late Pleistocene climates and vegetation patterns are presented because they shaped the environments that were encountered by the first peoples that entered the region. The dynamics involved in the readjustment of vegetation and land-forming processes from the climates of the last glacial cycle to those of the Holocene must have played an important role in shaping the adaptations of peoples entering the region 12,000 to 11,000 years ago. Patterns that the region has experienced during the last 12,000 years have occurred in previous interglacials. Even the magnitude of interstadial variation has some parallel during the Holocene, except that the climatic baseline from which fluctuations occurred was much warmer. Therefore, Holocene environmental shifts must have been impacting people in ways that we are only now beginning to understand.

Late Pleistocene

The Intermountain West from the Columbia Plateau in the north to the Mojave Desert in the south has been a land of extremes during the late Quaternary. Although most vegetation community transformations have been in response to climate change, dramatic movements of vegetation due to brushes with cordilleran ice sheets and isolated ice caps, catastrophic floods, formation of huge lakes in juxtaposition with extensive dune-fringed dry playas have typified the region. J. Harlan Bretz (1969) described massive, short-duration floods that swept much of the Plateau just about the time that people were first beginning to explore the region. It is ironic that these floods not only provide the Columbia Plateau with its distinguishing Pleistocene event, but were also responsible for creating the majority of the sites where Holocene pollen records have accumulated in the region. Unfortunately, these floods may have destroyed sites dating to full glacial and interstadial, or perhaps even, interglacial times. In addition, localities to the north in the Okanogan Highlands were glaciated. As a result, almost the only Pleistocene records that can be used to reconstruct vegetation and climates in the Plateau occur at its edges or outside the region entirely.

A growing body of research in the Columbia Plateau and the northern and western Great Basin is clarifying the relationships between past climate, lake and marsh history and vegetation change (Mehringer 1985, 1986). Only three long pollen records, that include more than just a Holocene record, are currently available: Summer (Cohen et al. 2000) and Owens (Smith and Bischoff 1997) Lakes from the southern portion of the region, and Carp Lake (Whitlock and Bartlein 1997) near Mount Adams in the north.

Carp Lake contains a pollen record spanning the last 125,000 years (Whitlock and Bartlein 1997). Although dating of the lower portion is still an issue, a general pattern of glacial/interglacial vegetation change is clear. Glacial episodes are clearly characterized by the expansion of sagebrush into those areas around the lake where pines are currently abundant (Figure 3.5). In addition, increased spruce and grass abundance also occurs. The sagebrush indicated is probably not big sagebrush, but an alpine variety, because the accompanying plant assemblage suggests a grassy, upper elevation sagebrush steppe with an admixture of spruce, probably occurring as isolated clumps as we see at upper tree-line today in the mountains. Climates were clearly much cooler and drier than those of today.

Intermountain climate links can be seen further south in the northern Great Basin. The Summer Lake Basin of south-central Oregon currently lies on the transition between the pine woodlands of the eastern slope of the Cascades and the sagebrush steppe of the northern Great Basin. Three pollen localities from the Summer Lake Basin span the period from ~250,000 to ~20,000 years ago (Cohen et al. 2000). Shifts in the dominance of mixed montañe conifers, cold steppe shrubs and desert scrub species record the varying effects of precipitation and temperature, and their changing seasonal influence.

Last Glacial Cycle (35,000-12,000 B.P.). At Carp Lake cool, dry alpine conditions are recorded after ~34,000 until ~27,600 years ago (Whitlock et al. 1997). They describe the immediate area (lying at ~714 m) as being dominated by open forest composed of a mixture of both higher and lower elevation species. Spruce and fir occurred in mesic locales, whereas drier, lower elevation woodlands were comprised of lodgepole and ponderosa pine, Douglas fir or western larch and grand fir. Between 27,600 and 14,100 years ago the area around Carp Lake was dominated by cold steppe composed of alpine sagebrush and grass. The presence of American bistort (*Polygonum bistorta*) indicates alpine meadow conditions. Copses of spruce occurred in and around the site. These conditions climaxed ~15,000 years ago, and ended abruptly ~12,000 years ago. As mentioned above, other records from the Plateau are absent either because they were destroyed by floods or because the areas to the north, west and east were glaciated. This author has observed late Pleistocene lake clays in the Dry Falls area of Washington with abundant willow leaf impressions suggesting the presence of riparian habitats adjacent to lakes near the ice margin in the northern Plateau. Density winds directed off the ice itself may have kept areas immediately adjacent to the ice margin extremely cold. During the glacial maximum global circulation patterns just south of the ice margin may have subjected the Columbia Basin to extremely cold and dry conditions (COHMAP 1988; Thompson et al. 1993). The modern rainshadow effect may have been exaggerated during the glacial maximum because even less Pacific storms than today were able to cross the Cascade and Sierra Mountains into the area.

Terrestrial conifer pollen evidence from the northern Great Basin indicates that from 34,000 to 30,000 years ago cooler, wetter conditions prevailed. This shift was heralded by an expansion of juniper and sagebrush. Increased regional abundance of grass together with retreat of saltbush and greasewood (*Sarcobatus* sp.) are indicated. A return to drier conditions at ~29,600 years ago is evidenced by a retreat of fir and spruce and an increase in saltbush (Cohen et al. 2000). A major drought from 28,000 to 26,000 years ago is flanked by regionally correlatable increased grass abundance. This appears in the woodrat midden pollen record at Pyramid Lake as well as in all three cores from Summer Lake (Wigand and Rhode 2002:Figures 2 and 3). The drought is reflected in lowered lake levels throughout the northwestern Great Basin (Negrini and Davis 1992).

After 24,500 years ago effectively moister climate, documented by renewed growth of pluvial Lake Lahontan, encouraged the spread of Utah juniper woodlands and allowed whitebark pine to intrude into scattered localities down to 1380 m (Wigand and Nowak 1992). By 20,000 years ago,

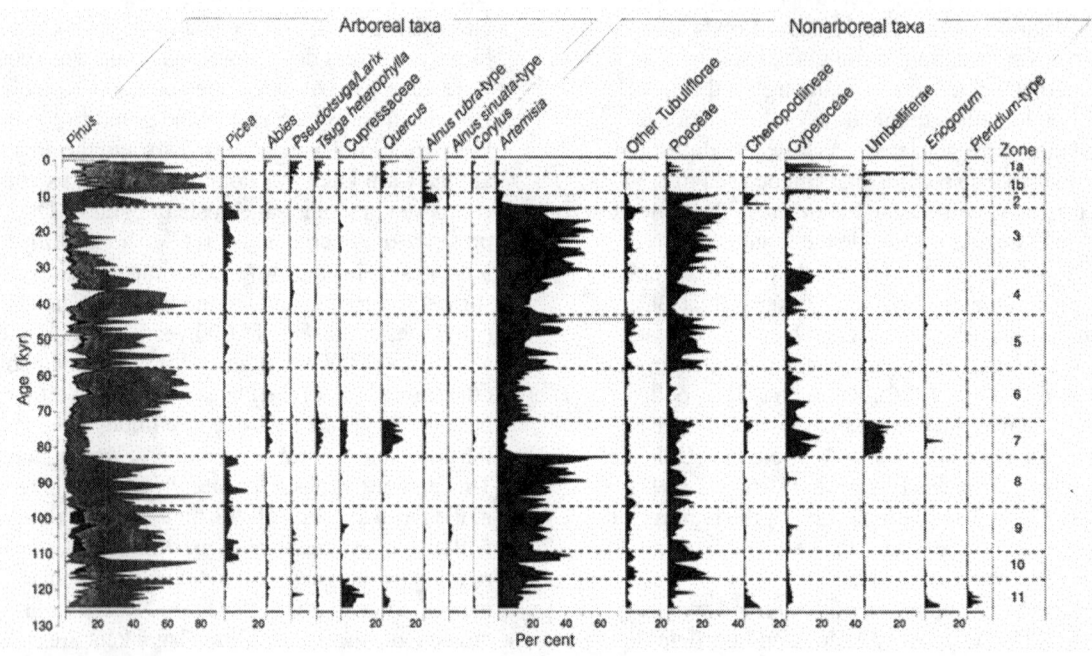

Figure 3.5 Pollen diagram of Carp Lake, Washington near Mount Adams from Whitlock and Bartlein 1997. Note increases of pine (*Pinus*) and sagebrush (*Artemisia*) pollen are out of phase and that spruce (*Picea*) and grass (Poaceae) pollen increase with sagebrush pollen.

the onset of much cooler, drier Glacial Maximum conditions led to a decline in Utah juniper and to expansion of mixed sagebrush and desert scrub communities (Nowak et al. 1994a). Whitebark pine appears over a thousand meters lower than its current elevational distribution ~23,000 and 12,000 years ago along the western margin of the Lahontan Basin during wetter periods leading up to and following, but not during, the glacial maximum (Nowak et al. 1994a, 1994b; Wigand and Nowak 1992). This provides an indication of just how cold and dry the glacial maximum may have been in northwestern Nevada.

Using vegetation dynamics from the northern Mojave Desert for the last glacial cycle as a model, and applying modern analogue temperature and precipitation requirements for limber pine and white fir, late Pleistocene climates for the Intermountain West can be extrapolated. Drought-resistant limber pine favors cool, dry continental conditions with rainfall distributed relatively evenly during the year and is drought tolerant. White fir favors cool, moist conditions often with a winter peak in rainfall. The wettest periods were those when white fir appeared in the woodrat midden record from the northern Mojave Desert, but limber pine disappeared. These appear to coincide with the onset (23,000-21,000 years ago) and demise (16,000-14,000 and 13,000-12,000 years ago), but not the climax, of the late glacial maximum. Conversely, based upon the absence of white fir in communities with abundant limber pine, periods of cold, dry continental climates (continental polar and arctic air mass predominance) occurred from 21,000 to 16,000 and again from 32,000 to 29,000 years ago.

Latest Pleistocene/Early Holocene (12,500-8000 B.P.).

This period was characterized by often incredibly diverse plant communities that retained plant species characteristic of the cooler conditions of the glacial period and saw the appearance of pioneering plant species that heralded the Holocene. At Carp Lake, Whitlock et al. (1997) find the area dominated by drier, but still grassy sagebrush steppe. Warmer, drier conditions are indicated by the appearance of saltbushes in the record, and the retreat of fir and spruce from mesic sites and its replacement by alder. The climbing pine pollen reflects both the regional expansion of pine and perhaps, to a certain extent, increased representation of long-distance pollen transport as local pollen production decreased as a due to drier conditions. The first 2,000 years (~13,000 to ~11,000 B.P.) of a pollen record from Creston Bog on the northern edge of the Plateau reflects a sagebrush steppe with high percentages of fir (*Abies*), spruce (*Picea*) and haploxylon pines (either western white pine [*Pinus monticola*] or whitebark pine [*Pinus albicaulis*]) (Mack et al. 1976). Mack et al. (1976) suggest that the pollen record reflects a vegetation mosaic comprised of both steppe and woodland elements much as is suggested by Whitlock et al. (1997) for Carp Lake during the glacial. Within this mosaic, copses of haploxylon pines, fir and spruce dotted a steppe dominated by sagebrush and grasses. Sedges (Cyperaceae) were clearly not an important element in this steppe so it was evidently a relatively dry steppe. Between 11,250 B.P. (the fall of Glacier Peak ash) and a radiocarbon date of 9,300 B.P. falling percentages of haploxylon pines, fir, spruce and sagebrush reflect a major change in local conditions to more xeric climates. Diploxylon pines (lodgepole pine [*Pinus contorta*] and ponderosa pine [*Pinus ponderosa*]) replace these conifers on the landscape probably as a dispersed woodland much as can be seen in the region today. This probably occurred by the end of the Younger Dryas (~10,200 B.P.). A brief increase in *Nuphar* (water lily), *Typha* (cattail), coupled with single-sample resurgences of spruce and fir might signal this period of much colder climate. By 9,400 years ago, however, vegetation and climate conditions that characterize the remainder of the record were established.

Mack et al. (1978b) record that by 9,500 B.P. many mesic plant species including willow, birch and a *Populus* (probably aspen) as well as grasses, sedges and pines had declined dramatically at Simpsons Flats in the San Poil River Valley north of the Plateau in the Okanogan uplands as well. An even earlier disappearance (by 10,000 B.P.) of mesic conifers (in this case fir) may be indicated at the bottom of the pollen record and paralleling the record at Creston Bog. By the fall of Mazama ash grass and sagebrush abundance that characterizes much of the remaining Holocene is established, pine is the exception (more below). At Waits Lake in the Colville River Valley sagebrush declines sharply by 10,000 B.P. as well (Mack et al. 1978c).

A pollen record further east at Big Meadow displays a sharp decline in sagebrush by the fall of Glacier Peak volcanic ash that parallels the decline at Creston Bog. Diploxylon pine (ponderosa and/or lodgepole pines) became more common than haploxylon pines, and replaced sagebrush as the dominant in the local vegetation (Mack et al. 1978a).

Mehringer (Johnson et al. 1994: text and Figure 14) suggests that his own ongoing research throughout the central and southern Plateau as well as pollen records from the northern Plateau, and the Okanogan (Mack et al. 1978a, 1978b, 1978c, 1978d, 1979, and 1983) indicate the following general sequence:
1) ~12,000 years ago sagebrush and grass dominance with spruce, fir, lodgepole and other pines appearing before 11,250 years ago (the fall of Glacier Peak volcanic ash);
2) ~11,250 to 10,250 years ago mixed conifers with occasional birch dominate; this period corresponds to the Younger Dryas Period of northern Europe, a period of much cooler climate;
3) ~10,000 years ago and following grass and/or sagebrush reasserts dominance at lower elevations; grassier conditions promoted the expansion of bison populations that were hunted by native peoples; to the east in the Rocky Mountains larch, Douglas fir, and lodgepole or ponderosa pines together with sagebrush and grass comprised dry interior forests and steppe.

In the northeast corner of the Plateau, Williams Lake Fen contains one of the more spectacular examples of the Younger Dryas event in western North America (Johnson et al. 1994). Mehringer indicates that between 11,200 and 10,200 years ago pine pollen (most probably from whitebark pine) becomes more abundant than at any other time in the late Pleistocene and Holocene (Figure 3.6). The bimodal nature of pine abundance during its expansion at that time corresponds to the two climatic events of the Younger Dryas described in paleoenvironmental records. In addition, shrub species, such as willow and birch and juniper become very abundant at the onset and decline of this two-part pine expansion.

This colder period also may be reflected in polygonal fractures in the Marmes site early floodplain deposits that Fryxell interpreted as "frost cracks" (Fryxell and Keel 1969) (Figure 3.7). Fryxell in Mack et al. (1976) suggests that most patterned ground in the Plateau formed after 13,000 following the withdrawal of ice along the northern margin. Because there is no patterned ground on surfaces younger than 8000 B.P. he assumed that its formation ceased by this time. Patterned ground is, however, abundant on surfaces dating between 11,200 B.P. and 8000 B.P. Further south, observations by the author indicates that throughout the northern Great Basin patterned ground occurs on surfaces just above the highest pluvial lake stands in the region, but not below them. This

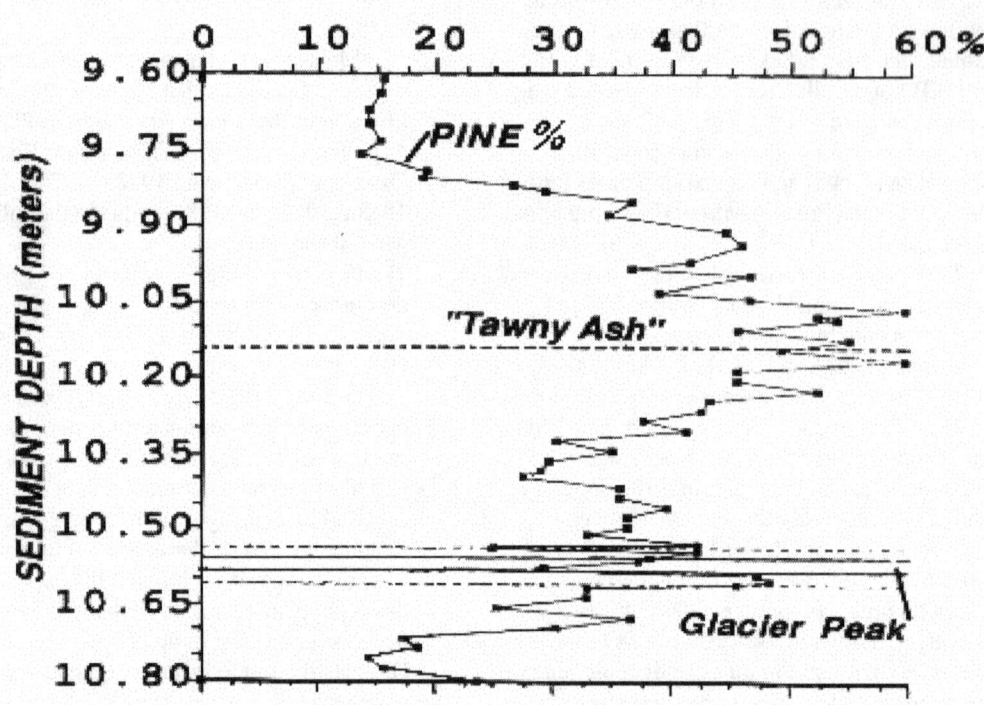

Figure 3.6 Pine pollen percentage between 11,250 B.P. (the age of Glacier Peak volcanic ash) and 10,200 years ago when pine pollen percentage resumes its lower levels (~ 9.75 meters). Modified from Johnson *et al.* 1994.

Figure 3.7 Photograph of remnant frost cracks in the Marmes floodplain area.

would suggest that these features formed during the transitional periods between stades and interstades when moisture was abundant, lake stands were highest, and annual temperatures may have had their greatest amplitude. By the time the pluvial lakes began falling conditions were no longer favorable for the formation of patterned ground. This period of time seems to have occurred slightly earlier in the south than in the north where average annual temperatures may have remained colder into the early Holocene.

Further south in the northern Great Basin, Mehringer (1985) notes several characteristics that tie many plant communities together at this time, especially at higher elevations and in the more xeric areas of the northern Great Basin. This includes:
1) a treeless sagebrush steppe often with abundant grass;
2) common occurrence of *Shepherdia canadensis* (L. Nutt. - russet buffaloberry) and occasionally juniper, and more rarely spruce and fir;
3) combinations of pollen types such as *Rumex-Oxyria, Bistorta, Polemonium, Eriogonum* and *Koenegia* that indicate sub-alpine or alpine plant communities;
4) abundance of the spores of cold climate plants such as *Selaginella densa, S. selaginoides, Botrychium*, and *Lycopodium*.

At lower elevations the Early Holocene upward expansion of sagebrush steppe vegetation is matched by expansion of saltbush-dominated desert scrub vegetation prior to the fall of Mazama Ash (Wigand and Rhode 2002:Figure 7). The climax of saltbush scrub vegetation before and after the fall of Mazama ash is clearly evident in the pollen records from the Warner and Chewaucan valleys analyzed by Hansen (1947) (Figure 3.8). Hansen's record shows what he suggests are cool, wet climate pines retreated as warm, dry climate pines expand. (Note: Hansen's exact identification of pines to the species level is controversial because it is based upon a mixture of pollen size and texture criteria. However, if one takes the counts at face value as large and small diploxylon pines, and a small haploxylon pine, Hansen's identifications appear to be consistent based upon the comparison of the Warner Valley and Chewaucan Marsh counts.) Grasses seem to be expanding reflecting the early Holocene intensification of summer rainfall that is noted throughout the Intermountain West. Gehr (1980) and Dugas (1998) record a series of radiocarbon dates between 9,600 and 7,400 years ago on molluscs from beach ridges and charcoal that indicate multiple pluvial Lake Malheur high stands including some within five meters of its overflow into the Snake River drainage at Malheur Gap. In the Fort Rock Basin, a wet episode close to the fall of Mazama ash is recorded by Mehringer and Cannon (1994). In any case, generally drier conditions resulted in the establishment of saltbush communities by the time that Mazama volcanic ash fell in the region - a pattern that matches pollen records in the Plateau.

As juniper was disappearing from lower elevation areas it occupied during the Pleistocene it expanded into new areas in the north. The presence of semi-arid juniper woodland in the northwestern Great Basin by ~8500 B.P. is evidenced by juniper pollen values in cores from McCoy Flat, and from Bicycle Pond in south-central Oregon (Wigand and Rhode 2002:Figures 5 and 7).

Middle Holocene (8000-5500 B.P.).

At Carp Lake, this period is characterized by ponderosa pine and oak woodland suggesting relatively dry forest conditions (Whitlock and Bartlein 1997). Throughout the central Columbia Plateau sagebrush steppe seems to predominate. Grassy, sagebrush steppe characterizes the pollen record from Wildcat Lake below Mazama volcanic ash in the pollen record (Johnson et al. 1994:Figure 9). However, it is much less grassy than it became during the last 4,400 years. In addition, saltbushes seem to have been much more abundant regionally, perhaps in the Pasco Basin. To the north in Creston Bog, pine became less abundant on the landscape.

Evidence within sediments on the Plateau indirectly suggests that vegetation cover was less dense than today. Reworked Mazama volcanic ash occurs in soil exposures throughout the region from the time of the initial eruption around 6730 B.P. (Hallett et al. 1997). After ~5400 B.P., the ash generally was no longer exposed to aeolian deflation because it was held in place by a denser vegetation cover that formed due to a return of more mesic conditions.

A comparison of the Carp Lake record with that of Hansen (1947) indicates that there is considerable similarity. Hansen's categories are slightly different and occasionally represent a mixing of climatic signals (e.g., the combined ChenoAm + Composite record), but the ponderosa pine in Hansen's composite record and the total pine record from Carp Lake are very similar as are the grass from both records. In total, these data indicate that major changes occurred prior to the fall of Mazama tephra.

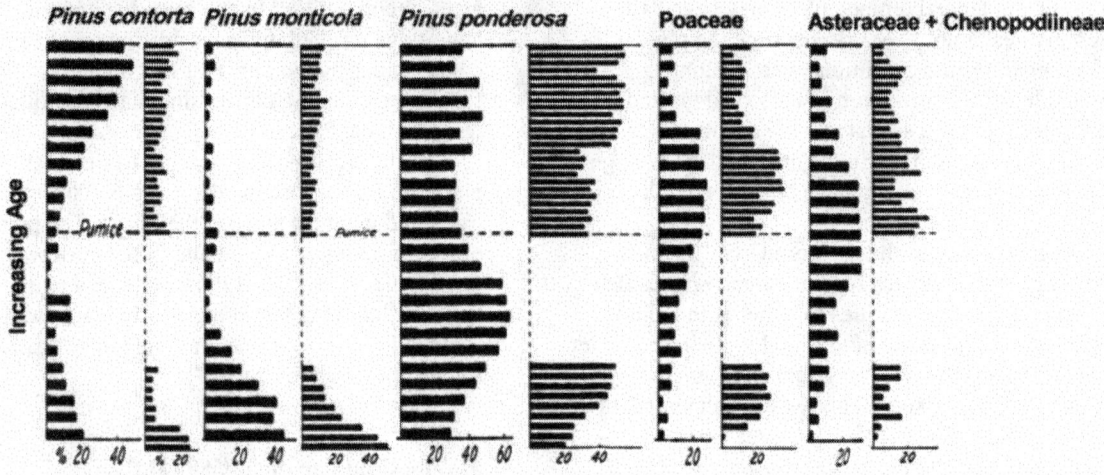

Figure 3.8 Hansen's record from the northern Great Basin although not radiocarbon dated (pumice is almost certainly Mazama ash at 6,730 B.P.) indicates early Holocene drying conditions. (Note: data are assembled from Hansen (1947); broad bars are Chewaucan Marsh pollen and narrow bars are Warner Valley pollen.)

South of the Plateau in the Steens Mountains, the Fish Lake pollen record indicates that sagebrush steppe became the dominant vegetation at elevations lying above the juniper woodland zone (Mehringer 1985). Sagebrush to grass pollen ratios from Fish Lake indicate that the climate that typified the period between 8,800 and 5,600 years ago was the driest to occur there during the Holocene (Mehringer 1986). The Wildhorse Lake record evidences the arrival of high-elevation sagebrush steppe by ~8000 B.P. (Mehringer 1986); it was to remain dominant at these elevations until about 3,800 years ago. In southern Oregon dry pluvial lake basins are indicated by dune deposits dated to this period including ones in the Catlow Valley at Skull Creek (Mehringer and Wigand 1987), the Fort Rock Valley (Mehringer and Cannon 1994), and the Malheur Valley (Dugas 1998).

The all-pervasive drought of the Middle Holocene is reflected, in part, by the dearth of paleobotanical records throughout the northern Great Basin at lower elevations. The almost continuous midden record which occurs in the Painted Hills on the west shore of Pyramid Lake and spans almost 34,000 years is interrupted for almost four thousand years (from 8,000 to 4,000 years ago) (Wigand and Nowak 1992). Juniper macrofossils disappear from woodrat middens lying up to 200 meters above the early and late Holocene lower elevational limits of juniper evidenced in the woodrat midden record. This may indicate that juniper retreated upward in elevation by at least 200 meters or more above its current distribution during the middle Holocene. LaMarche (1974) suggests that upper tree-line temperatures were considerably warmer during this period and that upper tree-lines extended upward.

A comparison of the Hidden Cave drought index with one generated from the pollen record of Kelso (1970) from Hogup Cave in northwestern Utah suggests that widespread drought characterized much of the northern Great Basin (Wigand and Rhode 2002:Figure 9). Pyramid Lake may have reached its lowest levels at this time (Born 1972), lakes and marshes throughout the region desiccated, and pollen of drought-tolerant salt desert species increased substantially (Mehringer and Wigand 1990; Wigand 1987).

At Diamond Pond south of Malheur Lake in south-central Oregon greasewood pollen reached levels of 80% of the terrestrial pollen where currently it comprises less than ~20% of the pollen record. These changes coincided with dramatically reduced evidence of the activities of Native American populations (Grayson 1993). Baumhoff and Heizer (1965) even suggested abandonment of large areas of the Great Basin by native peoples during this period. This is consistent with a decline in foraging resources and corroborates the severity of this drought (Aikens 1986).

Early Late Holocene (5500 to 2000 B.P.).

Beginning ~5,400 years ago, the extreme drought of the middle Holocene came to an end. Temperatures remained warm but pollen and macrofossil data indicate gradually increasing annual precipitation punctuated by periodic increases in rainfall abundance (Davis 1982; Mehringer 1986; Wigand 1987). A brief, extremely dramatic climatic event that can be correlated from the Columbia Plateau to the spring deposits of Ash Meadows in the northern Mojave Desert effectively signals the end of the middle Holocene in the Intermountain West.

Evidence for this pan-regional event begins in the north at Wildcat Lake in eastern Washington. At 5400 B.P., the deposition of re-worked Mazama ash in the lake sediments suddenly ends (Blinman et al. 1979). As noted above, this indicates that vegetation density on the landscape became great enough to prevent the re-suspension of the volcanic ash either by wind or water. This increase in vegetation density probably directly reflects a dramatic shift toward greater rainfall.

At Diamond Pond in south central Oregon there was a shift from dry climate pollen assemblages to much more mesic assemblages (Wigand 1987). Within just a few decades about 5400 B.P., pollen spectra which had been dominated by as much as 80% greasewood pollen are instead suddenly characterized by as much as 40% sagebrush pollen. Although there is a shift back towards drier climate during the next fourteen hundred years, the conditions that characterized the period before 5400 B.P. never re-occur. In the Tahoe Basin, montañe forests were so rapidly submerged beneath the rising cold waters of Lake Tahoe at 5500 B.P. that they had no opportunity to decay (Martin Rose, personal communication). At the same time mesic shrub species such as willow, birch and fir increase significantly in the pollen record of Little Valley on the east slope of the Sierra Nevada Mountains just east of Lake Tahoe (Wigand and Rhode 2002). An eight-thousand year bristlecone pine tree-ring record from Methuselah Walk in the White Mountains reveals this same climatic event which terminated a period of drought that may have lasted as much as 1,500 years (Graybill et al. 1994). This event is documented as far south as the northern Mojave Desert as renewed spring discharge evidenced by the growth of peats at Little Lake (Mehringer and Sheppard 1978), in Ash Meadows (Mehringer and Warren 1976), and at Lower Pahranagat Lake (Wigand 1997b). The pollen record from Wildcat Lake between 5400 and 4400 B.P. also indicates reduced abundance of saltbush pollen compared with the record below Mazama ash (Johnson et al. 1994:Figure 9). These data suggest that sagebrush may have begun replacing saltbushes in the Pasco Basin at this time.

The shift to wetter conditions after ~5,500 years ago initiated the re-expansion of woodlands and montañe forests. Along the northern margin of the Plateau and in the Okanogan Highlands pollen records record the initiation of pine expansion regionally (Mack et al. 1976; Mack et al. 1978a; Mack et al. 1978b; Mack et al. 1978c). Although pollen records from the Columbia Plateau (including the Carp Lake record, which has a low Holocene sample resolution) do not reveal the 5500 B.P. event, they clearly evidence the "Neoglacial" between 4,000 and 2,000 years ago. Whitlock et al. (1997) indicate that modern forest communities were established at this time. Ponderosa pine and Douglas fir became the local dominants. Alder, hazel, and other shrubs formed the understory. Grand fir, white pine and western hemlock grew on mesic sites. Oak dominated the lower elevation woodland. Grasslands seem to have become established at this time in the area of Carp Lake. Pine expansion reaches Holocene maximums at both Simpsons Flats and Waits Lake in the Okanogan Highlands at this time (Mack et al. 1978b; Mack et al. 1978c). At Big Meadow the highest values of spruce for the Holocene probably document the "Neoglacial" (Mack et al. 1978a).

Movement of juniper woodland into areas at elevations lower than where it is currently found in the northern Great Basin are recorded in Diamond Pond by 4000 B.P. The juniper pollen record and western juniper macrofossils from woodrat middens in Diamond Craters east of Diamond Pond both evidence this event (Mehringer and Wigand 1990; Wigand 1987). Except for two lapses, western juniper woodland remained near this lower elevational limit (~150 m below its current extent) until about 1,900 years ago (Mehringer and Wigand 1990; Wigand 1987). Broad "Neoglacial" expansion of pine in the woodlands on the northernmost boundary of the Great Basin is recorded in the Craddock Meadow pollen record north of Burns, Oregon (Wigand 1989). Together these data reflect a pattern of forest and woodland expansion that occurred throughout the entire northern and western Great Basin. These episodes are characterized not only by the expansion of both forests and woodlands, but also by a regional rise in water table, and a re-expansion of lakes and marshes (Grayson 1993; Wigand 1987). In the northwestern Great Basin the archaeology of this period is characterized by the marsh-adapted Lovelock culture (Baumhoff and Heizer 1965).

In the Columbia Plateau proper, Mehringer indicates that between 4,000 and 2,000 years ago sagebrush steppe retreated toward the most arid southeastern corner of the region (Johnson et al. 1994). The pollen record from Wildcat Lake between 4400 and 2400 B.P. has high values of grass pollen and much reduced sagebrush and saltbush abundance than previously (Johnson et al. 1994:Figure 9). These data suggest that sagebrush steppe was increasingly grassy and that replacement of saltbushes with sagebrush in the Pasco Basin was well advanced. Forests comprised of ponderosa and lodgepole pines, and larch or Douglas fir advanced along the northern and eastern edges of the Plateau. Spruce and fir invaded the formerly dry forests east and north of the Plateau. This corresponds to the record from Carp Lake. Chatters (1995:382) asserts that from 3900 to 2400 B.P., there is evidence for an increase in available moisture and a decline in temperature that persisted until at least 2200 B.P. (1995:387). Records from the forests at the eastern edge of the Plateau (e.g., Blue Lake, Idaho) indicate warm, moist conditions prior to 4000 B.P. (Smith 1983). Between 4,000 and 3,000 years ago cool moist conditions predominate. Following 3000 B.P. drying conditions suggest the replacement of dense, mixed conifer forest with ponderosa pine parkland (Smith 1983).

Conditions within the many deeper canyons of the Columbia Plateau between 4000 and 2000 B.P. must have become much more pleasant than during the Middle Holocene. Stream-flow and spring reliability must have increased significantly as it did in the Great Basin. Increased fire frequency, whether natural or encouraged by human intervention created a more complex mosaic of habitats where berry crops and game animals could proliferate.

In general, climates south of the Plateau in the northern Great Basin during the "Neoglacial" were cooler and significantly wetter, with winter precipitation dramatically increased with respect to summer precipitation (Davis 1982; Wigand 1987). Periodic increased abundance of grasses relative to sagebrush and saltbushes during this period indicate that the broad areas around Diamond Craters, which today are dominated by desert scrub vegetation, may have been characterized by a grassy, sagebrush steppe. A dramatic reduction in desert shrub vegetation also resulted from marshland expansion into areas previously dominated by greasewood and saltbushes (Mehringer and Wigand 1990; Wigand 1987). This pattern probably characterized much of the northern and western Great Basin.

Increased abundance of grass pollen coincident with woodland expansion also mirrors the presence of a vigorous herbaceous understory and of occasional fire episodes. Three very pronounced grass pollen increases at Diamond Pond between 4,000 and 2,000 years ago are closely tied to preceding charcoal events, evidence of dramatic local grass expansion after fire (Miller and Wigand 1991). Periodic, regional mega-droughts resulted in extensive fires that characterized both the lower and middle elevational distribution of juniper woodland in the Great Basin during the "Neoglacial" (Wigand 1987).

A comparison of moisture indexes generated for the Intermountain West suggests that the Neoglacial wet period appears to have been more pronounced in the northern Great Basin and the Pacific Northwest, rather than the southern Great Basin or southwestern United States. Only two (at ~3,700 and 2,700 years ago) of the three major wet periods seen in the Diamond Pond record appear in the Lower Pahranagat Lake (southern Nevada) record (Wigand and Rhode 2002:Figure 17). They are of considerably lesser magnitude at Lower Pahranagat Lake than at Diamond Pond (Wigand 1987:Figure 18). This may indicate that Pacific storm systems were primarily responsible for these periods of wetter climate, and that their impact was focused primarily to the north.

Reconstructed deposition rates of the Lower Pahranagat Lake record, together with dramatic changes in relative pollen values, indicate that at times the transition to wetter, or conversely drier, conditions often took less than a decade or two (Wigand and Rhode 2002:Figure 17). This may have been characteristic for the entire Intermountain West during the "Neoglacial". The local increase in effective precipitation needed to accomplish the observed changes in juniper abundance, based upon the difference in the minimal annual rainfall requirements between sagebrush and Utah juniper, must have been at least 10 to 20 mm per year at elevations around 1500 m. Reduced evaporation rates due to reduced mean annual temperature probably played a significant role in increasing effective precipitation during the "Neoglacial" as well.

Late Holocene (2,000 to Present).

Beginning ~1,900 years ago, Intermountain West climate generally became warmer and drier (Davis 1982; Wigand 1987; Wigand and Nowak 1992; Wigand and Rose 1990). Mehringer, in Johnson et al. (1994:Figure 9) indicates that the Wildcat Lake record between 2,400 and 600 years ago was at times more grassy than the "Neoglacial", and occasionally more saltbush-rich than during the period before the fall of Mazama ash. This suggests a period of great variability. However, it may hint at shifts in climate that characterize the central and southern Intermountain West as well. Lower pine values at Simpsons Flats and Waits Lake during this period suggest that annual precipitation values may have been generally lower (Mack et al. 1978b; Mack et al. 1978c). The retreat of spruce at the same time in the Big Meadow record seems to confirm this (Mack et al. 1978a). Significantly increased aquatic plant abundance at Waits Lake (including that of sedge, cat-tail, bayberry [*Myrica*] and birch) may suggest either: 1) expansion of the marshy margins of the lake into the lake as a result of drying, or 2) expansion into the surrounding forest margins as a result of flooding (Mack et al. 1978c). Similar expansions of sedge at Simpsons Flats may reflect either case as well. As indicated above, Smith (1983) sees the replacement of dense, mixed conifer forest with ponderosa pine woodland. This again suggests drying.

Juniper pollen values in the northwestern Great Basin declined dramatically with respect to shrubs and grasses during this period (Wigand 1987). In addition, the shift in the ratio of coarse to fine charcoal when compared with the changes in the dominant vegetation type at Diamond Pond, Oregon clearly reflect the change from juniper woodland to shrub steppe fuels (Wigand 1987).

At Fish Lake an increase in big sagebrush pollen relative to grass pollen reflects decreased grass in response to drier conditions and was probably characteristic of the upper sagebrush zones of the northern Great Basin (Mehringer 1985). An increase in desert scrub vegetation, indicated by increasing saltbush and greasewood pollen at Diamond Pond provides additional evidence for increasing local and regional aridity, particularly between 1900 and 1000 B.P. (Wigand 1987). Pollen and macrofossils of aquatic plant species at Diamond Pond indicate that water levels had dropped significantly since the "Neoglacial" (Wigand 1987).

An unpublished pollen record from Lead Lake in the Carson Sink records terrestrial and aquatic vegetation dynamics for the central Great Basin during the last 2,200 years. The major trend for the last two millennia has been the decline in greasewood while marsh species became more dominant. The decline of greasewood is matched by increases in pine, sagebrush, and grass (Wigand and Rhode 2002:Figure 11). These data document marsh expansion into areas previously occupied by saltbush communities. The predominance of emergent aquatic plants relative to littoral plant species indicates slightly deeper water conditions in marshes between 1,900 and 800 B.P. (Wigand and Rhode 2002:Figure 12). Deeper water at this time does not mean the climate was wetter. Instead, evidence from other localities suggests that although annual precipitation remained relatively unchanged a shift toward summer rainfall resulted in the persistence of deeper water into the summer when high evaporation rates would normally have shrunk the marsh or dried it out.

A rapid expansion in piñon pine distribution at this time was primarily triggered by increased rainfall throughout the summer as well. This encouraged seedling establishment (Wigand and Rhode 2002:Figures 13a and 13b). However, this expansion would probably not have been possible without the milder winters that followed the "Neoglacial" after 1,900 years ago. At Diamond Pond south-central Oregon an early post "Neoglacial" expansion of grass between 1,900 and 1,000 years ago coincides with piñon pine expansions seen at Lead Lake in central Nevada (Wigand 1987, 1997). Western juniper, which during most of the record from Diamond Pond had varied in concert with grass, did not increase in abundance during this period, indicating that grass alone was responding to summer-shifted rainfall as well.

That this expansion of grass was regional in nature is demonstrated by the dramatic increase of bison remains in archaeological sites of the northern Intermountain West. Radiocarbon dates plotted with standard deviations on bison remains from archaeological sites in the northern Great Basin (Marwitt 1973) and the Plateau of eastern Washington (Schroedl 1973) show remarkable coincidence with this episode of grass expansion (Figure 3.9). This correspondence suggests that the Columbia Plateau was also experiencing increasingly grassy conditions as well. This is most likely what is evidenced in the higher grass values for Wildcat Lake between 2,400 and 600 years ago.

Significant increase in juniper pollen values ~1000 B.P. in the northern Great Basin at Diamond Pond, coincident with local increased abundance of woodrat middens containing western juniper

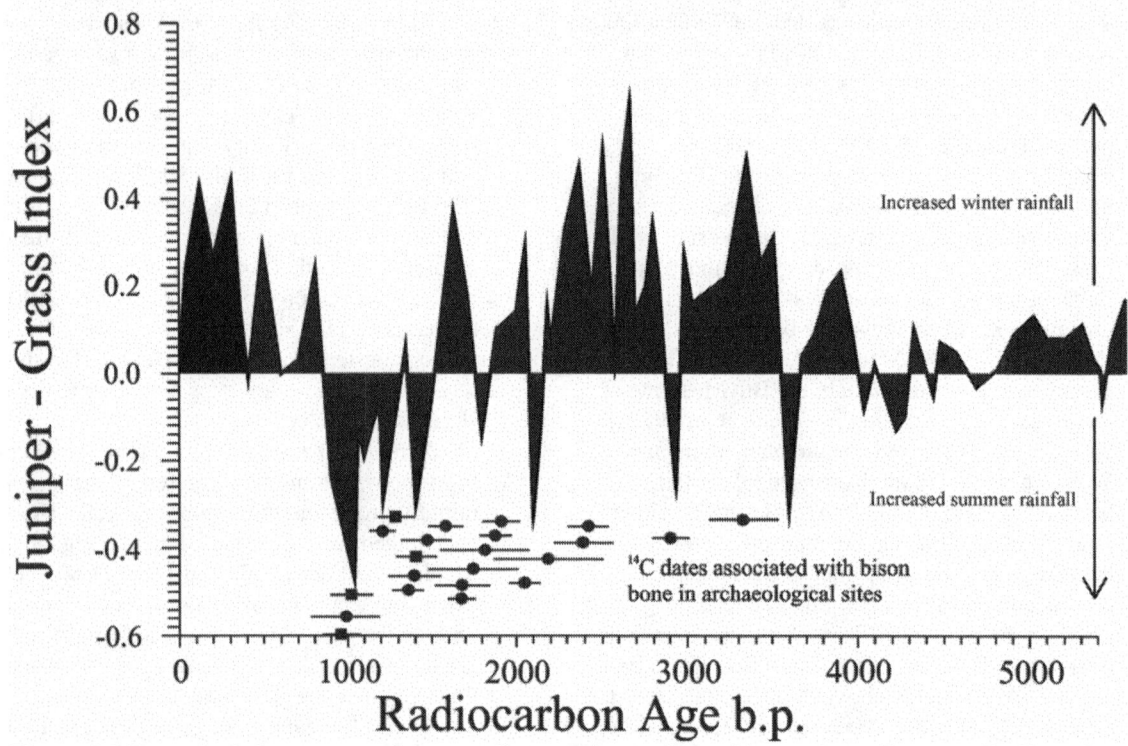

Figure 3.9 Drought index generated from pollen at Diamond Pond in southern Oregon. Radiocarbon dates on bison bone have been collected from the northeastern Great Basin and the Columbia Plateau.

macrofossils, evidences renewed expansion of woodland in the northern Great Basin and the end of grass abundance (Mehringer and Wigand 1990; Wigand 1987). This corresponds to increased large vs. small charcoal values indicating both a change in fuel type as well as more frequent fires as fuels accumulated in response to wetter climate (Wigand 1987; Wigand et al. 1995). Dugas (1998) indicates that higher lake stands may have occurred in the Malheur basin ~1,000 and ~800 years ago.

A severe drought in the middle of the first millennium is recorded by the expansion of saltbushes in the northern Great Basin, as well as by evidence that Diamond Pond reached some of its lowest levels since the middle Holocene (Wigand 1987). Re-expansion of marsh with the recession of Lead Lake evidences regional drying of the western Great Basin after 800 years ago (Wigand and Rhode 2002:Figures 12 and 13). Clear Lake, located 19 km northeast of the Marmes site, provided Bartholomew (1982) with a late prehistoric record of pollen, plant macrofossils, and algae. He concluded that a dry climatic interval began around 1,100 years ago and continued until about 600 years ago (Bartholomew 1982:70), roughly equivalent with the drought noted in the Great Basin record.

Charcoal evidence from the cores at Little Valley indicates that fire frequency increased considerably after about 600 years ago (Wigand and Rhode 2002:Figure 11). Destruction of the lower elevation woodlands appears as a sudden gap in the woodrat midden evidence of northern Nevada for this period (Wigand and Nowak 1992:Figures 14a and 14b). Stine (1990) sees extended droughts terminated about 900 B.P. and 600 B.P. by brief wet events, and following 350 years ago by the "Little Ice Age".

The "Little Ice Age," a pattern of stronger winter precipitation and cooler temperatures beginning 300 to 400 years ago, terminated the droughts of the middle portion of the millennium and initiated a gradual re-expansion of juniper woodland in the northern Great Basin (Mehringer and Wigand 1990). By the time Europeans first entered the area, climate-initiated re-expansion of Great Basin woodlands was well underway. In the northern Great Basin increased occurrence of pine in the pollen record at Lead Lake during the last 150 years corresponds to the post-"Little Ice Age" 2.5-fold increase in areal coverage of piñon pine recorded in stand establishment records (Tausch et al. 1981). The pollen record at Diamond Pond indicates increasing aridity since the end of the Little Ice Age. Juniper woodland retreated and sagebrush steppe advanced (Wigand 1987). Grasses seem to have become less abundant throughout the region. A decrease in water depth in Diamond Pond reflects a regional drop in water table in the Harney Basin (Mehringer and Wigand 1990; Wigand 1987). Little or no evidence for the "Little Ice Age" exists in the Columbia Basin. The one possible clue to its occurrence are the pollen per cm^3 in the Wildcat Lake record. A dramatic increase in pollen abundance that began after 400 B.P. and terminated before the impact of European settlement upon the lake became pronounced may reflect higher regional organic (and pollen) production in response to the wetter conditions that characterized the "Little Ice Age" (Davis et al. 1977).

The post "Little Ice Age" trend toward lower effective precipitation brought on by warmer mean annual temperatures combined with the spread of woodlands should have increased the potential for fire in the Great Basin. Despite increased drought stress on the trees comprising the semi-arid areas, the occurrence of fire decreased. This stands in contradiction to the prehistoric pattern of massive fires that typified periods when Great Basin woodlands were subjected to drought conditions (e.g., the end of the "Neoglacial") (Wigand 1987). In part, decreased fire frequency may have occurred because conditions were dry enough to keep the production of light fuels (i.e. grasses and forbs and even shrubs) low. On the other hand, Peter Skene Ogden noted abundant evidence of Native American-set fires in the Harney and Malheur lakes region during the middle 1820s (Davies et al. 1961). A decline in fire frequency as native populations were displaced and declined may have enabled the expansion of semi-arid woodlands at the time of European settlement. Livestock grazing may also have played a role in decreased fire frequency through the reduction of fine fuels (Miller and Wigand 1991). Grazing may also have played a role in juniper expansion through seed dissemination and encouragement of shrubs that provided nursery areas for juniper seedling establishment. The record from Wildcat Lake indicates at least two periods of intense erosion that reflects impact of grazing animals on groundcover and erosion rates (Davis et al. 1977).

Prehistorically, climate change was the primary factor affecting the expansion of semi-arid woodlands and steppe. Droughts and resultant disturbance phenomena such as insect infestations, disease and fire all contributed to periodic retreat of woodlands and expansion of shrub steppe. Since European arrival, both climate change and human action have significantly affected woodland/steppe distributions. Drier climates during this century have resulted in increased physiological stress on plant communities. Fire suppression practices and the use of insecticides have disrupted normal cycles of disturbance phenomena such as fire, insect infestation and disease. Resulting fuel build-up has increased the potential for intense, widespread fires.

Study Area Vegetation

Current vegetation in the area is shrub-steppe, referred to as the *Agropyron spicatum-Poa secunda* zone, associated with deep, loamy soils along the Snake River and its tributaries (Daubenmire 1970; Franklin and Dyrness 1973). The dominant plants are bluebunch wheatgrass (*Agropyron spicatum*) and bluegrass (*Poa secunda*). Upland settings support Idaho fescue (*Festuca idahoensis*) associated with bluebunch wheatgrass as major components of the *Agropyron-Festuca* association (Daubenmire 1970). Rabbitbrush (*Chrysothamnus nauseosus*) and other shrubs occur in both communities, but do not form a true shrub canopy. Rabbitbrush has a tendency to occur in disturbed contexts. Cheatgrass (*Bromus tectorum*), an introduced species, has replaced many of the native grasses, particularly in those areas used for cattle grazing. Other introduced species, such as teasel (*Dipsicus sylvetris*), mullein (*Vergascum spp.*), and horseweed (*Conyza spp.*) also are present.

Other plant species in the study area include sagebrush (*Artemisia tridentata*), Indian wildrye (*Elymus spp.*), and mustard (*Brassica spp.*). Willow (*Salix spp.*) and cottonwood (*Populus hasata*) occur along stream banks and moist habitats (Daubenmire 1942; Hitchcock and Cronquist 1981). Cat-tails (*Typha latifolia*) and tules (*Scirpus spp.*) are common along the Palouse River.

Many plant species occurring in the study area provided the Native American groups in the area with foods and materials for manufacturing mats and baskets. These include, but are not limited to, cat-tail and tule, many grass species, wild onion (*Allium spp.*), wild carrot (*Daucus spp.*), bitterroot (*Lewisia rediviva*), balsamroot (*Balsamorhiza spp.*), biscuitroot (*Lomatium spp.*), various lilies (*Liliaceae*), and a myriad of berries, and currants (Grossulariaceae) (Jorgensen 1980; Ray 1933; Spinden 1908). Most of the roots are available from early spring to early summer, whereas the berries are available from mid summer to early fall.

Study Area Faunal Resources

Although both mule deer (*Odocoileus hemionus*) and whitetail deer (*O. virginianus*) occur in the vicinity of the Marmes site today, mule deer are the most common ungulate resource (Asherin and Claar 1976:117). However, most deer move up to the major tributary canyons and/or to the higher elevations by late spring and do not return to the bottomlands until early winter (Asherin and Claar 1976:133). Gustafson (1972, 1996) reports that prehistoric archaeological sites in the area contain the remains of elk (*Cervus canadensis*) and pronghorn antelope (*Antilocapra americana*). Antelope, however, became extinct locally before the arrival of Euroamericans, and elk disappeared in eastern Washington by 1910 (Dalquest 1948). Dalquest (1948) also notes that bighorn sheep (*Ovis canadensis*) were present along the Snake and Columbia River canyons into the historic period; Lothson (1989) describes prehistoric procurement strategies for bighorn sheep based on archaeological sites located in the mid-Columbia River area. Deer, elk, antelope, and bison have all been reported in Snake River site assemblages.

Medium and small sized mammals of importance to Native American populations include a number of species: cottontail (*Sylvilagus nutalli*) and jackrabbit (*Lepus californicus*), marmot (*Marmota flaviventris avara*), muskrat (*Onadatra zibethicus osoyoosenis*), beaver (*Castor canadensis*), badger (*Taxidea taxus*), coyote (*Canis latrans*), and bobcat (*Lynx rufus pallescent*) (Dalquest 1948). Evidence of badger activity, namely burrowing, was observed during the Marmes Rockshelter excavations and is common in rockshelters throughout the region. A variety of mice, rats, and squirrels also occur in the area (Dalquest 1948; Gustafson 1972).

Avian species of importance include Canadian geese (*Branta canadensis*), mallards (*Anas platyrhynchos*), and other ducks. Grouse (*Pedioecetes phasianellus*) and sagehen (*Centrocerus urophasianus*) also were once common. Various birds of prey, including hawks, eagles, falcons, and owls, are present.

In addition to ungulate and plant resources, the Palouse and Snake Rivers provided several varieties of fish and other aquatic resources that were utilized extensively by local Native American groups. Anadromous species of note include Chinook (*Oncorhynchus tshawytscha*) and sockeye (*O. nerka*) salmon, steelhead trout (*Salmo gairdneri*), and lampreys (*Lampetra planieri*). Resident species, such as squawfish (*Ptychocheilus orogenisis*), sturgeon (*Ancipencer transmontanus*), whitefish (*Prosopium spp.*), and suckers (*Catostomus spp.*), also were exploited. Freshwater mussels (*Margaritifera falcata* and *Gonidea angulata*) also were available (Netboy 1980).

4
CULTURAL CONTEXT

Brent A. Hicks

This report cannot present a cultural overview as thoroughly or as well as any number of recent summary reports on Plateau archaeology. Of particular use to this study has been the Plateau volume of the Handbook of North American Indians (Walker ed. 1998) and Reid's Snake River Overview (ed. 1995).

Prehistory

In 1962, Daugherty outlined four periods in which he described a developmental progression within what he called the Northwest Riverine Areal Tradition, which included the lower Snake River area, a regional manifestation of a larger Intermontane Western Tradition (1962). The first three periods correlated with climatic periods: the Early period with postglacial anathermal climates (11,000-8000 B.P.); the Transitional period with hypsithermal climates (8000-4500 B.P.); the Developmental period began with the inception of essentially modern climatic conditions beginning ca. 4500 B.P.; and a Late period beginning about 2000 B.P. and ending at historic contact. Summarizing the Northwest Riverine Areal Tradition, Reid and Gallison (1995) described:

> Major trends included a shift toward riverine adaptations and increased use of storage facilities during the Transitional period, and the appearance of local art styles and elaborate burial practices during the Late period. The process of change was viewed as gradual and cumulative, with new traits appearing and then adhering to the body of previously existing customs and lifeways....culture along the lower Snake didn't become more complex with the passage of time; it just got bigger. [Reid and Gallison 1995:2.23-2.24]

Much of Daugherty's conceptualization of the Northwest Riverine Areal Tradition came from his involvement in and familiarity with the rapidly growing data base from the lower Snake River area (e.g., Daugherty 1956a, 1960; Fryxell and Daugherty 1962).

Leonhardy and D. Rice (1970), building on Daugherty's periodization, have provided the most commonly utilized cultural chronological framework for the Snake River drainage; this chronology is also generally cited throughout the southern Plateau region. The area described by Leonhardy and Rice (1970) consisted of three *districts* between the confluence of the Snake and Columbia Rivers and the vicinity of the Clearwater River, with the database weighted toward the upstream district (Reid and Gallison 1995:2.7). Their chronology was based on the analysis of assemblages from 19 sites, but relied in large part upon the findings from excavations at Windust Cave (H. Rice 1965), Marmes Rockshelter (Fryxell and Daugherty 1962; D. Rice 1969), and Granite Point (Leonhardy 1970).

Leonhardy and Rice (1970) proposed four periods and six phases as a basis for organizing archaeological collections recovered from the lower Snake River. Beginning with the oldest, the time periods included the Pioneer, Initial Snake River, Snake River, and Ethnographic. The Pioneer period encompassed the Windust Phase (8000-7000 B.C.) and the Cascade Phase (6000-3000 B.C.); the Initial Snake River period consisted of the Tucannon Phase (3000-500 B.C.); the Snake River period included the Harder Phase (500 B.C.-A.D. 1300) and the Piqunin Phase (A.D. 1300-1700); and the Ethnographic Period consisted of the Numipu Phase (A.D. 1700 to contact).

These phases were defined in terms of their formal content and restricted distributions. "While the Cascade Phase was believed to have evolved out of the earlier Windust Phase, no such continuity was assumed between the Cascade and Tucannon Phases" (Reid and Gallison 1995:2.26). The remaining phases, beginning with the Tucannon Phase, represented a second, distinct evolutionary continuum to Leonhardy and Rice (1970). During this second continuum, fish and plant resources became increasingly important in the economies of the people. It is uncertain whether the earlier and later continua are related or represent two separate manifestations. This represented a major departure from Daugherty's (1959, 1962) conceptualization of slow, cumulative, in place adaptation.

The Windust and Cascade Phases were considered to represent an evolutionary continuum of the first documented human occupation of the area (Leonhardy and Rice 1970:22-24). How the Windust populations correlate with the Clovis peoples, documented as being in the Wenatchee area soon after the 11,225 B.P. eruption of Glacier Peak (Mehringer and Foit 1990), is uncertain. The Windust people primarily were big game hunters but also exploited rabbits, beaver, and river mussels; artifacts associated with plant processing are rare in these assemblages. Diagnostic artifacts include shouldered, short bladed projectile points with straight stems and straight or concave bases. Stone tools were based on tabular flakes and blades rather than bifacial blanks or cores (Reid and Gallison 1995:2.26).

The Cascade Phase was divided into early and late subphases separated by the Mazama ashfall. Lanceolate (willow leaf-shaped) projectile points are considered diagnostic of the early subphase, while large side-notched Cold Springs points occur alongside generally smaller versions of the leaf-shaped points in the later subphase. Bense (1972) noted other differences between assemblages from the two subphases, and Reid and Gallison (1995:2.30) note that evidence from Hells Canyon and Hatwai may suggest that the two subphases may represent different adaptations. Atlatl weights are found for the first time, although the occurrence of bone and antler spurs in Windust assemblages implies an earlier use of this throwing tool. Plant processing tools associated with the use of seeds rather than roots is noted for the first time, as is evidence of the exploitation of salmon and steelhead (Reid and Gallison 1995:2.26-2.30). Bense (1972) draws a correlation between the subsistence base of Cascade Phase and ethnographic Nez Perce populations.

Tucannon Phase projectile points, described as stemmed or corner notched dart points with triangular blades, are easily distinguished from Cascade style points. Kennedy (1976) posits that the change in point style is a result of combined influences from the west, south and east. Lucas (1994) has sought to explain the change in assemblages from the Cascade Phase to the Tucannon Phase in terms of environmental pressure leading to an increase in the use of upland subsistence resources, particularly root crops as indicated by the presence of hopper mortars and pestles. Still, the increased exploitation of riverine resources such as river mussels and fish also is documented by anadromous fish remains, net weights, and bone shuttles indicating the manufacture of nets. Hammatt (1977) has suggested that the decrease in apparent Tucannon Phase sites can be attributed to aggradation of the Snake River at this time, with frequent floods scouring the floodplain and potentially affecting evidence of cultural use. Harder (1998) agrees with Leonhardy and Rice (1970) that the Tucannon Phase exhibits significant differences from the Cascade Phase. Harder's (1998) expanded definition of the Tucannon Phase largely disposes with the perception that there are fewer sites during this period, and the initial use of housepits (5050 ± 220 B.P. at Hatwai) in tandem with evidence of resource intensification marks the beginning of this phase.

The Harder Phase was presented in two subphases differentiated by an apparent change in point styles and different site types associated with the components assigned to the two subphases. The early subphase is marked by corner notched dart points and assemblages reflecting base camp sites, while the later subphase assemblages include increasing numbers of small corner and basally notched arrow points and all come from pithouse sites. Leonhardy and Rice (1970:14) believed that the Harder Phase marks the establishment of villages composed of multiple pithouses on the lower Snake River (Reid and Gallison 1995:2.33).

The Piqunin Phase is defined from a single site (45GA61) and exhibits variously notched forms of the Columbia Valley corner-notched and rectangular stemmed arrow points. Items such as twined basketry, bone awls, matting needles, and harpoon elements, rarely found in earlier phases, also were found here. Leonhardy (1975) later determined that a distinction between Harder and Piqunin Phases was probably not justified. The Numipu (or ethnographic) Phase is described from burials and material culture dating after A.D. 1700 and continues to the reservation period after the treaties of 1855 and 1863 (Reid and Gallison 1995:2.33-2.35).

Reid and Gallison (1995:2.35) note that, following the publication of Leonhardy and Rice's (1970) cultural chronology framework, archaeological manifestations of the individual phases were analyzed by a number of researchers:

Rice [1972] for Windust; Bense (1972) for Cascade; Kennedy (1976) [and Harder (1998)] for Tucannon; Brauner (1976) and Yent (1976) for Harder; Yent (1976) for Piqunin; and Adams (1972) for Numipu. Integrative studies of lithic technology (Muto 1976), archaeofaunas (Schroedl 1973, Lyman 1976), and archaeological stratigraphy (Hammatt 1977) were also completed within this framework. Most of the work was done at Washington State University under a New Archaeology paradigm that looked more to Chang (1967) and Clarke (1968) than to Binford (1968). The approach combined the culture-historical systematics of Willey and Phillips (1958) with a...systems theory perspective derived from Clarke's (1968) *Analytical Archaeology*. [Reid and Gallison 1995:2.35]

Reid and Gallison (1995:2.35-2.36) also note that soon after most of the above referenced works were completed, a shift in the orientation of field research to a Binfordian or neo-evolutionary perspective occurred. This perspective called for regional research designs and emphasized the record forming potential of long term demographic processes. Schalk and Cleveland (in Schalk 1983) applied this approach in developing a stadial model of plateau prehistory based on long term changes in settlement and subsistence patterns. The three principal adaptations described in their model were early and middle Holocene *broad spectrum foragers*, late Holocene *semisedentary foragers*, and protohistoric *equestrian foragers*.

This stage framework was presented as a processual advance on earlier culture-historical schemes, including the Leonhardy and Rice (1970) framework. The new approach emphasized the greater use of testing and surface collection data in the context of regional research designs, rather than the analysis of whole assemblages from large excavation blocks. Reports from the second wave of "new archaeologists" include Schalk (1980a) at McNary Dam; Burtchard et al. (1981) at McNary Reservoir; Miss and Cochran (1982) at Riparia; Schalk (1983a, b) at Lyons Ferry and Strawberry Island; and Thoms (1983) at McNary Reservoir. Behind this paradigm shift from Clarke to Binford was the changed archaeological mission from pre-reservoir salvage excavation to post-inundation monitoring and compliance studies. [Reid and Gallison 1995:2.36]

Ames et al. (1998) offered a cultural chronology for the Southern Plateau region that uses numerical designations (e.g., Period IA) rather than named subphases in establishing the chronology's periodization. Ames et al. (1998) chronology appears to mesh Leonhardy and Rice's chronology (including Leonhardy's [1975] revisions) and Sappington's (1994) chronology for the Clearwater River region. The Windust and Cascade Phases are collapsed into a single period (Period IB) similar to Sappington's Early Prehistoric Period and the timing of the boundaries between several phases is adjusted (Figure 4.1).

Only the Richey-Roberts Clovis Cache site, approximately 150 km northwest of Marmes Rockshelter, is included in the initial Paleo-Indian sub-period (Period IA), which dates between ca. 11,500-11,000 years ago in Ames et al. (1998) chronology. Leonhardy and Rice's (1970) classification did not include a typological unit for this time period. The Richey-Roberts site is "classic Clovis" rather than attributable to the "Western Fluted" materials (after Dixon 1999, in Ames 2000:7). Referred to as a "cache", the site's materials included a diverse assemblage of formed objects and debitage believed to be intentionally buried in a manner that suggests ceremonial activity, which Ames et al. (1998:103) suggest hints at an evolved socioreligious system for these early occupants.

The materials were placed on top of sediments rich in Glacier Peak ash, essentially dating them to immediately after these eruptions around 11,250 B.P. (Johnson et al. 1994). Surface finds of Clovis points are fairly rare throughout the Plateau although an unpublished paper from the mid-1980s by M. Avey compiled more than twenty confirmed surface observations of Clovis points in Washington.

Ames' et al. (1998) major change to Leonhardy and Rice (1970) is the combination of the Windust and Cascade phases into a single Period IB. This diminishes the significance of a transition in projectile point types from the stemmed lanceolate Windust points to the foliate Cascade points that diminish in size through time until they overlap with large side-notched (Cold Springs) points that defined the late Cascade sub-phase. In what seems a backward justification for these opposing period

Dates (B.P.)	Leonhardy & Rice 1970	Sappington 1994	Ames et. al 1998
200	Numipu	Protohistoric Period	Early Modern
300			
500			
700	Piq'unin	Late Prehistoric Period	Period III
1,000	Harder		
1,500			
2,000			
2,500			
3,000	Tucannon	Middle Prehistoric Period	Period II
3,500			
4,000			
4,500			
5,000			
5,500	Late Cascade		
6,000			
6,500			
6,730 (Mazama)		Early	
7,000	Early Cascade	Prehistoric Period	Period IB
7,500			
8,000			
8,500			
9,000			
9,500	Windust		
10,000			
10,500			
11,000			
11,500			Period IA
12,000			
12,500			

Figure 4.1 Comparison of Cultural Chronologies for the Southern Plateau.

definitions: Rice (1972) and Bense (1972) see no differences between the Windust and Cascade Phases other than these gradually changing, overlapping point styles, which would seem to argue for a single period; while Ames et al. (1998:106) describe shifts in settlement patterns and tool technologies around 9000 B.P. (the time assigned as the boundary between the Windust and Cascade Phases by Leonhardy and Rice [1970]), which would seem to be a better argument for two periods. The result of Ames' et al. (1998) compilation of Windust and Cascade Phases appears to favor the continuity observed by Bense and Rice and recognizes the gradual change in assemblages over this time period.

Willig and Aikens (1988) in looking at the larger Far West region (west of the Rocky Mountains) place both the Windust and Cascade Phases in the Western Stemmed Complex (ca. 11,000 to 7500 B.P.) and note that while "there is a certain diversity of regional styles,…all complexes share similarities in technology, typology, and implied settlement-subsistence patterns" (1988:4). Marmes Rockshelter deposits have the potential to offer information that may clarify the Windust-Cascade transition. Other sites from this period that are particularly important in describing the full complement of assemblages (and the subsistence variation they imply) from the region include Five-Mile Rapids at the east end of the Columbia River Gorge, Lind Coulee in the central basin, Granite Point on the Snake River above Marmes Rockshelter, and Hatwai on the Clearwater River in Idaho.

In general terms, Period I sites reflect low population densities and high levels of mobility. Windust projectile points exhibit wide bases relative to blade size, edge grinding of the stems, and can be highly variable in both size and shape due to reworking. The early Cascade foliate point style also occurs in Windust Phase assemblages and is similar in size to "classic" stemmed and/or indented base Windust points, but clearly represents a different hafting mechanism that may reflect initiation of darts thrown with atlatls or other throwing boards. Early on in this period lithic technology includes blade and flake manufacture from prepared, usually chert, cores. Other lithic tools of note include burins and small, girdled pebbles that may be bola stones or a kind of net weight. Later in Period I, large side-notched points occur and lithic manufacturing includes Levallois prepared core technique (associated with the Cascade Phase), with microblades and microblade cores found in some assemblages but not along the lower Snake River. Bone tool technology is well represented. Subsistence appears to focus on hunting of large and medium mammals and exploitation of plants is suggested by the presence of small milling stones and edge ground cobbles, particularly after ca. 9000 B.P. There is some evidence of salmon fishing, but no storage is indicated during this period. There are a few examples of structures built on the ground surface that may be temporary shelters, perhaps used only for sleeping (Ames et al. 1998; Ames 2000; Muto 1976). Ames (2000) suggests that the central Columbia Basin was abandoned at some time during Period I, but many fewer sites have been excavated in the central basin compared with the major river valleys and more work is needed to demonstrate this pattern.

Ames' et al. (1998) Period II represents the last of the Late Cascade subphase and the Tucannon Phase in Leonhardy and Rice's (1970) chronology. As discussed above, this period is marked by settlement and subsistence changes reflected in the elaboration of Period I tool assemblages and the gradual disappearance of certain Period I artifact types. In particular, the first pithouses appear, both along the rivers and in the southern uplands. Settlements appear small, with few houses radiocarbon dated to the same occupations despite substantial associated cultural deposits, and virtual abandonment of housepits by the end of the period (ca. 4000 B.P.). Alpowa, Hatwai, and Hatiupuh all have housepits with dates that place their use at the beginning of this period, and two houses at Hatwai have intriguing earlier dates. But there are breaks in dated housepits from different parts of the Plateau that raises questions about the timing, locations, and success of the earliest transition to semisedentism in the Plateau. Upland sites are documented in a range of environments indicating use of these areas for a wide range of activities. Mortars and pestles, some quite large, become regular parts of assemblages and are thought to indicate an increasing reliance on food plants; together with fish and medium-sized mammals they appear to form the basis of subsistence for this period (Ames et al. 1998; Ames 2000; Chatters 1989).

Cascade points diminish in size and are joined by smaller dart-sized projectile points of varying forms (i.e. straight and excurvate blades, convex, straight, and concave stems, rectangular and rounded shoulders, and side notching that approaches corner-notch in some examples). The latter (Tucannon points) often look like small versions of the large late Cascade side-notched points (e.g., Cold-Springs) but with the notching occurring further toward the base. Harder (1998) demonstrates that both the lanceolate and large side-notch Cascade points occur into dated Tucannon Phase occupations and it may be that some of the Tucannon Phase points are reworked from the larger side-notched forms, with excurvate blade shape a result of working down points with broken tips and the varied stem shapes and near corner-notch examples a result of working down points with damaged bases. At the same time, the chipped stone assemblages appear to reflect less refined manufacturing. But while formed stone tools are less well-made, the bone tool industry appears to flourish and includes large needles and decorative objects.

This summary description of Period II sites indicates increases in exploitation of resources used in Period I, in particular, roots and salmon. Because of the increase in exploitation of roots and salmon, and the beginnings of housepit occupation and some evidence of initial storage late in this period (Harder 1998), it might be argued that this Period can be seen as the beginning of what further elaborates into the "Ethnographic Plateau Pattern".

Ames' et al. (1998) Period III correlates with the last of the Tucannon Phase and all of the Harder and Piqunin Phases in the Leonhardy and Rice (1970) chronology. This period represents the full establishment of the settlement and subsistence pattern of winter-occupied pit house villages as part of a seasonal round that focused on the intensive exploitation and storage of salmon and roots (especially camas). This pattern continues into the ethnographic period with elaboration as the horse became widely available in the early 1700s, marking the end of this Period. Pithouses increase in number at settlements and their size varies. A large housepit village on Strawberry Island included over 100 housepits dating to two principal occupations of the site (Schalk 1983b). After ca. AD 500 mat lodges appear but do not replace housepits. Increased population is indicated and may be related to evidence of use of the central Columbia Basin as well as other areas of the Plateau (but see discussion of Reid 1991 below). Projectile points decrease in size and become almost exclusively basal and corner-notch forms indicative of the use of bow and arrow early in this period, but atlatls continue to be found in sites until ca. AD 1000. By the time of Leonhardy and Rice's (1970) Piqunin Phase points are quite small, almost delicate in appearance, and indicate considerable refinement in manufacturing skill. There is also a considerable presence of botanical artifacts and raw materials in the archaeological record during this period, including basketry, matting, and cordage of varying thickness (reflecting widespread applications). It is expected that botanical technology was used prior to this time period, but has not persisted in the archaeological record to this degree. Cemeteries associated with housepit villages appear ca. 2500 B.P. (Ames 2000; Ames et al. 1998).

Reid (1991:29-30) has suggested that a dry interval that began ca. AD 1000 may have influenced human populations in the lower Snake River area, possibly leading to greater use of upland sites during the winter at this time. While Leonhardy (1975) ultimately favored dropping the Piqunin Phase distinction, Reid notes that this drought coincides with the Harder/Piqunin phase transition and suggests that there may be "patterned relationships between severe regional droughts, settlement pattern shifts, and perhaps even changes in the organization of subsistence and material culture" (1991:31).

Ames' et al. (1998) Modern Period and Leonhardy and Rice's (1970) Numipu Phase both begin with arrival of the horse and constitute the ethnographic period through to establishment of reservations. This is a time of great transition in Native American lifeway, with Euroamerican diseases rapidly diminishing the population base and the rapid spread of the horse and access to trade goods and material types all impacting the archaeological record of this period. Despite information from early explorers and traders in the region, and accounts collected by ethnographers beginning after 1900, archaeologists still look at this period as an elaborated extension of the previous prehistoric pattern. There also is an enticement to view the late prehistoric period in terms of the ethnographic period information (ethnographic analogy), which has contributed to model-building (e.g., Gould 1977) but must be conducted with caution (e.g., Dunnell 1979; Kelly 1995).

Ethnographic Period

The Marmes site lies within the territory historically occupied by the Palouse Indians or *palu'uspam* (Sprague 1998:358). At the time of contact the Palouse were found from near the confluence of the Snake and Clearwater rivers to the confluence of the Snake and Columbia Rivers (Sprague 1998:352) (Figure 4.2). The Palouse were comprised of Upper, Middle, and Lower groups (Trafzer and Scheuerman 1986). Upper Palouse groups are reported to have lived near and among the Nez Perce Indians; the Palouse "were apparently dominant at Almota and shared…the historic villages of Penawawa, Wawaiwai, and to a lesser extent Alpowa" (Sprague 1998:352). The Palouse and Tucannon Rivers region was occupied by the Middle Palouse, and the Lower Palouse occupied the area from Fishhook Bend to the confluence of the Snake and Columbia Rivers where they shared territory with the Wanapam (Sprague 1998:352). Additional bordering groups included the Cayuse and Walla Walla to the south, the Couer d' Alene, Spokane, and Columbia-Sinkiuse to the north, and the Yakama to the west (Spier 1936; Ray 1936, 1939).

Epidemic diseases decimated the Palouse population after Euroamerican contact. Bartholomew (1982:69) has estimated the Palouse population to have been roughly 5,400 individuals around A.D. 1780. However, populations across the Plateau may have been considerably larger prior to this; Boyd (1985) has documented extensive population loss from a series of recurring epidemics beginning ca. 1770. Between 1805 and 1854, the Palouse Indians' population declined from 1,600 to about 500 individuals, according to Swanton (1952). Trafzer and Scheuerman (1986) attribute this decline largely to a measles epidemic in 1847. Additional casualties were suffered by the Palouse during hostilities with the U.S. Army after 1858. By 1910, only 82 Palouse Indians remained in the area (Swanton 1952). As a consequence, what little is known about the Palouse has been inferred from more detailed knowledge of the neighboring Nez Perce and their culture (e.g., Haines 1955; Schwede 1966, 1970; Spinden 1908). Today persons of Palouse descent are found on several reservations in eastern Washington and Oregon and in Idaho.

As a political entity, the Palouse Tribe is represented by the Confederated Tribes of the Colville Reservation as a result of an Executive Order (July 2, 1872) that defined the current reservation as the reservation for any Indians the Department of Interior saw fit to locate thereon. Many of the Upper Palouse, who had fought with young Chief Joseph's Nez Perce and were removed to Indian Territory following Chief Joseph's surrender, were placed on the Colville Reservation by Commissioner of Indian Affairs John Atkins and special agent W. H. Faulkner upon their return. Many of the Middle and Lower Palouse already had moved to the Colville Reservation with the remaining tribal leaders (including Kamiakin's sons) who had established residence there by invitation of Chief Moses (Ray 1975:12-13).

Like the Nez Perce, Wanapum, Walla Walla, and Yakama Indians, the Palouse were Sahaptin speakers, although their dialect differed from the others in having four additional vowels (Sprague 1998:352). The Palouse constituted bands, which lacked formal tribal organization but shared a common language, religion, and culture. Group composition typically was fluid, as bands shared hunting and root grounds, intermarried, and often lived and worked together. The Palouse retained their group cohesion through marriage and shared subsistence, religion, language, and customs, as did other groups (Ray 1936). Marriage was exogamous and kinship was bilateral. Patrilocal residence commonly was practiced, and land tenure or ownership of fishing and hunting grounds reportedly was communal. The Palouse participated in task groups for fishing, root gathering, hunting, and warfare with most of the neighboring groups.

The ethnographic Palouse bands, generally comprising related individuals, typically exploited a drainage system of tributary streams feeding the Snake River. Each settlement was an autonomous entity, and headmen were elected to their position. Winter settlements, the permanent residential sites of the Palouse, were located along the major rivers and at the confluences of streams. These areas provided access to fish, fuel for cooking and heating, construction materials, and protection from the winter cold and snow typical of more upland settings. Each settlement consisted of one or more types of structures, including mat and wood covered longhouses, conical mat lodges, semi-underground houses, sweat lodges, menstrual huts, and sudatories for the men (Ray 1975; Spinden 1908; Thwaites 1959, vol. 3). Historically, the most common winter house was a rectangular, multifamily, A-framed pole lodge covered with mats (Lewis 1906:185).

Most of the major settlements "were occupied seasonally but the three or four largest had populations, of varying size, the year round" (Ray 1975:189). In the spring, the mat and

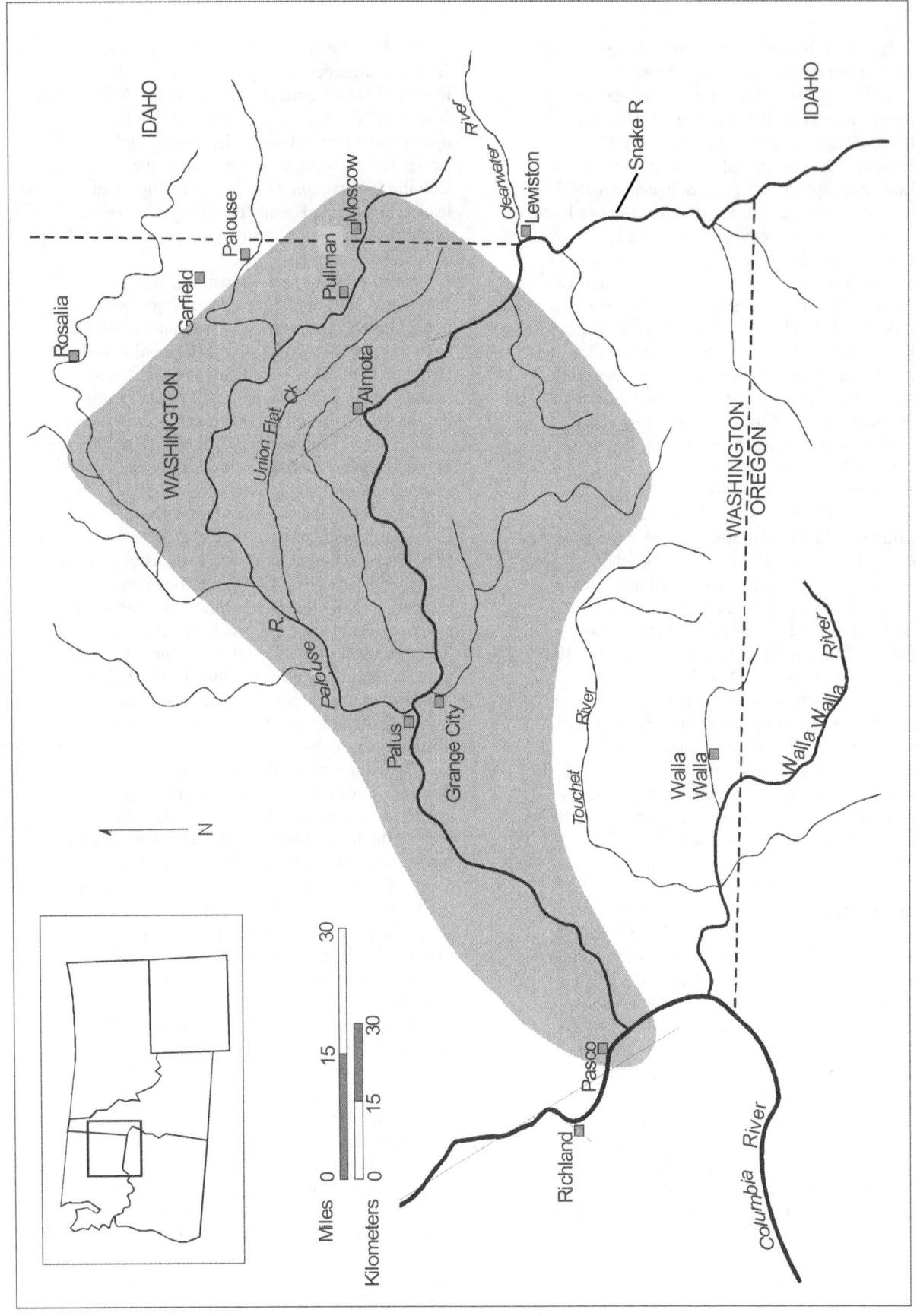

Figure 4.2 Palouse territory during the 19th century. Marmes Rockshelter is near Palus (from Sprague 1998:Figure 1).

board covered houses were dismantled and the poles and mats were carefully stored against the threat of damage from severe winds, floods and fire. Because the houses were essentially tied together, a "large house could be dismantled by two or three persons in a day's time, re-erected in two or three days" (Ray 1975:190). Lewis and Clark noted piles of these construction materials along this stretch of the Snake River during their trip downstream and reluctantly used some of the boards for fuel, having no other wood available (Thwaites 1959, vol. 3). Ray (1975:190) notes that not all of the permanent villages were occupied at the same time (although it is speculated that this may be a factor of depopulation caused by disease by the time of the recording of Palouse ethnographic lifeway).

> Conditions of the rivers, the distribution of game, and mere human caprice were factors involved in the shifts and preferences. Any family of the Palus tribe was privileged to reside at any village of its choice. The total territory was small and everyone knew one another; hence shifting from one village or village location to another was convenient and congenial. [Ray 1975:190]

Palus ("big rock in the river" [Ray 1975:193] or "what is standing up in the water" [Sprague 1998:358]) Village (45FR36), one of the two largest of the Palouse settlements, was located at the mouth of the Palouse River.

> At the time of Lewis and Clark (1805) and other early explorers, the settlement was situated on the west side of the Palouse River. Later, the east side was used, also.

> The population here was considerable, probably averaging twenty-five houses of six to ten occupants each. Lewis and Clarks's map shows more house symbols at this location than for any other Palus village and the explorers speak of it as a very great fishing place. At the time of Ross Cox (circa 1812) the town consisted of forty mat-covered houses ("tents"). Cox remained at the village for seven days, early in August, at which time "the inhabitants were busily employed in catching and drying salmon for their winter and spring stock." Cox's party purchased fifty horses from the townspeople. At the same period, Alexander Ross found this town to be the capital of the Palus tribe, the residence of the chief.

> Salmon were taken in great numbers near this town, both in the Snake River, by spearing and seining, and in the Palouse by spearing, dip-netting, weirs, and traps. Smaller fish were caught in small nets or with hook (gorget) and line. This was a convenient location for antelope and deer hunting in the plains to the north.

> This town was the most important commercial center of the tribe. The principal north-south Indian trail crossed the [Snake] at this point. [Ray 1975:193-194]

Several residential sites also existed on the Palouse River. A winter fishing village was located about two miles above the mouth on the west side of the river. This would be in the vicinity of a major bend in the river south of Marmes Rockshelter, in the area currently termed "Palouse Bay" because of the breadth of the inundated canyon at this point. *Claxo'pa* was a small fishing village on the west side of the river about four miles above the mouth, opposite a small stream. A fishing village, referred to as *A'patap*, was located just below Palouse Falls ("falling water"). "The fishing was exceptionally good...and the site was close to berrying grounds and hunting territory, and the important root-digging area of the western lobe of Palus territory" (Ray 1975:198).

Other settlements include *Maxmaxc*, located near Steptoe Butte; *Qainakpa*, located on the north bank of the Snake River, and *Wapnitna*, at the mouth of the Tucannon River (Trafzer and Scheuerman 1986). The cultural deposits at *Wapnitna* are known archaeologically as 45CO1, the Tucannon Site. A fish trap was maintained in the Tucannon River at this point (Ray 1975:194). Palouse Indians also occupied settlements located at Almota, Page (Fishhook Bend), and Ainsworth at the confluence of the Snake and Columbia Rivers (Sprague 1998:352).

The bulk of the tribe moved to temporary camps at least during the heights of the root-digging, berry-gathering and hunting seasons. These temporary camps

> were set up wherever convenience dictated and therefore were variously located from year to year, depending on crops, the availability of pasture for horses, the condition of water-holes, and other such factors. These temporary camps were occupied as long as economic productivity justified: from a few days to several weeks. The houses erected at these camps were ordinarily small, of conical or gable shape, with frameworks of light poles covered with mats of

tule, cattail or other rushes or grasses. Rush mats were particularly favored because they were very light in weight and could be rolled up conveniently for carrying by horseback. [Ray 1975:189]

Sometimes these dwellings were covered with hides or brush. Conical mat lodges became more common after the introduction of the horse and served as winter structures, often housing single families. Summer lean-tos constructed of poles and mats also were used (Mallory 1966). Menstrual huts were semisubterranean structures constructed over deep pits. Semisubterranean lodges occupied for many millennia prior to the use of mat covered structures may have only been used during the winter (Spinden 1908) (see Rice [1985] for further information on Plateau Indian structures).

The seasonal movements and settlement pattern of the Palouse were governed in large part by the availability of plant foods and to some extent by that of fish and game. Trafzer and Scheuerman (1986:7), citing Lewis and Clark, report that plants contributed a significant portion of the Palouse diet. Kous and camas were primary plant foods; kous was available in the spring, and camas was available from spring to fall, with major harvesting occurring in the late summer. Other plant foods of importance include bitterroot, salmon berries, serviceberries, chokecherries, huckleberries, currants, and sunflower (balsamroot arrowleaf) seeds. Plant food procurement and processing appears to have been accomplished by women, while the men engaged in hunting activities. Meat from deer, elk, antelope, mountain sheep, and small mammals most frequently was procured by the men. Bison, particularly after the acquisition of the horse, also were hunted (Spinden 1908).

Anadromous fish, such as chinook, sockeye, and silver salmon, as well as steelhead, probably were the most important fish. Resident fish, such as suckers, squaw fish, and sturgeon, also were used, as were river mussels and lampreys. Fish were dried for winter storage and consumption (Spinden 1908).

The material culture of the Palouse and other Sahaptin groups consisted in large part of food procurement, food processing, and hide-working implements. Flaked stone tools were manufactured from chert, basalt, and obsidian materials found in the local deposits or obtained through trade. Such artifacts include knives, projectile points, scrapers, drills, and stone net weights, all of which were made by males. Pestles, mauls, mortars and other pecked and ground stone tools were made and used by women. Antler, bone, shell, and wood provided materials used in manufacture of awls, needles, beads, wedges, and cooking utensils. Baskets, made by the women from Indian hemp, cattails, and tule, were used for storage and cooking containers (Spinden 1908).

Historically, the storage of plants, meat, and fish was practiced to sustain the Palouse through the winter months. Among the Nez Perce, Walker (1978) describes bitterroot and camas being placed in watertight baskets that were cached in grass- and bark-lined storage pits on well drained slopes. When Lewis and Clark passed through the area, they observed people at Palus Village storing fish in large pits dug into the river bank (Trafzer and Scheuerman 1986:3). There is archaeological evidence for prehistoric groups in the study area using rockshelters and rock overhangs as storage caches (Endacott 1992; Hicks and Morgenstein 1994; Mallory 1966). Among the Kalispel to the north, Smith (1986:194-195) describes pits of various sizes in which dried camas and fish were stored. The food stores were placed in the pit, which was covered by a layer of poles across the top. The poles were then covered with cedar bark, grass, and/or pine needles and finally were covered with a layer of earth to protect the food from moisture. Elevated platforms and storage houses also were constructed by some groups, such as the Sanpoil and Nespelem to the north (Ray 1933).

Historic Period

The first encounter with Euroamericans among the Palouse occurred on October 13, 1805 when Lewis and Clark (The Corps of Discovery) traveled through the area. Their journal entries note a village site at the mouth of the Palouse River associated with a picketed graveyard (Scheuerman and Trafzer 1980:4; Thwaites 1959, v.3:112) and described as "a large fishing establishment, where there are the scaffolds and timbers of several houses piled up against each other, and the meadow adjoining contains a number of holes, which seem to have been used as places of deposit for fish for a great length of time" (Sprague 1998:353). At this village, several medallions were given to Palouse headmen, including a silver peace medal that was subsequently recovered from a grave when the Palus Burial Site was excavated in 1964 in advance of its inundation by the reservoir behind Lower Monumental Dam.

Five years later, David Thompson of the Northwest Company camped at Palus on August 8, 1811. During the following year, members of the American Astoria Company, including David

McKenzie, John Clarke, and Ross Cox, visited the settlement (Trafzer and Scheuerman 1986:12-13). John Clarke returned to the settlement on June 30, 1813. Although previous visits had been friendly, a disagreement over a stolen silver goblet led Clarke to hang the offender, an incident that resulted in tension between whites and the Palouse for years to follow (Cox 1957:118-119).

On July 3, 1825, John Work camped at the mouth of the Palouse River and purchased two horses from the Palouse (Elliott 1914:88). Work, who was traveling with David Douglas, passed through the area again on July 19, 1826, and May 6, 1830 (Elliott 1909:297).

The first missionaries to arrive in the area in 1836 were the Protestant Marcus and Narcissa Whitman, who settled at Waiilatpu in Cayuse territory. Whitman, along with Henry H. Spaulding, visited the mouth of the Palouse River in 1839 in the hope of establishing a mission site closer to the Nez Perce Indians (Drury 1958:257). Spaulding decided the setting would not be conducive to a mission. After several serious epidemics of measles and smallpox, however, tensions among the missionaries at Waiilatpu and Indians continued to grow. Blaming the missionaries for the epidemics, the Cayuse killed the Whitmans and 11 other whites in November 1847. The Palouse were not involved in this event (Trafzer and Scheuerman 1986) and encounters between whites and the Palouse over the next few years appear unchanged.

Charles Wilkes passed by the area in 1841 in route from Waiilatpu to Fort Colville (Wilkes 1856, Vol. 4:466). Although little mention is made of the Palouse, he visited and noted his impression of Palouse Falls. In 1845 and 1846, Father Pierre Jean de Smet, a missionary among the Couer d'Alene Indians, briefly visited the Palouse (de Smet 1905, Vol. 2). During the 1846 visit, he traded powder and lead for the Indians' salmon before continuing his journey (de Smet 1905, Vol. 2:561).

Father de Smet's visit was followed in 1847 by the artist Paul Kane, who reported that the Palouse village at the mouth of the river consisted of 70 to 80 individuals (Kane 1925). Like de Smet, Kane, too, visited Palouse Falls, accompanied by the Palouse chief *Slo-ce-ac-cum* who Kane illustrated in watercolor during the visit.

During the next few years, several visitors observed patches of land under cultivation by the Palouse and noted that corn, wheat, and potatoes were being raised (Sprague 1998:353). Major John Owen noted the Palouse River in his journals in 1852 and 1857 (Owen 1927). On July 31, 1853, on the bank of the Snake River opposite the Palouse River, Lt. Rufus Sexton met a delegation of Nez Perce and Palouse Indians. While documenting the location and composition of the Indian tribes of Washington, George Gibbs traveled past several Palouse settlements in 1854. In his report to Captain George McClellan, Gibbs (1972) noted that the Indian headman displayed a medal given to his father by Lewis and Clark.

Isaac Stevens, Territorial Governor and Superintendent of Indian Tribes, negotiated a treaty with the Yakama Indians in 1855 to place them on a reservation. Participants listed in the treaty signing included the Palouse Indians, represented by Kamiakin and Kahlotus (Kappler 1904, Vol. 2:698, 702). Kamiakin was well-known and admired by Indian people of many tribes in the region. But when the treaty negotiators asked him to represent the Yakama Indians at the treaty signing, Kamiakin complained, saying that he shouldn't be signing for the Yakama because he was Palouse, a statement that was entered into the record of the proceedings.

Continued hostilities between whites and Indians in eastern Washington and Oregon led to U.S. military involvement in the mid-1850s. In 1858, troops from Fort Walla Walla, headed by Colonel E. J. Steptoe, hoped to halt these activities by more direct intervention. As a consequence, Steptoe recommended that a fort be constructed at the mouth of the Tucannon River. The fort would primarily function as a supply and support base for campaigns conducted to the north. Fort Taylor was constructed east of the Tucannon overlooking a ford used by military personnel to cross the Snake River. Colonel George Wright and a troop of soldiers arrived at the post on August 15; during the following days, the soldiers occasionally were fired upon by the Indians (Fletcher 1982; Manring 1975).

On August 25, the soldiers crossed the river and camped on the north bank before they moved downstream (Trafzer and Scheuerman 1986). The post was left in the hands of a small group of soldiers to continue its use as a supply base during the campaign. When the hostilities ended in October, the fort command was given to a Palouse headman (Fletcher 1982). At the end of the Indian campaigns in 1858, only a small number of Palouse continued to live at Palus. Most of the remaining Palouse people followed their leaders to the Colville Reservation at the invitation of Chief Moses.

> They could have chosen Coeur d'Alene, Umatilla, or Yakima but most of them wanted either to stay in their homeland, a wish that became more and more untenable, or to go to the Colville Reservation. All of their leaders and

chiefs, who were not too old or infirm to travel, moved to the Colville Reservation....[including the] sons of Chief Kamiakin... [Ray 1975:12-13]

In 1860, an observer wrote that Palouse Indians in two villages between the Tucannon River and Alpowa planted crops on islands and irrigated from tributary creeks (Sprague 1998:353). That same year "the Palouse [living at Palus] were reported to be reduced in numbers from war, disease, and starvation; to be poor, having few horses and cattle; and to be reliant upon fish and gardens for subsistence (Sprague 1998:356). A visit to Palus by an Indian Agent in 1897 described generally poor conditions, with about 50-100 Indians living in teepees and only a small plot of land under cultivation. By this time the residents were no longer catching sufficient salmon to support themselves because of the use of fish wheels and nets downstream (Sprague 1998:354).

The first permanent non-Indian settlers to the Columbia Plateau were miners, spurred by gold mining, who arrived after the conclusion of the Indian Wars. Isaac Kellogg constructed a cabin at the present location of Starbuck in 1860 and operated a ferry located at the mouth of the Tucannon River. In 1858 Edward L. Massey was granted the right by the territorial legislature to build a ferry at the mouth of the Palouse River, the first such undertaking on the Snake River (Steinberger 1897). This ferry provided an important link in the Mullan Military road, which was opened in the early 1860s and served as a busy route between Fort Walla Walla and all points north. A roadhouse was reported at the crossing (Beall 1917:87), but the reference does not say on which bank of the Snake River. The ferry was known as the Palouse Ferry until 1926, when it was renamed Lyons Ferry in honor of a former owner.

Miners were soon followed by cattle and sheep herders from the Willamette Valley who moved to the interior grasslands; the mining rush abated in the 1870s (McGregor 1982:10-12). Members of the Starbuck Grange built a large warehouse at the former site of Fort Taylor in 1875; the warehouse provided storage space for grain and freight carried by steamboats that began navigating this stretch of the Snake River two years earlier. Platted as Grange City by A. Simmons, it was abandoned by 1883 after the arrival of the railroad in 1881 (Fletcher 1982:88).

By the early 1880s, the most promising forest and humid prairies in the West had been settled, so population pressure and successive bountiful wheat crops from the Walla Walla area led to a major influx of immigrants to the Columbia Plateau (McGregor 1982:12-13). The McGregor brothers came to the region at about this time and began raising sheep in the Alkali Flat area (McGregor 1982:22). However, due to often violent relations between cattlemen and sheepmen in the mid-1880s, the McGregors moved their winter range for their sheep herds to the protected valleys of the lower Palouse and Snake Rivers, which were impractical for wheat farming. The McGregors maintained a winter camp for their herders at Palouse Falls (McGregor 1982:23,119).

Sheep herding in Palouse Canyon is evidenced by sheep dung within the large rockshelters. More recent use of the study area for cattle grazing is indicated by the many cattle trails, trail-terraced slopes, and fences. The Roland Marmes family maintained a farm just south of Marmes Rockshelter, and raised cattle in the canyon until they left prior to inundation of their settlement by the reservoir.

The Joso Trestle, spanning the confluence of the Palouse and Snake Rivers, was completed in 1914 (Sprague 1983:76) for the Union Pacific. The structure is named after the Joso siding, formerly located on the north side of the river where the Jaussaud Sheep Company owned land prior to the coming of the railroad (Stratton and Lindeman 1976:56). This structure still is in use today.

The reader is directed to Stratton and Lindeman (1976 and 1978) for a regional perspective on historic cultural resources.

5
STRATIGRAPHY AND SITE FORMATION PROCESSES

Gary Huckleberry, Carl E. Gustafson, and Shawn Gibson

This chapter provides a geological context to the archaeological remains at the Marmes site (45FR50) with an emphasis on the sedimentological record. The Marmes site is a deeply stratified archaeological site with evidence of over 10,000 years of cultural use in distinct levels of sediment and cultural debris. Given the rich stratigraphy at the Marmes site, perhaps it is not too surprising that a man trained in geology would be a principal investigator of the site. Roald Fryxell was keenly interested in the details of the Marmes site stratigraphy not only for providing a basic relative cultural chronology but also for taking into account the big picture of interactions between humans and a dynamic natural environment. Although Fryxell's death in 1974 preceded his final analysis of the Marmes site, many of his ideas regarding landscape evolution, paleoclimate, and human adaptation on the Columbia Plateau were shaped by his work at the site between 1962 and 1968 (Fryxell 1963a, 1963b, 1964, 1971; Fryxell and Daugherty 1962; Fryxell and Cook 1964). During the excavations at Marmes in 1962-65 and 1968, much stratigraphic data was collected including field descriptions and profiles, soil monoliths and acetate peels, and bulk samples of sediments to be analyzed in a soils laboratory. Although laboratory analysis was performed in 1968 on selected sediment samples, the data were never compiled, analyzed, and presented. This report attempts to achieve that goal.

The sedimentology of the Marmes site presented here is based on stratigraphic profiles and sediment data originally collected and partially analyzed by Fryxell and his students and colleagues. The primary objectives are to 1) characterize the physical and chemical properties of sediments at the Marmes site, 2) reconstruct the depositional and post-depositional environments at the site, and 3) assess the integrity of archaeological and faunal materials excavated at the site. It should be stated from the outset that this study is inherently limited by the fact that not all field and laboratory data were fully recovered during this study. Most of Fryxell's field notes on stratigraphy were not found, and some of the laboratory data could not be used due to a lack of provenience. Perhaps the greatest limitation to the study is that it is reported by someone who did not see the actual stratigraphy of the Marmes site. There is no substitution for direct observation of in situ stratigraphy. All information presented here is directly or indirectly archival, based on maps, profile descriptions, photographs, monoliths, and laboratory data. Although photographs, profile drawings, and monoliths help place the sedimentological data in their appropriate spatial context, some patterns in the stratigraphy are inevitably missed when viewed retrospectively. Likewise, sediment samples tend to be "bags of dirt" to the person who never saw them in place within the stratigraphy or played a role in sampling. Despite these handicaps, a considerable wealth of sedimentological data exists for the Marmes site, and many reasonable interpretations can be made regarding the depositional and post-depositional history of the site based on the archival record. The sedimentological information presented here provides a framework for characterizing, distinguishing, and interpreting the Marmes site stratigraphy and augments the other analyses. Together, these multiple lines of evidence can be used to formulate and test hypotheses regarding site formation, paleoclimate, and human activities at the Marmes site.

Geomorphic Context

The sedimentary record is the product of sediment production, transfer, and deposition within an evolving landscape. Although a detailed overview of the region's geomorphology is beyond the scope of this study, it is appropriate to briefly mention the geomorphic history of the Marmes site as it pertains to the deposition of sediments and formation of the site.

The Marmes site is composed of three landscape elements (Figure 5.1). One is the rockshelter, a shallow re-entrant formed into a basalt cliff facing south. The second element is the floodplain of the Palouse River, which at the site consists of a single early Holocene alluvial terrace (Marshall 1971) approximately 40 feet (12 m) below the rockshelter. The third element is the hillslope, herein referred to as the colluvial slope, connecting the rockshelter with the floodplain. This topography is the product of geological forces extending back at least 10 million years (see Chapter Two). The rockshelter itself is likely the product of the glacial-era Missoula floods.

After the last Missoula Flood, the lower Palouse River canyon bottom filled with alluvium from the Palouse River interfingered with fan alluvium and talus deposits derived from tributaries and adjacent hillslopes. Marshall (1971) identified six, post-glacial alluvial terraces between the Marmes site and the Snake River. The terraces are composed of stratified gravels, sand, silt, tephra, and weakly developed soils. The highest and oldest terrace (Terrace 1) forms the topographic bench below the Marmes Rockshelter. Outside the floodplain, most of the Holocene geological events are characterized by loess deposition and slopewash/colluviation on hillsides. Compared to geomorphic events of the Pleistocene, the Holocene has been relatively quiet in the lower Palouse River canyon.

Methods

Field and laboratory sedimentological data from the Marmes site were compiled from records housed at the Washington State University Museum of Anthropology. Records were searched to see what sediment samples were available or had already been analyzed. An inventory of sediment samples previously analyzed in 1968 was constructed (Table 5.1) and used to assess the representativeness of the sample from different parts of the site. Areas estimated to be underrepresented by laboratory data were identified, and sediment samples were selected from those areas for laboratory analysis. In all, 162 sediment samples were processed in 1968 and 1998 for sedimentological analysis.

Details of data collection and laboratory methods are presented below.

Field Methods

Most stratigraphic analyses and many of the profile drawings were done by Fryxell in the 1960s; all were conducted under his direction. Fryxell and crew members drew the profiles exposed in the rockshelter from 1962 through 1965. Madge G. Gleeson and Joan Brodhead exposed and drew nearly all of the original profiles on the floodplain in 1968. Fryxell approved or modified field drawings before acceptance. All site measurements were made in units of feet and tenths because only surplus surveying instruments were available to the field crew in the early 1960s. An arbitrary datum of 100 feet vertical elevation was established at the highest point in the rockshelter. All elevations presented herein are relative to that arbitrary elevation. Excavation units consisted of squares five feet on a side. To ensure uniformity, this system was followed throughout the entire period of excavation and is continued herein.

During the fieldwork at the Marmes site, stratigraphy was mapped and described using terminology of the U.S. Department of Agriculture (U.S.D.A. Soil Survey Staff 1951; see Appendix B). Sediments were collected in selected locations by stratigraphic level for subsequent laboratory analysis. Sediments also were collected through the use of acetate peels and soil monoliths (Smith and Moodie 1948). For the current study, laboratory analysis was performed on both bulk sediments and selected monoliths; several monoliths had been described during the original field project as well (Table 5.1). All bulk sediment samples stored since 1968 were air-dried and sealed in plastic bags.

Laboratory Analysis: 1968

Many of the sediment samples from the Marmes site were analyzed at the University of Idaho Soils Laboratory in 1968. Data from these analyses were recorded in simple tables with no explanation of the methods used. In an effort to identify and document the 1968 laboratory procedures, the data were presented to Paul McDaniel and Anita Falen of the Soils Department at the University of Idaho, and the most probable laboratory methods employed during the 1968 analyses were determined.

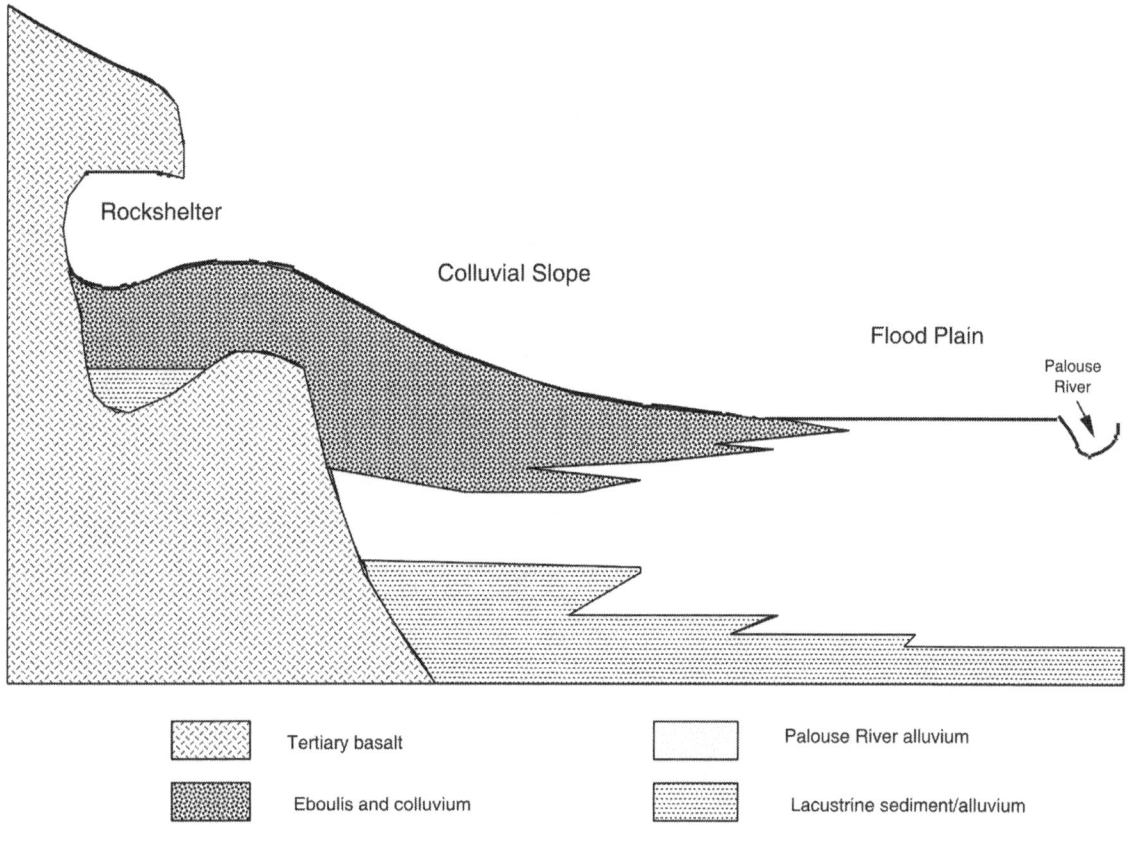

Figure 5.1 Schematic cross-section of the Marmes Site.

In 1968, it is likely that particle size fractions were analyzed through a combination of dry screening with nested sieves for the sand fractions and settling in a solution using the pipette or hydrometer method for the silt and clay fractions (Bouyoucos 1962; Gee and Bauder 1986; see Appendix C). A suite of chemical analyses were also performed on selected Marmes site sediments (see Appendix D) including:

- pH
- electrical conductivity
- exchangeable ions (Ca^{2+}, Mg^{2+}, Na^+, K^+, CO_3^{2-}, HCO_3^-, Cl^-, and SO_4^{2-})
- cation exchange capacity
- exchangeable sodium percentage
- calcium carbonate
- gypsum
- organic matter
- nitrogen
- phosphate

Many of the chemical tests conducted in 1968 are more useful for assessing soil fertility than for geoarchaeological studies (although, see Moody and Dort 1990). More commonly employed in geoarchaeological studies are analyses such as pH, and organic matter, calcium carbonate, and phosphate content which can provide insight into depositional environments, soil formation, and human activities (Stein 1985). Because most of the laboratory data are dependent upon the procedure used (i.e., they are measurement-dependent), it is important to present the laboratory procedures used in the Marmes Rockshelter study.

pH is a measure of hydrogen ion (H^+) activity and is used to estimate soil acidity or alkalinity. The measured pH values are highly dependent on how the analysis is performed (Bohn et al. 1985:219). For example, the ratio of soil and water, the composition of exchangeable cations, clay mineralogy, the composition and concentration of soluble salts, and the

Table 5.1 Inventory of sediment data for the Marmes site.

Landscape Component	Provenience	Sample Size (n)	Sieve	Hydrometer	Chemical	Descriptions
Rockshelter	N85/W25	30	yes	yes	yes	yes (Roald Fryxell 1964)
Colluvial Slope	Monolith A (N25/W20)	25	yes	no	yes	no
Colluvial Slope	Monolith B (N20/W20)	27	yes	no	yes	yes (Gary Huckleberry 1998)
Colluvial Slope/Flood Plain	Monolith G (N10/W20)	36	yes	yes	yes	no
Flood Plain	Monolith K (S75/W20)	17	yes	yes	no	yes (Joan Broadhead 1969)
Flood Plain	Monolith M (S120/W20)	21	no	no	no	yes (Claudine Weatherford 1969)
Flood Plain	Monolith O (S137/W20)	0	no	no	no	yes (Gary Huckleberry 1998)
Flood Plain	Monolith N (S139/W20)	6	yes	yes	no	yes (Sarah VanGalder 1998)

presence or absence of gypsum and carbonates can affect the measured pH (Richards 1954:18). For the Marmes site samples, a pH electrode was used on a sediment paste consisting of sediment and distilled water.

Electrical Conductivity (EC) is an indirect measure of total dissolved salts contained within a soil solution. It is measured with an electrode, and the amount of current that passes through the solution is proportional to the total concentration of dissolved salts. The concentration of the sediment solution (sediment:water ratio) can affect the solute concentration in the sample (Rhoades 1982a). EC measurements were determined using a Solu-bridge electrode on the soluble extract derived from the soil paste.

Exchangeable ions are ions that are easily exchanged between different soil surfaces like clays and organic matter (Bohn et al. 1985:154). They originate through the solubilization of salts and weathering of minerals. Two methods were used to analyze exchangeable ions in the Marmes site samples. In one, the supernatant derived from the soil:water paste used in the pH analysis was used to determine the amount of saturation extractable ions. In the other method, exchangeable cations were measured by treating the sediment with ammonium acetate resulting in the displacement of exchangeable cations into solution (Richards 1954:18-20). Ca^{2+}, HCO_3^- and Cl^- were determined using a titration procedure (Rhoades 1982b). Mg^{2+} was measured using spectrophotometry, and K^+ and Na^+ were measured using flame emission photometry. SO_4^- was calculated as the difference between the sum of the cations and the sum of HCO_3^- and Cl^-.

Cation-exchange capacity (CEC) is the ability of a soil to absorb and release exchangeable cations. There are several different ways of measuring CEC, some of which depend on the pH of the sediments (see Rhoades 1982b). The direct displacement method using Ba^{2+} as the index cation was employed to measure CEC for the Marmes site sediments.

Exchangeable sodium percentage (ESP) reflects the dominance of Na^+ amongst the exchangeable cations and is an important measure for assessing the alkalinity of soils. Soils with ESP > 15% are considered alkaline and have specific associated physical and chemical properties deleterious to plant growth (Bohn et al. 1985:242; Richards 1954). ESP for the Marmes site sediments was determined by dividing the exchangeable Na^+ content by the CEC.

Gypsum ($CaSO_4 \cdot 2H_2O$) is a soluble salt common to poorly leached soils and sediments derived from gypsiferous bedrock. Gypsum was extracted from the Marmes site sediments using distilled water and content was determined through a versenate titration.

Calcium carbonate ($CaCO_3$) is a salt that accumulates in moderately leached soils common to arid and semiarid environments. Whereas the pedogenic environment at the Marmes site is adequately moist to leach gypsum from the soil, there is inadequate rainfall to remove $CaCO_3$, and hence most natural soils in the Columbia Basin are calcic (Busacca 1989). Changes in the amount of $CaCO_3$ at the Marmes site may indicate episodes of increased exogenous dust flux, soil development, or previously elevated water tables. $CaCO_3$ was determined by acid neutralization and titration (Richards 1954:105).

Organic matter content may be used to infer previous biological activity, including that of humans. Organic matter consists of different materials composed of various amounts of carbon, hydrogen, oxygen, nitrogen, phosphorus, and sulfur, and the amount and type of organic matter varies between pedogenic and cultural environments (Stein 1992). Organic matter is estimated by measuring organic carbon and multiplying by an empirically-defined conversion factor. There are a variety of methods for determining organic carbon content (see Nelson and Sommers 1982). The semiquantitative Walkley-Black method (Walkley and Black 1934) was used to determine organic C content and then multiplied by the empirically-derived VanBemellen factor (1.724) to estimate organic matter content.

Nitrogen (N) is a basic component of soil organic matter and thus can be viewed as a proxy for previous biological activity. The Kjeldahl (wet oxidation) method (Bremner and Mulvaney 1982) was used for the Marmes site sediments.

Phosphorus (P) is also a common constituent of organic matter that can be used as a chemical tracer for previous human activity

(Eidt 1977). However, P also occurs in other inorganic forms, and one needs to be specific as to the type of phosphorus being analyzed in an archaeological context. Phosphorus generally occurs in soil as phosphate ion (PO_4^{3-}) bound to aluminum, iron, calcium, and organic compounds depending on the soil environment (Bohn et al. 1985:304). The type of phosphate measured depends on the type of extraction method used (see Olsen and Sommers 1982). Available P was measured on the Marmes site sediments using a bicarbonate extraction method and spectrophotometry with a molybdate blue solution.

Laboratory Analysis: 1998

During the 30 years since the Marmes Rockshelter excavations were completed, Washington State University (WSU) moved the collections from one storage building to another until they were moved into a permanent facility at WSU in 1984. For these investigations, it was first necessary to locate and assemble all of the stratigraphic drawings, field notes and documentation essential for reproducing accurate profiles suitable for use in detailing the provenience of archaeological features, artifacts and biological remains. At the same time, appropriate monoliths, sediment, and faunal samples were located. Once accomplished, the materials were gathered together in laboratories in College Hall at WSU. When needed, field notes and other documents were consulted to fill in missing data and to upgrade profile drawings that were partially complete or contained questionable information.

Various renditions of the stratigraphic drawings on Mylar, along with old blueprints and original field drawings were used to reconstruct the stratigraphic profiles for this study. Fryxell's original intent was to include as much detail as possible in the scale drawings, so any rocks or other objects greater than 2 cm in diameter are included. Except for a few trenches, most of the original drawings were made from walls of the individual 5 ft x 5 ft excavation squares. In the rockshelter, drawings were made over a period of at least four project or field season years. Not only did the stratigraphers have to match drawings from one square to the next, but also they had to match vertical sections that were drawn in one year with lower sections drawn in later years when deeper excavations were conducted. In addition, some of the walls had been drawn facing north or east whereas others were drawn facing south or west. To complete a profile during the current study, it was necessary to make reverse copies in order to match the majority of the original drawings from individual excavation squares. Sometimes as many as four or five profiles were superposed to produce a completed drawing. A total of nearly 60 profile drawings comprising 2,370 lineal feet of trench walls drawn at a scale of one-half inch to the foot were completed. These represent all of the stratigraphic profiles that could be located. All of the profiles were then scanned and copied onto compact disks.

Particle-size analysis was performed on sediments from the Marmes site in the spring of 1998 at the Washington State University Geoarchaeology Laboratory. Because of a paucity of laboratory data for deposits from the colluvial slope and floodplain, sieve and hydrometer analyses were performed in 1998 on sediments from the Monoliths G and K locality, and directly from Monolith N. The fine fractions were treated with hydrogen peroxide and dilute hydrochloric acid to remove organic matter and calcium carbonate, respectively. Re-examination of the sieved fractions from the rockshelter indicated that the samples were never fully disaggregated during the 1968 analysis. Consequently, most of the samples from the rockshelter control block (85N/25W) were resieved for this study after disaggregation with a mortar and pestle. Grinding force was limited in order to avoid grinding down primary grains. Measures of particle size mean, sorting, skewness, and kurtosis were determined using graphical methods (Boggs 1987:110-116). Finally, physical properties were described for Monoliths B and N in the laboratory as a proxy for descriptions of stratigraphy made in the field during excavations. Field and monolith descriptions are presented in Appendix E.

Rockshelter Stratigraphy

Archaeologically, caves and rockshelters provide a trove of information because they tend to be associated with well preserved organic materials and a fairly complete and protected depositional record (Butzer 1971; Farrand 1985). Decades of geoarchaeological research in caves and rockshelters have demonstrated that each is unique with its own bedrock lithology, internal geometry, orientation and history of biotic activity. Thus, it should come as no surprise that caves and rockshelters formed by limestone dissolution would likely be different in geometry and stratigraphy than those formed in volcanic rock. However, even adjacent caves and rockshelters of the same lithology and local environment can differ significantly due to a myriad of microenvironmental and biological factors. For example, the microclimate and sedimentation rate within nearby McGregor and Porcupine Caves, both formed in basalt, are different due to differences in the direction and orientation of their entrances (Hicks 1995). Climate, microclimate, bedrock lithology, chamber geometry, and biotic activity (including humans) all influence the types of sediments and their arrangement in caves and rockshelters. Despite a myriad of origins for cave and rockshelter sediment (Draper and Morgenstein 1993:B-65; Lau et al. 1997:508), all such deposits fall under the category of either endogenous or exogenous. Endogenous sediments include materials derived from the walls and ceilings by chemical and mechanical weathering, especially roof-collapse debris or e'boulis. Exogenous sediments originate outside the cave or rockshelter and may be blown, washed, or mass wasted into the interior through the main entrance or other possible passages. Any one deposit may consist exclusively of endogenous or exogenous sediments, or a mixture of the two. Depending on the lithology of the rock forming the cave or rockshelter, the size of particles generated endogenously and exogenously may differ significantly, thus enhancing the power of granulometry to distinguish sediment origins.

The Data (1968 and 1998)

Marmes Rockshelter is a shallow alcove extending approximately 25 feet (8 m) into a basalt cliff facing south. Prior to excavations, the entrance was approximately 40 feet (12 m) wide and over 10 feet (3 m) tall. Preliminary stratigraphic studies at Marmes Rockshelter in 1962 revealed eight major units. These were identified from lowest (deepest) to highest as units I through VIII (Fryxell and Daugherty 1962; Gustafson 1972; see Figures 1.7 and 1.9). Once identified, attempts were made to excavate according to these "natural" stratigraphic units. Archaeological excavations penetrated through over 21 feet (6 m) of rockshelter deposits without ever reaching bedrock (see Appendix B, 85N). By the end of summer 1964, cultural deposits spanning at least 10,000 years were recognized.

A stratigraphic control section was selected for the rockshelter at 85N/25W, which was deemed representative of the overall stratigraphy (see Figure 1.9). The control block attained a depth of about 13 feet (4 m) during the 1962-64 excavations and was described by Fryxell[1] (Appendix E). In 1968, the control block was extended to a depth of approximately 17 feet (5 m) depth. Based on how the depths were recorded on the sediment bags for 85N/25W (i.e., a change from depth below the rockshelter surface to elevation relative to site datum), it appears that the five sediment samples below 10.8 feet (3.3 m) were collected during the 1968 excavations. Also in 1968, a deeper test unit was placed at the rear of the cave (95N/20W) which extended even deeper into fine-textured lake sediments (Keel and Fryxell, 1969:15), the top of which is recorded at 15 feet (5 m) depth (84 feet site elevation).

Overall, the entire stratigraphic sequence is poorly sorted (graphical standard deviation 2.0 to 4.0 ϕ; Table 5.2) and dominated by various amounts of angular basaltic gravels, cobbles, and boulders in a matrix of very fine sand and silt. The amount of matrix and coarse clastics varies with depth (Figures 5.2 and 5.3). The upper 3 feet (1 m) is poorly sorted containing a mixture of angular gravels/cobbles, sand, and silt. At 3 to 4 feet (1 to 1.3 m) depth, there is a bimodal distribution with a dominance of > 4 mm diameter clasts and fine sand and silt. At 4 to 6 feet (1.3 to 2 m) depth, the deposits are relatively well-sorted and dominated by fine sand and silt. These deposits correspond with the pumicite layer and zone of reworked Mazama tephra (Fryxell and Daugherty 1962). Most of the deposits from 6 to 16.6 feet (2 to 5 m) have a distinct bimodal distribution of angular gravels/cobbles and fine sand and silt. Based on the particle size mean (Figure 5.4; Table 5.2), the coarsest sediments occur in the upper 3 feet (1 m) and below 7 feet (2.3 m); deposits between 3.7 and 5.9 feet (1.1 to 1.8 m) are relatively fine textured. It is worth mentioning that this vertical distribution of mean particle size matches fairly well with Fryxell's field measurements of rock (gravel and coarser) frequency with depth (Fryxell et al. 1968b:Figure 2) suggesting that rock frequency is a good proxy for relative changes in predominant particle sizes with depth.

---

[1] Field descriptions were found at the American Heritage Center, University of Wyoming.

Table 5.2 Particle-size statistics for rockshelter control block (85N/25W).

Depth (ft)	Graphic Mean (ϕ)	Graphic Standard Deviation[2]	Graphic Skewness	Graphic Kurtosis
.65 - 1.00	-0.50	2.76	28.84	13.24
1.0 - 1.3	0.07	2.50	22.86	14.45
1.3-1.6	-0.07	2.56	20.57	14.63
1.6-1.9	0.00	2.54	19.44	13.78
1.9 - 2.3	0.10	2.45	13.67	13.27
2.3 - 2.6	0.30	2.58	17.96	14.79
2.6 - 2.9	0.73	3.25	27.88	21.74
2.9 - 3.3	-0.30	3.52	25.45	26.39
3.3 - 3.7	0.87	3.74	29.28	29.69
3.7 - 4.0	1.90	2.88	-10.38	14.48
4.0 - 4.3	3.17	3.46	-27.42	24.76
4.3 - 4.6	3.33	3.24	-20.94	14.34
4.6 - 4.9	1.90	2.28	-7.54	2.82
4.9 - 5.15	1.80	2.28	-8.08	2.89
5.15-5.3	1.57	2.81	-8.49	14.10
5.3 - 5.6	2.60	-	-	-
5.6 - 5.9	1.77	2.04	-10.36	1.51
5.9 - 6.15	0.10	-	-	-
6.15 - 6.4	2.73	3.13	2.90	12.00
6.4 - 6.8	0.70	2.81	-28.97	16.39
6.8 - 7.2	-0.30	-	-	-
7.2 - 7.8	0.97	-	-	-
7.8 - 8.0	-0.53	-	-	-
8.0 - 8.6	-0.90	-	-	-
8.6 - 9.0	-0.63	-	-	-
10.8-11.3	-0.83	-	-	-
11.5-12.0	-0.50	-	-	-
12.1-12.3	-1.40	-	-	-
13.3-13.5	-0.57	-	-	-
15.5-16.6	1.67	-	-	-

[2] Graphic standard deviation, skewness, and kurtosis could not be calculated on samples with a large > 4 mm (-2 ϕ) fraction.

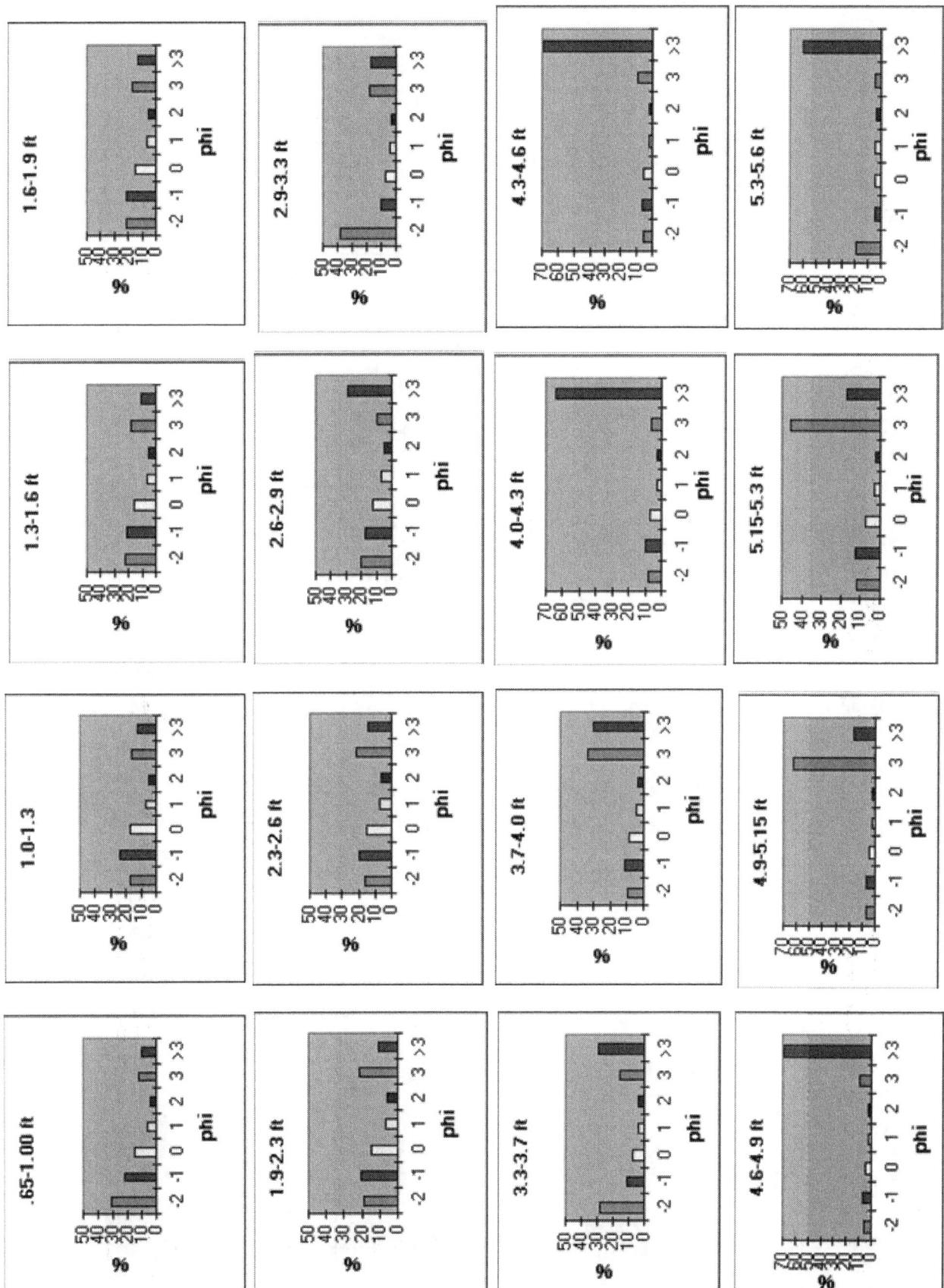

Figure 5.2 Particle-size histograms from rockshelter control block (85N/25W).

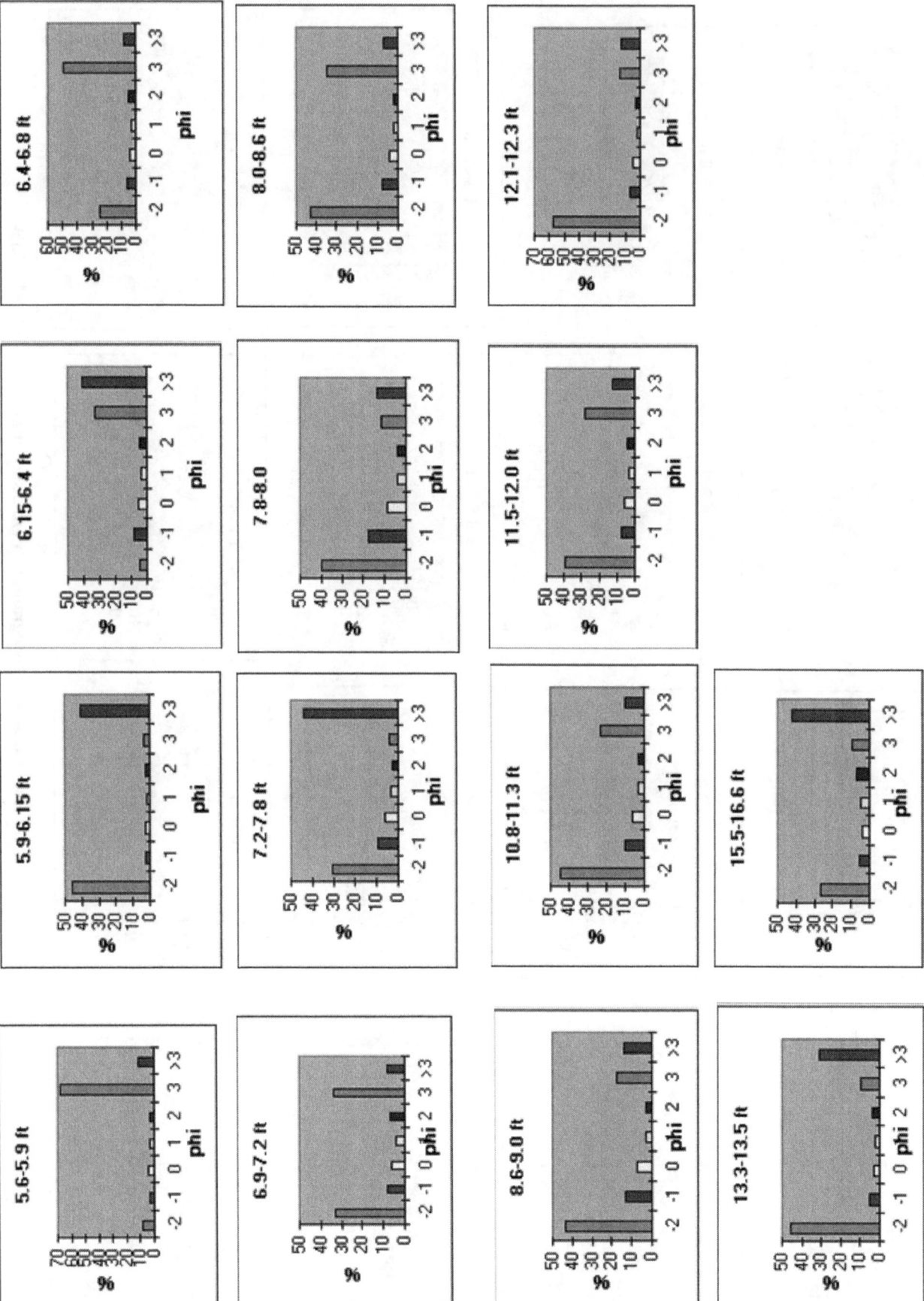

Figure 5.2 Particle-size histograms from rockshelter control block (85N/25W) (continued).

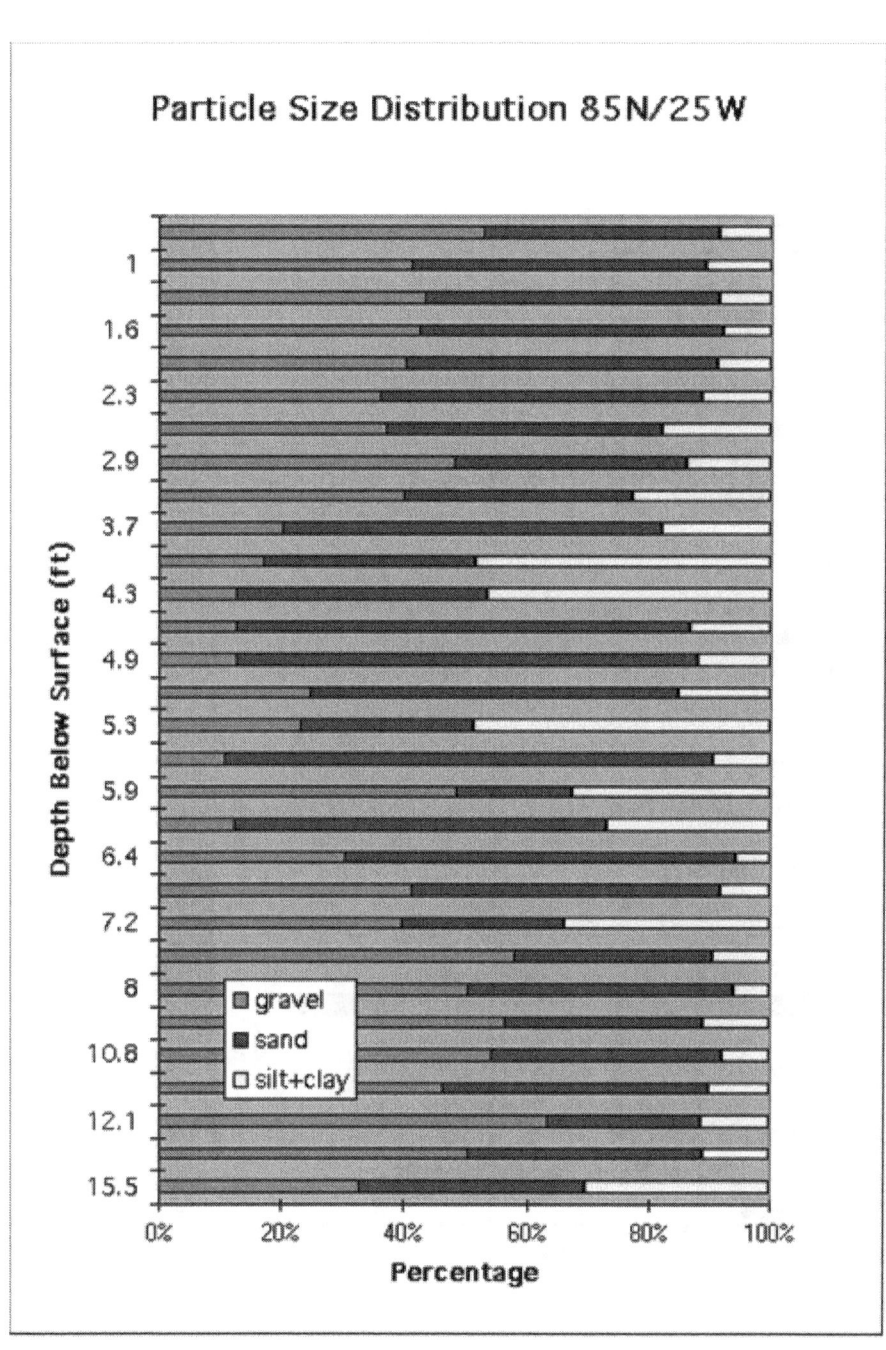

Figure 5.3 Vertical particle-size distribution from rockshelter control block (85N/25W).

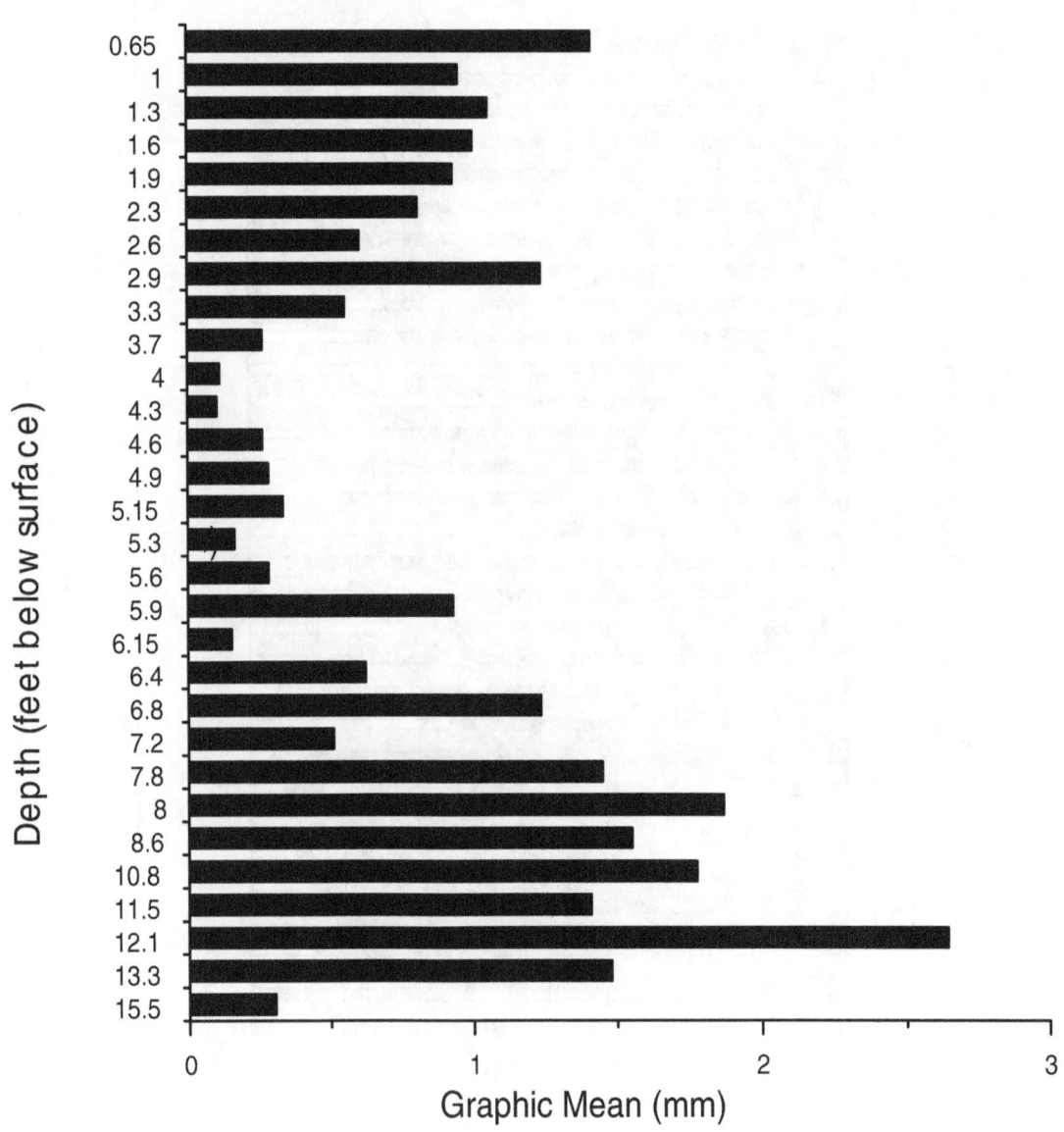

Figure 5.4 Vertical distribution of mean particle size (mm) at rockshelter control block (85N/25W).

Of the chemical analyses performed on the rockshelter sediments, most do not show any dramatic pattern with depth. pH values are neutral to slightly alkaline and generally highest (e.g., 7.5 to 8.3) below 5.6 feet (1.7 m) depth (Figure 5.5). pH values close to 8.0 usually reflect the presence of calcite, and indeed the calcium carbonate values are greatest at depth. Electrical conductivity values are less than 2 mmhos/cm indicating that there are few soluble salts in the rockshelter sediments, and like pH and calcium carbonate, the EC values are greatest at depth although the differences are negligible. Cation exchange capacity does not display any dominant trend with depth although sediments below 5.3 feet (1.6 m) have slightly higher values. Organic matter and nitrogen content is low with no trend with depth, although there is a relative spike in nitrogen content (.06%) at 2.3 to 2.6 feet (.7 to .8 m). Phosphorus values are all less than 12 parts per million and are lowest between 3.7 and 5.3 feet (1.1 to 1.6 m), somewhat mirroring the particle-size trend with depth. Gypsum content varies from none to 2 millequivalents per 100 grams with maximum values below 5.6 feet (1.7 m) depth.

Interpretations

The granulometric data clearly show different contributions of endogenous and exogenous sediments. Given the cryptocrystalline texture of the basalt forming the rockshelter, the mechanical breakdown and exfoliation of e'boulis produce mostly gravel and coarser clasts (i.e., sediments greater than 2 mm). In contrast, sediments blown into the rockshelter tend to be fine sand (.25 mm) and finer. There is the possibility of silt and clay held in suspension in sheetwash entering the rockshelter from the higher hillslope. Although observations during a visit to the rockshelter in 1998 by the author failed to identify any significant sheetwash component entering at the brow or sides of the rockshelter, sheetwash may have entered the rockshelter interior from the berm formerly located below the dripline at the rockshelter's entrance.

Three basic patterns are seen in the particle-size histograms (see Figure 5.2). One is a mixed assemblage with no prevailing grain size mode (i.e. very poorly sorted). This pattern is limited mainly to the upper 3 feet (1 m) of the stratigraphy and may indicate heavy contributions of anthropogenic sediment that contain a higher percentage of midrange grain sizes. It may also reflect the significant mixing of the stratigraphy observed near the surface (Fryxell and Daugherty 1962). The second pattern is a unimodal peak of fine textured sediments. This pattern is prevalent between 4.0 and 6.0 feet (1.2 to 1.8 m) depth and likely indicates the dominance of eolian sediment, and particularly tephra associated with the eruption of Mount Mazama, most recently dated at 6,730 ^{14}C years B.P. (Hallett et al. 1997). The third pattern is a stark bimodal distribution such as is evident at 3.3 to 3.7 feet (1.0 to 1.1 m), 5.9 to 6.15 feet (1.8 to 1.9 m), and the entire lower section at 6.4 to 16.6 feet (2.0 to 5.0 m) depth. This pattern records both e'boulis and fine-textured, exogenous sediment and could be produced by concurrent deposition of the two materials (Draper and Morgenstein 1993:B-3) or by predominantly e'boulis deposition and post-depositional leaching of silt and clay from higher levels, particularly from the tephra-rich zone. Post-depositional downward translocation of sediment has likely occurred in the rockshelter, but probably does not result in deep (> 1 m) displacement of fines. In fact, there is no uniform decrease in fines below the tephra-rich zone suggesting that it is likely the lower deposits contain a bimodal particle-size distribution produced by simultaneous deposition of e'boulis and eolian sediments. It is also possible that the fine sand, silt, and clay fractions in the lowest two samples below 13.3 feet (4.0 m) may be partly lacustrine/alluvial in origin (i.e. deposited during the last of the Missoula floods).

The amount of soil formation within the rockshelter is uncertain and likely to be spatially variable. The limiting factor is moisture, and the interiors of rockshelters receive less precipitation and runoff than areas located at the entrance. Consequently, soil development should progressively decrease towards the backwall away from the entrance. Indeed, natural soil formation at McGregor and Porcupine Caves was limited to the berms and generally absent within the chamber (Hicks and Morgenstein 1994:80-81; Hicks 1995:82). There is evidence of soil formation within part of the Marmes rockshelter based on Fryxell's field descriptions of weak structure near the surface and carbonate development at depth (Appendix E). Moreover, soil profile notes taken by Maynard Fosberg of the University of Idaho Soils Department on April 28, 1968 regarding 80-75 N (no west coordinate given) indicate the presence of a mollic epipedon at 0 to 14 inches (0 to 20 cm) depth, a cambic B horizon at 30 to 35 inches (56 to 76 cm) depth, and a buried argillic horizon at 61 to 87 inches (156 to 220 cm) situated immediately below Mazama tephra. Both of these locations are situated at or close to the drip line and former berm of the rockshelter, and thus it is likely that these sediments would have been subject to

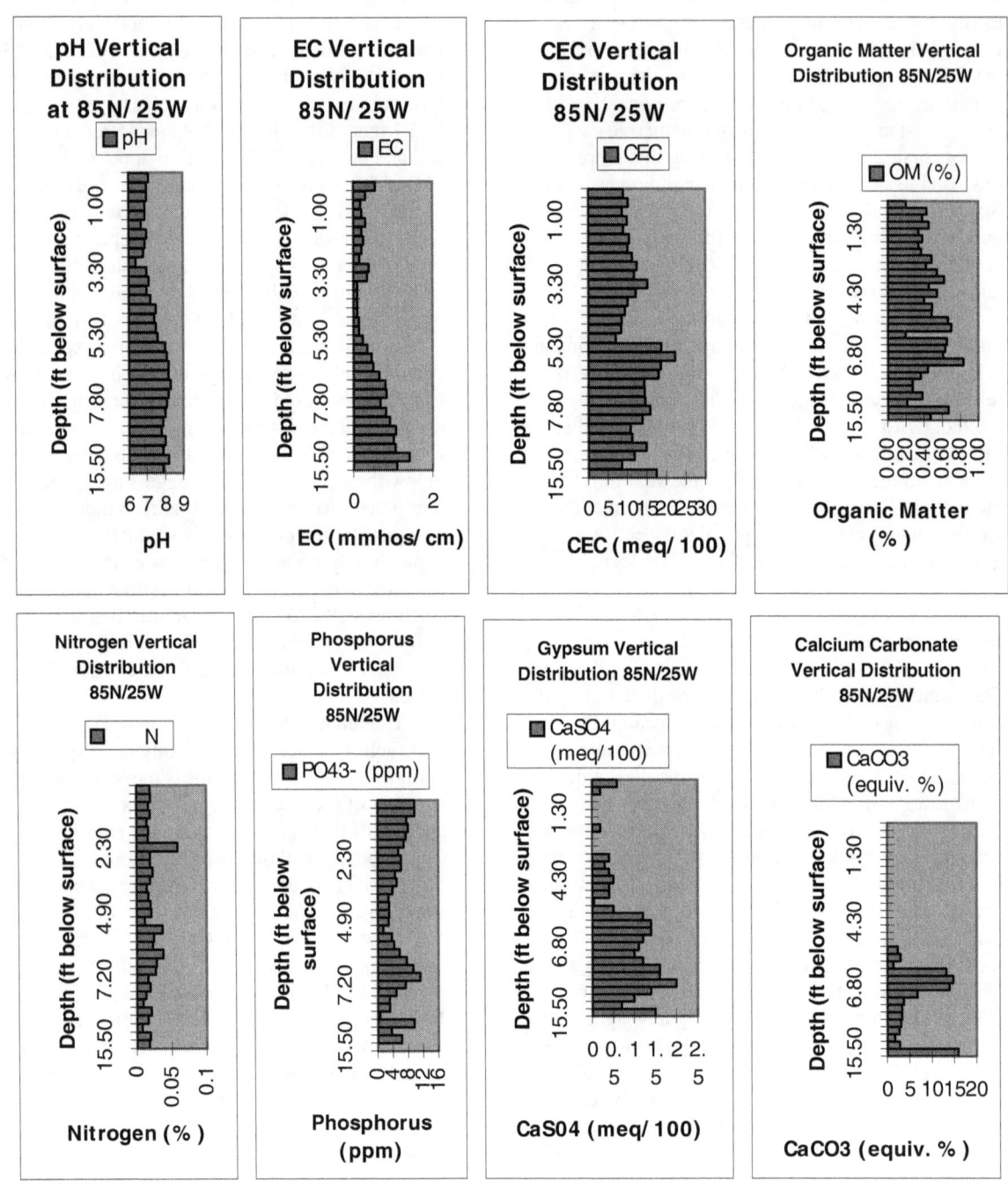

Figure 5.5 pH, electrical conductivity, cation exchange capacity, organic matter, nitrogen, phosphorus, and calcium carbonate data for rockshelter control block, 85N/25W.

water infiltration requisite for pedogenesis. Unfortunately, the laboratory data from 85N/25W do not clearly reveal these pedogenic horizons. As mentioned above, there is no clear indication of silt/clay translocation in the particle-size data; increases in the fine fractions tend to occur erratically with depth and are restricted to narrow zones atypical of pedogenesis. However, the increase in pH, EC, gypsum, and calcium carbonate at 85N/25W below 5.3 feet (1.6 m) more likely reflects pedogenesis in the form of leaching and precipitation of salts. If this is indeed the product of soil formation, then a surface of stability lies immediately above 5.3 feet depth at 85N/25W.

Organic components in the rockshelter stratigraphy may also indicate the presence of a buried soil. Erratic changes in organic matter, nitrogen, and phosphorus content within the rockshelter at 85N/25W are interpreted as reflecting variable exogenous biological inputs rather than pedogenic processes. Organics are probably introduced mainly by animals and humans entering the rockshelter and bringing with them an assortment of organic detritus. Of interest is a spike in nitrogen at 2.3 to 2.6 feet (.7 to .8 m) without a corresponding increase in organic matter content. Although a decrease in the ratio of organic matter (a proxy for organic carbon) to nitrogen may reflect a shift from shrub to grassland vegetation (Moody and Dort 1990:366), this would be an unlikely explanation for this example. It may, however, indicate some type of occupation layer. There are no similar spikes in organic matter or phosphorus throughout the entire sequence. There is, however, a simultaneous slight increase in organic matter, nitrogen, and phosphorus between 5.3 and 7.8 feet (1.6 to 2.4 m), beneath the tephra-dominated zone. This may correlate to the buried paleosol postulated to occur in Unit III and/or increased human activity in the rockshelter during this time. Fryxell and Daugherty (1962:Table III) describe Unit III as containing lenses of shell and bone midden, which would increase the organic matter, nitrogen, and phosphorus levels. Hence, the increased values of organic matter, nitrogen and phosphorus at 5.3 to 7.8 feet probably reflects both anthropogenic and pedogenic processes.

Based on the 1962 excavations in Marmes Rockshelter, Fryxell and Daugherty (1962:Table III) divided the rockshelter stratigraphy into eight major units which are well represented in the front half of the rockshelter. A ninth unit interpreted in the field as lacustrine sediment with rockfall was later exposed during the 1968 excavations. These presumed lake sediments predate the culture-bearing floodplain alluvium. Based on the laboratory data, it is possible to correlate specific depths in the control block samples to Fryxell and Daugherty's system (Table 5.3). Starting from the bottom, Unit I which consists of "coarse, heavy rockfall with little or no interstitial fill", does not correlate well with the samples collected for laboratory analysis, probably due to the fact that such deposits are difficult to sample for granulometry and likely were avoided. However, there are relatively high amounts of coarse clastics relative to fines between 8.6 and 12.3 feet (2.6 to 3.8 m) depth (see Figure 5.2), and these likely correlate to Unit I. Unit II contains "coarse rockfall with windblown sediment filling space between fragments" (Fryxell and Daugherty 1962:Table III). This correlates well with samples from 7.8 to 8.6 feet (2.4 to 2.6 m) depth that have a distinct bimodal particle-size distribution. Unit III, described as "windblown sediment with moderate amounts of rockfall and with interbedded lenses of shell and bone midden" (Fryxell and Daugherty 1962:Table III), correlates to samples at 5.9 to 7.8 feet (1.6 to 1.8 m) depth in a zone that is culturally and pedogenically modified (see above). In the field this unit was traceable laterally beneath the volcanic ash of Unit IV. As might be expected, this deposit is rich in $CaCO_3$. Bones and teeth from this layer are partially impregnated and coated with carbonates, which makes them considerably heavier and easy to distinguish from those from above and below.

Unit IV is the "undisturbed pumicite" or relatively pure Mazama tephra, and Unit V is "dominantly wind-blown sand with sparse rockfall, generally showing extreme mixing and including much pumicite..." (Fryxell and Daugherty 1962:Table III). It is difficult to discern Units IV and V from the laboratory data, but the two can both be correlated to samples from levels between 3.7 and 5.15 feet (1.1 to 1.6 m) depth which are dominated by fine sands and silts. Unit VI is a mixed zone with "moderate rockfall and windblown sediment" (Fryxell and Daugherty 1962:Table III). Such a deposit would have a bimodal particle size distribution and thus probably correlates to samples from 2.6 to 3.7 feet (0.8 to 1.1 m) depth. Unit VII is a complex mixture of "fire hearths, storage pits, organic debris, moderate rockfall and wind-blown sediment" (Fryxell and Daugherty 1962:Table III). This Unit correlates well with the very poorly sorted samples from .65 to 2.6 feet (.2 to .8 m) depth.

Table 5.3 Correlation of stratigraphic units I-VIII (Fryxell and Daugherty, 1962:Table III) with the rockshelter control block (85N/25W) based on particle-size data.

Unit	Generalized Description	Approximate Thickness at Control Block (feet)	Correlated Samples (Depth in Feet Below Surface)
VIII	Manure from domestic stock, organic debris; locally includes disturbed areas filled with debris from stratigraphically earlier deposits.	0.15 - 0.25	Not sampled
VII	Complex; interbedded fire hearths, storage pits, organic debris, moderate rockfall and windblown sediments.	1.0 - 2.0 ft	0.65 - 2.6
VI	complex; filled excavations, fire hearths, moderate rockfall and windblown sediments	0 - 1.15	2.6 - 3.7
V	Dominantly windblown sand with sparse rockfall, generally showing extreme mixing and including much pumicite, reworked during excavation of graves from layer of volcanic ash immediately below.	1.5 - 2.0	3.7 - 5.9[1]
IV	Undisturbed pumicite.	0.3 - 0.8	3.7 - 5.9[1]
III	Windblown sediment with moderate amounts of rockfall and with interbedded lenses of shell and bone midden.	1.0 - 1.5	5.9 - 7.8
II	Zone of transition; coarse rockfall with windblown sediment filling space between fragments.	0.25 - 0.5	7.8 - 8.6
I	Coarse, heavy rockfall with little or no interstitial fill; depth undetermined.	> 2.5	8.6 - 12.3

[1] Units IV and V could not be distinguished by particle-size.

Toward the front of the shelter, units VII, VI and V exhibited minimal mixing, and storage pits, fire hearths and burials had not disturbed the deposits beyond easy recognition. Such was not the case toward the rear and sides of the shelter (Figure 5.6). Here multiple storage pits and fire hearths were found occurring in complex cut-and-fill relationships. Occasional burials further complicated the situation. Attempting to trace deposits laterally in these upper units was not considered practical where such complexity occurred (Gustafson and Gibson 1998:14)

Unit VIII, described from field accounts as consisting of manure from domestic stock, other organic debris and modern artifacts, was not sampled for this analysis. Occasional pits and pockets had been excavated through this layer into older deposits bringing their material to the surface. Gustafson recalls one pit nearly six feet deep into which a cow carcass apparently had been dumped and then burned. However, for the most part, a relatively thin, uniform layer was present over the entire rockshelter surface. Much of this was removed or tramped down in preparation for establishing the site grid and beginning excavations (Gustafson and Gibson 1998:14).

An important process in the formation of the rockshelter stratigraphy and of direct relevance to the density of archaeological and faunal material is the rate of sedimentation. It is unlikely that the rate

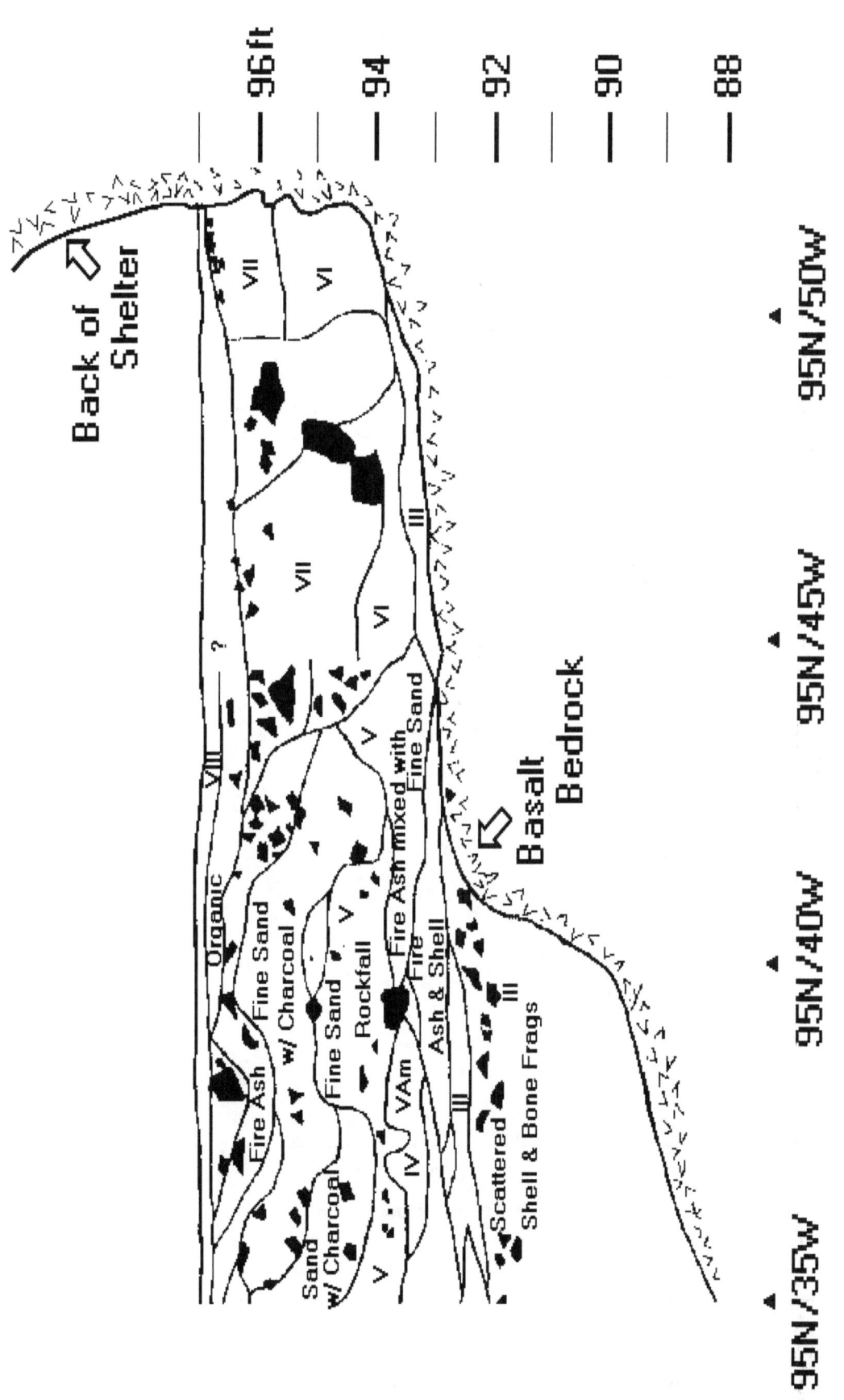

Figure 5.6 Typical disturbance in the sediments near the back of the rockshelter.

of rockshelter sedimentation would be constant during the dramatic climatic changes associated with the Pleistocene to Holocene transition. Changes in the amount of eboulis and eolian sediment within Marmes Rockshelter indicated that the sedimentation rate probably varied through time (Fryxell 1964). Also, variation in the amount of soluble salts and secondary silica on eboulis in McGregor Cave suggested to Hicks and Morgenstein (1994:D-73) that rockfall is episodic rather than constant. At the Marmes Rockshelter, the Mazama tephra and a suite of ^{14}C dates (Sheppard et al. 1987) provide the age-control necessary to assess general rates of sedimentation within the rockshelter (Figure 5.7). These rates, however, are only approximate since the rockshelter has experienced considerable stratigraphic mixing that may have moved the sampled materials from the location of their original associated deposits. Rockshelter stratigraphic depths and thicknesses are based on Units II-VIII provided by Fryxell and Daugherty (1962:Table III); median values were used to estimate deposit thickness. A thickness for Unit I was not provided by Fryxell and Daugherty (1962) because the bottom of the unit was never defined. It is estimated in this study based on the 1968 excavations at 95N/20W (see Appendix B, 95N) as the distance between the top of the underlying lacustrine/alluvial and the base of Unit II. Also, the limiting age of 11,200 ^{14}C years B.P. for the base of Unit I is based on an assumption that the unidentified tephra identified in the upper part of the lacustrine/alluvial sediments by Fryxell is Glacier Peak (Foit et al. 1992), although this has yet to be confirmed.

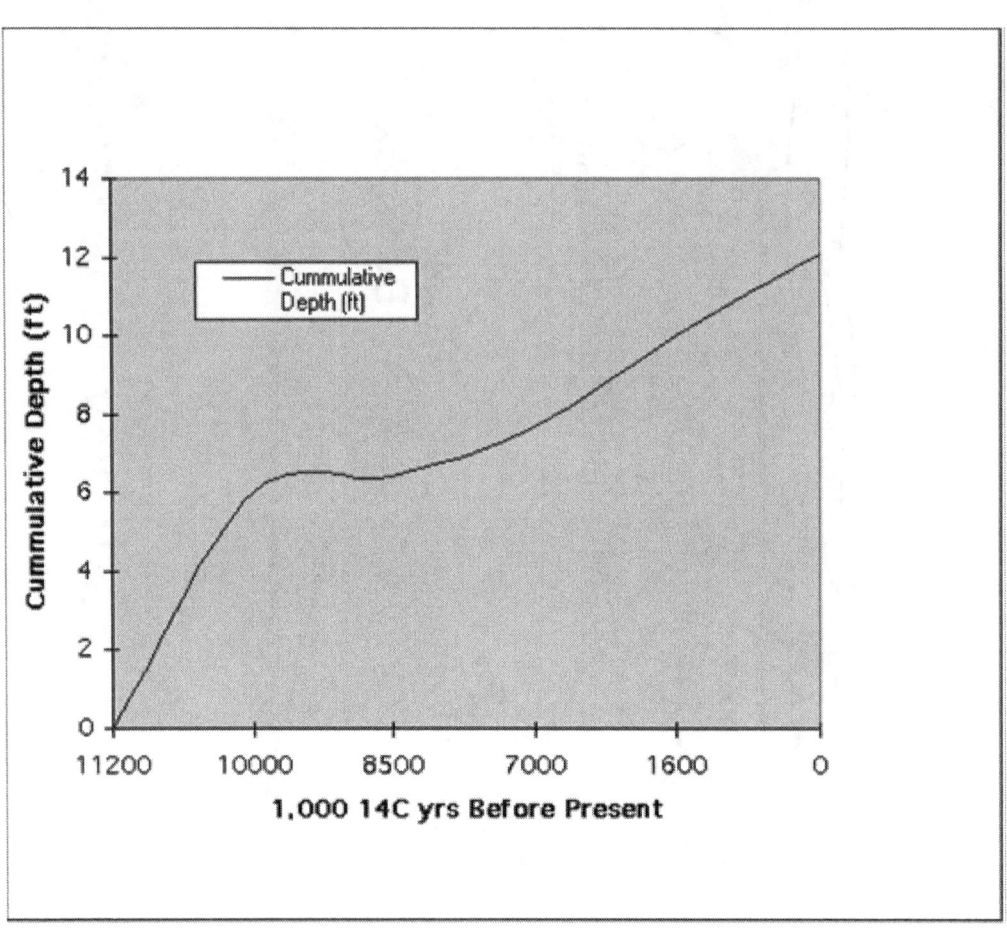

Figure 5.7 Cumulative rockshelter sedimentation rate.

The resulting estimates of depositional rates vary from .4 ft/1000 years (.1 m/1,000 years) for Unit II to approximately 5.0 ft/1,000 years (1.5 m/1,000 years) for Unit I (based on the revised thickness of Unit I)(Table 5.4; Figure 5.7). Even if the more conservative thickness estimate of 2.5+ feet is used for Unit I, the rate is still 2.5+ feet/1,000 years (0.8 m/1,000 years) or at least approximately twice that of the deposition rates for subsequent Holocene units. The mean depositional rate for Marmes Rockshelter calculated by Hicks (1995:152) was .6 m/1,000 years. In sum, rates of sedimentation were greatest in the Marmes Rockshelter at the close of the Pleistocene 11,200 to 10,000 ^{14}C years ago and relatively low from 9,500 to 7,500 ^{14}C years, the latter supporting the interpretation of a buried paleosol prior to the eruption of Mount Mazama. Sedimentation then increased again but at a rate less than that of the latest Pleistocene. For the most part, the rate of sedimentation in the Marmes rockshelter appears to be controlled by climate and vulcanism. Climate affects the production of coarse éboulis through freeze-thaw and eolian sedimentation through changes to regional vegetation (Fryxell 1964). Large volcanic eruptions (e.g., Mount Mazama) affect the amount of silt-sized tephra available for wind-transport.

Colluvial Slope Stratigraphy

Although artifactual and faunal materials recovered from the Marmes site were found in the rockshelter and floodplain, the colluvial slope connects the two areas, and represents an integral part of the entire site. Indeed, the production and transport of debris on the hillslopes around the Marmes site since the last of the Missoula floods is not only a reflection of previous climate and vegetation, it also represents an important process in the burial and preservation of archaeological materials in the floodplain. The term "colluvial" is defined as any geological process associated with slopes where gravity plays a predominant role in sediment movement (i.e. materials that are largely emplaced through mass wasting). In reality, colluvium also may contain sediment washed in by sheetwash runoff or blown in from distant sources. The sediments comprising the colluvial slope at the Marmes site derive from multiple origins including alluvial, eolian, and mass wasting sources. However, given slopes of up to 60 percent below the entrance to the rockshelter, gravity has played a significant role here, as indicated by the presence of large angular talus boulders (see Appendix B, 20W).

The Data

Colluvium at the Marmes site extends approximately 100 feet (31 m) away from the mouth of the rockshelter and interfingers with flood-plain sediments of the Palouse River. The primary sedimentological data from the colluvial slope comes from stratigraphic profiles, monolith descriptions, and laboratory data derived from bulk sampling of sediments adjacent to monoliths. Monoliths A and B (Figure 5.8), located 56 feet (17.1 m) from the brow of the rockshelter, are considered representative of the midslope colluvium (Figure 5.9). Monolith A extends six feet (1.8 m) below the surface, and Monolith B, located five feet (1.5 m) to the south of Monolith A, extends another seven feet (2.1 m) below the base of Monolith A and into buried floodplain sediments.

Monolith B was described in the laboratory in 1998 in order to verify and supplement stratigraphic profiles constructed in the field (Appendix E). It was selected because it includes both colluvium and flood-plain deposits. Overall, the sedimentology of Monolith B matches well with the stratigraphic profile. For example, the profile indicates krotavina in the upper part of Monolith B, and indeed the upper 1.7 feet (.5 m) of the monolith contains a structureless fill rich in volcanic ash as a result of disturbance by burrowing. The tephra presumably was reworked from higher in the column and correlates to the Mazama ash in stratigraphic profiles. Where not disturbed by krotavina, the monolith contains a soil with angular blocky structure and filaments of a white, noneffervescent salt. The entire monolith contains diffuse calcium carbonate as evidenced by effervescence when treated with dilute hydrochloric acid. The white noneffervescent salt is best expressed in the middle part of the monolith which contains large angular basalt clasts. Below the large clasts, the salts disappear and there is little sign of pedogenesis. No clear stratification could be identified in the lower part of the monolith although horizontal layering was recorded in the field (see Figure 5.8).

Particle-size data are derived from bulk samples collected adjacent to the monoliths. The results, however, are somewhat misleading in that they only record the silty matrix of the colluvium. Monolith A contains an almost uniform vertical particle-size distribution dominated by silt

Table 5.4 Estimates of rockshelter sedimentation rates.

Stratum	Age[1] (14C years B.P.)	Length of Time (14C years)	Median Thickness[2] (ft)	Deposition Rate (ft/1,000 14C years)	Deposition Rate (m/1,000 14C years)
Unit VI-VIII	1,600 - present	1,600	2.1	1.3	0.4
Unit IV-V	1,600-6,700	5,100	2.3	0.5	0.2
Unit III	7,000-8,000	1,000	1.3	1.3	0.4
Unit II	8,500-9,500	1,000	0.4	0.4	0.1
Unit I	10,000-11,000	1,000	2.5+	2.5+	0.8+
Unit I	10,000-11,250[3]	1,200	~6.0	~5.0	~1.5

[1] Ages based on Sheppard et al. (1987).
[2] Depths based on Fryxell and Daugherty (1962:Table III) except for Unit I which includes a second estimate based on 1968 excavations at 95N/20W.
[3] Revised age based on 1968 excavations into "lacustrine" sediments at 95N/20W and possible Glacier Peak tephra.

(Appendix C; Figure 5.10). This contrasts with the stratigraphic profile, which shows large angular cobbles and boulders (see Figure 5.8). Gravel fractions were not determined during the 1968 laboratory analysis (Appendix C; Figure 5.10), and it may be that the > 2 mm fraction was too coarse to sample (as was the case for the middle part of Monolith B). Hence, the true particle-size distribution is likely to range from strongly bimodal consisting mostly of angular cobbles and boulders in a silty matrix to unimodal consisting mostly of silt. Although clay represents a very small part of the total fraction in Monolith A, there is a slight increase at the surface and at the base, both of which may indicate slight pedogenic enrichment of fines. The increase in clay towards the base of Monolith A is matched by a similar slight increase at 0.5-1.5 feet (0.2-0.5 m) depth on Monolith B. This zone on Monolith B contains the angular blocky structure indicative of pedogenesis (Appendix E). This zone also underlies the primary Mazama tephra layer and correlates to the paleosol identified in Unit III of the rockshelter, as well as the paleosol underlying Mazama ash in the berms at McGregor and Porcupine Caves (Hicks and Morgenstein 1994:D-71; Hicks 1995:82). Also of interest is the greater variability and amount of sand in the lower part of Monolith B compared to the upper part of the monolith and all of Monolith A. This likely corresponds to the fluvial origin of sediment at the base of Monolith B which allows for more deposition of sand than in the eolian-dominant upper part of the stratigraphy.

The most intriguing aspect of the chemical data from the colluvial slope is the high alkalinity of the sediments. In the lower 3.5 feet (1.1 m) of Monolith A and upper 0.5 feet (0.2 m) of Monolith B, pH values are greater than 9 (Figures 5.11 and 5.12; Appendix D). Such high pH's are usually associated with poorly drained soils that are dominated by exchangeable sodium (Bohn et al. 1985:247; Richards 1954). Based on the exchangeable sodium percentage (ESP) values alone, the lower half of Monolith A and all but the lowermost part of Monolith B classify as sodic soils (i.e., they have ESP values > 15 percent). In fact, the upper part of Monolith B has ESP values of approximately 100 percent suggesting that all exchangeable sites in these

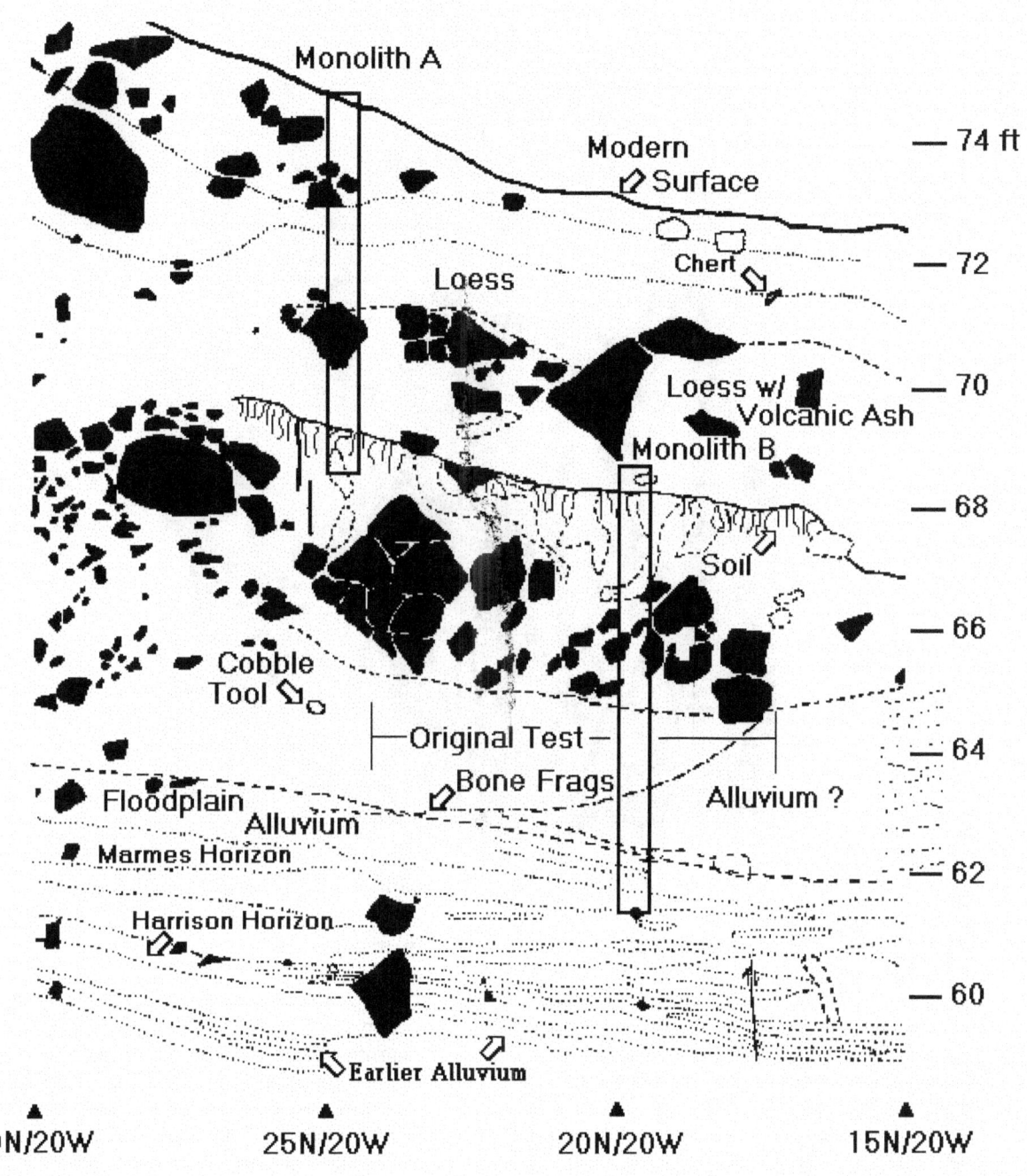

Figure 5.8 Stratigraphic profile of Monoliths A and B localities on the colluvial slope.

Figure 5.9 Photograph of one of the soil monoliths being prepared for collection.

sediments are occupied by sodium ion. Meanwhile, the electrical conductivity (EC) values in the upper 0.5 feet (0.2 m) of Monolith A are > 4 mmhos/cm indicating a high salt content sufficient to classify as a saline soil. Peak gypsum content also corresponds to the surface horizon, but no stratum contains both a high EC and ESP value to classify as a saline-sodic soil. Organic matter and nitrogen content displays a typical pattern of maximum value at the surface A horizon and decreasing values with depth. Calcium carbonate is present at most levels but is greatest in the lower levels of Monolith A above the Mazama tephra and in Monolith B below the bioturbated upper zone.

Interpretations

The depositional history at the Monolith A and B locality can be summed as low-energy floodplain deposition followed by pulses of colluvial deposition consisting of large talus blocks and reworked loess from the higher hillslope. Coeval with this deposition was eolian sedimentation of mostly silt. Eolian sedimentation appears to have been gradual with the exception of increased deposition during and immediately following the Mazama eruption. In contrast, colluvial deposition is much more periodic, marked by rapid deposition and then subsequent overall slope stability. Hillslope stability was sufficient immediately prior to Mazama time for the formation of a paleosol within the colluvium, and was sufficient to form

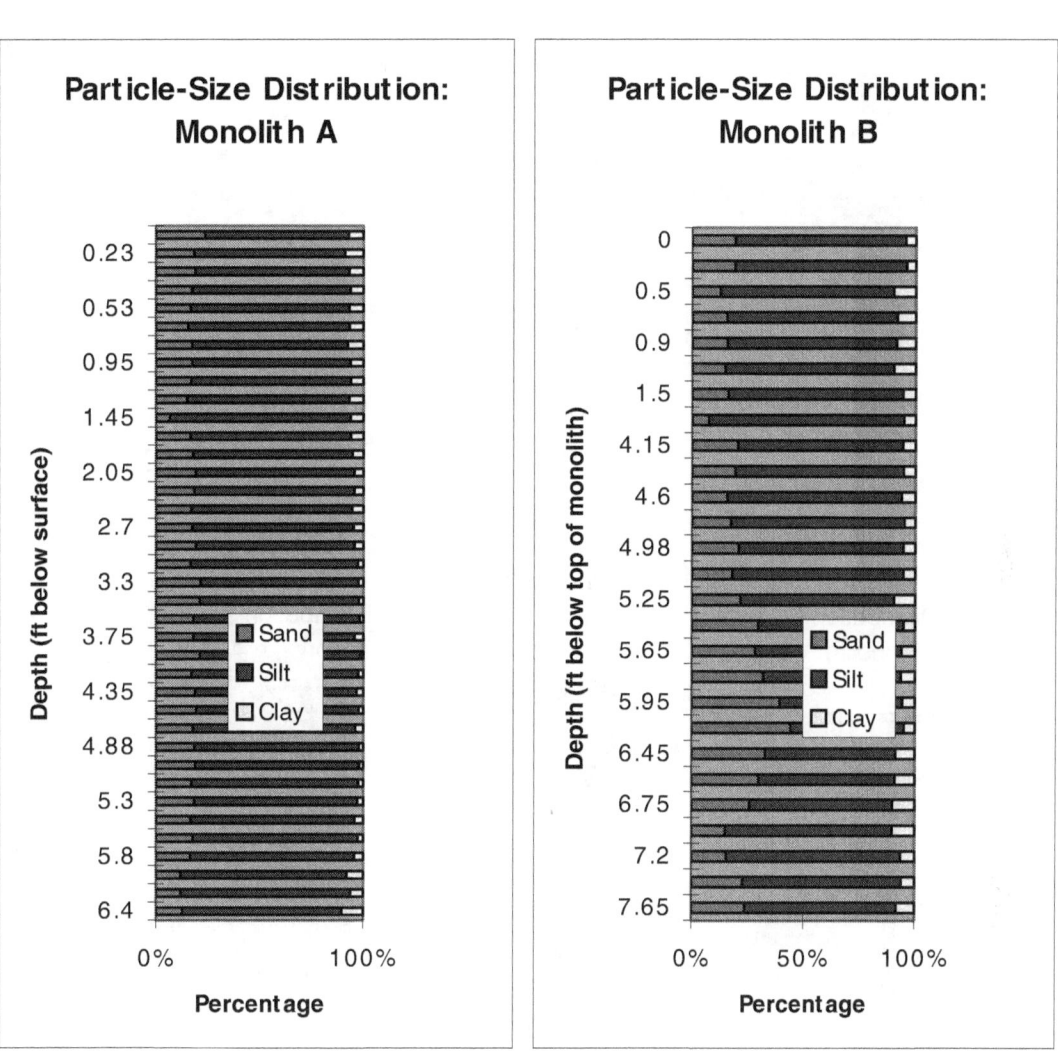

Figure 5.10 Vertical particle-size distribution from Monoliths A and B locality.

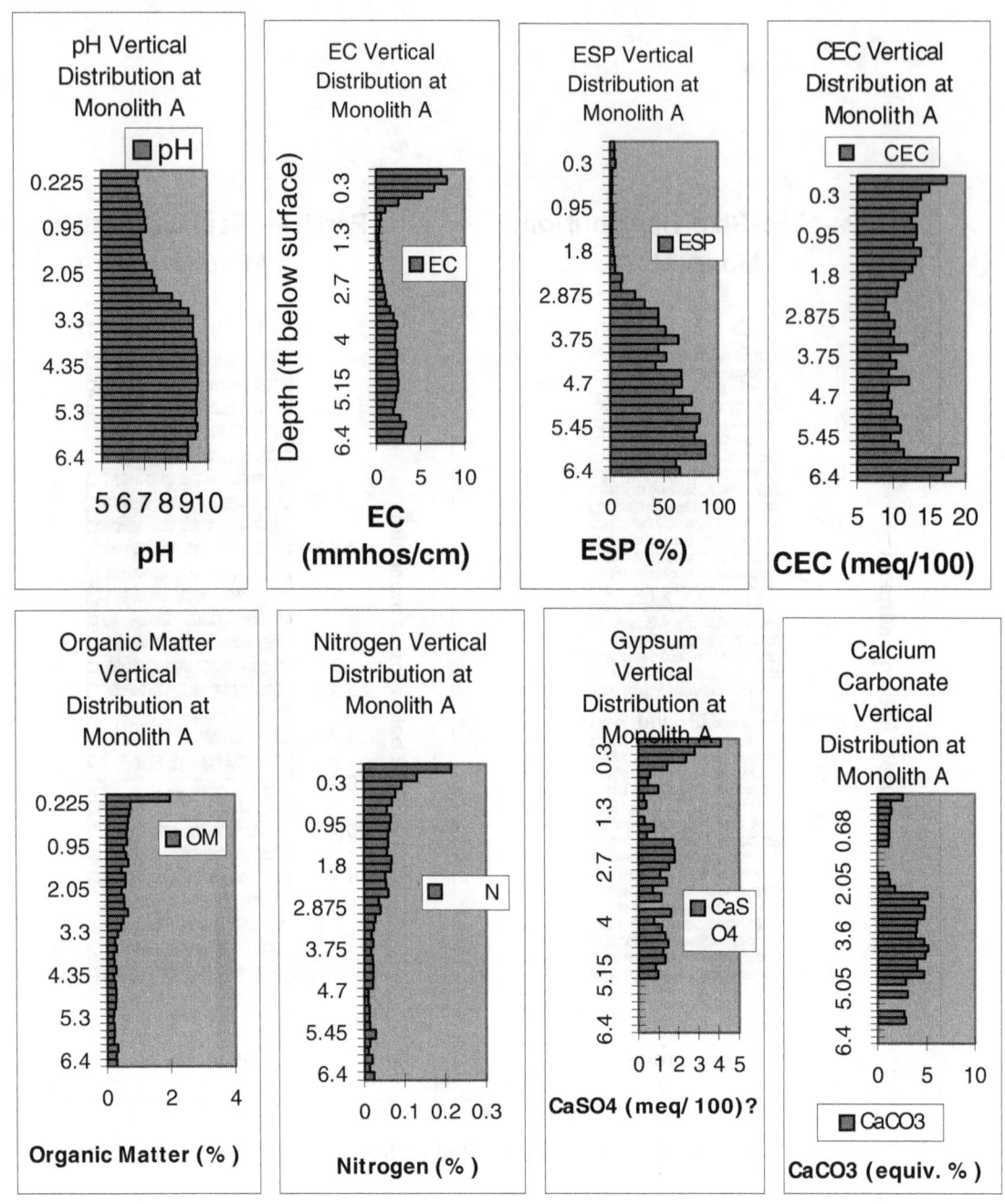

Figure 5.11 pH, electrical conductivity, exchangeable sodium percentage, cation exchange capacity, organic matter, nitrogen, and calcium carbonate measurements with depth for Monolith A locality.

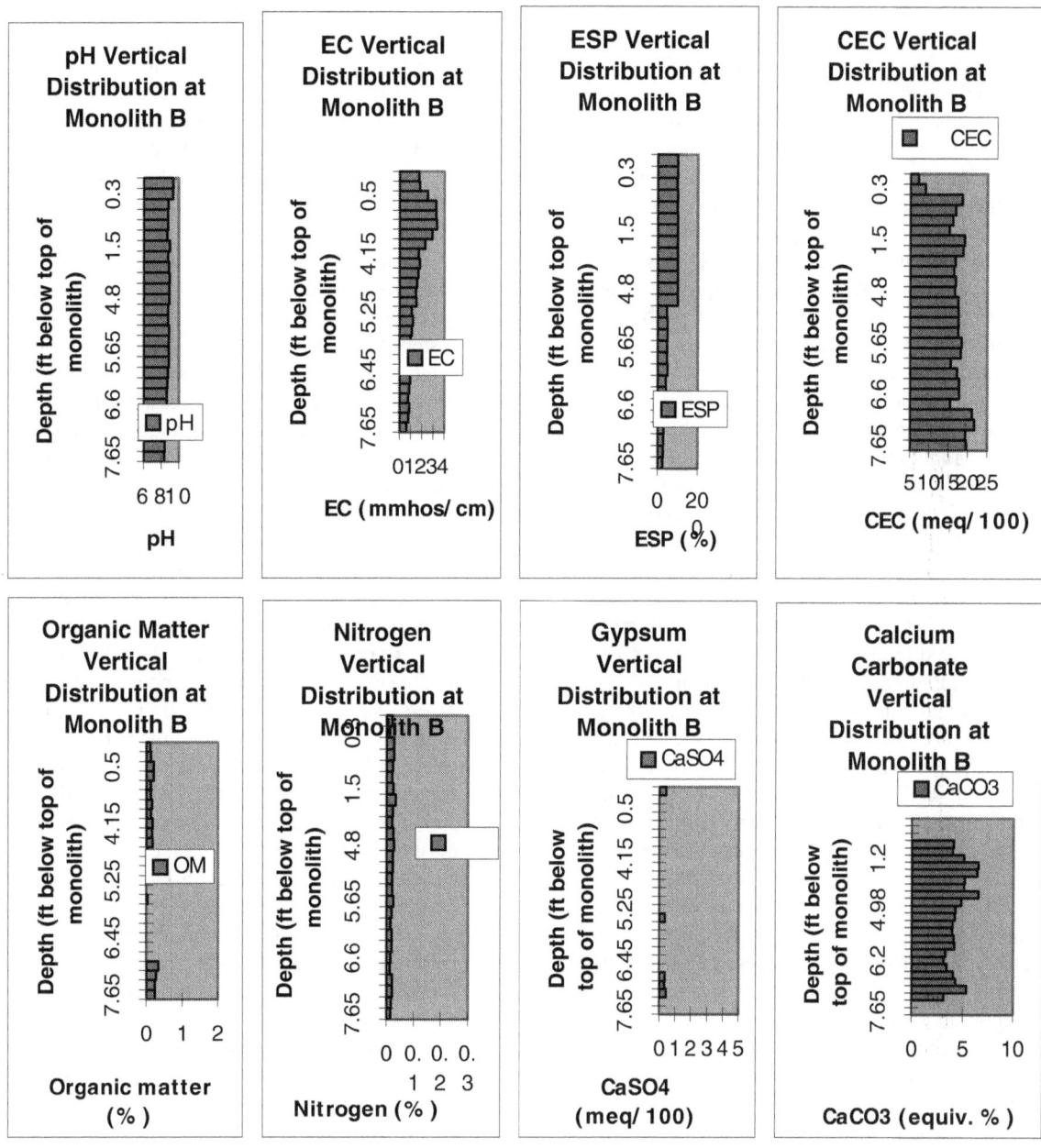

Figure 5.12 pH, electrical conductivity, exchangeable sodium percentage, cation exchange capacity, organic matter, nitrogen, gypsum, and calcium carbonate data for Monolith B locality.

the modern soil at the surface. It is uncertain how much time is represented by the paleosol with strong blocky structure and Stage I+ salt morphology but it probably involves over 1,000 years. The soil is shown as truncated in the profile (see Figure 5.8; Appendix B, 20W), but it is unknown whether this unconformity is real or the product of mechanical excavations in 1968.

Although stratigraphy at the Monolith A and B locality indicates that alluviation predates colluviation, the lake sediments exposed at the base of the rockshelter contained rockfall, probably e'boulis, and it is probable that the lake sediments outside the rockshelter underlying Palouse River alluvium but not exposed in excavations also contains rock material derived from the hillslopes. At the Monolith A and B locality, floodplain sedimentation was the dominant depositional process later followed by predominant colluviation. This begs the question as to what would cause colluvium to encroach upon the floodplain. Was it a change in climate? It is impossible to rule out climate change given that the lower stratigraphy represents at least the early Holocene and possibly the latest Pleistocene, a period of significant global and regional climate change (Mehringer 1996; Thompson et al. 1993). Moreover, the shift to a drier climate that characterizes the early to middle Holocene in the Columbia Basin is compatible with predictions of increased hillslope sediment yield based on geomorphic process-response models (e.g., Bull 1991). Specifically, reduced effective moisture results in reduced vegetative cover thus allowing for greater erosion on the upper hillslopes and deposition on the toe slopes. However, other factors have to be considered including the type of vegetation associated with the changing biogeography and how sediment is produced on the hillslopes. For example, the large talus blocks in the upper stratigraphy are generated by mechanical weathering and separation from the basalt cliff and hillslope. It is feasible that the increase in talus blocks in the upper stratigraphy near Monoliths A and B reflects greater production due to climatic shifts favorable for enhanced freeze-thaw. On the other hand, it could merely reflect an eventual exposure of a well-jointed seam in the basalt conducive for talus generation. Finally, changes in the predominance of alluviation versus colluviation at this one location may simply reflect shifts in the depositional loci of the hillslope and floodplain without any implied climate change. For example, the main channel of the Palouse River may have shifted south away from Marmes Rockshelter thus decreasing the frequency of alluvial deposition on the northern side of the floodplain. This would have increased the percentage of colluvial deposits preserved in the stratigraphic section without necessarily increasing the rate of colluviation.

Of possible paleoclimatic significance is the presence of alkaline and saline soils in a nonhyperarid, well-drained position on the landscape. There is nothing unusual with the presence of salts in soils; they are the byproducts of mineral weathering and can enter the solum through eolian sources. However, in most well-drained conditions, soluble salts are leached from soils except in hyperarid environments. In semiarid eastern Washington, most soils are leached of soluble salts but retain the less soluble salts such as calcium carbonate. Nonetheless, alkaline soils are known in the Palouse region as implied by Alkali Flat Creek, the next major tributary on the north side of the Snake River upstream from the Palouse River. The distribution of alkaline soils in the Palouse region, however, is spotty and usually linked to localized impeded drainage (Peterson 1961). In upland contexts, they are commonly found either on bedrock benches with shallow loess cover or on hillslopes covered with thick loess where subsurface, relatively impermeable layers divert seepage to the surface. In both cases, sodium-saturated waters are concentrated near the surface (due to either bedrock or some pedogenic aquiclude) resulting in soils with high ESP values and characteristic natric soil morphologies (e.g., massive eluvial horizon over a prismatic, clay-enriched natric horizon) (U.S.D.A. Soil Survey Staff 1975:28). Neither of these hydrologic settings apply to the colluvial slope at the Marmes site, nor do the soils display well-developed natric properties. There is no evidence of limited hydraulic conductivity in the stratigraphy such as an impenetrable subsurface horizon or blue, gray, green, and orange mottled colors reflective of oxidation/reduction and fluctuating water tables. However, soils with > 50% silt can have low permeability (Richards 1954:31), and the matrix of the colluvium typically has silt fractions greater than 60%. Moreover, layering of different silty layers, such as occurs in loess with tephra, can reduce the hydraulic conductivity, and it is conceivable that the Mazama tephra and the underlying truncated paleosol on the colluvial slope play a role in limiting infiltration.

Saline soils are defined as soils with high soluble salt (e.g., primarily chlorides and sulfates) content but not dominated by sodium ion. Such soils are not common to upland sites in the Palouse. For example, most of the alkaline soils studied by Peterson (1961) in eastern Washington had EC values below 4 mmhos/cm. This is interesting since the conventional model is that most alkaline soils evolve from saline soils. However, the surface soil in the vicinity of Monolith A contains EC values greater than 5 mmhos/cm which meets the requirements of a saline soil. Of more significance is the vertical distribution of the EC values: they are at maximum near the surface and decrease with depth. This suggests previous shallow, saturated conditions whereby salts were emplaced at the surface through evapotranspiration and capillary rise (Gerasimov and Glazovskaya 1965:105; Richards 1954:4). Again, for this to occur, the silty matrix of the colluvium must inhibit infiltration and leaching.

The conventional model for the genesis of alkaline soils requires that there first be an accumulation of soluble salts near the surface due to incomplete leaching and an abundance of Na^+. This has to be followed by a period of increased leaching such that some of the soil solutes are shifted downward, either preferentially leaving Na^+ behind, or, as suggested by Peterson (1961) for the Palouse region, preferentially transporting Na^+ along the upper contact of some aquiclude that eventually emerges at the surface. It is possible that much of the sodium ion in the colluvial sediments at the Marmes site is derived from upslope sources, however the presence of saline soil at the colluvial slope surface suggests that an extralocal source of sodium ion is not required. Once Na^+ dominates the exchange complexes of the soil, then concomitant physical changes occur such as the dispersion and eluviation of clay and formation of the natric horizon (Gerasimov and Glazovskaya 1965: 105-110; Peterson 1961). Peterson (1961:214-215) postulated that the shift from the accumulation of soluble salts to their redistribution and concentration of Na^+ might be driven by Holocene climate change. Specifically, the period of reduced effective moisture during the middle Holocene ca. 5,000-7,000 years ago (Hansen 1947) may have been a period of salt concentration that was followed by preferential leaching and lateral movement of the sodium downslope during the subsequent moister period of the late Holocene. At the Marmes site, sediments affected by alkalization both pre- and post-date the Mazama eruption ca. 6,730 ^{14}C years ago. It is uncertain from the laboratory data and profiles whether or not the alkalization process occurred before and/or after the Mazama eruption. Because soluble salts are presently accumulating at the surface, it appears that conditions necessary for the first step in the alkalization process need not be too different from today's climate. In fact, the accumulation of soluble salts and subsequent differential leaching of exchangeable ions may only require small shifts in climate that cross some type of moisture threshold between saline and alkali soil formation. Although somewhat conjectural in lieu of detailed pedologic study, it nonetheless suggests that Holocene climate change need not be dramatic to have significant effects in soil formation at the Marmes site. Because salinization and alkalization have soil morphological and chemical consequences that affect plant growth, it is plausible that these small magnitude changes in climate during the Holocene also affected the composition and distribution of plant communities at the Marmes site. However, the overall vegetation in the lower Palouse River area (as indicated by pollen from Wildcat Lake) has remained a grassland-sagebrush steppe since 10,000 years ago (Mehringer 1996).

Floodplain Stratigraphy

Colluvium gradually grades into the highest alluvial terrace of the Palouse River (see Figure 1.11), or what Marshall (1971) mapped as Terrace 1. This terrace was exposed in several cutbanks along the lower Palouse River prior to downstream damming and inundation of the canyon. Marshall (1971:27-30) described the general stratigraphy of the terrace as consisting of basal deposits of boulders, cobbles, and gravels of basaltic and nonbasaltic lithologies overlain by predominantly silty sediment. Mazama tephra was identified as overlying the gravels and usually located at the base of the silty upper deposits. At the Marmes site, Marshall (1971) reported that the Mazama ash lies unconformably over a calcic B horizon, and he interpreted much of the upper silty sediment and associated tephra as eolian in origin.

Because the water to be impounded behind Lower Monumental Dam was scheduled to flood the site in less than eight months at the beginning of fieldwork in 1968, Fryxell and other supervisory personnel elected to sacrifice the alluvial/eolian silts and sand capping Terrace 1. Most of the overlying deposits were removed by bulldozer (Fryxell and Keel 1969; Keel and Fryxell 1969) allowing a concentration of effort on the lower alluvial deposits. Excavations in the floodplain extended much deeper than what was exposed in stream cutbanks. The north-south stratigraphic trench at 20W extended onto the floodplain to 140S and exposed both low-energy, horizontally stratified, fine sand and silty alluvium and higher energy, crudely bedded, gravelly and coarser alluvium (Appendix B, 20W). The horizontally stratified, overbank deposits contained several organic zones and were rich in cultural remains including large elk, artifacts, and human bones (mostly charred skull fragments). Fryxell (personal communication to Gustafson) concluded that the darker organic zones were weakly developed soils that could form in a few decades and thus were surfaces of human activity. Fryxell called these darker colored, coalescing organic zones incipient "A" horizons. He named the two uppermost (A1 and A2) the Marmes horizon after the Marmes family who owned the land (Figure 5.13). These were separated from the next lower three A horizons by sterile alluvial silts. Layers A3, A4, and A5 were named the Harrison horizon after the owners of the adjacent land. Artifacts and charred animal bones were found in both horizons, but only the Marmes horizon contained human bone fragments. Figure 5.14 depicts the surface of both the Marmes and the Harrison horizons. In both cases, there is a rise of about five feet (1.5 m) from south to north indicating that the main Palouse River channel was to the south and providing a minimum estimate of the depth of large floods during formation of the floodplain. The predominant silt-size of the alluvium suggests that the velocity was not great as the water spilled over the banks of the channel.

In addition to the fine-textured alluvium with organic horizons, a gravelly paleochannel of the Palouse River was encountered trending in an east-west direction with its north bank near the zero N/S grid line (Figure 5.15). As seen in the figure, major excavations ceased at about the same line. Apparently it was assumed that this channel represented the stream responsible for the culture-bearing and other pre-Mazama overbank deposits. Fryxell and Keel (1969) reported:

> Exposures provided by bulldozing of the overburden, particularly extension of the original north-south trench and the main cross-trench, both sectioned and exposed in plan view portions of relict Palouse River channels. Reconstruction of these channels confirmed earlier interpretations, and limited the area of potentially valuable culture-bearing deposits to an area of about 300X400 feet. Such reconstruction also confirmed interpretation that early inhabitants of the site had occupied a small floodplain bench between the former channel of the Palouse and the rockshelter. [Fryxell and Keel 1969:40]

The north edge of the channel is fairly well defined in the stratigraphic profiles, but the southern edge is apparent only in a few profiles near the 30W and 20W walls of the extended stratigraphic trench. However, gravelly channel deposits, possibly part of this paleochannel or other depositional events, extend southward along the stratigraphic trench (Appendix B, 20W).

The Data

The compiled stratigraphic profiles and sedimentological data add further insight into the channel and overbank deposits at the Marmes site. In addition to the base of Monolith B, descriptive and laboratory data for the floodplain were available from Monoliths G, K, M, N, and O, which extend progressively to the south along 20W (Figures 5.16-5.18). Fine-textured alluvial deposits containing the organic horizons dominate the lower stratigraphy north of 85S and are cut and overlain by the aforementioned gravelly paleochannel. Overlying the gravelly channel deposits is a fine-textured, silty unit interpreted in the field as loess. Monolith G and the lower part of Monolith K record the fine-textured deposits; the upper part of Monolith K, all of Monolith M, and the lower part of Monolith O record the coarse-textured channel deposits. The upper part of Monolith O and all of Monolith N record the capping, silty deposits.

Field descriptions are not available for stratigraphy in the vicinity of Monolith G (Figure 5.16), but profiles indicate that there are a series of low-energy, well-sorted, horizontally stratified, sandy and silty deposits overlain by the distal ends of bouldery hillslope colluvium. Monolith G transects the fine-textured sediments including the organic Marmes and Harrison Horizons (Keel and Fryxell 1969; see Figure 5.16). Silt is the dominant grain size (Appendix C; Figure 5.19) although there is more sand present than in the colluvial or rockshelter deposits. There are no significant spikes in clay content which supports field observations that pedogenesis was limited to organic enrichment and A-horizon formation. pH is high (8.0 to 8.7) and electrical conductivity low (< 1 mmho/cm) in all of the samples (Appendix D; Figure 5.20). Exchangeable sodium percentage is considerably less than in the colluvial deposits but still high enough to classify the sediments as sodic (i.e. > 15 percent), at least at depths from 1.1 to 5.0 feet (0.3 to 1.5 m) below the top of the monolith. Cation exchange capacity is relatively uniform with depth suggesting no significant change in the quantity of exchange sites such as clays and organic matter. This is supported by the organic matter and nitrogen values that are generally low with only a few abrupt increases. Gypsum is present in the

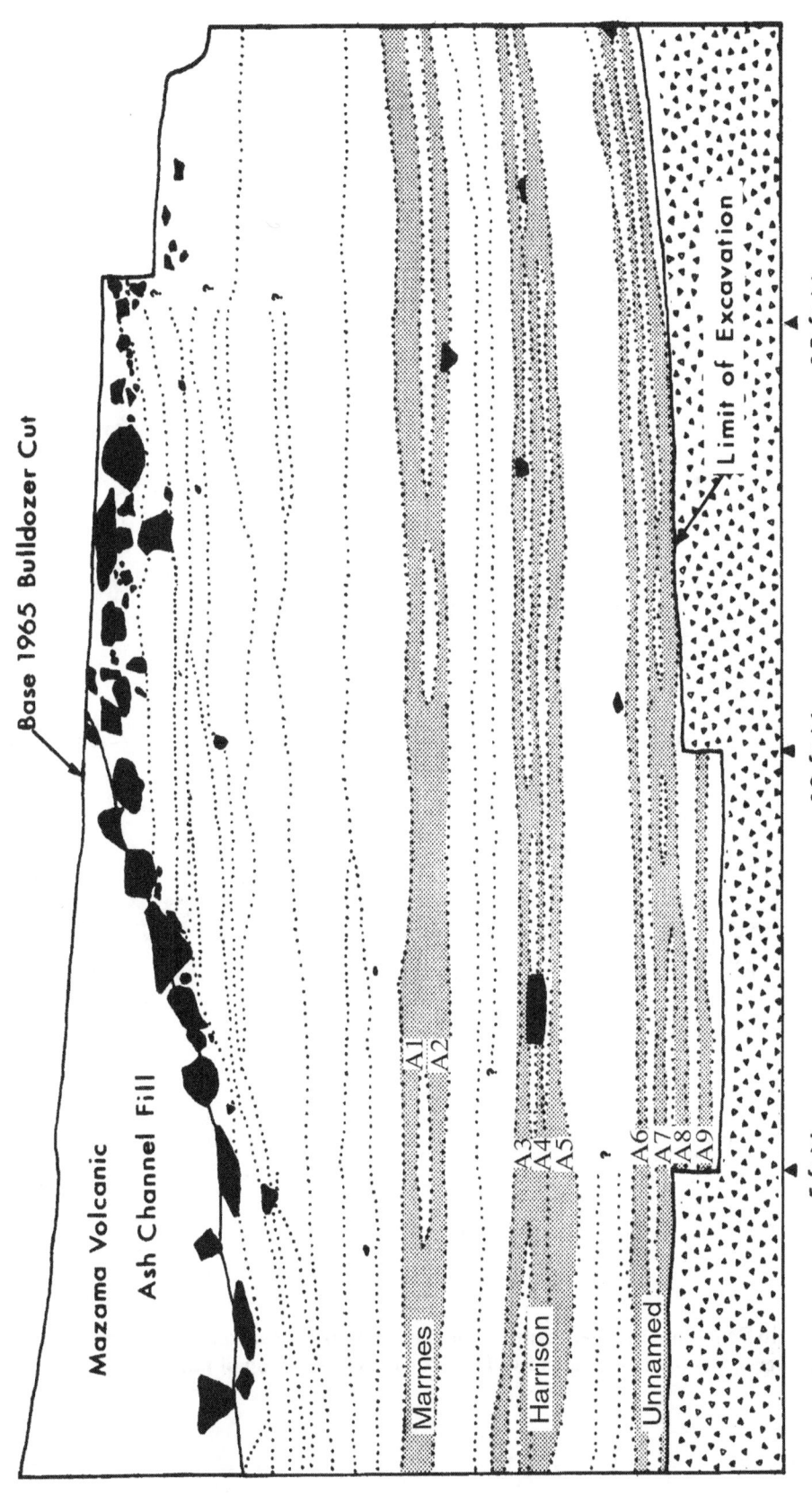

Figure 5.13 Stratigraphic diagram of stream overbank deposits showing the position of buried A horizons in relation to an abandoned stream channel filled with Mazama tephra (from Gustafson 1972).

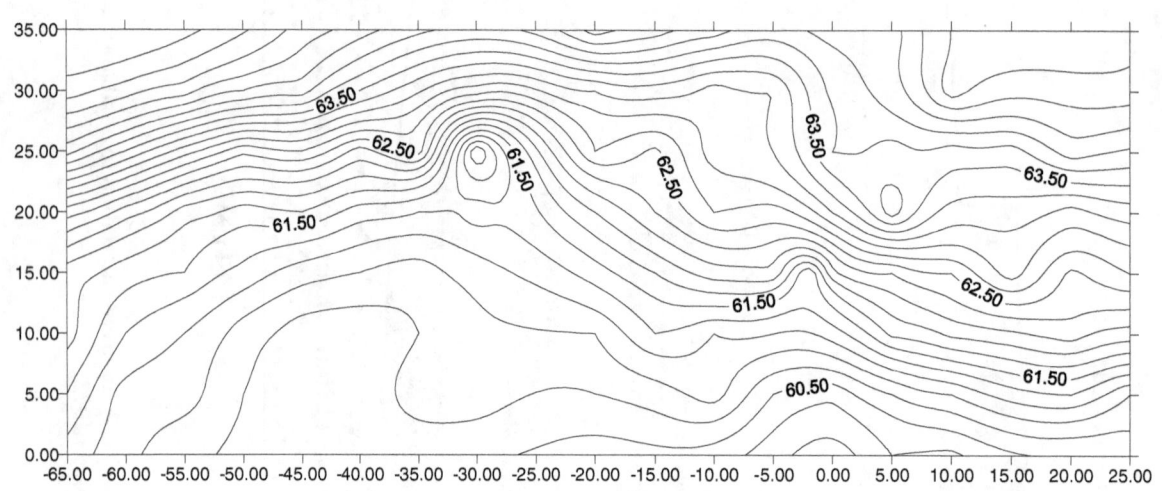

MARMES SURFACE

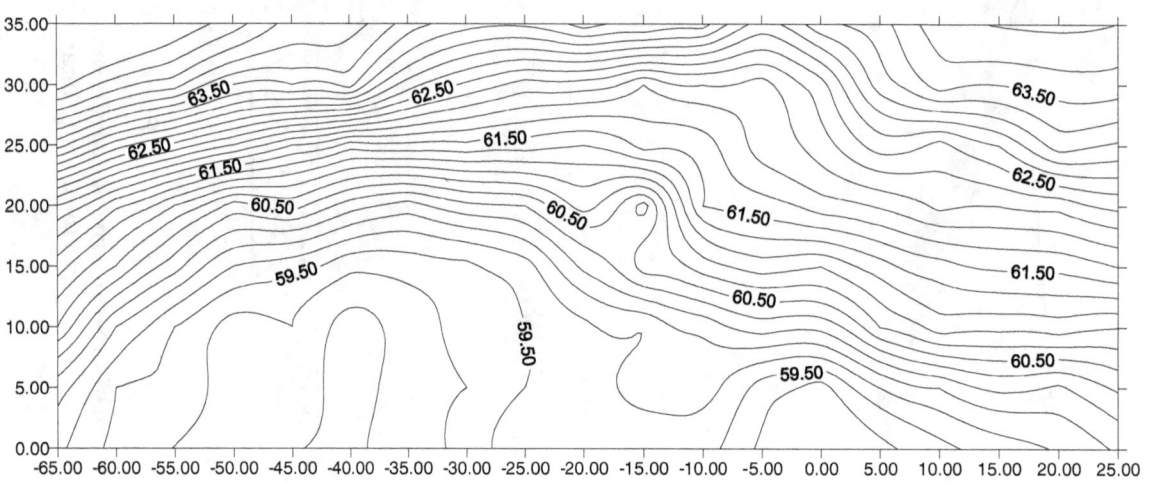

HARRISON SURFACE

Figure 5.14 Main floodplain excavation area showing contours of Marmes and Harrison surfaces. Contour interval is two feet. North is toward top of page.

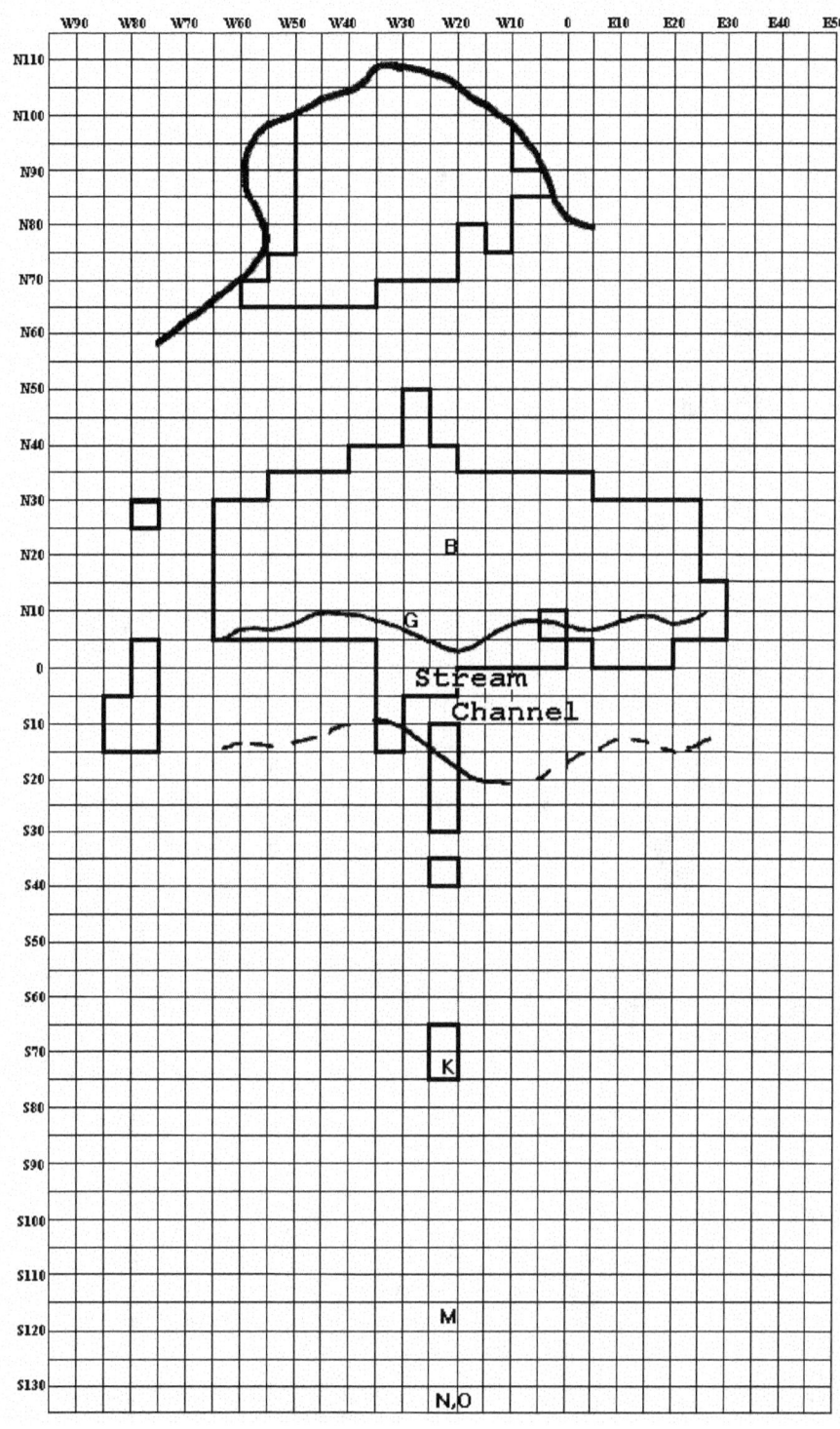

Figure 5.15 Excavation grid showing relationship of stream channel to excavated area (dark outlines). Bold letters designate monolith locations.

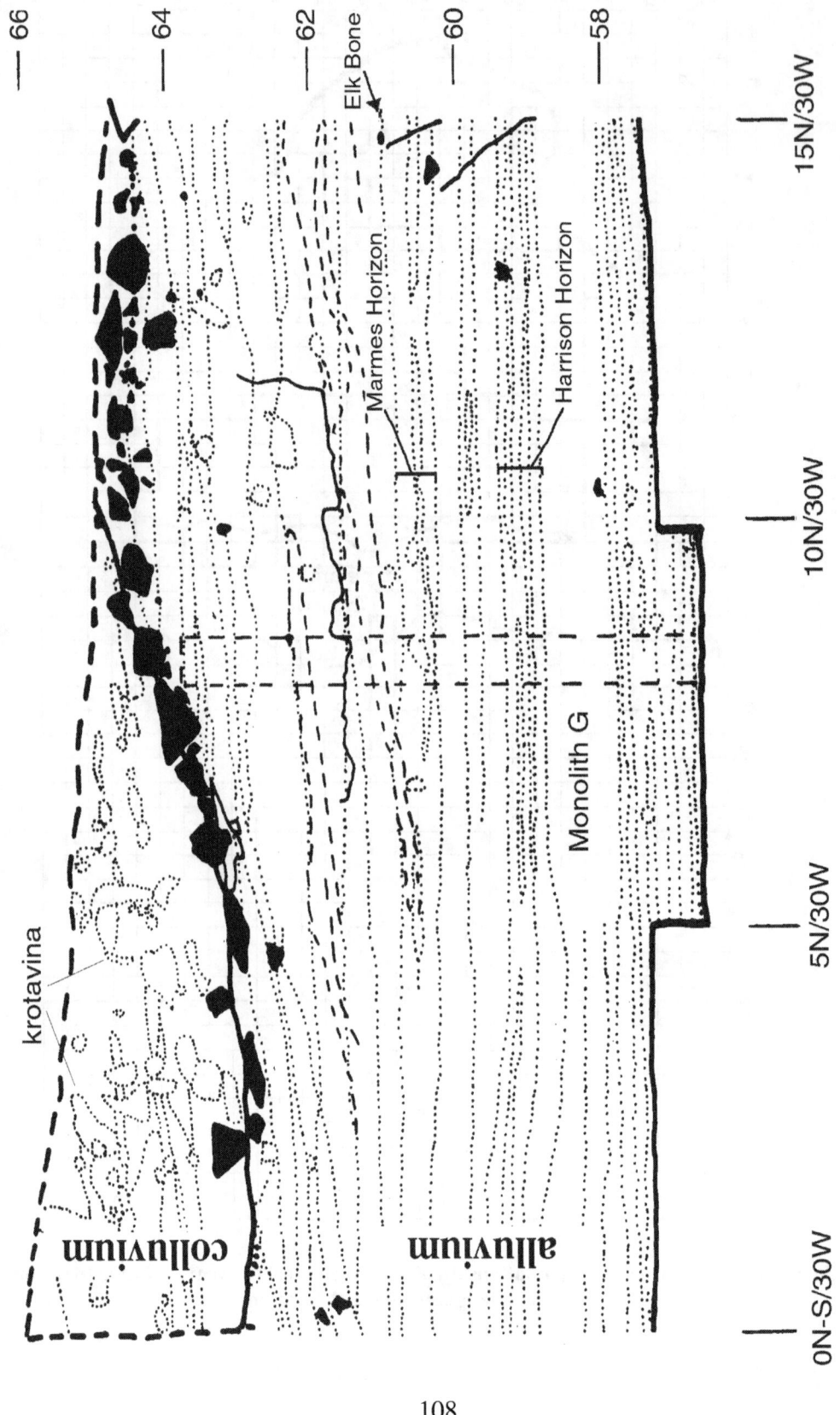

Figure 5.16 Stratigraphic profile at Monolith G locality at the colluvial slope-floodplain transition.

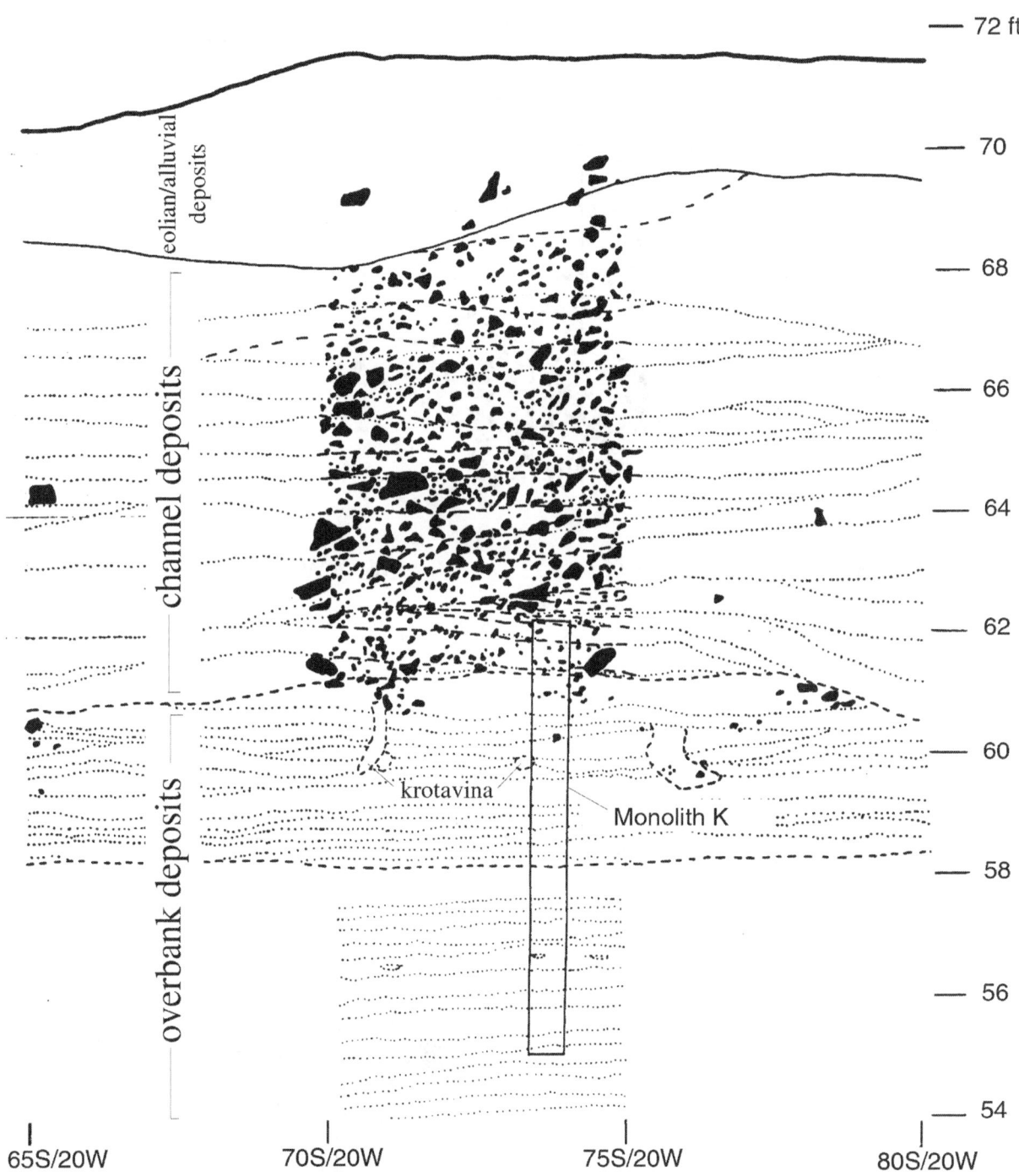

Figure 5.17 Stratigraphic profile of Monolith K locality. Unit 70S was excavated for stratigraphic purposes and thus its profile reflects greater detail than the profiles of the adjacent units.

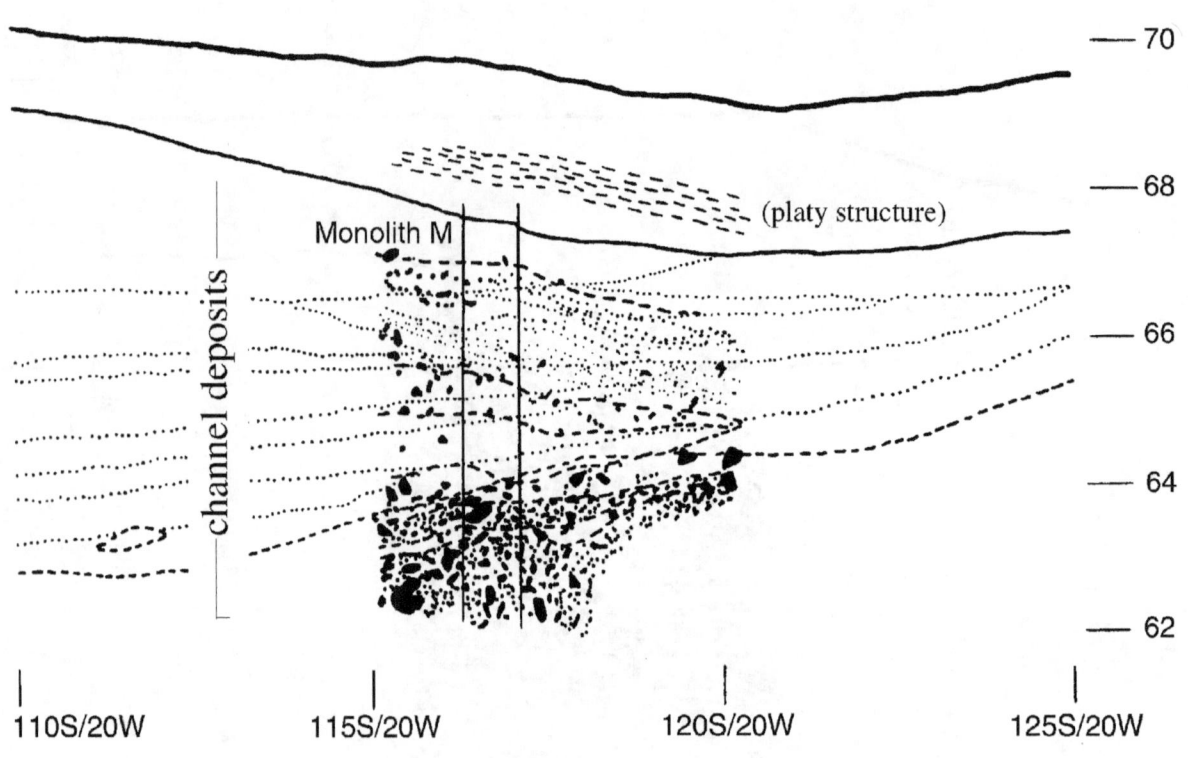
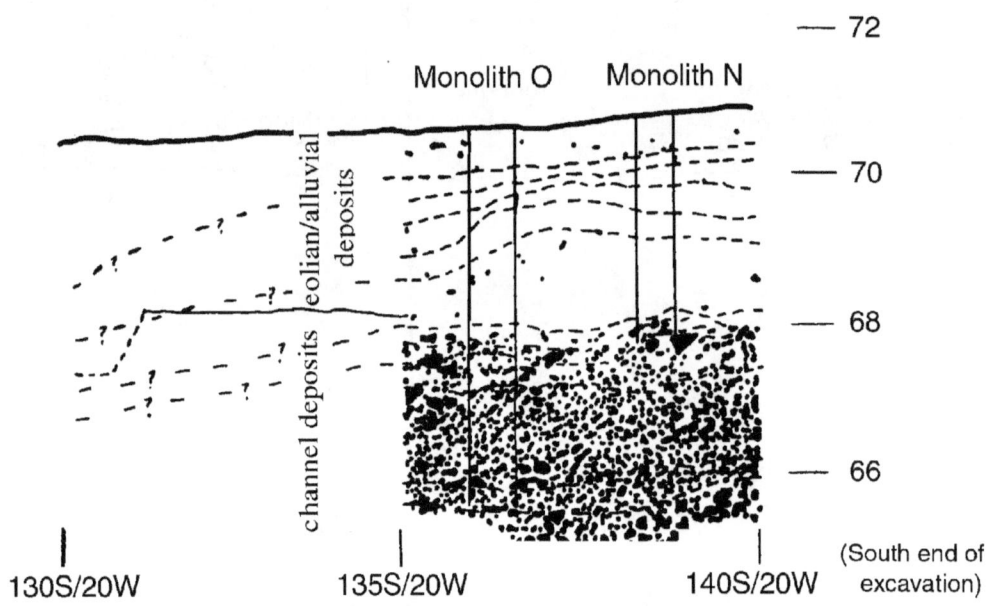

Figure 5.18 Stratigraphic profile of Monolith M, N, and O localities.

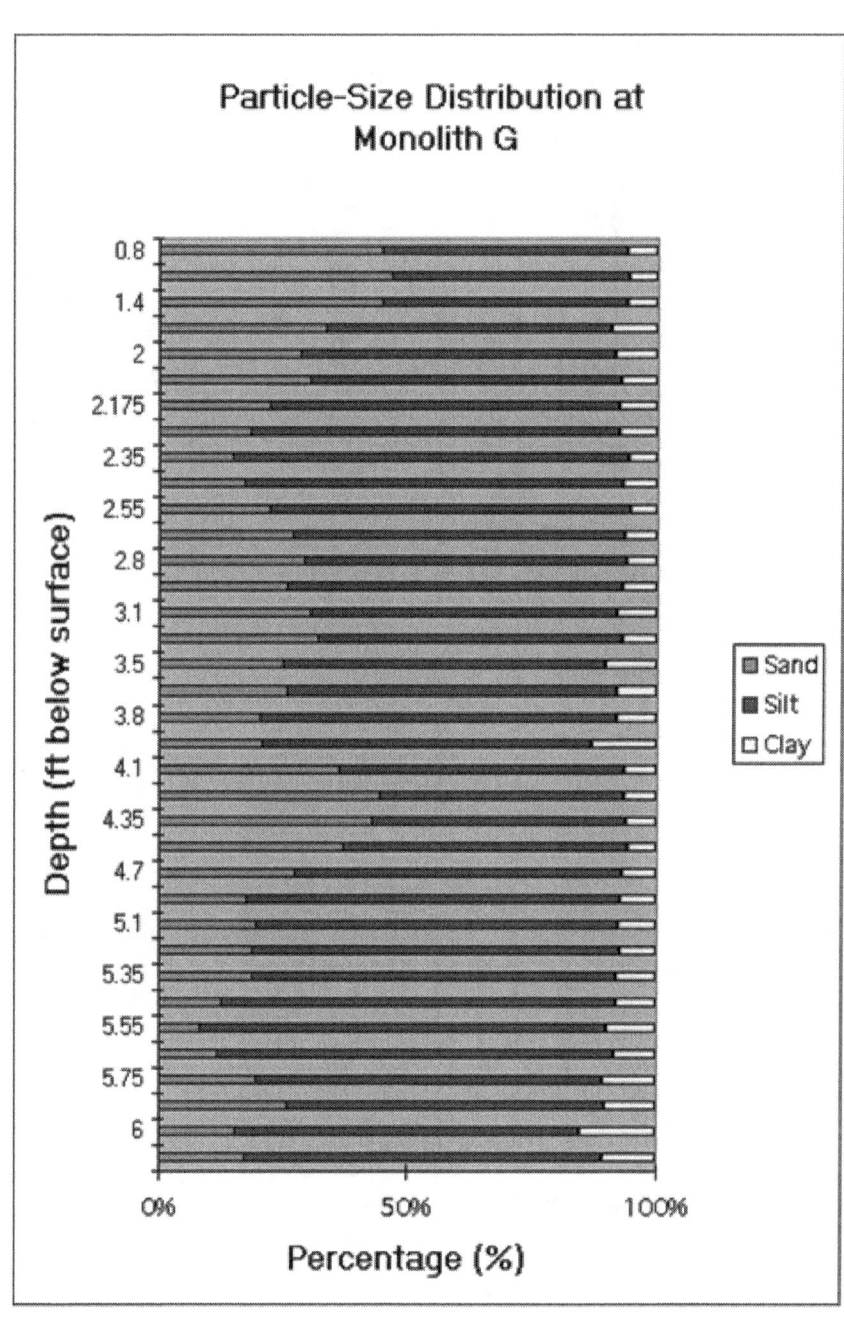

Figure 5.19 Vertical particle-size distribution from Monolith G locality.

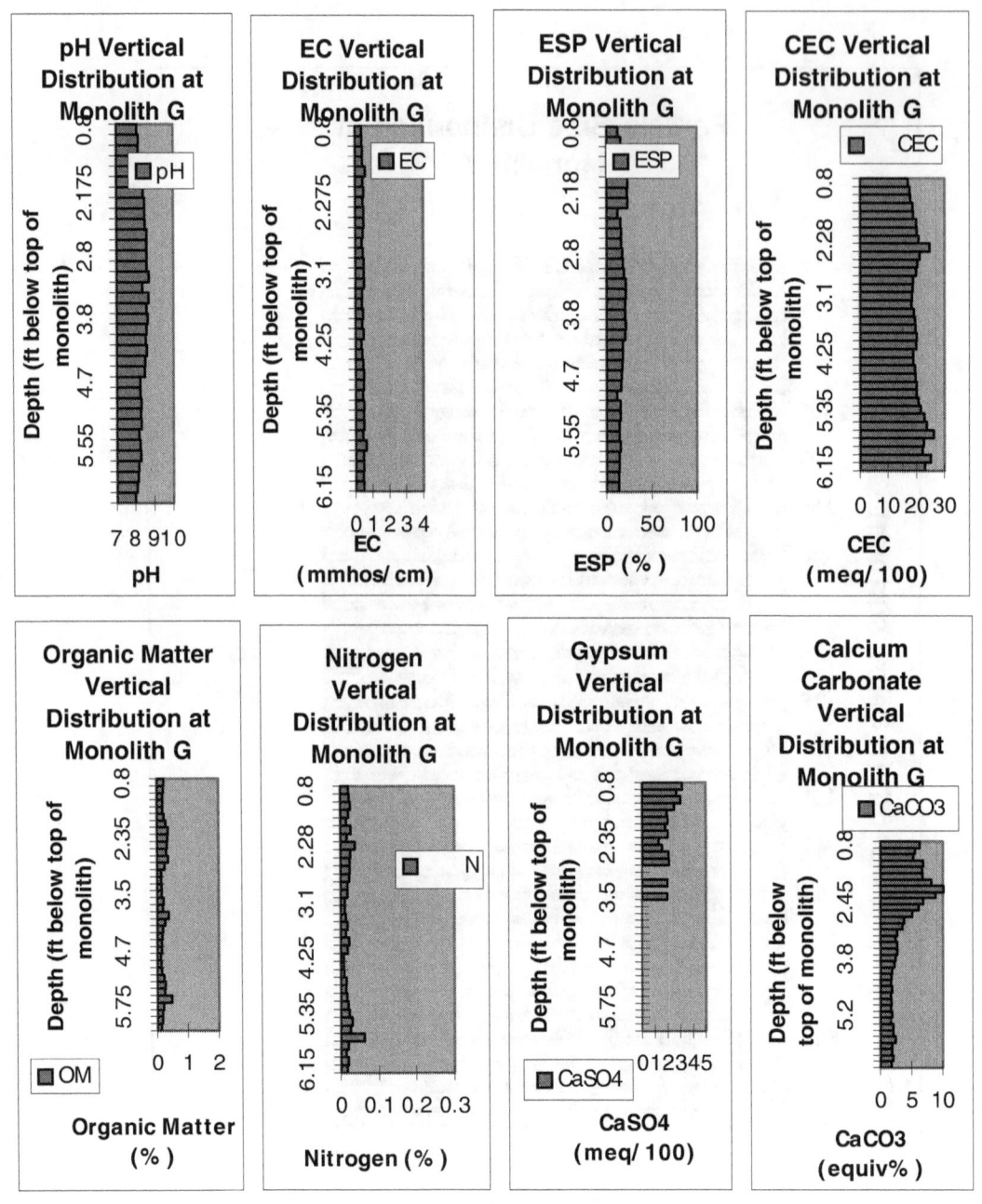

Figure 5.20 pH, electrical conductivity, exchangeable sodium percentage, cation exchange capacity, organic matter, nitrogen, gypsum, and calcium carbonate data for Monolith G locality.

In the vicinity of Monolith K, stratigraphic descriptions were made in the field in 1968 by Joan Brodhead and Madge Gleeson at 70-75S/20W (see Figure 5.17) and were recovered during this study (Appendix E). These are summarized below:

Surface overlain by .5 ft [.2 m] of compacted backdirt.

Natural surface to 3.0 ft [.9 m] depth: silt and reworked ash; very few rocks; pedogenesis including a 1 ft [.3 m] thick organic A horizon and a coarse, angular blocky to prismatic structure B; white salts (carbonate?).

68.4-66.6 ft above site datum (a.s.d.): angular to subangular basalt pebbles in a granule and sand matrix.

66.6-65 ft a.s.d.: less angular (than above) basalt pebbles in a granule and sand matrix.

65.0-62.0 ft a.s.d.: several gravel deposits; primarily subangular to subrounded cobbles and pebbles in a sandy silt matrix.

62.0-61.0 ft a.s.d.: fewer cobbles (than above).

61.0-57.0 ft a.s.d.: silt and sandy silt; few pebbles; several weakly to strongly developed organic A horizons.

Granulometric data are available for the Monolith K locality (Appendix C; Figure 5.21), but the exact elevations of the samples could not be determined. Nonetheless, all of the samples were collected between 58 and 62 feet site elevation and provide a glimpse of particle-size variations with depth. Gravel is prominent in the upper 1 ft where the monolith extends into coarse-textured alluvium (see Figure 5.17) but gravel also occurs in places at greater depth (e.g., Samples 13, 17, and 20) (Figure 5.21). Again, like in Monolith G, sand is more abundant relative to the rockshelter or colluvial slope, and there is no significant increase in clay with depth.

Field descriptions of stratigraphy at the Monolith M (117S/20W) locality by Claudine Weatherford in 1969 were also recovered (Appendix E) and are summarized below:

0-1 ft [0-.3 m] below surface: Plowed A horizon; thin platy structure in places.

1.0-3.0 ft [.3-.9 m] below surface: soil with A, A/B, and B horizons; A horizon is dark and developed into very fine sand and silt; A/B horizon is structureless; B horizon is stained and darker than overlying A/B and developed into fine sand, silt, gravels, and subangular cobbles and pebbles; coarse, angular blocky structure.

3.0-6.0 ft [.9-1.8 m] below surface; subangular and subrounded pebbles and fine cobbles in matrix of granules, sand and silt; well sorted in places; most of the pebbles and cobbles show "dense" deposits of silica on the under side.

Monoliths N and O, which are located 3 feet (0.9 m) apart at the far southern end of the project area, were both described in 1998 (Appendix E), and particle-size analysis was performed on the finer textured sediments on Monolith N (Appendix C; Figure 5.22). Monolith N contains a 0.4 feet (0.1 m) thick dark brown A horizon overlying a 1.7 foot (0.5 m) thick structural B horizon. The A and B horizons each have one thin zone of strong platy structure, but the prevailing structure of the B horizon is strong prismatic to angular blocky structure. The parent material is largely sand with some silt and a few gravels that in turn overlies a layer of well-rounded cobbles and gravels. The sand fractions increase with depth toward the cobbly layer. The entire solum is noncalcareous.

Monolith O consists of 2.1 feet (.6 m) of fine sand and silt overlying well-rounded cobbles and gravels in a sandy matrix. Within these deposits is a moderately developed soil consisting of a 0.7-foot (0.2 m) thick organic A horizon overlying a 1.4-foot (0.4 m) thick structural B horizon. Only the A horizon has a zone of strong platy structure; the B horizon contains weak angular blocky structure. The entire solum is noneffervescent, but it was not possible to apply acid to the white salts located on the bottom sides of the cobbles due to the resin used in the preservation of the monolith.

Interpretations

Excavations on Terrace I revealed three main lithostratigraphic units. At the base is a fine-textured, horizontally stratified alluvial deposit containing weakly developed, organic soils. This is overlain by coarse-textured, channel gravels and cobbles. At the very top is a layer of mostly sandy material that represents a mixture of alluvium and loess.

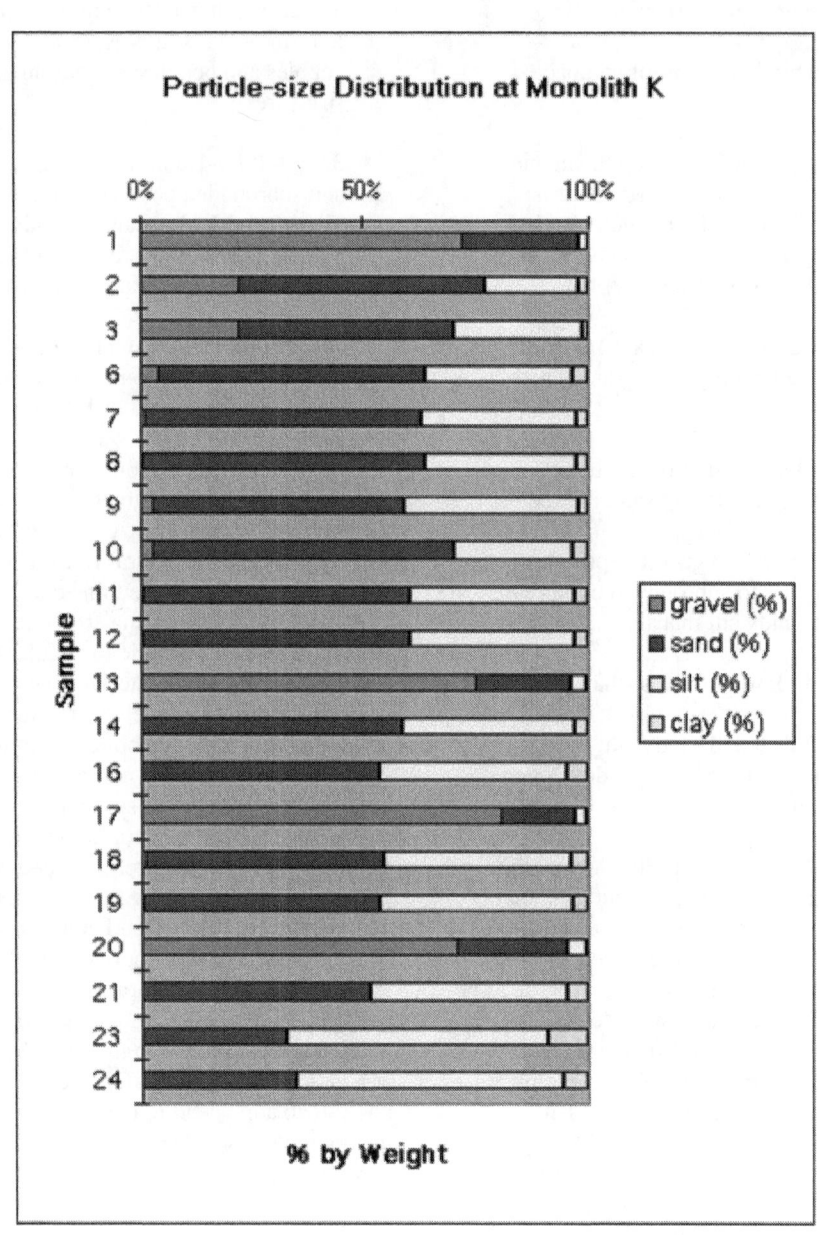

Figure 5.21 Vertical particle-size distribution from Monoloth K locality, at approximately 58-62 feet of site elevation.

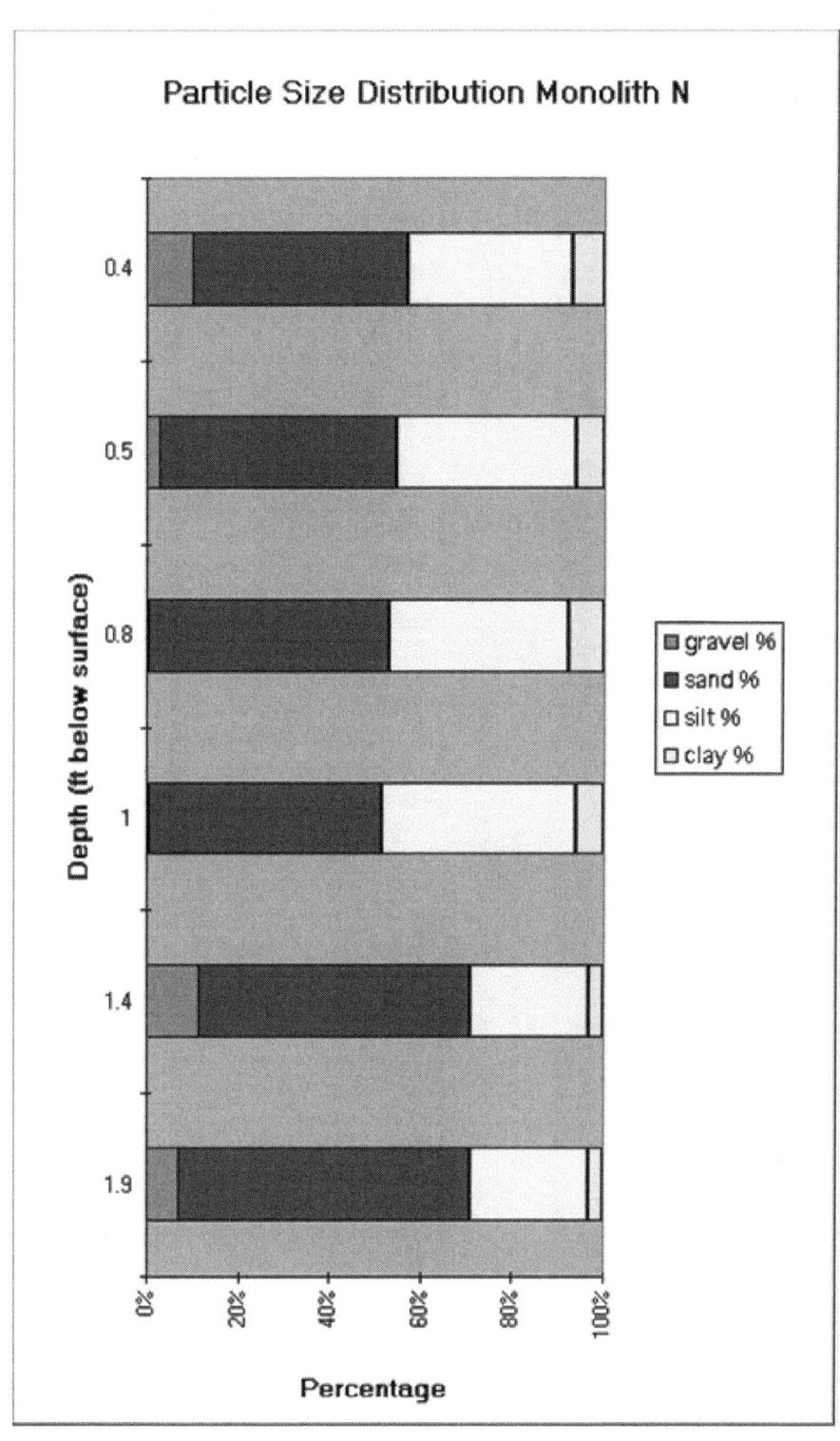

Figure 5.22 Vertical particle-size distribution from Monolith N locality.

The lower, horizontally stratified, silty/sandy alluvium formed mostly by vertical accretion of Terrace 1 by the Palouse River. These deposits represent a fluvial environment situated away from the main channel in the transition from natural levee to flood basin (Reinick and Singh 1980:289-298). Here, sediments settle out of suspension in a relatively low-energy environment conducive for burial and preservation of cultural materials. Overbank deposition was episodic as evidenced by the formation of weakly developed A horizons (e.g., the Marmes and Harrison Horizons) (Gustafson 1972). The organic content in these horizons is not great, but may be higher than indicated using the Walkley-Black method which may not completely oxidize all of the organic matter. The time required for the formation of these horizons varies within a time frame of less than 1,000 years (Birkeland 1984). It is reasonable that these thin, incipient A horizons at the Marmes site formed in less than 100 years, as is supported by radiocarbon dates from the Marmes and Harrison Horizons (Sheppard et al. 1987). Such soils are cumulic (i.e., they form as the floodplain slowly aggrades). In fact, these types of soils are in a sense both geological and pedological because they contain both depositional and soil-forming traits (Ferring 1992). To form, there has to be a balance between the amount of sedimentation and humus formation. When rates of aggradation exceed humification, the soil becomes buried. Because deposition on the terrace was spatially discontinuous, the A horizon splits and divides when viewed in cross-section. There is very little other evidence for soil formation in this alluvium. Although the exchangeable sodium percentage values are >15 percent for the overbank deposits at Monolith G, it is difficult to argue that this represents soil formation. At these depths, the sodium is more likely emplaced through groundwater processes.

The depositional environment changed on Terrace I as evidenced by the more gravelly channel stratigraphy above the finer-textured vertical accretion deposits (Appendix B, 20W). As previously mentioned, Fryxell and Keel (1969) interpreted the paleochannel of the Palouse River encountered near 0N/S (see Figure 5.15) as being contemporaneous with the overbank deposits and associated cultural materials. However, reconstructions based on the stratigraphic profiles compiled in 1998 indicate that the paleochannel was not the one responsible for the deposition of the Marmes Horizon and earlier overbank deposits. The paleochannel clearly cuts through and is filled into the older floodplain deposits (Figure 5.23).

Thus, the Palouse River channel that was active at the time of deposition of the culture-bearing deposits must be located somewhere farther to the south. This means that there may be an even larger area of potential archaeological significance that has not been investigated provided that it was not eroded by the paleochannel of the Palouse River. This would depend on how the paleochannel formed and arrived at the 0N/S vicinity. If the paleochannel at 0N/S entered through a sudden channel avulsion (i.e., a short-cut across Terrace 1), then earlier overbank deposits located to the south may have been preserved. On the other hand, if the channel entered via channel migration and lateral accretion across the terrace, then the lateral shifting of the main channel eroded previous low energy overbank deposits and possible archaeological materials. Lateral accretion of the paleochannel is evident in the stratigraphy exposed along 20W (Appendix B, 20W) but this may have occurred later as the paleochannel retreated southward. If so, then older cultural deposits coeval with the Marmes and Harrison Horizons may still be preserved at depth.

Channel processes dominated the floodplain below the Marmes rockshelter until the Palouse River downcut creating Terrace 1. After that, deposition on the terrace included colluviation from adjacent hillslopes and loess sedimentation. The uppermost lithological unit that overlies the fluvial gravels likely consists of both alluvium and loess. Sand fractions are relatively high and increase with depth at Monolith O, a pattern atypical of loess but typical of overbank alluvium. The lower part of this lithological unit likely represents a layer of channel and levee sands deposited just prior to incision by the Palouse River that has subsequently been pedogenically mixed with loess. This surface deposit eventually grades into more loess-dominant material toward the hillslope.

Soil formation in the Terrace 1 surface deposit includes development of a moderately thick (probably mollic) A horizon overlying a well-structured B horizon. At depth the B horizon becomes calcareous with Stage I salt development in the gravels. Although there are no chemical data available, anecdotal evidence suggests that the surface soil might be sodic. This evidence includes prismatic structure and the presence of white salts (Appendix E). Moreover, the Monolith M locality descriptions indicate a relatively light colored and structureless "A/B" horizon in the near surface which could be a possible eluvial (E) horizon such as is a common with sodic soils in the region (Peterson 1961). Moreover, the presence of secondary silica as patinas on the undersides of the

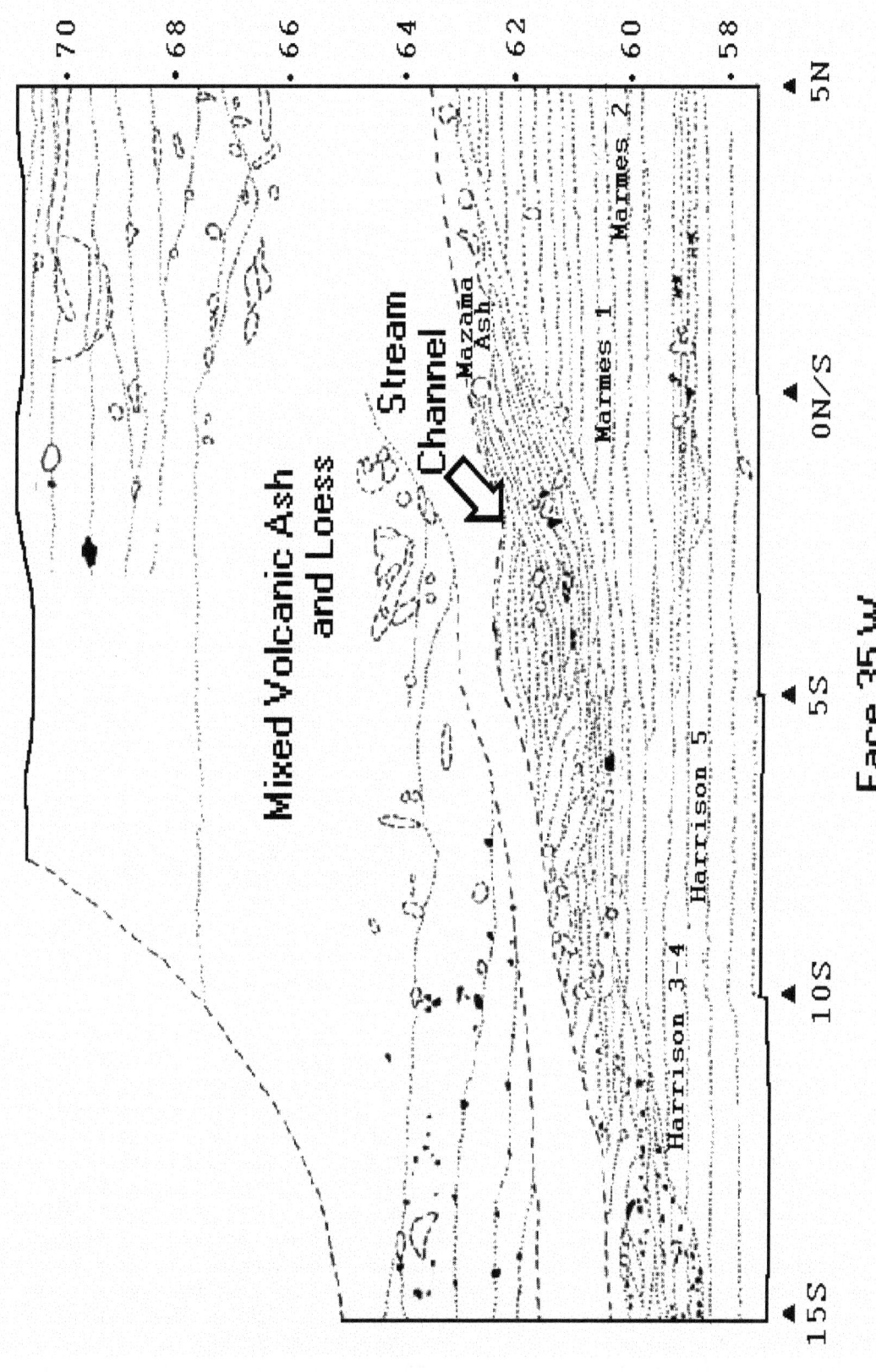

Figure 5.23 Cross section of stream channel illustrating cut-and-fill relationship to Marmes and Harrison horizons.

stones at Monolith M also suggests alkaline conditions which favor silica solubility (Chadwick et al. 1987). However, such soils were not identified along the lower Palouse River in the late 1950s by Peterson (1961:3) whereas they were identified along Alkali Flat Creek. At this stage of the investigation, the alkalinity of the surface soil on Terrace 1 prior to dam construction and any paleoenvironmental inferences are speculative. It should be noted that Marshall (1971) identified a buried calcic paleosol beneath Mazama tephra in Terrace 1 below Marmes Rockshelter. Although not traceable on the stratigraphic drawings (Appendix B, 20W), it is probable that this pre-Mazama paleosol correlates to that on the colluvial slope and in the rockshelter.

The alluvium of Terrace 1 is latest Pleistocene and earliest Holocene in age. Four radiocarbon ages from the Marmes and Harrison Horizons range from approximately 9,500 to 10,500 ^{14}C years B.P. and correlate in time with Unit I in the rockshelter (Sheppard et al. 1987; see Tables 5.3 and 5.4). There is no good age control for the gravelly channel deposits or the capping sandy alluvial/loess deposit. At several different localities in the lower Palouse River floodplain, Marshall (1971) observed tephra presumed to be Mazama directly overlying Terrace 1 gravels. Hence, it is probable that the gravelly alluvial channel deposits in front of the Marmes rockshelter date between 9,500 and 6,700 ^{14}C years B.P. and thus correlate in time with Units II and III in the rockshelter. Accordingly, most of the surface deposits on the terrace overlie Mazama tephra and post-date 6,700 ^{14}C years B.P. thus correlating in time with Units V through VIII in the rockshelter.

The alluvial chronology at the Marmes site can be correlated to other sites along the Snake and Columbia River. For example, Hammatt (1977) recognizes an early Holocene terrace composed of alluvium dated approximately 8,000-10,000 ^{14}C years B.P. along the lower Snake River between Lewiston and Lower Granite Dam (approximately 50 km upstream from the mouth of the Palouse River). In fact, Hammatt's (1977:72) descriptions of the early Holocene alluvium are remarkably similar to descriptions of the overbank deposits at the Marmes site. Other pre-Mazama ashfall, early Holocene terraces capped by eolian sands and silts have been identified along the middle Columbia River (Chatters and Hoover 1992; Huckleberry, Stafford, and Chatters 1998). Such regional correlations in alluvial stratigraphy strongly imply climatic mechanisms for hydrologic change and terrace formation in central and eastern Washington during the early and middle Holocene (Chatters and Hoover 1992; Hammatt 1977). Although the details of Holocene climate change for the inland Northwest are still being refined, there is general agreement that the middle Holocene (ca. 4,000-8,000 years B.P.) was a period of reduced effective moisture (Chatters and Hoover 1992; Hansen 1947; Mehringer 1996). Consequently, it is reasonable to link the formation of Terrace 1 (i.e. aggradation and subsequent downcutting of the lower Palouse River) with a shift to reduced fluvial discharge. However, the dynamics of sediment production and transport within a hydrologic basin are complex, and a simple model of "dry equals downcutting" and "wet equals aggradation" is overly simplistic. More detailed alluvial chronological work along streams with contrasting catchment area is needed before a hydroclimatological model can be established linking alluvial terraces and climate in the Columbia Basin.

Site Formation Processes at the Marmes Site

The archaeological record at the Marmes site is the product of both natural and cultural processes. Clearly, geological processes have played the dominant roles in forming the site stratigraphy. In the rockshelter, rockfall (eboulis) and eolian sedimentation are prominent. On the colluvial slope, mass wasting prevails. On the floodplain, channel and overbank alluviation are dominant depositional processes. The degree to which previous human activity is recognized and preserved in the natural stratigraphy obviously depends on the nature of the activity but also on the geological environment. For example, artifacts are more likely to be buried in place in low energy depositional environments where there is less potential for artifact or sediment entrainment and transport. It is thus no surprise that archaeological features and faunal materials from the Marmes site were recovered from rockshelter deposits and weakly developed soils in overbank alluvial deposits, the two depositional environments with the least potential for lateral displacement by geological processes. In contrast, colluvial slopes are characterized by pulses of high-energy talus deposition and intervening episodes of lower energy slopewash. Consequently, there is ample opportunity for cultural materials to be displaced downslope, although the degree of disruption decreases towards the distal end of the colluvial slope. Likewise, channel environments in floodplains are known for gravel bar migration and fluvial reworking of materials and usually have little archaeological potential (Waters 1992:138). An exception was the Granite Point Site where archaeological materials were found relatively in situ in fluvial cobble deposits (Leonhardy 1970 referenced in Hammatt 1977:144).

It is important to consider how geological processes may have obscured possible past human activity at the Marmes site and vicinity. For example, now that a pre-Clovis model for the peopling of the Americas is gaining favor with the general acceptance of the Monte Verde Site in Chile, one has to seriously consider what effect the last of the Missoula floods might have had on the preservation and visibility of possible archaeological sites. If such sites were present, there remains little chance of their preservation in areas affected by high-energy flow such as along the lower Palouse River. Less spectacular floods also have affected the distribution of known archaeological sites in the area. For example, along the lower Snake River two cycles of alluvial aggradation 10,000 to 8,000 and 4,000 to 2,500 years B.P., have made conditions unfavorable for the in situ preservation and visibility of Windust and Tucannon phase archaeological sites (Hammatt 1977). Consequently, the paucity of such sites in the lower Snake River area is likely to reflect geological obfuscation more than any reduced levels of cultural activity. At the Marmes site, the situation is different. Windust-age materials actually have a good chance of being preserved at depth in the overbank alluvial deposits and associated organic soils that date approximately 9,000 to 11,000 ^{14}C years B.P., although such cultural materials also have a good chance of being partly eroded by the subsequent channel erosion. Late Cascade and younger materials also have a good chance of being preserved in the surface sandy deposits that post-date the Mazama ashfall. However, the same cannot be said for the early Cascade materials that predate the Mazama ashfall and correlate in time to the gravelly channel deposits; such materials would have little chance of being preserved in the terrace below the rockshelter. In comparison, the rockshelter has a relatively continuous depositional history of éboulis and loess amenable for preservation of archaeological materials ranging from Windust to historic age.

If the geological environment is conducive to the preservation of cultural resources, there are still post-depositional processes that can affect the integrity of those resources. These include freezing and thawing, bioturbation, soil formation, and subsequent human activities. For example, although the Marmes rockshelter is well suited for preserving archaeological materials, a host of post-depositional disturbance processes have disrupted the stratigraphy more than on the colluvial slope and floodplain. The most obvious evidence of disturbance comes from the upper layers of the rockshelter, particularly near the back of the cave, where there are multiple krotavina and intrusive pits (see Figure 5.6). These disturbances have mixed the sediments resulting in multimodal particle-size distributions (see Figure 5.2) versus the more common bimodal éboulis and loess pattern. It is probable that artifacts and faunal material have also been vertically and horizontally displaced.

Another less obvious but probable disturbance process in the rockshelter is cryoturbation. Repeated freezing and thawing has the effect of lifting artifacts upward depending on the depth of burial, shape of the artifact, and heterogeneity of materials with different thermal properties in the sedimentary matrix (Wood and Johnson 1978). The amount of cryoturbation experienced in the Marmes rockshelter is difficult to define. No ice-wedge casts or similar freeze-thaw structures were observed within the rockshelter, but given present climate which experiences frequent diurnal cycles of freezing and thawing, and the regional paleoenvironmental record of colder temperatures in the past, it is reasonable to assume that any material exposed near the surface for an extended period of time has been displaced from its original position. This is especially true for archaeological materials located on the colluvial slope where freeze-thaw and gravity play important roles in stratigraphic mixing and downslope displacement of deposits. Although the degree of displacement decreases towards the distal end of the colluvial slope where slope diminishes, cultural materials from hillslope contexts, especially those in environments prone to freezing and thawing, have to be viewed as occurring in secondary contexts. Archaeological materials in the floodplain are also prone to freeze-thaw processes as well as greater potential for root disturbances given the presumably greater amount of vegetation close to the river. Other possible post-depositional disturbance processes include salt translocation and precipitation, which can play a role in weathering and, to a lesser degree, movement of artifacts.

In sum, archaeological and faunal materials have a good chance of being preserved in the Marmes rockshelter and in overbank flood deposits, whereas colluvium and alluvial channel deposits have a lower potential for containing relatively intact archaeological materials. Although the preservation potential for archaeological materials in the rockshelter was probably time-invariant, it varied in the floodplain and was minimal during a period of high-energy deposition of gravels between 9,000 and 6,700 ^{14}C years B.P. All cultural materials in all geomorphic contexts at the Marmes site have been affected by post-depositional processes, the most prominent being human excavation, faunal burrowing, and cryoturbation. Although these latter processes can displace and degrade archaeological materials, they do not preclude analysis of cultural chronology and interpretation of human activities at the Marmes site.

Summary and Conclusions

Stratigraphy provides a context for interpreting the cultural and paleoenvironmental record at the Marmes site. Through the compilation of stratigraphic profiles and sedimentological data, this report addresses 1) the physical and chemical properties of the sediments at the Marmes site; 2) depositional and post-depositional environments; and 3) the integrity of archaeological and faunal materials (i.e., the degree to which they have been modified by natural processes).

The Marmes site contains three landscape elements: rockshelter, colluvial slope, and floodplain. Each has its own distinct assemblage of deposits that have formed since the last of the Missoula floods approximately 13,000 years ago. The rockshelter contains basaltic roof collapse blocks (eboulis) mixed with windblown silts. These sediments have been modified since deposition by freeze-thaw, pedogenic, and cultural processes. The main period of soil formation (as evidenced by salt translocation and field descriptions of structural development) appears to have occurred within the rockshelter during the early Holocene prior to the eruption of Mt. Mazama approximately 6,700 years ago. This corresponds to a time of reduced sedimentation in the rockshelter based on an evaluation of the radiocarbon chronology (Sheppard et al. 1987). The rate of sedimentation in the rockshelter was maximum prior to 10,000 ^{14}C years B.P. and then decreased substantially until approximately 7,000 ^{14}C years B.P. when it began to increase again but at a lesser rate than in the Pleistocene (see Figure 5.7). This is somewhat different from Fryxell's (Fryxell et al. 1968b) estimate of relatively high rockfall (eboulis) frequency extending from approximately 12,000 years B.P. to as recently as 8,000 years B.P. Fryxell's (1964) interpretation that increased rockfall during the latest Pleistocene and earliest Holocene reflected colder conditions more amenable to mechanical weathering is reasonable, but other factors are also possible including changes in the exposure of well-jointed bedrock in the ceiling and the rate of eolian sedimentation which if increased can reduce the density of rock spalls in the sedimentary matrix without decreasing the rate of rockfall. The period of reduced sedimentation and soil development in the rockshelter corresponds to soils developed into early Holocene alluvium along the lower Snake and Palouse Rivers that are estimated to have formed 8,000 to 6,700 years B.P. (Hammatt 1977; Marshall 1971). The calcic soils record a period of landscape stability during a period of relatively dry conditions.

The colluvial slope below the Marmes rockshelter contains an assemblage of basaltic rock debris and silty loess that traces down and interfingers with floodplain deposits of the Palouse River. The colluvium contains a buried paleosol capped by Mazama tephra that is traceable into the rockshelter and down into the floodplain and can be correlated with other early Holocene soils in the region. Of climatic significance within the colluvial deposits are saline and sodic soils. Given the well-drained position of the colluvial slope, it is believed that saline soils formed during relatively dry periods of the Holocene whereas the alkaline soils formed during subsequent shifts to more mesic conditions and greater effective leaching. The frequency and chronology of these pedogenic events is uncertain, as is the magnitude of climate shifts necessary to bring about such changes. The fact that saline soils are presently forming at the surface suggests modern climate is adequate for the accumulation of salts. The fact that sodic soils occur at depth but within Holocene sediments suggests that conditions previously were more moist than today, or possibly cooler allowing for more effective moisture and greater leaching. It is possible that the effective moisture threshold between the two conditions was crossed several times in the Holocene and need not require large magnitude changes in climate.

Below the colluvial slope at the Marmes site is an early Holocene alluvial terrace previously mapped by Marshall (1971) as Terrace 1. The terrace contains both low-energy, silty and sandy overbank and high-energy, gravelly channel deposits. Between 10,500 and 9,500 ^{14}C years ago, the area experienced relatively low-energy, overbank sedimentation during which weakly developed, organic soils formed in the floodplain. These include the Marmes and Harrison Horizons, which represent paleosurfaces with good potential for preservation of in situ archaeological and faunal materials. Sometime between 9,500 and 6,700 ^{14}C years ago, the main channel of the Palouse River encroached northward either through avulsion or migration towards the rockshelter and eroded some of the earlier, overbank deposits. The floodplain continued to aggrade until sometime following the deposition of Mazama tephra approximately 6,700 ^{14}C years ago when the Palouse River downcut forming the terrace. Subsequent deposition on the terrace has been limited to eolian sediments and alluvial fan and colluvial sediments towards the margins of the floodplain. This terrace correlates in time and lithology to other terraces along the lower Snake and middle Columbia Rivers and indicates a regional lowering of base level, ostensibly due to climate change, prior to the eruption of Mount Mazama.

All cultural materials from the Marmes site have been modified to some degree by post-depositional processes. However the amount of modification varies from relatively low in the rockshelter and overbank alluvial silts and sands of the floodplain to high on the colluvial slope and channel gravels of the floodplain. Whereas in the floodplain there is likely to be geological processes biasing the visibility and preservation of different aged cultural horizons, the rockshelter has a continuous depositional sequence from the latest Pleistocene through the Holocene that allows for a more realistic measure of past human activity.

The Marmes site with its artifactual, faunal, and sedimentological record provides a wealth of information regarding late Quaternary environmental change and human activity in the Columbia Basin. Much of this information could not have been gleaned from the site if not for the "big picture" approach taken by Roald Fryxell and Richard Daugherty over thirty years ago. Specifically, the spatial and temporal variation in geological processes that have helped to shape the archaeological record at the Marmes site would have been difficult to define without the combined stratigraphic record from the rockshelter, colluvial slope, and floodplain. All three are linked physically as best expressed by the Mazama ash which can be traced from floodplain to rockshelter. Moreover, all three are linked culturally for people are not limited to specific areas like rockshelters, hillslopes, or floodplains but instead utilize all these areas and more. Human activities occur at the scale of the landscape. By excavating and recording the stratigraphy of these three landscape components, much more can be said about how the site formed and how people used it over the last 12,000 years.

6
FEATURES

Brent A. Hicks

In addition to the human remains features, fire hearths, storage pits, and concentrations of specific kinds of cultural materials were recorded as features at the Marmes site. As was noted in Chapter One, much more attention was given to identifying features in the floodplain excavations in 1968 than for any of the excavations in the rockshelter. Of the 52 floodplain features reported in Fryxell and Keel (1969:38), many were represented by only a single item (e.g., a tool) or a small number of items (e.g., several lithic flakes or bone fragments), but no hearths or pit features were identified.

In contrast, only a small number of features represented by concentrations of faunal or botanical materials were identified in the rockshelter either by the excavators or Rice from his review of the records, and no lithic concentrations were featured. Such concentrations certainly existed. Review of the field records for this study found various hints of the presence of such concentrations and in one rare instance a notation in the field notes speculated that a concentration of shell found in a cleft in the bedrock was a single-event dump. In addition, stratigraphic Unit III in the rockshelter is a 'living floor' consisting of high densities of shell, faunal bone, and lithics. That such dense materials, deposited within an occupation area as a direct result of activities related to that occupation, would not result in concentrations of material that could be considered features is unlikely. However, assigning, defining, or refining feature status of spatially-related accumulations of cultural material based on records reviewed during this study was not possible (see Human Remains chapter for exceptions); in general, the field records do not provide the necessary information. Specific provenience relationships of the different items recovered from features defined in the field, or detailed descriptions of feature contents are very rarely included in the field records. Rice (1969:49), referring to the rockshelter deposits, cautions that "The numbers of features reported should in all cases be considered minimal." This is especially true of the 1962-1964 excavations that encountered concentrations of shell and botanical materials, the latter probably related to storage pit features. But moderate density lithic and faunal bone concentrations also must be underrepresented as defined features based on comparisons of the level bag counts from various excavation units within the same stratigraphic level across the rockshelter.

It appears that "features" were not recognized for a number of reasons: the relative inexperience of excavators working in complex cave deposits, often with little oversight from an experienced archaeologist or graduate student; the extension of the material concentration from or into an adjacent unit (which may not have been excavated in the same year); the general lack of significance given to botanical materials related to storage pit features, especially the recognition of storage pit feature materials displaced from their original provenience by the site's occupants digging later storage pit features; and, disturbance and spreading of material concentrations by burrowing rodents. We echo Rice's caution and suggest that excavation of the same volume as that removed from Marmes Rockshelter using contemporary techniques and research foci would have documented at least two times the number of features that Rice was able to delineate from the records.

Rice (1969) conducted a thorough review of the field records in developing "The Features" section of the preliminary report. However, he chose to define only three broad categories of features for his presentation: 1) hearths, fire pits, and storage pits; 2) miscellaneous stone features; and, 3) miscellaneous faunal and botanical features. Within these three categories Rice numbered the 127 features he identified consecutively within the eight stratigraphic units. Therefore, for the first category of feature types, eight features are numbered as "1" since there is at least one feature in each of the stratigraphic units. To avoid confusion in the current presentation, all of the features were renumbered consecutively beginning with the deepest (oldest) stratigraphic unit. Within each stratigraphic unit, feature numbers were applied based on the date the feature was identified in the field. Thus, Feature 1 (F1) in the new numbering system is the first feature identified by excavators within stratigraphic Unit I and Feature 124 is the last feature identified within stratigraphic Unit VIII. Also, where multiple, superimposed pit features were given a single feature number designation by Rice, alphabetical extensions were added to further designate each pit feature in the complex. Three features were dropped from Rice's feature list that upon review appear to fail the criteria of a "feature". These include: a "Broken cobble with ochre stain on One End" (Rice 1969:73) for which no other information could be found to place it in a context with other materials and is considered only an artifact; "One Bear Claw" (1969:75) found in a level bag and which may have been displaced from a burial by rodent activity; and, a "Yew Wood Needle" (1969:76) which without other contextual information also is just an artifact.

For the purposes of the current study, Rice's feature type categories are further broken out with hearths and fire pits separated from storage pits. This was done to aid in discussion and graphical depiction of the different functions these kinds of features represent. While it is certainly arguable that hearths and fire pits may represent some different functional uses as well, the interchangeable use of both terms in the field notes, and the lack of specific information for most that would support the application by excavators of one term over another, makes asserting a functional difference herein arbitrary. It is rarely made clear in the field records whether excavators recognized that the fire-related feature they were excavating had been dug into or built on top of the rockshelter floor. It appears that Rice did not attempt to authenticate the feature designations as applied in the field (hearth or fire pit) since, in one case, a fire pit in Unit II includes a comment of "no well defined outline" (1969:50) which would seem to be a major factor in differentiating between a fire pit and a hearth. This study also does not authenticate the validity of the two terms as the lack of descriptive information in the field records would prevent this from being applied consistently for all such features. In general, this is the approach of this study for all of Rice's descriptive information on the features.

Rockshelter Features

Prior to the period associated with stratigraphic Unit VII, nearly all of the features found in the rockshelter, other than burials, are fire-related features and likely represent activities related to occupation of the rockshelter (Figure 6.1). Unit VII features also are dominated by fire-related features but also include numerous storage pits. The small number of Unit VIII features identified by Rice include only a single "recent" fire pit (1969:73); the rest of the features are a result of historic disturbance, generally related to pothunting activities. The overall number of features increases greatly from the early to the late stratigraphic units. In addition, there is a gradual horizontal movement of cultural features from the front of the rockshelter in the earliest strata towards the interior of the rockshelter through time.

Hearths/Fire Pits

Rice (1969:49-73) lists 93 hearth and fire pit features in the rockshelter (see data table in Appendix F). One of these consists of two superimposed fire pits (F105) and another is described as a "fire pit complex" with five superimposed pits in the same feature area bringing the total number of fire-related features to 98. Other hearths and fire pits also may represent multi-use events but were not described as such in the field notes. Only three of these feature descriptions make mention of fire-cracked or fire-blackened rocks; most of the features were defined by the presence of fire-reddened earth, charcoal, ash, or oxidized soil. Only one fire-related feature is described as having been made in a ring of rocks.

The hearth feature count also must include the cremation hearth feature complex found in Units I-II (see Chapter Seven). This hearth complex was noted to have "A series of small hearth areas" with "rings of rock and rock piles…mammal bone and chipping detritus…and a wide variety of artifacts including projectile points" (Rice 1969:85) interspersed throughout in addition to the many burned human bone fragments. The only other published statement on the contents of the cremation hearth are from Krantz's study of the human remains where he states that "A number of non-human bones have been separated out of this mass, but the overwhelming majority was and is human" (1979:169). Unfortunately, Krantz does not describe the area represented by the 'mass' or the methods used in addressing it; whether the 'separation' referred to involved identifying and cataloging all contents of the hearth complex or if the point was solely to retrieve the human

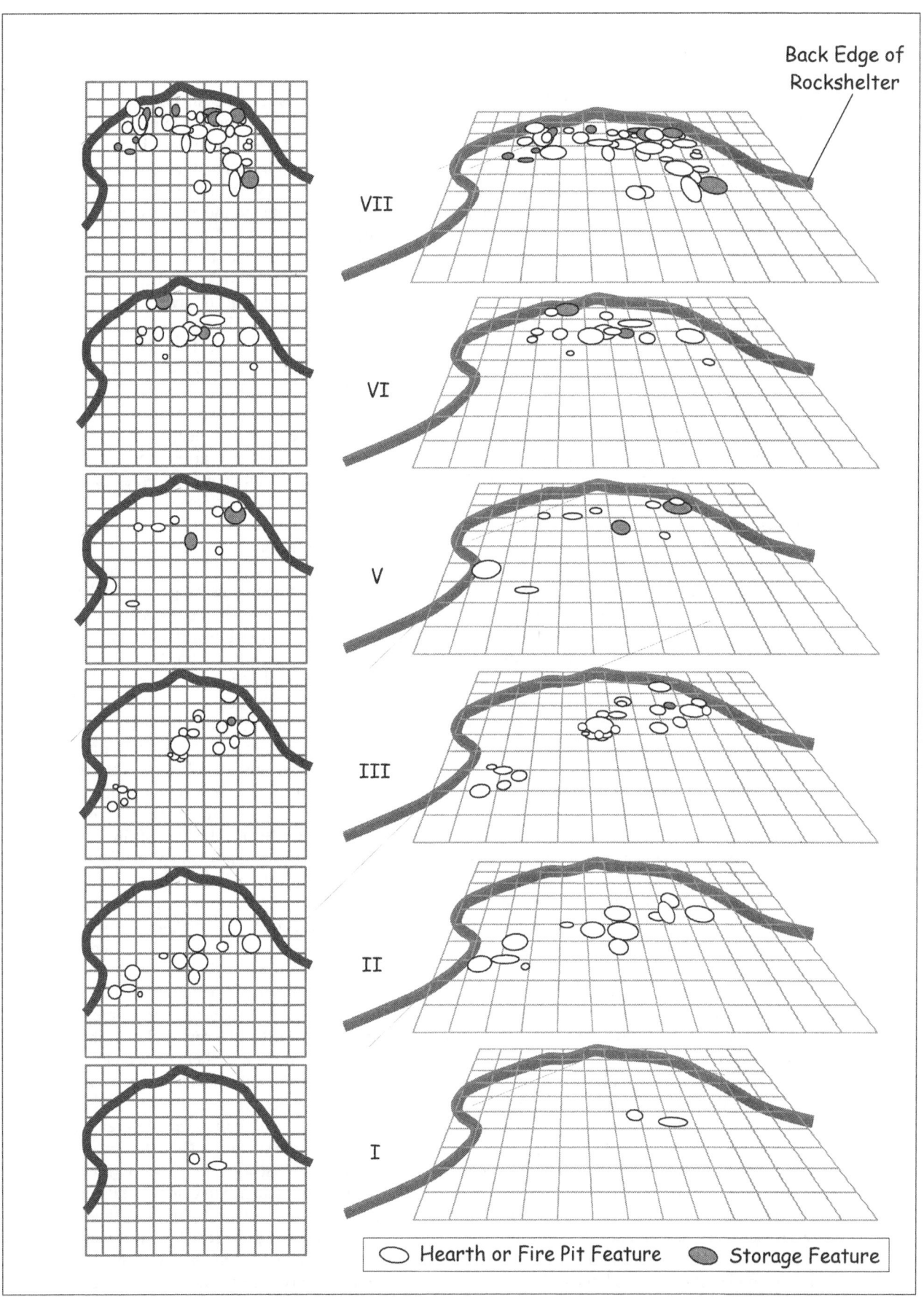

Figure 6.1 Plan and oblique views of the features by stratigraphic unit in Marmes Rockshelter.

remains that Chatters sorted and glued together (1979:169). No records have been found that address the contents of the hearth complex other than the excavation records and it appears that no post-field analysis of the feature (or any features) was conducted other than that which led to Krantz's article on the human remains. Thus, the 'mass' Krantz refers to remains largely unidentified except for what Rice (1969) gleaned from the field records.

The cremation hearth complex occurred over a horizontal area greater than 200 square feet and vertically through more than three feet of the rockshelter deposits in some excavation units. The presence of mammal bone, lithic tools and debitage in association with a hearth feature would be expected in an occupation site and would not be construed as representing ritualistic activities as the presence of the burned human bone may indicate. It seems clear the cremation hearth feature complex served multiple purposes for the rockshelter occupants, in addition to the burning of human bone, over a long period of time and are probably related to the day-to-day occupation of the rockshelter during the initial period(s) of use. Rice did not attempt to enumerate individual hearth areas within the feature complex, probably because the detail is not available in the field records to do so. This is unfortunate since more careful delineation of hearth areas and the bone types found associated with each may have helped ascertain whether certain areas or particular hearths within the hearth complex were more associated with reduction of human bone. It may be that the degree of rodent disturbance, the very high amount of rooffall rock in the Unit I deposits, and the vertical emphasis of the excavations prevented further feature delineation and identification of concentrations of animal versus human bone. For this study, the information on the contents and internal patterning of the cremation hearth complex is too limited to offer a determination on whether the deposits represent cremation or cannibalism.

All but two of the fire-related features identified in stratigraphic Unit I occur within the cremation hearth complex area close to the west wall of the rockshelter and at or outside of where the dripline would have been at that time. The other two hearth features identified in Unit I occur just behind the current dripline of the rockshelter which probably would have been within the roof cover of the shelter at that time. This appears to indicate that the bulk of the cultural activities within the rockshelter occurred in the western part of the rockshelter near the mouth. However, there are concentrations of lithic artifacts in the center-east portion of the rockshelter around the other two hearth features. It may be that there was a distinct separation of certain activities between the area of the cremation hearth and the other concentration of hearth features with lithic materials.

By the time period associated with stratigraphic Unit II, fire-related features are dispersed throughout much more of the middle and eastern part of the rockshelter. This change may reflect abandonment (or greatly reduced use) of the floodplain as an activity area after the time period associated with Unit I. This pattern was not tested however, in that only minimal excavations were conducted in the floodplain deposits above the Marmes horizon. A separation of activities between the cremation hearth area (west) and the balance of the other hearths and concentrations of lithic artifacts and debitage (center-east) also is present in Unit II. The exception is several hammerstones recovered from the approximate inside perimeter of the cremation hearth. Verification is impossible, but given the very small size of the human bone fragments in the cremation hearth, these hammerstones may have a direct functional relationship with the cremation hearth, which is the largest cultural feature in the rockshelter at this time.

This trend of activities moving generally back from the mouth of, and to the east within the rockshelter continues through stratigraphic Unit V when the first fire-related features are found in the west-interior. By Unit VII, numerous fire features are found in the west interior portion of the rockshelter. Despite the lack of fire features in the west interior area until Unit V, beginning with the cremation hearth in Unit I, fire features are found at the west edge of the mouth of the rockshelter, but not the east edge, through Unit V, including fairly dense use in Unit III. This shows a continued selection of this area for fire-related activities after the cremation hearth feature ceases to be used for that purpose. Fire-related features are never found this far forward in the rockshelter at the east end of the mouth of the rockshelter; this may indicate something about the development of the dripline berm (see below).

Features are found against the east wall near the rear of the shelter in Unit III, but are not found against the west wall at the rear until Unit VI. Some of this patterning can be explained by the erosion of the rockshelter through time as the rear wall and the dripline recede. The natural piling of rooffall rock would not be an impediment to cultural activities unless rocks too large to move occur and these appear to have been rare. If erosion is the sole factor for the spread of features within the shelter through time, then it is apparent that the east interior walls did not erode as much as the west interior walls. Other factors may include wind patterns within the rockshelter and how those may have changed as the berm developed. It may be that during Unit II through Unit IV occupations, prevailing winds made the eastern interior part of the rockshelter better for fire-related activities, but as the berm grew in height the same benefit was afforded the western interior area as well.

There is a great increase in the number of fire-related features from Unit I through Unit III. No features are attributed to Unit IV, then a relatively small number in Unit V, a small increase in Unit VI, then a great number in Unit VII. Assuming that the presence and numbers of such features are a proxy for the amount of cultural use of the rockshelter, the drop off in Unit IV and the slow increase after this is important for site interpretation. The break in the presence of fire-related features in Unit IV must be related to the Mazama ashfall and implies decreased activities within the rockshelter despite the presence of lithics typical of occupation activities (see Chapter Eight).

The contents of the fire-related features do not offer much to understanding this pattern. The amounts and diversity of cultural materials found in the features varies within and between each stratigraphic unit. Unit II may have the most cultural material associated with each feature overall, but given the variability in record keeping this pattern can not be asserted to a degree that sets Unit II apart from the other units. Bone and shell were observed in most fire-related features in all stratigraphic units. Lithic tools are most numerous in Unit II, Unit VI, and Unit VII features, but Unit III features have a high incidence of projectile points or point fragments. Overall, the tool types associated with fire-related features follow the patterns described in Chapter Eight for lithic tools in the stratigraphic units as a whole. The presence of lithic debitage does not necessarily follow with the presence of tools. For example, Unit VI and Unit VII features have many tools but there is almost no mention of debitage. Shell beads are first mentioned in descriptions of fire-related features in Unit V and are found consistently, although in small numbers, throughout the overlying strata. All other materials are found in too low numbers to be demonstrative. At least two of the Unit VII fire-pits appear to have been made in storage pit features as their bases were found to be grass-lined and matting fragments were recovered, both of which are commonly associated with storage pit construction (see below).

From the earliest use of the rockshelter until the time period associated with Unit V, the area at the west end of the rockshelter mouth outside the current dripline consistently exhibits fire-related features while the east end never does (although the west end received more excavation than the east end). One explanation may be related to the shelter afforded by the rockshelter's location. With the rockshelter facing southeast, sunlight would reach the shelter mouth relatively early in the day, and first at the west end of the shelter mouth, but would be shadowed by late afternoon. At colder times of the year, these two events could make this portion of the rockshelter a place to gather, perhaps around warming fires. This concentration of features also may indicate where the principal entry point to the rockshelter was located. Even today, trails accessing rockshelters throughout the canyon (and the plateau in general) enter rockshelters at one of the ends. Generally, this is because the berms tend to be tallest in the middle. But even at Porcupine Cave, which has no berm, the entry trails arrive from either end rather than in the middle. At Marmes Rockshelter, berm development may have been slower at the west end making this the natural entry point.

The presence of fire-related features at the mouth of the rockshelter consistently until Unit V, and then none after that, may indicate when the development of the berm (and perhaps recession of the cave roof at the mouth) restricted use of that area. Following the Mazama ashfall, the sediment available upslope of the rockshelter mouth would have been increased and deposition along the rockshelter dripline may have increased as a result, creating or adding to a berm at the front of the rockshelter's floor; review of the stratigraphic profiles for units along the dripline (Appendix B) appear to indicate an increase in the rate of accumulation of the berm through time. This is a time of relatively low rooffall rock exfoliation as the climate had become warm and dry (see Chapter Three). A prominent berm would have inhibited use of the mouth of the rockshelter for certain cultural activities, including those involving prepared features. In Unit VI, fire-related features move further back into the interior of the rockshelter. However, in Unit VII a handful of these features are again found towards the front of the rockshelter. In this case, it may be that the density of features in the interior of the rockshelter forces the use of the front of the site again.

Storage Pit Features

Storage, particularly of foodstuffs, is important for understanding settlement and subsistence patterns. Pit features in habitation sites, implying their use for storage, are found well back into the Plateau archaeological record, but prior to ca. 2500-3000 B.P. only in small numbers. The archaeological record also includes examples of the storage of durable items in pits. A small storage site (45FR444b) located approximately a kilometer from the Marmes site within the Palouse Canyon was found to contain two well-formed hand mauls in a storage pit built against the base of a bedrock wall (Hicks 1996:41). The storage of durable items is recorded in the oral history of Plateau tribes as well. Items such as pestles, stone bowls, durable utensils and other artifacts not prone to deterioration that were needed at each village or campsite regularly used in the course of a seasonal round could be placed in a pit and left for the next time the site was occupied (Adeline Fredin, personal communication, 1998). However, the presence of numerous pit features in rockshelters is generally associated with storage of food resources.

Documentation of storage pit feature structural materials and stored remains is lacking for the Marmes Rockshelter deposits, although both the stratigraphic profiles and some description in the field notes indicate that at least grass linings were present allowing the identification of superimposed and interbedded features, sometimes together with fire pits (Figure 6.2). Rice assigns numbers to 15 storage pit features at Marmes Rockshelter, three of which are multiple, superimposed storage pit features. In addition, he includes a "small pit containing bone" (1969:57) and a "shallow pit" (1969:70), both with provenience information only (no contents remarks), in his "Hearth, Fire Pit, and Storage Pit Features" category implying that he considers them also to have been pits used for storage. Two of the multiple, superimposed storage pit features include two pits while the third (F100) is made up of five superimposed features. To this accounting we can add the two features listed as fire pits that apparently were built in storage pit features. This totals 25 storage pit features. By far, the majority of the identified storage pit features occurred in stratigraphic Unit VII; the only pits in Unit VIII are considered recent pothunter holes, the deepest of which intrudes to Unit IV. One storage pit feature and the small pit containing bone occur in Unit III, two storage pit features occur in Unit V, and two others, including a "stone filled storage pit" occur in Unit VI (Rice 1969:61).

A review of the stratigraphic profiles of the upper matrix for the rockshelter interior suggest that 25 storage pit features probably is a fairly accurate accounting, although storage pits and fire pits are not differentiated in the profiles and most of the depressions referenced in the field records were filled with sediment that included charcoal. The picture is also confused by the pothunter holes that intrude from the surface. However, this total of 25 storage pit features probably does not represent partial pit feature remnants, which can be difficult to recognize during excavation, particularly with a vertical excavation focus. Previous research in Palouse Canyon rockshelters (Hicks and Morgenstein 1994; Hicks 1995, 1996) indicates that storage pits were reused as well as rebuilt within older pit features (leading to superimposed pit features such as F100a-e). When pit features were reconstructed or new features were prepared, they sometimes disturbed parts (and probably sometimes all) of previously constructed features. When excavated, these disturbed features often can be identified as pit walls truncated by a subsequent pit feature. Each instance of storage pit preparation might include some cleanout and reuse of pit materials used in the previous storage event(s), such as grass lining, matting and cordage fragments and perhaps even trace amounts of the materials that had been stored. In the Marmes Rockshelter storage pit features, grass lining was often noted at the base of the pits, particularly in Unit VII. Below this, only one storage pit feature (in Unit V) was noted as having a grass lining. Feature 112 appeared to have been lined with rocks averaging two inches in diameter.

It is impossible to ascertain how many partial storage pit features may have been encountered by the excavators at the Marmes site. As noted in Chapter One, there is a startling lack of botanical materials in the site's collection, which must reflect the collecting directives rather than the lack of such materials in the site. Excavations in other rockshelters in Palouse Canyon that contained storage pits all observed considerable amounts of botanical materials associated with the use of the pits (e.g., Draper and Morgenstein 1993; Endacott 1992; Hicks and Morgenstein 1994; Hicks 1995, 1996; Mallory 1966). A model of rockshelter storage pit construction and use developed out of some of those previous excavations (Hicks and Morgenstein 1994; Hicks 1995, 1996). Borrowing heavily from Hicks and Morgenstein (1994) and Blukis Onat, Hicks, and Stump (1996) the model is presented below as a descriptor of what likely existed in Marmes Rockshelter.

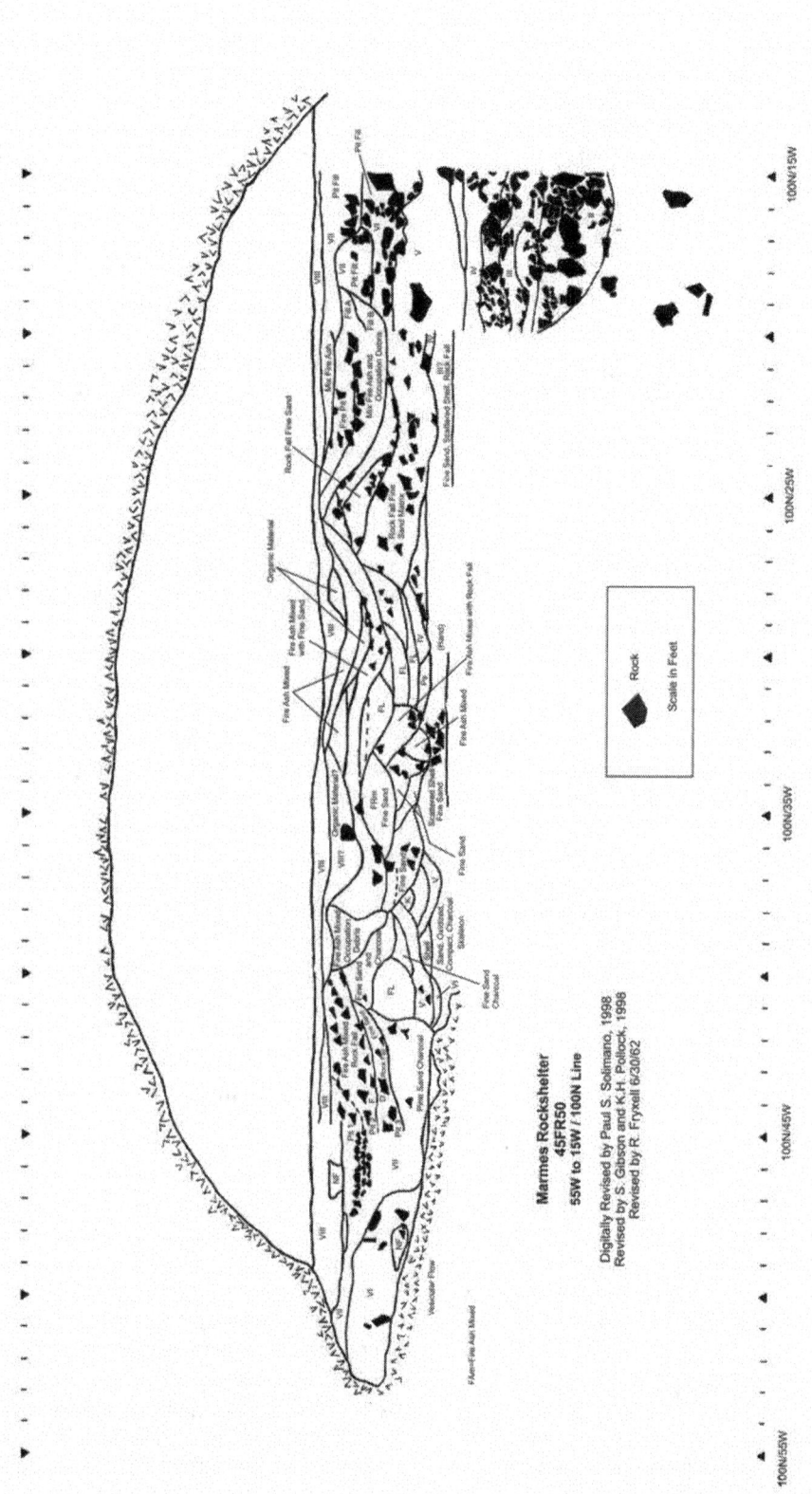

Figure 6.2 Stratigraphic profile showing the complexity of feature relationships near the back of the rockshelter.

At McGregor Cave (45FR201), two surface depressions thought to be storage pit features were completely excavated by BOAS, Inc. staff under the direction of the author using a horizontal rather than vertical focus. Feature layers were removed in their entirety from the multiple excavations units that made up each block that encompassed the area of each pit feature. While both of the features excavated were found to have been emptied of stored resources they were largely intact. A third pit feature, partly truncated by construction of one of the surface-visible pit features, was also encountered that was not visible on the surface. Because of the care taken during excavation of the features, individual construction layers and events were discernable and recovered, allowing a thorough description of the construction procedures used in these features.

Both of the intact pit features were bowl-shaped with flat floors measuring ca. 1 m x 80 cm. The floors and sidewalls were lined with overlapping layers of grasses. During recovery, these liners could be removed in sections that appeared to have been bundled, varying in size from what amounted to a large handful to a small armful for an adult. These grass bundles were often still cohesive from their original collection, as suggested by the intertwining of the base of the stalks and in some cases, the roots. It appeared that these bundles were placed in the pits in their approximate intended position and then spread to form a layer. This pattern was also evident in the horizontal layers of grass interpreted as pit covers.

Construction of a new storage pit feature in the rockshelter deposits is hypothesized to have begun by removing rooffall rocks until a circular or ovoid depression of the desired depth was attained (Figure 6.3). In one feature, small boulder-sized rocks, generally with a long, flat surface, were set in at the base of the excavation as pit wall liners. Excavations of other rockshelter pit features suggests that the use of facing rocks such as these is not the common method, but it has been observed elsewhere. The floor within the perimeter of these liner rocks consisted of pebbles with a blanket of loess and volcanic ash that probably resulted when the larger rocks were removed during construction of the feature; some of the fine-grained sediments undoubtedly settled there during the excavation of the features as well.

Overlying the floor of both pits was a layer of large-stalk grasses that continued up onto the rocks forming the side walls of the pits, thus forming a continuous pit liner that rose approximately one-third the distance from the floor to the current surface. In one of the features, this basal pit liner was made up of several layers of grasses and small matting fragments, with the second layer placed higher on the walls of the pit, thus thickening the layer and extending it further toward the surface. Over this basal layer of grasses was a framework of criss-crossed sticks of both roughly-broken willow limb wood and split cedar, all of a relatively consistent length (Figure 6.4). This frame of sticks may have served as a platform, built up higher at the edges with a shallow depression in the middle, to hold the stored resource off the ground to aid in preservation.

Over the stick platform was another layer of grasses that also continued up onto the pit side walls forming another complete pit liner. In one feature, overlying the grass layer on top of the stick platform, a large grass mat fragment was recovered that covered nearly all of floor of the feature at this depth. Similarly, a large tule mat fragment was recovered from an equivalent level in the other pit feature. On top of this mat layer was another thick layer of grass that was not part of a pit lining (it did not continue onto the side walls of the pit) and so was dubbed a pit cover. In one feature, this layer was made up of two large bundles and one smaller bundle of small-stalk grasses spread out to form a layer that covered the entire feature. That the pits would be intact, and the pit cover still in place flat atop the feature, but with no stored resources remaining within the feature, suggested that the resource was easily removable and the cover dropped back into place.

All deposits overlying the pit covers were attributable to cultural activities unrelated to these features and natural deposition. On top of the pit covers was loose rooffall rock with redeposited ash and loess and a thin (ca. 5 cm) layer of unstructured grasses, small clumps of grasses, and scattered small fragments of matting and cordage all mixed together, lying below the loose rooffall rock and sediment visible on the surface. This unstructured layer of mixed grasses and textile fragments is hypothesized to be redeposited from the cleaning out and reconstruction of neighboring storage pit features and redeposition of windblown plant fragments that have filtered down from the surface since the last use of the site.

A third storage pit feature with no surface depression, was found during excavation of one of the surface-evident pit features. This feature represented the remnants of an abandoned storage pit that predated the surface-visible pit feature. It is clear from its orientation that it was encountered during construction of the latter feature, and was

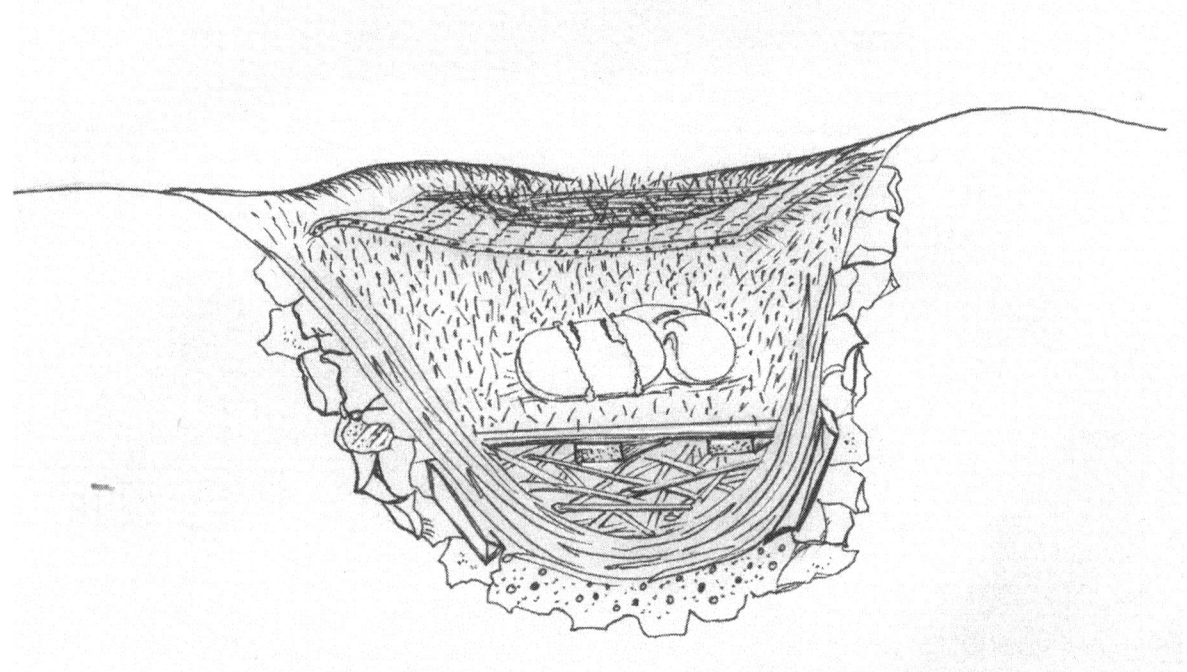

Figure 6.3 Drawing of typical storage pit feature in cross-section.

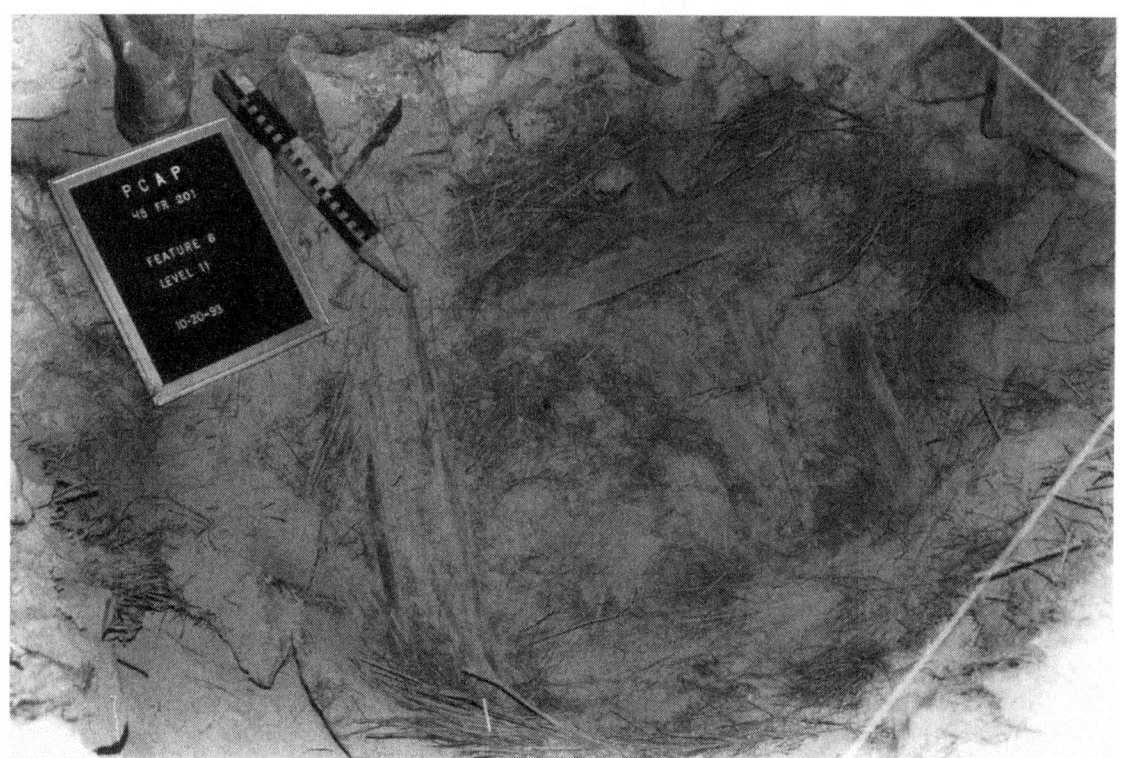

Figure 6.4 Stick "platform" in a storage pit at McGregor Cave (Hicks and Morgenstein 1995: Figure 3.42).

just pushed aside to make room for the new feature. In pushing the older feature aside, the thick pit liner on the side where the new feature was being constructed was rolled and folded over into the interior of the older feature, trapping a number of rocks and sticks, and a unique, carved wooden artifact interpreted as having been used for pounding camas or for shaping camas into loaves for storage (after Spinden 1908 and Walker 1973).

There is evidence from several excavated rockshelter storage sites (e.g., McGregor Cave, Porcupine Cave, Squirt Cave) other than Marmes Rockshelter that storage pit features were reused, at the least by constructing a new feature within the depression of a previously abandoned feature. Reuse of storage features could have been an annual activity that likely required some housekeeping. In particular, these storage features apparently are periodically cleaned out by purposeful burning of the pits' structural botanical materials, particularly the grasses. This burning out of the features may have been required to kill bacteria and mold that resulted from the previous storage event before storing new resources in it. This would require only a quick burn, and could be done so as to preserve much of the feature materials for reuse. Badly burned or highly fragmented materials may have been pulled out of the pit and replaced. This is archaeologically testable in that replaced materials likely would end up on the floor of the rockshelter.

Sediment analysis of both feature and non-feature deposits in McGregor Cave found granular grass charcoal to be a common constituent of the anthrosol identified in the sediment profile next to the excavated storage pit features (Hicks and Morgenstein 1994). In addition, portions of the pit liner in one of the storage pit features had been burnt. Also, pieces of burnt and charred wood and mat and cordage fragments were found throughout the upper deposits of McGregor Cave, which have been interpreted as cleanout from previous uses of Feature 5 and other adjacent pit features. Burned material of this kind was also observed in the grass liners of excavated features in nearby Squirt Cave.

Based on the observed structure of these pit features, in particular the shape of the framework of sticks, and that the stored resource was removed with little disruption to the pit covers, it was suggested that the stored items were placed in these features in bundles or 'packages'. No matting fragments and only a few cordage items were recovered from outside the feature-related deposits, but were common within the feature deposits. This suggests a correlation between the function of the matting and cordage, and storage activities in this site. This is significant, in that it indicates that the site's users brought matting and cordage to this site for use in the features, with even small fragments of mats having been included in construction of the pit features.

Evidence for specific uses of the cordage is less evident though, since they weren't found in an intact context. The longer, thicker cords may have been used for tying up bundles of stored resources. The recovery of several longer cords that rebounded to loops of roughly equivalent size (10 x 15 cm) once they were freed from the rockshelter deposits suggests that they were originally wrapped or tied in this shape when the grass was still green and pliable and apparently stiffened into this shape as they dried. Two of these looped cords are still knotted assuring the circumference of the loop. No evidence was found that cordage was used in any other context in these rockshelter sites than in association with storage activities.

The relatively consistent shape and size of the looped cordage artifacts implies a bundle of that size and shape which raises implications for the kinds of resources that may have been stored in such bundles. That a cord was tied or wrapped around something suggests that a wrapping material (e.g., matting, basketry, animal hide, etc.) was included as part of the bundle, since a cord would not be effective at tying most materials in this shape. The sizes of the bundles somewhat restricts the resource type that could have been contained in them. It is unlikely the bundles held large mammal remains. Butchered elk haunches have been observed in two rockshelter storage pits in the Palouse Canyon (45FR54 and 45FR404), neither associated with looped cordage of this kind or with long cordage fragments that may have been looped at one time. Instead, smaller mammals, fish, or plant resources are suggested as the contents of stored bundles.

It is difficult to determine the contents of the storage pit features at the Marmes site. Rice (1969) summarized the information available on storage features in the field records together with hearth and fire-pit features; this information is presented as a table in Appendix F. Often, no information on contents was provided by the excavators. Many pit features were found to have cultural materials that may have sifted into the pits from the surrounding matrix (e.g., shell, bone fragments, lithic flakes, lithic and bone tools). The field records generally offer little interpretive information about the context of these materials within the pit features; for example, whether these materials were concentrated in a fashion that suggested they were stored items, or scattered throughout the feature fill matrix. This may have been considered appropriate, however, since rodent activity could have spread stored items within a feature. Several pit features had rocks within the pits on top of grass lining that formed their base, in two cases these included river-rounded cobbles, and while questions remain about their primacy, they would be less susceptible to movement from rodent activity. With these reservations stated, the contents of the storage pit features are summarized below.

Of the 25 distinct storage pit features identified in the rockshelter, contents (other than structural elements or rockfall) are indicated in the field records for only ten. Bone and shell fragments were common in these ten features, but in six of the features the material contents described in the field notes and in Rice's (1969) feature summaries are similar to the content of the matrix of the stratigraphic unit in which they occur and may be the result of infilling of the features after abandonment. Some of these features may have been used to store mammal remains or shellfish, but the available information is too limited to offer such a conclusion.

Stored resources are interpreted as remaining in just four features, all from Unit VII. The lower one and/or two of the five superimposed storage pits that make up Feature 100 contained much bone, and Feature 112 contained a number of large pieces of shell; in both cases, the pit matrix did not include other materials that would indicate infilling of the feature after abandonment. The relatively large volume of bone in Feature 100, and the rarity of large fragments of shell (such as those in Feature 112) in the rockshelter matrix contributed to the interpretation that these materials represent the remnants of stored food resources. Feature 100 also included five projectile points, which may have been intentionally placed in one of the five superimposed pits that make up this feature.

In addition to the bone found in Feature 87, material that occurred in the Unit VII matrix outside of features also was described, but in this case the relatively high volume of bone found appears to indicate this was the stored resource. Bones, shell, "vegetable matter" and a bone awl were found within Feature 99, the description of which appears to indicate little disturbance to the feature. It is not clear if the botanical material referenced is the pit's structural material or remnants of a stored plant resource. Neither the bones or the shell are further described, but they may be the remnants of a stored resource. The available information on feature contents is sparse and the presence of stored resources in these features is not certain.

Despite the uncertainties of interpreting which storage pit features retained their stored contents based on the available records, a low proportion of features with stored resources is not unexpected. It seems unlikely that stored resources would be abandoned or forgotten. It is expected that the aboriginal occupants of the area had a good understanding of the potential of storage, which resources would last the longest, and how best to prepare them to be available when needed. That Marmes would have been used for storage seems unquestionable given the relative distances between the Palus Village site at the mouth of the Palouse River and other documented storage rockshelters that are as much as several miles further away or in difficult to access locations. While storage pit features have been recorded in talus slopes and against the base of marginally overhung bedrock walls (Hicks and Morgenstein 1994; Hicks 1995), Marmes would have been close, easy to access, and offered a better protective environment than those other storage locales.

Initially archaeologists believed that these rockshelters served as good storage locations as a dry environment. But data collected in the Palouse Canyon (Hicks and Morgenstein 1994; Hicks 1995) indicates that this assumption is incorrect. Morgenstein initiated the burial of paired sets of Ambrose temperature and relative humidity cells at 10, 25, and 50 centimeters below the surface within five rockshelters of differing aspect. These measurement cells were then recovered after one year in the ground. The temperatures and relative humidity obtained by this method are mean values representing the entire year of data collection at each site. Maximum and minimum values are not obtained by this method. The premise is that the effective temperature and moisture content of the soil affects weathering reactions (which is relevant for hydration studies) and also affects the preservation of food stocks that might have been stored in these shelters.

The resulting data indicate that the available soil moisture is much greater than originally thought and increases with depth in all of the sites (presumably due to air drying effects near the surface). A small rockshelter with a northern exposure had the highest moisture reading of 95-96% relative humidity. The other smaller rockshelters had 87-92%, 93-95%, and 57-69% relative humidity while McGregor Cave provided readings of 54-72%. Soil temperature also was relatively low, and considerably below the open air mean temperature for the area. It was suggested from this data that preservation of organics in these sites is due to a combination of relatively cool temperatures (particularly during the fall and winter months when most resources are being stored) and low moisture flux (Hicks 1995).

As noted above, it seems it would be unlikely to find unrecouped resources in very many storage pit features. In sites where storage pit features are found to retain stored materials, a disruption of the seasonal round may have occurred with those who stored the resources. Storage of most of the resources would be limited to several seasons before deterioration rendered them inedible, or before rodents or insects found them (although prairie sage leaves, used by northern Columbia Basin tribes as an insect repellent in bedding [Turner 1979:177], was found in a storage pit cover in McGregor Cave [Hicks and Morgenstein 1994]). Where residents were prevented from returning for the resources for a year, the abandonment of the feature may have been voluntary, and new features were dug rather than attempt to clean out a tainted one. Complicating matters may be the inability of groups to return to stored resources in the early historic era, perhaps when battles with the U.S. Army didn't go well (such as when upper Palouse Tribe members were shipped to Oklahoma following their defeat). In addition, Wilfong (1990) noted that white "volunteers" intentionally disrupted stored plant foods to pressure the Indians into moving to reservations.

Floodplain Features

Fifty-two features were recorded in the floodplain during the 1968 excavations (Fryxell and Keel 1969:38), many of which were individual artifacts or only a small number of items of cultural material. However, except for a few examples (e.g., the owl-foot artifact), this study's review of the available records offers little to substantiate such fine feature definitions. In nearly every case where field records for material defined as features could be located, those records offered little information that subsequent researchers could use to interpret either the 'feature' itself, or how the defining of the 'feature' contributes to interpretation of the site. In some cases it is clear that excavators were uncertain in their identification of the complex, often ephemeral stratigraphic layers that Fryxell observed in the floodplain deposits. The field notes include many statements by excavators about uncertainty of whether their level was in one of the cultural horizons or in the thin bands of sterile overbank deposits between them. This may have led them to be overly cautious about potential relationships between cultural items. It also may be that the very low recovery rate of lithic or bone artifacts from the floodplain deposits (n=96), despite the careful hand excavation of 394 cubic yards of sediment (Fryxell and Keel 1969:38), may have led excavators to overemphasize the potential importance of individual items.

In an attempt to describe feature areas that offer something to interpretation of the use of the floodplain portion of the site, this study redefined the floodplain features based on the assumption that accumulations of material found in the same general locations may be due to related cultural activities. Beginning with the faunal analysis in the first year of the current study, four areas within the floodplain excavation identified as having concentrations of cultural material were defined as areas of specific interest for analysis and interpretation. The boundaries of these concentrations were defined roughly from a plan drawing of the floodplain units that showed the location of individual items that had been recovered (see Figure 1.19). These were defined as feature areas and numbered 1 through 4:

1. The northeast quadrant where Arctic fox remains and the owl foot were found and lithic flakes were scattered about among some large boulders;
2. The north-central area near where the original floodplain discovery of human and elk bone was made in 1965;
3. The south-central area where semiarticulated remains of a large elk were found; and,
4. The southwest quadrant among a concentration of smaller bone fragments

The initial review of the field records indicated that lumping of concentrations in adjacent units in this manner appeared justified since the materials recovered from feature areas 1 and 4 came almost entirely from the Harrison horizon, while the bulk of the materials from feature areas 2 and 3 were contained within the Marmes horizon. These two cultural horizons date close together in time, and Fryxell treated them as a single 'component' when viewed against the rockshelter stratum units. However, they were separated by thin bands of overbank flood silts indicating some passage of time between them that appears to correlate with some shifting of activity areas.

The following discussion of the contents of the four floodplain feature areas relies on the original field records from the excavation units associated with each feature area. The faunal contents of the bulk samples from the floodplain that were sorted for the current study are discussed in Chapter Ten and will not be reiterated here. Likewise, analysis of the specific material remains from the floodplain feature areas was presented in the results of previous chapters to the extent that the items described in the field records are still present in the site's collection.

Northeast Feature Area

The northeast floodplain feature area occupies a roughly triangular area between excavation units N5/E0 to the south and west, N30/E5 to the north and west, and N30/E15 to the north and east, an area of approximately 250 square feet (ca. 23 m^2). The dominant features of the floodplain in front of the rockshelter were several large boulders (see Figure 1.15) around which the cultural material in this feature area was found (Figure 1.19). These boulders may have served as an attraction to activities that took place in front of the rockshelter. They may have served as shelter from the sun and wind or as blinds, keeping in mind that the cooler climate during the time period when the floodplain was being used would have resulted in considerably more vegetation than at present. Certain surfaces and edges of the boulders also may have served for working on tools or other resources.

The most prominent cultural materials from the northeast feature area are the faunal bone fragments. The owl claw artifact and the arctic fox skull fragment have been described in Chapter Nine and Chapter Ten, respectively. But the field records show that some 54 other individually identified faunal bone and tooth fragments also were collected from these units and probably many more since nearly every level resulted in a "miscellaneous bone fragments" jar as well. In addition, more than 125 pieces of lithic debitage, 17 tools, and some possible ochre are individually identified in the field records as having been collected from the excavation units in this feature area. Charcoal and shell jars also were collected from most of the levels from these units.

Most of the bone fragments collected were too small for species identification, but deer and red fox, as well as unidentified medium- and small-sized mammals were represented. Deer, canine, and rodent teeth fragments also were observed. Slightly more of these faunal items were recovered from the Marmes horizon than the Harrison horizon, although all three bone tools, a bone needle in two pieces, a bone point, and a worked bone fragment, came from the Harrison horizon.

The opposite was true for the lithic materials with many more coming from the Harrison horizon than the Marmes horizon. At least 118 of the more than 125 pieces of lithic debitage came from the Harrison horizon, including 84 chert flakes (often referred to as "chips" in the field records), 11 obsidian flakes, one basalt flake, one quartzite flake, two chert "chunks" that probably are shatter, 19 flakes unidentified to material type, and a single reference to having collected "various chipping detritus" from one level. Only seven similar lithic items were recovered from the Marmes horizon, including three chert flakes, one basalt flake, and three flakes of unidentified material. The same disparity in representation is seen in the lithic tools recovered from this feature area. Only two tools, a chert scraper/graver and a possible cobble chopper, came from the Marmes horizon. The Harrison horizon produced 12 lithic tools, including three projectile point fragments: two "Lind Coulee" (Windust) point bases, one of basalt and one of cryptocrystaline silicate, and a chert lanceolate point base. The other nine tools include a chert scraper, a chert graver tip, three chert utilized flakes, an obsidian utilized flake, two utilized flakes of unstated material type, and a core fragment of unknown material. In addition, the two scrapers that were part of the owl-foot feature were both chert. This latter feature was probably a ceremonial item, perhaps the contents of a medicine bag/bundle (Gustafson 1972). One level record also notes the presence of possible ochre.

North-central Feature Area

The north-central floodplain feature area occurred within grid units N25-30 and W25-30, encompassing an area of ca. 40 square feet (ca. 3.7 m^2). Cultural materials were collected from this feature area at least three different times. This feature area included the concentration of cultural materials first observed in the floodplain deposits by Fryxell during excavation of the bulldozer trench and described as "a dark oval concentration of bone, some of which was burned" (Fryxell and Keel 1969:10). Fryxell collected an unspecified number of items from this location and took them to Pullman, where Gustafson identified them as "the remains of elk, deer, and – most importantly – man" (Fryxell and Keel 1969:10). In April 1968, Fryxell and graduate students collected additional materials from a few feet north of the 1965 finds while attempting to verify that the initial discovery represented in situ deposits. Again, it is uncertain how many items were collected during that work, but "more of the elk as well as rabbit, fish, and antelope (as well as) additional fragments of the charred human remains, some pieces articulating with the material found in 1965" (Fryxell and Keel 1969:10) were identified. Sieving of the sediments excavated at that time also produced a fragmented bone projectile point (Fryxell and Keel 1969:11). In addition, prior to removal of the floodplain overburden deposits by bulldozer, the backdirt from the 1965 bulldozing was sifted "to recover any additional human remains which might have been stripped from the trench in 1965. Skeletal material of the original find was recovered, as were 13 artifacts" (Fryxell and Keel 1969:29). While it is likely that the human remains were associated with the Marmes I feature from this area, it can't be asserted that the 13 artifacts all came from within the north-central feature area. Interestingly, while no lithic projectile points were recovered from the Marmes horizon by the controlled excavations, 8 of the 13 lithic tools recovered from the sifted backdirt accumulated by the 1965 bulldozer work were point fragments and likely all came from the Marmes horizon or higher deposits.

The full-scale excavations that followed were documented by field records that provide an accounting of the materials found in the summer of 1968. That documentation included a plan drawing of the feature area made after the materials found "lying on and partly in the top Marmes A Horizon" (Marmes Field Catalog Form, 7/26/68) were fully exposed (Figure 6.5). The plan drawing shows more bone fragments than are indicated in the field notes and catalog forms for these units. From the different records, it is apparent that more than 17 fragments of bone from the hind quarters of the butchered elk, a basalt flake, a piece of burned worked bone, a possible basalt chopper, and a hammerstone were collected from the Marmes horizon at this location. The field notes indicate that unspecified amounts of small bone fragments and charcoal also were associated with this feature area. Immediately below this feature, a canine jaw and tooth fragments and a chalcedony scraper were collected from the Harrison horizon in these units. Further description of the elk remains, described by Gustafson as larger than contemporary elk (personal communication, 1998), is presented in Chapter Ten. Caulk (1988) also identified an elk bone in his analysis sample from the Marmes horizon.

South-central Feature Area

This feature area occurred in three units, N15/W30, N15/W35, and N20/W35, but the largest portion of the feature was in unit N15/W35; the total area was approximately 40 ft^2 (ca. 3.7 m^2). This feature area represents much of the balance of the large elk that was found about ten feet northeast on top of the Marmes A1 horizon (Figure 6.6) with some elements extending up into the silt layer above the cultural horizon.

Thirty-seven individual bones or bone fragments are listed in the field catalog for the Marmes horizon in the units associated with this feature area, and the field notes indicate that many other miscellaneous burned and unburned bone fragments also were observed. Shell fragments also were collected from several levels. The lithics recovered were limited to a single tool (see below), three rounded river cobbles and seven basalt rocks, two of which are described in the notes as "non indigenous" (Marmes field catalog, 7/24/68). The field notes describe this as the apparent butchering area of the elk:

> "the remains present are those of the axial skeleton minus the head and (undecipherable word) vertebrae. The position indicates that the head, quarters were removed and consumed elsewhere, while the rib cage, etc were left since they would not have been very important as a food source or for raw material for tools" (Marmes field notes, 10/10/68).

However, only a single lithic tool, a rounded pebble with possible flaking, was recovered from the Marmes horizon in the units associated with the feature. Instead, two potential butchering tools, a bifacially retouched chalcedony trimming flake and a chert scraper retouched on both sides were found in the Harrison horizon below the feature. The Harrison horizon levels also produced a bone needle, "two flat, smooth mano-like rocks" and miscellaneous bone fragments including "many fish vertebrae" (Marmes field catalog, 7/26/68, 8/6/68, and 8/20/68).

Southwest Feature Area

While much of the floodplain excavation area was found to include small bone fragments and charcoal, particularly in the Marmes Horizon, the southwest feature area had a higher density of these small bone fragments and charcoal, and more so in the Harrison horizon than the Marmes horizon. In addition, two adjacent units within this feature area found bone fragments and charcoal below an A16 horizon, more than three feet below the lowest Harrison (A5) horizon. The area of this scatter encompasses an area of ca. 150 ft^2 (ca. 14 m^2) in

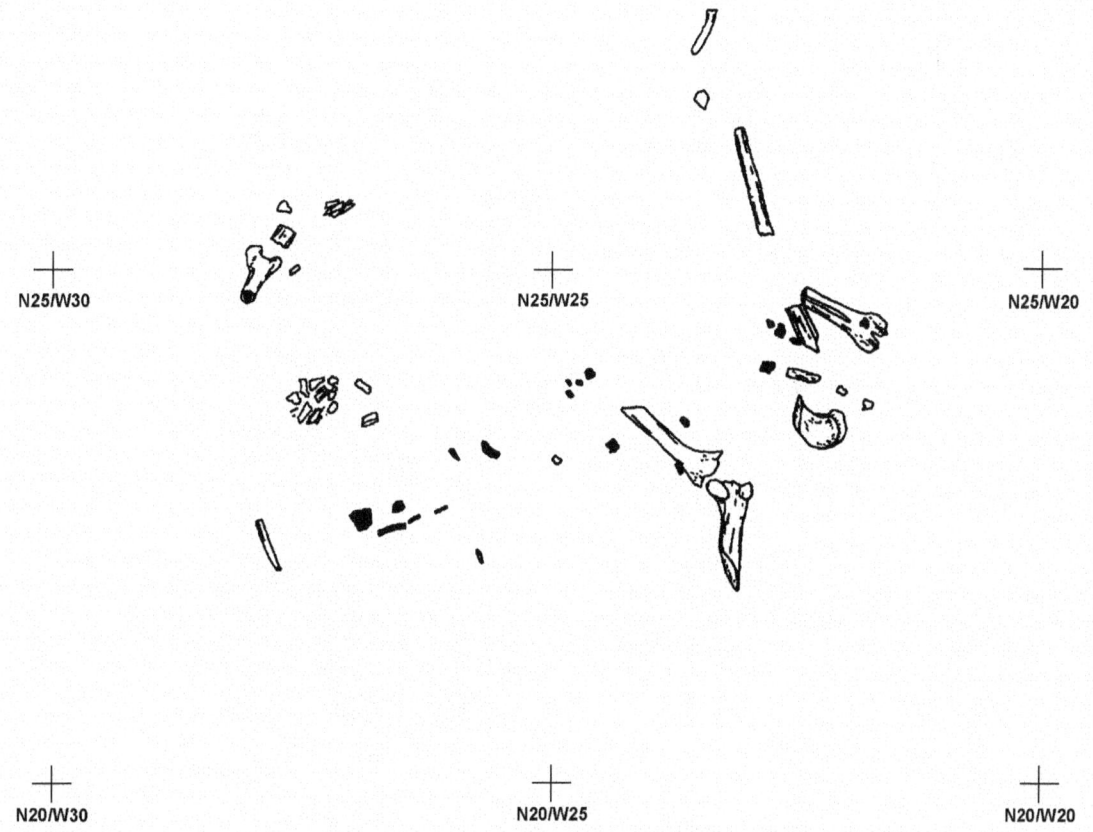

Figure 6.5 Diagram of hind limb bones of a large elk from the northern portion of the Marmes floodplain excavation (from Gustafson and Wegener 1998).

the southwest corner of the floodplain excavation area; no units were excavated south and west of this feature area. Within this area the density of the bone fragments and charcoal varies with greater densities generally occurring in the westernmost units (see Figure 1.19). Elk and other mammal bone fragments and some lithics occur in both the Marmes and Harrison horizons. More bone fragments were found in the Marmes horizon than the Harrison horizon, while a few more lithic items were recovered from the Harrison horizon than the Marmes horizon. This was not true of lithic tools, however, as a chert projectile point, a chert end scraper, a chert beaked scraper, and a possible hammerstone all were recovered from the Marmes horizon in this feature area. A worked flake and a small obsidian graver or drill were recovered from the Harrison horizon. This pattern was not typical in the floodplain where the total number of lithic tools recovered from the Harrison horizon (n=40) outnumbered those from the Marmes horizon (n=22). A Windust style projectile point also was recovered from the Harrison horizon in a unit adjacent to this feature area.

Another area of interest in the floodplain was a shell midden feature west of the main excavation block that was revealed by the excavation of a bulldozer trench (for additional access) early in the summer of 1968. Excavation in this area of the "western trench" is only reported in a few sentences in Fryxell and Keel (1969) and concentrated on pre-Mazama ash deposits that included a concentration of shell. Fryxell and Keel may have been referring to this area when presenting the "Cascade Component" (1969:31-32) in their final progress report but the items listed in their table actually came from throughout the floodplain excavation area. A single sample of shell from the western trench area produced a conventional radiocarbon date of 7980 ± 300 B.P. (Sheppard et al. 1987:122).

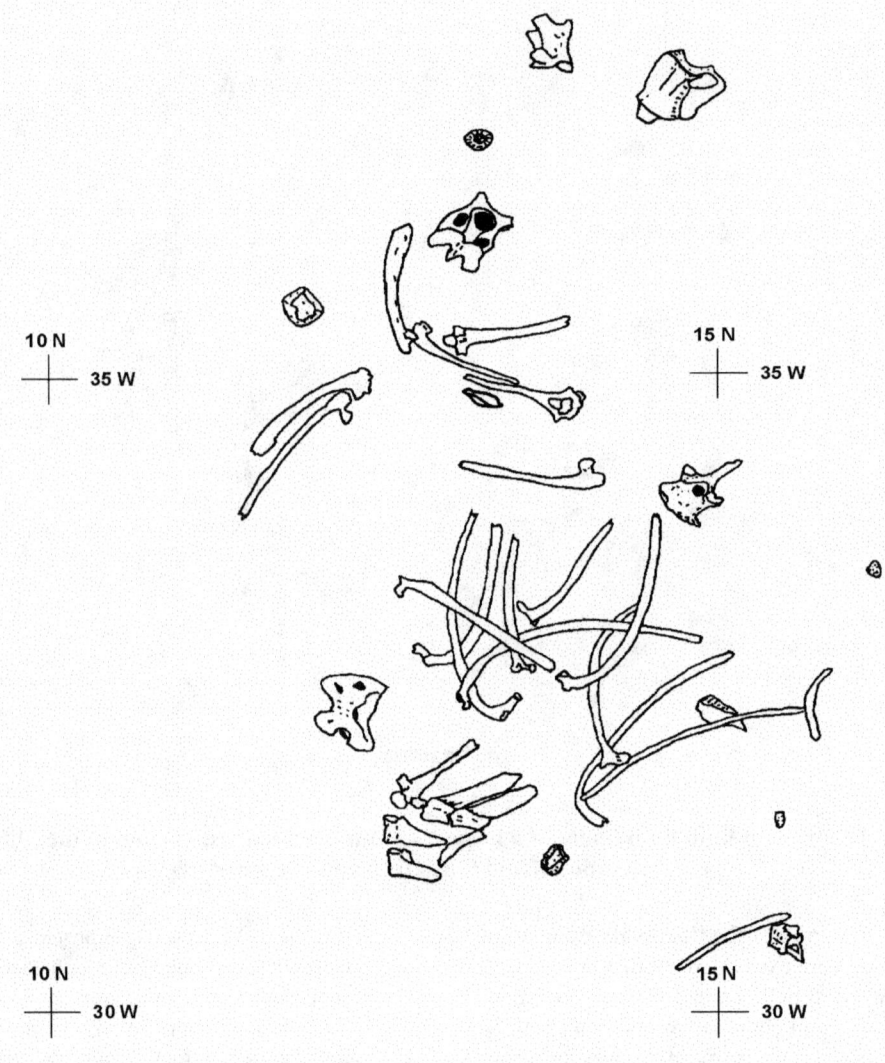

Figure 6.6 Diagram of vertebrae and ribs of the large elk from the south-central portion of the Marmes floodplain excavation.

7
HUMAN REMAINS

Brent A. Hicks

A number of researchers have analyzed the collected human remains from the Marmes site. Krantz (1979) and Breschini (1975, 1979) have published the results of their analyses. Krantz also prepared the "Examination of Human Skeletal Remains" section of the 1969 Final Progress Report by Fryxell and Keel. Tadeusz Bielicki, a physical anthropologist visiting from Poland at the time that the initial human remains were exposed in the floodplain, examined portions of those remains (Marmes I) and briefly described them for the 1968 American Antiquity article (Fryxell et al. 1968b). Fryxell and Daugherty (1962) described the discovery and treatment of the human remains in the field during the 1962 field season. Rice summarized the burial feature information from the rockshelter deposits from the field records for the preliminary report to the National Park Service (1969).

In addition to the published accounts, students at Washington State University accessed the human remains from the site for study, which included opening some of the plaster encased burial features prior to Breschini's comprehensive effort for his thesis project (1975). Unfortunately, other than Breschini, only one student's work with the human remains was documented to any degree. Krantz (1979:169) notes that James Chatters sorted and glued together the matching cranial fragments from the rockshelter cremation hearth (Burial 23) while an undergraduate at WSU. Chatters prepared a color-coded key on a sheet of graph paper that correlated with colored adhesive dots he had placed on the portions of the crania he was able to reconstruct; that key was found in the Marmes site records at WSU. Chatters also has obtained loans of human remains from the collection within the last decade but there are no records associated with any recent analysis in the site records. Because the current study did not include any new analysis of the human remains or grave goods, the following is based entirely on the extant record that is the published accounts and the field records.

Krantz (1979) considered the human remains from the Marmes site as occurring in three groupings: "those from the main shelter deposit, those from the "cremation hearth" within the shelter, and those found in the floodplain deposits in front of the shelter" (1979:159). However, his decision on how to group the remains surely was in part because he was addressing the latter two groupings in his article and needed to separate the cremation hearth remains from the balance of the remains found in the rockshelter. Consistent with the balance of this report, this chapter addresses the rockshelter and floodplain human remains in two groups.

Methods

The current study includes no new analysis of human remains or identified grave goods. As such, this discussion of the methods employed by previous analysts of those materials is limited to what is presented in the few published accounts of analysis of the human remains from the site. Those accounts include little discussion of the analysts' particular approaches to the analyses or their goals or expectations based on the state of physical anthropological research at the time of their studies.

The total number of human remains features identified by the excavations at the Marmes site has been revised based on this research. A minimum of 38 individuals are represented in the human remains features found in the rockshelter and floodplain areas of the site. This is a "minimum number of individuals" count as this study took a conservative approach in determining what constituted the presence of a human remains "feature". The term "human remains features" includes all concentrations of human remains. This term is differentiated from "burial features" because not all of the human remains found at the site were buried; all burial features are human remains features. Only where the site records noted the presence of a pit, or indicated that the human remains were "intrusive" into a particular stratum, or that the remains were associated with such a quantity of cultural items (e.g., ochre, Olivella shell beads, formed tools) that interment is a reasonable explanation, was a collection of human remains considered a "burial." Only 14 individuals could be confidently described as burials.

Individual or small numbers of human bones, without additional context to suggest they represented only the last remnants of a burial feature or a person that may have died in the cave and was not interred, were not considered a human remains feature for this study. The site records commonly cite the discovery of individual bones or fragments, even several within a level, without mentioning the presence of a pit or other evidence that the bones (either complete or partial) appeared to have been intentionally buried. Further, only rarely do the records ascribe such finds to nearby human remains features. These isolated bones were considered the result of the extent of disturbance, both cultural and natural, to the upper strata in the site. These activities, including storage pit excavation by later occupants and rodent burrowing, had scattered burial elements from their original locations throughout the site's strata.

Rockshelter

Human remains became a focus of the Marmes Rockshelter investigations within a few weeks of the initiation of excavations at the site in 1962. Eleven individuals were located that year, three stratigraphically below pumicite believed to be from Mount Mazama, the eruption of which at that time was thought to date to ca. 6,500 B.P. (Fryxell and Daugherty 1962:16). Fryxell and Daugherty estimated the older burials to date between 6,000 and 8,000 B.P. and noted a paucity of previously reported altithermal burials (1962:16, 20). This, and the promise of good stratigraphy that could be used to understand the stratigraphy at the many other sites on the lower Snake River that Daugherty believed he would be excavating in the years prior to their immersion in the reservoir behind Lower Monumental Dam, led to additional excavations in the next two years.

The burials encountered in the first year's investigations in Marmes Rockshelter were 'difficult' to excavate. The "human bones were found to be extremely friable, sometimes disintegrating from the simple pressure of a paintbrush used during cleaning for identification" (Fryxell and Daugherty 1962:21). In spite of their fragile condition, the human bones were fairly easy to distinguish from the considerable amount of faunal bone in the deposits because most of the faunal bone had been cracked to remove the marrow. This generally allowed the excavators to recognize human bones soon after uncovering them. In addition, disturbance of the normal stratigraphic layers, a result of the digging of burial pits by the site's inhabitants could often be recognized before encountering the human remains. Because of the amount of time and effort it would have taken to excavate such fragile features in the field, a number of the burial features were encased in plaster casts, ostensibly for analysis in the laboratory at WSU. When a burial was recognized

> "...no effort was made to uncover the skeleton beyond determining its position and lateral extent. When this had been done the minimum possible block of matrix was marked out, exposed bones were covered with soil, and the block of matrix isolated as a pedestal. Surgical crinoline then was cut to convenient working lengths, usually between two or three feet long, and plaster mixed....The upper surface and sides of the isolated burial block were covered with cloth strips dipped in the plaster as rapidly as possible to fix them in place before the plaster began to set; any plaster left over then was smeared over the cloth surface until smooth." [Fryxell and Daugherty 1962:23-24] [Figure 7.1]

> "Subsequent layers were applied as rapidly as drying of the preceding coat would permit, until a cast three to five layers in thickness had been constructed. Before the final coats of plaster on the upper surface and sides had been applied, the bottom of the cast was begun by gradually undercutting the pedestal a small portion at a time, replacing dirt removed with a 2x4 or 1x6 plank, by tunneling through the remaining pedestal to insert additional wood flooring, and then by removing the remaining supporting dirt and plastering the bottom in the same manner as had been done with the sides and top. When completed, casts were allowed to remain in position until the last week of the field season." [Fryxell and Daugherty 1962:23-24] [Figure 7.2]

Figure 7.1 Photograph of feature being encased in plaster.

Figure 7.2 Photograph of completed plaster cast.

Removal of the casts, each weighing 600 pounds or more, required up to six crewmembers working the completed cast by hand onto a "wood sledge or platform….[that] was slowly dragged to the entrance of the rockshelter…." and slid down the talus slope into a flatbed truck and transported to Pullman (Fryxell and Daugherty 1962:24).

Daugherty numbered the burial features in the field in the order they were encountered. All subsequent researchers used those burial designations in their reporting. Fryxell and Daugherty report that 11 burials were encountered in 1962: "…three individuals, including two adults and an infant, were found four feet beneath the surface in positions stratigraphically lower than the Mazama pumicite; the skeletons of seven adults and one infant lay above it" (Fryxell and Daugherty 1962:16). Burial 9, however, is reported with two designations (Burials 9a and 9b) and is mapped as two separate locations in the same excavation unit in Figure 4 of the 1962 report (ibid:12). According to the field notes, three concentrations of human remains were found at this location, all initially subsumed under the designation Burial 9. The field notes appear to indicate that the initial identification of the skeletal elements in the three concentrations suggested that it was a single dispersed burial with a visible pit outline that intruded through Units V and IV to the top of Unit III (Breschini 1979:131-2). In the end, two of these concentrations were placed in separate casts and designated as separate features while the third concentration was excavated and collected in the field. Clearly it was ascertained in the field that at least two individuals were represented in Burial 9, a conclusion supported by Breschini from his analysis of Burials 9a and 9b and the remains from the third concentration that were collected in level bags (1979:132, 134). At the same time, a third cast was made in the same vicinity that was not considered part of Burial 9, but was not given a separate number. Breschini (1979) refers to this feature as the 'small, unnumbered cast' and he included it in his 1975 thesis study. In addition, examination of the records for the current study indicates two distinct burials have been designated as Burial 5 (see below). As such, fourteen concentrations of human remains actually were recovered in 1962 with eleven being removed in plaster casts. Burial 6 was placed in a partial cast in 1962 to protect the bones exposed in a sidewall, but was excavated the following year when the adjacent unit was excavated (Breschini 1979:116). Burial 11 was excavated in the field.

Other burials recovered in subsequent years of the project also were encased in plaster casts, but there is disagreement in the available records about how many and which burial features they were. Rice, working from the available records and by questioning available excavators, could only identify "at least nine…burials reported were encased in plaster casts" (1969:77). His tally does not include Burials 2, 5, and 19, all of which Breschini included in his analysis, and Rice's report lists Burial 9 as a single interment. Conversely, Rice does include Burial 6 as in a cast even though this feature was excavated in the field in 1963. This confusion is likely the result of a photograph of Burial 6 partly encased in plaster in 1962, a feature that was completely excavated the following year.

A better accounting of the features placed in casts comes from Breschini's tally developed from examination of the records and what he found in storage at WSU in 1975. Breschini's study included description of Burials 1 through 10 (including 9a and 9b) from 1962, Burials 19 and 20 from 1963, and the small unnumbered cast designated S.C. in his Table 1 (1979:117). This totals 14 and with Burial 6 having been excavated in the field after being placed in a partial cast, the total number of burials placed in casts and transported to Pullman would appear to be 13, a number affirmed by Breschini (1979:116). The details of the discrepancies in the total number of burial features from the sites are presented below in the discussion of the individual burial features.

Discussion of the methodology of laboratory analyses of the human remains from the rockshelter is limited to Krantz's description of the cremation hearth remains (1979:159-174) and Breschini's 1975 thesis study; this report references Breschini's 1979 article in the journal Northwest Anthropological Research Notes (1979:111-157) that was derived from his thesis. Krantz's (1979) discussion of analysis methods is limited to a description of how the human remains were separated from the mass of other cultural materials recovered from the cremation hearth feature. Because virtually all of the bone in the feature was extensively burned, discriminating between human and animal bone was often difficult. Adding to this difficulty was the fact that all of the bone fragments were small, the largest identified skull fragment being just 64 mm in length (1979:169). Attempts to separate the cranial fragments by lots based on differences in the degree of burning proved somewhat successful, and when correlated with the apparent ages of non-overlapping elements from the individuals represented led to the conclusion of the estimated number of individuals recovered from the feature's contents. Those conclusions are tentative however, "as the ages are lumped as juvenile or adult with no finer distinctions, and also because each skull was not necessarily burned to the same degree in all its parts. Evidence of a second act of burning further complicated matters" (Krantz 1979:169). As noted above, the work of sorting and gluing together the matching cranial bones was conducted by Chatters.

Breschini found analysis of the plaster-encased rockshelter burial features

"extremely difficult (due to) the exceedingly deteriorated condition of the bones. I excavated these burials using an air hose into which a glass tube from an eyedropper was inserted. Because most of the sediment around the burials was reworked pumicite and other windblown deposits, and had dried thoroughly since removal from the rockshelter, a very weak stream of air usually was sufficient to expose the bones. In many cases, however, the very weak air stream was also enough to cause the bones to crumble, and they had to be treated with a preservative as soon as they were exposed....When individual bones had been exposed and stabilized, they were photographed in situ and were removed, sorted, and set aside for later measurement and analysis.

All sediment contained in the burial casts was screened through a 1/8 in. mesh screen, and...the midden constituents...placed in separate bags for later analysis, and the sediment was discarded after screening." [1979:117-118]

Breschini kept detailed notes of his work and took approximately 400 photographs of the eight casts that he excavated. He did not attempt to conduct detailed analysis of the artifacts recovered from the burial features but described and took measurements of specific items.

"During analysis of the burials, techniques of both archaeology and physical anthropology were used. The archaeological data were used to determine burial practices, burial orientations, temporal positions, and artifact associations....The physical data were used for the determination of age, sex, and whenever possible to attempt to determine the physical attributes of the Marmes individuals.

The ages of the skeletons were estimated through cranial suture closure, tooth eruption, epiphyseal fusion, and the condition of the pubic symphysis....The sexes were determined through relative width of the sciatic notch, length of the pubis (when present), femur head diameter, mandibular angle, and cranial and post-cranial morphology. The pre-auricular groove was examined, when available, to try to determine if the individuals had ever been pregnant....Measurements were taken (of bones that were intact)....Nonmetric analysis of the skeletal material was also attempted, both to record pathologies, if present, and to give data on bones that could not be measured." [1979:118-119]

This report will not report the results of the measurements; interested readers are referred to Breschini's data tables (see 1979:120-122).

Breschini examined the original field notes and photographs from the years that burials were removed in plaster casts (1962 and 1963). These records provided information on provenience, stratigraphy, depths of burials and associated cultural material, and on any skeletal remains that were not included in the casts. He also sought information concerning the casts previously opened in the lab but found little documentation of those excavations (1979:119). Krantz told Breschini that "several of the casts were excavated in the past, and did not contain skeletal material" (1979:146).

Floodplain

The first human remains found in the floodplain were within a feature of concentrated bone, observed in 1965 by Fryxell during excavation of a bulldozer trench for his geological investigations. Subsequent investigations by Fryxell and graduate students in April 1968 located additional human remains in the bone feature and confirmed that the remains were in situ, leading to emergency salvage efforts beginning in June 1968. The floodplain excavations focused on the Marmes and Harrison horizons, but human remains only were found in association with the Marmes horizon. By September, fragments of four human skeletons had been recovered from this stratum and were numbered consecutively Marmes I, Marmes II, Marmes III, and Marmes IV. Recovery included a number of elements assigned to Marmes I that were sifted from the backdirt pile created during the excavation of the trench by the bulldozer in 1965. The age of the Marmes horizon, and therefore the period of deposition of the floodplain human remains, was established by two radiocarbon dates of 9820 ± 300 B.P. (W-2209) and 9970 ± 110 B.P. (Y-2481) (Sheppard et al. 1987:122).

The importance of the floodplain human remains (i.e. their antiquity) was known from the time of their first discovery. Fryxell's investigation of the stratigraphy in the floodplain was prompted by his knowledge of the antiquity of the rockshelter's deepest cultural stratum and his desire to trace the rockshelter strata into the floodplain deposits for interpretation of the geochronology of other locations along the Snake River. Because of the awareness of the importance of any human remains present, great care was taken throughout the emergency salvage excavations in the floodplain to identify human remains and to excavate them with attention to detail.

Little information on specific methods of excavation of the human remains features found in the floodplain is available in any of the published reports. Fryxell and Keel (1969:33) describe the use of the five-foot strips of 1 mm mesh screen bags in the vicinity of Marmes I find, but it is not clear whether sediments associated with the feature were included in any of those bags. They also state that the Marmes horizon sediments were troweled and features were drawn and photographed (1969:41), but no further information on how the human remains features were excavated is provided. A series of photographs taken during the latter stages of the exposure of the Marmes I skull fragments show Grover Krantz (Figure 7.3) and Henry Irwin working with several different hand tools to make the recovery, and using small, wooden 'posts' to hold the exposed bones in place while the rest of the skull fragments were being freed from the surrounding sediments. Photographs of the other floodplain skull fragments being recovered indicate similar care was taken with those.

Results of Analysis of Rockshelter Remains

The following discussion of the human remains recovered from within the rockshelter is organized from the earliest features to the most recent based on their locations in the stratigraphic units defined by Fryxell (Fryxell and Daugherty 1962) (Figure 7.4). This presentation will not incorporate technical descriptions of the remains beyond that needed for demonstration of descriptive points regarding the burial features. Technical descriptions are available in the Krantz's (1979) and Breschini's (1979) primary reports. Of the burial features recovered from the rockshelter, Breschini's analysis (1975, 1979) provides the best information for 14 while Krantz (1979) provides the only reported analysis of the cremation hearth remains (Burial 23). For the remaining rockshelter burial features, Rice's (1969) feature descriptions are cited along with any information culled from the available field records. Breschini's summary table (1979:117) has been modified to include all rockshelter burials (Table 7.1).

Unit I/II

The cremation hearth, referred to as Burial 23 by Rice (1969:85), represents the only human remains associated with the earliest cultural stratum in the rockshelter.

Burial 23. Rice describes this feature as "a series of small hearth areas each containing ash, charcoal and charred human bone; rings of rock and rock piles are present; shell, mammal bone and chipping detritus are interspersed throughout this cremation hearth complex" (1969:85) that covered a ca. 238 ft^2 area. In addition, a wide variety of artifacts including Windust projectile points were recovered from the area of the feature although their association with the human remains is problematic (1969:85). While faunal bone was identified, the overwhelming majority of the bones and identifiable bone fragments were human (Krantz 1979:169).

Krantz notes that "The human remains from the "cremation hearth" are small fragments of mostly burned bone. Over 1000 cranial pieces have been counted after all possible restorations, and the postcranial total probably far exceeds this amount" (1979:169). While at least six individuals are represented, the fine breakage

Figure 7.3 Photograph of Grover Krantz excavating the Marmes I cranium.

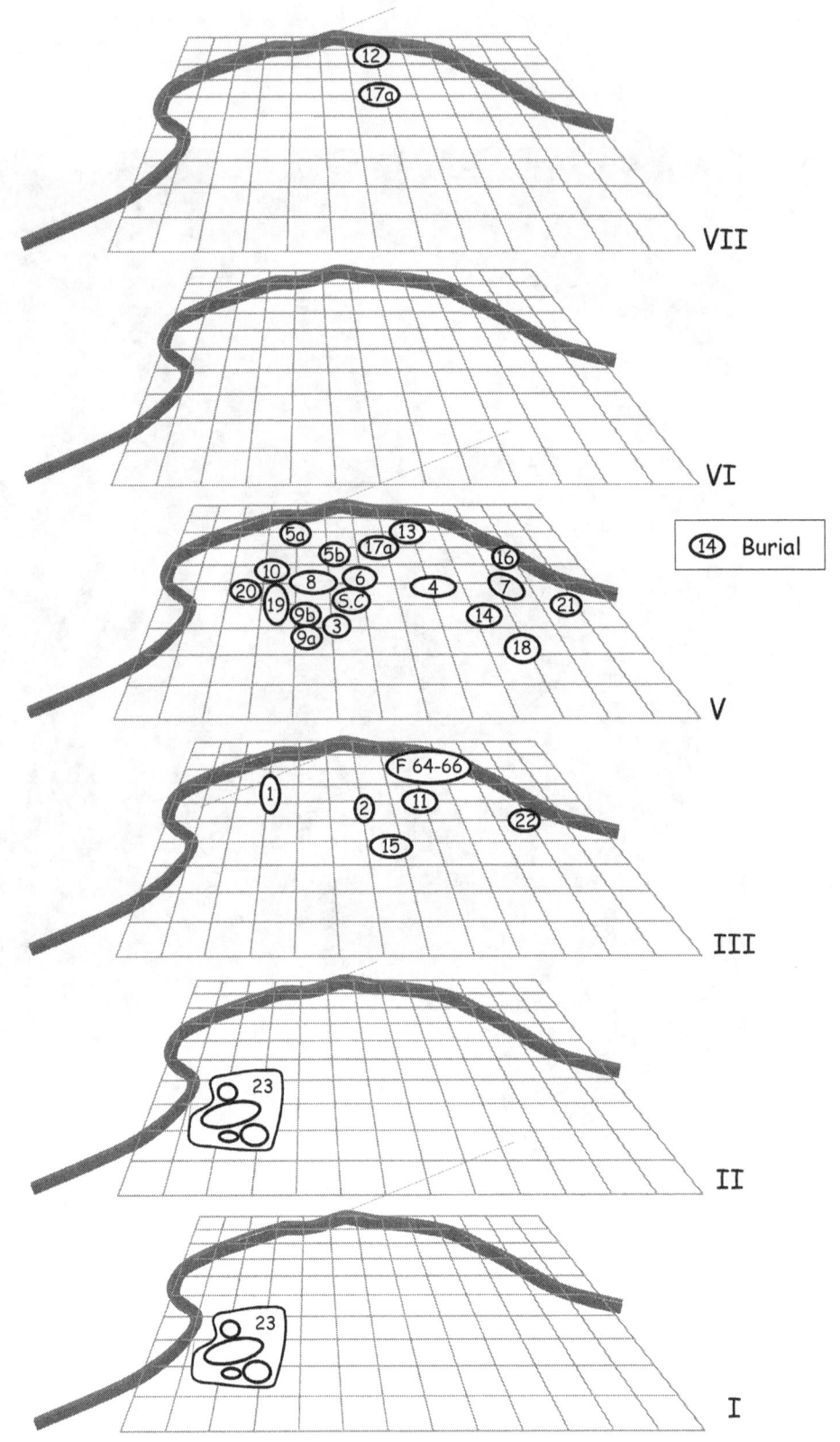

Figure 7.4 Location of burials in Marmes Rockshelter by stratigraphic unit.

Table 7.1 Human remains feature summary (adapted from Breschini 1979).

Feat. #	Age	Sex		Soil Unit	Burial Type	Burial Position	Side	Vertebral Orientation	Orbital Orientation	Ochre	Artifacts Associated
1	30-40	M?	35-40W/93-99N	III	not buried	none	face down	W-E	?	very little	9 olivella, little ochre
2	?	?	25-30W/95-100N	III	not buried	?	?	?	?	none?	none
3	16-18	F?	30-35W/85-90N	V	secondary	?	none	W-E	North	large amounts	80+ olivella, much ochre
4	adult	?	20-25W/90-95N	V	primary	semi-flexed	left	W-E	South	Yes	olivella, projectile point, ochre
5a	?	?	39.5W/101N	V	?	?	?	?	?	?	olivella beads
5b	child	?	30-35W/85-90N	V	?	?	?	?	?	?	olivella beads
6	adult	?	27-32W/90-95N	V	primary	semi-flexed	back	E-W	South	Yes	18+ olivella, projectile point, ochre.
7	adult	F	10-15W/90-95N	V	primary	semi-flexed	right	NNW-SSE	NE	large amounts	50+ olivella, 150+ obsidian flakes, 2 proj. pts., drill, bone awl, much ochre
8	35-50	M?	35-40W/93-96N	V	?	?	?	?	?	very little	4 olivella, proj. pt, scraper, little ochre
9A	?	?	35-40W/84-88N	V	?	?	?	?	?	?	olivella, 2 bear teeth
9B	6-8	?	35-40W/85-90N	V	primary	semi-flexed	left	W-E	South	very little	3 antler shaft fragments, 7 olivella, blade, little ochre
10	young adult	F	40-45W/90-95N	V	primary	semi-flexed	back?	W-E?	?	very little	7 olivella, little ochre
11	?	?	20-25W/95-100N	III	?	?	?	?	?	?	?
12	?	?	25-27W/103-106N	VII	?	?	?	?	?	?	projectile point, 2 olivella, graphite bead, worked chert fragment, bone pendant
13	?	?	20-25W/100-105N	V	?	?	?	?	?	?	?
14	adult?	?	15-20W/85-90N	V	?	?	?	NW-SE	?	?	?
15	?	?	21-26W/87-89N	III	?	?	?	?	?	?	?
16	adult	M	12-15W/95-99N	V	?	?	?	?	?	yes	olivella, 2 proj. points/2 frags, chopper, 2 knife frags, pendant, milling stone, stone/shaft polisher, ochre
17a	older child	?	10-15W/95-100N	V	not buried	?	?	?	?	?	grinding pallet
17b	?	?	28.8W/98.8N	VII	interred?	?	?	?	?	?	cradleboard hoop
18	?	?	10-13W/83-85N	V	?	?	?	?	?	?	olivella
19	?	?	39-44W/85-91N	V	interred	flexed?	?	?	?	Yes	60+ olivella, 5 projectile points, 2 atlatl weights, dart fragment, ochre
20	adult	?	43-47W/85-90N	V	interred	?	?	?	?	?	2 projectile points
21	?	?	0-5W/85-90N	V	?	?	?	?	?	?	?
22?	?	?	?	III?	?	?	?	?	?	?	?
23	6+ individuals	?	35-52W/70-84N	I/II	?	?	?	?	?	?	?
S.C	child	?	34-40W/85-90N	V	?	?	left	?	?	Yes	2 olivella beads, ochre
64-6	?	?	15-20W/100-105N	III	?	?	?	?	?	?	?

(the longest skull fragment is 64 mm) prevented further definition or assigning of cranial fragments to postcranial elements as none of these bones were complete enough to yield any measurements or observations as to their morphology (1979:169, 173). If the estimate of six individuals is correct, then about 2/3 of the skeletal bone was missing when Krantz conducted his analysis. Several hundred fragments, assignable generally as adult or juvenile based on size were tentatively attributable to certain individuals based on the degree of burning although evidence of secondary burning complicated the task. The adult skull fragments are not so completely burned as are the juvenile elements. Krantz concluded that there were at least three adults and three children (roughly between ages eight and 14) represented and he designated them Hearth I through Hearth VI, beginning with the largest adult (1979:169-172).

Krantz noted that one identifiable human bone, an apparent femur fragment, "has two clearly marked parallel cuts on it at right angles to the length of the shaft" and several additional apparent cuts roughly parallel to the first (1979:173). "From each of the major cuts the surface bone has broken away on the same side in a thin layer, so that the second cut was made inside the bone flake scar from the first cut. These could easily be taken for butchering marks except for the fact that no other such cuts have been found. Of course, the fragmentary nature of the remains may well have obscured other such cutting marks" (1979:173).

Krantz also observed that in a number of cases where pieces of bone had been glued back together the color and degree of burning was not continuous across the break, indicating that some burning occurred after some of the breakage. "In a few cases, a well burned piece fitted into some less burned pieces in such a way that it could not have originally been separated without breaking up the whole region. What this shows is that the bones were first burned, apparently whole, then…broken into many small pieces and a few of them subjected to fire a second time. While this whole operation was carried out in a small area of only a few square yards at most, a majority of the bone fragments were at some time transported elsewhere" (1979:174). Krantz suggests cannibalism as one activity that could have created this feature, but his statement is not supported. Krantz does not describe any specific features on the bones as indicative of cannibalistic activities. Nor does he make comparisons between the condition of these bones and observations made in other studies of human remains where cannibalism was concluded. Offering a conclusion of cannibalism for burned human remains should be backed up by rigorous study. Cremation as a method of treatment of the dead is known worldwide through time and is the more elegant inference in the absence of specific data that could suggest otherwise.

Unit III

Five human remains features are attributable to use of the rockshelter during the time period corresponding with Unit III. Burials 1 and 2 were found on the surface of this stratigraphic unit and were not interred in a pit. Burials 11 and 15 were interred in pits. No information could be located for Burial 22 and the information presented for this feature is considered unverified. However, a collection of human remains designated Feature 64-6 likely represents a fifth human remain feature. Burials 1 and 2 were two of the four burials moved to Pullman in casts but that had been partially excavated by 1975 when Breschini began his study of the Marmes burial casts (Breschini 1979:116).

Burial 1. According to field notes, this skeleton consisted of skull fragments, ribs, vertebrae, and various limb bones. The skeleton was found lying face down on a dense shell and bone layer immediately underlying the dense Mazama ash layer that is Unit IV. "The positioning of the bones suggested that the individual was not buried, but rather had died in the rockshelter and was subsequently covered by windblown sediments" (Breschini 1979:138). Despite many of the bones being disarticulated and somewhat scattered, perhaps due to its not being interred in a pit, one arm was described as being under the rib cage. While skull fragments from this skeleton were described in the field notes, they were not incorporated into the plaster cast as they were scattered over a large area, and Breschini was unable to locate the skull fragments for his study (Breschini 1979:137-8). Excavation of the plaster cast prior to 1975 resulted in the removal of a few of the bones but this did not impact Breschini's interpretation of the feature (Breschini 1979:137).

Within the plaster cast the sediments underlying the skeleton matched the field description of a dense midden of shell and split animal bone, with densely-packed rooffall rock in the bottom of the cast. Within these midden materials, varying amounts of cultural material occurred, including charcoal, hackberry seeds, fish vertebrae, and lithics, including a basalt projectile point and a few obsidian flakes. The sediments above the 'bed' consisted of reworked Mazama ash with few midden constituents.

Immediately around the skeletal remains, nine olivella beads were found along with a very small amount of red ochre (Breschini 1979:138).

Despite the absence of the skull and the pelvis, Breschini was able to conclude, based largely on the mandible and teeth, that the individual was a male about 35 years of age.

Burial 2. When located in the field, this burial feature consisted of hand bones and a few other bones identified as ribs and teeth. The feature was encased in plaster and transported to Pullman. A laboratory excavation of the cast was begun sometime prior to Breschini's investigation in 1975 and apparently removed the upper half of the cast although no documentation of that work has been found. Breschini excavated the remainder of the cast but found no human remains (Breschini 1979:140).

Fryxell and Daugherty (1962:16-20) described Burial 2 as fragmented and the bones as jumbled. Similar to Burial 1, these remains were found directly atop the Unit III shell and bone midden and did not appear to have been interred. Breschini found "large amounts of shell and split animal bones, some charcoal and snail shells, a few lithic materials, and one basalt projectile point fragment" in the remainder of the cast but could not be sure that they were in association with the skeletal material (1979:140). Breschini also notes that the field notes mention that a few olivella beads were found in the immediate area of the human hand (Breschini 1979:140). Rice (1969:78) notes that two projectile point fragments, a crude biface, and a cobble spall scraper were recovered in association with the human remains. Rice also refers to this feature as being "adult" (1969:78).

Burial 11. A definite interment, the skeletal remains of this feature consisted of only the skull and teeth of an infant buried in the Unit III cultural midden and overlain by the midden materials and the Mazama ash layer. The feature was excavated in the field but salvage of the skeleton was impossible because of the poorly preserved condition of the few delicate bone fragments remaining" (Fryxell and Daugherty 1962:17). The burial was found to have "five matched but crudely flaked basalt blades placed with it" (Fryxell and Daugherty 1962:17). Rice also notes that olivella beads and a projectile point were recovered in association with the feature (1969:81).

Burial 15. When encountered in the field in 1964, this burial feature was described as the "oldest burial to date" based solely on its apparent stratigraphic position (Marmes field notes, 7/15/64). Subsequent examination revealed a burial pit dug through Unit II into Unit I and including fill from Unit II. The skeletal remains consisted of a skull, vertebrae, ribs and long bones found in association with duck bones and a cryptocrystalline core, all lying on three sections of matting (1969:82). An additional unidentified bone found in an adjacent unit two months later was attributed to this feature and was put in the Burial 15 level bag (Marmes field notes, 9/8/64).

Rice also notes that shell from the burial area was radiocarbon dated but may have incorrectly cited the resulting date (7870 ± 110 B.P.) and sample number (WSU-211). Sheppard et al. (1987:122) indicates that WSU-211 was collected from Unit I and returned a date of $10,750 \pm 300$ B.P. Rice may have meant sample WSU-210, collected from Unit III, that returned a date of 7870 ± 300 B.P. (Sheppard et al. 1987:122). However the provenience for this sample as reported by Sheppard et al. (1987) (unit N87.5/W40, depth = 89.7') does not match the provenience of Burial 15 as reported by Rice (unit N87-89/W21-26, depth = 87.6-88.07'). Sheppard et al. (1987:123) note that sample WSU-210 was initially incorrectly reported as being from stratum Unit II, but that error would appear to have no bearing on Rice's reference to WSU-211. Rice may be incorrect that shell from this feature was dated. Another possible source of confusion may be that a second concentration of human bone was excavated the same week as Burial 15 (see Feature 64-6 below) and the minimal field records kept in 1964 do not readily distinguish between the two events. However, Rice does not report on this second burial feature and its provenience also does not match any of the dated contexts reported in the American Antiquity article (Sheppard et al. 1987).

Feature 64-6. This concentration of human bone fragments is only recorded in the unit excavator's field notes. It was never identified as a burial 'feature' in the minimal records from the field and unlike other similar situations Rice did not give the concentration a burial number during his research for the 1969 report. The field staff may have attributed these remains to Burial 13, which was excavated from an adjacent unit the year before, but this is not stated. This is somewhat doubtful though as there was little continuity of excavators in the first three years' excavations and the 1964 participants may not have been aware of the location of Burial 13. The concentration is described only as "Several human bone fragments, in aluminum foil, no artifacts" (Marmes field notes, 7/15/64). No unit drawings have been found but very few such graphics were made in the field in 1964.

Burial 22. Rice (1969) represents the only reporting of this burial feature, the provenience for which places it approximately one foot below Burial 16 in the same excavation unit. The field records were consulted for the current report and while we found the field notes corresponding to the day when the provenience Rice lists for this feature was excavated, there is no description of any bones being found. Rice was unable to provide any description of the feature but lists the skeletal remains as tarsals, phalanges, and a metacarpal. Rice also reports that the remains were recovered in probable association with two bone point fragments, and a drill or graver (1969:84). The field notes for this provenience list two different artifacts as being recovered from this location. It is unknown what records Rice used to report Burial 22 leaving this feature's presence and description unverified.

Unit V

Unit V by far contained the greatest amount of human remains of any of the stratum. Nine primary burials, one secondary burial, and eight features with human remains of uncertain burial status (i.e. no information available in the field records) were located in this stratum; however, see also the discussion of Burial 17a in Unit VII below. Three of these features are shown in Breschini's Table 1 (1979:117) as associated with Stratum IV, however it is clear from the text that he was in agreement with Rice (1969) that they were intrusive into or through Unit IV from Unit V. It should be noted that the decision in 1968 to remove by backhoe the upper strata in the west portion of the rockshelter to enable the concentration of excavation below the Mazama ash layer (Unit IV) may have unknowingly removed additional human remains.

Burial 3. The field notes describe this as a small feature that consisted of skull fragments, one tooth and arm bones. This feature was encased in plaster and transported to Pullman. Breschini found this cast to be in the best condition of all the burial casts and contained many more elements than were reported in the field. He also concluded that the feature was a secondary interment: "The crushed skull was situated at the east end of the cast, and the bones were arranged within an area of about 25 x 55 cm, extending to the west of the skull. Not only was this area too small in which to bury an adult body, but the arrangement of the bones would be impossible for an articulated body" (1979:124). The skeletal material lay on a layer of undisturbed pumicite. The skeletal material had been interred with ca. 80 olivella shell beads with no apparent orientation, suggesting that "they were not a part of an item of clothing which had disintegrated, but rather that they were burial goods of some other type" (1979:124). In addition, a 1- to 2-cm layer of red ochre lay over a large portion of the skeletal material, both in powder form and as small pebbles.

The Burial 3 cast also contained some 200-300 seeds that Rice (1972:159-160) tentatively identified as probably chokecherry but Gustafson suggested might be hackberry (Breschini 1979:124-6). Breschini considered this number of seeds to be relatively large compared to other burial feature casts he excavated (1979:124). Mastroguiseppe (see Chapter Thirteen) emphasizes the consistent presence of hackberry in the rockshelter deposits but discounts the significance of the minimal presence of native cherry pit fragments. While there were other midden materials in the cast of Burial 3, including shell, charcoal, 34 'undistinctive' lithic flakes, "relatively few" animal bones, and rock (Breschini 1979:124-6), given the large number of seeds and the formality of preparation of this feature, it is reasonable to conclude that they were intentionally included in the burial pit.

Rice notes that in addition to the items excavated from the cast by Breschini, other olivella beads, a utilized flake, a medial fragment of a projectile point, and a 'chokecherry' pit also were recovered in the field in association with the feature (Rice 1969:78). Rice further notes that there were "circular charcoal deposits" in the feature (1969:78) but there is no further information on these deposits.

Breschini concluded that Burial 3 consisted of an almost complete individual between 16 and 18 years of age and probably female. He noticed no pathologies other than greater tooth wear than expected for a person this age. Breschini suggested that greater tooth wear may have been a result of contamination of food resources by the Mazama ash (1979:127). Because the burial feature retained so many of the skeletal elements despite being a secondary burial, Breschini suggested a primary burial may have been disturbed by a later pit excavation (burial or storage) and then given a ceremonial reburial (1979:127).

Burial 4. This feature was transported to Pullman as a plaster cast and excavated in the laboratory in the fall of 1970. The only records Breschini located regarding that work were photographs that show the skeletal remains were those of a semi-flexed adult of indeterminate sex. One photograph showed a projectile point adjacent to the upper end of a femur. The field notes describe a burial interred in pure pumicite with olivella beads and red ochre in association. The notes also suggest that a scatter of rocks above the burial may have represented a four-foot long cairn (Breschini 1979:144; Rice 1969:78). Rice also notes the recovery of red ochre and olivella beads in association with this feature, as well as an atlatl weight and four basalt projectile points (1969:79). Rice included photographs of the feature encased in plaster in the 1969 report. No other information is available for this feature.

Burial 5. There is some discrepancy in the reporting of this feature that suggests that two different features are being referenced. Breschini refers to the field notes description of the feature consisting of "small limb bones" with measurements that indicate they are from an infant or very young child. The burial was associated with "great quantities" of olivella beads, some shell, and a few faunal bones (1979:144). The field notes reference unit 30-35W/85-90N, the same designation for this burial number given in the report of the 1962 field season (Fryxell and Daugherty 1962:12). Whereas, Rice (1969) reports the provenience of Burial 5 as N101/W39.5 at a depth of 94.01' and describes the skeletal remains as "skull fragments and tooth" (1969:79). Breschini notes that bones were found at the location given by Rice but they were not encased in plaster (1979:144). Rice does not indicate that his Burial 5 was encased in plaster although he provides no descriptive information for his Burial 5 (1969:79). None of the other burial features in Rice's report are from excavation unit 30-35W/85-90N or match the description of the infant long bones. Thus, it appears that two different human remains features are being referenced and one of the features was not assigned a burial feature number in the field. Rice's feature is noted as Burial 5a and Breschini's as Burial 5b for the purposes of this report.

Little information is available for either of the Burial 5 features. As noted above, Rice was not able to find any descriptive information on the context of the skull fragments and tooth other than that an unspecified number of olivella shell beads were in association (1969:79). Breschini was not able to find a cast for Burial 5 at WSU. He found photographs of a cast shown at the site in the area denoted in Fryxell and Daugherty's (1962) report. Breschini concludes that the cast probably was excavated in the laboratory sometime before his study began in 1975 but there are no records of that work. He speculates that the cast may have been found to not have human remains; that the bones may have been removed in the field prior to the feature being encased in plaster (1979:144-6).

Burial 6. This feature was found to extend into the sidewall of the stratigraphic control block in 1962 and was encased in plaster to hold its position as Daugherty and Fryxell did not want to excavate the control unit. In 1963, the control block was excavated as a unit and when the burial was encountered, the entire feature was excavated in the field. The skeletal remains, observed in photographs and two plan drawings from the 1963 field records, and described by Rice as "Skull, vertebrae, ribs, sternum, pelvis, femur, tibia, fibula" (1969:79), were lying with legs flexed on top of Stratum Unit III, the burial pit having been excavated through Unit IV from Unit V. The burial was of an adult and had red ochre, a leaf-shaped basalt projectile point, and up to 18 olivella beads in association. No other information is available for this feature. Breschini (1975:66) cites the orientation as westerly.

Burial 7. The field excavators first encountered "red ochre and approximately 150 obsidian flakes, some up to 2 in. in diameter" as they came down on the feature ultimately designated as Burial 7 (Breschini 1979:140). These materials, a 'leaf-shaped point', and olivella beads were found to be associated with a layer of large rocks thought to be a possible burial cairn under which a skeleton was located. A human sternum, two scrapers, two projectile points, and a drill were found nearby and are suspected to have been displaced from the burial by rodent activity of which there was considerable evidence in this excavation unit. Breschini's excavation of the resulting cast support a conclusion of the association of these materials with the burial feature in that many of the bones of the thoracic area were found to be missing. Rice notes the association with this burial of the olivella beads and the lithic tools noted above as well as a bone awl (1969:80). After excavating the feature only enough to determine the general orientation of the skeleton, the feature was encased in plaster and transported to Pullman.

Excavation of the cast in the laboratory began late in 1962 and Breschini speculates that this was probably the first of the burial casts to be opened in the lab (1979:141). There are no records of the initial laboratory excavation of this cast, but slides dated December 1962 show only a portion of the skull exposed. A photograph in Kirk (1970:19) shows the cast in a partial state of excavation, which matches the cast's condition when Breschini began his work in 1975. "The major bones had been exposed, and partially treated with vinylite resin as a preservative, but apparently no bones had been removed. It is likely, however, that some of the olivella beads had been taken out" (Breschini 1979:141).

Breschini's excavation of the cast confirmed that the burial was interred in a pit dug into the Unit IV ash layer. Overall, the feature was intact but most of the bones were too fragmentary for measurement. Many olivella beads were found with the skeleton as well as a bone shaft fragment and several obsidian flakes in the area of the skull, which would have been directly below the possible burial cairn described in the field notes. A "layer of red ochre extended from the scapula to the ilium" part of which Breschini treated with a resin and removed intact (Breschini 1979:141). Based on the field records, Rice described "An ochre deposit one inch thick" in association with this burial feature (1969:80). These two descriptions appear to indicate a great amount of ochre in association with this burial since the feature was only minimally excavated in the field, but it is the field records that form the basis of Rice's observations. It is expected that Rice would have been unaware of the amount of ochre in the cast that Breschini describes from his work in 1975.

Additional "midden constituents within burial cast 7 consisted of mussel shell, small amounts of split animal bone, hackberry seeds, some charcoal, two fish vertebrae, about a dozen small lithic flakes, and several snail shells" (Breschini 1979:141). It is not clear if the 'midden' underlies the burial and is from stratum Unit III or if it is pit fill from Unit V. Both strata were found to contain these culturally-derived materials, although in considerably different densities.

Breschini concluded that the skeleton was that of an adult, probably a female, of average development. He was not able to ascertain the individual's age beyond adult (1979:143). In addition, Breschini identified two phalanges from an infant or young child, their presence he ascribes to rodent activity. This, he notes, "was not unusual….As almost every cast contained at least one infant bone" (1979:141). Two teeth found in the cast, exhibiting widely varying wear, also were noted as probably not associated with the burial and intrusive due to rodent activity (1979:143).

Burial 8. Based on the field records, Rice described this burial as "broken up and scattered, and in a state of disarticulation" (1969:80). While the major portion of the feature was encased in plaster, the leg bones were found in an adjacent excavation unit and were collected in level bags. Construction of the plaster cast was not very successful. Perhaps this was because the feature retained little integrity, was in a pit filled with mixed volcanic ash, some rooffall rock, and other "midden constituents", and lay directly on pure ash (stratum Unit IV). When Breschini found the cast in 1975 "The sides were eroded at the lower edges, and large quantities of sediment had escaped. It is likely that some of the bones were displaced, and some may have been lost through the openings at the sides and bottom" (1979:128). As a result, Breschini's excavation of the cast found only scattered parts of a skeleton with little apparent organization and no indication of mortuary customs, such as the presence of olivella shell or red ochre, as was common in the other interred remains in the rockshelter (although Rice notes that eight olivella shell beads were recovered in the field [1969:80]).

Breschini postulated that despite the clear presence of a pit feature the individual "died in the rockshelter and was covered by natural accumulations of sediment" (1979:130). That accumulation included rock, shell, split animal bone, lithic materials, small seeds or pits, and some small snail shells (probably naturally occurring), all materials associated with the Unit V cultural deposits. Rice also notes that a scraper and two projectile points were recovered in the field in association with the feature (1969:80). A "fibrous mat was found immediately north of the burial, and was described as…lying immediately above the layer of pure pumicite" (from Marmes field notes, in Breschini 1979:128). While the mat's vertical position is the same as the base of the burial feature, its association with the feature is unclear and it was not encased in the cast.

Breschini suggests that the amount of rock in the feature could support a conclusion of natural infilling of the pit feature if increased rooffall rock at the time of deposition of Unit V corresponds with a post Altithermal cooler period. However, Fryxell and Daugherty's (1962:14) summary of the stratigraphic units for the rockshelter describes Unit V as having sparse rockfall, a point noted by Breschini as well (1979:128). [But see Chapter Fourteen for this study's conclusions] It is difficult to discount the feature as an interment given the presence of a pit feature with no other apparent feature contents (e.g., cache of shell or faunal resources) than human bones, and infilled with sediments and cultural debris that correlate with the stratum associated with the time period of the interment (Unit V). The presence of large amounts of rooffall rock could be attributable to disturbance of the feature by later occupants at a time when rock deposition had increased. But Breschini discounts that the scattered nature of the remains could be due to subsequent pit excavation by the site's occupants based on his description of a fairly intact pit feature in the cast, an observation made by the field excavators as well. Rodent activity, mentioned in his descriptions of other Unit V burial features, could be a contributing factor but the degree of scattering would seem too great to attribute as the only cause.

Breschini describes the skeleton as that of an adult male between 35 and 50 years of age. He describes the individual as "very large and powerfully built, but within the normal range of variation" (1979:131) based on the size of the tibia and fibula fragments. The skull had a "very thickened frontal region" and based on an "inexact method of estimating the cranial index....also appeared to be longheaded" (1979:131).

Burial 9a. Burial 9a represents the easternmost of three concentrations of human remains initially dubbed Burial 9. Burials 9a and 9b (the central concentration) were determined to represent two individuals and these features were encased separately in plaster casts. The westernmost concentration, which was excavated in the field, represents the posterior portion of Burial 9b. Rice does not differentiate between these concentrations in his minimal description of Burial 9 even though the 1962 report differentiates between Features 9a and 9b (1969:80).

Burial 9a appears to have been largely excavated in the field. Breschini states that the feature contained "olivella beads, scattered vertebrae, some ochre, teeth, and fragmentary human bones...." as well as two bear teeth (1979:146). The human bones appeared to "consist of scattered vertebrae, what appeared to be arm bones, and possibly some pelvic bones" and teeth (1979:132). After the field crew removed many of the bones observed in the feature, the area was encased in plaster. But when Breschini found the cast in 1975, only the bottom half of the cast remained and it was empty, apparently having been previously excavated. No records have been found of that excavation and it may be that this was one of the casts that was found to not contain skeletal material. Breschini did not locate the excavated feature contents and provides no other information on the human remains other than that they represent a distinct individual from Burial 9b (1979:134).

Burial 9b. As noted above, this feature was collected in two concentrations, one in a plaster cast, the other excavated in the field. These two concentrations represented the upper and lower portions, respectively, of a child. The lower portion is described in the field notes as articulated leg bones in a flexed position; these remains have not been analyzed but the field notes indicate a polished bone shaft was found in association. The field notes indicate the burial was interred in a pit excavated through stratum Units V and IV to the top of Unit III, a description confirmed by Breschini from his excavation of the cast (Breschini 1979:132).

Unlike with Burial 9a, no bones were removed from the Burial 9b feature area prior to the area being encased in plaster, but a blade of petrified wood is noted in association with the burial in the field notes that Breschini did not find in the cast. Breschini found the Burial 9b cast to contain the thoracic and cranial regions of a child. Based on analysis of the teeth, Breschini estimated the age of the child as between six and eight years of age. No obvious pathologies were noted and he was not able to determine the sex of the child (1979:134). The pit fill sediments included three antler shaft fragments, seven olivella shell and a small amount of red ochre, all associated with the burial. The sediments surrounding the pit contained a few small lithic flakes, a few seeds, and only a few shell and animal bones. The sediments underlying the burial contained large amounts of material commonly associated with the Unit III midden deposit, especially shell and some split animal bone (1979:132).

Burial 10. This burial feature, encased in plaster in the field, had become inverted by the time Breschini excavated it in 1975. The burial was interred in a pit intrusive into stratum Unit III and that appeared to have been capped by a large basalt rock. While the field notes indicated the presence of a portion of a skull in this burial, only the posterior portion of a skeleton was found in the previously unopened cast (Breschini 1979:134).

The burial "consisted of the articulated pelvis, legs, and feet of an adult skeleton...but all of the bones were fragmentary and very deteriorated" (1979:134). Despite this, Breschini determined that the remains were that of a female, approximately 20 years of age, who probably had been pregnant at some time in her life based on the irregularity of and pitting on the pre-auricular groove of the os coxae (1979:136). The feature rested on faunal materials associated with stratum Unit III (e.g., dense shell deposits with split animal bone) that included hackberry seeds, a small amount of charcoal, and five lithic artifacts that were not associated with the burial feature. Within the burial pit, seven olivella shell beads and a small amount of red ochre were found. A number of infant bones also were found within the burial feature but Breschini discounted that they had been interred with the adult female since they "were scattered throughout the cast, and included some of the smallest bones of the infant body" (1979:137). He cited rodent activity as the likely reason for their presence.

Burial 13. Rice (1969:82) reports only the minimal reference to Burial 13 that is contained in the field notes: "Burial 13 comprised of several teeth assorted bones. Apparently disturbed, bones fragmented, scattered" (Marmes field notes, 7/16/63). No artifacts are mentioned in association with the human remains and there is no information on whether the remains were interred in a pit or were left on a surface. No drawing of the unit or level were found in the records from the site. The additional notes on the excavation level describe typical Unit V sediments and contents.

Burial 14. Reporting of this feature suffers from a lack of information as well. The only records of this feature that could be found were a plan drawing of the human remains and a photograph of the feature encased in plaster; no corresponding field notes or feature description was found and Breschini makes no reference to a burial cast associated with a Burial 14. Rice (1969:82) was unable to find any information on what bones were found or any descriptive information on the feature. His only information is the provenience of the feature and it is apparent that his source for this information was the photograph he presents as Figure 13 in the 1969 report. Unfortunately, that photograph may suggest that there are two features identified as Burial 14.

The readerboard in the photograph shown in Rice lists the date as July 24, 1963 and the provenience as "Unit IV, Datum – 93.10, 85-90 N, 15-20 W" (1969:105, Figure 13). Other than the stratigraphic unit designation, this matches all the information Rice presents on the feature. But the plan drawing of Burial #14 found in the "1963 Field Notes" binder at WSU is dated August 22, 1963 and lists the provenience as 85-90N 0-5W and the elevation as 92.34-92.09 feet. The later date of the plan drawing and its listing of a lower elevation than the photograph, and Rice's reporting of the stratigraphic unit as Unit V rather than Unit IV all is explainable if the feature had been excavated in the field subsequent to the picture of the feature encased in plaster cast. This does not explain the different designation of the west grid point for the excavation unit but this may have been an error on the part of the photographer or the feature drawing's artist. This seems a better explanation than the presence of two burial features with the same designation given the lack of documentation.

The plan drawing is fairly well rendered allowing description of the orientation of the burial and identification of a number of elements. It appears that only the upper portion of a skeleton is present, oriented NW/SE with the cranium to the northwest. Twenty-seven discernible bones and fragments are illustrated and a number are identifiable to element. They are six vertebrae, at least five rib fragments, four long bone fragments, a single carpal, and skull fragments. Given their location within the feature, the long bones are probably of the arm(s) and the vertebrae are likely thoracic. The skull appears crushed in the drawing and includes at least a portion of a mandible. Assuming the scale of the drawing is accurate, it appears to be a small adult of undetermined sex.

Burial 16. This feature was excavated in 1964. It was first spotted in the sidewall of an adjacent unit, then excavated in whole when that unit was taken down. The field notes relate little information about the feature. They note that a "scetch of the burial" was made, and both color and black and white photographs were taken, and that one photograph was taken "with most bones exposed" (Marmes field notes 1964:20). Another photograph was taken "after bones on that level were removed and scapulas clavicles ect. were exposed. Bones were preserved w/ elmer's glue and wrapped in toilet paper and foil" (Marmes field notes 1964:23). Walter Birkby was on site and "From the pubic symphasis he was able to sex & age the burial. It's a male in its 20's" (Marmes field notes 1964:23). The burial was in a pit excavated into Unit IV from Unit V.

Rice noted that teeth, phalanges, the skull, a femur, humerus, ulna, scapula, and the left humerus were the bones encountered. He had no data on the burial feature but lists many artifacts recovered in association, including olivella shell beads, red ochre, two projectile points, two projectile point fragments, a chopper, two basalt knife fragments, a stone pendant, a milling stone, and an abrasive stone/shaft polisher (1969:83).

Burial 18. This feature is described only in the field notes and with a poorly rendered plan drawing of the excavation unit level directly above the level on which this minimal burial feature rested. The feature drawing indicates that the skeletal remains, consisting only of skull and long bone fragments, protrude up from stratum Unit IV into a level excavated as Unit V. These two raised areas are shown as being approximately one foot apart with the west concentration including two long bones broken into five or six pieces and the east concentration including one additional long bone fragment and a number of skull fragments. The feature was first found when a shovel full of dirt from the area of the east concentration was found to include a clod of dirt "that had most of a skull cap clinging to it." The field notes conclude that "It was evident that it was not a whole burial" and that other "bone chips" were found in the level (Marmes field notes, 8/6/63). The next level produced olivella shell beads assumed to be "associated with the bones in the level above" (Marmes field notes, 8/7/63). There is no description of a burial pit and no accounting of the number of bone fragments or shell beads.

Burial 19. This burial feature was designated as "burial northeast" in the field in 1963; Rice provided the Burial 19 designation during his research for the 1969 report. Both Burials 19 and 20 were found near where Burial 5 was located in 1962. According to the field notes, 'Burial NE' apparently was placed in a cast but after most of the feature contents were excavated in the field. Breschini could not locate a cast associated with this feature in 1975, but his review of the field records indicate that the cast likely would have included "only a few foot bones, and possibly the lower legs of a scattered burial" (1979:146). Notations and several sketches in the 1963 field records indicate the remains were interred in a pit that was dug through stratum Unit IV and into Unit III.

Rice lists the recovered bones as a "skull, ribs, femurs, tibia" of an adult; the field notes indicate that a femur, tibia, and ulna were in a flexed position. Rice notes that a single child femur also was recovered adjacent to the adult femur and charred human bones also were observed. He also indicates that a "charcoal lens containing burned Olivella shell beads and red ochre was found" (1969:83-84), but it is unclear if the charred human bones were associated with the lens. At least sixty olivella shell beads, five projectile points, two atlatl weights, and a dart shaft fragment were recovered in association with the feature. Rice also lists choke cherry pits as having been recovered from the feature (1969:84) as reported in the field notes; as noted above, these may be hackberry seeds that were misidentified in the field. The skeleton was definitely interred as the field notes describe considerable differences between the sediments around the burial and the sediment in the rest of the excavation unit at these levels.

Burial 20. Excavated from the same unit as Burial 19, Burial 20 was originally designated "burial northwest" in the field. As with Burial 19, the field notes and a plan drawing of the excavation unit show "Burial NW" in a cast, but Breschini was unable to locate a cast associated with this feature. He concludes that it probably was "opened in the laboratory and classed as a "non-burial, but no data could be located concerning its laboratory excavation or contents" (1979:146). He notes that the field records indicate that a skull fragment was known to have been encased in the cast. Rice's description lists the recovered bones as "skull, radius, ulna" of an adult in a definite burial pit intrusive into Unit III from Unit V. Two projectile points and choke cherry (or hackberry) pits also were found (1969:84).

Burial 21. This feature was excavated in 1963 and reported in the field notes as containing only a partial skeleton including ribs, vertebrae, pelvis, arm and leg bones, and phalanges, but no skull. The feature is described as very disturbed and no other feature contents are reported. The bones were treated with 'resin' prior to being boxed up for shipment to the lab. The field notes say that a sketch was drawn but no such drawing was found in the site's records at WSU.

Small Unnumbered Cast. "S.C." is the designation that Breschini gave to the small unnumbered cast he found in the lab in 1975. This feature was recovered in 1962 just north of Burial 9a and also was quite close to Burial 19 recovered in 1963. The field notes describe the contents of the cast as a crushed skull and teeth. Breschini's excavation of the cast revealed "a crushed skull that was in most places little more than orange colored soil, and no parts could be recovered intact" (1979:143). Based on several teeth found in the cast, Breschini concluded the skull was that of a young child. Two olivella beads and moderate amounts of red ochre also were present (1979:143).

Unit VII

Two burial features are attributable to stratum Unit VII in the rockshelter.

Burial 12. This was the first burial encountered in the 1963 field season. The feature was described as disturbed with most of the bones broken and no longer articulated, and with shell and animal bone mixed in the feature sediments. The field notes mention the presence of the skull, an ulna, phalanges, and a tarsal. Small fragments of human bone, including a "burned finger bone" also were found in the next two levels (Marmes field notes, 7/12-

7/16/63). Associated artifacts include a projectile point, two olivella beads, a graphite bead, and a worked chert fragment that Rice refers to as a "crude biface" (1969:81). Rice also includes a bone pendant in association although there is no mention of this artifact in the field notes.

Burial 17. There may be two Burial 17 features. A burial identified as Burial 17 in the field notes from the 1964 season describes a burial feature in N95-100/W10-15 and sitting atop stratum Unit III (now Burial 17a); this feature is likely associated with Unit V. Rice reports a Burial 17 at N98.8/W28.8 in stratum Unit VII (now Burial 17b); the only other information offered is that wood fragments were found in association that Rice believed are the remnants of a hoop from a cradleboard (1969:83) (see Figure 13.1 f in Chapter Thirteen). Despite the minimal information available, these two descriptions lead to a conclusion that two different burial features were described.

The field notes describe Burial 17a as "not a regular interment burial. Only the skull, left femur & one other long bone frag…were in close association. Several other human bones & a variety of animal bones were scattered about. Walt Birkby look at it afternoon and identified the vertebrea & phlange as belonging to a late adolecent" (Marmes field notes, 8/11-8/13/64). A left humerus, scapula fragments, and teeth also are mentioned. A small grinding pallet was found in association.

Results of Analysis of Floodplain Remains

Four concentrations of human remains were located in the floodplain, all associated with the Marmes horizon (Fryxell and Keel 1969). Krantz provides the only published analysis of these remains in his Northwest Anthropological Research Notes article (1979:159-174). The four individuals represented by these features were numbered Marmes I, Marmes II, Marmes III, and Marmes IV in the order in which they were encountered in the field. All highly fragmentary, Marmes I includes some postcranial elements, the other three are represented only by cranial fragments. All occur on old ground surfaces, not in pits.

Marmes I. As noted above, Tadeusz Bielicki briefly described the portion of Marmes I that had been found by June 1968 for an American Antiquity article the same year (Fryxell et al. 1968b), then wrote up his analysis of those fragments upon his return to Poland. Krantz includes Bielicki's submittal in its entirety in the 1979 article (Krantz 1979:160-162). Bielicki placed the age of Marmes I as probably in the late teens or early twenties but was unable to establish the sex of the individual noting only that it was suggestive of a rather small braincase. The skull measurements possible given its fragmentary nature indicate the individual was a fully modern homo sapien with the "rather prominent, flaring cheek bones" expected of a person of Mongoloid stock (Krantz 1979:161).

Bielicki cautioned that his inferences had to be viewed as "purely presumptive - pending further possible discoveries of other parts of the cranium" (Krantz 1979:161-162). And despite the subsequent discovery of "about 20 pieces of cranial vault, an incus bone, many tooth fragments, a large piece of the chin region, and about 100 tiny fragments of bone which cannot be identified but most likely are from this skull" (1979:162), Krantz's analysis of all of the Marmes I fragments fully confirmed Bielicki's observations (1979:162). In particular, Krantz noted the presence of a shovel-shaped incisor, an indicator of a "Mongoloid type of face", and his measurements yielded a cranial index of 84 thus barring a description of longheaded (1979:162-163). Based mostly on his impressions, Krantz regarded Marmes I as female. A number of postcranial bones also were available to Krantz including pieces of ribs, vertebrae, a carpal, and possible long bone fragments although all were so fragmentary that no measurements or observations could be made. However, he noted that their presence "indicated that at least a large part of a human body, not just the head, was left at this spot…" (1979:163). A number of small fragments of two burned antler rods, originally called bone rods (Fryxell et al. 1968b), were recovered in possible association with the Marmes I remains.

Additional human remains, described as "Skeletal material of the original find…" (Fryxell and Keel 1969:29), were recovered while sifting the backdirt pile created by the bulldozer in 1965. While it can't be certain that all of the human remains recovered in this way came from the Marmes I feature area, since this was the only human remains feature exposed in the floodplain at the time that the sifting was done, and because the results of the floodplain excavations found only four discrete human remains features, it is likely the recovered human bone fragments did come from the Marmes I human remains. However, it is not known if these remains were included with the Marmes I remains collected in situ.

Marmes II. Marmes II consisted of "…the greater part of the frontal squama, in pieces, and parts of four upper permanent teeth" of a child's skull (Krantz 1979:163, 167). Almost all of the outer table of bone was missing from the frontal fragments, making restoration difficult and measurements impossible. Four pieces of tooth were identifiable as from an incisor, a canine, a premolar, and a first molar.

"All but the molar were clearly unworn on the occlusal surfaces, while the molar showed slight wear. The roots of the molar appeared somewhat developed judging from the dentine thickness in the neck region, but the other teeth evidently had little root development. An age of about six years may be inferred from the teeth and is fully in accord with the frontal development." [Krantz 1979:167]

Krantz also noted that a lateral margin from the incisor fragment clearly showed shoveling morphology (1979:167). Little else could be offered regarding the Marmes II remains.

Marmes III. The remains of this feature and about 500 pounds of surrounding matrix were brought back to WSU for excavation. The identifiable bones were found to be seven fragments of a skullcap without any of the frontal bone. In addition, an upper left medial incisor was found in dirt inside the skull and hundreds of tiny bone fragments (<2 mm in size) were found in the surrounding sediments. Krantz concluded the skull fragment represented a male of similar age to Marmes I who had sustained and survived an injury to the left parietal. Krantz was able to estimate a cranial index range of 76-80 despite the lack of the frontal. The tooth fragment had a strongly shoveled appearance.

Marmes IV. Marmes IV was a minimal portion of a skull, consisting of only three fragments of vault, each one square inch (ca. 2.5 cm) or smaller. While none of the fragments overlap with the crania from Marmes I, II or III, the fragments were considered representative of a distinct individual based on their recovery from an excavation unit somewhat north of the others and "their condition of preservation (not burned but very strong)….The bones are clearly of an adult, but no clear age or sex determination" was possible (Krantz 1979:169).

Conclusions

There were at least 38 individuals encountered as concentrations of human remains during the excavations at the Marmes site. Fourteen concentrations within the rockshelter are identifiable as burials or intentional interments. Burial features were conservatively defined by this study. Many of the other concentrations of human remains found in the rockshelter probably also were interred, but later disturbances scattered the bones and made it hard for excavators to identify pits, or the excavators just failed to note such evidence in the site records. In some cases, minimal concentrations of human remains (e.g., a humerus, scapula fragments, and teeth [Human Remains Feature 17a]) are found with possible grave goods (e.g., Olivella shell, formed tools), and probably are the minimal remnants of interments. But without more evidence in the records from the excavations, an assertion that such concentrations represent burials cannot be made. Especially since there are human remains features clearly documented to not have been interments.

A minimum of six individuals are represented in the highly fragmented human remains in the cremation hearth in Stratum I/II in the Rockshelter (Krantz 1979). However, given that there were over 1,000 cranial fragments left over after all possible restorations had been made, it is likely that more than six individuals were cremated there. Four individuals were identified in the minimal human remains found in the floodplain excavation area. Additional human remains may exist in areas of the floodplain that were not excavated. Fourteen additional individuals are the minimum number attributable to other concentrations of human remains within the rockshelter.

In the rockshelter, five human remains features predate the Mazama ash fall (Stratum IV), not including the individuals represented in the cremation feature. A sixth, Human Remains Feature 22, also probably dates to before the Mazama ash fall based on the elevation of the recovery location, but there is very little information available to substantiate its vertical provenience. Two of the pre-Mazama human remains features are definite interments and included grave goods. The remaining 12 burial features post-date the Mazama ash fall and all included grave goods, though less than half included red ochre. Olivella shell beads, formed tools, and other cultural items were found in association with a number of the human remains features that were not considered burials. They may be grave goods if the features were interments, or they may have been items on their person at the time of death in the rockshelter. For example, Human Remains Feature 1 was lying face down on the surface of Stratum III and was covered with Mazama ash (Stratum IV); these remains were definitely not in a burial pit. A basalt projectile point, nine Olivella shell beads, and a few obsidian flakes were found in association with this feature. The shell beads may have represented a bracelet or part of a necklace, while the point and flakes could have been part of a tool kit carried in a pocket or pouch.

8
LITHIC DEBITAGE AND FORMED TOOLS

Terry L. Ozbun, Daniel O. Stueber, Maureen Zehendner, and John L. Fagan

The lithic artifact assemblage of tools and debitage recovered from the 1960s excavations at the Marmes site (45FR50) was analyzed by Archaeological Investigations Northwest, Inc. (AINW), under contract terms specified by the Confederated Tribes of the Colville Reservation (CCT). A total of 10,793 artifacts, including 9,230 pieces of the debitage and 1,563 tools were analyzed for information on reduction technology and function. In addition, twenty of the obsidian artifacts were selected from the assemblage and submitted to Northwest Research Obsidian Studies Laboratory for X-ray fluorescence analysis to determine volcanic glass trace element composition and geological sources. Also, twenty-nine artifacts and four soil samples were analyzed for possible blood residues.

This analysis encountered a number of difficulties related to the excavation methodology. Sediments excavated from the floodplain in the 1968 field season were water-screened through 1-mm plastic mesh, resulting in recovery of much smaller artifacts by comparison with the 1962-1964 excavations. The 1968 excavations in the rockshelter were dry-screened with ¼-inch (6-mm) wire mesh. Unfortunately, the database for the 1968 rockshelter artifacts does not retain correlations with the stratigraphic layers defined in the initial work, although since the sediments above Mazama ash were removed for this work, they presumably represent Strata I-IV. Pit features associated with occupations of the shelter during the various periods represented by the stratigraphic layers intruded on earlier deposits and caused some degree of mixing. Some of the pit features contained human remains, and lithic materials from these features have been excluded from this analysis. Some items are reportedly missing from the original collections and some items in the collection have incorrect or incomplete provenience information (see Chapter Two). The analysis data presented here includes all of the lithic artifacts (debitage and tools) extant in the collections delivered to AINW without any sampling. Proveniences listed here are those provided with the collections according to the CCT's revised database (see Chapter Two, Appendix A) and information on the artifact bags.

All of the 10,973 lithic artifacts in the collection delivered to AINW were analyzed for technological attributes. These data, along with a list of technological abbreviations used in the AINW database descriptions are presented in Appendixes G and H of this report. The results of AINW analyses are further described in the following text, tables and figures. The analytical discussions in the body of this report emphasize materials securely identified with the major stratigraphic units defined for the rockshelter and the floodplain. These ten Marmes site subassemblages are materials associated with the Harrison and Marmes Horizons on the floodplain in front of the rockshelter and Stratum Units I-VIII within the rockshelter. These ten subassemblages exclude a large number of the artifacts in the collection due to a lack of stratigraphic provenience information in the site's records (Table 8.1).

Excluding about one-half of the tools and one-third of the debitage from the ten designated subassemblages due to poor provenience information was of great concern. Many of the projectile points recovered, such as those shown in photographs (Figure 8.1) found in the site records, are no longer in the collection. AINW attempted to assign provenience to these artifacts (see Chapter Two) but ultimately determined that strata could not be confidently assigned without additional information. Perhaps a thorough and painstaking study of field notes, previous reports, and stratigraphic information would allow assignment of some additional artifacts to strata, but this task is outside of the scope of this analysis. Technological analysis data for the artifacts excluded from the ten subassemblages is provided in Appendix H of this report so it is available if others want to make a detailed study of provenience for any of them in the future.

Since the stratigraphically designated subassemblages from the site are so important for culture-historical modeling, and since a lot of the previously published discussions of the site rely on secondary sources of information for the stratigraphic correlations, a conservative approach to artifact stratum assignment was

Table 8.1 Excavation contexts for analyzed lithic assemblage.

EXCAVATION AREA *	WSU** STRAT UNIT	SUBASSEM-BLAGES DISCUSSED IN TEXT	TOOLS	DEBITAGE	TOTALS
FLOODPLAIN	NONE (1968)		17	197	214
FLOODPLAIN	HARRISON	HARRISON	9	315	324
FLOODPLAIN	HARRISON 3&4	HARRISON	0	284	284
FLOODPLAIN	HARRISON 5	HARRISON	0	4	4
FLOODPLAIN	MARMES	MARMES	2	64	66
FLOODPLAIN	MARMES 1	MARMES	0	1	1
FLOODPLAIN	MARMES 1&2	MARMES	0	19	19
FLOODPLAIN	MARMES A	MARMES	0	202	202
FLOODPLAIN	MARMES II	MARMES	0	2	2
FLOODPLAIN	MARMES SURFACE	MARMES	0	11	11
FLOODPLAIN	SPLIT		0	5	5
ROCKSHELTER	NONE (1962-64)		598	2,544	3,142
ROCKSHELTER	NONE (1968)		155	680	835
ROCKSHELTER	I	STRAT I	64	321	385
ROCKSHELTER	I & II		19	20	39
ROCKSHELTER	II	STRAT II	128	364	492
ROCKSHELTER	II & III		16	43	59
ROCKSHELTER	III	STRAT III	191	896	1,087
ROCKSHELTER	IV	STRAT IV	17	150	167
ROCKSHELTER	IV & V		0	10	10
ROCKSHELTER	V	STRAT V	79	859	938
ROCKSHELTER	V & VI		1	13	14
ROCKSHELTER	VI	STRAT VI	85	710	795
ROCKSHELTER	VI & VII		2	5	7
ROCKSHELTER	VII	STRAT VII	145	1,289	1,434
ROCKSHELTER	VII & VIII		0	16	16
ROCKSHELTER	VIII	STRAT VIII	25	198	223
UNKNOWN	UNKNOWN		10	8	18
TOTALS			1,563	9,230	10,793

* Floodplain materials from less than 60 North in excavation grid; Rockshelter materials from 60 North, or greater, in excavation grid.

** From Washington State University database "Comments" and "Level/Strata" fields.

Figure 8.1 Digital enhancement of photographic prints of projectile points from the Marmes site (found in Marmes site records at WSU Museum of Anthropology archives). The quality of the original prints is poor, but the size and variability of the point collection originally excavated from the site is demonstrated.

selected. That is, although many artifacts are excluded, the artifacts within the subassemblages have clear associations based on field assignments preserved in the curation records. Selection of these subassemblages for more detailed contextual analysis and discussion may help to clarify the sequence of culture-historical developments represented by the Marmes site deposits.

Despite the problems in assignment of specific provenience designations for many of the artifacts, an attempt was made to use the excavation grid unit data to define possible activity areas represented by the ten subassemblages within their associated strata. For this purpose, the distribution of lithic debitage and tools were plotted on the block excavation grid, along with information on the locations of features. Some patterns emerged, and these are discussed and illustrated below. Potential problems with interpretation of these patterns involve the nature of the samples included. Many of the grid units within the excavation blocks for each stratum did not include any artifacts. Whether these negative results represent absence of sediments from the stratum of a grid unit, absence of artifacts from the stratum within a grid unit, loss of provenience data, or some other vagary of data transformation is unknown. Certainly, the presence of intrusive pit features plays a role in the distributional patterns for strata that underlie them.

Lithic analysis methods were chosen to address research issues identified in the contract with the CCT. These research issues involve technological patterns evident at the site especially with regard to tool production, use, breakage patterns, and rejuvenation, as well as functional and morphological classifications. Intrasite patterning of these artifact and assemblage attributes is also addressed. Multiple lines of evidence were collected during the lithic analyses to address these research issues.

Methods of Analyses

AINW approaches lithic analyses by modeling reduction sequences or, more broadly, *chaînes opératoires* (chain of operations) of technological systems. A *chaîne opératoire* is a succession of " . . . actions and mental processes required in the manufacture of an artifact and in its maintenance . . . (Sellet 1993:106)." It involves modeling lithic technological systems from raw material procurement, through tool manufacture, use, maintenance, and disposal. The models reference objects (artifacts), technical gestures (actions), and technical knowledge (decisions) as elements of cultural systems (Sellet 1993:107). This approach was developed mainly through experimental flintknapping and stone tool use. The experimental products and byproducts were closely compared with archaeological materials (Crabtree 1982). The AINW approach is unique in some particular classificatory schemes and definitions, but is an outgrowth of French structuralist and American processualist schools in anthropology as applied to stone tool studies.

Several scales or units of measure are applied in combination for the analyses. These include attribute, item, and aggregate or assemblage scales (Steffen et al. 1998). The pivotal unit is the item (flake or tool) type. For debitage, the flake type identifies both reduction stage and reduction method (for example late stage percussion bifacial thinning flake). The term tool is used in a broad sense meaning any artifact formed or modified by removal of mass through flaking, pecking, abrasion, or use. Tools are classified by functional types that represent a reduction stage function (for example core, blank, or preform) or a use function identified from wear patterns or form or both (for example hammerstone, knife, or projectile point). Attribute measurements are taken for the individual items (flakes and tools) so that they can be evaluated relative to each other and to the item scale. Attributes measured include raw material composition, cortex type, remnant surface type, use wear evidence, thermal alteration evidence, breakage patterns, rejuvenation evidence, and size. The aggregate scale is measured through evaluation of the quantitative relationships between the various attribute and item unit measurements within the assemblage. These are used to characterize variability within and between subassemblages or portions of the assemblage (as sorted by provenience for example).

The AINW analysis classification categories emphasize variability related to bifacial reduction and bifacial tools (Appendix G). This tendency is common in North American lithic analysis and has been criticized as a "biface bias" potentially masking evidence for non-bifacial reduction (Steffen et al. 1998:133). To compensate for the "biface bias", particular classification categories have been added for various strategies of core reduction and tool maintenance (for example multidirectional, unifacial, blade, bipolar, radial break, and burin). Pecked and ground stone tool industries are also described with specific classifications following the *chaîne opératoire* concept (cf. Schneider 1996). Although these classificatory schemes are not as well-developed as the bifacial reduction schemes, they capture a broader range of variability. AINW's emphasis on replicative experimentation also allows discovery of additional previously undescribed variability (*àla* Crabtree 1982; Flenniken 1981).

This analysis includes an assessment of the projectile points from the site and correlates the projectile point classes defined by Rice (1969) with identifiable point styles (historical types) recognized from the Lower Snake River region. The resulting point style classifications should be viewed with caution as they cannot describe the potential technological variability in the outcome of biface reduction. Reliance on ascribed styles can lead to poor interpretations of the intended function(s) of a tool. Indeed, only 24 of the 35 classes of projectile points defined by Rice (1969) are attributable to recognized styles from the region. While the historical types are presented in the discussion of this analysis (see *Conclusions* of this chapter), functional distinctions (e.g., dart, arrow) are considered more useful for interpreting site use through time.

Raw Materials

Lithic material composition was identified during the analysis according to attributes macroscopically visible. The identifying attributes included color, texture, opacity, and mineral type (Appendix G). Eleven different lithic materials were identified during analysis of the Marmes site lithic assemblage (Appendix H). These consist of igneous rocks (basalt [BAS], granite [GRA], rhyolite [RHO], and obsidian [OBS]), metamorphic rocks (marble [MAR], quartzite [QTZ]), and sedimentary rocks (cryptocrystalline silicate [CCS], petrified wood [PET], quartz [QUA], vein quartz [VQU], and sandstone [SAN]). The CCS materials were predominant (n=6,668 or 62% of the assemblage) followed by basalt (n=3,280 or 30% of the assemblage) while the other materials were represented in minor amounts (596 obsidian, 194 petrified wood, 22 quartzite, 17 rhyolite, seven granite, three sandstone, one massive quartz, one vein quartz, and one marble). During the lithic debitage analysis, three concoidal flakes of bone were identified (Marmes Horizon, Stratum VI, and Stratum VII). These may also represent bone tool manufacture or they may be byproducts of breaking bone to process it for grease.

The CCS materials identified at the site include many varieties of cryptocrystalline and microcrystalline quartz such as chert, chalcedony, and jasper. A similar array of CCS materials outcrop in the lower Snake River basin, many exposed in the canyons of the Snake River and its major tributaries, especially in Hells Canyon upstream from the mouth of the Palouse River (Reid 1997a). The CCS materials and other toolstones also occur as alluvial gravels throughout the drainage system (Reid 1997a). Although some of the source geological formations have been described, the materials vary in visual attributes, especially in grain-size, color, and opacity, such that it is not possible to attribute particular sources. Heat treatment and heat damage also contribute to the variability in color and luster. Petrified wood (PET) is a form of CCS identifiable from fossil wood grain replacement structures sometimes preserved in the material. Petrified wood occurs in interbed deposits between Columbia River Basalt group flows and elsewhere in the region (Reid 1997a).

Quartz (QUA) occurs in massive and monocrystalline forms. Locally, these rarely occur in sizes large enough for toolstones. Quartzite (QTZ) is a metamorphosed sandstone with a sugary-texture and is very hard. Quartzite occurs as alluvial gravel in the area (Reid 1997b). Marble (MAR) is a metamorphosed limestone, rare in the region, but associated with ancient marine rocks such as occur in the Blue Mountains. Sandstone (SAN) is also a rare rock type for the area since it is commonly associated with marine formations.

Basalt (BAS) was identified on the basis of macroscopic characteristics and the name basalt was used to classify a variety of fine-grained, highly siliceous, dark gray and dark greenish gray volcanic toolstones including andesite. Petrographic and geochemical analyses of toolstones in the region suggest that despite the widespread availability of basalts, the generally smaller lava flows forming andesites and dacites were selected for prehistoric toolstones (Bakewell 1993; Reid 1997a:69-70). These toolstones generally outcrop in the highlands surrounding the Snake River drainage, but are incorporated into the low elevation alluvial gravels including those near the Marmes site through geological processes. Rhyolite (RHO) is sometimes difficult to distinguish from andesite and basalt but is generally lighter in color, often flow banded, and has larger phenocrysts. Granite (GRA) or similar igneous rocks are easily distinguished by their granular texture. Granite is poorly suited to well-controlled flintknapping, but useful for some stone tool purposes.

Obsidian (OBS) is a rhyolitic volcanic glass highly valued for its flakability and edge sharpness. Geochemical data from a sample of obsidian artifacts show that it was imported to the Marmes site from non-local sources located to the south in the Blue Mountains of northeastern Oregon and Timber Butte in an adjacent area of Idaho. Local obsidian sources are unknown, although a brown glassy basalt toolstone identified in the Marmes assemblage is geochemically similar to some sources in Washington (and unlike the Beeler Ridge glassy basalt from the Joseph Uplands [Hughes 1993]) (Craig Skinner, personal communication 2000).

Reduction Technology

A detailed lithic analysis of the recovered tools and debitage was conducted to provide baseline technological information needed to model reduction technologies represented in the assemblage (Appendix H). Stone tools and debitage were examined for technological information regarding production sequences; methods of resharpening, reworking, and recycling; use of heat treatment; and selection of raw materials. Debitage was tabulated by size, raw material type, reduction technology, and reduction stage. Presence or absence of, and type of, cortex was noted. Evidence of heat damage (likely from post-depositional alteration of the assemblage) was distinguished from evidence of heat treatment (likely part of the prehistoric technological process). Appendix G lists the abbreviations used in technological analyses and defines the specialized terms used.

The manufacture, use, and maintenance of stone tools requires several steps which leave evidence on the debitage and tools in the form of flake scars and other marks. Flake scars and other technological variables were noted to determine if there were patterns represented in the data which may reflect cultural choices made by prehistoric tool producers and users. Choices made as to specific reduction strategies, methods of flake removal, and stage at which heat treatment is done, for example, are based on cultural preferences which are passed on from one person to another and reflect behavioral subsystems which are part and parcel of larger more complex cultural systems. A wide range of choice is possible in each step involving stone tool production, use, and rejuvenation. Clusters of specific attributes reflect choices or selections which represent learned cultural preferences and patterned human behavior and are indicative of specific cultural groups or traditions.

The lithic debitage was organized into categories that reflected technological information. Sets of attributes characteristic of specific reduction techniques and strategies were used to identify and classify the debitage. The most important attributes used to distinguish between reduction strategies include: flake size, amount of cortex, location of cortex, platform thickness, platform preparation methods, shape of flake, curvature of flake, and dorsal scar morphology. These were not recorded individually, but contributed to the definition of types (Appendix G).

Artifacts were identified as debitage or tools and then assigned to more specific categories related to attributes of their manufacture and use (Appendix G). Other attributes were measured separately, including size, amount of cortex, thermal alterations, and presence of certain traits such as remnant ventral surface, flake scar patterns or evidence of use-wear. The debitage, a byproduct of tool manufacture, consists of flakes, some of which were diagnostic of particular technologies or stages of reduction. The diagnostic debitage includes burin spalls, core reduction flakes, percussion bifacial thinning flakes, and pressure flakes. These classes are distinguished on the basis of technological and morphological attributes defined from flintknapping experimentation and attributes common in the lithic technological literature (Crabtree 1982; Flenniken 1981; Holmes 1919; Titmus 1985). The tools were assigned to technological and functional classes on the basis of similar criteria. Tool classes include production rejects such as cores, blanks, and preforms, and discarded implements such as projectile points, scrapers, and flake tools.

Use-wear

All of the identified tools in the analyzed assemblage were examined for evidence of use-wear. Potential use-wear attributes identified and described for this analysis include polish, rounding, microflaking, and breakage patterns (Appendix H). These were identified by microscopic examination with a hand lens and an Olympus SZ40 stereoscope at magnifications of 20-120X. The characteristics of the identified attributes were described and evaluated on the basis of comparisons with published information (Boldurian and Hubinsky 1994; Keeley 1980; Odell and Odell-Vereecken 1980; Semenov 1964; Vaughan 1985), comparisons with other prehistoric artifacts bearing traces of use-wear, and comparisons with experimentally used lithic tools. The use-wear literature and experience in making and using stone tools indicate that several factors can influence the formation and appearance of the attributes analyzed. These factors include the composition of the lithic tool (e.g., obsidian, CCS, basalt), the methods used to manufacture the lithic tool (e.g., percussion flaking, pressure flaking, abrasion), the composition of the worked material (e.g., meat, hide, wood, antler, bone, plants), the nature of the action applied (e.g., scraping, shaving, slicing, sawing, chopping), and incidental or post-depositional affects (e.g., trampling and weathering). Composition, manufacture, and post-depositional effects are attributes independent of the use-wear traces and can be evaluated for each artifact to determine each factor's potential influence on tool surface appearance. Additional contextual information is derived from comparison with other artifacts in the assemblage and the technological data obtained from them.

Polish and associated rounding was identified on several of the artifacts examined. The polish consists of localized areas of artifact surfaces that reflect light more brightly than other areas (Vaughan 1985:13) because they are worn smooth. Polish may be formed by chemical and mechanical weathering of artifacts in their depositional environment, which usually results in a relatively homogeneous distribution of the polish across artifact surfaces. Polish from use-wear is generally confined to certain working edges or surfaces.

Some edges were crushed as evident from patterns of flake scars (often very small "microscars") with step and hinge terminations. Such microflaking can occur as a result of forces exerted on the edges of tools during use, but use-wear microflaking is virtually indistinguishable from microflaking caused by other means such as abrasion during flintknapping and trampling (Flenniken and Haggerty 1979). For the purpose of this analysis, microflaking is defined as small flake scars produced as a result of lithic tool contact with other materials but not from intentional removal of individual flakes during flintknapping (i.e., may include shearing or flintknapping abrasion but not pressure flaking). The tool manufacture context is important in identifying use-wear. Microflaking patterns considered most likely related to tool use are associated with a uniform edge, have overlapping scars, and exhibit hinge or step terminations.

The location and type of breaks on lithic artifacts can be indicative of tool use. Certain breakage patterns (such as burinations initiated at the tip or base) reflect impacts from use of bifacial tools as projectile points, while other breakage patterns (such as bending fractures across the proximal section) are more indicative of use of bifacial tools as knives (Flenniken 1991; Woods 1987). Some breaks (such as perverse fractures) are generally characteristic of manufacturing errors and can serve to help distinguish byproducts of lithic reduction from tools broken in use (Johnson 1979) especially when other attributes indicative of stage of reduction and edge maintenance are taken into consideration.

Some of the observed use-wear was associated with residues adhering to the artifacts. These residues commonly appeared as fibers and resinous masses, sometimes in combination. These residues were distinguished from clearly modern residues such as clothing fibers and labeling substances (observed but not noted). Although the materials represented were not specifically identified, most of the fibers resembled bits of sinew. The resinous masses could represent a variety of materials including blood and other animal protein materials or plant compounds.

Obsidian X-ray Fluorescence

A sample of obsidian artifacts was submitted to Northwest Research Obsidian Studies Laboratory for geochemical sourcing (Appendix I). Energy dispersive X-ray fluorescence (XRF) analysis was employed to estimate the concentrations of various trace elements for each specimen. Obsidian is a volcanic glass composed mostly of silica, often with a number of trace elements such as zinc (ZN), lead (Pb), thorium (Th), rubidium (Rb), strontium (Sr), yttrium (Y), zirconium (Zr), niobium (Nb), barium (Ba), lanthanum (La), cerium (Ce), titanium (Ti), manganese (Mn), and iron oxide

(Fe_2O_3). Iron oxide is generally the most abundant, and is measured in weight percent. The other elements are measured in parts per million (ppm). The varying proportions of these elements are used as a geochemical "fingerprint" to determine from which volcanic source a given obsidian artifact originated.

Only one previous study of artifacts of the Marmes site lithic assemblage has included obsidian X-ray fluorescence (XRF) analysis (Hess 1997). Hess submitted two artifacts from Stratum Unit I and nine artifacts from Stratum Unit III of the rockshelter to BioSystems Analysis, Inc. One of these (a Windust type projectile point from Stratum Unit III) was attributed to the Whitewater Ridge geochemical source and the others were assigned to three unknown geochemical types (Hess 1997:Table A.10). The three unknown geochemical types (Unknowns 6, 7, and 8) had distinctive trace element compositions, but did not match the geochemistry of any geological sources in the laboratory's database. Two of the three unknown geochemical types have been subsequently identified by Craig Skinner of Northwest Research Obsidian Studies Laboratory in connection with the current study. Unknown 6 has been identified as the Indian Creek geochemical source and Unknown 7 has been identified as the Gregory Creek geochemical source. Unknown 8 remains unidentified. The revised data from this previous study are incorporated into the results for this report.

The current study includes an additional 20 artifacts selected from the rockshelter deposits and submitted to Northwest Research Obsidian Studies Laboratory (Appendix I). Although a fairly large number of obsidian artifacts (n=596) are represented in the collections from the Marmes site, the majority of these are too small for reliable analysis by the XRF technique. The sample was selected among artifacts (tools and debitage) greater than 9 mm in diameter and at least 2 mm thick (less than 200 potential candidates). The obsidian artifacts large enough for analysis were further selected on the basis of provenience information to complement the previous study (Hess 1997) that represented the earliest deposits in the rockshelter. That is, artifacts were selected from strata that were underrepresented in the previous study. Samples from the floodplain were excluded since it was thought that the stratigraphic and temporal relationships between the rockshelter and the floodplain were more tenuous than within the rockshelter. In retrospect, it might have been useful to have included some samples from the floodplain. Many of the potential samples from the rockshelter were from unidentified strata and therefore were excluded from consideration. The remaining pool of potential samples included only a few more than the 20 samples ultimately selected. The final selections were based on distributional considerations and an attempt to represent the broadest variety of phenotypes (variation in color, opacity, and flow banding). The resulting selection included seven projectile points and 13 pieces of debitage. Every stratum unit within the rockshelter is represented except II and IV (the ash layer).

Blood Residue Analysis

A number of lithic artifacts were selected for blood residue analysis using the cross-over immunoelectrophoresis (CIEP) method to gather additional information regarding subsistence targets in specific strata in the site. The results also can offer information to interpretation of tool functions. The CIEP method used for this study is based on techniques developed by the Royal Canadian Mounted Police Serology Laboratory in Toronto, Ontario (Culliford 1971; Newman 1990; Williams 1990). The CIEP technique uses the immune (antibody-antigen) reaction, the principle that all animals produce immunoglobulin proteins (antibodies) that recognize and bind with foreign proteins (antigens) as part of the body's defense system. The ability of antibodies to precipitate antigens out of solution is the basis of CIEP analysis (Newman 1990:56). Extracts of protein residues from artifacts or soil samples in an ammonia solution are tested against antisera from known animals. The solutions are placed on a gel substrate and exposed to an electric current, which causes the proteins to flow together. An immune reaction between the extract and the antiserum causes a precipitate to form, which is visible after being stained. CIEP indicates the presence or absence of a particular antigen, and is not designed as a quantitative test. While other types of immunoassay have been used effectively to analyze blood protein residues under various conditions, the CIEP test is particularly suitable in that it is sensitive (able to detect protein in concentrations of about two parts per million), does not require expensive or bulky equipment, is relatively fast (about 48 hours per test), and can easily and efficiently accommodate multiple samples (Newman 1990:52).

This analysis was conducted with full knowledge of the criticisms of residue analysis studies (e.g., Eisele 1994). Blood residue analysis using the CIEP method has been used for over 35 years in archaeological and other scientific (e.g., medical, legal) applications with considerable success, including for studies of several archaeological sites near Marmes Rockshelter in the Palouse River canyon (Hicks 1995, 1996; Hicks and Morgenstein 1994). Several factors can raise the level of confidence in the applicability of this method of residue analysis including careful handling of artifacts, routine testing of soils collected from the vicinity of the tested items, routine assays of antibodies, and the use of appropriate controls in the lab. The latter cannot be overemphasized in establishing the confidence in results of particular analyses. It is acknowledged that several of these factors are beyond the control of this analysis. It is unknown to what extent the tested artifacts have been handled in the years since their recovery in the site. In addition, the provenience of the control sediment samples used in this analysis are not as precise as is desirable. After analysis of a tool produces a positive result (i.e. identification of a residue to its probable species), a sample of sediment from the same location is tested to verify that the identified residue doesn't occur throughout the sediments in which the tool was found. If that test produces the same result, the residue on the tool may have resulted from being in the location of a butchering event, for example, rather than from the use of the tool to conduct the butchering. For this analysis, while control soil samples were located in the collection that approximate the provenience of the tested items, it cannot be shown that they were collected in immediate proximity to the item to be tested. This weakens the level of confidence in any positive results and caution in the use of such results for interpretation of cultural activities is appropriate.

Twenty-nine artifacts and four soil samples from the Marmes Rockshelter were analyzed for possible blood residues. Standard analysis procedures begin with extracting the residues from the artifacts and soil samples in a 5% ammonia solution. As some of the tools had evidence of handling in the form of accession numbers, the relatively less disturbed side or portion of each tool was extracted whenever possible. The artifact and soil sample extracts were then tested against a non-immune serum (NIS), in this case a solution prepared from goat serum and a 5% ammonia solution. This is done to determine if there are any contaminants or extraneous proteins that might give false positive results. The NIS is not an antiserum and the specimens should not react to it. If a reaction does occur, the extract solutions are mixed with an equal volume of a 1% solution of a non-ionic detergent to increase chemical bonding specificity and are run again. If the specimens still react to the NIS after the addition of the non-ionic detergent, any reactions of those specimens to the antisera are discounted. None of the artifact or soil sample extracts analyzed for this project reacted to the non-immune serum.

The 29 artifacts included four cobble tools (hammerstones), 11 flake tools, 10 projectile points, two scrapers, one graver, and one flaked pebble tool. Raw materials represented by the tools include obsidian (2), basalt (15), and CCS (12). The four soil samples represented control samples that had been selected from stratigraphic units that contained tools that produced positive reactions to certain antisera. The extracts from the artifact samples were tested against bear, bovine, deer, chicken, rabbit, sheep, trout, and human antisera, as well as against a non-immune serum as a control against false positives (Table 8.2). The antisera are forensic-grade and were obtained from commercial laboratories.

Results of Analyses

Lithic Technology

Floodplain Harrison Horizon. The Harrison Horizon archaeological assemblage was found in deeply buried floodplain alluvium associated with three closely related organic layers (incipient A soil horizons [A3, A4, A5]). Charcoal from these deposits has been radiocarbon dated to $10,130 \pm 300$ ^{14}C years BP and 9840 ± 300 ^{14}C years BP (Sheppard et al. 1987:Table 1). The age and stratigraphic relationships of these deposits suggest that they are approximately contemporaneous with the Stratum Unit I materials from the rockshelter and represent the Windust Phase occupation of the site. The assemblage was excavated from the floodplain in 1968 using water-screening with 1-mm mesh to recover artifacts. The Harrison Horizon assemblage includes 603 pieces

Table 8.2 List of species that react to the antisera used in the Marmes site blood residue analysis.

ANTISERUM	REACTS WITH:
BEAR	Family Ursidae: black bear, brown bear, grizzly
BOVINE	Family Bovidae: domestic cow, bison
CHICKEN	Order Galliformes, Order Anseriformes, Order Columbiformes
DEER	Family Cervidae: white-tail and mule deer, elk, moose
HUMAN	Order Primates: humans, apes, monkeys
RABBIT	Family Leporidae: rabbit, jackrabbit
SHEEP	Subfamily Ovidae: domestic sheep, bighorn sheep
SHEEP	Order Artiodactyla: ovids, less strongly with other bovids, cervids, antilocaprids
TROUT	Genus Oncorhynchus: salmon, steelhead, rainbow trout

of lithic debitage and nine stone tools (Tables 8.3 and 8.4). The majority of the debitage is small and fragmentary, reflective of the small screen mesh used in its archaeological recovery and possibly some sorting and damage related to post-depositional alluvial processes and burning (Table 8.3).

CCS materials predominate (69%) and were mainly obtained from bedrock geological sources (37 of 42 pieces of cortical debitage have primary geological cortex). A small piece of tabular CCS material was tested for use as a core (Artifact 4833) and about one-quarter of the diagnostic CCS debitage was produced in percussion core reduction (Figure 8.2), suggesting that some CCS raw material was obtained nearby and reduced through early stages of tool manufacture. Evidence for heat treatment of the CCS material is limited to one observation of differential luster (on an early-stage pressure flake) and one observation of differential color on a late-stage core reduction flake. Percussion bifacial thinning is represented by nearly one-quarter of the CCS diagnostic debitage, and a dart-sized bifacial blank was broken in manufacture (Artifact 3461). Pressure bifacial reduction accounts for over one-half of the diagnostic debitage, representing both bifacial tool finishing and maintenance. Four recovered CCS flake tools and one informal scraper indicate other processing activities in addition to CCS tool manufacture. Two of the flake tools (Artifacts 3471 and 3486) were made expediently from locally available alluvial CCS cobbles.

The obsidian artifacts (24%) from the Harrison Horizon consist entirely of debitage. The technologically diagnostic debitage represents late-stage percussion bifacial thinning and bifacial pressure reduction. The ratio of late-stage pressure flakes to early-stage pressure flakes is over 3:1 indicating a heavy emphasis on resharpening and repair of worn or broken bifacial tools, probably projectile points. The character of the obsidian debitage suggests that it was imported as finished bifacial tool forms, many probably near the end of their use-lives (Figure 8.2).

Only a small amount of basalt (7%) is represented. The eight pieces of diagnostic debitage are associated with percussion core reduction and early-stage percussion bifacial thinning (Figure 8.2). At least some of the basalt was obtained from local alluvial gravels as indicated by incipient cone cortex on one flake and on the cobble chopper tool (Artifact 7021). One basalt stemmed Windust point base with impact damage is also represented (Artifact 3060). The point base exhibits characteristic collateral pressure flaking scars on the stem and a square facet at the proximal end, probably a remnant platform surface from the parent flake blank.

The analyzed lithic artifacts are concentrated in two areas, perhaps representing activity loci. A high density of debitage and tools was identified in grid units N25/E5-10 (n=314) in the northeastern portion of the excavation block near some large boulders (Figure 8.3). A broad array of CCS debitage and the tabular CCS core were found in this area. The broken CCS bifacial blank was recovered about 10 feet away to the southwest. The obsidian debitage in these two grid units includes 42 of the 47 late-stage pressure flakes in the Harrison Horizon subassemblage, probably indicating a bifacial tool resharpening event(s). In addition to the flintknapping activities represented in this locus,

Table 8.3 Summary of debitage analysis data for the floodplain Harrison Horizon assemblage.

COMPOSITION	BASALT/ RHYOLITE		CCS/QUARTZ/ PETRIFIED WOOD		OBSIDIAN		QUARTZITE	TOTAL
TECHNOLOGY								
bipolar								0
core percussion, early	1	13%	15	11%				16
core percussion, late	6	75%	18	13%				24
biface percussion, early	1	13%	27	20%				28
biface percussion, late			4	3%	5	7%		9
pressure, early			28	21%	15	22%		43
pressure, late			42	31%	47	70%		89
SUBTOTALS	8	100%	134	100%	67	100%		209
undetermined percussion	27		97		4			128
undetermined	2		147		74			223
thermal	3		40					43
TOTAL	40		418		145		0	603
OPERATION								
platform preparation			1		1			2
bulb removal								0
alternate			3					3
edge preparation								0
grinding or abrasion								0
overshot			2					2
margin removal			3					3
notch								0
rejuvenation								0
radial break								0
remnant ventral surface			1		5			6
CORTEX								
primary geological			37					37
incipient cone	1		5					6
THERMAL								
heat damage	3		58					61
differential luster			1					1
differential color			1					1
SIZE CLASS								
2 (1.6 - 3.2 mm)	1		7		29			37
3 (3.2 - 6.4 mm)	13		245		84			342
4 (6.4 - 12.7 mm)	15		96		30			141
5 (12.7 - 25.4 mm)	9		50		2			61
6 (25.4 - 50.8 mm)	2		16					18
7 (>50.8 mm)			4					4
TOTAL	40		418		145		0	603

Table 8.4 Summary of tool analysis data for the floodplain Harrison Horizon assemblage.

	BASALT/ RHYOLITE	CCS/QUARTZ/ PETRIFIED WOOD	OBSIDIAN	OTHER	TOTALS
CORES		1			1
primary geological cortex		1			
BLANKS		1			1
PREFORMS					0
PROJECTILE POINTS	1				1
complete/proximal	1				
FLAKE/BIFACIAL TOOLS		4			4
CHOPPER	1				1
SCRAPER		1			1
TOTALS	2	7	0	0	9

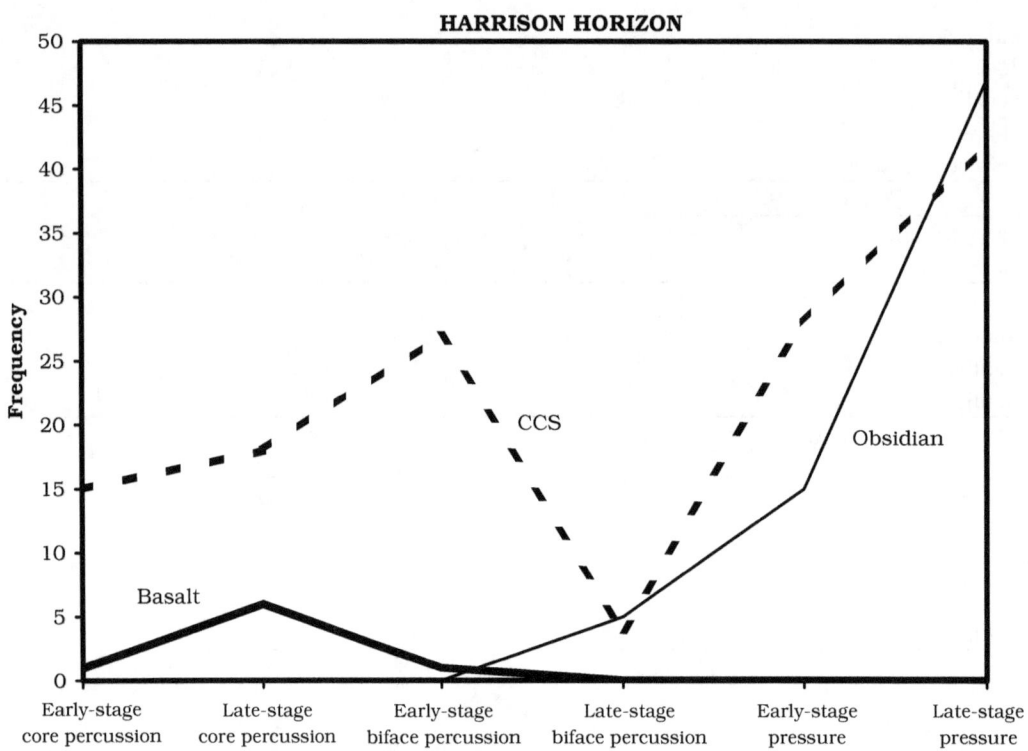

Figure 8.2 Frequency distribution of stage-diagnostic debitage from the floodplain Harrison Horizon assemblage, reduction trajectory from left to right.

tools found in these units and immediately adjacent units include the CCS scraper and three of the CCS flake tools, reflecting processing activities, perhaps with expedient tools made on that spot. The second Harrison Horizon high density locus is represented by grid unit N10/W35 in the south central portion of the excavation block (Figure 8.3). This grid unit contains 118 pieces of debitage, mainly CCS pressure flakes, representing another event(s) of bifacial tool resharpening.

Floodplain Marmes Horizon. The Marmes Horizon archaeological assemblage was also recovered from deeply buried floodplain alluvium overlying the Harrison Horizon deposits. The Marmes Horizon materials were found in association with two organic layers (incipient A soil horizons [A1 and A2]). A radiocarbon AMS sample from elk bone found in A1 produced a date of 9870 ± 50 ^{14}C years BP (Hicks 1998:156). Mussel shell fragments from the Marmes Horizon have been radiocarbon dated to 9970 ± 110 ^{14}C years BP and 9820 ± 300 ^{14}C years BP (Sheppard et al. 1987:Table 1). The radiocarbon age and stratigraphic relationships of these deposits suggest that they are slightly younger than the Stratum Unit I materials from the rockshelter but also represent the Windust Phase occupation of the site. The assemblage was excavated from the floodplain in 1968 using water-screening with 1-mm mesh to recover artifacts. The Marmes Horizon assemblage includes 299 pieces of lithic debitage and two stone tools (Tables 8.5 and 8.6). Most of the debitage is small and fragmentary, reflective of the small screen mesh used to recover it from the site and possibly some sorting and damage related to post-depositional alluvial processes and burning (Table 8.5).

The CCS materials (all debitage) again predominate but by a somewhat smaller margin (59%) than in the Harrison Horizon. Only two CCS flakes exhibit cortex, and these are of the incipient cone (alluvial) type rather than the primary geological (bedrock) type. The Marmes Horizon diagnostic CCS debitage is similar to the Harrison Horizon debitage in character and proportions of the various types and reduction stages represented (Figure 8.4). However, the relatively greater number of early-stage pressure flakes indicated more emphasis on original tool manufacture than on maintenance. Very limited evidence for heat treatment of the CCS material was observed (differential luster on three early-stage pressure flakes and one undiagnostic percussion flake). As with the Harrison materials, the relatively small representation of late-stage percussion bifacial thinning debitage may indicate that bifacial tools were made from thin linear flake blanks that required little systematic percussion thinning.

Basalt comprises 37% of the Marmes Horizon assemblage, although most of the artifacts are flake fragments. Only 13 pieces of diagnostic debitage and 2 tools are represented. The debitage is mainly associated with percussion core reduction (Figure 8.4). Four pieces of debitage and one chopper tool (Artifact 4491) have primary geological (bedrock) cortex. One piece of debitage and one cobble tool (Artifact 11685) have incipient cone (alluvial) cortex. Three early-stage basalt pressure flakes were identified, probably representing manufacture of projectile points. The obsidian artifacts (4%) from the Marmes Horizon consist entirely of debitage. The technologically diagnostic debitage (n=6) is all attributed to late-stage pressure bifacial reduction indicating a heavy emphasis on resharpening and repair of worn or broken bifacial tools, probably projectile points imported to the site near the end of their use-lives (Figure 8.4).

Although the debitage and tools were sparse in the Marmes Horizon layers a few minor concentrations of flakes were identified in the vicinity of a cluster of elk ribs and vertebrae (59 flakes in grid units N5-10/W35 [grid unit datum in northwest corner]) and near a cluster of fox bones (90 flakes in grid units N25/E5-15) (see Figure 1.19) (Figure 8.5). Both debitage clusters were technologically similar to one another and similar to the overall Marmes Horizon subassemblage pattern in proportions of the represented debitage types and compositions. The basalt cobble tool and chopper are nearest the fox bones on the eastern (grid unit N15/E20) and northern (grid unit N30/E5) peripheries of the fox bone cluster, respectively (Figure 8.5).

Rockshelter Stratum Unit I. The Stratum Unit I archaeological assemblage is from the deepest stratigraphic layer within the rockshelter containing artifacts. The thick Stratum I sediments are primarily composed of coarse rockfall from the roof of the shelter which overlie ca. 13,000 year old Missoula Flood deposits (Huckleberry, Gustafson, and Gibson 1998:96). Mussel shell from Stratum I has been radiocarbon dated to 10,810 ± 300 ^{14}C years BP, 10,750 ± 300, and 10,475 ± 300 ^{14}C years BP (Sheppard et al. 1987:Table 1). The artifacts

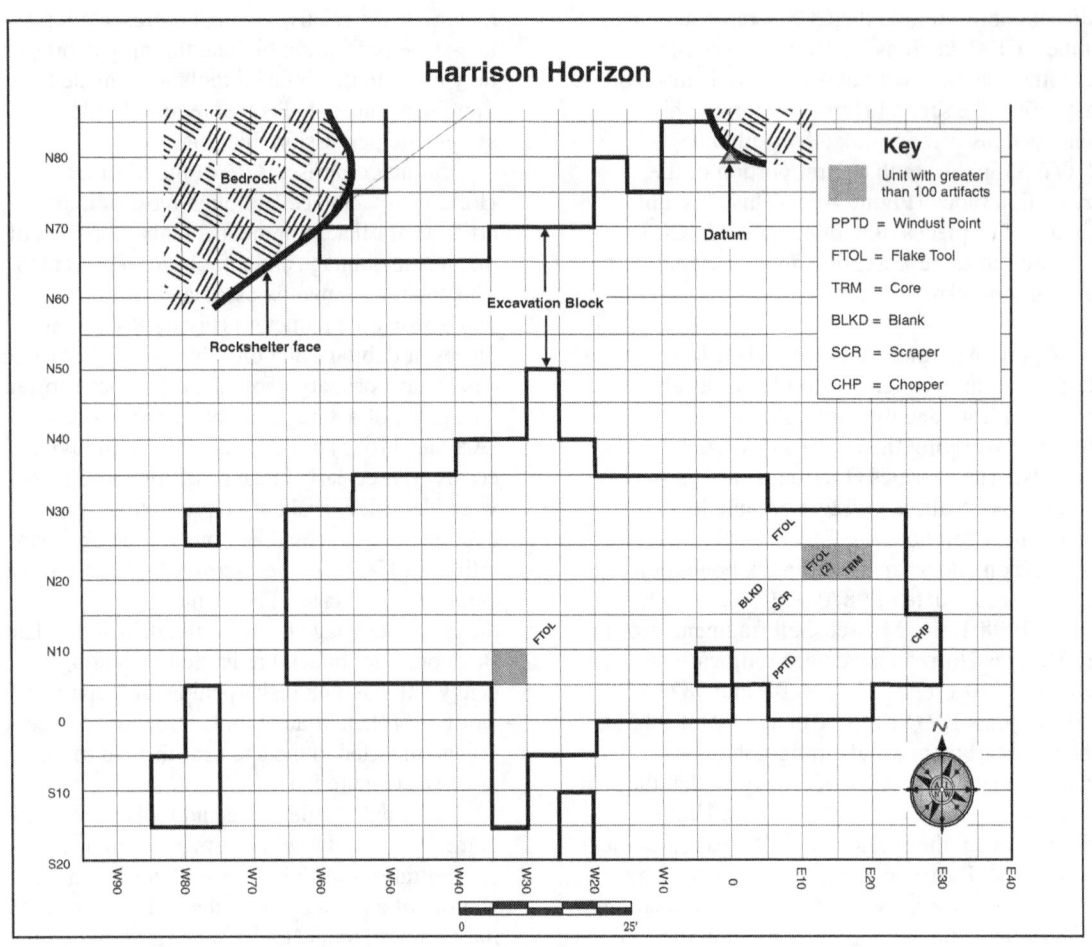

Figure 8.3 Spatial distribution of high-density lithic artifact deposits and stone tools associated with the floodplain Harrison Horizon assemblage.

from Stratum I (and II) were the primary materials used to define the Windust Phase of the lower Snake River region (Rice 1972). The assemblage was excavated using dry-screening with ¼-inch mesh to recover artifacts. The Stratum I assemblage includes 321 pieces of lithic debitage and 64 stone tools (Tables 8.7 and 8.8). Most of the debitage is large and unbroken, probably because of the large screen mesh used to recover it and the relatively good preservation within the shelter. The incidental effects of fire, probably associated with human activity within the shelter are indicated by heat damage to about 21% of the CCS debitage (Table 8.7). The extreme difference in size distributions between materials recovered in 1-mm mesh on the floodplain and materials recovered in ¼-inch (6-mm) mesh in the rockshelter make comparisons difficult. It is likely that pressure flaking debitage and other commonly small artifacts are significantly underrepresented in the rockshelter assemblages by comparison with the floodplain assemblages.

Artifacts of CCS composition are the most abundant (76%) in the Stratum I assemblage. Primary geological cortex is the most common type of cortex in the debitage (18 of 27) and occurs on three of the four CCS cores as well. One of the cores (Lot #4827-01) resulted from bipolar reduction of a biface. About one-third of the diagnostic CCS debitage was produced in core reduction (Figure 8.6), suggesting that some CCS raw material was obtained nearby and reduced through the early stages of tool manufacture in the rockshelter. Over one-half of the diagnostic debitage was produced in percussion bifacial thinning with a nearly even split between early-stage and late-stage.

Table 8.5 Summary of debitage analysis data for the floodplain Marmes Horizon assemblage.

COMPOSITION	BASALT/ RHYOLITE		CCS/QUARTZ/ PETRIFIED WOOD		OBSIDIAN		QUARTZITE	TOTAL
TECHNOLOGY								
bipolar								0
Core percussion, early	2	15%	1	2%				3
Core percussion, late	8	62%	5	9%				13
biface percussion, early			3	6%				3
biface percussion, late			1	2%				0
pressure, early	3	23%	29	55%				32
pressure, late			14	26%	6	100%		20
SUBTOTALS	13	100%	53	100%	6	100%		72
undetermined percussion	67		69		5			141
undetermined	26		52					78
thermal	3		5					8
TOTAL	109		179		11		0	299
OPERATION								
platform preparation								0
Bulb removal								0
alternate								0
Edge preparation								0
grinding or abrasion								0
overshot								0
margin removal			1					1
notch								0
rejuvenation								0
radial break								0
remnant ventral surface			1					1
CORTEX								
primary geological	4							4
incipient cone	1		2					3
THERMAL								
Heat damage	4		22					26
differential luster			4					4
differential color								0
SIZE CLASS								
2 (1.6 - 3.2 mm)	13		11		1			25
3 (3.2 - 6.4 mm)	51		124		7			182
4 (6.4 - 12.7 mm)	32		35		3			70
5 (12.7 - 25.4 mm)	6		7					13
6 (25.4 - 50.8 mm)	7		2					9
7 (>50.8 mm)								0
TOTAL	109		179		11		0	299

Table 8.6 Summary of tool analysis data for the floodplain Marmes Horizon assemblage.

	BASALT/ RHYOLITE	CCS/QUARTZ/ PETRIFIED WOOD	OBSIDIAN	OTHER	TOTALS
CORES					0
BLANKS					0
PREFORMS					0
PROJECTILE POINTS					0
FLAKE/BIFACIAL TOOLS					0
COBBLE TOOL	1				1
CHOPPER	1				1
TOTALS	2	0	0	0	2

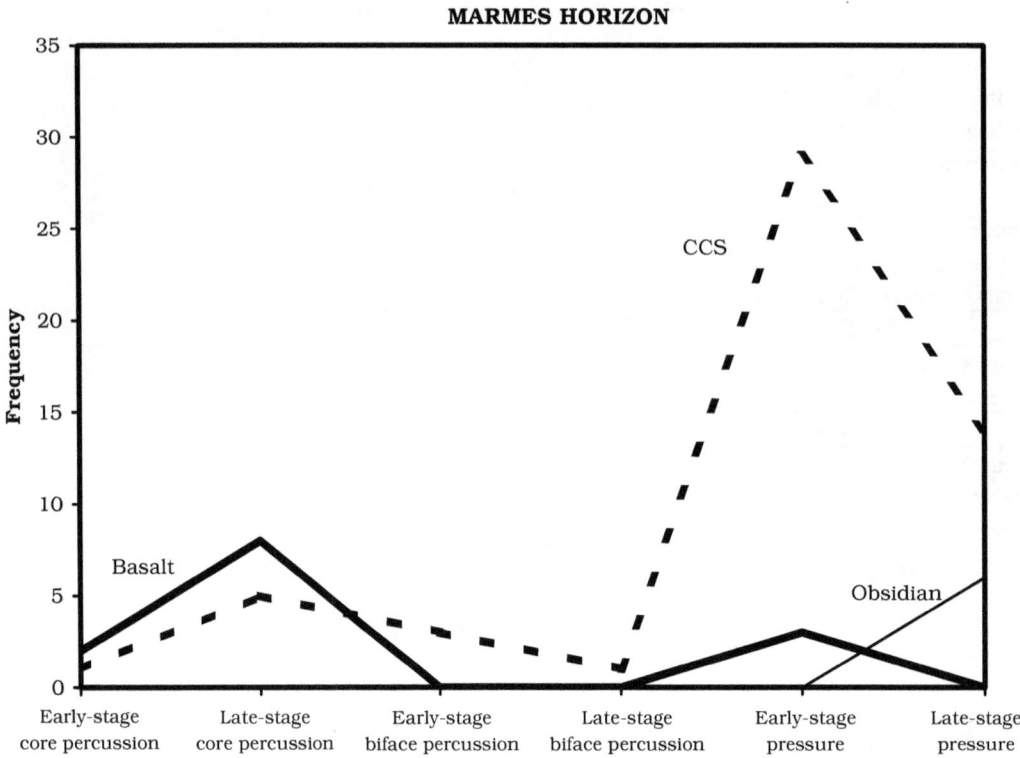

Figure 8.4 Frequency distribution of stage-diagnostic debitage from the floodplain Marmes Horizon assemblage, reduction trajectory from left to right.

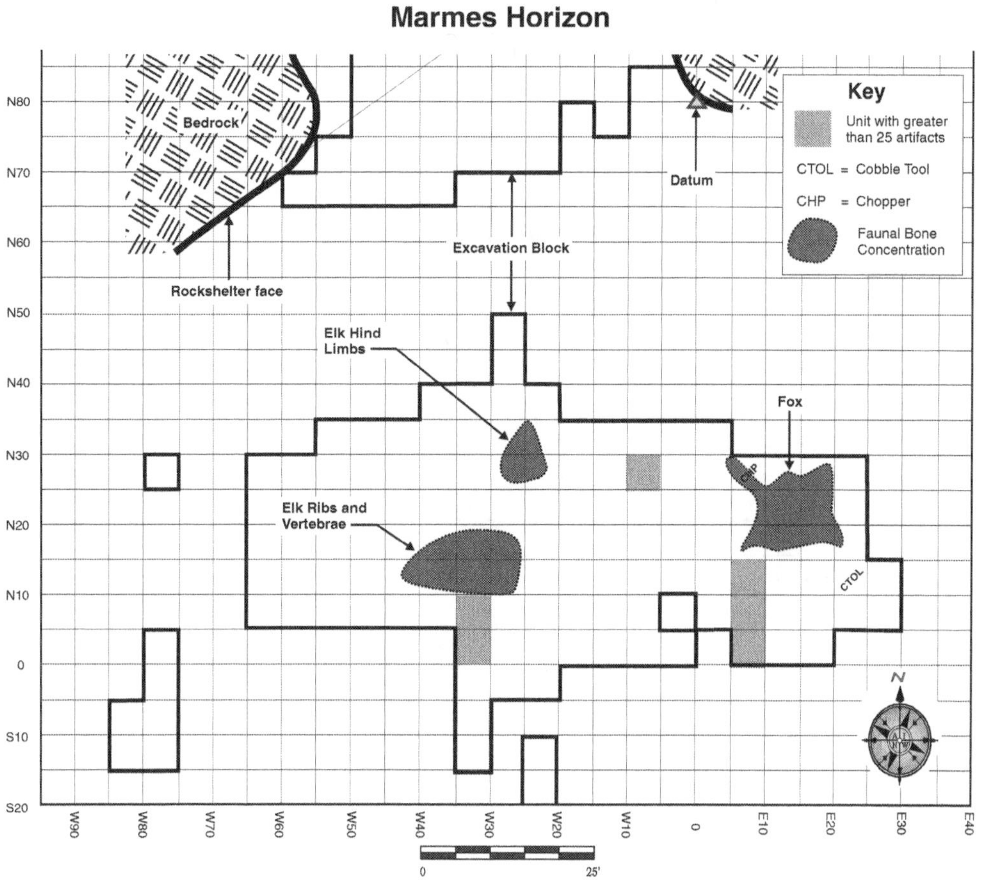

Figure 8.5 Spatial association of high-density lithic artifact deposits and stone tools with faunal remains from the floodplain Marmes Horizon assemblage.

The late-stage percussion bifacial thinning debitage reflects systematic thinning of blanks.

Evidence for heat treatment of the CCS material is sparse, but associated with early-stage percussion bifacial thinning debitage and one small preform. Remnant ventral surfaces were identified on one early-stage percussion bifacial thinning flake, three early-stage pressure flakes, two preforms, and one projectile point. The heat treatment and remnant ventral surface data suggest that both flake blanks and bifacial blanks were heat-treated to improve their flakability before the later stages of percussion and pressure flaking to finished tool forms. Pressure bifacial reduction flakes comprise only 21% of the diagnostic debitage, but may be underrepresented in comparison with the percussion categories due to poor recovery of small artifacts in the shelter. Other evidence for the final stages of CCS bifacial tool production and maintenance is identifiable from the tools themselves. Three of the CCS projectile points exhibit flake scars from resharpening.

Basalt tools and debitage account for 22% of the Stratum I lithic assemblage. The diagnostic debitage is primarily associated with core reduction from bedrock (10 flakes with primary geological cortex) and alluvial gravel (four flakes with incipient cone cortex) sources (Figure 8.6). The alluvial cobbles were used to produce expedient flake tools and choppers as indicated by incipient cone cortex on the identified tools. Larger bedrock pieces might have been preferred for production of flake blanks suitable for further reduction into larger bifacial tools. The relatively small amount of basalt bifacial reduction debitage and the single dart-sized preform broken in manufacture may indicate that much of the initial manufacturing for basalt projectile points occurred elsewhere.

Table 8.7 Summary of debitage analysis data for the rockshelter Stratum I assemblage.

COMPOSITION	BASALT/ RHYOLITE		CCS/QUARTZ/ PETRIFIED WOOD		OBSIDIAN		QUARTZITE	TOTAL
TECHNOLOGY								
bipolar								0
core percussion, early	14	29%	23	14%				37
core percussion, late	27	56%	27	16%				54
biface percussion, early	4	8%	43	26%				47
biface percussion, late	1	2%	40	24%	2	40%		43
pressure, early			19	11%	2	40%		21
pressure, late	2	4%	16	10%	1	20%		19
SUBTOTALS	48	100%	168	100%	5	100%		221
undetermined percussion	22		68					90
undetermined								0
thermal	3		7					10
TOTAL	73		243		5			321
OPERATION								
platform preparation								0
bulb removal								0
alternate			6					6
edge preparation			3					3
grinding or abrasion								0
overshot								0
margin removal								0
notch								0
rejuvenation								0
radial break			1					1
Remnant ventral surface			4		2			6
CORTEX								
Primary geological	10		18					28
incipient cone	4		9					13
THERMAL								
heat damage	1		51					52
differential luster			5					5
differential color								0
SIZE CLASS								
2 (1.6 – 3.2 mm)								0
3 (3.2 – 6.4 mm)								0
4 (6.4 - 12.7 mm)	10		88		3			101
5 (12.7 - 25.4 mm)	34		118		2			154
6 (25.4 - 50.8 mm)	24		34					58
7 (>50.8 mm)	5		3					8
TOTAL	73		243		5		0	321

Table 8.8 Summary of tool analysis data for the rockshelter Stratum I assemblage.

	BASALT/ RHYOLITE	CCS/QUARTZ/ PETRIFIED WOOD	OBSIDIAN	OTHER	TOTALS
CORES			4		4
bipolar		1			
primary geological cortex		2			
BLANKS			3		3
heat damage		1			
PREFORMS	1		3		4
remnant ventral surface		2			
heat damage		2			
heat treatment		1			
PROJECTILE POINTS	3		8	1	12
remnant ventral surface		1			
heat damage					
use wear	2	1			
resharpening	1	3			
complete/proximal	1	5	1		
other fragments	2	3			
FLAKE/BIFACIAL TOOLS	4		29		33
radial break	1				
ABRADER				1	1
CHOPPER	1				1
HAMMERSTONES	3				3
PEBBLE TOOL	1				1
SCRAPERS			2		2
TOTALS	13		49	1	1	64

The obsidian artifacts may be underrepresented because of their small size. Only five obsidian flakes and one obsidian tool are associated with the Stratum I assemblage. The technologically diagnostic debitage represents late-stage percussion bifacial thinning and bifacial pressure reduction and the tool is a dart-sized point base (Figure 8.6). These artifacts probably represent repair and replacement of worn and broken hunting equipment.

The 12 finished bifacial tools in the Stratum I assemblage are all classified as projectile points. These include six proximal (base) fragments, two medial (midsection) fragments, and three distal (tip) fragments from dart-sized points. Projectile point fragments are often associated with debris from hunting equipment repair (Flenniken 1991:185-190). The dart-sized points are both stemmed (Windust type, n=4) and lanceolate (Cascade type, n=2) in outline form. The stemmed points exhibit heavily ground basal margins. One complete side-notched arrow point was also identified, but the arrow point context is noted as disturbed in the catalog and on the bag, so the artifact may be intrusive to Stratum I.

Other finished tools include a large number of flake tools, some with visible fibrous (Artifact 4822-01 and Artifact 14501) and resinous (Artifact 4238 and Artifact 17953) residues adhering to the working edges. One of the basalt flake tools (Artifact 14177) is radially broken, an intentional reduction technique used to create

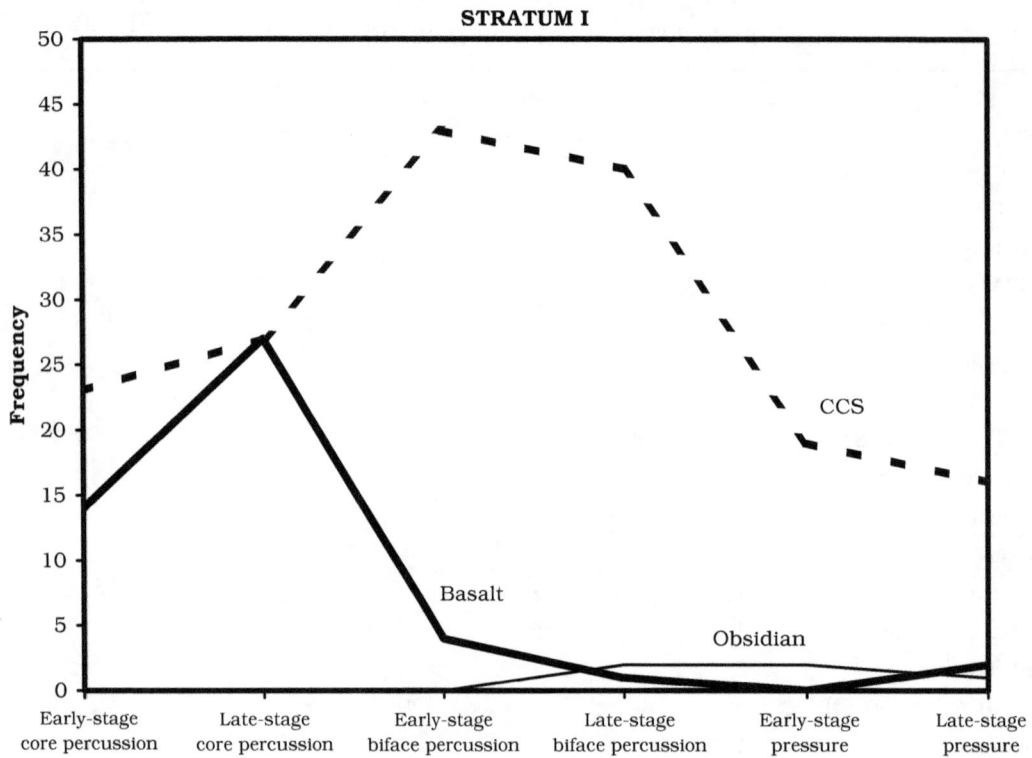

Figure 8.6 Frequency distribution of stage-diagnostic debitage from the rockshelter Stratum I assemblage, reduction trajectory from left to right.

strong square edges for heavy processing tasks such as bone or wood working. A single basalt pebble tool (Artifact 8293) also exhibits a fibrous residue. The sandstone abrader (Artifact 4260) has scratches from use in shaping hard materials. The hammerstones are basalt alluvial cobbles with battering from undetermined uses.

The Stratum I lithic subassemblage is from a relatively small area (25 feet square) near the center of the excavation block behind two Stratum I hearth features located near the shelter's dripline (Rice 1969:49, Figure 52). All three of the basalt hammerstones were found in units adjacent to a fire hearth feature (Figure 8.7). The highest concentration of tools and debitage is in the middle area in Grid Unit bifacial thinning flakes. Adjacent to another fire hearth in Grid Units N80-85/W25-35 (Figure 8.7), significantly high proportions of CCS late-stage percussion bifacial thinning and pressure flakes occur, indicating a CCS bifacial tool finishing area. At the rear of the shelter (Grid Units N90-100/W20-25), a concentration of basalt core reduction debitage with primary geological cortex was identified (Figure 8.7).

Rockshelter Stratum Unit II. The Stratum Unit II archaeological assemblage is from rockshelter sediments overlying Stratum I. Stratum II consists of a thin layer of rockfall from the roof of the shelter mixed with windblown sands and silts (Huckleberry, Gustafson, and Gibson 1998:63). Mussel shell from the Stratum I/II transition zone has been radiocarbon dated to 9540 ± 300 ^{14}C years BP and 9010 ± 300 ^{14}C years BP (Sheppard et al. 1987:Table 1). Charcoal from the Stratum I/II transition zone was dated to 9200 ± 110 ^{14}C years BP and 8700 ± 300 ^{14}C years BP. Mussel shell from Stratum II produced an additional radiocarbon date of 8525 ± 100 ^{14}C years BP. The artifacts from Stratum II were also used (along with Stratum I) as the primary materials defining the Windust Phase of the lower Snake

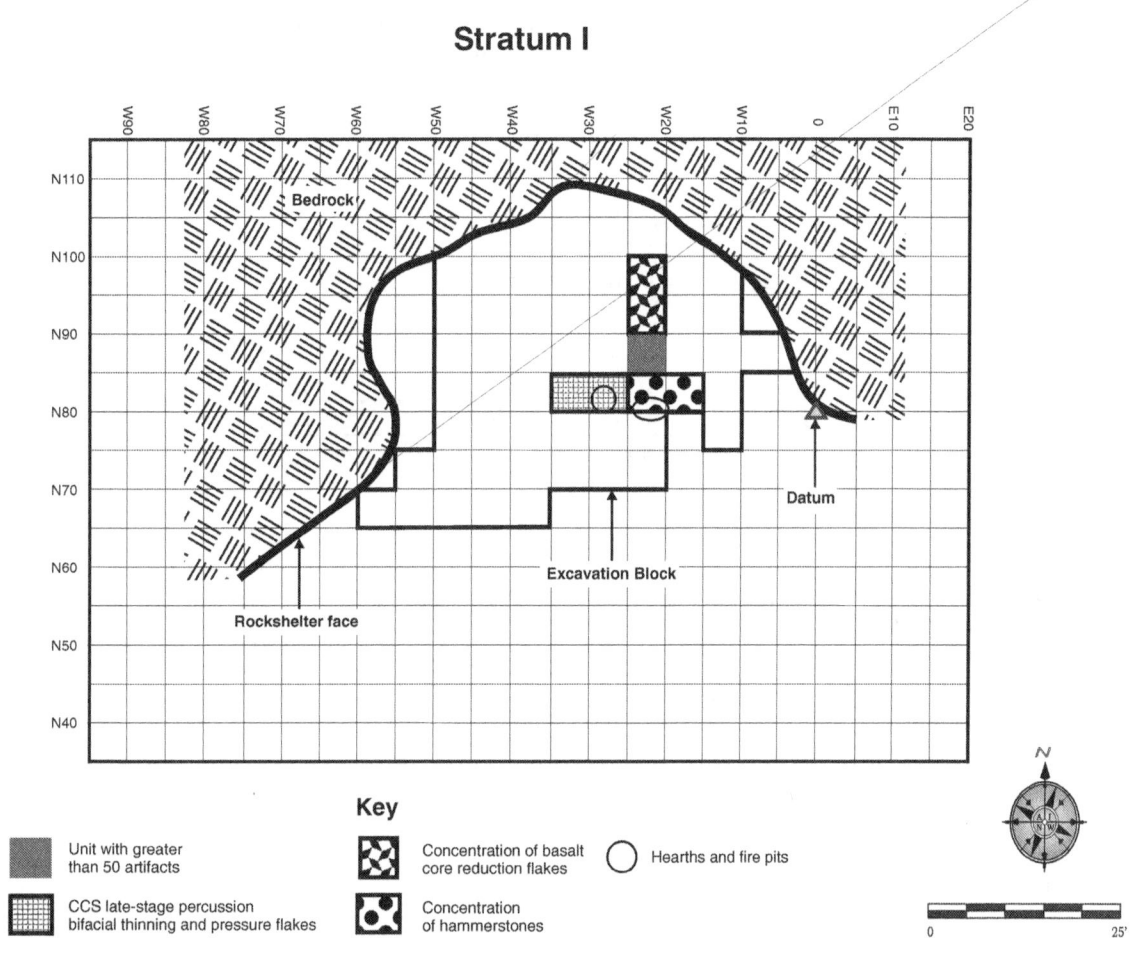

Figure 8.7 Spatial association of high-density artifact deposits and features from the rockshelter Stratum I assemblage.

River region (Rice 1972). The Stratum II assemblage includes 364 pieces of lithic debitage and 128 stone tools (Tables 8.9 and 8.10). The incidental effects of fire, probably associated with human activity within the shelter are indicated by heat damage to about 19% of the CCS debitage (Table 8.9).

Sixty-one percent of the Stratum II assemblage consists of CCS artifacts. Incipient cone cortex is the most common type of cortex in the debitage (25 of 42 cortical flakes) although primary geological cortex is more common in the tools (8 of 12 with cortex). Forty percent of the diagnostic CCS debitage was produced in percussion core reduction (Figure 8.8), suggesting that some CCS raw material was obtained nearby and reduced through the early stages of tool manufacture in the rockshelter. The seven CCS cores exhibit scars from multidirectional reduction and are small (15 to 63 mm in maximum dimension). About one-half of the diagnostic debitage was produced in percussion bifacial thinning, mainly in the early-stage with relatively little late-stage systematic percussion bifacial thinning. Seven of the early-stage percussion bifacial thinning flakes have alternate flaking operation characteristics indicating beveling for square edges as occur on flake blanks. Remnant ventral surfaces were identified on three early-stage percussion bifacial thinning flakes, two early-stage pressure flakes, four blanks, three preforms, and five projectile points. These remnant ventral surfaces indicate reduction from large flake blanks for bifacial

Table 8.9 Summary of debitage analysis data for the rockshelter Stratum II assemblage.

COMPOSITION	BASALT/ RHYOLITE		CCS/QUARTZ/ PETRIFIED WOOD		OBSIDIAN		QUARTZITE	TOTAL
TECHNOLOGY								
bipolar								0
core percussion, early	26	23%	34	20%				60
core percussion, late	63	55%	34	20%				97
biface percussion, early	11	10%	57	34%				68
biface percussion, late	12	11%	25	15%	4	57%		41
pressure, early	1	1%	4	2%	1	14%		5
pressure, late	1	1%	13	8%	2	29%		16
SUBTOTALS	114	100%	167	100%	7	100%		288
undetermined percussion	17		42		1			60
undetermined								0
thermal	1		15					16
TOTAL	132		224		8			364
OPERATION								
platform preparation			2					2
bulb removal								0
alternate	1		8					9
edge preparation								0
grinding or abrasion								0
overshot			2					2
margin removal					1			1
notch								0
rejuvenation	3		1					4
radial break								0
remnant ventral surface			5					5
CORTEX								
primary geological	11		17					28
incipient cone	28		25					53
THERMAL								
heat damage	2		42					44
differential luster			12					12
differential color								0
SIZE CLASS								
2 (1.6 - 3.2 mm)								0
3 (3.2 - 6.4 mm)	1				1			2
4 (6.4 - 12.7 mm)	12		44		5			61
5 (12.7 - 25.4 mm)	51		132		2			185
6 (25.4 - 50.8 mm)	56		48					104
7 (>50.8 mm)	12							12
TOTAL	132		224		8		0	364

Table 8.10 Summary of tool analysis data for the rockshelter Stratum II assemblage.

	BASALT/ RHYOLITE	CCS/QUARTZ/ PETRIFIED WOOD	OBSIDIAN	OTHER	TOTALS
CORES		2	7		9
incipient cone cortex	2	2			
primary geological cortex		1			
BLANKS		3	5		8
incipient cone cortex		1			
remnant ventral surface	2	4			
heat damage		1			
PREFORMS			5		5
primary geological cortex		1			
remnant ventral surface		3			
heat damage		3			
PROJECTILE POINTS		9	21		30
remnant ventral surface	1	5			
heat damage		5			
heat treatment		1			
use wear	6	2			
reworking		2			
resharpening	5	9			
serrated edges		1			
complete/proximal	5	12			
other fragments	4	9			
FLAKE/BIFACIAL TOOLS		9	35		1	45
radial break		4			
square edge	1				
graver		1			
BOLA		1				1
CHOPPERS		4	1		2	7
UNDETERMINED		1				1
HAMMERSTONES		14				14
KNIFE			1			1
MANO		1				1
PEBBLE TOOLS		3				3
SCRAPERS			3			3
TOTALS		47	78	0	3	128

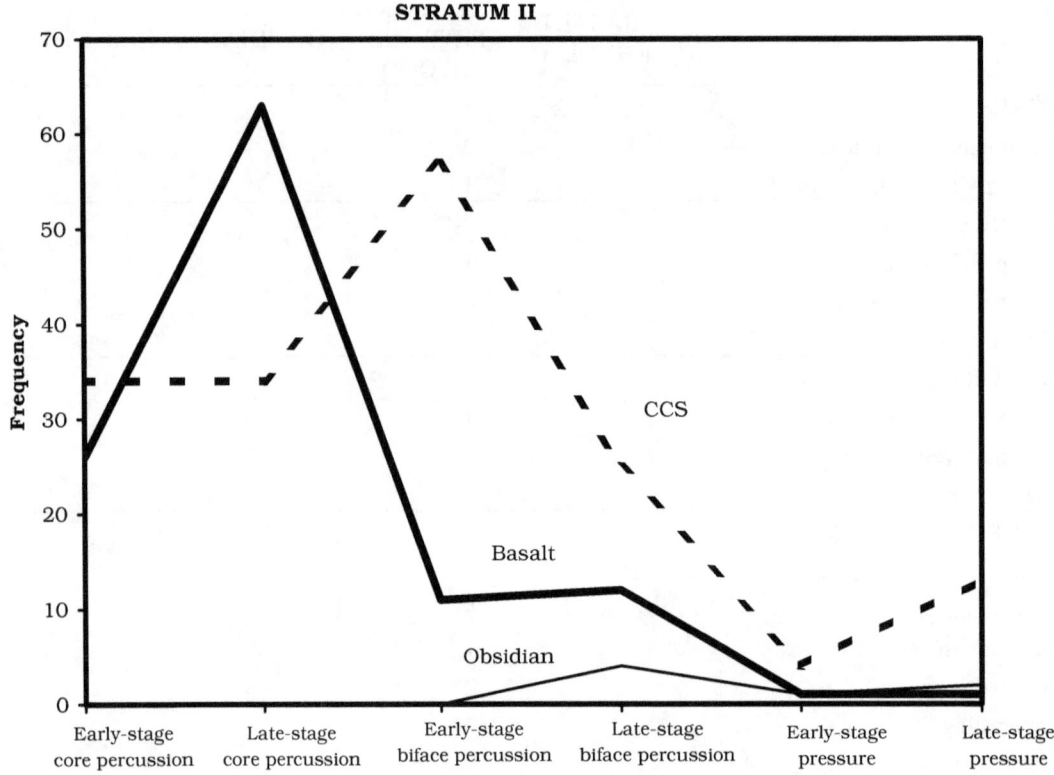

Figure 8.8 Frequency distribution of stage-diagnostic debitage from the rockshelter Stratum II assemblage, reduction trajectory from left to right.

tools. Evidence for heat treatment of the CCS material is clearly associated with early-stage percussion bifacial thinning debitage. The heat treatment and remnant ventral surface data suggest that flake blanks were heat treated to improve their flakability before the later stages of percussion and pressure flaking to finished tool forms.

Pressure bifacial reduction flakes comprise only 10% of the diagnostic debitage, but may be underrepresented in comparison with the percussion categories due to poor recovery of small artifacts in the shelter. Most of the CCS projectile points exhibit flake scars from resharpening or reworking and indicate extensive rejuvenation activity.

Basalt tools and debitage account for 36% of the Stratum II lithic assemblage. The diagnostic debitage is primarily associated with core reduction from alluvial gravels (28 flakes with incipient cone cortex) and bedrock materials (11 flakes with primary geological cortex) (Figure 8.8). The alluvial cobbles were used to produce expedient flake tools and large tools such as choppers as indicated by incipient cone cortex on the identified tools. Flakes or prepared blanks derived from larger bedrock pieces of fine-grained basalt or andesite from highland sources might have been imported for bifacial tools. Two of the three basalt bifacial blanks have remnant ventral surfaces from their original flake blanks. The relatively small amount of basalt bifacial reduction debitage and small number of blanks may indicate that much of the initial manufacturing for basalt projectile points occurred elsewhere. The relatively greater number of basalt projectile points indicates that many of these tools entered the site in the later stages of the manufacturing and use-life trajectory.

The obsidian artifacts consist of only eight flakes and may be underrepresented because of their small size relative to the archaeological recovery techniques. The technologically diagnostic debitage is from late-stage percussion bifacial thinning and bifacial pressure reduction and probably represents repair and replacement of worn and broken hunting equipment (Figure 8.8).

The 30 projectile points in the Stratum II assemblage include eight complete dart-sized points and a variety of fragmentary pieces typical of sites associated with game animal butchery and hunting equipment repair (Flenniken 1991:185-190). The points are both stemmed (Windust type, n=7) and lanceolate (Cascade type, n=12) in outline form. Some of the stemmed points are well designed for durability with biconvex cross-sections and long-sections and others are expediently made from curved flakes with mainly unifacial pressure flaking (Artifact 3773). The lanceolate points are commonly broken straight across the base at bending fractures one to two centimeters from the proximal end.

The large number of flake tools include some with visible fiber residues (Artifact 4567-01, Artifact 5294-01, Artifact 5300-01, and Artifact 5300-02). Four of the CCS flake tools exhibit radial breaks used to create strong square edges for heavy processing tasks such as bone or wood working. One bifacial tool (Artifact 13117-02) has a graver tip, probably used for incising bone or wood. Fourteen basalt hammerstones with battering wear patterns probably reflect a variety of activities including flintknapping, pecking, bone splitting, and food processing. Choppers may have been used for similar purposes. The grooved basalt stone identified as a bola stone (Artifact 3636) may represent a poorly understood hunting technology or a net weight.

The lithic artifacts recovered in Stratum II were within and behind a series of hearths and firepits aligned diagonally from southwest to northeast across the mouth of the shelter (Figure 8.9). Debitage was concentrated in three units with a similar diagonal orientation (Grid Units N80-85/W40-45, N90-95/W25-30, and N95-100/W20-25). Each of these grid units contained 50 or more pieces of debitage, but none exhibited distinctive debitage types or compositions relative to the complete subassemblage. Relatively high proportions of CCS percussion bifacial thinning flakes were recovered from two lower density grid units (N85-95/W15-20) in the eastern portion of the shelter, perhaps representing an activity area or specific flintknapping event (Figure 8.9). Nine hammerstones were concentrated in two grid units (N80-85/W35-45), possibly associated with a small hearth (Feature 3) (Rice 1969:50, Figure 53) (Figure 8.9).

Rockshelter Stratum Unit III. The archaeological assemblage from Stratum Unit III of the rockshelter was found in a deposit of windblown sands and silts with lenses of bone and shell (Huckleberry, Gustafson, and Gibson 1998:63). Five radiocarbon dates on mussel shell range from 6200 ± 475 ^{14}C years BP to 7870 ± 300 ^{14}C years BP (Sheppard et al. 1987:Table 1). The Early Cascade Subphase of the lower Snake River region was defined primarily on the basis of materials from this deposit (Leonhardy and Rice 1970). Artifacts include 896 pieces of lithic debitage and 191 stone tools (Tables 8.11 and 8.12). The incidental effects of fire, probably associated with human activity within the shelter are indicated by heat damage to about 30% of the CCS debitage (Table 8.11).

Basalt artifacts represent nearly one-half of the Stratum III assemblage, marking the zenith of a trend in increasing basalt use through the early occupations of the site. The diagnostic basalt debitage is overwhelmingly associated with core reduction from alluvial gravels (117 of 126 cortical flakes have incipient cone cortex) (Figure 8.10). The alluvial cobbles were used to produce expedient flake tools and choppers as well as flake blanks for bifacial tools (16 of 54 early-stage percussion bifacial thinning flakes have incipient cone cortex as does the only percussion bifacial blank fragment [Artifact 4147]). However, the relatively small amount of basalt bifacial reduction debitage and small number of blanks may indicate that much of the initial manufacturing for basalt projectile points occurred elsewhere. The large number of basalt projectile points indicates that many of these tools entered the site in the later stages of the manufacturing and use-life trajectory.

Despite the ascendancy of basalt, the CCS artifacts are still numerically superior (52% of the assemblage). Both alluvial gravels and primary geological deposits of CCS material were used for toolstone. As with Stratum II, 40% of the diagnostic CCS debitage was produced in percussion core reduction, but proportionately more of it was for the late-stage (Figure 8.10). Eleven CCS cores range from 14 mm to 50 mm in maximum dimension reflecting the size of the exhausted raw material. Most of the cores were reduced multidirectionally, but four have bipolar reduction attributes (including bipolar reduction of an alluvial pebble [Artifact

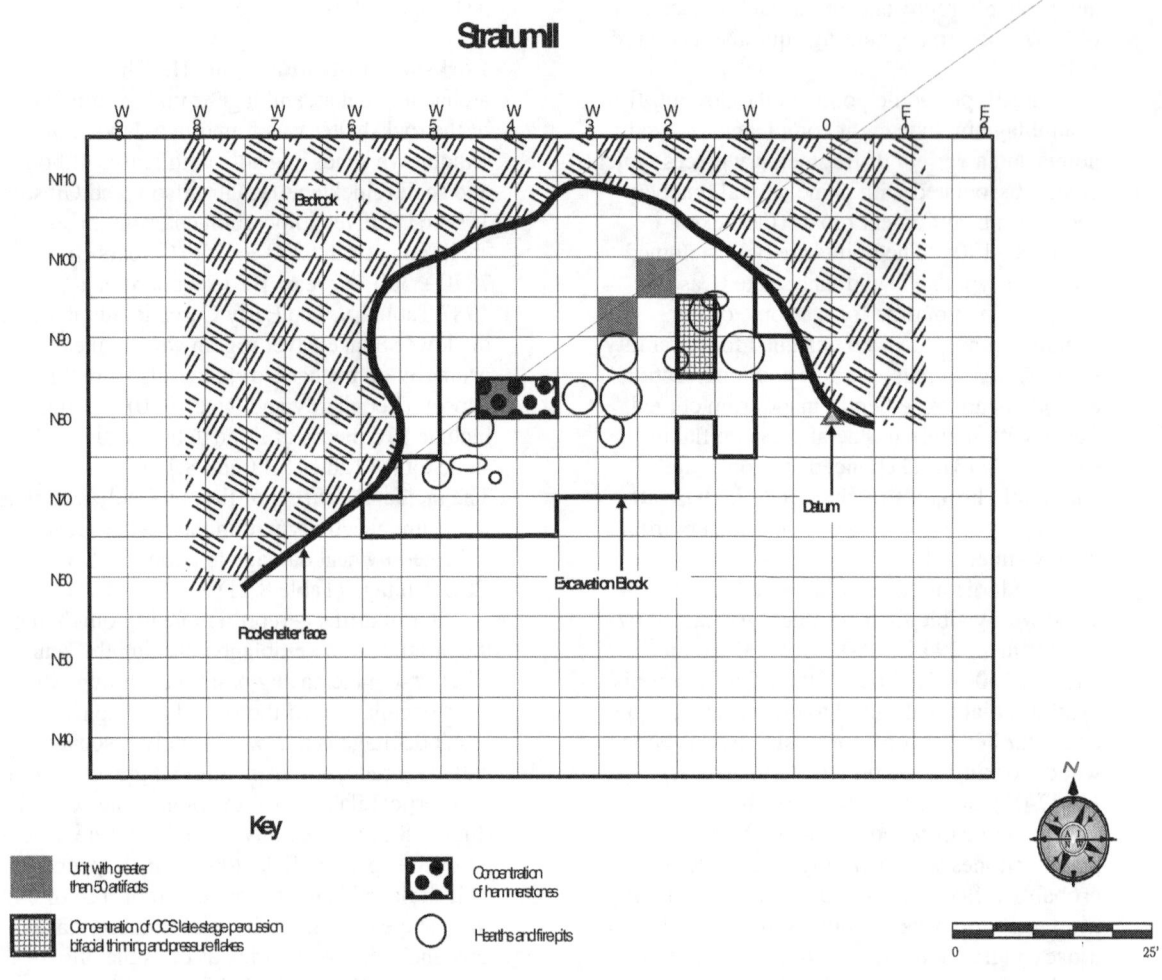

Figure 8.9 Spatial association of high-density artifact deposits and features from the rockshelter Stratum II assemblage.

10643], a flake tool [Artifact 15436-01] and a dart point [Artifact 15424-02]). About one-half of the diagnostic debitage was produced in percussion bifacial thinning, mainly in the early-stage with relatively little late-stage systematic percussion bifacial thinning. Twelve of the early-stage percussion bifacial thinning flakes have alternate flaking operation characteristics indicating beveling for square edges to prepare them for further bifacial reduction. Remnant ventral surfaces were identified on five early-stage percussion bifacial thinning flakes, three blanks, four preforms, and three projectile points, indicating reduction from flake blanks for bifacial tools. Evidence for heat treatment of the CCS material corresponds with early-stage percussion bifacial thinning debitage in 28 of 36 observations of differential luster and differential color (Table 8.11). Only 9% of the diagnostic debitage was classified as pressure bifacial reduction flakes, probably due to poor recovery of small artifacts in the shelter.

The pattern for the obsidian artifacts is generally consistent with the underlying materials from Strata I and II (Figure 8.10). Only 20 obsidian flakes and 1 tool were recovered. The obsidian debitage is from percussion bifacial thinning and bifacial pressure reduction (Table 8.11). The obsidian tool, however, is an early-stage core reduction flake with primary geological cortex and polish on one lateral margin (Artifact 4185, sourced to Indian Creek).

Table 8.11 Summary of debitage analysis data for the rockshelter Stratum III assemblage.

COMPOSITION	BASALT/ RHYOLITE		CCS/QUARTZ/ PETRIFIED WOOD		OBSIDIAN		QUARTZITE		TOTAL
TECHNOLOGY									
bipolar									0
core percussion, early	55	16%	50	15%					105
core percussion, late	190	54%	82	25%			2	67%	272
biface percussion, early	54	15%	118	35%	3	16%	1	33%	172
biface percussion, late	45	13%	55	16%	5	26%			105
pressure, early	1	0%	10	3%	6	32%			16
pressure, late	4	1%	19	6%	5	26%			28
SUBTOTALS	349	100%	334	100%	19	100%	3	100%	705
undetermined percussion	41		108		1				150
undetermined thermal	11		30						41
TOTAL	401		472		20		3		896
OPERATION									
platform preparation	2								2
bulb removal			2						2
alternate	6		12		1				19
edge preparation									0
grinding or abrasion									0
overshot	1		1						2
margin removal			1						1
notch									0
rejuvenation	14		1						15
radial break									0
remnant ventral surface			5		1				6
CORTEX									
primary geological	9		27						36
incipient cone	117		35						152
THERMAL									
heat damage	12		140						152
differential luster			30						30
differential color			6						6
SIZE CLASS									
2 (1.6 - 3.2 mm)									0
3 (3.2 - 6.4 mm)			4						4
4 (6.4 - 12.7 mm)	24		92		9				125
5 (12.7 - 25.4 mm)	166		283		8				457
6 (25.4 - 50.8 mm)	177		87		3		2		269
7 (>50.8 mm)	34		6				1		41
TOTAL	401		472		20		3		896

Table 8.12 Summary of tool analysis data for the rockshelter Stratum III assemblage.

	BASALT/ RHYOLITE	CCS/QUARTZ/ PETRIFIED WOOD	OBSIDIAN	OTHER	TOTALS
CORES	2	11			13
bipolar		4			
incipient cone cortex	1	2			
primary geological cortex		2			
BLANKS	1	8			9
incipient cone cortex	1				
remnant ventral surface		3			
heat damage		3			
PREFORMS		4			4
primary geological cortex		1			
remnant ventral surface		4			
heat damage		2			
PROJECTILE POINTS	18	16			34
remnant ventral surface	2	3			
heat damage		4			
use wear	9	1			
resharpening	9	1			
serrated edges	1				
complete/proximal	13	9			
other fragments	5	7			
FLAKE/BIFACIAL TOOLS	28	55	1		84
burin		2			
radial break	1	5			
square edge	1	1			
COBBLE TOOLS	2				2
CHOPPERS	10			1	11
DRILL		1			1
EDGE GROUND COBBLE	1				1
HAMMERSTONES	18				18
PEBBLE TOOLS	8	1			9
SCRAPERS	3	2			5
TOTALS	91	98	1	1	191

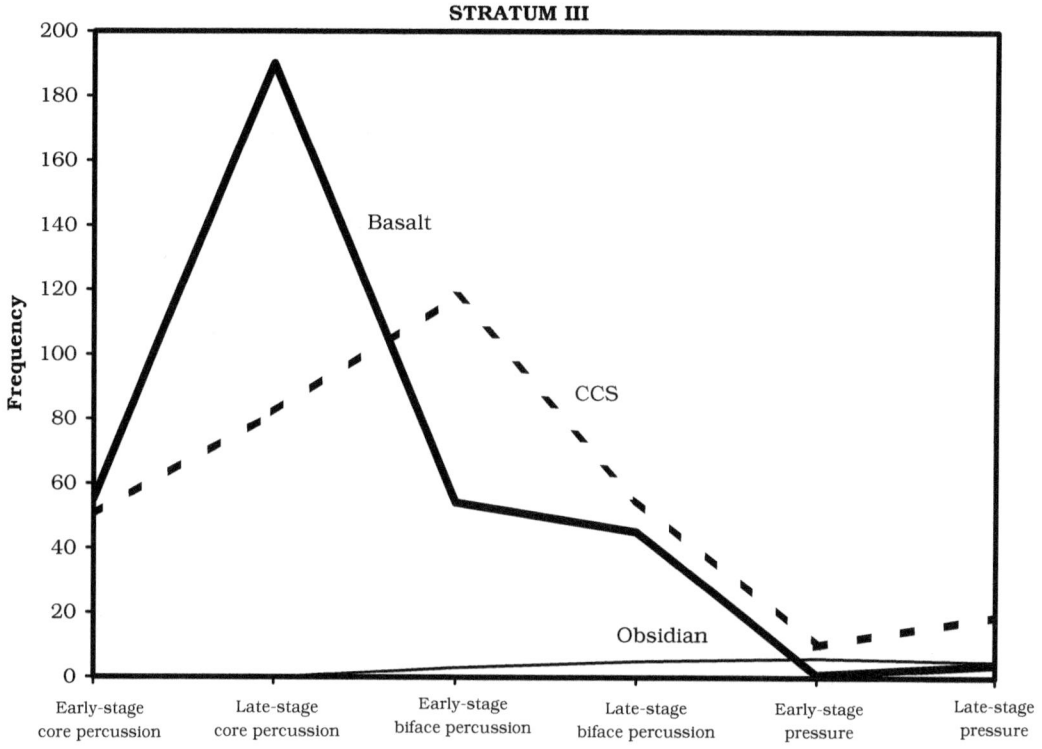

Figure 8.10 Frequency distribution of stage-diagnostic debitage from the rockshelter Stratum III assemblage, reduction trajectory from left to right.

The 34 projectile points associated with Stratum III include 18 lanceolate (Cascade type), three stemmed (Windust type), and two side-notched forms (notches low on the base, within 3 mm of the proximal end). The ratio of complete and proximal fragments to other fragments is high, possibly indicating more hunting equipment maintenance than in the earlier occupations. The lanceolate points are primarily made from basalt (12 basalt, six CCS) and are relatively small due to extensive rejuvenation. The most extensively resharpened Cascade points are relatively narrow and thick such that the margin edge angles become blunted. Serrations on the edges of these points appear to be a design measure used to resharpen these blunted edges just prior to the end of their use-life exhaustion (Artifact 3025).

Flake tools are numerous and predominantly of CCS composition (Table 8.12). Flake tools with square edges (including burins and radially broken pieces) are well-represented, indicating bone or wood working. Hammerstones and choppers are also abundant, probably reflecting food processing and a variety of other possible activities. Pebble tools with one or two flake scars are fairly numerous and may represent a processing tool for shredding sinew or some other fiber or possibly net weights. Only one edge-ground cobble was identified in the Stratum III assemblage (Artifact 6611).

The lithic subassemblage recovered from Stratum III is from the area containing 21 hearth and fire pit features in the central and northeastern part of the shelter (Figure 8.11). Within this area, several debitage concentrations were identified, but these did not exhibit distinctive variability in technological attributes, except as noted below in association with the choppers. And, with the exception of hammerstones and possibly choppers, neither did the tools seem to cluster into functionally related spatial groupings. A high concentration of 15 hammerstones in Grid Unit N80-85/W40-45 is west of a dense cluster of intersecting features (Figure 8.11). The hammerstones are all basalt with incipient cone cortex, and mainly unmodified except from use that produced battering on the otherwise smooth cortical surfaces. These may represent a cache of

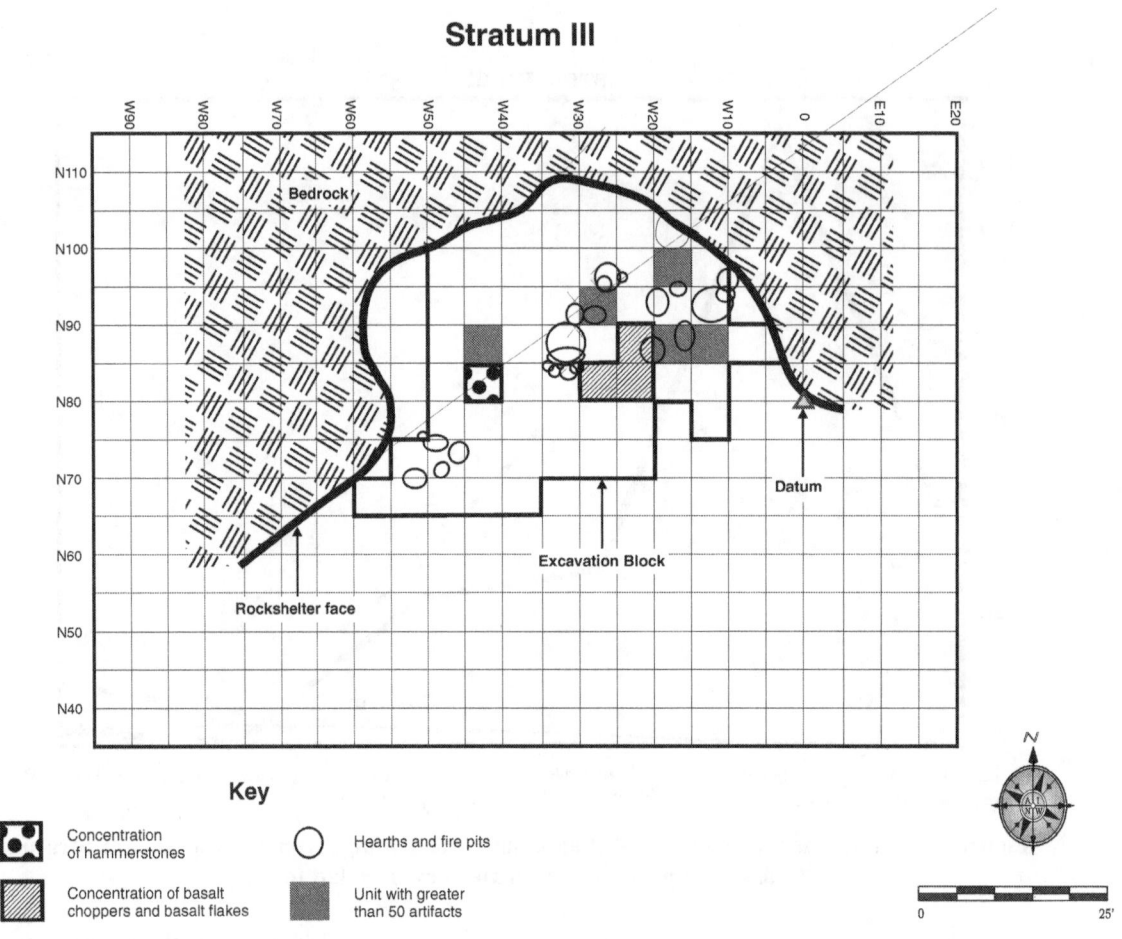

Figure 8.11 Spatial association of high-density artifact deposits and features from the rockshelter Stratum III assemblage.

hammers or an activity area, possibly for processing animal bones. The choppers are less densely concentrated, but six were found in three contiguous grid units (N80-85/W20-30 and N85-90/W20-25) at the mouth of the shelter near a hearth feature (Figure 8.11). These six choppers are all made from basalt alluvial cobbles and are associated with a concentration of basalt late-stage percussion core reduction flakes (in Grid Unit N85-90/W20-25), possibly resulting from resharpening the working edges or bits of the choppers.

Rockshelter Stratum Unit IV. Rockshelter Stratum Unit IV consists of volcanic ash identified as a primary airfall deposit from the climactic eruption of Mt. Mazama, approximately 6,730 radiocarbon years ago (Hallett et al. 1997; Huckleberry, Gustafson, and Gibson 1998; Sheppard et al. 1987). The ash layer was deposited over a short period, perhaps only a few days, and lacks occupation surfaces. Artifacts found within it are likely intrusive from deposits above and below the ash. Only 150 pieces of lithic debitage and 17 stone tools are associated with Stratum IV (Tables 8.13 and 8.14). These most closely resemble the assemblage from the overlying Stratum V in technological character and spatial distribution (Figure 8.12).

Rockshelter Stratum Unit V. The archaeological assemblage from Stratum Unit V of the rockshelter is associated with a thick layer of redeposited volcanic ash and loess (Huckleberry, Gustafson, and Gibson 1998:63-65). A single radiocarbon date on mussel shell produced an age of 4250 ± 300 ^{14}C years BP, but

Table 8.13 Summary of debitage analysis data for the rockshelter Stratum IV assemblage.

COMPOSITION	BASALT/ RHYOLITE		CCS/QUARTZ/ PETRIFIED WOOD		OBSIDIAN		QUARTZITE		TOTAL
TECHNOLOGY									
bipolar									0
core percussion, early	15	28%	8	11%			1	100%	24
core percussion, late	21	40%	7	10%					28
biface percussion, early	4	8%	27	39%					31
biface percussion, late	9	17%	18	26%	3	50%			30
pressure, early	1	2%	3	4%	3	50%			6
pressure, late	3	6%	7	10%					10
SUBTOTALS	53	100%	70	100%	6	100%	1	100%	130
undetermined percussion	3		15						18
undetermined thermal			2						2
TOTAL	56		87		6		1		150
OPERATION									
platform preparation									0
bulb removal									0
alternate			6						6
edge preparation									0
grinding or abrasion									0
overshot									0
margin removal			1						1
notch									0
rejuvenation	3								3
radial break									0
remnant ventral surface			3		3				6
CORTEX									
primary geological	10		3						13
incipient cone	17		6				1		24
THERMAL									
heat damage			16						16
differential luster			5						5
differential color									0
SIZE CLASS									
2 (1.6 - 3.2 mm)									0
3 (3.2 - 6.4 mm)									0
4 (6.4 - 12.7 mm)	10		28		2				40
5 (12.7 - 25.4 mm)	19		41		3				63
6 (25.4 - 50.8 mm)	25		18		1		1		45
7 (>50.8 mm)	2								2
TOTAL	56		87		6		1		150

Table 8.14 Summary of tool analysis data for the rockshelter Stratum IV assemblage.

	BASALT/ RHYOLITE	CCS/QUARTZ/ PETRIFIED WOOD	OBSIDIAN	OTHER	TOTALS
CORES		1	2		3
incipient cone cortex	1				
BLANKS			1		1
heat damage		1			
PREFORMS			1		1
remnant ventral surface		1			
PROJECTILE POINTS					0
FLAKE/BIFACIAL TOOLS	3	5		1	9
radial break	1				
square edge	1				
CHOPPERS	2				2
HAMMERSTONE	1				1
TOTALS	7	9	0	1	17

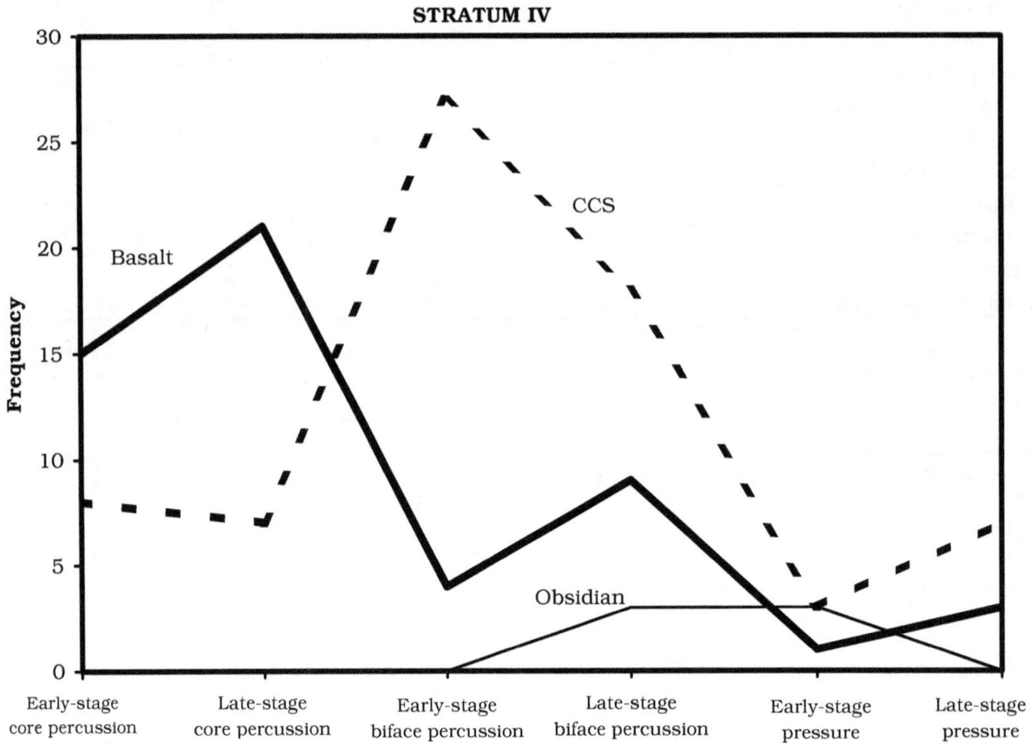

STRATUM IV

Figure 8.12 Frequency distribution of stage-diagnostic debitage from the rockshelter Stratum IV assemblage, reduction trajectory from left to right.

the context of the dated material is unsure (Sheppard et al. 1987:Table 1, 124). Despite the young radiocarbon date, archaeological materials from Stratum V have been correlated with the Late Cascade Subphase of the lower Snake River region (7000 B.P. to 4500 B.P.) (Ames 1988: 328; Leonhardy and Rice 1970:6). The Stratum V deposit is associated with seven human burials and may represent a period of ceremonial rather than residential activity at the site (Fryxell and Daugherty 1962:28). Intrusive storage pits from the overlying strata may have caused some mixing of the deposits in Stratum V.

The Stratum V assemblage was excavated using dry-screening with ¼-inch mesh to recover artifacts including 859 pieces of lithic debitage and 79 stone tools (Tables 8.15 and 8.16). Heat damage has affected about 24% of the CCS debitage (Table 8.15).

As in Stratum III, basalt artifacts represent nearly one-half of the Stratum V assemblage. Sixty-five percent of the diagnostic basalt debitage is associated with core reduction (Figure 8.13). The basalt raw materials were derived from alluvial gravels (55 of 99 cortical flakes have incipient cone cortex) and bedrock materials (44 or 99 cortical flakes have primary geological cortex). The alluvial cobbles were used to produce some of the flake blanks for bifacial tools (5 of 41 early-stage percussion bifacial thinning flakes have incipient cone cortex). Most of the basalt reduction appears to be associated with the more expedient tool production for flake tools and choppers. Some rejuvenation of worn or broken basalt projectile points is indicated by the pressure flakes recovered (Table 8.15).

CCS artifacts comprise 46% of the Stratum V assemblage of stone tools and lithic debitage. Cortex from alluvial gravels is more common than primary geological cortex in the debitage, but all three of the cortical cores and the single bifacial blank with cortex are from CCS bedrock materials. Only about 19% of the diagnostic CCS debitage was produced by percussion core reduction (Figure 8.13). Seven CCS cores range from 17 mm to 48 mm in maximum dimension reflecting the size of the exhausted raw material. Most of the cores were reduced multidirectionally, but one represents bipolar reduction of a biface as a means of recycling the material (Artifact 13006-02). More than one-half of the diagnostic debitage was produced in percussion bifacial thinning with early-stage and late-stage fairly evenly represented. The higher

representation of late-stage thinning may indicate a shift to use of broader blanks for projectile points. Eleven of the early-stage percussion bifacial thinning flakes and two early-stage pressure flakes have alternate flaking operation characteristics indicating beveling for square edges to prepare them for further bifacial reduction. Remnant ventral surfaces were identified on three early-stage percussion bifacial thinning flakes, two pressure flakes, two blanks, and two projectile points, indicating reduction from flake blanks for bifacial tools. Evidence for heat treatment of the CCS material corresponds with early-stage percussion bifacial thinning debitage in 13 of 17 observations of differential luster and differential color (Table 8.15). Only 13% of the CCS diagnostic debitage was classified as pressure bifacial reduction flakes, probably due to poor recovery of small artifacts in the shelter.

The obsidian artifacts consist of 50 flakes, 38 of which are diagnostic of reduction stages (Figure 8.13). Three pieces of obsidian have primary geological cortex and six have remnant ventral surfaces, both indicators of relatively early stages of reduction. Percussion bifacial reduction is predominantly of the late-stage (Table 8.15). Thinning debitage includes one more early-stage piece than late-stage pieces and bifacial pressure reduction is predominantly of the late-stage (Table 8.15).

Projectile points associated with Stratum V include three lanceolate (Cascade type) and four side-notched (Cold Springs type) forms (e.g., Figure 8.14). The high ratio of complete and proximal fragments to other fragments (8:4) again reflects hunting equipment maintenance. The lanceolate points are all broken near the base and one base fragment exhibits a burination scar perhaps created when it was pried from the haft for replacement (Artifact 3398).

Flake tools of CCS and basalt are numerous and some have square working edges (Table 8.16). Basalt hammerstones and choppers are also abundant, probably reflecting food processing and a variety of other possible activities. Other tools are also similar to those found in the pre-Mazama deposits (Table 8.16).

The Stratum V lithic subassemblage is most concentrated at the rear of the shelter in the vicinity of a cluster of features including a large grass-lined storage pit with rocks at the bottom (Feature 7, Grid Unit N95-100/W14-20) (Rice 1969:60; Figure 55). Tools in this area (N95-100/W10-20 and N100-105/W15-20) include a

Table 8.15 Summary of debitage analysis data for the rockshelter Stratum V assemblage.

COMPOSITION	BASALT/ RHYOLITE		CCS/QUARTZ/ PETRIFIED WOOD		OBSIDIAN		QUARTZITE		TOTAL
TECHNOLOGY									
bipolar									0
core percussion, early	69	21%	21	7%			1	50%	91
core percussion, late	143	44%	35	12%					178
biface percussion, early	41	13%	85	30%	13	34%	1	50%	127
biface percussion, late	55	17%	74	26%	12	32%			141
pressure, early	3	1%	32	11%	3	8%			35
pressure, late	12	4%	34	12%	10	26%			56
SUBTOTALS	323	100%	281	100%	38	100%	2	100%	644
undetermined percussion	75		92		12				179
undetermined			25						25
thermal	11								11
TOTAL	409		398		50		2		859
OPERATION									
platform preparation									0
bulb removal			2						2
alternate	2		13		1				16
edge preparation	1		2						3
grinding or abrasion									0
overshot					1				1
margin removal			4						4
notch									0
rejuvenation	8		2		1				11
radial break									0
remnant ventral surface			5		6				11
CORTEX									
primary geological	44		13		3				60
incipient cone	55		25				1		81
THERMAL									
heat damage	16		97						113
differential luster			17						17
differential color			1						1
SIZE CLASS									
2 (1.6 - 3.2 mm)									0
3 (3.2 - 6.4 mm)	6		5		5				16
4 (6.4 - 12.7 mm)	78		151		21				250
5 (12.7 - 25.4 mm)	187		208		14				409
6 (25.4 - 50.8 mm)	124		34		10		2		170
7 (>50.8 mm)	14								14
TOTAL	409		398		50		2		859

Table 8.16 Summary of tool analysis data for the rockshelter Stratum V assemblage.

	BASALT/ RHYOLITE	CCS/QUARTZ/ PETRIFIED WOOD	OBSIDIAN	OTHER	TOTALS
CORES	1	7			8
bipolar		1			
incipient cone cortex	1				
primary geological cortex		3			
BLANKS	1	3			4
primary geological cortex		1			
remnant ventral surface	1	2			
heat damage		2			
heat treatment		1			
PREFORMS	1	1			2
heat damage		1			
PROJECTILE POINTS	3	9			12
remnant ventral surface		2			
heat damage		2			
heat treatment		1			
reworking	1				
resharpening	2	3			
complete/proximal	3	5			
other fragments		4			
FLAKE/BIFACIAL TOOLS	13	17			30
radial break		1			
square edge	1	1			
COBBLE TOOL	1				1
ANVIL	1				1
CHOPPERS	8			1	9
DRILL	1	1			2
HAMMERSTONES	7				7
PEBBLE TOOLS	3				3
TOTALS	40	38	0	1	79

concentration of four hammerstones, four pebble/cobble tools, and six choppers (Figure 8.15).

All but two of these tools are made from basalt alluvial gravels. The debitage in this area exhibits a high density but is not unusual except for a slightly elevated percentage (39%) of basalt alluvial gravel and bedrock sources in nearly even proportions. Two of the cores represent bipolar recycling of a projectile point in one case (Artifact 14735), and a flake (Artifact 13099). Nearly two-thirds of the diagnostic debitage is associated with percussion bifacial thinning, mainly in the early stage. Alternate flaking operation attributes representing edge beveling were observed on 15 of 120 CCS early-stage percussion bifacial thinning flakes. Evidence for heat treatment was observed on 45 of 120 CCS early-stage percussion bifacial thinning flakes, early-stage core reduction debitage, perhaps

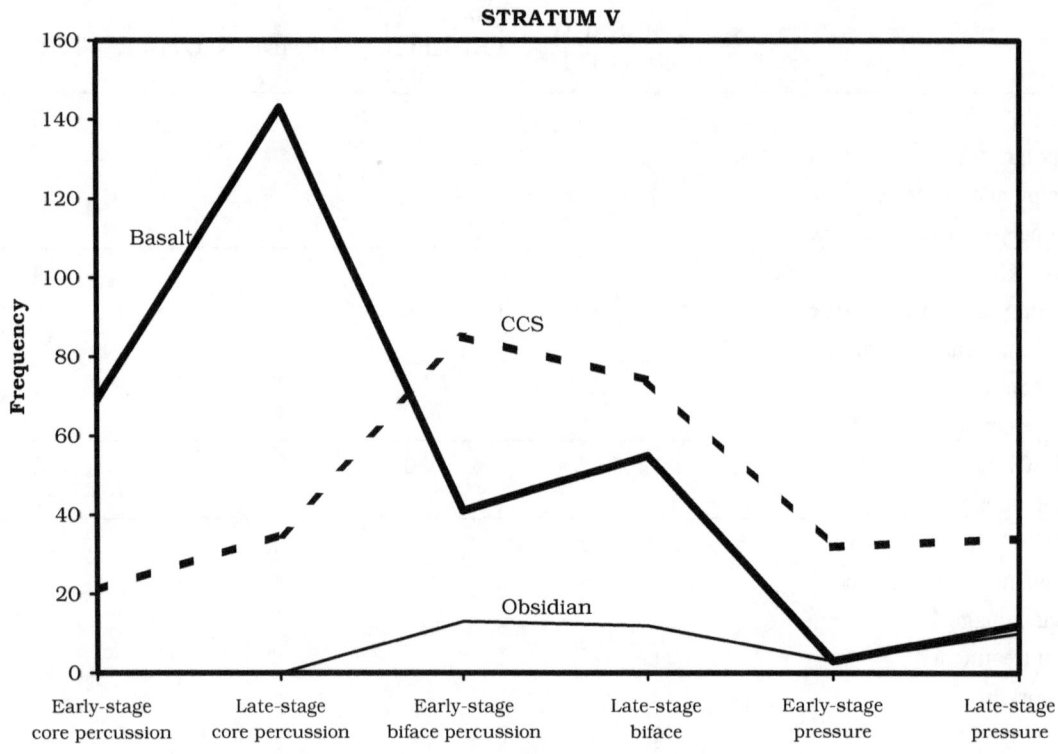

Figure 8.13 Frequency distribution of stage-diagnostic debitage from the rockshelter Stratum V assemblage, reduction trajectory from left to right.

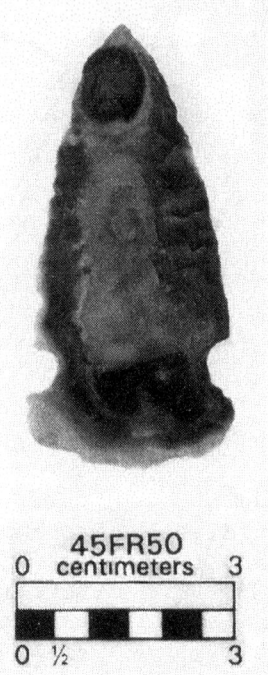

Figure 8.14 Example of a Cold Springs side-notched point (Inv. No. 2768) (print found in Marmes site files in WSU Museum of Anthropology archives).

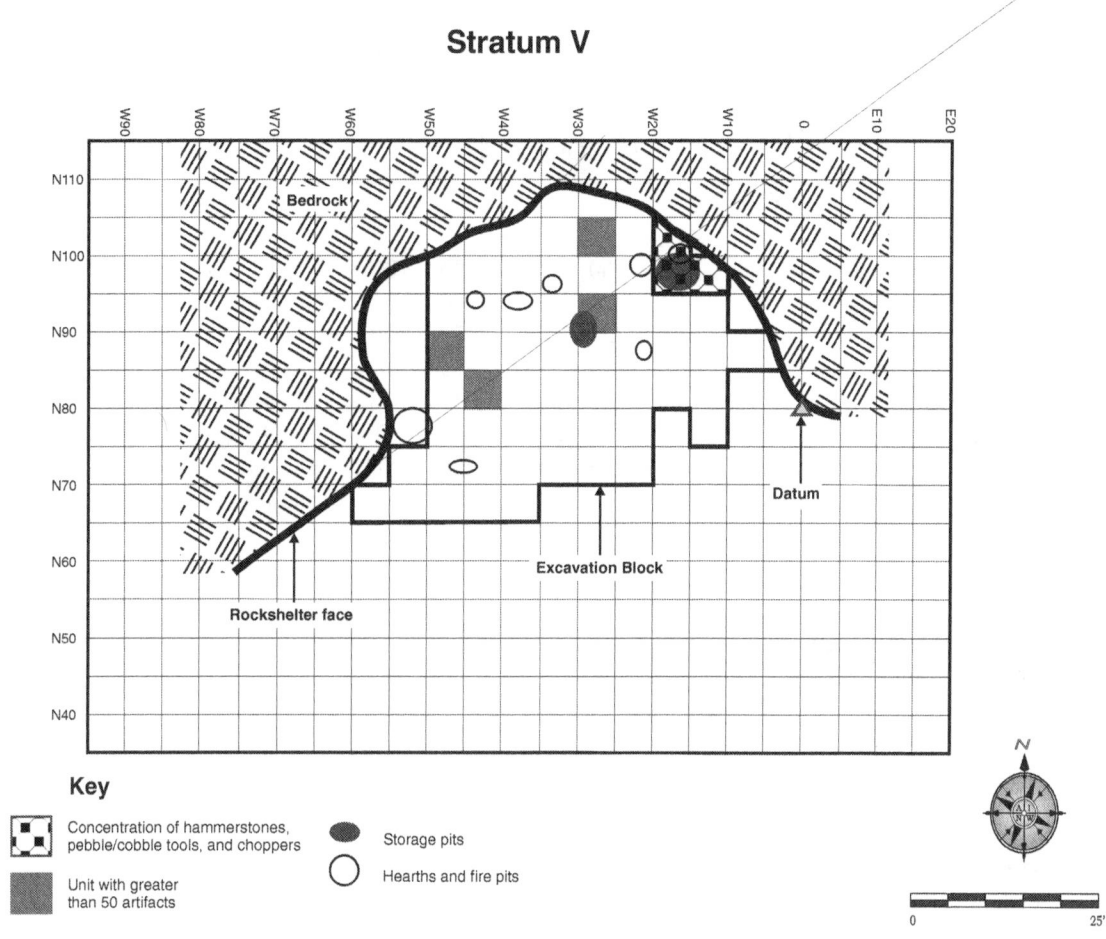

Figure 8.15 Spatial association of high-density artifact deposits and features from the rockshelter Stratum V assemblage.

related to production of the basalt tools. Activities in this area may have been related to food processing for storage, or the expedient tools may have been collected from the shelter area for use in construction of the storage pit.

Rockshelter Stratum Unit VI. The sediments of Stratum Unit VI in the shelter consist of windblown loess and roof fall rock mixed and interrupted by pits and fire hearths (Huckleberry, Gustafson, and Gibson 1998:63-65). Two radiocarbon dates on bone and charcoal produced ages of 1940 ± 70 ^{14}C years BP and 1300 ± 300 ^{14}C years BP, respectively (Sheppard et al. 1987:Table 1). Although these radiocarbon dates are too young, the Stratum VI assemblage was used, in part, to define the Tucannon Phase (2500 B.P. to 5500 B.P.) of the lower Snake River region, possibly one reason that the phase was described as discontinuous with the previous phases (Leonhardy and Rice 1970:11). Artifacts include 710 pieces of lithic debitage and 85 stone tools (Tables 8.17 and 8.18). Heat damage has affected about 25% of the CCS debitage (Table 8.17).

The CCS artifacts (n=464) significantly outnumber basalt artifacts (n=287) and are 58% of the entire Stratum VI lithic assemblage. Core reduction accounts for only 22% of the diagnostic CCS debitage (Figure 8.16), although 10 cores were recovered (15 mm to 51 mm in maximum dimension). The cores and core reduction debitage exhibit cortex from both, and only on a few flakes from other stages of reduction. Remnant ventral surfaces were also observed on a few of the flakes from heat-treated percussion-shaped bifaces and one heat-treated bifacial blank fragment (Artifact 10260-02).

Table 8.17 Summary of debitage analysis data for the rockshelter Stratum VI assemblage.

COMPOSITION	BASALT/ RHYOLITE		CCS/QUARTZ/ PETRIFIED WOOD		OBSIDIAN		QUARTZITE		TOTAL
TECHNOLOGY									
bipolar					1	3%			1
core percussion, early	60	27%	35	12%			1	100%	96
core percussion, late	126	56%	30	10%					156
biface percussion, early	22	10%	120	40%	1	3%			143
biface percussion, late	14	6%	63	21%	12	32%			89
pressure, early	2	1%	30	10%	7	19%			39
pressure, late	1	0%	21	7%	16	43%			38
SUBTOTALS	225	100%	299	100%	37	100%	1	100%	562
undetermined percussion	28		87		1				116
undetermined			25						25
thermal	7								7
TOTAL	260		411		38		1		710
OPERATION									
platform preparation	2								2
bulb removal									0
alternate			15						15
edge preparation									0
grinding or abrasion									0
overshot			4						4
margin removal									0
notch									0
rejuvenation	4		1		3				8
radial break									0
remnant ventral surface			9		2				11
CORTEX									
primary geological	16		26						42
incipient cone	70		19						89
THERMAL									
heat damage	2		102						104
differential luster			52						52
differential color			1						1
SIZE CLASS									
2 (1.6 - 3.2 mm)									0
3 (3.2 - 6.4 mm)	1		2						3
4 (6.4 - 12.7 mm)	22		115		27				164
5 (12.7 - 25.4 mm)	90		228		10				328
6 (25.4 - 50.8 mm)	112		66		1		1		180
7 (>50.8 mm)	35								35
TOTAL	260		411		38		1		710

Table 8.18 Summary of tool analysis data for the rockshelter Stratum VI assemblage.

	BASALT/ RHYOLITE	CCS/QUARTZ/ PETRIFIED WOOD	OBSIDIAN	OTHER	TOTALS
CORES		10			10
bipolar		2			
incipient cone cortex		2			
primary geological cortex		2			
BLANKS		9			9
incipient cone cortex		1			
primary geological cortex		1			
remnant ventral surface		2			
heat treatment		2			
PREFORMS		7			7
remnant ventral surface		5			
heat damage		2			
heat treatment		3			
PROJECTILE POINTS	2	6	3		11
remnant ventral surface		3			
heat damage		3			
heat treatment		1			
use wear	1				
resharpening			1		
complete/proximal	1	2	1		
other fragments	1	4	2		
FLAKE/BIFACIAL TOOLS	16	18	2		36
burin	1	1			
radial break	3				
square edge	3				
ANVIL	1				1
CHOPPERS	2	1			3
HAMMERSTONES	5				5
PEBBLE TOOL	1				1
SCRAPERS		2			2
TOTALS	27	53	5	0	85

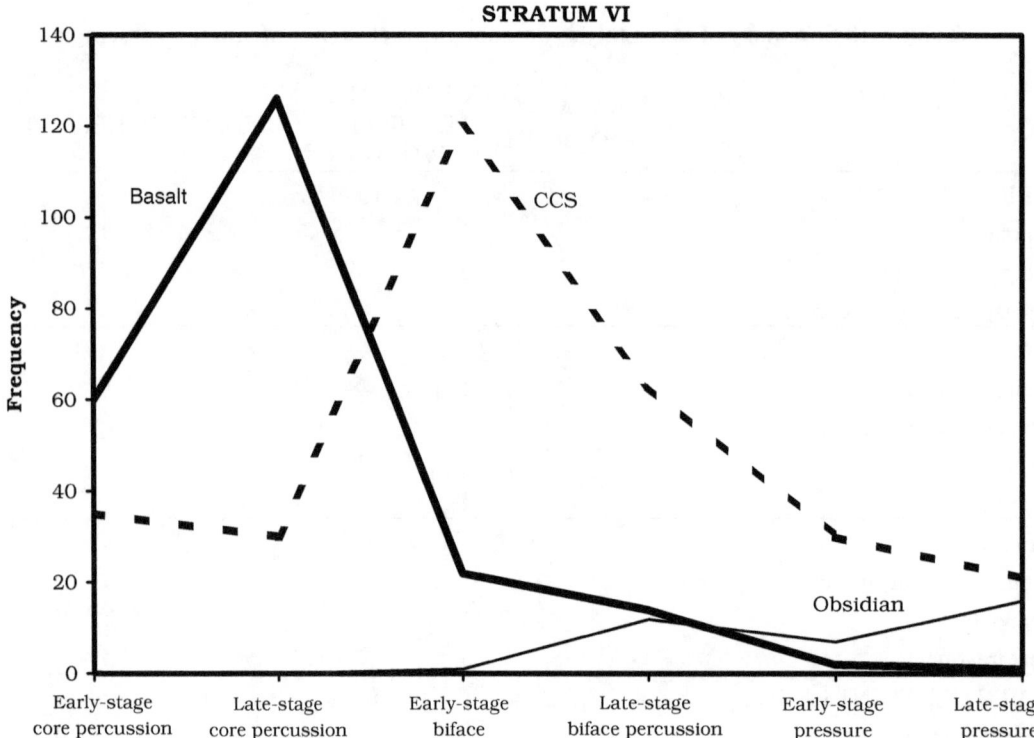

Figure 8.16 Frequency distribution of stage-diagnostic debitage from the rockshelter Stratum VI assemblage, reduction trajectory from left to right.

These attributes suggest that the CCS raw material was commonly heat-treated as flake blanks prior to bifacial reduction. Blanks and preforms of petrified wood were also heat-treated, but as tabular pieces rather than flake blanks. Bifacial pressure flaking debitage is poorly represented in comparison with the pressure flaked bifacial tools (preforms and points).

The basalt tools and debitage from Stratum VI comprise about 36% of the assemblage. Nearly all of the basalt is associated with expedient core reduction technologies, mainly for production of flake tools (Figure 8.16). Some limited use of basalt for bifacial reduction is indicated by a small amount of the diagnostic debitage (Table 8.17) and two projectile points (Table 8.18).

Obsidian artifacts in Stratum VI consist of 38 flakes and 5 tools. The diagnostic obsidian flakes are classified mainly to the late-stages of percussion bifacial thinning and bifacial pressure reduction (Figure 8.16). The tools are projectile point fragments and flake tools made on late-stage percussion bifacial thinning flakes.

Eleven projectile points associated with Stratum VI are extremely fragmentary and include a variety of possible types. One basalt base fragment (Artifact 3109) is characteristic of the side-notched Cold Springs type and one CCS base (Artifact 14325) fragment has a remnant platform surface and outline form suggestive of the lanceolate Cascade type. One CCS distal point fragment (Artifact 8812) includes all but the base and appears to be a corner-notched arrow point broken in manufacture. The remaining fragments are small pieces of points, mainly of dart-size.

Other tools include a relatively large number of flake tools, with square working edges on many of the basalt specimens (Table 8.18). A few of the larger tools such as choppers and hammerstones are also present, mainly in basalt.

More than half of the tools and debitage recovered from Stratum VI are from grid units on the western edge of the excavations within the shelter (W40-50). This higher concentration of artifacts, however, does not appear to correspond with any significant differentiation in technological character. That is, similar types and proportions of tools and debitage are represented in the western high-density area as in the rest of the shelter. One possible activity area at the rear of the shelter (Grid Unit N100-105/W40-50) consists of a high density of debitage (n=117), three choppers, a pebble tool, and a hammerstone just west of three features including Feature 4, a stone filled storage pit (Rice 1969:61, Figure 56). This association of stones in the storage pit is similar to one described for a storage pit in Stratum V.

Rockshelter Stratum Unit VII. Stratum Unit VII in the rockshelter was described as a complex layer of interwoven cultural features such as fire hearths and storage pits along with organic debris formed in windblown loess and roof fall rock (Fryxell and Daugherty 1962:Table III). This material was deposited relatively recently, as indicated by four radiocarbon dates that range from 1600 ± 100 ^{14}C years BP to 660 ± 75 ^{14}C years BP (Sheppard et al. 1987:Table 1). The archaeological deposits have been attributed to the Harder and Piqunin Phases of the lower Snake River region (Leonhardy and Rice 1970). The Stratum VII assemblage contains the largest number of artifacts among the 10 site subassemblages described here and along with the cultural features, represent intensive activity during this period of rockshelter occupation. Stratum VII was excavated using dry-screening with ¼-inch mesh to recover 1,289 pieces of lithic debitage and 145 stone tools (Tables 8.19 and 8.20). The incidental effects of fire, probably associated with human activity within the shelter are indicated by heat damage to about 26% of the CCS debitage (Table 8.19).

The CCS materials constitute fully three-quarters of the artifacts. Core reduction accounts for only 22% of the diagnostic CCS debitage (Figure 8.17), although 13 multidirectional cores were recovered (13 mm to 60 mm in maximum dimension). The cores and core reduction debitage exhibit cortex from both alluvial gravel and bedrock materials in nearly even proportions. About one-half of the diagnostic debitage is associated with percussion bifacial thinning, mainly in the early stage. Alternate flaking operation attributes representing edge beveling were observed on 20 of 218 CCS early-stage percussion bifacial thinning flakes. Evidence for heat treatment was observed on 75 of 218 CCS early-stage percussion bifacial thinning flakes and on 41 of 111 early-stage pressure bifacial reduction flakes. Remnant ventral surfaces were identified on 10 early-stage percussion bifacial thinning flakes and nine early-stage pressure bifacial reduction flakes. The CCS blanks and preforms also exhibit remnant ventral surfaces and evidence of heat-treatment (Table 8.20). These attributes suggest that the CCS raw material was commonly heat-treated as flake blanks prior to bifacial reduction by either percussion or pressure or both. Blanks and preforms of petrified wood were also heat-treated, but as tabular pieces rather than flake blanks. Bifacial pressure flaking debitage is poorly represented in comparison with the pressure flaked bifacial tools (preforms and points).

The basalt tools and debitage from Stratum VII comprise about 21% of the assemblage. Nearly all of the basalt is associated with expedient technologies, mainly for flake tools and choppers made from alluvial cobbles. Some limited use of basalt for bifacial reduction is indicated by a small amount of the diagnostic debitage (Table 8.19) and twelve bifacial tools (blanks, preforms, and projectile points, Table 8.20).

The obsidian artifacts are a meager 3% of the Stratum VII assemblage. The technologically diagnostic obsidian debitage includes a few core reduction flakes, but mainly represents percussion bifacial thinning and, to a lesser degree, bifacial pressure reduction (Figure 8.17). Both primary geological and incipient cone cortex are present on the debitage. The tools include a blank and a preform, two projectile points, and a flake tool. The character of the obsidian debitage suggests that it was imported in a variety of forms and reduction stages.

The 24 projectile points include a variety of nearly complete dart-sized points (n=12) and arrow-sized points (n=6), and some undetermined pieces. The dart-sized points include corner-notched (n=5), side-notched (n=2) and stemmed (n=1) forms. One of the dart-sized point bases has characteristics of a Cascade point (Artifact 14039-02).

The arrow points include basally notched (n=3), corner-notched (n=1), and stemmed (n=2) forms. The forms resemble illustrated types associated with Tucannon, Harder, and Piqunin Phases (Leonhardy and Rice 1970). The diversity of point sizes and forms reflects use of different weapons systems (dart/atlatl and bow/arrow), reworking and resharpening activities associated with hunting equipment, and some mixing of deposits.

Table 8.19 Summary of debitage analysis data for the rockshelter Stratum VII assemblage.

COMPOSITION	BASALT/ RHYOLITE		CCS/QUARTZ/ PETRIFIED WOOD		OBSIDIAN		QUARTZITE		TOTAL
TECHNOLOGY									
bipolar									0
core percussion, early	62	28%	78	11%	1	2%			141
core percussion, late	117	53%	79	11%	2	5%	1	50%	199
biface percussion, early	14	6%	218	31%	16	39%	1	50%	249
biface percussion, late	21	10%	136	20%	9	22%			166
pressure, early	3	1%	111	16%	5	12%			119
pressure, late	4	2%	73	11%	8	20%			85
SUBTOTALS	221	100%	695	100%	41	100%	2	100%	959
undetermined percussion	27		211		4				242
undetermined									0
thermal	12		76						88
TOTAL	260		982		45		2		1289
OPERATION									
platform preparation	2								2
bulb removal			1		1				2
alternate	1		29		1				31
edge preparation									0
grinding or abrasion					1				1
overshot			4						4
margin removal			5						5
notch			2						2
rejuvenation	5				1				6
radial break					2				2
remnant ventral surface			20		9				29
CORTEX									
primary geological	29		49		2				80
incipient cone	55		49		3				107
THERMAL									
heat damage	11		257						268
differential luster			121						121
differential color			3						3
SIZE CLASS									
2 (1.6 - 3.2 mm)									0
3 (3.2 - 6.4 mm)	2		4		1				7
4 (6.4 - 12.7 mm)	19		381		17				417
5 (12.7 - 25.4 mm)	117		489		22		2		630
6 (25.4 - 50.8 mm)	93		107		5				205
7 (>50.8 mm)	29		1						30
TOTAL	260		982		45		2		1289

Table 8.20 Summary of tool analysis data for the rockshelter Stratum VII assemblage.

	BASALT/ RHYOLITE	CCS/QUARTZ/ PETRIFIED WOOD	OBSIDIAN	OTHER	TOTALS
CORES		13			13
incipient cone cortex		3			
primary geological cortex		3			
BLANKS	2	16	1		19
incipient cone cortex	2				
primary geological cortex		3			
remnant ventral surface	1	3			
heat damage		6			
heat treatment		4			
PREFORMS	2	12	1		15
incipient cone cortex		1			
primary geological cortex		1			
remnant ventral surface	1	8	1		
heat damage		4			
heat treatment		3			
PROJECTILE POINTS	5	17	2		24
incipient cone cortex		1			
primary geological cortex		1			
remnant ventral surface	1	6	2		
heat damage		7			
use wear	2				
reworking		3			
resharpening	2	2			
complete/proximal	3	8	1		
other fragments	2	9	1		
FLAKE/BIFACIAL TOOLS	14	35	1	1	51
radial break		5			
square edge	3	1			
CHOPPERS	6			1	7
DRILL		1			1
EDGE GROUND COBBLE				1	1
HAMMERSTONES	8			1	9
PEBBLE TOOL	1				1
SCRAPERS		4			4
TOTALS	38	98	5	4	145

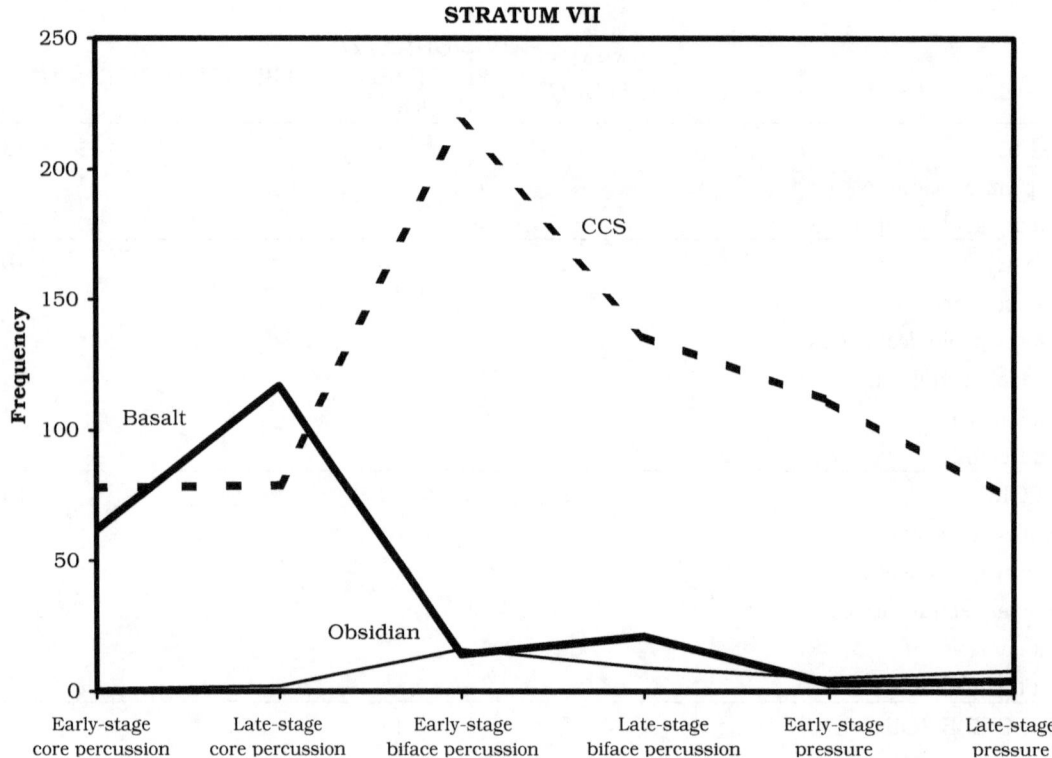

Figure 8.17. Frequency distribution of stage-diagnostic debitage from the rockshelter Stratum VII assemblage, reduction trajectory from left to right.

Other tools include a relatively large number of flake tools, many with square working edges (Table 8.20). Choppers and hammerstones are also well represented.

The lithic subassemblage from Stratum VII was recovered mainly from concentrations at the rear of the shelter in a distribution that matches the features (especially storage pits) identified for the stratum (Rice 1969:64-72, Figure 57). The grid units with high densities of debitage (n=108 – 272) and tools (n=5 – 24) occur against the northcentral (N100-105/W20-30) and northwestern (N90-100/W45-50 and N100-105/W40-45) walls of the shelter in association with the storage pits (Figure 8.18). The technological attributes of the two artifact clusters and the rest of the subassemblage appear fairly homogeneous. Some slight variation was noted for the northcentral cluster where the proportions of CCS early-stage bifacial percussion thinning (52% of diagnostic) and early-stage pressure flakes (23%) were higher and the overall percentage of basalt debitage (31%) was slightly higher than the whole subassemblage. This variation, along with the cores (n=5), blanks (n=4), and preforms (n=4) may suggest the location of a flintknapping area for the production of bifacial tools or a disposal area for the byproducts of such production.

Rockshelter Stratum Unit VIII. The upper surface of the rockshelter was covered in a layer of domestic livestock manure when excavations began. It contained archaeological materials from disturbances to the underlying strata and was designated as Stratum VIII (Fryxell and Daugherty 1962:Table III). A modern radiocarbon date was obtained from charcoal found on the surface (Sheppard et al. 1987:Table 1). Stratum VIII has been correlated to the historic period and ethnographic Numipu (Nez Perce) Phase of the lower Snake River region (Fryxell and Daugherty 1962; Leonhardy and Rice 1970; Sheppard et al. 1987). Some of Stratum VIII was removed from the site prior to setting the excavation grid. The assemblage contains a relatively small number of artifacts recovered in dry-screening with ¼-inch mesh.

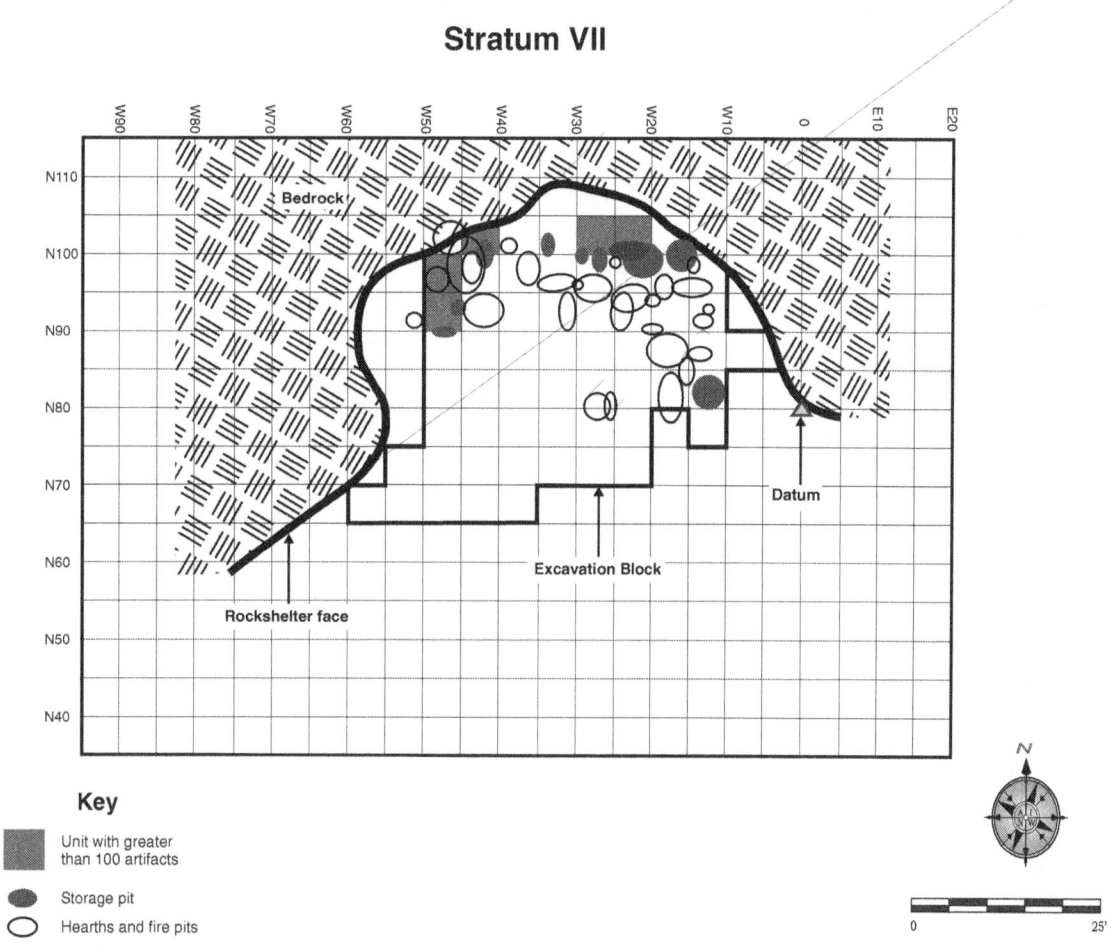

Figure 8.18 Spatial association of high-density artifact deposits and features from the rockshelter Stratum VII assemblage.

The artifacts include 198 pieces of lithic debitage and 25 stone tools (Tables 8.21 and 8.22). Attributes of these materials are similar to those described above for Stratum VI and Stratum VII. One anomalous aspect of the Stratum VIII assemblage is a high number of obsidian percussion bifacial thinning flakes (Figure 8.19). The spatial distribution of tools and debitage is similar to that identified for Stratum VII, although in much lower densities.

Obsidian Sourcing

The results of the previous (Hess 1997) and current (Appendix I) XRF studies of 31 obsidian artifacts from the Marmes site indicate that five distinct geochemical types are represented (Table 8.23). These geochemical types are Indian Creek (n=19 or 61% of the sample), Whitewater Ridge (n=7 or 23% of the sample), Timber Butte (n=2 or 6% of the sample), Gregory Creek (n=2 or 6% of the sample), and Unknown 8 (n=1 or 3% of the sample). The Gregory Creek source, only recently located, was identified as Unknown 7 in Hess' (1997) analysis of two Cascade Phase points from Marmes Rockshelter. The obsidian sources at Indian Creek, Whitewater Ridge, and Timber Butte are described in a report from Northwest Research Obsidian Studies Laboratory (Appendix I). The Gregory Creek source is located on the Malheur River ca. 50 km above its confluence with the Snake River (Figure 8.20). Unknown 8 has yet to be correlated with a geological source location.

Table 8.21 Summary of debitage analysis data for the rockshelter Stratum VIII assemblage.

COMPOSITION	BASALT/ RHYOLITE		CCS/QUARTZ/ PETRIFIED WOOD		OBSIDIAN		QUARTZITE	TOTAL
TECHNOLOGY								
bipolar								0
core percussion, early	10	40%	10	11%	2	9%		22
core percussion, late	12	48%	17	18%	3	13%		32
biface percussion, early	3	12%	36	38%	15	65%		54
biface percussion, late			10	11%	1	4%		11
pressure, early			15	16%	1	4%		16
pressure, late			7	7%	1	4%		8
SUBTOTALS	25	100%	95	100%	23	100%		143
undetermined percussion	3		36					39
undetermined								0
thermal	3		13					16
TOTAL	31		144		23		0	198
OPERATION								
platform preparation								0
bulb removal								0
alternate			3					3
edge preparation								0
grinding or abrasion								0
overshot			1		1			2
margin removal					1			1
notch								0
rejuvenation								0
radial break								0
remnant ventral surface			4		5			9
CORTEX								
primary geological	7		8		6			21
incipient cone	10		6					16
THERMAL								
heat damage	3		38					41
differential luster			15					15
differential color			1					1
SIZE CLASS								
2 (1.6 - 3.2 mm)								0
3 (3.2 - 6.4 mm)								0
4 (6.4 - 12.7 mm)	2		36		2			40
5 (12.7 - 25.4 mm)	4		83		17			104
6 (25.4 - 50.8 mm)	18		23		4			45
7 (>50.8 mm)	7		2					9
TOTAL	31		144		23		0	198

Table 8.22 Summary of tool analysis data for the rockshelter Stratum VIII assemblage.

	BASALT/ RHYOLITE	CCS/QUARTZ/ PETRIFIED WOOD	OBSIDIAN	OTHER	TOTALS
CORES		3		1	4
incipient cone cortex			1		
primary geological cortex	1				
BLANKS		3			3
primary geological cortex	1				
remnant ventral surface	1				
PREFORMS					0
PROJECTILE POINTS		3			3
heat damage	1				
heat treatment	1				
other fragments	3				
FLAKE/BIFACIAL TOOLS	3	9		1	13
radial break	1				
square edge	1				
HAMMERSTONE	1				1
PEBBLE TOOL	1				1
TOTALS	5	18	0	2	25

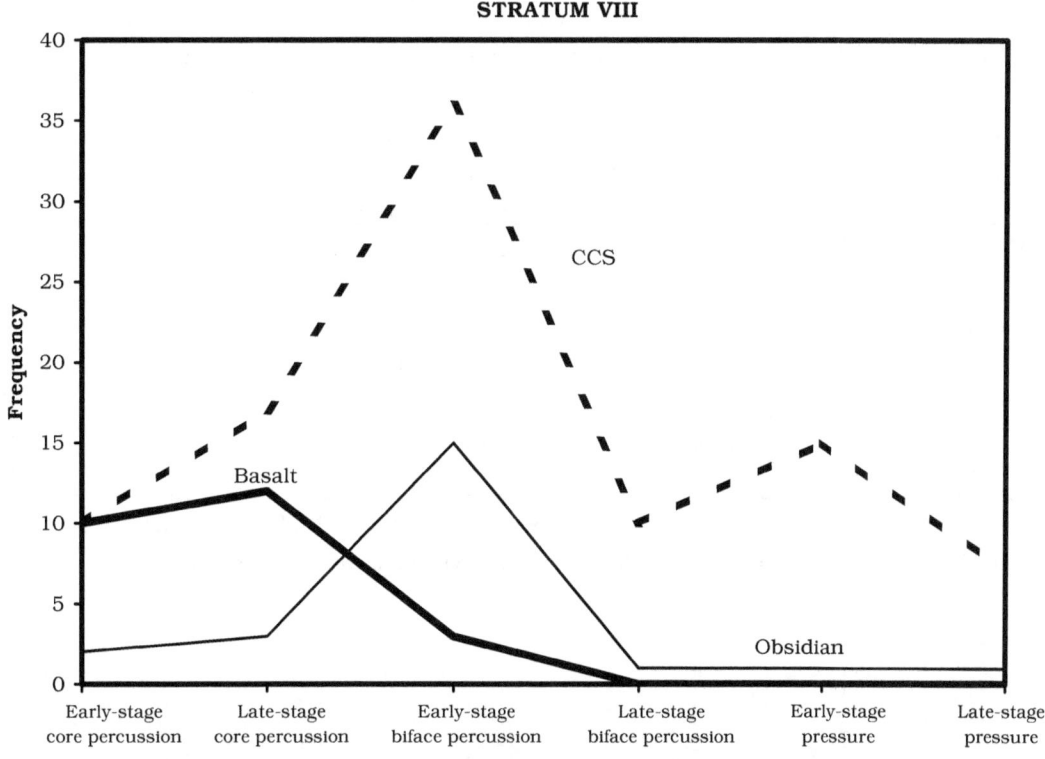

Figure 8.19 Frequency distribution of stage-diagnostic debitage from the rockshelter Stratum VIII assemblage, reduction trajectory from left to right.

Table 8.23 List of obsidian sourcing samples from the Marmes site.

STRATUM	LOT/SPEC #	ARTIFACT TYPE	GEOCHEMICAL SOURCE	OBSIDIAN XRF LAB *
I	4684	Projectile Point**	Gregory Creek	BIOSYS
I	4688	Projectile Point**	Gregory Creek	BIOSYS
I	4818	Early-stage pressure flake	Indian Creek	NWR
I	10683	Late-stage biface percussion flake	Indian Creek	NWR
III	2818	Projectile point, dart-sized	Whitewater Ridge	NWR
III	3672	Windust projectile point	Whitewater Ridge	BIOSYS
III	4185	Flake tool	Indian Creek	BIOSYS
III	4426	Early-stage biface percussion flake	Indian Creek	BIOSYS
III	5273-A	Late-stage pressure flake	Indian Creek	BIOSYS
III	5273-B	Early-stage pressure flake	Unknown 8	BIOSYS
III	5468	Late-stage pressure flake	Indian Creek	BIOSYS
III	7649	Late-stage biface percussion flake	Indian Creek	BIOSYS
III	8756	Early-stage pressure flake	Indian Creek	BIOSYS
III	13398	Early-stage pressure flake	Indian Creek	BIOSYS
IV/V	3767	Projectile point, dart-sized	Timber Butte	NWR
V	15238	Late-stage pressure flake	Indian Creek	NWR
V	13011	Late-stage biface percussion flake	Indian Creek	NWR
V	14360	Early-stage biface percussion flake	Indian Creek	NWR
VI	3230	Projectile point, dart-sized	Whitewater Ridge	NWR
VI	14093-1	Late-stage biface percussion flake	Whitewater Ridge	NWR
VI	13073	Late-stage biface percussion flake	Indian Creek	NWR
VI	5447	Late-stage biface percussion flake	Indian Creek	NWR
VI	12578	Late-stage biface percussion flake	Whitewater Ridge	NWR
VI	8200	Late-stage biface percussion flake	Whitewater Ridge	NWR
VI/VII	3066	Projectile point, dart-sized	Indian Creek	NWR
VII	3098	Projectile point, arrow-sized	Whitewater Ridge	NWR
VII	2898	Projectile point, dart-sized	Timber Butte	NWR
VII	11584	Early-stage core percussion flake	Indian Creek	NWR
VII	10736-1	Early-stage biface percussion flake	Indian Creek	NWR
VII/VIII	2822	Projectile point, dart-sized	Indian Creek	NWR
VIII	14796	Late-stage core percussion flake	Indian Creek	NWR

*NWR = Northwest Research Obsidian Studies Laboratory; BIOSYS = BioSystems Analysis Inc.,
**Analysis data from Hess 1997:Table A10, not included with collection delivered to AINW.

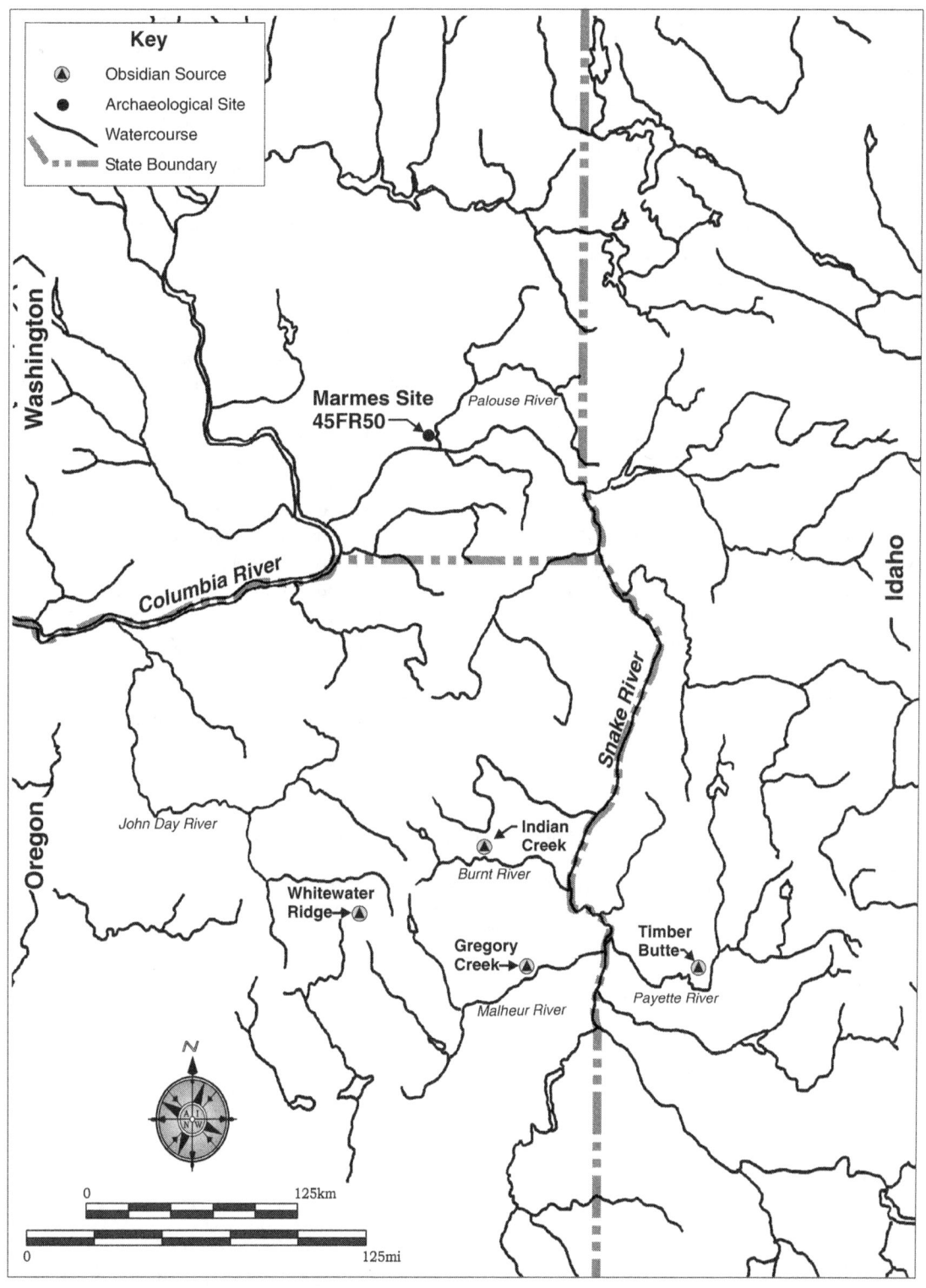

Figure 8.20 Location of obsidian sources identified in XRF analysis of obsidian artifacts from the Marmes site.

river drainages were "important avenues of obsidian transport (Hess 1997:266)." The Indian Creek, Gregory Creek, and Timber Butte sources are located upstream from the Marmes site on highlands above tributaries of the Snake River. Whitewater Ridge is in the divide between the John Day River drainage (a Columbia River tributary) and the Malheur River drainage (a Snake River tributary), but previous studies indicate that Whitewater Ridge obsidian is most abundantly distributed along the John Day River. If Whitewater Ridge obsidian was transported downstream along the John Day River drainage, then it could have arrived at the Marmes site by upstream travel on the Columbia and Snake Rivers.

There is little evidence of change in obsidian source use through time at the Marmes site. Indian Creek appears to be the predominant source throughout the history of site occupation. Deviations in this pattern might be represented by the occurrence of both Gregory Creek samples in Stratum Unit I where they equal the number of Indian Creek samples (Table 8.24). In Stratum Unit VI, the Whitewater Ridge samples (n=4) actually outnumber the Indian Creek samples (n=2). In general, the diversity of sources appears to decrease through time, however the XRF sample is quite small and may not be representative for each period of site use.

Obsidian may have been procured directly by the Marmes people from the geological sources or obtained through trade and exchange relationships with other groups. Imported obsidian arrived at the Marmes site most commonly in the form of bifacial tools. Obsidian debitage used in the XRF analyses is predominately classified as late-stage percussion bifacial thinning flakes and late-stage pressure bifacial reduction flakes derived from maintenance of bifacial tools. The obsidian tools analyzed by XRF are all bifacial projectile points representing reworked and broken or exhausted forms.

All four of the identified sources are located in highland areas south of the Marmes site. These southern obsidian toolstone sources do not necessarily indicate the predominant mobility or trade orientation of Marmes site people since all of the major regional sources of obsidian are located south of the site (Figure 8.20). Although the identified obsidian sources are located long distances from the site, they are probably the closest available sources. The nearest source is Indian Creek, located on Dooley Mountain near the Burnt River approximately 220 km (137 mi)

Table 8.24 Distribution of obsidian samples from the Marmes site.

STRATUM	INDIAN CREEK	WHITEWATER RIDGE	TIMBER BUTTE	GREGORY CREEK	UNKNOWN 8	TOTAL
I	2			2		4
II						0
III	7	2			1	10
IV						0
IV/V			1			1
V	3					3
VI	2	4				6
VI/VII	1					1
VII	2	1	1			4
VII/VIII	1					1
VIII	1					1
TOTALS	19	7	2	2	1	31

from the Marmes site. The most distant source is Timber Butte located near a tributary of the Payette River about 320 km (199 mi) south-southeast of the Marmes site. At intermediate distances are Whitewater Ridge, about 280 km (174 mi) from Marmes near the headwaters of the John Day River, and Gregory Creek, approximately 300 km (186 mi) from the Marmes site near the Malheur River. These direct linear distances considerably underestimate the probable prehistoric transportation distances required to carry obsidian from the geological sources to the Marmes site through the rugged and mountainous terrain that divides them.

Aboriginal trails followed by early Euroamerican explorers and later Oregon Trail emigrants through the Blue Mountains to the Columbia River roughly bisect the highland country containing these obsidian sources. Although these overland routes were probably also used, previous analyses of prehistoric obsidian distribution patterns indicate that these attributes probably reflect extensive use of the obsidian bifaces imported to the Marmes site. Only three of the analyzed flakes represent initial stages of reduction, and these are all from the latest periods of site occupation (Stratum Units VII and VIII) (Table 8.24).

During AINW's technological analysis of the Marmes collection, it was noted that a few pieces of brown and orange-brown glassy volcanic material was present. This material resembles the Beeler Ridge glassy basalt (tachylyte) described for the Joseph Upland area (Hughes 1993; Reid 1997a:70). One piece of the brown material from the Marmes site was informally analyzed by Northwest Research Obsidian Studies Laboratory. It did not match the trace element chemistry of Beeler Ridge, but did resemble the chemistry of tachylytes from other archaeological sites in eastern Washington (Craig Skinner, personal communication 2000).

Blood Residue Analysis

There were eight positive reactions to three antisera on seven artifact samples (Table 8.25). Extracts from two tools tested positive to the deer antiserum: a CCS flake tool (Catalog #8034-1) and a CCS scraper (Catalog #8098). Both tools were from Stratigraphic Unit I. Deer residue may indicate the presence of deer, elk, or moose. The control soil sample from Stratigraphic Unit I (Catalog #17348) was tested against deer antiserum and produced a positive results as well indicating that this result may be due to widespread presence of *Cervidae* residues in the Unit I deposits.

Extracts from a basalt hammerstone (Catalog #2351) from Stratigraphic Unit III tested positive to the chicken antiserum, which may indicate any number of birds that occurred in the site vicinity (e.g., quail, pheasant, ducks, geese, swan). The control soil sample from Stratigraphic Unit III (Catalog #4180) tested negative to the chicken antiserum, indicating the tool itself may have come in contact with bird residue.

Extracts from five tools from Stratigraphic Units I, III, V, and VI tested positive to the trout antiserum, which can indicate the presence of salmon and steelhead in addition to trout. Two of these tools were from Stratigraphic Unit I, a CCS scraper (Catalog #8098) and a CCS flake tool (Catalog #17953); a basalt projectile point (Catalog #3014) was from Stratigraphic Unit III; a CCS projectile point (Catalog #3328) was from Stratigraphic Unit V; and a basalt flake tool (Catalog #14739) was from Stratigraphic Unit VI. The control soil samples from Stratigraphic Units I, V, and VI tested positive to the trout antiserum suggesting that fish residues may be widespread in these deposits, especially given the varied tool types exhibiting this residue. The soil sample from Stratigraphic Unit III tested negative to the trout antiserum.

As noted above, the CCS scraper from Stratigraphic Unit I (Catalog #8098) tested positive to both the deer and trout antisera, as did the soil sample from Stratigraphic Unit I. Positive reactions were confirmed by repeat analysis. It should be noted that the negative results from testing against the selected antisera do not preclude the possibility of any of the specimens retaining residues from other animals.

Summary of Analyses

The analyzed lithic assemblage from the Marmes site includes 1,563 tools and 9,230 pieces of debitage. The assemblage has been divided into a series of artificial units based on stratigraphy and excavation unit and level information from excavations on the floodplain and in the rockshelter. The technological profiles of these analytical units or subassem-blages has been discussed in previous sections of this report and reflect the results of the analysis of the debitage and tools associated with each of these units. The subassemblages from the recognized

Table 8.25 Blood residue analysis comparative results.

RAL #	STRAT. UNIT	CATALOG #	TYPE OF ANTISERUM								
			Bear	Bovine	Deer	Chicken	Rabbit	Sheep	Trout	Human	NIS
1	III	2351	-	-	-	+	-	-	-	-	-
2	VII	2472	-	-	-	-	-	-	-	-	-
3	II	2833	-	-	-	-	-	-	-	-	-
4	III	3014	-	-	-	-	-	-	+	-	-
5	VII	3052	-	-	-	-	-	-	-	-	-
6	VII	3097	-	-	-	-	-	-	-	-	-
7	V	3117	-	-	-	-	-	-	-	-	-
8	II	3169	-	-	-	-	-	-	-	-	-
9	II	3171	-	-	-	-	-	-	-	-	-
10	V	3328	-	-	-	-	-	-	+	-	-
11	III	3649	-	-	-	-	-	-	-	-	-
12	III	3788	-	-	-	-	-	-	-	-	-
13	III	3791	-	-	-	-	-	-	-	-	-
14	III	3792	-	-	-	-	-	-	-	-	-
15	VI	5460	-	-	-	-	-	-	-	-	-
16	III	6221	-	-	-	-	-	-	-	-	-
17	I	6921-1	-	-	-	-	-	-	-	-	-
18	I	8034-1	-	-	+	-	-	-	-	-	-
19	I	8098	-	-	+	-	-	-	+	-	-
20	N/A	10078	-	-	-	-	-	-	-	-	-
21	N/A	10083	-	-	-	-	-	-	-	-	-
22	V	11224	-	-	-	-	-	-	-	-	-
23	VI	12749	-	-	-	-	-	-	-	-	-
24	II	13166	-	-	-	-	-	-	-	-	-
25	VI	14739	-	-	-	-	-	-	+	-	-
26	III	17923	-	-	-	-	-	-	-	-	-
27	II	17945	-	-	-	-	-	-	-	-	-
28	I	17953	-	-	-	-	-	-	+	-	-
29	III	18046	-	-	-	-	-	-	-	-	-
A	III	4180			-				-		
B	V	12217							+		-
C	VI	14741							+		-
D	I	17348			+				+		-
GEL #			1660	1661	1657	1665	1662	1659	1663	1657	1660
GEL #					1658	1689			1689		
GEL #					1689						
REPEAT GEL #			1668	1668	1670	1669	1671	1667	1669		
REPEAT GEL #					1690				1690		

Key: + = Positive; - = Negative; N/A = Not Applicable

stratigraphic units in the rockshelter have been used to define cultural phases for the lower Snake River region (Leonhardy and Rice 1970; Rice 1972). In the following section of this report, the salient features resulting from technological analysis of the lithic artifacts will be discussed and compared for each of these cultural phases.

The Marmes site assemblage has been divided into five cultural/temporal phases. From earliest to most recent, these are the Windust, Cascade, Tucannon, Harder, and Numipu phases. Our technological analysis has identified some major differences and some similarities or continuities between these assemblages that are noted below.

The Windust Phase

The Windust Phase at the Marmes site is represented by the Harrison and Marmes horizons from the floodplain, and Stratum Units I and II from the rockshelter. As a whole, the Windust debitage reflects a predominance in the use of CCS (67%) for both the floodplain and the rockshelter subassemblages, followed by basalt (22%), while obsidian accounts for 11% of the debitage. Raw materials used for the production of tools from the Windust assemblage also show a predominance in the use of CCS (66%), followed by basalt (32%). Although the use of the fine screen for recovery of artifacts from the floodplain has greatly affected the debitage profile resulting in a much higher count and percentage of obsidian from the floodplain, there are other differences that apparently are not related to sampling differences between the floodplain and the rockshelter.

The technologically diagnostic debitage from the Windust Phase deposits for all raw materials reflects a similar profile for percussion core reduction, percussion bifacial thinning, and pressure flaking. There is a slightly greater abundance of percussion core reduction debitage (39%), however, pressure flakes (31%) and percussion bifacial thinning flakes (30%), each account for approximately one-third of the diagnostic debitage. Thus, there appears to be relatively equal representation for percussion core reduction, percussion biface reduction, and pressure flaking within the debitage assemblage from the Windust Phase deposits.

There appears to have been distinctive and differential use of the floodplain and the rockshelter during the Windust Phase. For example, while 57% of the debitage is from the floodplain and 43% is from the rockshelter, 95% of the tools are from the rockshelter and only 5% are from the floodplain. The tools from the floodplain include a core, a blank, a Windust projectile point fragment, four flake/bifacial tools, two choppers, a cobble tool, and a scraper. These tools reflect activities associated with flintknapping for the production of bifaces, presumably projectile points to replace ones broken in use, and the manufacture and/or processing of other materials for food and perishable tools. By contrast, the Windust Phase tools from the rockshelter reflect a much wider range of activities. Flintknapping tools (cores, blanks, and preforms) account for 17% of the tools from the rockshelter, while projectile points account for 22% of the tools. Flake/bifacial tools account for 17% of the tool assemblage from the Windust Phase rockshelter deposits and are interpreted as reflecting tools used to make other tools from wood, bone, antler, and hide. Hammerstones account for 9% of the tools and are interpreted as pounding tools for processing food stuffs and for flintknapping.

Types of tools present in the Windust Phase rockshelter deposits, but in low frequencies or as single items include an abrader, a bola stone, a knife, a mano, four pebble tools, eight choppers, and five scrapers. For the most part, these miscellaneous tools suggest that a wide range of food preparation and procurement tasks were conducted in the rockshelter.

Overall, the Windust Phase deposits reflect use of the site as a base for food procurement and for food processing. The lithic technology also reflects an emphasis on the manufacture and production of tools from organic materials (probably wood, antler, and bone) that would be consistent with a habitation site. There is also ample evidence of stone tool production, but with an emphasis on the maintenance and production of bifacial edges through percussion bifacial thinning and pressure flaking as indicated in the CCS and obsidian debitage. Late-stage percussion core reduction of basalt is also notable, and moderate amounts of late-stage percussion core reduction CCS debitage suggest that flake blanks were produced for use as flake tools as well as for the production of biface blanks and preforms.

The Early Cascade Phase

The Early Cascade Phase is represented by Stratum Unit III from the rockshelter. As with the Windust Phase debitage, use of CCS during the Early Cascade Phase predominates at 53%, however, there is a marked increase in the use of basalt that consists of 45% of the debitage. The raw materials selected for the production of tools reflects a similar relationship, with CCS accounting for 51% and basalt 48% of the Early Cascade Phase tools. By contrast, basalt in the Windust Phase assemblage accounted for only 32% of the tools.

The technological profile for the Early Cascade Phase debitage for all raw materials reflects a preponderance of percussion core reduction. Over half of the diagnostic debitage was produced by percussion core reduction (54%), while 39% was produced by percussion biface thinning. Only 6% of the diagnostic debitage was produced by pressure flaking, however, we suspect that pressure flakes are underrepresented in the assemblage from the rockshelter due to screen size used during the excavations. A comparison with the Windust Phase diagnostic debitage from the rockshelter with the diagnostic debitage from the Early Cascade Phase deposits indicates that both assemblages are very similar. The Windust Phase diagnostic debitage profile for the rockshelter assemblage reveals a lower percentage of percussion core reduction flakes (49%), the same percentage of percussion biface thinning flakes (39%), but with a higher percentage of pressure flakes (12%). Thus, the overall diagnostic debitage profiles for the debitage from the rockshelter for both the Windust Phase and the Early Cascade Phase deposits are virtually indistinguishable.

Tools from the Early Cascade Phase assemblage include an abundance of flake/bifacial tools (44%), some of which are interpreted as tools used to produce other tools from wood, bone, antler, or hide. Projectile points account for 18% of the tool assemblage, while cores, blanks, and preforms account for 14% of the tools. Hammerstones represent 9% of the tool assemblage, while choppers account for 6% of the tools. The cores, blanks, and preforms reflect flintknapping activities, while the hammerstones represent both food processing and flintknapping activities. Other tools present in low numbers or represented by a single item include an edge ground cobble, a drill, nine pebble tools, and five scrapers.

The Early Cascade Phase assemblage reflects an occupation by people using basalt and CCS for the production of flake blanks for projectile points and flake tools. The tools suggest that a wide range of domestic and economic activities were conducted at the site, with an emphasis on the production of tools from wood, bone, antler, or hide, as well as food processing. Projectile points represent tools that have been maintained and rejuvenated, most of which have been broken in use, and presumably discarded at the site and replaced with new points. The edge ground cobble, Cascade points, and choppers are commonly associated with Cascade Phase deposits, however, the notched pebble tools described above, appear to be a specialized type of tool used for the processing of sinew or other fibers. The Early Cascade Phase assemblage reflects a habitation site where a wide range of domestic tasks were performed.

The Late Cascade Phase

The Late Cascade Phase is represented by Stratum Unit V from the rockshelter at the Marmes site. The lithic assemblage from the Late Cascade Phase deposits is similar to that of the Early Cascade Phase deposits. There are minor differences in the percentages of raw materials represented in the debitage. The Late Cascade Phase debitage assemblage is comprised of 48% basalt, 46% CCS, and 6% obsidian, while the raw materials selected for tools include basalt at 51%, and CCS at 48%. This represents a continuation of the trend from Windust through the Early Cascade Phase of increasing proportional use of basalt over CCS in both the debitage and tool assemblages.

The technological profile for the Late Cascade Phase assemblage differs from the Early Cascade Phase assemblage with both percussion core reduction and percussion bifacial thinning representing 42% of the diagnostic debitage. The Early Cascade Phase, on the other hand, was characterized by a predominance of percussion core reduction debitage.

Like the Early Cascade Phase tool assemblage, the Late Cascade Phase tools include a broad range of tool types associated with flintknapping, maintenance of projectile points, and the processing of food products and organic materials for tools. The category "flake/bifacial tools" accounts for 38% of the tool assemblage from the Late Cascade Phase assemblage. These tools represent a wide range of items interpreted as tools used in the production of other tools from wood, bone, antler, or hide. Tools associated with flintknapping, cores, blanks, and preforms, account for 18% of the tool assemblage, while choppers comprise 11%. The Late Cascade tool assemblage also includes an abrader, two drills, seven hammerstones, and three notched pebbles.

The lithic assemblage from the Late Cascade Phase at the Marmes site represents a wide range of tool production and maintenance and food processing activities. As with the Early Cascade Phase assemblage, basalt and CCS cores were used to produce flake blanks for flake tools and projectile points. Broken points were discarded at the site, and presumably replaced with new points; and wood, bone, and antler was fashioned into tools. Foods and fibers were also processed at the site during Late Cascade Phase times.

The Tucannon Phase

At the Marmes rockshelter, Stratum Unit VI has been assigned to the Tucannon Phase. Unlike the Late Cascade Phase debitage, the CCS debitage from the Tucannon Phase at 58% greatly outnumbers the basalt debitage (37%), reversing a prior trend of increasing basalt use from the Windust through the Lake Cascade Phases; obsidian remains relatively low at 5%. Raw materials used for the production of tools reflect a similar distribution with CCS accounting for 62%, basalt for 32%, and obsidian 6% of the tools.

The technological profile for the diagnostic debitage from the Tucannon Phase deposits generally mirrors that of the Late Cascade Phase debitage. While 45% of the debitage was produced by percussion core reduction, 41% resulted from percussion bifacial thinning, and 14% from pressure flaking. The major difference in the debitage between the Tucannon Phase and Late Cascade Phase occurs in the distribution of the diagnostic obsidian debitage. While the Tucannon Phase exhibits an abundance of late-stage pressure flakes that represent 43% of the diagnostic debitage, and late-stage percussion bifacial thinning at 32%, the Late Cascade Phase obsidian debitage reflected more of an emphasis on both early- and late-stages of percussion bifacial thinning (34% and 32% respectively).

The Tucannon Phase tool assemblage differs from the preceding Late Cascade Phase tool assemblage, by having more types of tools represented. For example, the Tucannon Phase flintknapping tools account for 31% of the tool assemblage, and all of the cores, blanks, and preforms are made from CCS raw materials. Tucannon Phase projectile points account for 13% of the tools, and flake/bifacial tools account for just over 42% of the tools. Other miscellaneous tools assigned to the Tucannon Phase assemblage include an anvil, three choppers, five hammerstones, a flaked pebble, and two scrapers. As with the previous Late Cascade Phase, the Tucannon Phase tool assemblage reflects a wide range of tool manufacturing and economic activities including the processing of food and organic raw materials. During the Tucannon Phase, the Marmes rockshelter served as a habitation site where people procured and processed foods and made tools and other products from wood, bone, antler, and hides.

The Harder Phase

At the Marmes rockshelter, Stratum Unit VII has been assigned to the Harder Phase. Like the Tucannon Phase debitage, the CCS debitage from the Harder Phase, at 76% greatly outnumbers the basalt debitage (20%), while obsidian remains relatively low at 3%. Raw materials used for the production of tools reflect a similar distribution with CCS accounting for 68%, basalt for 26%, and obsidian 3% of the tools.

The technological profile for the diagnostic debitage from the Harder Phase deposits is quite similar to that of the Tucannon Phase debitage. While 45% of the debitage was produced by percussion core reduction, 43% resulted from percussion bifacial thinning, and 21% from pressure flaking in the Harder Phase diagnostic debitage assemblage. The major difference in the diagnostic debitage between the Harder Phase and the Tucannon Phase is evident in the distribution of the obsidian debitage, although this is only 3% of all debitage. While the Tucannon Phase exhibits an abundance of late-stage pressure flakes that represent 43% of the diagnostic debitage, and late-stage percussion bifacial thinning at 32%, the Harder Phase obsidian debitage reflects much more of an emphasis on percussion bifacial thinning on both early- and late-stages with 39% and 20% respectively.

The Harder Phase tool assemblage differs from the preceding Tucannon Phase tool assemblage, by having fewer basalt tools, slightly more CCS tools, fewer obsidian tools, but more tools from "other" types of lithic raw materials. The percentage of flintknapping tools at 32% in the Harder Phase is similar to the Tucannon Phase. However, the flintknapping tools in the Harder Phase assemblage reflect the use of CCS, basalt, and obsidian, whereas these tools in the Tucannon Phase reflected only the use of CCS. Harder Phase projectile points account for 17% of the tools (slightly more than for the Tucannon Phase), while Harder Phase flake/bifacial tools account for 35% of the tools, slightly less than for the Tucannon Phase. Other miscellaneous tools assigned to the Harder Phase assemblage include seven choppers, a drill, an edge ground cobble, nine hammerstones, a flaked pebble, and four scrapers. As with the previous Tucannon Phase, the Harder Phase tool assemblage reflects a wide range of tool manufacturing and economic activities including the processing of food and organic raw materials. The Harder Phase lithic assemblage at the Marmes rockshelter reflects use of the site for processing food and the manufacture of wood, bone, and antler tools. The location also served as a place for processing hides and fibers and is interpreted as a Harder Phase habitation site.

The Numipu Phase

Stratum Unit VIII at the Marmes rockshelter has been assigned to the Numipu Phase. The debitage from this most recent cultural deposit is characterized by a predominance of CCS (73%), with 16% basalt, and 12% obsidian. This distribution reflects a marked increase in the amount of obsidian over the previous Harder Phase debitage assemblage. Lithic raw materials selected for the production of tools are heavily weighted toward CCS accounting for 72% and basalt accounting for 20% of the tools. There were no obsidian tools in the Numipu Phase assemblage in spite of the relatively high percentage of obsidian debitage.

The technological profile for the diagnostic debitage from the Numipu Phase deposits is quite similar to that of the Harder Phase debitage. While 38% of the debitage was produced by percussion core reduction, 45% resulted from percussion bifacial thinning, and 17% from pressure flaking in the Numipu Phase diagnostic debitage assemblage. The distribution of diagnostic debitage by raw material between the Harder Phase and the Numipu Phase is very similar for basalt with an emphasis in both assemblages on percussion core reduction. On the other hand, the CCS in the Numipu Phase assemblage exhibits a preponderance of early-stage percussion bifacial thinning flakes (38%) and late-stage percussion core reduction flakes. The obsidian exhibits a similar pattern with 65% reflecting early-stage percussion bifacial thinning flakes and 13% late-stage percussion core reduction flakes.

The Numipu Phase tool assemblage is relatively small, however, 72% of the tools are made from CCS while 20% are basalt. There are no obsidian tools. While 52% of the tools are flake/bifacial tools, 28% are flintknapping tools (cores and blanks). Projectile points account for 12% of the tool assemblage and there is one flaked pebble and a hammerstone. Unlike previous assemblages, the range of tools is quite narrow, but does reflect limited flintknapping activities, the maintenance of projectile points, and the manufacture of tools from wood, bone, or antler. The hammerstone may reflect limited amounts of food processing and flintknapping, and the flaked pebble may represent a specialized fiber processing tool.

The Numipu Phase lithic assemblage at the Marmes rockshelter reflects a very limited use of the site. Much of the deposits attributable to the Numipu Phase may have been removed from the rockshelter prior to the archaeological excavation; this is certain for those units excavated during the 1968 effort.

Conclusions

Culture-History

Probably more than any other site on the Columbia Plateau, information obtained from the Marmes site has influenced archaeologists' perceptions of regional prehistory. The model of a culture-historical sequence for the lower Snake River region that remains a standard reference 30 years after it was published (Leonhardy and Rice 1970) is based primarily on tool assemblage data from the Marmes Rockshelter. Reanalysis of the lithic artifact collections from the site confirms the general outline of the model and adds some data for the debitage, certain tool types, and lithic raw material sources. Correlation of the artifacts with the stratigraphic associations also show that much of the provenience data from the original excavations has been lost or was never assigned to some excavated materials. The existing provenience data used in this report indicate that many of the diagnostic artifact types cross-cut stratigraphic and temporal boundaries for the assemblages and periods that they characterize. The reasons for this probably include both mixing of the archaeological deposits and survival of technological traditions through long periods of time.

The original investigators at the Marmes site described the prehistoric occupations as essentially continuous from about ten thousand years ago to the present (Fryxell and Keel 1969:viii). Although the occupations may have been continuous, the level of activity as measured by the number of lithic artifacts recovered from the rockshelter, reflect significant fluctuations in the intensity of the occupations. The artifact distribution data (Table 8.1) and previously reported information on faunal material density and cultural features (particularly fire hearths) suggest very intensive occupations during the Cascade Phase from about 7900 to about 6200 radiocarbon years B.P. and again during the Harder Phase from about 1600 to about 700 radiocarbon years B.P. The flurry of activity during these periods appears to be responsible, in part, for mixing the deposits. Earlier and later uses of the site surrounding both of these intensive periods of occupation produced fewer artifacts, some of which appear to be intermingled with artifacts from the periods of intensive occupation.

Technologies change through time, but the speed and direction of technological change depends on factors such as environmental conditions, population movements, and historical events. The evolution of technologies at the Marmes site appears to be deliberate and conservative. That is, while there are differences between successive occupations, they seem gradual and incremental. For example, the Windust Phase assemblages include stemmed projectile points commonly made on CCS materials and the Cascade Phase assemblages include lanceolate projectile points commonly made from basalt materials. However, both point types made on both materials occur in the deposits of each respective age (Table 8.26). Furthermore, the breakage patterns, rejuvenation techniques, and choices regarding what elements to discard are similar. Moreover, the other tool types and technologies associated with the projectile points of each period are also similar. What distinguishes one period from another is simply the relative frequency of certain artifact types and traits. Uncertainty regarding the degree to which the artifacts representing these traits are mixed between deposits and their frequencies masked by periodic flourishes in activity are the major obstacles to understanding technological change at the site.

Some more subtle aspects of the lithic technologies appear to change within the periods representing the putative cultural phases or style zones. For example, the earliest Windust phase assemblage at the site from Stratum I in the rockshelter shows more emphasis on the later stages of percussion bifacial thinning (Figure 8.6), suggesting a biface-based technology akin to Clovis and other early Paleoindian technologies where extensive and systematic bifacial thinning are important elements of the reduction technologies (Wilke et al. 1991). The slightly later Windust materials from the floodplain (Harrison [Figure 8.2] and Marmes [Figure 8.4] Horizons) and Stratum II [Figure 8.8] in the rockshelter contain proportionately fewer late-stage percussion bifacial thinning flakes indicating a shift to thin, linear flake blanks for the production of projectile points. These blanks are produced by different core reduction techniques and do not require systematic percussion thinning. This pattern of reduction extends from the later Windust occupations and into the early Cascade occupation represented by Stratum III in the rockshelter. However, after the eruption of Mt. Mazama, a brief return to use of more extensive percussion bifacial thinning is indicated for the late Cascade materials from Stratum V (Figure 8.13).

Another underlying trend in the lithic technologies involves proportional use of different lithic material compositions. Basalt (or andesite) and CCS materials predominate and obsidian is present in small amounts throughout the site occupations (Figure 8.21). CCS materials are overwhelmingly predominant in the earliest occupations, but the use of basalt increases steadily (apart from the Stratum IV Mazama ash layer) through the late Cascade occupation reaching equal importance, and then declining again thereafter (Figure 8.21). The heaviest use of basalt materials appears to be linked with production of basalt projectile points. In this respect, the Windust and Cascade assemblages both contribute to a trend that suggests continuity. This trend has been previously noted by other researchers (Reid 1995:Figure 80).

As noted above, the projectile point types represented in the subassemblages from the rockshelter strata vary generally according to the developed chronologies. Only one typeable projectile point was identified for the floodplain

215

Table 8.26 Seriation of Marmes site projectile point styles using Rice's classes correlated with identified historical types (adapted from Ames 2002).

Rice's types	Marmes strata								Historical types
	I	II	III	IV	V	VI	VII	VIII	
1.35							3		
1.34						3	3		
1.33						11	11		Columbia Valley[1]
1.32						12	44	5	Wallula[2]
1.31					2	3	3		
1.30						4	9		
1.29						1			Cascade
1.28					2	2	3		Tucannon[3]
1.27						2	1		
1.26						8	3		Tucannon
1.25					3	6	4		Tucannon
1.24			2		8	2	6		Snake River Corner-notched[4]
1.23					2	12	2		Cascade
1.22					5	4			Snake River Corner-notched
1.21					2	1			
1.20			1		1				
1.19					35	7			Cold Springs
1.18			1			1			Cascade
1.17		1	30		9	2			Cascade
1.16		1	4		1	1			
1.15		1	6						Cascade
1.14		2	10			1			Cascade
1.13			3		2	1			Cascade
1.12		4	2						
1.11		2	17						Windust
1.10			3						Windust
1.09		1	1						
1.08		1							Cascade
1.07		3							Windust
1.06	1	1	1						Tucannon
1.05	1	2	2						Cascade
1.04	1	1	1						Windust
1.03	1	4	4						
1.02		8	1						Windust
1.01	2	10	2						Windust

[1] corner-notched arrow points
[2] basally-notched arrow points
[3] stemmed dart points
[4] dart points

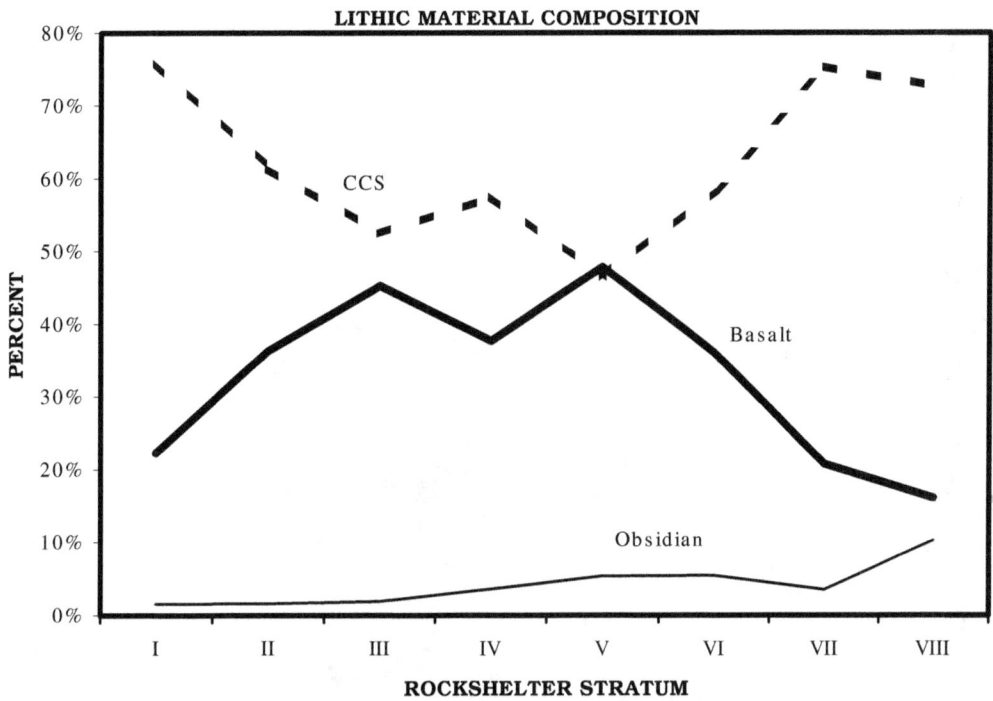

Figure 8.21 Proportional representation of the major lithic material composition types in the rockshelter strata.

subassemblages, a basalt Windust point in the Harrison Horizon. For this analysis, the projectile points were classified as Windust (n=15; e.g., Figure 8.24 a), Cascade (n=37; e.g., Figure 8.23 and 6.24 b), Cold Springs (n=9; e.g., Figure 8.24 c), Dart Points (n=8), and Arrow points (n=4). Comparison of the number of projectile points available for this analysis with the numbers of points in Rice's classification (see Table 8.26) indicates the number of points that have gone missing from the Marmes collection (see also Figure 8.1). The first three types are common archaeological designations for the region, the last two represent combined type designations, collapsed because of their individual low frequencies. Note that the terms dart point and arrow point have been used elsewhere in this report to identify technological classes of projectile points and fragments of projectile points.

Here the terms Arrow Point and Dart Point refer to specific typological classes of typologically complete specimens. The Dart Points include two stemmed "Tucannon" types from Stratum VII and six Snake River corner-notched types from Stratum VI (n=1) and Stratum VII (n=5).

The Arrow Points include three small basally-notched (Wallula) and one corner-notched (Columbia Valley) types, all from Stratum VII. A frequency distribution of the typologically distinctive projectile points identified within the subassemblages shows each type modality peak in the rockshelter stratum predicted by the Leonhardy and Rice (1970) chronology (Figure 8.22). The distributions around these modes, however, suggest that some types were long-lived, most notably the Cascade type. Cascade points occur in every stratum with typeable points. Also, Cascade points outnumber the Windust points in Stratum II (7 Windust and 12 Cascade). Again, the degree of mixing among the archaeological deposits and its contribution to the pattern of projectile point distribution is uncertain.

Index Fossils and Tool Functions

Certain types of artifacts, particularly distinctive projectile point forms, have acquired the status of temporal diagnostics based on their relative stratigraphic position within the deposits of the Marmes site. Although some of the contextual problems associated with use of these types as index fossils have been resolved by repeated

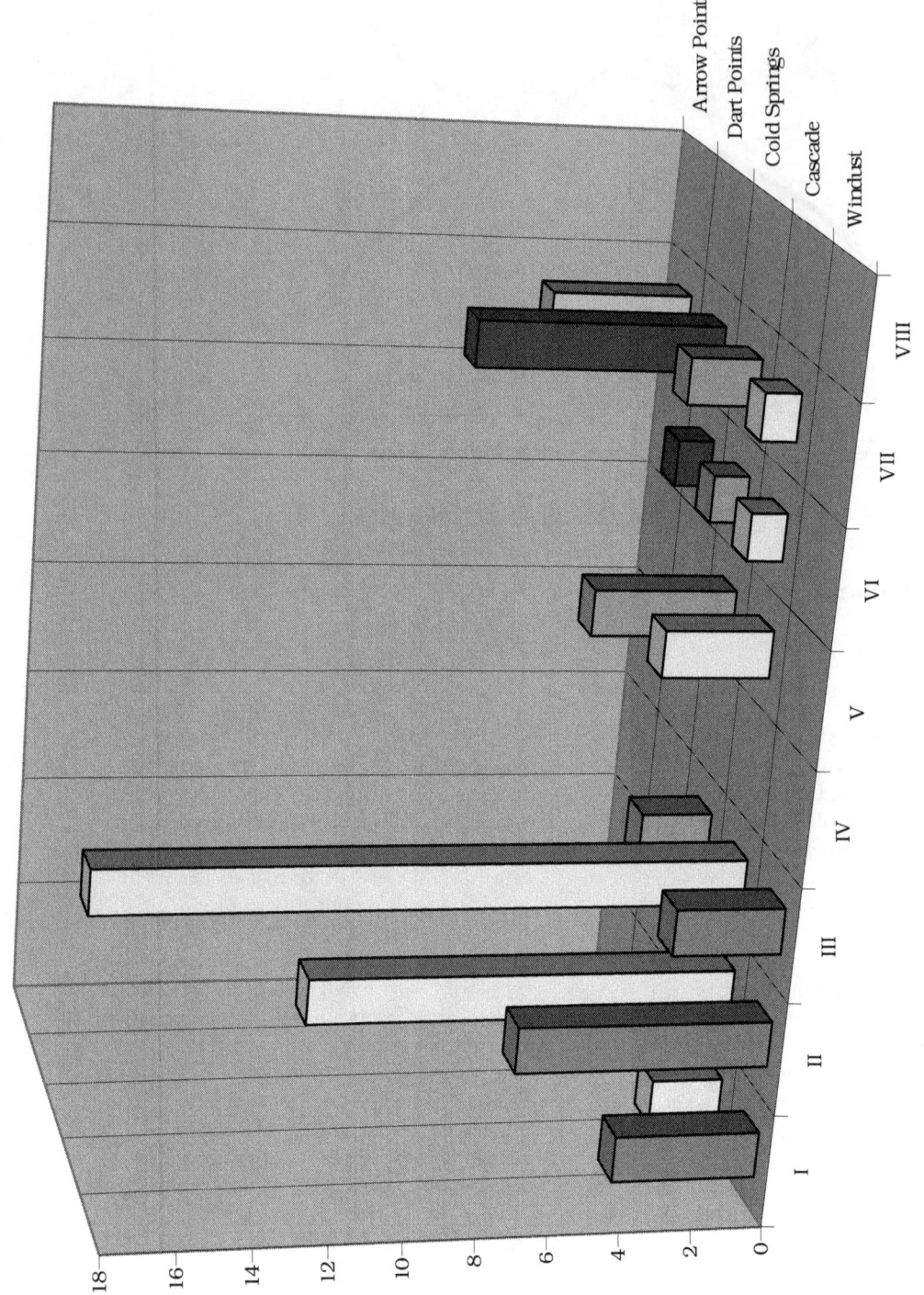

Figure 8.22 Frequency distribution of projectile point types among rockshelter strata. Dart points are stemmed and corner notched. Arrow points are basally notched and corner notched.

associations with well-dated deposits at other sites, the nature of the technologies and functions they represent are still poorly understood. This report emphasizes those rtifacts that could be associated with specific locations within the site and certain stratigraphic layers. Many of the important types of artifacts recovered from the site occur in greater numbers. Although the provenience information is lacking, technological relationships and patterns can provide a departure for further discussion.

The general function of projectile points is implicitly understood as piercing tips for spears, darts, and arrows used in animal hunting and warfare. More specific associations of certain kinds of projectile points with various propulsion systems and hunting or warfare strategies are poorly known. Some researchers have suggested that large lanceolate points (like Cascade, Figure 8.23) are associated with thrusting spears while large notched points (like Cold Springs) are associated with dart and atlatl systems (Flenniken and Wilke 1989). The impact fractures associated with Windust, Cascade, and Cold Springs points are all similar in location and type suggesting high velocity impacts (bends at the base and burinations of the margins and faces) (Figure 8.24). The recovery of artifacts identified as stone atlatl weights (Stratum Unit V, Fryxell and Daugherty 1962:Table IV) and a bone atlatl spur (Stratum I, II, or III, Root and Gustafson 1999:143) in the early-period deposits suggest use of Windust, Cascade, or Cold Springs points with a dart and atlatl system. Polish and other use-wear patterns observed on many of the points suggest they were used for other purposes in addition to projectile tips.

Patterns of Windust and Cascade projectile point breakage and reworking observed in the Marmes assemblage are similar to those described elsewhere for Paleoindian projectile points. This pattern involves common breakage near the base of the points at the top of the haft area and reworking to configure a new base on the larger distal portion. This pattern has been described for Clovis points from Oregon (Ozbun et al. 1998) and Angostura points from Texas (Collins 1993:90). Five grooved basalt pebbles were identified as bola stones (Artifact 2676, 3469, 3636 [Stratum II], 4507, and 5894) (seven described for Strata II and III by Rice 1969:38). These artifacts were made from alluvial pebbles shaped by pecking and finished by grinding. The groove is formed by pecking in a straight line around the long-axis of the pebble and accentuated by pecking to reduce the surface outside the lip of the groove. The pecking extends over the entire surface in the finished specimens. Grinding or abrasion was used to smooth the pecked surface. Three of the artifacts appear to be unfinished due to breakage during pecking or rejection of the piece following initial preparation of the groove. A bola is a cord or set of cords tied together with a loop at one end and with weights attached to the opposite ends, used for throwing at animals to entangle them during hunting. Ethnographic description of this weapon has been made for game animal hunting by South American Indians (Cressman 1960:60; Wormington 1964:161), and bone bola weights have been described for use by California Indians in hunting birds (Kelly 1978:418; Levy 1978:491). Similar artifacts were found in the early deposits at The Dalles Roadcut site in association with bones from large birds (Cressman 1960:60). The bola function was originally attributed to these artifacts based on comparison of their form to the South American ethnographic examples. Aikens (1993:Figure 3.10) has suggested they were more likely used as net weights. They are reported as an ancient artifact type that disappears from the archaeological record in the Plateau after the early Cascade Phase.

Another hallmark artifact type associated with the early Cascade Phase is the edge-ground cobble, two of which were identified in the collection (Artifact 6611 and 7979 [Stratum III]) (Rice 1969:36-37 reports a total of 10, from Strata II, III, V, and VII). These are relatively flat alluvial cobbles with facets developed along an edge from pecking and abrasion. They have been interpreted as hammerstones for percussion blade core reduction (Crabtree and Swanson 1968; Leonhardy and Rice 1970:9). Although some do show battering from use as hammerstones, the battering does not appear to be related to the facet. Polish on the facet of one (Artifact 6611) suggests use for some other purpose. Similar artifacts from The Dalles Roadcut site were identified as "side polishers" (Cressman 1960:57, Figure 51a) and may have been used to "polish" animal hides (Frison 1991:367).

A variety of flake tools and bifacial tools were identified as having square working edges (Figure 8.25 b-e, i). These include radial breaks, burins, and bending fractures on flakes and bifaces. The square or obtuse edges of these tools often exhibited use-wear or damage. Experimentation has shown that such edges are

Figure 8.23 Reduction sequence for Cascade projectile points, as illustrated by artifacts from the Marmes site.

220

Figure 8.24 Windust (a=Inv. No. 5779, d=#6215), Cascade (b=#5934, e=#3266), and Cold Springs (c=#3114, F=#3113) points illustrating impact fracture patterns.

useful for working hard materials such as bone, antler, or wood (Crabtree 1974).

Gravers are stone tools " . . . generally made by pressure flaking and intentionally designed to have a functional point or points. It is generally assumed that gravers are used to incise or form organic materials and soft stone (Crabtree 1982:37)." Although only a few gravers were identified in the collection, one exhibits a distinctive form that has been described in association with Paleoindian assemblages. This graver (Artifact 10078) is a flake tool with three graver tips (and possibly a fourth that has been broken off and reduced by pressure flaking) and resembles the type described as a compass graver (Tomenchuk and Storck 1997). Compass gravers are interpreted as tools used to engrave circles, cut disks, and bore holes in soft (organic) materials, possibly for decorative or symbolic purposes. The proposed function involves use of one point as a pilot and another as scribe turned in a rotary motion (Tomenchuk and Storck 1997). Alternatively, multiple points may have been used to incise evenly spaced parallel lines.

Twenty-four artifacts identified as pebble tools in this analysis of the Marmes site lithic assemblage are primarily small, irregularly shaped, basalt alluvial pebbles with a single flake scar at one margin (Figure 8.25 h). The flake scar originates near the margin on one face of the pebble and extends across the opposite face, usually less than halfway to the opposite margin.

Figure 8.25 Lithic tools from the Marmes site (a=Inv. No. 3162, b=#9716-01, c=#2904, d=#5908, e=#5699-01, f=#5482, g=#13883, h=#13166, I=#2907).

The contact point (ring crack) and negative bulb of the scar appear to be modified such that the area forms a small notch at the edge of the pebble. The notch is usually worn so that the margin of the tool within the notch area is rounded or polished. Fiber resembling sinew and other unidentified residues were observed in the notch area and negative flake scar on some of the artifacts. Two of these pebble tools (Artifact 7869 and 10087) have two flake scars on opposite margins and resemble net weights or sinkers. The others do not appear to be configured for such use and their function may involve processing fiber through the notch.

Twelve bipolar cores were identified in the collection (Figure 8.25 f, Appendix H). The bipolar reduction was used primarily for recycling flakes or broken tools (Figure 8.25 g), probably to derive flakes for other uses. The bipolar technology occurs in several identified strata in the rockshelter, but most commonly in Stratum III (n=4).

Seventy-two choppers were identified in the collection from the Marmes site (Appendix H). These are primarily basalt cobbles with unifacially and bifacially percussion flaked bits that exhibit battering, rounding, polish, and microflaking wear from heavy use. Many exhibit evidence of resharpening at the bit and rejuvenation. Twenty-six basalt rejuvenation flakes from resharpening chopper bits were also identified. The choppers occur most abundantly in strata associated with the Cascade Phase (Stratum III [n=11] and Stratum V [n=9]) as do the rejuvenation flakes (Stratum III [n=9] and Stratum V [n=3]). These tools are designed for heavy processing tasks, but the exact nature of this function is unknown. Various functions have been proposed for these types of tools including use as cores (the "choppers" being simply byproducts of manufacture), wood working tools, bone marrow extraction tools (Cleveland 1977), fish processing tools (Thoms 1983), pecking tools for production and maintenance of ground stone equipment (Dodd 1979), and multiple functions (Schalk 1983). Deer and bison blood residues identified by immunological analysis on choppers from prehistoric sites on the Middle Fork John Day River indicate an association with animal processing there (Ozbun et al. 1997:88, 143).

One hundred fifteen artifacts were identified as hammerstones. Nearly all of these are basalt alluvial cobbles unmodified except by battering from use. Some appear to have functioned as flintknapping hammers or pecking stones, but most have battering that probably is associated with other heavy processing tasks. Similar tools have been described as pounding stones presumably for use in processing foods or other organic materials (Cressman 1960). One similar artifact from a site on the Middle Fork John Day River retained residue of bison blood on its surface (Ozbun et al. 1997:143). These artifacts are abundant throughout the deposits at the Marmes site in the rockshelter (absent from the floodplain), but are most common in Stratum III where they are associated with abundant faunal material and in Stratum VII in association with processing and storage features.

The Cascade Technique

Investigators working at Washington State University in the 1970s proposed a model to characterize a core reduction technique used during the Cascade Phase and earlier (Bense 1972; Leonhardy et al. 1971; Muto 1976). The Cascade Technique model describes core reduction by a particular method that resembles the Old World Middle Paleolithic Levallois technologies of core preparation to derive flake blanks and blades of particular distinctive forms for use as tools (Muto 1976:Figures 10-20). In the Cascade Technique model the core is reduced to create a convex face with a raised ridge through the middle formed by terminations of flake scars from two opposite margins. A "primary" flake is removed along the ridge much like a channel flake from a fluted point. After some additional core preparation, two "A blades" are removed along the ridges created by the "primary" flake scar and using the same platform surface as the "primary" flake. Subsequent trapezoidal cross-section "B blades" are removed from the same core face, platform, and direction. Additional "C blades" are removed following core maintenance by corner flake (crested blade) removal. The model suggests that the various flakes and blades were used to make knives, Windust points, Cascade points, end scrapers, concave bit scrapers, and crescents (Muto 1976:Figure 49). The exhausted cores were used for choppers. Although it is often mentioned in the literature (e.g., Ames et al. 1998:103), little additional research has been conducted to confirm or refine the model.

The model was developed on the basis of data from Wexpusnime (45GA61) and Granite Point (45WT41), but elements of the Cascade

Technique were said to be present at other sites including the Marmes site in the floodplain Harrison Horizon and Marmes Horizon assemblages (Muto 1976:142). This assertion could not be confirmed during AINW's analysis of these assemblages. Neither distinctive debitage, nor cores of the types described by the Cascade Technique model could be identified in the assemblages from the floodplain or the rockshelter at the Marmes site.

Distinctive aspects of the reduction technology for Cascade points that were identified included preferred use of linear flakes as manufacturing blanks and retention of a portion of the flake blank's platform (remnant platform surface) at the base of the finished points. These attributes have been previously described for Cascade points from other sites (Butler 1961:28; Nelson 1969:19-20; Ozbun and Fagan 1998:32) and have implications for both core reduction technology and hafting technology. The Marmes site lithic analysis data presented here suggest that early-stage core reduction for lanceolate points occurred mainly elsewhere, but probably included multidirectional reduction to produce large linear flakes. Retention of the remnant platform surface or bending break surface at the base of Cascade points may reflect socketted hafting (Ozbun and Fagan 1998:32).

Gravels and Travels

The mobility and settlement patterns of early Cascade Phase peoples of the lower Snake River region has been a issue of recent debate among archaeological researchers (Andrefsky 1995; Reid 1997b). The debate has centered on the relation of Cascade Phase sites to toolstone source locations. Information obtained from analysis of the Marmes site lithic artifact collection bears on this debate.

Obsidian sourcing data for Stratum III of the rockshelter (n=10) indicate that at least three obsidian sources were used during the early Cascade Phase occupation (Tables 8.23 and 8.24). These three sources are Indian Creek (n=7), Whitewater Ridge (n=2), and Unknown 8 (n=1). The two known sources associated with the early Cascade Phase are the closest of the four known sources identified for the site and possibly the closest available obsidian of good quality. Still, both occur more than 220 km (137 miles) south of the Marmes site in high elevation areas. These areas could have been accessed by travelling through the connecting drainage systems or on overland trails across the Blue Mountains (Figure 8.20). The obsidian entered the Marmes site during the early Cascade Phase primarily in the form of bifacial tools that were reworked, resharpened, or discarded there. It is likely that these tools were used at other (intermediate) locations prior to their importation, use, maintenance, and disposal at the Marmes site. Some were rejuvenated and returned to service for further use, perhaps at locations more distant from the geological sources.

One lanceolate projectile point (Artifact 3162) recovered from the rockshelter by the 1968 excavations was identified by AINW as made from CCS material that resembles Knife River Flint (Figure 8.25 a). The Knife River Flint quarries are located in Dunn County, North Dakota, approximately 1,300 km (806 miles) east of the Marmes site. Knife River Flint materials have been previously identified at Paleoindian sites over a broad area of North America reaching into western Montana, but not as far west as the Marmes site (Root 1992:Figure 2.6). The tentative identification of the composition of the Marmes site artifact as Knife River Flint is based on AINW's experimental flintknapping experience with the material, visiting the quarries and source area, comparative collections of Knife River Flint, and published descriptions of macroscopic characteristics (Ahler 1986:3). Although Artifact 3162 shares some attributes of both Cascade (lanceolate form) and Windust (ground margins) point types, it most closely matches type descriptions for Agate Basin points from Wyoming (Bradley 1982:194-195). If these assessments of the raw material source and artifact type are correct, the artifact represents a long-distance connection with contemporary (circa 10,000 B.P.) groups of the northwestern high plains.

The sources of lithic raw materials used at the Marmes site during the early Cascade Phase occupation can also be evaluated based on cortex remaining on flakes and tools. Incipient cone cortex constitutes 83% of 270 cortex observations on all flakes and tools of all compositions from Stratum III to indicate that alluvial gravels were used for tool-making at the site. These alluvial gravels were available from nearby Palouse and Snake River gravel deposits. Basalt and granite compositions are most strongly associated with incipient cone cortex, while CCS (including petrified wood) artifacts are more evenly split between incipient cone cortex (n=41) and primary geological cortex (n=36) (both observations of cortex on obsidian artifacts are primary geological). Alluvial gravel use seems to be associated most strongly with the expedient technologies for production and maintenance of flake tools and choppers. The more elaborate bifacial technologies also employed alluvial gravel materials as indicated by incipient cone cortex on 16 of 17 basalt cortical early-stage percussion bifacial thinning flakes, three of four cortical CCS early-stage percussion bifacial thinning flakes, and one basalt bifacial blank. However, the relatively small amount of bifacial reduction debitage and manufacturing byproducts compared to the large number of bifacial projectile points indicate that most of the bifacial tools were made elsewhere although discarded at the Marmes site (Tables 8.11 and 8.12). This relationship between bifacial tools and bifacial debitage may indicate that the raw materials preferred for manufacturing bifacial tools were not available in the local area.

The relationships of toolstone availability and projectile point manufacture and use are complex at the Marmes site. Most of the projectile points discarded during the early Cascade Phase appear to have been used and maintained over long use-lives and were designed to endure rigorous duty. Some, however, are clearly expedient, following the general form and size of the others but lacking in symmetry and execution of design. The latter may represent the learning attempts of children or simple opportunistic use of available resources and time.

Site Function

The analyzed lithic assemblage contains debitage and tools reflecting use of the site for a variety of purposes throughout the history of its occupation. The lithic artifacts represent tool manufacturing and maintenance (of stone tools and tools of hard organic materials such as bone, antler, and wood), food procurement (projectile points, bola stones, net weights), and processing of food and other materials, probably including hides and fiber. These combinations of activities characterize a residential camp. Given its location in the relatively warm and protected canyon, high density of artifacts in the natural shelter, and the heavy wear on the tools, the site may represent winter occupations. This general pattern appears to be repeated through each period of occupation.

Some researchers have suggested that the numerous burials associated with Stratum V in the rockshelter may indicate that the site functioned as a cemetery during the late Cascade Phase and that artifacts associated with the deposit are grave goods (Fryxell and Daugherty 1962:28). The abundance and variety of utilitarian artifacts identified in this analysis of Stratum V lithic artifacts seem contrary to this notion. Special ceremonial use of the shelter may be represented during some portion of the period, but does not appear to characterize its entire duration.

Bone working and wood working tools such as abraders, drills, gravers, burins, radial breaks, and square edge flake tools are best represented in Stratum III (n=11 in the Early Cascade subassemblage) and Stratum VII (n=10 in the Harder subassemblage), although they also occur in every other rockshelter subassemblage. This frequency distribution does not correspond with the distribution of bone tools which are most abundant in Strata I, II (n=17 for the Windust materials) IV, and V (n=13 for the Late Cascade materials)(see Table 9.12 in Chapter Nine). Although bone tool manufacturing and bone tool use or deposition may not correspond with one another, it seems likely that the low frequencies of these stone and bone tools may not accurately reflect the relative abundance of associated activity during any particular period.

Use of the site as a base for hunting forays might be reflected in the frequency of projectile points and butchery or animal processing tools. Projectile points are most abundant in the Stratum II (n=30, Late Windust), Stratum III (n=34, Early Cascade), and Stratum VII (n=24, Harder) subassemblages. Flake tools, choppers, hammerstones, and scrapers are also most abundant in these subassemblages. The high frequencies of these tools and faunal remains found in these subassemblages suggest intensive hunting activities during these periods.

The proportions of lithic tools and debitage vary through time in some aspects that may reflect change in site function. For example, in the rockshelter the ratio of debitage to tools is relatively low for the Windust and Early Cascade period subassemblages by comparison with the later subassemblages (Figure 8.26). This change may reflect more intensive tool use, tool exhaustion, and tool disposal during the early periods and more intensive tool manufacturing during the later periods. Likewise, the percentage of tools that are byproducts of manufacturing (cores, blanks, and preforms) also

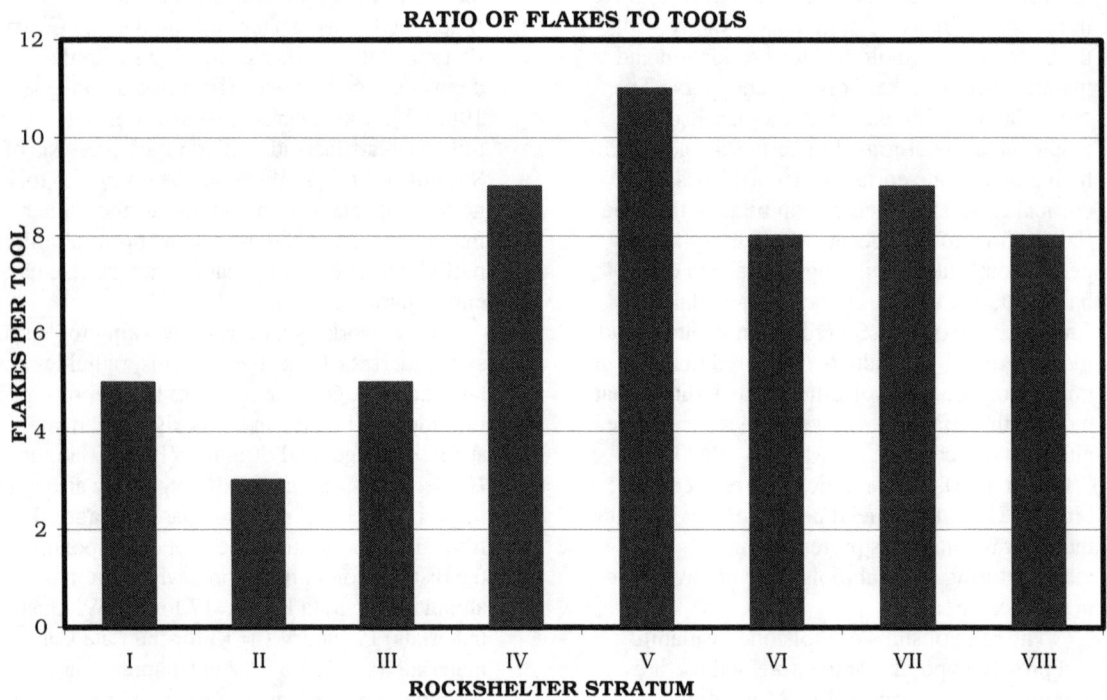

Figure 8.26 Ratio of flakes to tools for rockshelter strata.

increases for the later periods (Figure 8.27). The cores, blanks, and preforms are pieces broken and rejected during the tool manufacturing process and their increase corresponds with the increased debitage for the later periods. One could infer from this pattern that Windust and Early Cascade people generally arrived at the site with stone tools sufficient for planned activities while later people more often manufactured stone tools while occupying the site. In this sense, the site functioned, in part, as a stone tool production workshop during the later periods of occupation.

Comparisons With Other Sites

The Marmes site is unique in the lower Snake River region for its long duration of prehistoric use and preservation of stratigraphically layered archaeological deposits from many periods of occupation. Comparisons of the Marmes site technological attributes with other sites in the region containing lithic assemblages of similar ages provide some greater context for the discussion.

The issue of Cascade Technique core reduction is served by comparison with materials from Granite Point (45WT41) and Wexpusnime (45GA61), where the early stages of reduction are much better represented. The Marmes data suggest that a core reduction technique that produced linear flake blanks for further reduction into Cascade points, however, a "Levallois-like" strategy, was not apparent. Additional analysis of the Granite Point and Wexpusnime assemblages would be necessary to evaluate the nature of the Cascade core reduction technique.

The Granite Point assemblages were also used to assert a distinct separation of Windust and Cascade Phases that was blurred by mixing at Marmes and Windust Caves (45FR46) (Rice 1972:25). More recently, however, the Windust Phase assemblage at Granite Point has been described as containing both Windust and Cascade points (Reid and Gallison 1995:76). At the Hatwai site (10NP143), near the mouth of the Clearwater, a clear division of Windust and Cascade projectile points has been described (Sappington 1996:145). The Windust points from Hatwai also exhibit the near-base breakage

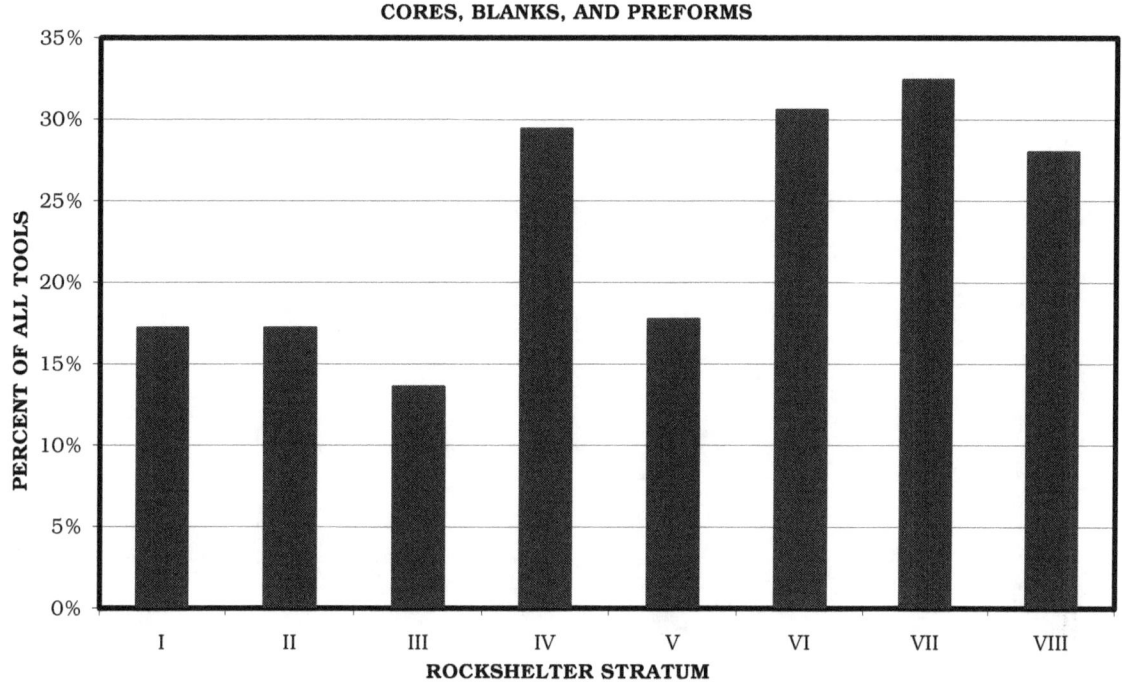

Figure 8.27 Percentage of lithic tools representing flaked stone manufacturing byproducts for rockshelter strata.

pattern identified at Marmes (Ames et al. 1981:Figure 15). Better separation of Early and Late Cascade Phase projectile points has also been described for Bernard Creek Rockshelter (10IH483) in Hells Canyon where 21 Cold Springs points all stratigraphically overlie the Early Cascade Phase materials (Reid and Gallison 1995:94). As with the Marmes assemblages, the plant food processing tools are absent or very rare from the Bernard Creek assemblage, and the site was apparently used as a camp for hunting and fishing (Reid and Gallison 1995:94).

Although microblade and similar technologies have not been described for the Lower Snake region, they have been found in early deposits at sites to the west (Ames et al. 1998:106). During analysis of the Marmes assemblage, cores and small linear flakes were examined for evidence of microblade production or use. Although a few small blade-like flakes were identified, there does not seem to be a microblade technology represented at the Marmes site.

The ubiquitous hammerstones and choppers in the Marmes assemblages are difficult to assess functionally. They seem to be associated with animal food processing at Marmes and other sites on the Columbia Plateau, as described above. Similar hammerstones are referred to as battered cobbles, pounders or pounding stones in many reports, but the materials to be pounded are unknown. Additional study of these tool types including immunological analysis would be helpful in assigning tool functions.

Overall, the Marmes site lithic assemblage represents a significant body of data and information about early Holocene to late prehistoric occupation of the Lower Snake River region of the Columbia Plateau. The lithic debitage and tools reflect a focus on use of imported tools and local raw materials for hunting and fishing as well as other food and materials processing as part of a seasonal round.

9
MODIFIED BONE AND ANTLER

Matthew J. Root and Carl E. Gustafson

Bone tools from the Marmes site have been previously described (e.g., Caulk 1988; Rice 1969, 1972), but the entire collection has not been described in one place with comprehensive descriptive data. This chapter presents information for all the modified bone artifacts inventoried from the collection at the Museum of Anthropology, Washington State University, Pullman. The purpose of this section is to describe the manufacturing technology, morphology, possible functions, and archaeological associations of the bone tools from Marmes. Changes through time are briefly described and the floodplain collection is compared with that from the rockshelter. Comparisons of tool morphology and function with other selected sites is presented.

The sample analyzed consists of 108 pieces that are included under 86 inventory numbers, as reported by Collins and Andrefsky (1995:Table 16). Several inventory numbers contain more than one piece of the same artifact. Many pieces were once glued together, but have broken apart in storage, and these were occasionally inventoried as more than one specimen. Cross-mended pieces within a single inventory number are counted as one artifact for this analysis. Therefore, the 108 pieces counted in the Marmes inventory consist of 91 individual artifacts. These 91 specimens consist of one bear tooth and 90 bone or antler artifacts. Rice (1969) described 93 bone artifacts, though it is clear that some of those specimens are now missing. Conversely, some of the artifacts described below were apparently not included in Rice's study.

Methods and Techniques

In 1994 and 1995, Washington State University inventoried the materials from the Marmes site that were in the collection housed at the Museum of Anthropology. During this inventory, lab technicians separated pieces that they thought might be bone tools or ornaments from pieces of unmodified bone. All bone and antler with questionable modification were also classified as modified (Collins and Andrefsky 1995). This analysis indicates that some of those questionable specimens are not modified, other than by pre-Columbian breakage from butchering or processing bone for foods such as bone grease. These items are identified by taxon and element and tallied, but are not further analyzed.

As defined here, modified bone includes all objects of bone or antler that exhibit humanly induced modification, either from shaping (e.g., carving, chopping, grinding) or use to modify other objects. We therefore do not include bone modified only by butchering, post-depositional weathering (Bonnichsen and Will 1980:9-10), or by trampling, gnawing, or digestion (Binford 1981:35-81). All tooth, bone or antler tools, ornaments, other non-utilitarian objects, and tool manufacturing debris are included as modified bone. Each piece of modified bone was identified to taxon and element where possible, and then analyzed under low magnifications for traces of manufacture, use, and the presence of mineral residues. Because antler is a kind of bone, it is sometimes difficult to distinguish it from other bone material. Also, recognizable bone artifacts are usually modified to such an extent that the kind of bone and the taxon and size of animal from which it came often cannot be determined.

Sometimes bone and tooth artifacts are little modified, if modified at all, and uses can only be inferred from associated artifacts or features, or by ethnographic analogy. For example, Rice (1969) reported two bear canines from Marmes Rockshelter. One is grooved around the root, but the other is unaltered. Both came from a human burial, and therefore, both are offerings. Two metapodials from a large canid and one from a large bear were recovered from the bison kill site, 45GA17 (Schroedl 1973). One canid metapodial had lines incised perpendicular to the shaft and had a hole drilled near the distal end. Neither the second canid bone nor that of the bear exhibited any modifications. They were considered to be artifacts because of their association, and because they were the only such bones recovered from the site. Highly modified bones, such as needles, small bone points, or bones ground flat are extremely difficult to identify, and often different bones from different kinds of animals can be used to make the same types of artifacts. Personnel at the Marmes site in 1968 discussed the kinds of bone that might have been used to make the tiny bone needles found there. Artiodactyl ribs, metapodials, long bone splinters, and scapulas were among the possibilities suggested. The dense shaft bone from rabbits or birds also seemed likely candidates. Flenniken (1978) used rabbit bones to replicate the bone needles from the Marmes and Lind Coulee sites. Frison and Craig (1982) believed that bison scapulas -- found with bone needles at the Agate Basin site, Wyoming -- were a better source of material than ribs or long bones.

We know of no nondestructive way to make unquestionable identifications of highly modified bone artifacts. Even bone splinters or flat pieces without diagnostic landmarks are difficult to identify. If part of the inner, cancellous tissue is intact, then one can have some idea of the thickness of the original piece, but most often this less resistant material has been cut or ground away leaving only compact bone. We can only provide our best judgment as to the kind of bone used in the manufacture of the artifacts discussed below.

Manufacturing Technology and Function

All pieces were examined under reflected light at magnifications ranging between 10x and 70x with a Leica Stereozoom binocular microscope and a Leica CLS 100 high intensity fiber-optic lamp. Microscopic examination sought to detect traces of manufacture and use-wear. Tool classifications were based on use-wear patterns, tool morphology, and ethnographic descriptions of tool uses. Types and intensities of manufacture and use-wear were recorded. All use-wear analysis was conducted at Plateau and Plains Research, Pullman, Washington.

The intensity of wear traces was recorded as pronounced, moderate, light, or absent. *Pronounced* traces are visible under the high intensity lamp without magnification. *Moderate* wear can be resolved between 10x and 40x and *light* wear can be resolved between 40x and 70x. If no wear is visible at 70x, it is recorded as *absent*. Wear types are listed and defined in Table 9.1; most types can result from either manufacture or use. In addition to the wear types listed below, we noted the presence of groove-and-snap and groove-and-splinter manufacturing techniques. Groove-and-snap is a technique of transversely breaking bone or antler. A groove is cut or sawed with a stone tool around the perimeter of a bone, but it is not cut all the way through. This scores the bone, which can then be snapped into two pieces along the groove. Traces of the cut and the roughly broken interior of the piece remain on each segment that is produced (Clark 1971:161). Cuts can also be made on opposite sides, but not around the entire circumference of the bone. The groove-and-splinter technique is similar, but it is used to divide bones longitudinally. A groove is cut lengthwise along the grain of the bone, and the bone is split or wedged apart (Clark 1971:115-117). Bifaces and burins are excellent stone tools for cutting grooves to make bone tools (Frison and Bradley 1980:130; Frison and Zeimens 1980; Root et al. 1999). Bipolar or other types of wedges can be placed in the groove and lightly hammered to split the bone (LeBlanc 1992).

Use-Life Classes

Durable goods used in a cultural system pass through the stages of raw material procurement, manufacture, use, and discard or loss. Tools are sometimes recycled, refurbished, laterally cycled, or scavenged (Schiffer 1987). Bone and stone tools are particularly amenable to analysis using such flow models. Therefore, all bone tools were classified according to their inferred position in this life-history sequence. Following Schiffer, these are called "use-life" classes. Use-life data can be used to study patterns of tool manufacture, use, and discard. Distinguishing manufacturing debris from tools discarded after breakage and those discarded or cached while still complete is critical for studies of site formation, site function, and settlement systems.

Five use-life classes were defined to indicate the general position in the use-life of a bone tool in the sequence of raw material selection, manufacture, use, and discard when it entered archaeological context (Moore 1985). Following Schiffer, the first two use-life classes include procurement and manufacture, the second two include use, maintenance, and recycling, and the fifth includes manufacturing debris. *Use-life class 1 tools* are complete and unbroken blanks, preforms, or other technologically unfinished items; these have the potential to be further reduced into usable tools. *Use-life class 2 tools* are those that broke or were otherwise rejected during manufacture, usually because of breakage. These are no longer suitable for continued manufacture. Unfinished items potentially include pieces of cut or split bone that could be made into implements such as awls or points. *Use-life class 3 tools* are technologically finished items that are not broken or worn out, and still have potential utility for their original function. Nearly complete tools, such as awls with snapped tips that could easily be resharpened, are also placed in this class. The presence of such tools indicates that they were cached on the site for potential future use, were lost, or were discarded while still functional. *Use-life class 4 tools* are technologically finished, but were broken during use, resharpening, or recycling;

Table 9.1 Definitions of wear types recorded in the modified bone collection from the Marmes site.

Wear Types	Definition (adapted from Moore 1985:64-66)
Grinding	The alteration of the shape or surface of a bone by abrasion. Diagnostic traces consist of shallow short striations, usually oriented parallel or subparallel to each other. Most grinding wear observed in the Marmes collection is from manufacture and resharpening.
Carving	The removal of small parings or strips of bone or antler. Few tools evidence this type of wear, probably because grinding during the later stages of manufacture has removed evidence of early shaping by carving.
Polish	Alteration of the surface by repeated rubbing against (e.g., scraping, piercing) a nonabrasive or very finely abrasive material. Diagnostic traces consist of highly reflective surfaces without visible microtopography (at magnifications up to 70x).
Drilling	The formation of a hole or socket with a tool used with a rotary motion, and perhaps an abrasive such as fine sand. Interior surfaces sometimes retain striations around the circumference of the hole. Holes can be smoothed or polished without remaining striations. Holes are usually drilled from both sides, indicated by biconical cross sections and holes from opposite sides that do not meet exactly.
Attrition	This is the removal of bone or antler through use which results in rounded working elements, or rough surfaces caused by the removal of small bits of bone. This type of wear is common on antler billets used in percussion flaking and on pressure-flakers.
Flaking	This wear is analogous to conchoidal flaking of stone. Small flakes of bone are removed, either due to pressure or percussion against resistant materials.

or were exhausted from repeated use or resharpening. These tools are inferred to have little or no utility remaining for their original function. *Use-life class 5* objects include debris or by-products created during manufacture (Moore 1985:41-42). Tools are assigned to use-life classes based on analysis of production technology and tool function. One tool has an indeterminate use-life class because of recent fracture (either during excavation or curation). Use-life classification is not applicable to unmodified specimens.

Finally, we recorded the presence or absence of burning and the presence of any mineral residues. Several tools have small black spots that are dendritic under low magnification; these are probably postdepositional manganese residues. Several tools are stained with a red mineral residue, and this is probably red ochre (hematite). Red ochre was recorded on cobbles from Marmes that were probably used to grind and prepare pigment. Forty-eight pieces of iron oxide were recovered from the rockshelter from all geologic units but Mazama ash (Stratum IV) and from the uppermost historic deposits (Stratum VIII) (Rice 1969:37, 39; 1972:120).

Description of the Modified Bone Collection: Morphological and Functional Analyses

The analyzed sample includes a total of 91 specimens. There are 78 tools, ornaments, and pieces of manufacturing debris. There are also 13 pieces of unmodified bone that were originally inventoried as modified. The 78 pieces of modified bone were placed into 15 classes that were defined by morphology and sometimes by inferred function. There are also two classes of manufacturing debris. Some of these classes reflect morphological variability of functionally similar implements, such as splinter awls and split metapodial awls. This variability results from selection of different elements as raw materials and different manufacturing technologies. Other classes, such as needles, are extremely homogeneous. Several classes, such as bone rods are of unknown function and are classified only by size and shape. We call these morpho-functional classes, and these are listed by use-life classes in Table 9.2. Unmodified debris was identified to taxon and element, if possible, measured but not further analyzed.

Bone Awls (n=18)

The awls were made from pieces of long bone or metapodials, most of which are from deer-sized

Table 9.2 Modified bone and antler by morpho-functional and use-life classes.

Morpho-Functional Class	Use-Life Classes					Total	
	Broken, Unfinished (2)	Finished, Complete, Usable (3)	Finished, nfs[1] (3/4)	Finished, Broken, Not Usable (4)	Manufacturing Debris (5)	(n)	(%)
Split metapodial awls	0	5	0	6	0	11	14.1
Splinter awls	0	7	0	0	0	7	9.0
Eyed needles	4	1	0	2	0	7	9.0
Atlatl spur	0	0	0	1	0	1	1.3
Splitting wedge	0	1	0	0	0	1	1.3
Spatulate tool	0	1	0	2	0	3	3.8
Narrow-diameter pointed tools (pins)	0	1	0	13	0	14	17.9
Bone/antler rods	0	3	0	9	0	12	15.4
Pointed tools, nfs	0	2	1	2	0	5	6.4
Pendants	0	2	0	0	0	2	2.6
Incised ornaments	0	2	0	3	0	5	6.4
Bone tubes	0	1	0	0	0	1	1.3
Drilled, polished bone, nfs	0	0	0	1		1	1.3
Polished bone, nfs	0	0	0	3		3	3.8
Pigment-stained bone	0	0	0	2		2	2.6
Groove-and-snap debris	0	0	0	0	1	1	1.3
Debris, nfs	0	0	0	0	2	2	2.6
Total	2	23	1.75	40	3	78	100

[1] nfs = not further specified

animals. The working ends of these tools have ovoid to circular cross sections and were shaped by grinding to fine points. Two types of awls occur: split metapodial awls and splinter awls. Rice (1969:40) reports 26 awls from Marmes, though only 18 are reported here. Some of this discrepancy may be due to differences in classification. For example, a tool from the Windust deposits that was previously called a "thin flat needle" (Rice 1972:120, Figure 34b) is classified as a resharpened awl in this report. Several specimens are, however, now missing from the Marmes collection (Collins and Andrefsky 1995). For example, Rice (1969:40) reports one awl made from a horn core from a pronghorn, but no such specimen is now in the collection. Rice (1972:120) also reported seven awls from the Windust phase deposits, though only one now remains. The single awl illustrated from the Windust deposits by Rice (1972:Figure 34a) is no longer in the collection. The awls remaining in the Marmes collection, their geologic associations, and measurements are listed in Table 9.3.

Split Metapodial Awls (n = 11). These tools were made from metatarsals or metacarpals that were longitudinally split. One is from a deer to elk-sized animal, eight are from deer-sized animals, and two are from deer. One has a remnant of a longitudinal groove-and-splinter break, suggesting that this was the technique employed to split the bone shafts. Where the proximal ends remain, they are rounded. Subsequent to splitting the bone, the tools were further shaped by grinding as indicated by pronounced to moderate intensity, sub-parallel striations that run parallel to the long axis of the tools. Working elements are sharply pointed distal tips that are highly polished; the polish diminishes in intensity away from the tip. All striations have been obliterated by the polish at the distal tips. Pronounced to moderate polish continues over the tool shafts, and this may have resulted from hand-held use. One awl has an eye in the proximal end. This hole is 3.7 mm in diameter, and was formed by drilling from both sides, as indicated by its biconical cross section. Such holes can be quickly drilled with a stone tool. The surface of the eye is obscured by secondary carbonates. It is likely that a line was strung though the hole, perhaps so that its user could loop it around the wrist. The distal tip of this tool is broken off, which perhaps led to its discard.

Table 9.3 Measurements of split metapodial and splinter awls.

Invent. No.	Geol. Unit	Morpho-Func. Class	Condition	Length (mm)	Width (mm)	Thick (mm)	Weight (g)	Figure
9248	V	metapodial awl	distal	28.9	5.65	3.65	0.4	9.3, b
10058	III	metapodial awl	complete	70.8	7.15	3.6	1.6	9.3, g
10342	-	metapodial awl	proximal	55.55	7.6	6.05	2.4	9.3, f
10876	-	metapodial awl	distal	60.8	13.65	4.7	3.7	9.1, u
10877	III	metapodial awl	distal	89.2	12.4	6.25	1.6	9.3, e
10878	-	metapodial awl	complete	122.85	8.3	4.15	3.8	9.1, y
10879	V	metapodial awl	complete	81.7	10.1	4.75	4.4	9.1, w
10882	-	metapodial awl	distal	70.8	10.7	3.25	2.8	9.1, v
10887	V	metapodial awl	distal	89.3	12.4	6.25	6.2	9.2, i
10888	I/II	metapodial awl	complete	94.15	9.0	5.2	4.8	9.1, x
10889	-	metapodial awl	distal	65.3	9.2	4.25	2.4	9.2, g
3479	Harrison	splinter awl	complete	36.65	5.0	1.7	0.4	9.3, a
10056	II	splinter awl	complete	31.1	5.9	6.1	0.7	9.3, j
10880	VII	splinter awl	nearly complete	39.95	10.3	3.7	1.7	9.3, i
10883	-	splinter awl	complete	54.2	3.7	1.4	0.4	-
10890	-	splinter awl	complete	51.75	8.4	1.85	0.9	9.1, t
10892	-	splinter awl	complete	54.05	4.05	3.5	1.2	-
10892	-	splinter awl	nearly complete	41.25	4.75	2.8	0.3	-

Use-wear at distal tips suggests repeated use to perforate soft materials, such as hides or plants. A limited series of experiments indicated that awls used to perforate hides and to enlarge holes in hides produced polish on the tips, but also produced striations that were oriented diagonal to the long axis of the awl. The same experiments indicated that awls used in basketry work produced pronounced polish on tips, but did not produce striations (Blomgren 1996:108-113; Bullock 1992:81-111). The tips of Marmes awls are polished, but lack striations, suggesting that they were used in basket working. Bone awls were probably used to split basketry materials into strips and pass stitching elements through the basket (e.g., Smith 1988:168-170). The lack of striations on the Marmes artifacts, however, does not preclude their use to punch holes in hides during sewing of items such as clothing and robes. If tools were used to work with fibers subsequent to hide working, striations might be obliterated, as indicated by the erasure of manufacturing striations on awl tips. The presence of bone needles in the floodplain deposits suggests use of awls in hide working (see below). Awls may also have been used to punch holes in hides during processing so that they could be staked and stretched over the ground or on wooden frames (e.g., Ray 1933:94-95, Plate 5). Ray (1933:35-44) reports that Sanpoil and Nespelem women used the same bone awls in both hide-working and basket making, and the Marmes awls may have been multifunctional as well.

The split metapodial awls require a modest amount of time to manufacture, and were apparently used for long periods. Intermittent resharpening is also required to maintain sharply pointed distal tips. One awl has a ground distal tip, but polish is absent on the tip. The shaft of this awl is highly polished and blackened, perhaps from fire hardening. The grinding at the tip cross-cuts the polish on the shaft indicating that it formed subsequent to the polish and suggesting that it is from resharpening. The tool was not used long enough subsequent to resharpening for use polish to develop again. These types of awls were sometimes placed in a handle, another indicator of long use-life tools (Ray 1933:35). A split metapodial awl from nearby Squirt Cave had resin adhering on its proximal end, suggesting that it was hafted (Endacott 1992). No such evidence of hafting was present on the Marmes specimens, however, and hafting is only a possibility. Of the 11 split metapodial awls, six were broken prehistorically and this probably led to their discard. Five of the awls, however, are complete, and some are still long and pointed with lots of use-life apparently remaining. These specimens may well have been stored at the site for anticipated future use, though they were never retrieved and became buried in site deposits. Many awls are of unknown provenience, but split

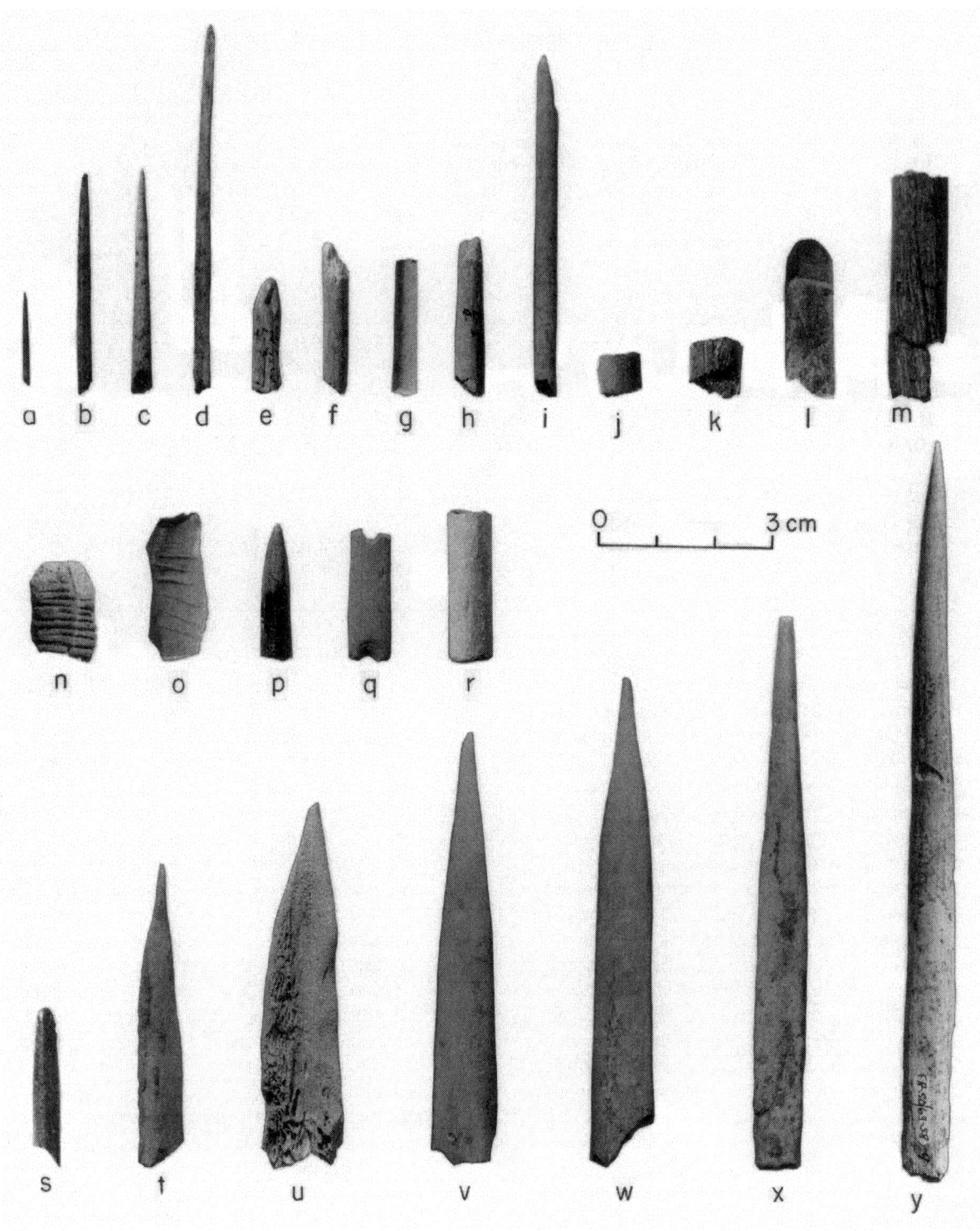

Figure 9.1 Photograph of bone and antler artifacts.

234

Figure 9.2 Photograph of bone and antler artifacts.

235

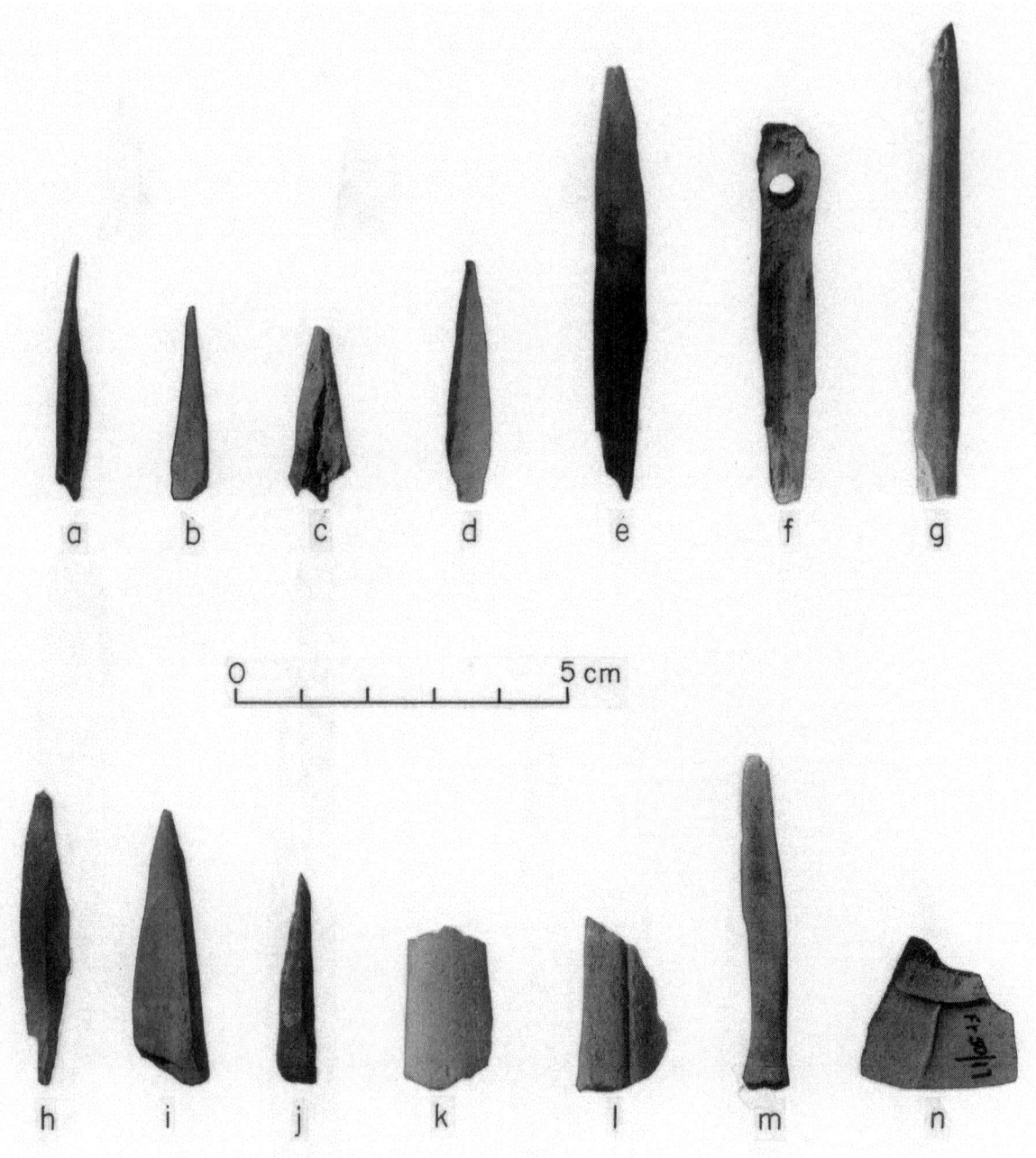

Figure 9.3 Photograph of bone artifacts and unmodified bone items.

metapodial awls were recovered throughout rockshelter deposits. These utilitarian implements are not time-diagnostic, and occur in sites of all ages across the Plateau (e.g., Endacott 1992:92-95; Nakonechny 1998:207; Nance 1966:29-30; Nelson 1969).

Splinter awls (n=7). These seven awls were made on small, pointed splinters of bone, with their shape determined by the original shape of the splinter. The defining attributes of these tools include an unpatterned form (that is, the splinter blanks were minimally modified) and a pointed end that displays moderate to pronounced intensity polish. The working ends of three of these awls are formed by the pointed end of the splinter without subsequent modification of the point. Thus, the tools are pieces of bone, either broken during butchering or broken to produce pointed splinters, that were selected for tool use without subsequent alteration. Three of these were made on the metapodials of deer-sized animals. Four were made on long bone fragments, two from deer-sized animals, one from a deer/elk sized animal, and one from a fox-sized animal. Functionally, these tools are similar to the split metapodial awls, though the working elements of the splinter awls do not display well-developed use-wear. Polish use-wear on two specimens is moderate, that is it is visible only under low magnification, suggesting a short period of use. In contrast to the patterned split metapodial awls, all of the splinter awls are complete and still usable. These simple and quickly made implements were likely used for short periods and may have been discarded when the task at hand was completed, rather that stored for future use.

Eyed Needles and Needle Preforms (n=7). Some of the most widely known artifacts from the Marmes site are the exquisite eyed needles recovered from the Windust phase deposits. The needles have been previously reported by several researchers (Caulk 1988; Flenniken 1978; Fryxell et al. 1968a; Kirk and Daugherty 1978:36). Rice (1969:41, Figure 48d) reported three needles, one each from Strata II, III, and VI. One of the specimens from either Stratum III or IV was described as eyed. As mentioned above in the discussion of awls, one of the specimens from Stratum II is actually a fairly large awl, not a needle. Later, Rice (1972:120-123) did not report any eyed needles from the Windust phase deposits, only the awl which was called a "thin, flat needle." All of the eyed needles from the Marmes site are from the Harrison horizon in the floodplain, which are Windust phase deposits. Therefore the eyed needle reported by Rice (1969) from later deposits, may have had the provenience misreported. Caulk (1988:104) sorted additional needles from water-screened samples and lists eight needle fragments from six proveniences in the Harrison horizon. Two of these fragments refit, making a total of seven needles or needle fragments from the site, all from the Windust phase, Harrison horizon. Two are eyed needles and one is a distal tip but is probably from an eyed specimen (Figure 9.4 a, b, d [also Figure 9.1 a]; Figure 9.5). Four are uneyed preforms that broke in manufacture (Figure 9.4 c).

The needles are all of similar morphology, and only slightly larger in diameter than modern sewing needles. The Marmes needles are exemplified by their extremely small diameters, their polished shafts, and of course, the eyes in the two finished specimens.

There are four needle preforms and three finished needles. Two preforms are proximal fragments and are 1.7 and 1.9 mm in maximum diameter. A preform midfragment is 2.0 mm in maximum diameter. The last preform is a tapering, distal fragment that lacks the distinctive use-polish characteristic of finished specimens, and is 1.3 mm in maximum diameter. The three finished specimens range from 1.3 to 1.5 mm in maximum diameter. The single complete needle (Inv. No. 3454) consists of three cross-mended fragments with recent breaks, and is 51.9 mm long. The proximal break is glued. It is likely that these breaks occurred inadvertently during screening or during transport and storage of water-screened samples. Hence this specimen is coded as complete and usable to reflect its condition when it entered the archaeological record some 10,000 radiocarbon years ago. The eye of this complete needle is circular and is 0.4 mm in diameter. The other eyed needle is missing its distal tip (Inv. No. 3451). The break is old, suggesting this fracture led to the discard of the needle. The eye of this needle is slightly elongate, and measures 0.93 by 0.72 mm. This needle is considerably shorter than the other specimen. During use, eyes of bone needles tend to break and proximal ends are reshaped and new eyes are made. The short condition of this specimen may be due to repeated breakage and refurbishing. All needles are slightly ovate in cross section; measurements are provided in Table 9.4.

Flenniken (1978) conducted an experimental replication of eyed needle manufacture using the shorter specimen (Inv. No. 3451) and a preform (Inv. No. 3511) from Marmes and two eyed needles from Lind Coulee as archaeological controls. Dried jackrabbit long bone was used in the replication, though the Marmes specimens were probably made on pieces of metapodials. Frison and Craig (1982:165-168) proposed that the Folsom eyed needles from Agate Basin were made from strips cut from bison scapula by the groove-and-splinter technique. Flenniken splintered the rabbit long bones, and carved the splinters with a flake knife into rough shape. The preforms were then ground on sandstone into a smooth and symmetrical form with an ovoid cross section. The resulting preforms were about twice the diameter of the finished tools. The proximal end was then flattened slightly to facilitate purchase of a thin stone drill tip. Eyes were drilled from both sides, resulting in a biconical cross section and holes 0.5 mm in diameter. The drilling of thicker preforms resulted in less breakage, and Flenniken broke only one of five preforms during drilling. The

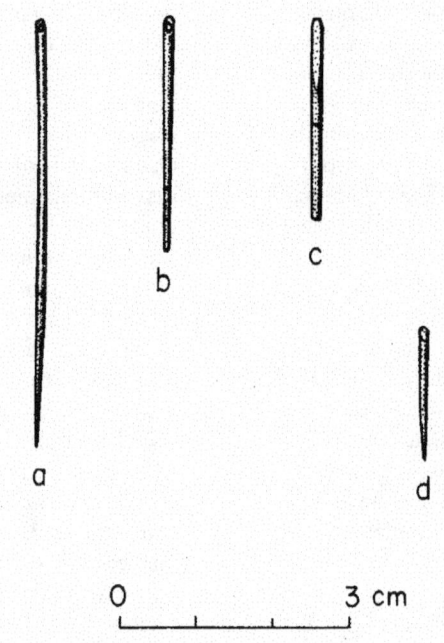

Figure 9.4 (a, b) eyed needles [Inv. Nos. 3454, 3451] (c) a needle preform [Inv. No. 3511], and (d) a distal needle tip [Inv. No. 3452] from the Harrison horizons (drawings by Sarah Moore).

Figure 9.5 Photograph of bone needles.

Table 9.4 Measurements and proveniences of bone needles from the Marmes site.

Inventory Number	Catalog Number	Condition	Excavation Unit	Level	Length (mm)	Width (mm)	Thick. (mm)	Weight (g)
3511	68.59	preform, proximal	25N/15W	Harrison 3-4/1	25.6	1.7	1.5	0.1
3545	68.0	preform, medial	10N/45W	Harrison, 4th A-horizon	11.7	2.0	1.2	<0.05
3543	68.0	preform, proximal	15N/0W	Harrison 3-4	7.1	1.9	0.7	<0.05
3543	68.0	preform, distal	15N/0W	Harrison 3-4	13.4	1.3	0.7	<0.05
3452	68.49	finished, distal	20N/10E	Harrison 3-4	15.75	1.3	1.2	<0.05
3451	68.27	eyed, proximal	15N/45W	Harrison 4/1, A-horizon	29.0	1.5	1.0	0.1
3454	-	eyed, complete	10N/10W	Harrison 3, 4, 5	51.9	1.3	1.3	0.1

drilling of thicker preforms is also indicated by the Marmes specimens, which are thicker than the finished tools. After drilling was complete, the preforms were again ground to bring the needles to final form and thickness, with finely tapering and pointed distal ends. The needles were ground longitudinally, creating small grooves in the sandstone abrader. All four preforms from the Marmes site exhibit tiny striations that are oriented parallel to their long axes. This suggests that the Marmes needles were also made using grooved abraders, in the manner described by Flenniken. The finished needles from the Marmes site do not exhibit striations, but are highly polished from use. These needles are also blackened, perhaps from fire-hardening during manufacture. The Marmes site needle with an elongate eye may have had the eye made by scratching back-and forth with a thin drill tip, pointed flake, or similar implement. If it was formed by rotary drilling, it should be round. Jeb Taylor, an accomplished preindustrial craftsman, has successfully made eyed needles of this size and form in this way rather than by rotary drilling (Jeb Taylor, Personal Communication, February 10, 1999).

Although the manufacturing technology of the needles is impressive, their presence indicates that the Marmes inhabitants wore tailored clothing of leather or possibly plant fibers, such as leggings, shirts, and dresses (e.g., Ray 1933:45-48). These needles are too delicate to have been used like modern steel needles to perforate leather. Instead, awls were likely used to puncture small holes, through which the needle, threaded with sinew or fiber, could then pass. Although gender equality was traditionally present among Plateau people, division of labor was certainly present. One traditional women's role was to dress hides and to sew clothing (Ackerman 1998:519-520). Though any analogies between ethnographically documented Plateau peoples and Marmes site society are speculative, division of labor by gender and age characterize all hunter-gatherer societies. Therefore, the presence of needles and awls indicates sewing of tailored clothing, and also suggests entire families or extended family groups resided at the site during the earliest occupations.

The needles from Marmes formed a part of the original definition of the Windust phase (Leonhardy and Rice 1970; Reid and Gallison 1996). Eyed needles were also found at the Lind Coulee site. Two needles with intact eyes and pieces of at least nine other needles or needle blanks were recovered at Lind Coulee, and these are similar to the Marmes site specimens (Irwin and Moody 1978:93-104,135). Eyed needles have also been found at other early Paleoindian sites in North America, most notably at Folsom components on the Northern Plains. One nearly complete needle and 16 needle fragments were found in the Folsom levels at the Agate Basin site, Wyoming (Frison and Craig 1982). These are similar to the Marmes specimens in diameter and eye configuration. Eighteen broken needles, only one of which retained an intact eye, were recovered from the Lindenmeier site, Colorado. The diameters of these range from 0.8 mm to 2.5 mm (Wilmsen and Roberts 1984:129-133), and these are similar to the Marmes site specimens. An eyed needle was also recovered from the Agate Basin component at the Hell Gap site, Wyoming, but dimensions were not given (Irwin-Williams et al. 1973). A midsection of a needle with a polished shaft was found at the Hanson site, another Folsom camp in Wyoming. Though the proximal end was not found, this specimen was similar in size and cross section to the Folsom needles at Agate Basin and Lindenmeier, suggesting that the Hanson needle was also eyed (Frison and Bradley 1980:103, Figure 69). Eyed bone needles were also recovered from the early deposits at the Namu site, though these are thicker and date later in prehistory than the Marmes site specimens (Carlson 1996:96, Figure 19).

Atlatl Spur or Hook (n=1). A single broken atlatl spur was recovered (Inv. No 10893), but there is some uncertainty over its provenience. Rice (1969:42) originally reported that this specimen was from Stratum III (early Cascade phase), but later (Rice 1972:123, Figure 35b) included it with the Windust collection (Strata I and II). The piece was made from a piece of a metapodial by grinding, and the surface consists of several striated facets. This piece is angled in long section (36 degrees) and is D-shaped in cross section. The flat side would have been attached with mastic or sinew to the atlatl board. One end is broken off, presumably the tip that protruded from the atlatl and engaged the end of the dart mainshaft. This is similar in size and shape to bone artifacts also identified as atlatl hooks from Component 1 (9,000-10,000 B.P.) at the Granite Point site (Leonhardy 1970:95-96, Figure 33s) and the Period 2 deposits (5,000-6,000 B.P.) at Namu on the British Columbia coast (Carlson 1996:95-100). The tool is 22.9 mm long, but may have been about 30 mm long prior to breakage. It is 5.6 mm thick and 5.2 mm wide (Figure 9.1 p).

Splitting Wedge (n=1). Rice (1969:42) originally reported three antler splitting wedges, one from Stratum V and two from Stratum III. The single splitting wedge that remains in the collection today (Inv. No. 9226) was illustrated by Rice (1969:Figure 49a), but was later illustrated and identified as coming from the Windust phase deposits (Rice 1972:123, Figure 36a). Based on data in Gustafson (1972:154-183), we have also determined that this artifact is from the Windust phase deposits in the rockshelter (Stratum I/II), and therefore, the original geologic association reported by Rice must be in error. This artifact is elk-sized, but is poorly preserved and the distal end has broken off since its original illustration. Prior to recent breakage, the tool was 183 mm long, and it is 35.8 mm wide and 27.0 mm thick. The working element is single beveled, but is weathered and disintegrating and no use-wear observations are possible. The tool has been shaped by grinding and the working edge may have been formed by chopping or carving. The proximal end is badly deteriorated, and was previously described as splintered from pounding (Rice 1972: 123). The proximal end is broken, but this may have resulted from postdepositional weathering, not pre-Columbian pounding (Figure 9.6 c). This tool is similar in size and shape to other implements identified as wedges (e.g., Aikens 1986:Figure 4.9; Endacott 1992:95-96; Nakonechny 1998:83, Figure 86c; Nelson 1969:404-405). Ray (1933:42-43) described the use of "heavy" pieces of elk antler to split logs and as wedges in felling trees. These were driven with stone and wooden mauls. The tool from Marmes may have been used in a similar manner, though it is too poorly preserved to confirm through low power use-wear analysis. This tool is coded as complete and still usable because breakage appears to be postdepositional.

Spatulate Tools (n=3). These tools are characterized by rounded distal working elements with pronounced to moderate polish (Table 9.5). Each of these was made from the metapodials of deer or deer-sized animals. Use-wear suggests prolonged contact with relatively soft materials, such as hides or plant fibers. None of these can be matched with previously described artifacts (Rice 1969, 1972). The largest of these tools is a distal fragment with a recent break (Inv. No. 3289). It was made on a split metapodial and shaped by grinding and subsequently polished. A small tool from the Windust deposits (Stratum II) was cut with a groove-and-snap fracture on its proximal end and had the distal end shaped by carving and grinding. The third spatulate tool is a poorly preserved distal fragment that is blackened, possibly by postdepositional manganese stains. The rounded distal tip has moderate use polish. The functions of these tools are unknown, but they may have been burnishing tools. Similar tools from Cayuse phase deposits at Sunset Creek (A.D. 1-1800) are called scrapers and fleshing implements (Nelson 1969:400-401), but there is no evidence that the implements from the Marmes site were used for those functions. Though the uses of the Marmes tools are unknown, they are similar to bone tools used as quill-flatteners. Quills used for embroidery were often softened in the mouth and then flattened between the fingernails. Lowie (1954:68-71) does, however, illustrate an Arapaho spatulate bone tool that was used for flattening quills, as well as for painting the embroidery designs on skins.

Narrow-Diameter Bone Pins (n=14). These tools have circular to slightly oval cross sections, taper to a sharp cylindrical point, have highly polished shafts and points, and are generally 3-5 mm in maximum

Figure 9.6 Photograph of bone artifacts and unmodified bone item (c).

Table 9.5 Measurements and geologic contexts of spatulate bone tools.

Inventory No.	Geologic Unit	Condition	Length (mm)	Width (mm)	Thickness (mm)	Weight (g)	Figure
3289	-	distal	70.85	14.45	6.3	4.8	9.6, a
5874	II	complete	35.5	6.65	5.35	1.2	9.3, d
11965	VII	distal	29.6	6.0	5.7	0.8	-

diameter. Most of these are from the Windust phase deposits, and it is possible that the three specimens from later deposits (two from Stratum III and one from Stratum IV) were displaced upward through the many postdepositional disturbances documented at the site (see Chapter Five). Four of these were reported by Rice (1969:41) as small needles or pins (Category 42). Two, however, were reported from Stratum VII, though none in the present collection are from those proveniences. Only one was reported from the Windust phase (Rice 1972:120). It is possible that the midfragments of shafts were classified as awls in the previous studies. These tools are, however, much narrower than awls with much more delicate tips. It is unlikely that these were implements for perforating hides or plant fibers.

Measurements and geologic associations are listed in Table 9.6. Most of these bone objects are small fragments, and it is difficult to reconstruct the morphology of complete specimens. One of the larger distal fragments (Inv. No 9243) and the largest proximal piece (inv. no 10881), suggest that some of these may have been ca. 10 cm long when complete. It is also possible that some were substantially shorter, and these objects are not necessarily functionally homogeneous. Most of these are made on pieces of metapodials from deer sized animals, though two may be antler (Inv. Nos. 9236, 10348). The distal tips taper very gradually to form sharp points. Proximal ends are rounded. Several of these are badly weathered and do not retain evidence of manufacture or use-wear. The well-preserved specimens were shaped by grinding, as indicated by subparallel striations. These usually run parallel to the long axis of the implement. The striations on three, however, spiral around the shaft, downward toward the distal tip. This indicates that these pins were rotated against the abrader, perhaps to aid in forming the symmetrical, gradually tapering points. The shafts of the unweathered specimens are polished, and all are undecorated. Pointed tips display polish that becomes more intense toward the end, suggesting that these were pushed through soft materials, likely through pre-existing holes. One of these (Inv. No. 10891) is triangular to diamond-shaped in cross section, but has been shaped by grinding and is otherwise similar to the cylindrical items.

Table 9.6 Measurements and geologic contexts of narrow-diameter bone pins.

Inventory Number	Geologic Unit	Condition	Length (mm)	Width (mm)	Thick. (mm)	Weight (g)	Figure
3491	Harrison	medial	37.3	2.7	2.6	0.2	9.1, b
9236	III	medial	21.8	5.1	4.7	0.6	9.1, h
9243	II	distal	38.5	3.5	3.5	0.4	9.1, c
9250	I/II	medial	40.25	5.3	4.7	0.9	-
9250	I/II	medial	24.5	4.3	3.4	0.3	-
9540	I/II	distal	26.6	3.6	3.5	0.3	-
9542	III	proximal	19.05	4.7	4.4	0.3	9.1, e
9579	II/III	medial	22.9	4.2	4.1	0.4	9.1, g
10334	IV	distal	39.0	4.6	3.7	0.5	-
10348	I/II	medial	25.65	4.05	3.8	0.4	9.1, f
10881	-	proximal	56.7	3.9	3.7	1.1	9.1, i
10886	I/II	proximal	21.2	4.0	3.9	0.4	9.1, s
10891	-	complete	62.8	3.8	3.9	0.5	9.1, d
15164	I/II	proximal	27.5	5.05	5.0	0.5	-

The function(s) of these objects is uncertain. They are much smaller in diameter than bone projectile points used in big-game terrestrial hunting. For example, bone points recovered from the Folsom component at the Agate Basin site are over 10 mm in diameter, over twice that of the Marmes site specimens of similar age. The tools from the Marmes site are probably too thin to be projectile points and, as mentioned above, are probably not awls. We suggest several possible functions. First, these may be barbs on compound fish spears or leisters. Barbs on leister heads used by the Shuswap for large fish are about 75 mm long and just over 5 mm in diameter, similar to the Marmes site specimens. Barbs on composite hooks and leisters for smaller fish are similar in diameter, but are only 3-4 mm long (Hewes 1998). These are also similar to the bone barbs attached to wooden side barbs on composite salmon spears that are illustrated and described by Ray (1933:60-61), which according to the description were about 10 cm long. Another possibility is that these were pins for fastening clothing or robes. Similar artifacts were recovered from the latest horizons at Sunset Creek, though most are decorated with incised lines (Nelson 1969:390-393).

Antler or Bone Rods (n=12). These are long solid pieces of antler (or possibly dense bone) that were carved, ground, and polished into rods that are circular to ovoid in cross section. The longest of these is 412 mm, though most are smaller fragments. These are distinguished from the bone pins described above by their larger diameters and longer forms. The rods are about 10 mm in diameter (except for small tapering distal fragments, or about twice the diameter of the bone pins. Eleven of the rods are antler or probable antler; one is either antler or large mammal long bone (Inv. No. 11654). Four of these were associated with burials in Stratum V. Two others also are from Stratum V, which is the layer above Mazama ash that included numerous burials, and may have been associated with burials. Therefore, this class of artifacts may be associated with burial rituals. Four small burned fragments were found adjacent to one of the concentrations of human skeletal material (Marmes I) found in the floodplain, but these remains were not buried in a pit and the association is not certain. Measurements and recovery context of these artifacts are summarized in Table 9.7. Rice (1969:40, Figure 43b) describes two of these artifacts as long, slender shafts and illustrates one (Inv. No 6410) from Burial 9B.

A group of small burned antler rod fragments are from the Windust floodplain deposits (N10/E5) with provenience listed as "bone points" from "base of east wall" (Inv. Nos. 3840, 3841). Another burned rod made up of seven small cross-mended pieces (Inv. No. 3540) is also labeled as "base of east wall," but the horizontal provenience is missing; it is probably from the same unit. Fryxell et al. (1968b) illustrated two of these pieces (Inv. Nos. 3540 and 3481) and did not illustrate two others. They also stated that although only two of the four directly refit, all were probably part of the same rod. The total length of the recovered fragments was 110 mm, and it was suggested that the original length of the complete rod may have been twice that length. None of the specimens in the present collection refit and the total length of pieces reported here is 82.9 mm. Furthermore, two of the pieces reported by Fryxell and others had worked ends, whereas only one of the present pieces is so modified. Thus, at least one piece originally reported is now missing, and one other small fragment has apparently been identified and added. Of the pieces reported here, the end fragment is narrower than the medial pieces, indicating that the rod gradually tapered toward the end. The end fragment was grooved-and-snapped by cutting almost all the way through the piece around its entire circumference. The snapped end was then ground, creating several facets, but not obliterating the original snap. Fryxell and his colleagues suggested that this artifact was probably a projectile point or foreshaft; we return to the function of these antler rods below.

The tapering end of a burned antler(?) rod was also recovered from the Windust deposits in the rockshelter (unit 85N 40W, Inv. No 5119). This was previously classified as an awl (Rice 1969), but is identical in size and morphology to the other rods. These small fragments from the Windust deposits all display moderate to pronounced striations from grinding during manufacture. The shafts of all are polished. The distal end of an unburned rod was recovered from Stratum III, and is similar to the piece described above. These may have been included as category 44 artifacts (bits and pieces of polished and striated bone) by Rice (1969:41), though this is uncertain. (These are Category 32 in Rice [1972]).

The remaining rods are associated with burials, or are from Stratum V, which contains numerous burials. As such, these artifacts may be associated with burial rites. The thickest rod, 18.45 mm by 9.0 mm (Inv. No. 11654), may be either a split long bone or split antler. This artifact was inventoried as eight pieces, but fragments cross-mend and it is counted as one piece in this analysis. As reconstructed, this artifact was 282 mm long. The piece is ovate in cross

Table 9.7 Measurements and associations of antler or bone rods.

Inventory Number	Burial Assoc.	Geologic Unit	Condition	Length (mm)	Width (mm)	Thick. (mm)	Weight (g)	Figure
3480	Marmes I?	Marmes	margin fragment	9.7	9.4	5.75	0.5	9.1, j
3480	Marmes I?	Marmes	margin fragment	7.1	7.65	5.0	0.3	9.1, k
3481	Marmes I?	Marmes	end fragment	27.9	9.5	8.5	2	9.1, l
3540	Marmes I?	Marmes	medial	39.2	10.15	7.75	3.9	9.1, m
5119		II	medial	18.65	8.5	6.0	0.6	9.2, c
6408	9B	V	complete	412	11.8	9.9	32.2	-
6409	9B	V	complete	355	9.8	5.7	16.4	-
6410	9B	V	distal	278.7	9.1	8.5	16.1	-
6411	7	V	medial	105.1	7.8	5.3	4.3	-
7772		III	distal	26.5	6.9	6.0	0.6	9.2, d
9237		V	distal	15.6	5.8	5.7	0.3	9.2, a
11654		V	complete	282	18.45	9.0	32.7	-

section with rounded ends. It is highly weathered and fragmented, and much of the original surface is gone. The tool was probably brought to final form by grinding and polishing, though weathering precludes any certain statements about manufacturing or use-wear.

The remaining rods are smaller in diameter, ranging from 7.8 mm to 11.8 mm in maximum diameters. Small, tapering distal fragments have smaller diameters, but were likely part of larger artifacts like the complete specimens. The two complete rods from Burial 9B are 412 mm and 355 mm long, and a pointed distal fragment from the same burial is 278.7 mm long. These rods are ovate in cross section, generally consisting of one evenly ground and polished semicircular side with the opposite side roughened and flat. It is possible that these pieces were purposely split after manufacture of a cylindrical form. The rough areas, however, are sometimes present at different positions around the circumferences of the artifacts. The ground and polished surfaces sometimes are present around the entire circumference on small portions of the shafts. Therefore, the rough, uneven surfaces may be from post-depositional weathering and erosion of the original surfaces. Two of the rods (Inv. Nos. 6408 and 6410) have faint red stains present below coatings of preservative (applied in the field?). These stains are likely remnants of red ochre that was applied prior to burial. The midshaft fragment from Burial 7 is weathered, but retains an evenly ground surface around its entire circumference. This item retains a surface polish visible under low magnification (moderate intensity).

The functions of osseous rods have long been a matter of archaeological investigation and speculation. Suggested functions include projectile point foreshafts, projectile points, handles for composite pressure flakers, and levers used in hafting large Clovis points for use as saws. Most of the Marmes specimens are much narrower than the rods of ivory, antler, or mammoth bone recovered from early Paleoindian sites. Those larger specimens from other sites range from 8-30 mm in maximum width, though only three are less than 10 mm (Lyman et al. 1998:Table 1). Furthermore, the Marmes specimens are pointed, not beveled as are most other early Paleoindian pieces. The thickest Marmes piece, which was recovered from above Mazama ash (Inv. No. 11654), is similar in size and cross-sectional shape to the early Paleoindian artifacts, but it is not beveled. Two other sites in Washington contained bone rods. Two specimens similar to those at Marmes were recovered from the Lind Coulee site, where they were called beveled bone shafts. A third beveled bone shaft is much thicker than the Marmes site artifacts (Irwin and Moody 1978:84-88). The beveled mammoth bone rods from the Richey-Roberts Clovis site, East Wenatchee, (Mehringer 1988, 1989b) are also generally much wider and thicker than the Marmes specimens.

Bone rods recovered from Clovis contexts have several suggested functions. Based on studies of Clovis points and bone rods from the Anzick site, Lahren and Bonnichsen (1974) proposed that they were foreshafts for Clovis points. Later researchers have noted problems with Lahren and Bonnichsen's specific hafting method (e.g., Lyman et al. 1998). The exact method of hafting proposed by Lahren and Bonnichsen is unlikely because of the limited penetration afforded the projectile tip. Slight modifications to the hafting proposed by Lahren and Bonnichsen, however, may allow deeper penetration and killing effectiveness. Foreshafts remain a possible function for at least some of these artifacts. The artifacts from the Richey-Roberts site, however, were not recovered in associations with the Clovis points that would indicate points were hafted to them at the time of their burial (Mehringer 1989a). Wilke et al. (1991) suggested that the bone rods from Anzick were handles for composite pressure flakers. This is a viable proposal, but it is difficult to confirm without recovery of the entire tool. Several researchers have also suggested that some bone or ivory rods were projectile points (e.g., Frison and Zeimens 1980; Stanford 1991), though Frison (1982) cautioned that the function of the ivory rod from the Sheaman site and other similar objects remained open to question. The Marmes specimens have distal points, and a projectile point function is possible. Lyman et al. (1998:897) note that projectile points are generally shorter than the early Paleoindian rods, the longest of which is 281 mm. Therefore, at least the Marmes specimens from Burial 9B are probably too long to have functioned as projectile tips.

Echoing Frison's statement for the Sheaman site, the function of the Marmes bone rods remains open to question. The burned fragments from the floodplain and similar fragments from other proveniences may have been parts of projectile points. The function of these small pieces, however, will likely remain uncertain. The long, pointed rods from the Stratum V burials and other Stratum V contexts may not be parts of utilitarian implements. We will never know why these pieces were interred with the dead. These have no obvious utilitarian use, and we regard these simply as grave goods and offer no further speculation as to their possible functions. The antler rod from Burial 7 may have been called an awl in the previous description of Marmes site (Rice 1969:79-80). Red ochre, *Olivella* shell beads, and several stone tools were also associated with two small fragmentary antler rods (Inv. Nos. 6411 and 7772). The long antler rods (Inv. Nos. 6408, 6409, and 6410) were associated with one polished antler point, shell beads, red ochre, several stone tools, and two bear canines (Rice 1969:80-81) (see below under Pendants/Ornaments). The presence of artifacts associated with burials is important for the study of the social structures of the people who buried their dead in the rockshelter, as well as for Pre-Columbian Plateau societies in general. Such an analysis, however, is beyond the scope of this descriptive report.

Pointed Bone Tools (n=5). These tools are defined by pointed distal ends that are cylindrical to ovate in cross section. These tools are functionally indeterminate and are of several different morphologies. They may also have served several functions. We use the term pointed tools as a morphological description, and do not imply that these necessarily functioned to tip projectiles, though that is a possibility. Measurements and geologic contexts of these are listed in Table 9.8.

These tools are highly variable in morphology, and therefore we describe them individually. Specimen number 3473 is a single beveled, highly polished piece of antler that is ovate in cross section. It is nearly complete, but has three recent longitudinal breaks that have been glued together. The tool was cut to length and then shaped and finished by grinding and polishing. The thin lateral edge adjacent to the bevel retains pronounced striations, probably from cutting or sawing. The only possible use-wear is a small snap break at the distal tip. The tool is of unknown function, but may have been a projectile tip.

Specimen 5053 is a split and polished metapodial. This specimen has the label "FR36/379" written in ink on it and is catalog number 62.379 (indicating a 1962 excavation date). Site 45FR36 is Palus Village located immediately west of mouth of the Palouse River. Palus Village excavations were conducted in 1962 along with excavations at Marmes Rockshelter (Fryxell and Daugherty 1962). This artifact is apparently from Palus Village, not Marmes Rockshelter. The tool is coated with preservative, making use-wear observations difficult. It was made from a metapodial and was shaped by grinding. The tool is now broken longitudinally and the proximal end is also snapped off. Both fractures cross-cut striations from grinding, indicating that the breaks occurred after manufacture, probably during use. The face opposite the longitudinal break is ground and polished to a flat surface. The tip is broken from a recent fracture. The breaks may have been caused from use as a projectile point, but this is uncertain.

A small pointed piece of antler (Inv. No. 5135) was shaped by grinding. It displays a recent break that is covered with glue, suggesting another piece of this tool was once present in the collection. The tool is weathered and use-wear, if once present, is not preserved. The tool is morphologically unlike awls. Function is indeterminate, but it may have been a projectile point, or possibly the tip of an antler rod such as those from Stratum V. The piece was

Table 9.8 Measurements and geologic context of pointed bone tools.

Inventory Number	Geologic Unit	Condition	Length (mm)	Width (mm)	Thickness (mm)	Weight (g)	Figure
3473	Harrison/ Marmes	complete	38.0	9.0	4.8	1.9	-
5035	45FR36?	distal	48.0	8.2	5.3	2.3	-
5135	VI	distal	18.2	8.4	5.8	0.4	9.2, b
5863	Harrison	complete	90.1	10.2	6.2	5.9	9.2, h
10885	III	distal	44.45	8.05	6.3	2.1	9.2, f

recovered from Stratum VI and may have been displaced upward from Stratum V, but it does not cross-mend with the other antler rods.

A complete point made from a deer or elk split metapodial (Inv. No. 5863) was recovered from Feature 1A in the Harrison horizon. This complete tool does not appear to be included in Rice's analysis, but he does state (1972:120) that two tools classified as awls, are possible "fragments of bone points." The tool was fashioned by grinding and polishing. The proximal end displays two pronounced facets, each of which are marked by pronounced parallel striations that angle almost perpendicular to the tool's long axis. The proximal end is evenly rounded and was shaped by carving and grinding. The tool is burned, possibly fire-hardened. The distal tip is broken off by a recent fracture, but the remaining portion displays pronounced polish that obscures all striations from manufacture and decreases in intensity away from the tip. This wear pattern is similar to that displayed on awls. The striated proximal facets may have been for hafting by creating more surface area for mastics to adhere. No mastic residues, however, were observed. It is possible that this was a projectile point, but its function is uncertain.

Specimen 10885 was apparently previously tallied as an awl (Rice 1969). This tool is subtriangular in cross section and has grooves that run down two of the three faces. The tip was removed by an ancient fracture, but there are recent proximal and longitudianl breaks (the longitudial breaks are glued). The entire surface of the tool is polished and no striations are present, but it was probably shaped by grinding. Polish is most intense at the distal point, suggesting that this was used to pierce soft materials. It is unlike the other awls in shape, especially the presence of longitudinal grooves. As with the other points, function is uncertain, but it may have been a projectile point.

Ornaments and Polished Bone

This group of artifacts includes pendants, incised bone, a bone tube, and other miscellaneous pieces of drilled, polished, or pigment-stained bone. Many of these are of unknown function, but none appear to be utilitarian implements. Measurements and geologic associations are listed in Table 9.9.

Pendants (n=2)

One drilled pendant that was made from the metapodial of a deer-sized animal was recovered from Burial 12 (Inv. No 10343). The Museum of Anthropology lists no provenience for this artifact, but Rice (1969:42, 81, Figure 50g) lists its provenience as Stratum VI/VII (103N/25W, 94.0-94.75 feet). This pendant is plano-convex in cross section and has a small fracture along one edge. All surfaces are ground and polished. The convex side has five parallel rows of five dots per row, a sixth row of four dots, and a seventh row of two dots. The dots are small, shallow semihemispherical holes formed by drilling, and range from 0.78 to 0.91 mm in diameter. Rice (1969:42) describes the pattern as seven rows of five dots. It is possible that some dots were removed along the fractured edge, but this is an old break and clearly occurred prior to archaeological recovery. The opposite flat side has two arcs of six dots each that form a V-shape, with a line of four dots that bisects the V. Below this pattern are two parallel rows of dots, one of four dots and the other of five dots. This pattern was previously described as five parallel rows of five dots each (Rice 1969:42). The eye is 1.9 mm in diameter and is biconical in cross section, indicating that it was drilled from both sides.

Two bear canine teeth (*Ursus* cf. *americana*) were associated with Burial 9. An upper right canine was included in the sample of modified bone, and it has a groove ground around the circumference of the

Table 9.9 Measurements and geologic contexts of ornaments and polished bone.

Inventory number	Geologic Unit	Morpho-Functional Class	Condition	Length (mm)	Width (mm)	Thick (mm)	Weight (g)	Figure
10343	VI/VII, Burial 12	decorated pendant	complete	47.3	10.6	2.4	1.5	-
11931	V, Burial 9	bear canine pendant	complete	62.9	18.0	12.2	8.7	-
7789	II	incised bone	medial	16.9	11.4	5.5	0.7	9.1, n
9222	III, Burial 2	incised bone	nearly complete	32.15	14.0	3.3	1.7	-
10344	-	incised bone	complete	15.2	10.4	3.2	1	9.1, o
10349	-	incised bone	proximal	11.5	7.05	6.0	0.3	-
10351	-	incised bone	interior fragment	21.9	10.55	2.25	0.4	-
10346	VI/VII	bone tube	complete	25.8	8.0	7.1	0.5	9.1, r
10341	-	drilled, polished	end	22.4	7.25	2.4	0.5	9.1, q
9233	II	polished bone	proximal	19.25	6.2	3.05	0.3	-
10347	-	polished bone	medial	42.65	8.95	4.6	0.8	9.2, e
10340	-	polished bone	medial	23.6	13.8	3.05	0.9	9.3, k
6828	V	pigment-stained antler	interior fragment	31.95	16.0	6.0	2.2	-
6844	III	pigment-stained antler	interior fragment	26.0	9.0	5.55	0.8	-

root (Inv. No. 11931). The other canine was inventoried as an unmodified grave good (Inv. No. 11932). The groove is 0.8 mm wide and there are striations present adjacent to the groove that also encircle the root of the tooth. The groove was probably made to facilitate tying the tooth to a cord (Gustafson 1972:107). It may have been worn as a pendant.

Incised Bone (n=5)

These artifacts are small fragments of bone incised with parallel lines. The incised lines are narrow and U-shaped in cross section. They were probably made with the edges of thin flakes, or with tools such as gravers or burins. One piece from Stratum II is a long bone fragment that has nine lines incised around its circumference (Inv. No. 7789). The bone is longitudinally split, but originally the artifact may have included the entire circumference of the long bone. This is a medial fragment, broken on both ends through incisions indicating that originally it was also longer, possibly with more lines. The incisions are U-shaped in cross sections and each incision is 0.6 mm wide at the surface and 0.6 mm deep.

A long bone shaft fragment from a deer-size animal has ten parallel lines incised across its surface (Inv. No. 9222). This artifact was found associated with Burial 2, below Mazama ash. This artifact was not previously listed as being associated with Burial 2, but was apparently reported as a possible tubular bead (Rice 1969:41, 78). The piece is poorly preserved, but is nearly complete. Apparently a small fragment was selected for decoration. Ten parallel incisions run perpendicular to the long axis; each is about 0.5 mm wide.

A spirally fractured long bone fragment from a deer-sized animal has six subparallel lines incised across the outer bone surface (Inv. No. 10344). The exterior bone surface is highly polished, though the concave interior surface is not. All spiral fracture surfaces are polished, indicating that this piece is complete. Incisions are 0.4-0.6 mm wide and 0.5 mm deep. A burned, highly polished, proximal fragment of a small mammal long bone has two parallel lines that spiral around the shaft, but it is uncertain whether these were incised (Inv. No. 10349) (this was identified by Rice [1969:41] as a rodent bone). Another highly polished long bone fragment of a small mammal has three incised lines at one end, and is broken through the last preserved incision (Inv. No. 10351).

Though the uses to which these incised pieces were put are uncertain, they are likely not utilitarian. Incised pieces of bone were commonly used as gaming pieces on the Plateau (e.g., Ray 1933:156), and the Marmes site pieces may have had similar uses. Incised bones from Sunset Creek were identified as gambling bones (Nelson 1969:298-299, 408-409), though most of those pieces are larger and most have modified ends that are rounded or pointed. Aikens (1986:62) also identifies incised bone gaming pieces from the Alderdale village site along the Middle Columbia, though again these are highly modified and patterned pieces.

Bone Tube (n=1)

A medial piece of a bird long bone shaft was made into a hollow tube by grooving and snapping each end (Inv. No. 10346). The bone was cut almost all of the way through prior to snapping. A red mineral residue remains in the rough cut marks; and this is likely red ochre (hematite). This may have been a tubular bead or some other kind of ornament or decoration.

Rice (1969:42) lists three tubular bone beads, but only one is included in the present collection. The provenience of the piece analyzed here is not listed, but Rice lists two pieces from Stratum VII and one piece from Stratum VI. Therefore, this piece must be from one of those late strata (measurements indicate that this piece was included in Rice's analysis).

Drilled, Polished Bone (n=1)

This artifact is probably a metapodial fragment. It was apparently listed by Rice (1969:41) as a piece of polished and striated bone. This piece was shaped to a flattened ovate cross section by grinding; all surfaces are also highly polished. There is one intact end that has a small notch that was produced by bifacial pressure flaking in a technique analogous to notching a projectile point. A single notching flake was pressed off of one surface, then a second flake was removed from the opposite surface using the first flake removal as a platform. The end opposite the notch bears the remnant of a biconical drilled hole; the piece is broken through this hole. (This piece is listed in the inventory as having holes drilled in both ends, but one is clearly a notch, not a drilled hole.) The function of this piece is unknown, but it may have been an ornament or a bangle.

Polished Bone (n=3)

These artifacts are small highly polished fragments of larger pieces, but are too fragmentary to place in a more specific class. One is a small long bone fragment (Inv. No. 9233) that may be the proximal end of an awl. A second is plano-convex in cross section with pronounced subparallel striations that run along the long axis of the piece (Inv. No. 10340). The piece has snap fractures on both ends, but one fracture is polished indicating that this is the original end of the object. This may be the proximal end of a spatulate bone tool. The third artifact is a distal piece of a rib from a deer-sized or smaller animal (Inv. No. 10347). The distal end tapers and has been carved, but the distal tip was removed by an old fracture. This may be a piece of an awl, but this is now indeterminate.

Pigment-Stained Antler (n=2)

These two pieces are probably antler, and each is an interior fragment with all edges formed by dry breaks. The exterior surfaces are stained with a red mineral, most likely red ochre (hematite). The artifacts were probably shaped by grinding, and the red ochre is burnished on to the surface. One fragment is from Stratum V and one is from Stratum III, but both may be from the same artifact. The Stratum V burials were dug into Mazama ash and in one case into Stratum III below the ash, therefore some mixing of these units is expected. These may be pieces of larger rods such as artifact 11654 (Table 9.7), and may have originally been part of a Stratum V burial.

Manufacturing Debris (n=3)

Three pieces of manufacturing debris were recovered. Two of these are long bone fragments from deer-sized animals that have pronounced striations from grinding but are otherwise unmodified. These pieces are probably residue from on-site manufacture of bone tools, analogous to flake debris from stone tool manufacture. The third piece of debris is the proximal end of a humerus from a large bird. The shaft end has the remains of two cuts made on opposite sides that went partially through the shaft. The bone was then snapped through the cut. Though the cut does not go around the entire circumference, this is groove-and-snap debitage from the production of bird bone artifacts such as tubular beads. Measurements and geologic contexts of debris are listed in Table 9.10.

Unmodified Bone

Thirteen pieces of unmodified bone were originally inventoried as modified. These likely represent food remains discarded after processing or consumption, though some may be natural inclusions to the site. Two bird feathers were recovered from Stratum VIII, the modern surface sediments in the rockshelter.

Table 9.10 Measurements and geologic contexts of bone manufacturing debris.

Inventory Number	Geologic Unit	Condition	Morpho-Functional Class	Length (mm)	Width (mm)	Thickness (mm)	Weight (g)
9232	II	interior fragment	indeterminate debris	31.2	9.45	4.2	1.5
9235	I	interior fragment	indeterminate debris	42.2	7.05	1.5	0.5
16292	V	complete	groove-and-snap	39.8	12.2	7.7	2.2

These are clearly modern and are undoubtedly from residents during the period of excavations. Taxon identifications, geologic associations, and measurements are listed in Table 9.11. A bison costal cartilage (Inv. No. 10332) was listed by Rice as an uncertain bone tool (Figure 9.6 b). The piece is thickly coated with a preservative, but we noted no modifications of any kind and list it here as unmodified.

Other Bone Artifacts From Marmes

This section presents other modified bone or antler artifacts not listed as modified bone in the Museum of Anthropology, Washington State University inventory or not listed in the inventory, but described in previous reports on the Marmes rockshelter.

An articulated owl foot (Inv. No. 5780, great horned owl or snowy owl) was found sandwiched between two modified flakes of chert or chalcedony in the floodplain excavations (Figure 9.7). Something must have held the bones together such that they remained articulated. They were the only articulated bones found at the site. The flakes and foot bones may have been bound together, perhaps in a medicine bundle or similar kind of package. This suggests that these artifacts were of special importance to their original owner (Gustafson 1972:108).

The following artifacts were listed by Rice (1969:40-43), but were not in the inventory. An awl made from the horn of a pronghorn (antelope) was listed, but this artifact is now apparently missing. A beaver incisor from Stratum II is not among the modified bone, but it may have been inventoried as unmodified. A "chipping tine" from Stratum II was described, but no similar artifact from Stratum II is now present. Other pieces of modified bone included in previous studies may also be missing from the inventory, but individual artifacts cannot be identified.

Changes Through Time

This section compares the modified bone collections from different geologic units/strata in the site. Strata I and II are grouped because they were difficult to distinguish during excavation, and many artifacts from Stratum II have probably moved down into Stratum I. These geologic units were, however, easily distinguished in profiles. These units are equivalent in time to the Marmes and Harrison horizons in the floodplain. We list these separately to compare artifacts on the floodplain with those in the rockshelter. Stratum IV is Mazama ash and Stratum V is the overlying mixed ash and loess. Stratum IV, of course, does not have any occupation surfaces and Stratum V includes numerous burials that were dug into the Mazama ash. It has been suggested that Stratum V was a period when the rockshelter was not residentially occupied, but was used only as a burial site. As such, all Stratum IV and V artifacts likely would be grave goods or related to burial ceremonies (Fryxell and Daugherty 1962). Strata VI, VII, and VIII are combined because these deposits are often extremely mixed. Strata I and II and the floodplain deposits are assigned to the Windust phase. Stratum III is of early Cascade phase affiliation. Stratum V dates to the Tucannon phase. Strata VI, VII, and VIII date from the Harder, Piqúnin, and Numipu phases (Leonhardy and Rice 1970).

Tool morpho-functional classes are listed by geologic unit in Table 9.12. The column for Stratum II/III includes a bone pin of uncertain association. Eighteen of the 78 (23 percent) pieces of modified bone are of unknown provenience and cannot be used in comparative analyses. Therefore, only 60 modified bone artifacts are listed in Table 9.12. Just over one-half (32 of 60) of the bone artifacts of known provenience come from the Windust phase deposits. Only six come from the upper cultural horizons.

Eyed needles and needle blanks only occur in the floodplain Windust phase deposits. At least one needle was complete when it entered the

Table 9.11 Measurements and geologic context of unmodified bone.

Inventory Number	Geologic Unit	Taxon	Element	Length (mm)	Width (mm)	Thick (mm)	Weight (g)	Figure
5034	-	bird	coracoid	41.7	11.7	9.4	0.8	-
8946	III	turtle	carapace	23.1	22	8.9	1.6	9.3, n
9234	II	deer-size	rib?	25.8	9	3.3	0.2	9.3, c
9251	I/II	deer/elk	long bone	42.8	6.95	3.25	0.8	9.3, h
9536	I/II	deer	metacarpal, vestigal	49.0	6.15	3.5	0.6	9.3, m
9538	I/II	deer-size	humerus	52.5	18.25	3.45	3.4	-
9538	I/II	deer-size	metapodial	41.0	12.2	6.6	2.6	-
10165	Harrison	fox	femur	93.6	9.4	7.3	5.1	-
10332	-	*B. bison*	costal cartilage (rib)	132.0	21.0	9	8.9	9.6, b
10345	-	deer-size	metapodial	25.7	13.6	5	1.5	9.3, l
10350	I	bird	tibio-tarsus	56.1	5.6	4.2	0.4	-
15148	VIII	bird	2 feathers	-	-	-	0.1	-
16761	VI/VII	deer/elk	long bone	33.4	20.6	3.1	2.2	-

Table 9.12 Bone tool morpho-functional classes by geologic unit.

	I, II		Floodplain		II/III		III		IV, V		VI-VIII		Total
Tool Class	n	%	n	%	n	%	n	%	n	%	n	%	n
Awls	2	11.8	1	6.7	0	0.0	2	25.0	3	23.0	1	16.7	9
Eyed needle	0	0.0	7	46.7	0	0.0	0	0.0	0	0.0	0	0.0	7
Atlatl spur	1	5.9	0	0.0	0	0.0	0	0.0	0	0.0	0	0.0	1
Splitting wedge	1	5.9	0	0.0	0	0.0	0	0.0	0	0.0	0	0.0	1
Spatulate tool	1	5.9	0	0.0	0	0.0	0	0.0	0	0.0	1	16.7	2
Narrow-diameter pins	7	41.2	1	6.7	1	100	2	25.0	1	7.7	0	0.0	12
Bone/Antler rods	1	5.9	4	26.7	0	0.0	1	12.5	6	46.2	0	0.0	12
Pointed bone tools	0	0.0	2	13.3	0	0.0	1	12.5	0	0.0	2	33.3	5
Ornaments, polished bone	2	11.8	0	0.0	0	0.0	2	25.0	2	15.4	2	33.3	8
Manufacturing debris	2	11.8	0	0.0	0	0.0	0	0.0	1	7.7	0	0.0	3
Total	17	100	15	100	1	100	8	100	13	100	6	100	60

archaeological record. It is unlikely that this was discarded, and it was more likely lost. Small items are rarely picked up and discarded in secondary trash dumps (Schiffer 1987), suggesting the needles and needle blanks were lost or discarded near the places that they were used or made. The floodplain may have been an area where people sewed clothing. The presence of needle blanks also suggests that tool manufacture took place on the floodplain. The Windust deposits contain a single broken atlatl spur, suggesting weapon repair. This also indicates that at least some Windust phase points were atlatl darts. A splitting wedge from the rockshelter Windust phase deposits suggests woodworking was linked to those occupations. Bone pins are concentrated in the Windust phase deposits, where 8 of 12 were recovered.

Antler/bone rods are strongly associated with burials. The four antler rod fragments from the floodplain may be part of the same artifact, and all are associated with a concentration of human remains there. Six of the rods are associated with Stratum V burials. Single fragments were recovered from the Windust and early Cascade rockshelter deposits. Thus ten of 12 rods may have burial associations. Discussions of the functions of bone rods should take

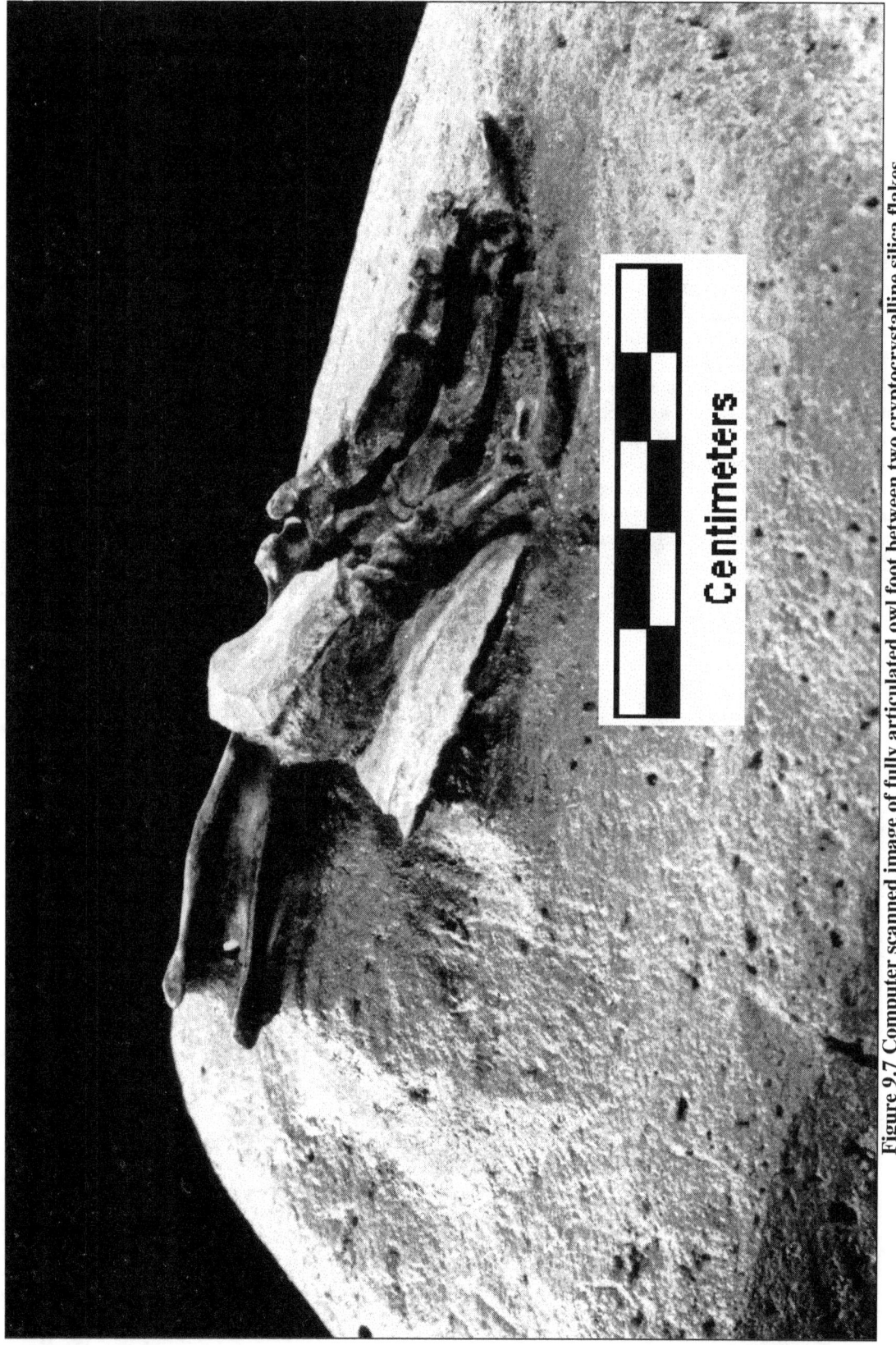

Figure 9.7 Computer scanned image of fully articulated owl foot between two cryptocrystalline silica flakes.

such associations into account, and it is possible that these are not utilitarian artifacts. The larger osseous rods from early Paleoindian sites are also strongly associated with caches (or offerings?), and the Anzick specimens may be associated with a burial. Some rods are, however, from camp sites. Context, as well as form, should be considered in determining the functions of such artifacts.

Spatulate tools occur in the earliest and the latest deposits. One is from the Windust deposits (Stratum II) and one was from Stratum VII. No bone pins were recovered from the Stratum VI-VIII deposits. Pointed bone tools, at least some of which may be projectile points, occur in the floodplain and in early Cascade (Stratum III) and late deposits (Stratum VI-VIII) in the rockshelter. Ornaments and polished bone also occur throughout the rockshelter deposits. Both of the pendants, however, are associated with Stratum V burials. Bone awls are common and occur in all geologic units, suggesting that basketry or hide working were important activities throughout the occupational history of the site. The awls in Stratum IV/V suggest that these utilitarian implements were interred with their owners.

Comparison of the Windust Phase Floodplain and Rockshelter Deposits

Only 32 pieces of modified bone from the Windust phase deposits retain provenience. There are 17 specimens from the rockshelter and 15 items from the floodplain (though four pieces are probably from the same artifact). The most striking contrast between these areas is the presence of needles and needle blanks only in the floodplain (see Table 9.4).

Most of the rockshelter deposits were screened with ¼-inch mesh (ca. 5.6-mm square openings), and this may account for the lack of needles. The presence of awls in both areas suggests that hide or basket working was carried out in both parts of the site. The needles are small and are probably primary refuse, implying that people sewed and made clothing on the floodplain. Narrow diameter pins occur principally in the rockshelter, though the function of those artifacts is uncertain. If they are barbs from fishing spears, their occurrence in the rockshelter may reflect tool repair. Most of the pins are small broken pieces and were discarded in an unusable state. The tools that are certainly utilitarian from the rockshelter are complete and still usable. These include two awls, an antler wedge, and a spatulate tool. These implements may have been stored in the rockshelter, but for unknown reasons were not retrieved by their owners.

10
FAUNAL REMAINS

Carl E. Gustafson and Robert M. Wegener

The Marmes site is situated within the *Agropyron spicatum-Poa secunda* bunchgrass zone of Daubenmire (1970) and the soils are generally shallow and stony with occasional gravel bars, rockslides and basalt bedrock outcrops. This bleak landscape with its impoverished flora and fauna belies the former productivity of the region. In addition to the early historic mosaic of native vegetation types, communities in various stages of plant succession and natural disturbances added even greater diversity to the prehistoric environments (Daubenmire 1970; Franklin and Dyrness 1973). When climatic fluctuations through time are applied to the environmental formula, the potential variety of biotic communities increases tremendously. Marmes Rockshelter provides an excellent opportunity to infer the effects of such diversity on human adaptations and on plant and animal populations.

Current Research

This study addresses only non-human animal bone other than fish. The Marmes floodplain was treated as a single component representing only the earliest period at the site. The study concentrated on the Marmes floodplain area because it was least understood in terms of both the fauna and the cultural material.

Sampling Population

This study focused primarily on the water-screened, bulk samples that had not been sorted previously. During excavations on the floodplain, recognizable artifacts and larger pieces of bone or shell were mapped *in situ* prior to removal. However, most material was too small (i.e., < 15 mm) to separate easily from the carbonate rootlet casts and small stone clasts. Therefore, all sediments were removed and water-screened through 1-mm mesh. A tag with provenience was attached to each sample before it was set aside to dry before packaging. Some of these "bulk samples" were sorted in the field laboratory, but time was limited and coupled with the goal of the 1968 excavations of removing as much material as possible with reasonable controls and documentation resulted in the accumulation of 1,198 unsorted floodplain bulk samples. These were returned to storage at Washington State University (WSU) where they remained until the present project was initiated in January 1998.

The entire Marmes assemblage was re-inventoried in the early 1990s (Collins and Andrefsky 1995) and a new database created for the collection with each specimen or sample assigned a unique inventory number in the database. Information extracted from this database indicated that 164 of the 1,198 floodplain bulk samples available for analysis lacked excavation unit designations. These 164 samples without provenience were not considered for analysis, so the resulting number of floodplain bulk samples considered for analysis totaled 1,034. Each bulk sample was located in one of 355 boxes stored in the WSU storage facility; these boxes also contained 642 other floodplain sample and items bags. Not all samples from a particular unit were found together because in 1968 bulk samples were boxed as they arrived daily in the field laboratory. Thus, material from different units and different levels is included in each of the 355 boxes containing material from the Marmes site. This situation made the retrieval of samples to be analyzed in this study an exceedingly cumbersome task.

Sample Selection

The intent of this faunal study was to analyze about 300 of the 1,034 floodplain bulk samples from known locations. Initially, the floodplain sample focused on four areas of particular cultural interest as defined by Hicks (see Chapter Two, *Sampling*):

1. The northeast quadrant of the excavation where Arctic fox and pine marten remains were associated with flaking debris amidst several large boulders,
2. The area surrounding the owl's foot first reported by Gustafson (1972:Figure 6.1),
3. The south-central area where semi-articulated remains of a large elk were found,
4. The southwest quadrant among a concentration of smaller bone fragments.

Some units selected contained only material from the Harrison horizons, others only from the Marmes horizon or from silts above or below. Rather than remove about two hundred ca. 50-pound boxes to satisfy a random sampling technique, we chose to sort and analyze all bulk samples that were in the boxes containing samples selected from one of the four subareas above. This procedure resulted in the selection of 221 bulk samples from 51 boxes and 18 field sorted samples.

Figure 10.1 shows the horizontal coverage resulting from the selection of samples in this study, but not all vertical excavation levels were accounted for in each of the 68 sampled excavation units in this study or the 60 sampled excavation units analyzed by Caulk (1988:Appendix A). Table 10.1 provides a summary of sampled floodplain excavation units and strata for both studies. The Marmes horizons were the most extensively excavated so this analysis focused first on 49 units yielding material from the Marmes horizons (31 with only Marmes, plus 18 containing both Marmes and Harrison remains), and then on materials from the 32 excavation units containing samples from the Harrison horizons (14 with only Harrison, plus the 18 with both Marmes and Harrison remains), 18 containing material from both the Marmes and Harrison horizons (these add to both the Marmes and Harrison figures as shown above), and other strata (n = 5) in the time remaining. Only five units are shaded as "Other strata" in Figure 10.1 in order to simplify the figure. The seven units not identified as containing samples analyzed from strata other than the Marmes and Harrison horizons include: 10N/20-30W, 10N/40-45W, 10N/60-65W, and 25N/25W. Twelve additional units are not identified in Figure 10.1 as containing samples from strata other than the Marmes and Harrison horizons because they already were included with the Marmes and/or Harrison units; these remains were from units 5N/15E, 10N/20W, 10N/25W, 10N/30W, 10N/40W, 10N/45W, 10N/55W, 10N/60W, 10N/65W, 25N/25W, 30N/10E and 30N/15E. Several of the latter units contained remains from two of the "other strata". This resulted in the 18 total (three "Above", five "Between", and ten "Below") depicted in Table 10.1.

Table 10.1 lists strata sampled in the 1998 study. More than 500 additional bulk samples would be required to examine the materials collected from each stratum and/or level excavated in the 68 units sampled. This was beyond the scope of the present study.

The 1998 sample contained 21% of the previously unanalyzed bulk samples, but relatively few field-sorted samples. Table 10.2 shows the number and percentage of sampled floodplain excavation units and strata in this study and those presented in Caulk (1988:Appendix B). Caulk focused solely on field-sorted samples (1988:23-25). Caulk analyzed 75 field-sorted samples from 60 floodplain excavation units, and except for a single sample, focused on the Marmes and Harrison horizons. The 1998 and Caulk's sample are integrated in Figure 10.2. Combined, the two samples represent 85 (54%) of the 158 units excavated on the Marmes Floodplain.

Unanalyzed Faunal Materials and Bulk Samples

Most of the Marmes floodplain faunal and bulk samples remain unexamined. Appendix J lists the inventory number, box number, and provenience for samples analyzed in this study. Unanalyzed presorted faunal samples containing mammal remains (n = 512) are listed in Appendix K. Appendix L lists samples coded as bird remains (n = 26). Appendix M lists 104 samples not assigned currently to a vertebrate class as well as unanalyzed reptile and amphibian bone because it was coded as "other bone" and re-curated together during the NAGPRA inventory (Collins and Andrefsky 1995:80). The appendices are organized by excavation unit and should prove useful during future analyses conducted on the Marmes fauna.

Sample Processing and Coding

Individual specimens were removed from the bulk samples using 10X magnifying lamps, forceps, and brushes. The separated material was subdivided into 11 categories, each of which was given a decimal extension of the original WSU bag inventory number. For example, mammal bone from inventory number 1734 was labeled as 1734.1, fish bone as 1734.7, plant remains as 1734.9 and so on. For future reference, the inventory number extensions are:

.1	Mammal bone
.2	Amphibian bone
.3	Reptile bone
.4	Bird bone
.5	Insect remains
.6	"Mica-laden pitch"
.7	Fish bone
.8	Mollusk shell
.9	Plant remains
.10	Stone artifacts
.11	Other

The category "mica-laden pitch" refers to small, very thin flakes of black, somewhat flexible material with even smaller flecks of what appear to be mica

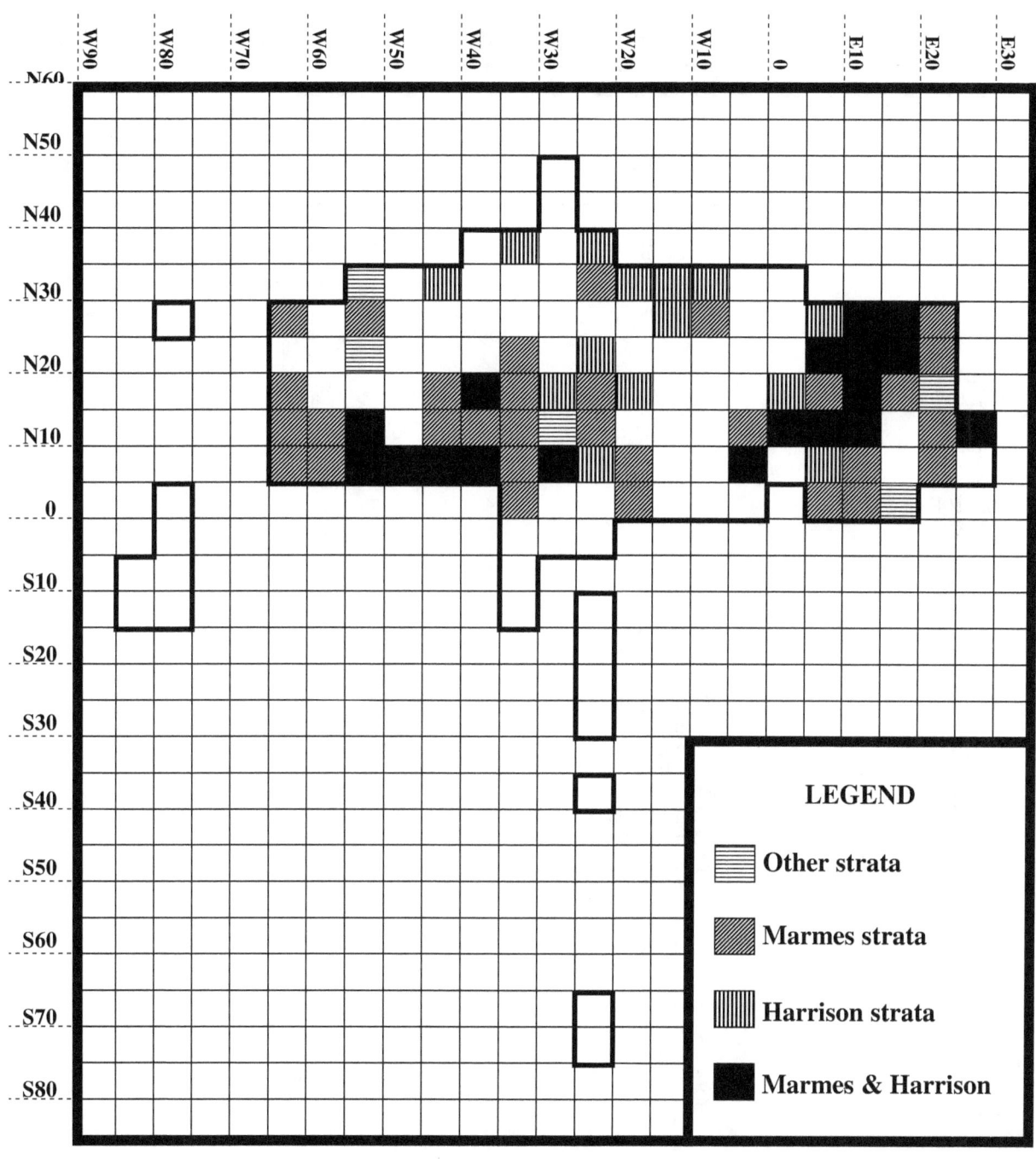

Figure 10.1 Plan map showing Marmes (45FR50) floodplain excavation units and strata that contained bulk samples analyzed in this study.

Table 10.1 Summary of Marmes (45FR50) floodplain units sampled in this study (1998) and from data presented in Caulk (1988:Appendix B).

Unit	Silts Above Marmes Horizons	Marmes Horizons	Silts Between Marmes & Harrison Horizons	Harrison Horizons	Silts Below Harrison Horizons
0N/30W	--	Caulk	--	--	--
5N/5E	--	1998	--	Caulk	--
5N/10E	--	1998	--	Caulk	--
5N/15E	--	--	1998	--	--
5N/5W	--	Caulk	--	--	--
5N/20W	--	Caulk/1998	--	Caulk	--
5N/30W	--	Caulk	--	Caulk	--
5N/35W	--	Caulk/1998	--	--	--
10N/5E	--	Caulk	--	Caulk/1998	--
10N/10E	--	1998	--	Caulk	--
10N/20E	--	1998	--	Caulk	--
10N/0W	--	--	--	Caulk	--
10N/5W	--	--	--	1998	--
10N/10	--	--	--	Caulk	--
10N/20	1998	Caulk/1998	--	Caulk	--
10N/25	1998	--	1998	1998	--
10N/30	--	1998	--	1998	1998
10N/35	--	Caulk/1998	--	Caulk	--
10N/40	--	1998	--	Caulk/1998	1998
10N/45	--	Caulk/1998	--	Caulk/1998	1998
10N/50	--	1998	--	1998	--
10N/55	--	1998	--	Caulk/1998	1998
10N/60	--	Caulk/1998	--	--	1998
10N/65	--	Caulk/1998	--	Caulk	1998
15N/5E	--	Caulk/1998	--	1998	--
15N/10E	--	1998	--	Caulk/1998	--
15N/20E	--	1998	--	--	--
15N/25E	--	1998	--	1998	--
15N/0W	--	1998	--	Caulk/1998	--
15N/5W	--	1998	--	--	--
15N/10	--	Caulk	--	--	--
15N/25	--	1998	--	--	--
15N/30	--	--	--	Caulk	1998
15N/35	--	Caulk/1998	--	Caulk	--
15N/40	--	1998	--	--	--
15N/45	--	1998	--	--	--
15N/55	--	1998	--	Caulk/1998	--
15N/60	--	1998	--	Caulk	--
15N/65	--	1998	--	--	--
20N/5E	--	1998	--	--	--
20N/10E	--	Caulk/1998	--	Caulk/1998	--
20N/15E	--	1998	--	Caulk	--
20N/20E	--	Caulk/1998	--	--	--
20N/0W	--	--	--	1998	--
20N/5W	--	Caulk	--	--	--

Table 10.1 Summary of Marmes (45FR50) floodplain units sampled in this study (1998) and from data presented in Caulk (1988:Appendix B).

Unit	Silts Above Marmes Horizons	Marmes Horizons	Silts Between Marmes & Harrison Horizons	Harrison Horizons	Silts Below Harrison Horizons
20N/20W	--	--	--	Caulk/1998	--
20N/25W	--	1998	Caulk	Caulk	--
20N/30W	--	--	--	Caulk/1998	--
20N/35W	--	1998	--	--	--
20N/40W	--	1998	--	1998	--
20N/45W	--	1998	--	--	--
20N/50W	--	--	--	Caulk	--
20N/55W	--	Caulk	--	--	--
20N/65W	--	Caulk/1998	--	--	--
25N/5E	--	Caulk/1998	--	1998	--
25N/10E	--	1998	--	Caulk/1998	--
25N/15E	--	1998	--	Caulk/1998	--
25N/20E	--	1998	--	Caulk	--
25N/0W	--	Caulk	--	--	--
25N/25W	--	Caulk	1998	Caulk/1998	1998
25N/35W	--	1998	--	Caulk	--
25N/50W	--	Caulk	--	--	--
25N/55W	1998	Caulk	--	Caulk	1998
30N/5E	--	--	--	Caulk/1998	--
30N/10E	--	1998	1998	1998	--
30N/15E	--	1998	1998	Caulk/1998	--
30N/20E	--	1998	--	--	--
30N/10W	--	Caulk/1998	--	Caulk	--
30N/15W	--	--	--	1998	--
30N/25W	--	Caulk	--	--	--
30N/50W	--	Caulk	--	--	--
30N/55W	--	Caulk/1998	--	--	--
30N/65W	--	1998	--	--	--
35N/5W	--	--	--	Caulk	--
35N/10W	--	Caulk	--	Caulk/1998	--
35N/15W	--	Caulk	--	Caulk/1998	--
35N/20W	--	--	--	Caulk/1998	--
35N/25W	--	1998	--	--	--
35N/45W	--	--	--	Caulk	--
35N/55W	--	--	--	--	1998
40N/10W	--	--	--	Caulk	--
40N/25W	--	--	--	1998	--
40N/35W	--	--	--	1998	--
10S/35W	--	--	--	Caulk	--
25S/25W	--	--	--	Caulk	--

Table 10.2 Number and percentage of sampled floodplain excavation units (n) and levels by strata from this study and Caulk (1988:Appendix B).

STRATUM	PROJECT					
	Caulk (1988)		This Study		Total	
	n	%	n	%	n	%
Silts above Marmes Horizons	--	--	3	(3.0)	3	(1.7)
Marmes Horizons	30	(40.0)	49	(49.5)	79	(45.4)
Silts between Marmes & Harrison Horizons	1	(1.3)	5	(5.1)	6	(3.4)
Harrison Horizons	44	(58.7)	32	(32.3)	76	(43.7)
Silts below Harrison Horizons	--	--	10	(10.1)	10	(5.7)
TOTALS[1]	75	(100.0)	99	(100.0)	174	(99.9)

[1] Percentage totals do not equal 100.0 due to rounding.

embedded on one surface. Caulk (1988:25) referred to these flecks as "mica impregnated pitch". Although we are not certain, we think these flecks may have flaked off the lining of the black, rubber hoses used at the water screen. No such material was found in the sediment samples that were not water-screened.

Specimen Identification

Whenever possible, identifications were made by direct comparison with known specimens housed in the Comparative Faunal Laboratory or borrowed from the Conner Museum, departments of Anthropology and Zoology, at Washington State University. Guides to identification referred to include Blair et al. 1968, Dalquest 1948, Ingles 1965, Stebbins 1954, and Stebbins 1966. Zweifel 1994 proved especially useful in identifying isolated teeth, to family, genus and species. Bones of amphibians, reptiles and mammals are addressed by this study. Another researcher analyzed the fish remains (see Chapter Eleven). No bird bones were identified from the 1998 floodplain sample; however, some were noted by Gustafson (1972) and Caulk (1988).

Relatively large amounts of bone fragments, shell fragments and other remains had been recovered from those samples sorted previously in the field, and it was assumed the remaining floodplain bulk samples also would produce substantial quantities of faunal remains. However, this was not the case, suggesting that those sorted in 1968 were selected for their noticeable high content of organic remains. In addition, cultural items of note were bagged separately by the excavators, and in some cases by the crewmembers overseeing the water screen; such separately bagged items may be among the missing and non-provenienced bags and are not represented in the floodplain sample from this analysis. Together, the 221 bulk samples processed totaled 252.2 kg (>550 lb) of sediment, which yielded 0.83 kg (<2 lb) of pertinent specimens. Here, a specimen is defined as any bone or tooth, or fragment thereof. The remaining 251.8 kg (>550 lb) of sediment consisted primarily of silt and tiny (< 3 mm) carbonate rootlet casts interspersed with occasional sand and gravel. Approximately 60,000 tiny (e.g., < 4 mm maximum dimension) faunal specimens were separated from these bulk samples and 34 taxa identified representing three vertebrate classes and eight orders.

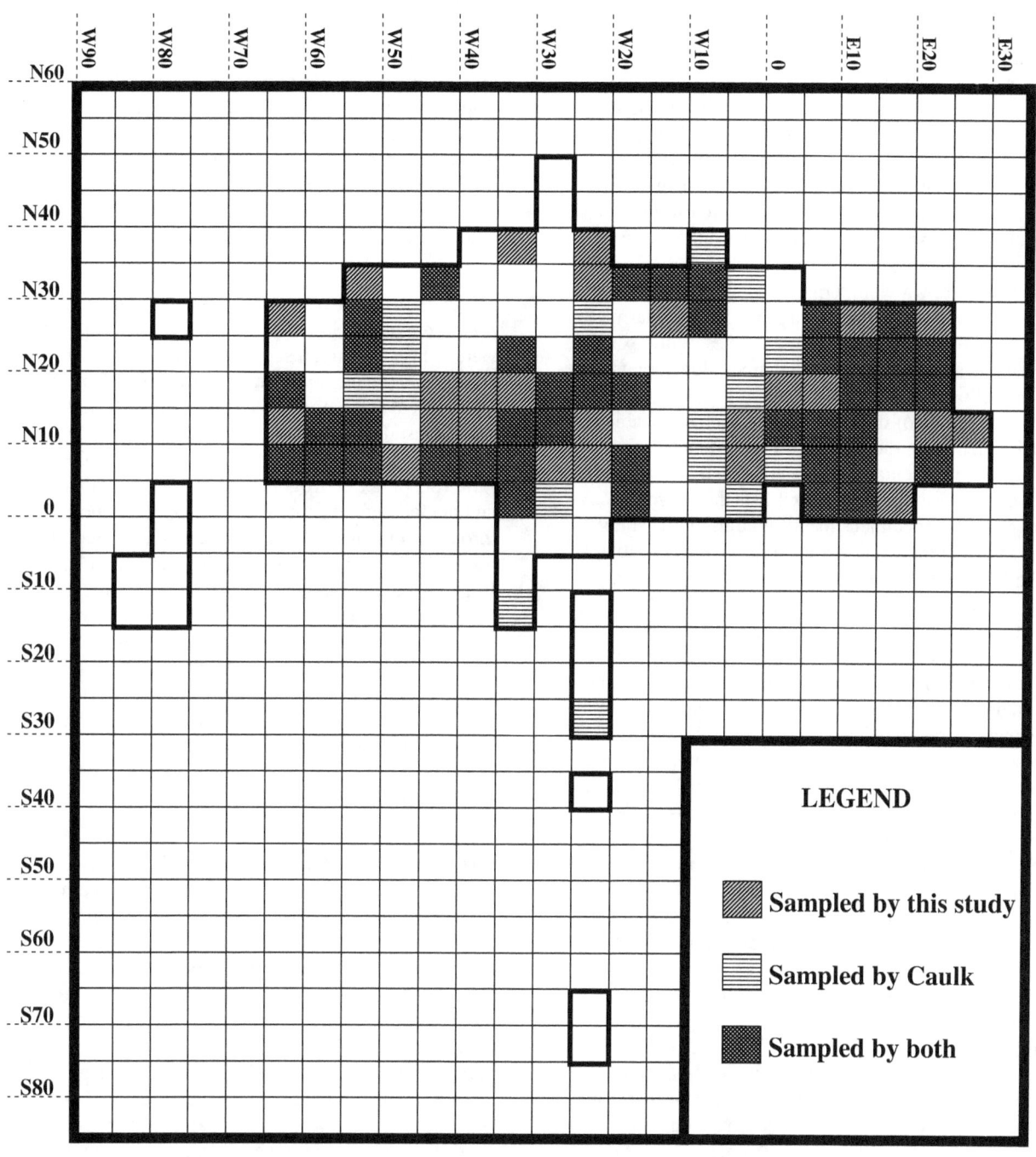

Figure 10.2 Plan map showing locations of Marmes (45FR50) floodplain excavation units sampled by this study and by Caulk (1988).

The Marmes Fauna

The discussion in this section is based on the identifications made during this study, and those presented in Gustafson (1972) and Caulk (1988). It provides a discussion for each taxon identified that: 1) describes each identified taxon and its habitat preference(s); 2) identifies the number of specimens assigned to that taxon and their distribution at the site; and, 3) presents the known ethnographic use(s), if any, of each taxon. Intended to be a synthesis, 75 taxa representing four vertebrate classes and 16 orders are discussed, which in turn, represent all non-human terrestrial vertebrate taxa documented at the Marmes site. The Scientific and common names of all taxa are given in Table 10.3. Scientific names alone are used in the following headings.

The results of this study are presented in Table 10.4 which gives the Number of Identified Specimens (NISP) and percentages for each taxon by excavated floodplain stratum. Appendix J lists the provenience and taxon for specimens identified during this study. All counts in this section are based on NISP. Table 10.5 lists the NISP and percentages—per-geologic unit and taxon—from "relatively undisturbed" contexts in the Marmes Rockshelter as reported in Gustafson (1972:Table 5.1). Identifications made on specimens from the Marmes and Harrison floodplain horizons given in Caulk (1988:Table 2) are listed by taxon and stratum in Table 10.6. In the following presentation, the numbers following the taxon are the total NISP assigned to that taxon. These NISP values represent a summation of the identifications made during this study and those made by Gustafson (1972) and Caulk (1988). Table 10.7 lists this information by taxon and study, and it includes all identifications made to date for non-human vertebrate taxa found at the Marmes Site. Reference to these summary tables also will be made later in this report.

Much of the material for this section is borrowed freely from Gustafson (1972: 57-83) and is used with his permission. This material is supplemented to include new data from the Marmes floodplain and information from Caulk (1988). Taxa identified from the Marmes Rockshelter and adjacent floodplain are described briefly with relevant statements on natural history and habitat preference. Information concerning cultural implications also is included where deemed appropriate. Numerous other taxa are present today along the course of the Lower Snake River, but only those recovered from sediments also including cultural material are included in this discussion.

Where species determination was not possible, the name of the species living in the region today has been appended. Thus, the conservative approach of Ziegler (1965), Smith (1965), and others has been followed. In all cases where this approach was used, the diagnostic skeletal features fall within the expected range for the extant species; in no case does this approach alter the interpretations to follow.

Class Amphibia

Order Anura (Frogs and Toads). *Hyla* sp. and *Rana pipiens* (NISP = 12): Frogs are common near water. They are given separate treatment here only because they have not been identified often from sites in eastern Washington and adjacent Idaho (Caulk 1988; Gustafson 1990). Seven bones from a small frog were found. These were assigned to the Genus *Hyla* (peeping tree frogs) based solely on small size and the fact that peeping frogs are very common throughout the northwest. The second anuran (five bones) was the size of *Rana pipiens* (common leopard frog). Although the toad, *Bufo borealis*, is about the same size, its skeleton is more robust; therefore, we have assigned the specimens from Marmes tentatively to the species *R. pipiens* based on the more gracile nature of the skeleton. Both taxa occur in the vicinity today. There is no indication that these amphibians were consumed by people at the Marmes site.

Class Reptilia

Order Squamata (Lizards and Snakes). As indicated in footnote 1 of Table 10.7, reptile values represent the number of units containing reptile remains and are not included in the NISP total calculations. NISP values included below are from the current study and are given here as an indicator of the large number of snake vertebrae present in the sample.

Sceloporus (in 1 unit): A single, blackened mandible from a small lizard tentatively is identified as belonging to the genus *Sceloporus* (fence or sage-brush lizards). As with the frogs, this is a tenuous identification based on insufficient diagnostic material. We ascribe no cultural significance to it at this time.

Indeterminate Snakes (in 13 units): With the exception of a few rattlesnake fangs, vertebrae are the only bones that could be attributed to snakes. Complete precloacal vertebrae could be

Table 10.3 Scientific and common names for all taxa from the Marmes site (45FR50).

AMPHIBIANS

cf. *Hyla* sp.	peeping tree frogs
cf. *Rana pipiens*	common leopard frog

REPTILES

Unidentified snakes	unidentified snakes
cf. *Pituophis* sp.	bull or gopher snake
cf. *Thamnophis* sp.	garter snake
cf. *Crotalus* sp.	rattlesnake
cf. *Sceloporus* sp.	fence or sagebrush lizards

BIRDS

Bucephala clangula	common goldeneye duck
Ptychoramphus aleuticum	Cassin's auklet
Empidonax sp.	flycatcher
Pica pica	magpie
Hylocichla sp.	olive-backed thrush
Sialia sp.	bluebird

MAMMALS

Indeterminate Medium Mammal	rabbit size
Unid. Lagomorph	rabbits and hares
Sylvilagus sp.	cottontail rabbits
Lepus sp.	jackrabbits
Marmota cf. *flaviventris*	yellow-bellied marmot
Indeterminate Small Mammal	Ground squirrel-size and smaller
Spermophilus sp.	ground squirrels
Spermophilus cf. *columbianus*	Columbian ground squirrel
Spermophilus cf. *townsendii*	Townsend ground squirrel
Spermophilus cf. *washingtoni*	Washington ground squirrel
Cricetidae-size	mice, rats, and voles
Thomomys cf. *talpoides*	northern pocket gopher
Perognathus cf. *parvus*	Great Basin pocket mouse
Castor canadensis	beaver
Reithrodontomys sp.	meadow mouse
Peromyscus cf. *maniculatus*	deer mouse
Onychomys cf. *leucogaster*	northern grasshopper mouse
Neotoma cf. *cinerea*	bushy-tailed wood rat
cf. *Clethrionomys* sp. (?)	red-backed mouse
cf. *Phenacomys* sp. (?)	heather vole
Microtus sp.	voles
Microtus montanus	montane vole
Microtus pennsylvannicus	meadow vole
Microtus oeconomus	tundra vole
Lagurus curtatus	sagebrush vole
cf. *Ondatra* sp.	muskrat
Canis sp.	dog, coyote, wolf
Canis latrans	coyote
Canis familiaris	domestic dog
Canis lupus	wolf
Unid. fox.	indeterminate fox
Alopex lagopus	arctic fox
Vulpes cf. *fulva*	red fox
Ursus sp.	bear
Mustella-size	weasel-sized
Martes americana	pine marten
Mustela frenata	long-tailed weasel
Taxidea cf. *taxus*	badger
Mephitis mephitis	striped skunk
Vulpes/Lynx-size	fox/lynx size
cf. *Lynx rufus*	bobcat
Lynx canadensis	Canadian lynx
Medium/Large Artiodactyl	deer-to-elk-size animals
Cervus canadensis	elk/wapiti
Odocoileus sp.	deer
Antilocapra americana	pronghorn
Bos taurus	domestic cattle

Table 10.4 Number (n) of identified specimens (NISP), percentages and totals, from the 1998 study of the Marmes floodplain (45FR50) fauna by stratigraphic unit and taxon.

TAXON	Silts Above Marmes Horizons		Marmes Horizons		Silts Between Marmes & Harrison Horizons		Harrison Horizons		Silts Below Harrison Horizons		NISP TOTAL	
	n	%	n	%	n	%	n	%	n	%	n	%
AMPHIBIANS												
cf. *Hyla* sp.	--	--	7	(1.1)	--	--	--	--	--	--	7	(.8)
cf. *Rana pipiens*	--	--	4	(.6)	--	--	--	--	--	--	4	(.5)
REPTILES[1]												
Unidentified snakes	--	--	10	--	--	--	3	--	--	--	13	--
cf. *Pituophis* sp.	--	--	12	--	--	--	1	--	--	--	13	--
cf. *Thamnophis* sp.	--	--	8	--	1	--	7	--	--	--	16	--
cf. *Crotalus* sp.	1	--	10	--	1	--	2	--	--	--	14	--
cf. *Sceloporus* sp.	--	--	1	--	--	--	--	--	--	--	1	--
MAMMALS												
Sylvilagus sp.	--	--	10	(1.6)	2	(10.5)	4	(3.1)	--	--	16	(1.9)
Lepus-size	--	--	1	(.2)	--	--	--	--	--	--	1	(.1)
Lepus sp.	--	--	9	(1.4)	--	--	19	(15.0)	--	--	28	(3.3)
Marmota cf. *flaviventris*	--	--	5	(.8)	--	--	5	(3.9)	1	(1.5)	11	(1.3)
Spermophilus sp.	--	--	6	(1.0)	--	--	2	(1.6)	--	--	8	(.9)
Spermophilus cf. *columbianus*	--	--	4	(.6)	--	--	5	(3.9)	--	--	9	(1.1)
Spermophilus cf. *townsendii*	--	--	2	(.3)	--	--	--	--	2	(3.0)	4	(.5)
Cricetidae-size	--	--	61	(9.7)	--	--	8	(6.3)	4	(6.0)	73	(8.6)
Thomomys cf. *talpoides*	--	--	136	(21.7)	5	(26.3)	12	(9.4)	4	(6.0)	157	(18.5)
cf. *Perognathus* sp.	5	(55.6)	16	(2.5)	2	(10.5)	13	(10.2)	11	(16.4)	47	(5.5)

Table 10.4 Number (n) of identified specimens (NISP), percentages and totals, from the 1998 study of the Marmes floodplain (45FR50) fauna by stratigraphic unit and taxon.

TAXON	STRATUM										NISP TOTAL	
	Silts Above Marmes Horizons		Marmes Horizons		Silts Between Marmes & Harrison Horizons		Harrison Horizons		Silts Below Harrison Horizons			
	n	%	n	%	n	%	n	%	n	%	n	%
Perognathus cf. *parvus*	4	(44.4)	48	(7.6)	3	(15.8)	12	(9.4)	8	(11.9)	75	(8.8)
Reithrodontomys sp.	--	--	3	(.5)	1	(5.3)	1	(.8)	--	--	5	(.6)
Peromyscus sp.	--	--	26	(4.1)	--	--	2	(1.6)	1	(1.5)	29	(3.4)
Peromyscus cf. *maniculatus*	--	--	5	(.8)	--	--	--	--	--	--	5	(.6)
Onychomys cf. *leucogaster*	--	--	2	(.3)	--	--	--	--	--	--	2	(.2)
Neotoma cf. *cinerea*	--	--	2	(.3)	--	--	--	--	14	(20.9)	16	(1.9)
cf. *Clethrionomys* sp. (?)	--	--	--	--	--	--	1	(.8)	--	--	1	(.1)
cf. *Phenacomys* sp. (?)	--	--	3	(.5)	--	--	--	--	--	--	3	(.4)
cf. *Microtus* sp.	--	--	3	(.5)	--	--	--	--	--	--	3	(.4)
Microtus sp.	--	--	83	(13.2)	4	(21.1)	22	(17.3)	12	(17.9)	121	(14.2)
cf. *Ondatra* sp.	--	--	1	(.2)	--	--	--	--	--	--	1	(.1)
Canid-size	--	--	--	--	--	--	2	(1.6)	7	(10.4)	9	(1.1)
Vulpes cf. *fulva*	--	--	1	(.2)	--	--	14	(11.0)	--	--	15	(1.8)
Mustella-size	--	--	--	--	--	--	1	(.8)	--	--	1	(.1)
Taxidea cf. *taxus*	--	--	1	(.2)	--	--	--	--	--	--	1	(.1)
Vulpes/Lynx-size	--	--	--	--	--	--	--	--	1	(1.5)	1	(.1)
cf. *Lynx rufus*	--	--	--	--	--	--	2	(1.6)	2	(3.0)	4	(.5)
Medium/Large Artiodactyl	--	--	188	(29.9)	2	(10.5)	2	(1.6)	--	--	192	(22.6)
Cervus canadensis	--	--	1	(.2)	--	--	--	--	--	--	1	(.1)
TOTALS[2]	9	(100.0)	628	(100.0)	19	(100.0)	127	(100.0)	67	(100.0)	850	(100.1)

[1]Reptile values represent number of units containing reptile remains and are not included in the NISP total calculations.
[2]Percentage totals do not equal 100.0 due to rounding.

Table 10.5 Number (n) of identified specimens (NISP), number and percentages, per-geologic unit and taxon from Marmes rockshelter (45FR50). Adapted from Gustafson (1972:Table 5.1).

^{14}C yrs. B.P. (Approx.)	4,000		6,850		7,500		10,000		NISP TOTAL	
Geologic Unit	VI-VIII		IV-V		III		I-II			
TAXON	n	%	n	%	n	%	n	%	n	%
Odocoileus sp.	26	(10.8)	49	(19.5)	108	(26.4)	22	(19.6)	205	(20.2)
Antilocapra Americana	31	(12.9)	65	(25.9)	79	(19.3)	14	(12.5)	189	(18.7)
Cervus Canadensis	2	(0.8)	8	(3.2)	36	(8.8)	5	(4.5)	51	(5.0)
Bos Taurus	48	(19.9)	--	--	--	--	--	--	48	(4.7)
Indeterminate Large Mammal	55	(22.8)	70	(27.9)	123	(30.1)	41	(36.6)	289	(28.5)
Subtotal	162	(67.2)	192	(76.5)	346	(84.6)	82	(73.2)	782	(77.2)
Sylvilagus sp.	14	(5.8)	12	(4.8)	10	(2.4)	2	(1.8)	38	(3.8)
Lepus sp.	2	(0.8)	3	(1.2)	5	(1.2)	3	(2.7)	13	(1.3)
Canis cf. latrans	3	(1.2)	5	(2.0)	20	(4.9)	2	(1.8)	30	(3.0)
Marmota cf. flaviventris	5	(2.1)	1	(.4)	2	(.5)	1	(.9)	9	(.9)
Ondatra zibethica	2	(0.8)	1	(.4)	4	(1.0)	2	(1.8)	9	(.9)
Indeterminate Medium Mammal	8	(3.3)	14	(5.6)	12	(2.9)	3	(2.7)	37	(3.7)
Subtotal	34	(14.1)	36	(14.3)	53	(13.0)	13	(11.6)	136	(13.4)
Spermophilus cf. washingtoni	22	(9.1)	6	(2.4)	2	(.5)	--	--	30	(3.0)
Thomomys talpoides	3	(1.2)	4	(2.4)	1	(.2)	3	(2.7)	11	(1.1)
Neotoma cinerea	4	(1.7)	4	(1.6)	2	(.5)	3	(2.7)	13	(1.3)
Peromyscus maniculatus	--	--	--	(1.6)	--	--	1	(.9)	1	(.1)
Indeterminate Small Mammal	16	(6.6)	9	(3.6)	5	(1.2)	10	(8.9)	40	(3.9)
Subtotal	45	(18.7)	23	(9.2)	10	(2.4)	17	(15.2)	95	(9.4)
Grand Total	241	(100.0)	251	(100.1)	409	(99.9)	112	(100.1)	1013	(100.0)

Table 10.6 Number (n) of identified terrestrial taxa reported by Caulk (1988: Table 2) from the Marmes and Harrison Horizons of the Marmes floodplain.

TAXON	HORIZONS					
	Marmes		Harrison		NISP TOTAL	
	n	%	n	%	n	%
AMPHIBIANS						
cf. *Rana pipiens*	--	--	1	(.5)	1	(.4)
REPTILES[1]						
cf. *Pituophis* sp.	6	--	7	--	13	--
cf. *Thamnophis* sp.	5	--	9	--	14	--
cf. *Crotalus* sp.	1	--	1	--	1	--
BIRDS						
Bucephala clangula	--	--	3	(1.6)	3	(1.2)
Ptychoramphus aleuticum	--	--	2	(1.1)	2	(.8)
Empidonax sp.	--	--	1	(.5)	1	(.4)
Pica pica	--	--	1	(.5)	1	(.4)
Hylocichla sp.	--	--	2	(1.1)	2	(.8)
Sialia sp.	--	--	3	(1.6)	3	(1.2)
MAMMALS						
Unidentified Lagomorph	--	--	4	(2.2)	4	(1.6)
Lepus sp.	1	(1.6)	8	(4.3)	9	(3.6)
Marmota sp.	--	--	10	(5.4)	10	(4.0)
Thomomys cf. *talpoides*	30	(47.6)	26	(14.1)	56	(22.9)
Perognathus cf. *parvus*	6	(9.5)	17	(9.2)	23	(9.3)
Castor canadensis	--	--	1	(.5)	1	(.4)
Peromyscus sp.	12	(19.0)	21	(11.4)	33	(13.5)
Neotoma sp.	1	(1.6)	12	(6.5)	13	(5.3)
Microtus sp.	--	--	4	(2.2)	4	(1.6)
Microtus montanus	--	--	1	(.5)	1	(.4)
Microtus pennsylvannicus	--	--	3	(1.6)	3	(1.2)
Microtus oeconomus	1	(1.6)	--	--	1	(.4)
Lagurus curtatus	3	(4.8)	3	(1.6)	6	(2.4)
Canid	1	(1.6)	6	(3.2)	7	(2.8)
Canis sp.	2	(3.2)	1	(.5)	3	(1.2)
Canis latrans	1	(1.6)	1	(.5)	2	(.8)
Canis familiaris	--	--	1	(.5)	1	(.4)
Unid. fox.	--	--	28	(15.1)	28	(11.4)
Alopex lagopus	--	--	6	(3.2)	6	(2.4)
Vulpes cf. *fulva*	--	--	5	(2.7)	5	(2.0)
Martes americana	--	--	3	(1.6)	3	(1.2)
Mustela frenata	--	--	3	(1.6)	3	(1.2)
Lynx canadensis	1	(1.6)	--	--	1	(.4)
Cervus canadensis	1	(1.6)	--	--	1	(.4)
Odocoileus sp.	2	(3.2)	4	(2.2)	6	(2.4)
Antilocapra americana	1	(1.6)	1	(.5)	2	(.8)
TOTALS	63	(100.1)	182	(99.6)	245	(99.6)

[1]Reptile values represent number of units containing reptile remains and are not included in the NISP total.

Table 10.7 Number (n) of identified specimens (NISP) summary, percentages and totals from this Marmes floodplain study, Caulk's (1988) Marmes floodplain study, and Gustafson's (1972) Marmes rockshelter research.

Taxon	Gustafson (1972) n	%	Caulk (1988) n	%	This Study n	%	NISP Total n	%
AMPHIBIANS								
cf. *Hyla* sp.	--	--	--	--	7	(.8)	7	(.3)
cf. *Rana pipiens*	--	--	1	(.4)	4	(.5)	5	(.2)
REPTILES[1]								
Unidentified snakes	--	--	--	--	13	--	13	--
cf. *Pituophis* sp.	--	--	13	--	13	--	26	--
cf. *Thamnophis* sp.	--	--	14	--	16	--	30	--
cf. *Crotalus* sp.	--	--	1	--	14	--	15	--
cf. *Sceloporus* sp.	--	--	--	--	1	--	1	--
BIRDS								
Bucephala clangula	--	--	3	(1.2)	--	--	3	(.1)
Ptychoramphus aleuticum	--	--	2	(.8)	--	--	2	(.1)
Empidonax sp.	--	--	1	(.4)	--	--	1	(<.1)
Pica pica	--	--	1	(.4)	--	--	1	(<.1)
Hylocichla sp.	--	--	2	(.8)	--	--	2	(.1)
Sialia sp.	--	--	3	(1.2)	--	--	3	(.1)
MAMMALS								
Indeterminate Medium Mammal	37	(3.7)	--	--	1	(.1)	38	(1.8)
Unid. Lagomorph	--	--	4	(1.6)	--	--	4	(.2)
Sylvilagus sp.	38	(3.8)	--	--	16	(1.9)	54	(2.6)
Lepus sp.	13	(1.3)	9	93.6	28	(3.3)	50	(2.4)
Marmota cf. *flaviventris*	9	(.9)	10	(4.0)	11	(1.3)	30	(1.4)
Spermophilus sp.	--	--	--	--	8	(.9)	8	(.4)
Spermophilus cf. *columbianus*	--	--	--	--	9	(1.1)	9	(.4)
Spermophilus cf. *townsendii*	--	--	--	--	4	(.5)	4	(.2)
Spermophilus cf. *washingtoni*	30	(3.0)	--	--	--	--	30	(1.4)
Cricetidae-size	40	(3.9)	--	--	73	(8.6)	113	(5.3)
Thomomys cf. *talpoides*	11	(1.1)	56	(22.9)	157	(18.5)	224	(10.6)
cf. *Perognathus* sp.	--	--	--	--	47	(5.5)	47	(2.2)
Perognathus cf. *parvus*	--	--	23	(9.3)	75	(8.8)	98	(4.6)
Castor canadensis	--	--	1	(.4)	--	--	1	(<.1)
Reithrodontomys sp.	--	--	--	--	5	(.6)	5	(.2)
Peromyscus sp.	--	--	33	(13.5)	29	(3.4)	62	(2.9)
Peromyscus cf. *maniculatus*	1	(.1)	--	--	5	(.6)	6	(.3)
Onychomys cf. *leucogaster*	--	--	--	--	2	(.2)	2	(.1)
Neotoma sp.	--	--	13	(5.3)	--	--	13	(.6)

Table 10.7 Number (n) of identified specimens (NISP) summary, percentages and totals from this Marmes floodplain study, Caulk's (1988) Marmes floodplain study, and Gustafson's (1972) Marmes rockshelter research.

Taxon	Study							
	Gustafson (1972)		Caulk (1988)		This Study		NISP Total	
	n	%	n	%	n	%	n	%
Neotoma cf. *cinerea*	13	(1.3)	--	--	16	(1.9)	29	(1.4)
cf. *Clethrionomys* sp. (?)	--	--	--	--	1	(.1)	1	(<.1)
cf. *Phenacomys* sp. (?)	--	--	--	--	3	(.4)	3	(.1)
cf. *Microtus* sp.	--	--	--	--	3	(.4)	3	(.1)
Microtus sp.	--	--	4	(1.6)	121	(14.2)	125	(5.9)
Microtus montanus	--	--	1	(.4)	--	--	1	(<.1)
Microtus pennsylvannicus	--	--	3	(1.2)	--	--	3	(.1)
Microtus oeconomus	--	--	1	(.4)	--	--	1	(<.1)
Lagurus curtatus	--	--	6	(2.4)	--	--	6	(.3)
cf. *Ondatra* sp.	9	(.9)	--	--	1	(.1)	10	(.5)
Canid-size	--	--	7	(2.8)	9	(1.1)	16	(.8)
Canis sp.	--	--	3	(1.2)	--	--	3	(.1)
Canis latrans	30	(3.0)	2	(.8)	--	--	32	(1.5)
Canis familiaris	--	--	1	(.4)	--	--	1	(<.1)
Canis lupus	1	(.1)	--	--	--	--	1	(<.1)
Unid. fox.	--	--	28	(11.4)	--	--	28	(1.3)
Alopex lagopus	1	(.1)	6	(2.4)	--	--	7	(.3)
Vulpes cf. *fulva*	--	--	5	(2.0)	15	(1.8)	20	(.9)
Ursus sp.	2	(.2)	--	--	--	--	2	(.1)
Mustella-size	--	--	--	--	1	(.1)	1	(<.1)
Martes americana	1	(.1)	3	(1.2)	--	--	4	(.2)
Mustela frenata	--	--	3	(1.2)	--	--	3	(.1)
Taxidea cf. *taxus*	--	--	--	--	1	(.1)	1	(<.1)
Mephitis mephitis	2	(.2)	--	--	--	--	2	(.1)
Vulpes/Lynx-size	--	--	--	--	1	(.1)	1	(<.1)
cf. *Lynx rufus*	--	--	--	--	4	(.5)	4	(.2)
Lynx canadensis	--	--	1	(.4)	--	--	1	(<.1)
Medium/Large Artiodactyl	289	(28.5)	--	--	192	(22.6)	481	(22.7)
Cervus canadensis	51	(5.0)	1	(.4)	1	(.1)	53	(2.5)
Odocoileus sp.	205	(20.2)	6	(2.4)	--	--	211	(10.0)
Antilocapra americana	189	(18.7)	2	(.8)	--	--	191	(9.0)
Bos taurus	48	(4.7)	--	--	--	--	48	(2.2)
Total[2]	1020	(100.1)	245	(99.6)	850	(100.1)	2115	(>98.6)

[1] Reptile values represent number of units containing reptile remains and are not included in the NISP total calculations.
[2] Percentage totals do not equal 100.0 due to rounding.

267

assigned with some confidence to one of the three genera discussed below. Damaged and postcloacal vertebrae were simply identified as "snake". Vertebrae from all snakes comprised the largest category of identified specimens: 501 (37%) of the total NISP of 1,359. When snake vertebrae were present, there usually was more than one. As many as 35-40 were found in a single sample. No attempts were made to plot their concentrations in the site, but finding so many vertebrae together suggests at least the presence of semiarticulated to nearly complete skeletons. Snake skeletons consist mainly of vertebrae and ribs, so the presence of vertebrae in large numbers probably does not represent many individuals.

Only three genera of snakes are included below. Other genera exist in the Pacific Northwest (Blair et al. 1968; Nussbaum et al. 1983; Stebbins 1966), but they are rare or seldom found in floodplain or arid steppe environments. We have been conservative in assigning generic identifications to only the common forms realizing that such assignments based on isolated vertebrae are subject to revision if more reliable remains become available.

Pituophis (in 26 units): Today there is only one species of bullsnake or gopher snake. At least 11 subspecies of *Pituophis melanoleucus* occupy many habitats throughout the United States. About 40% of the vertebrae identified had been burned; 60% were unaltered by fire.

Thamnophis (in 30 units): Garter snakes are common from the east coast to the west coast of the lower 48 United States and from Canada to Costa Rica (Nussbaum et al. 1983). There are many species and subspecies, but most prefer wet areas or are amphibious. Almost all (>99%) of the vertebrae assigned to this taxon were burned.

Crotalus (in 15 units): Only one species of rattlesnake (*Crotalus viridis*) inhabits eastern Washington today. Unlike garter snakes, both bullsnakes and rattlesnakes routinely enter rodent burrows in search of food. Their remains may be intrusive into deeper deposits, which could account for the smaller proportion of burned bones for the latter genera. Only about 30% of the rattlesnake bones had been burned.

Class Aves

Bird remains were present, but not abundant, in the rockshelter deposits excavated from 1962 – 1964, but Gustafson was not yet experienced enough to undertake their analysis. Mr. Sanford Leffler—an ornithologist—identified a sample of bird bones from the rockshelter. These are listed in Gustafson (1972:100-102).

> The sample is small, and the same species are represented in the deepest sediments as were known from the region about Marmes Rockshelter in historic times. This list of birds should not be considered exhaustive, because it represents only a sample of the total bird bones found in the site (Gustafson 1972:102).

Those birds identified by Leffler are from the rockshelter and are not discussed below. Only the owl (*Bubo* or *Nyctea*) and the sharptailed grouse (*Pediocetes phasianellus*) were identified from the early floodplain deposits studied by Gustafson (1972). The latter and those identified by Caulk (1988) are included in the following discussion.

Order Anseriformes. *Bucephala clangula* (NISP = 3): The common goldeneye duck is a winter resident of eastern Washington (Jewett et al. 1953). Thus, its presence suggests a winter death. If the bones are food refuse, it may indicate human occupancy in winter.

Order Passeriformes. *Empidonax* sp. (NISP = 1): Flycatchers occur widely in Washington occupying a variety of habitats from coniferous forest to brush (Jewett et al. 1953; Scott 1983). Because of their broad distribution, they have little value for environmental interpretation.

Pica pica (NISP = 1): The black-billed magpie is a common inhabitant throughout eastern Washington (Jewett, et al. 1953). Its significance at the Marmes site is not known.

Hylocichla cf. *ustulata* (NISP = 2): Olive-backed thrushes occur in a variety of wooded or brushy habitats in eastern Washington (Scott 1983). Its presence on the Marmes floodplain is not remarkable.

Sialia sp. (NISP = 3): Bluebirds also are common in woodlands or brushy areas (Scott 1983). We attach no particular significance to their occurrence.

Order Charadriiformes. *Ptychoramphus aleuticum* (NISP 2?): Caulk (1988:Table 2) records two elements of *Ptychoramphus aleuticum* (Cassin's auklet) from the Harrison horizon on the Marmes floodplain, but no further reference to this taxon could be found in his data set or in the discussions. For consistency, this species has been included in all tables incorporating Caulk's identifications. Cassin's auklet is a coastal form found from Alaska to Baja, California (Blair et al. 1968). If its presence in the early floodplain deposits could be documented, it could have special significance.

Order Galliformes. *Pedioecetes phasianellus* (NISP = 1): Sharp-tailed grouse once were common throughout the interior of western North America (Blair et al. 1968). They inhabit grasslands, brush and thickets (Jewett et al. 1953; Peterson 1961).

Order Strigiformes. *Bubo* or *Nyctea* (NISP = 1): A fully articulated foot of either the great horned owl (*Bubo virginianus*) or the snowy owl (*Nyctea scandiaca*) was recovered from the early floodplain deposits. The great horned owl ranges from coniferous forest to open grassland and steppe. The snowy owl is primarily an Arctic bird, but it sometimes occurs in great numbers on open areas far to the south (Peterson 1961; Blair et al. 1968). Like the goldeneye (*Bucephala clangula*), presence of the snowy owl could indicate a winter occurrence. Unfortunately it was preserved in the field with white glue in its sediment matrix, so it cannot be observed in detail.

The owl foot is "sandwiched" between two flakes of modified cryptocrystalline silica. It obviously is an artifact and is discussed more fully in Gustafson (1972:108-110) and in the section on bone tools.

Class Mammalia

Order Chiroptera. *Antrozous pallidus* (NISP = 1): A single mandible of the pallid bat was recovered from within the rockshelter. Dalquest (1948) records this bat only from localities along the Columbia and Snake rivers of eastern Washington. Because these bats prefer crevices or caves in basalt cliffs for their daytime hiding places, the presence of the pallid bat in rockshelter sediments is not surprising. There is no indication that this species ever was important in the diet or ceremonies of the aboriginal occupants of the Marmes site.

Order Lagomorpha. *Sylvilagus* sp. (NISP = 54): All specimens complete enough for identification are indistinguishable from *Sylvilagus nuttallii,* the common cottontail in the Columbia Basin today. The species present in archaeological context is consistently larger than the pigmy rabbit, *Sylvilagus idahoensis*, reported to be present in the sagebrush region of the Columbia Plateau (Dalquest 1948). The eastern cottontail, *S. floridanus*, has been introduced in the eastern part of the Columbia Plateau since 1925 (Dalquest 1941); therefore remains from sediments older than this date cannot belong to this species. Because there are other species of cottontails that conceivably could have been present in eastern Washington in former times, a determination of species of *Sylvilagus* will be deferred until suitable specimens are recovered.

Cottontails played a prominent role in the diet of early people in eastern Washington. Their remains are invariably broken and often are charred. Little environmental interpretation can be gained from either of the Leporidae found in our archaeological sites because both genera contain members that exhibit a wide range of environmental preferences.

Lepus sp. (NISP 50): Skeletal remains of rabbits in Washington archaeological sites are highly fragmentary, presumably because of the delicate nature of the bones and the porous condition of the skull. The supraorbital processes, reported to be useful for species identification (Ingles 1965), are not present on any material collected here thus far. John A. White (Personal communication 1965) finds no satisfactory method for identifying species of either *Lepus* or *Sylvilagus* on the basis of tooth characteristics alone. Because the sample used in this study consists only of isolated teeth, fragments of mandibles and maxillae, and long bones, the determination has been left at the generic level.

The two species of *Lepus* that are present on the Columbia Plateau are *townsendii* and *californicus*. *L. californicus* appears to have radiated northward from Oregon during the past 100 years or less (Couch 1927), so *L. townsendii* probably was the common species during at least the last 4,000 years. During the warmer, drier Altithermal period (8,000 to 4,000 years ago), *L. californicus* or a related species may have been the common species because of its greater adaptation to aridity. *L. townsendii* is somewhat larger than *L. californicus*, but all available sample specimens fall within the expected range of size variation for either of these species.

L. towsendii formerly ranged over the sagebrush and grassland areas of most of eastern Washington. With the invasion and subsequent spread of *L. californicus*, the range of *townsendii* has been restricted to the more grasslands of the area. *L. californicus*, which is restricted to arid, sagebrush regions, is apparently a better competitor in this environment, and certainly would appear to be a good ecological indicator of arid conditions. Obviously, both species can survive in arid environments, but in the case of *L. townsendii* the arid region appears to be marginal to the optimum conditions for survival. With the exception of the snowshoe hare (*Lepus americanus*), all species of jackrabbits in the New World are indicative of rather open tundra, desert, or grassland environments.

Their charred and fragmentary remains in archaeological sites imply the importance of jackrabbits as food and rabbits are one of the most important small animal food resources in the ethnographic literature. The Nez Perce held large communal drives, chasing rabbits into nets and corrals (Spinden 1908:214). The Sanpoil-Nespelem killed large numbers of rabbits for their meat and hides (Ray 1932:87), and rabbits were the most important small game taken by the Southern Okanogan (Post 1938:24).

Order Rodentia. *Marmota* cf. *flaviventris* (NISP = 30): The yellow bellied marmot is, and apparently has been, a common resident of the talus slopes near Marmes Rockshelter throughout the last 10,000 years. Bones and teeth of this animal were recovered from all major stratigraphic units of the rockshelter as well as from the culture-bearing sediments on the adjacent floodplain. Long bones recovered show no evidence of burning or breaking to obtain marrow, nor any recognizable signs of butchery. Therefore, there is no indication that this species represented a food source for the primitive hunters.

There is some possibility that the marmot remains from the lowermost sediments (interpreted as having been deposited during a cooler period) could belong to a different species, such as *M. caligata*. However, the size and general pattern of dentition of the Marmes specimens are entirely compatible with *M. flaviventris*.

Because marmots are highly efficient burrowers, and because there is no indication of burning or butchery, this species is considered to be intrusive. Certainly marmots are large enough to have provided an effective energy resource for humans, but there is no evidence that this taxon was a major food source at the Marmes site.
Spermophilus cf. *washingtoni* (NISP = 30): Because of the fragmentary nature of the skulls recovered, and the lack of a large series of comparative specimens, this species has not been successfully separated from the closely related *S. townsendii* (NISP = 4). According to Dalquest (1948), there is no overlap in the ranges of these two species. *S. townsendii* is restricted to a small area "...of the Yakima Valley from Ellensburg...south to the Columbia at Kennewick (W. W. D.)" (Dalquest 1948:268), whereas *S. washingtoni* is found only to the north and east of that area. *S. washingtoni* is the only species known to inhabit the region about Marmes Rockshelter today.

The niches of the two species are very similar. Both inhabit sage brush areas, especially where the sage brush is interspersed with grassland. "Burrows, nests, habitats, and food of this species [*S. washingtoni*] seem identical to those of *townsendii*" (Dalquest 1948:272). Because of the close similarity between the two species, identification to species is not necessary for most environmental interpretations.

Nez Perce and Sanpoil-Nespelem hunters pursued ground squirrels (Walker 1973:55; Ray 1933:87). Post's Southern Okanogan informants recalled burning the fur off these animals before cleaning and roasting them (1938:24). Unfortunately, due to the severe fragmentation encountered in the floodplain samples, we were only able to assign molariform teeth and tooth fragments to this genus. Thus, we were unable to assess any patterning in the burning and/or breakage of the postcranial remains of these animals.
Spermophilus columbianus (NISP = 9): This species is considerably larger than either *S. townsendii* or *S. washingtoni*; therefore, their remains are easy to distinguish on the basis of size alone. *S. columbianus* was the only ground squirrel encountered at 45WT41 near Wawawai, Washington. *Spermophilus* cf. *washingtoni* was the only ground squirrel recovered from the remainder of the sites to the west.

Excavation of stratified sites near the boundary of the ranges of these two species would be expected to disclose alternating occurrences of these two species paralleling shifting climates through time. Because *Spermophilus* cf. *washingtoni* was found at 45GA17 near Penewawa, Washington, the boundary between these two species must lie somewhere between there and Wawawai to the west. Unfortunately, most of the archaeological sites in this region now lie beneath backwater from the Little Goose Dam, and recovery of data needed for more precise determination of range boundaries will be difficult from sediments remaining exposed.

All ground squirrels are adept burrowers, and many undoubtedly are intrusive in archaeological sites. Unless one can demonstrate conclusively the surface for each burrow or a primary deposition by means such as owl pellets, the stratigraphic position of ground squirrel remains (or those of other burrowers) should not be used for short-term paleoenvironmental interpretations. As is the case with the marmots, there is no indication from skeletal remains that ground squirrels were an important food source for humans at this locality.

Cricetidae-size (NISP = 113): The postcranial remains of Cricetidae-size (e.g., mice, rats, and voles) animals are the fifth most abundant taxon at the site (Table 10.7) and fourth most abundant taxon present in the floodplain sediments (Table 10.8). The small size (e.g., cortical bone fragments < 2 mm in thickness) of nearly all the "unidentifiable" bones and bone fragments sorted during this study indicates that the remains of these small animals dominate the floodplain assemblage.

Thomomys talpoides (NISP = 224): *Thomomys talpoides* is the only species of pocket gopher known to inhabit the Columbia Plateau. Because the skulls of pocket gophers are thicker and stronger than those of the ground squirrels, cranial remains of the gophers are more complete than those of the squirrels. There is no question that these remains belong to the same species of pocket gopher that inhabits the area today.

Like ground squirrels, pocket gophers burrow deeply and usually cannot be relied on for environmental interpretation because even their modern burrows are intrusive into the deepest sediments of most archaeological sites. Again one must prove beyond doubt that the bones of this species represent a primary deposition within a stratigraphic unit rather than a later intrusion through burrowing into deeper sediments. Such proof cannot be demonstrated adequately with present data from any of our archaeological sites.

The highly fragmented nature of specimens identified as *T. talpoides* makes assessment of their cultural significance ambiguous. However, it is known that the Kutenai " . . . did not despise rodents. Gophers are extremely numerous in the summer and very easy to shoot. As summer progresses they get very fat and are considered excellent food" (Turney-High 1941:41). However, as Lyman (1976:36) correctly emphasizes, it is unclear whether Turney-High is referring to gophers (*Thomomys* sp.) or ground squirrels (*Spermophilus* sp.).

Perognathus parvus (NISP = 98): Remains of *Perognathus* are much more common in the sediments of the floodplain than in those of the rockshelter itself. This may be because these animals did not frequent the rockshelter often, or it may mean simply that the method of fine screening employed with the floodplain sediments provided a far greater recovery of these small bones. Future testing with the fine screening method will be necessary to evaluate this observation

Great Basin pocket mice are restricted mostly to arid areas wherever they are found. Davis (1939), however, states that they occur commonly in tall wet grass along a stream and in a marsh at Crane Creek in southwestern Idaho. This species is not known to occur in forested regions. Remains of *Perognathus* are found in the disturbed sediments of Marmes Rockshelter and throughout the floodplain sediments. No evidence for cultural significance was recovered for this species.

Castor canadensis (NISP = 1): Caulk (1988) found a beaver skull fragment and an incisor from the Harrison horizon on the Marmes floodplain. Beaver incisors apparently were used both as ornaments and tools. In some instances (e.g., at the Ozette Village Site Complex on the Olympic Peninsula) they also were used as gaming pieces much as dice are used today for gambling. Probably some of the decorated beaver teeth found in inland sites also represent these gaming pieces (Culin 1907). The Coeur d'Alene used beaver incisors for incising wood, bone, antler, and soft stone (Teit 1930:43). It is also known that the Nez Perce considered the beaver a great delicacy (Spinden 1908:207) and that the Southern Okanogan also hunted beavers (Post 1938:24).

Beavers were widely distributed in Washington wherever adequate water and nearby trees on which to feed exist. Because they occur under very diverse climatic conditions, they are of little value for climatic interpretation. However, their presence in a site indicates that there was a continuous availability of fresh water where they were living. Thus, beaver remains can be valuable environmental indicators in arid regions where no permanent water existed in recent time. Their presence in deeper sediments obviously means that permanent water was available at an earlier time.

Reithrodontomys sp. (NISP = 5): The western harvest mouse, *Reithrodontomys megalotis*, is the only species of this genus that currently inhabits

Table 10.8 Number (n) of identified specimens (NISP), percentages and totals, from the Marmes floodplain (45FR50) by stratigraphic unit and taxon including data presented in Caulk (1988:Table 2).

TAXON	A n	A %	B n	B %	C n	C %	D n	D %	E n	E %	NISP TOTAL n	NISP TOTAL %
AMPHIBIANS												
cf. *Hyla* sp.	--	--	7	(1.1)	--	--	--	--	--	--	7	(.6)
cf. *Rana pipiens*	--	--	4	(.6)	--	--	1	(.3)	--	--	5	(.5)
REPTILES[2]												
Unidentified snakes	--	--	10	--	--	--	3	--	--	--	13	--
cf. *Pituophis* sp.	--	--	18	--	--	--	8	--	--	--	26	--
cf. *Thamnophis* sp.	--	--	13	--	1	--	14	--	--	--	28	--
cf. *Crotalus* sp.	1	--	11	--	1	--	3	--	--	--	16	--
cf. *Sceloporus* sp.	--	--	1	--	--	--	--	--	--	--	1	--
BIRDS												
Bucephala clangula	--	--	--	--	--	--	3	(1.0)	--	--	3	(.3)
Ptychoramphus aleuticum	--	--	--	--	--	--	2	(.6)	--	--	2	(.2)
Empidonax sp.	--	--	--	--	--	--	1	(.3)	--	--	1	(.1)
Pica pica	--	--	--	--	--	--	1	(.3)	--	--	1	(.1)
Hylocichla sp.	--	--	--	--	--	--	2	(.6)	--	--	2	(.2)
Sialia sp.	--	--	--	--	--	--	3	(1.0)	--	--	3	(.3)
MAMMALS												
Unid. Lagomorph	--	--	1	(.1)	--	--	4	(1.3)	--	--	5	(.5)
Sylvilagus sp.	--	--	10	(1.4)	2	(10.5)	4	(1.3)	--	--	16	(1.5)
Lepus sp.	--	--	10	(1.4)	--	--	27	(8.7)	--	--	37	(3.4)
Marmota cf. *flaviventris*	--	--	5	(.7)	2	(10.5)	15	(4.9)	1	(1.6)	20	(1.9)
Spermophilus sp.	--	--	6	(.9)	--	--	2	(.6)	--	--	8	(.7)
Spermophilus cf. *columbianus*	--	--	4	(.6)	--	--	5	(1.6)	--	--	9	(.8)
Spermophilus cf. *townsendii*	--	--	2	(.3)	--	--	--	--	2	(3.1)	4	(.4)
Cricetidae-size	--	--	61	(8.8)	--	--	8	(2.6)	4	(6.3)	73	(6.7)
Thomomys cf. *talpoides*	--	--	166	(24.0)	5	(26.3)	38	(12.3)	4	(6.3)	213	(19.5)
cf. *Perognathus* sp.	5	(55.6)	16	(2.3)	2	(10.5)	13	(4.2)	11	(17.2)	47	(4.3)
Perognathus cf. *parvus*	4	(44.4)	54	(7.8)	3	(15.8)	29	(9.4)	8	(12.5)	98	(8.9)
Castor canadensis	--	--	--	--	--	--	1	(.3)	--	--	1	(.1)
Reithrodontomys sp.	--	--	3	(.4)	1	(5.3)	1	(.3)	--	--	5	(.5)
Peromyscus sp.	--	--	38	(5.5)	--	--	23	(7.4)	1	(1.6)	62	(5.7)
Peromyscus cf. *maniculatus*	--	--	5	(.7)	--	--	--	--	--	--	5	(.5)
Onychomys cf. *leucogaster*	--	--	2	(.3)	--	--	--	--	--	--	2	(.2)

Table 10.8 Number (n) of identified specimens (NISP), percentages and totals, from the Marmes floodplain (45FR50) by stratigraphic unit and taxon including data presented in Caulk (1988:Table 2).

TAXON	STRATUM[1]										NISP TOTAL	
	A		B		C		D		E			
	n	%	n	%	n	%	n	%	n	%	n	%
Neotoma sp.	--	--	1	(.1)	--	--	12	(3.9)	--	--	13	(1.2)
Neotoma cf. *cinerea*	--	--	2	(.3)	--	--	--	--	14	(21.9)	16	(1.5)
cf. *Clethrionomys* sp. (?)	--	--	--	--	--	--	1	(.3)	--	--	1	(.1)
cf. *Phenacomys* sp. (?)	--	--	3	(.4)	--	--	--	--	--	--	3	(.3)
cf. *Microtus* sp.	--	--	3	(.4)	--	--	--	--	1	(1.6)	4	(.4)
Microtus sp.	--	--	86	(12.4)	4	(21.1)	26	(8.4)	12	(18.8)	128	(11.7)
Microtus montanus	--	--	--	--	--	--	1	(.3)	--	--	1	(.1)
Microtus pennsylvannicus	--	--	--	--	--	--	3	(1.0)	--	--	3	(.3)
Microtus oeconomus	--	--	1	(.1)	--	--	--	--	--	--	1	(.1)
Lagurus curtatus	--	--	1	(.1)	--	--	3	(1.0)	--	--	4	(.4)
cf. *Ondatra* sp.	--	--	1	(.1)	--	--	--	--	--	--	1	(.1)
Canid	--	--	2	(.3)	--	--	8	(2.6)	7	(10.9)	16	(1.5)
Canis sp.	--	--	1	(.1)	--	--	1	(.3)	--	--	2	(.2)
Canis latrans	--	--	--	--	--	--	1	(.3)	--	--	1	(.1)
Canis familiaris	--	--	--	--	--	--	1	(.3)	--	--	1	(.1)
Unid fox.	--	--	--	--	--	--	28	(9.1)	--	--	28	(2.6)
Alopex lagopus	--	--	--	--	--	--	6	(1.9)	--	--	6	(.5)
Vulpes cf. *fulva*	--	--	1	(.1)	--	--	19	(6.1)	--	--	20	(1.8)
Mustella-size	--	--	--	--	--	--	1	(.3)	--	--	1	(.1)
Martes americana	--	--	--	--	--	--	3	(1.0)	--	--	3	(.3)
Mustela frenata	--	--	--	--	--	--	3	(1.0)	--	--	3	(.3)
Taxidea cf. *taxus*	--	--	1	(.1)	--	--	--	--	--	--	1	(.1)
Vulpes/Lynx-size	--	--	--	--	--	--	--	--	1	(1.6)	1	(.1)
cf. *Lynx rufus*	--	--	--	--	--	--	2	(.6)	2	(3.1)	4	(.4)
Lynx canadensis	--	--	1	(.1)	--	--	--	--	--	--	1	(.1)
Medium/Large Artiodactyl	--	--	188	(27.2)	2	(10.5)	2	(.6)	--	--	192	(17.5)
Cervus canadensis[3]	--	--	1	(.1)	--	--	--	--	--	--	1	(.1)
Odocoileus sp.	--	--	2	(.3)	--	--	4	(1.3)	--	--	6	(.5)
Antilocapra americana	--	--	1	(.1)	--	--	1	(.3)	--	--	2	(.2)
TOTALS[4]	9	(100.0)	690	(99.9)	19	(100.0)	309	(99.7)	68	(100.0)	1095	(100.6)

[1]Stratum designations equal, "A" strata above Marmes Horizon, "B" Marmes Horizon, "C" strata between Marmes and Harrison horizons, "D" Harrison Horizon, and "E" strata below Harrison Horizon.

[2]Reptile values represent number of units containing reptile remains and are not included in the NISP total calculations.

[3]Identified by Caulk (1988)

[4]Percentage totals do not equal 100.0 due to rounding.

southeastern Washington (Dalquest 1948). This grassland and open desert dweller is considerably smaller than *Peromyscus* sp. and *Onychomys* sp., thus the molariform teeth of *Reithrodontomys* sp. can be identified based on their size.

Peromyscus maniculatus (NISP = 6): Deer mice, *Peromyscus maniculatus*, undoubtedly are the most ubiquitous and least informative rodent found in Washington archaeological sites. Hall and Kelson (1959) show the distribution of the many subspecies to include nearly every conceivable habitat from rain forests to desert. Its presence in a site is almost useless from an interpretative standpoint.

Neotoma cf. *cinerea* (NISP = 29): Wood rats are distributed over most of North America. The single species found in Washington occurs throughout the state except along the seacoast of the Olympic Peninsula and Puget Sound (Dalquest 1948). Because it occurs in nearly every environment where talus slopes or rockslides are present, this species too is a poor indicator of environment. However, because of its curious habit of collecting a great variety of materials, ranging from jewelry to bones of other animals, it can be an important element of disturbance in many sites. Where wood rats are common, the excavator must be watchful of possible caches of bones and other materials placed there by these animals.

Onychomys cf. *leucogaster* (NISP = 2): Today, the carnivorous northern grasshopper mouse stalks its prey throughout the low-lying valleys and prairies of southeastern Washington (Dalquest 1948). *O. leucogaster* is the only species of these genus in the region surrounding the Marmes site and its restriction to low, arid environments may make it a valuable environmental indicator. Northern grasshopper mouse remains, isolated molariform teeth, have only been found in the Marmes floodplain horizons.

cf. *Cleithrionomys* sp. (NISP = 1): Southern red-backed voles, *Cleithrionomys gapperi*, are the only species of genus that currently occupies eastern Washington. These small crepuscular animals inhabit cool, moist forest, bogs, and swamps and may be valuable indicators of cool, more mesic environments (Dalquest 1948; Ingles 1965). Members of this genus possess rooted molars with a prismatic crown pattern. This trait readily distinguishes them from their ubiquitous microtine relatives that possess high crowned, rootless molars. However, the identification of this taxon is based on the presence of a single lower third molar and remains tentative.

cf. *Phenacomys* sp. (NISP = 3): As with *Cleithrionomys* sp., this genus is potentially represented by isolated molars retrieved from the Marmes horizons of the floodplain. The Heather vole, *Phenacomys intermedius*, is the only species of this genus that currently occupies eastern Washington. Today, Heather voles inhabit the boreal forests of the Cascade Range and higher mountains of extreme northeastern and southeastern Washington (Dalquest 1948). Their proclivity for these environments potentially makes them important environmental indicators. However, given the paucity of remains assigned to this genus, more remains with more reliable features would be necessary to confirm the identifications.

Microtus cf. *montanus* (NISP = 1): The skulls and mandibles of *Microtus* sp. found in most archaeological sites that we have examined are generally fragmentary and often the teeth are absent. It is very difficult to distinguish this genus from the related *Lagurus* sp. unless teeth are present in their alveoli. All specimens labeled "*Microtus* or *Lagurus*" have been kept for possible future identifications.

As has been pointed out by Ingles (1965), species of *Microtus* can be distinguished by the size and shape of the incisive foramina. Where these are preserved in specimens from sites included in the report, the foramina most closely resemble those of *Microtus montanus,* the montane meadow vole. Hence, all specimens tentatively have been referred to this species.

Microtus montanus occurs throughout the Blue Mountains in southeastern Washington, over most of the Palouse Hills in the vicinity of Colfax and Pullman, and along the valley bottoms of the Snake and Columbia rivers. Suitable environments probably have existed at least locally along the Snake River and its tributaries for the entire span of human presence in this region.

Microtus cf. *pennsylvannicus* (NISP = 3): Meadow voles (*M. pennsylvannicus*) have the widest distribution of any vole in North America, ranging from arctic tundra to the central United States. They are the dominant vole in the North American taiga biome and prefer wet meadows; however, they can inhabit virtually any grassland habitat. The only habitats these creatures avoid are deep forests and high dry grasslands (Banfield 1974; Getz 1985; Hoffman and Koeppl 1985; Johnson and Johnson 1982).

Microtus cf. *oeconomus* (NISP = 1): Of all the species of *Microtus* sp., only the tundra vole (*M. oeconomus*) and the singing vole (*M. miurus*) occupy arctic tundra. Caulk (1988:35) identified tundra vole dentition from the Marmes horizons of the floodplain. The tundra vole is found in the mesic meadows of the low arctic and in the taiga and alpine meadows in the southern portion of its North American range (Banfield 1974; Hoffman and Koeppl 1985).

Lagurus curtatus (NISP = 6): This is the only species of sagebrush vole found in the United States. As stated above, it is often difficult to distinguish from the closely related *Microtus*. Dalquest (1948:359) states that the sagebrush vole prefers "Upland areas of low sagebrush with sparse grass." Thus *Lagurus* should be a good indicator of the relatively arid conditions usually associated with sagebrush. Ingles (1965) also includes the bunchgrass area of eastern Washington within the range of *Lagurus*. We have found specimens of *Lagurus* from sediments in 45WT41 (approximately five miles south of Wawawai), in area C, a portion of the site which consists of dune sand. In the nearby 45WT41 areas A and B, located on floodplain deposits, only *Microtus* was present. Thus, it may be that *Lagurus* is restricted to well-drained, somewhat atypical soils within the bunchgrass steppe and river bottoms of extreme southeastern Washington. It also is possible that these remains may have been deposited in the regurgitated pellets of far-ranging hawks or owls. Until more is known about its distribution, little can be said for its use as an environmental indicator, although certainly one would not expect to find *Lagurus* in a region dominated by tall shrubs or forests.

Ondatra zibethica (NISP = 10): The muskrat, like the beaver, is an aquatic rodent and necessarily requires permanent fresh water for survival, although muskrats are known to occupy a marine habitat in Puget South and the San Juan Islands (Dalquest 1948). Their presence, therefore, indicates only that water was nearby. There are some reports that muskrats were eaten by preindustrial people, but muskrats occur in so few numbers in archaeological sites of eastern Washington that their presence may be simply the results of predation by owls or coyotes. The Sanpoil-Nespelem exploited muskrats (Ray 1932:86) and the Flathead utilized the muskrats for their hide and occasionally meat (Turney-High 1937:115). The Cour d'Alene (Teit 1930:96) and Kutenai (Turney-High 1941:41) also hunted muskrats for their hide.

Geological evidence at the Marmes site shows that the necessary stream was nearby throughout the time that people occupied the site. Consequently, the muskrat provides no information of environmental consequence except to confirm the presence of permanent water nearby.

Order Carnivora. Canid-size (NISP = 16), *Canis* sp. (NISP = 3): Severely fragmented molariform tooth fragments from Canid-size animals are relatively common throughout the floodplain sediments. These specimens lacked the essential diagnostic features to assign them to a particular genus, but their size suggested they possibly represented coyote, dog, or wolf-sized animals. Three additional specimens compare favorably in size and shape with coyote, dog, or possibly wolf-sized carnivores and were assigned to the *Canis* sp. category.

Canis latrans: Coyotes range over the entire state of Washington except in some regions above timberline and in some of the densest forests (Dalquest 1948). As a result, they are of little value as environmental indicators. Their remains occur in most archaeological sites and certainly could have been a source of food for early humans. Coyote bones often are not broken in the manner characteristic for food bones in this area; however, they seldom are charred. Hence, they may represent animals that died of natural causes. Indian legends commonly record "coyote" as a legendary being whose mystical powers were highly respected, and because of its special role many Indians of the recent past would not consider eating a coyote (Walker 1971).

However, the ethnographic record is ambiguous concerning the use of coyotes and wolves for food and fur. Nez Perce captured coyotes and wolves with deadfalls (Spinden 1908:214) and the Sanpoil-Nespelem (Ray 1932:86, 90) considered the coyote to be especially good food and they utilized wolf hides for robes and blankets, whereas, the Southern Okanogan kept young coyotes and wolves as pets (Post 1938:34). A larger sample of prehistoric postcranial elements is required to investigate the potential use of these animals for their food and fur.

Canis familiaris (NISP = 1): Caulk (1988:37) tentatively identified a distal radius from the Harrison horizons as belonging to a medium-sized dog. Continued analysis of the floodplain fauna may provide additional information on the nature of New World dogs during the early-Holocene. The Nez Perce, Sanpoil-Nespelem,

Southern Okanogan and Flathead did not eat dogs. Rather, they were kept as pets and proved useful during deer hunting (Spinden 1908:207; Ray 1932:90; Post 1938:34; Turney-High 1937:104).

Canis lupus (NISP = 1): According to Dalquest (1948), "Wolves occurred in western, northeastern, and southeastern Washington. They seem not to have occurred on the Columbian Plateau". There apparently are no recent records for wolves in Washington other than in timbered areas; however, in other regions of North America, wolves are known to have occurred on the plains and deserts (Young and Goldman 1944). Certainly wolves may have roamed over the Columbia Basin in pre-Columbian time. A wolf metatarsal dated at ca. 2,000 years B. P. was recovered from a bison kill site (45GA17). The only other specimen confidently identified as wolf is a first upper molar from Unit II at Marmes Rockshelter.

Vulpes fulva (NISP = 20): According to Dalquest (1948), the red fox of Washington is an alpine animal living at or slightly below timberline in the Cascades, Blue Mountains, and Okanogan Highlands. Ingles (1965) states that colonies of red foxes have been reported at lower elevations in the Sacramento Valley of California, although he believes these to be the introduced Midwestern form which regularly lives at lower elevations. The Midwestern subspecies of red fox also have been introduced successfully into the lowland regions of Puget Sound.

Isolated teeth of the gray fox (*Urocyon cinereoargenteus*) could possibly be confused with those of a small red fox. However, teeth in the sample collection lack the supplementary tubercle on M_1, and the size and curvature of the canines are more like those of *Vulpes* than *Urocyon*. Therefore, specimens tentatively have been referred to the genus *Vulpes*. This possible discrepancy can be resolved only when more skeletal material is available. Either the red fox or the gray fox could have been present on the steppe region or in the nearby streamside thickets. An articulated right frontal and parietal from floodplain sediments at Marmes Rockshelter are the only charred specimens found to date. The cultural significance of this skull fragment is not known.

Alopex lagopus (NISP = 7): The Arctic fox previously was not known to have inhabited the state of Washington. At present, this species is restricted in North America to the tundra and ice flows of Alaska and Canada. A single skull fragment consisting of the charred right parietal and articulated frontal was recovered from the Harrison Horizon at Marmes Rockshelter. The shape, size, and location of the coronal suture; the general contour of the skull fragment; and the location and extent of the temporal ridges match exactly with skulls of the Arctic fox housed in the Charles R. Conner Museum and the skull of an Arctic fox from St. Paul Island, Alaska. None of the external or internal features of this skull fragment compare favorably with a series of red fox skulls from Washington and Alaska or with the northern race of Arctic fox from Point Barrow, Alaska. Caulk (1988:38) found additional Arctic fox remains in his sample including, a distal fibula, left and right proximal radii and a distal radius. Thus far, Arctic fox remains have only been found in the northeastern portion of the Marmes Floodplain excavations.

A skull fragment attributable to the red fox, and consisting of the same skull parts (i.e., charred right parietal and frontal), also was recovered from the Harrison Horizon. This skull fragment resembles the red fox in all respects and is clearly distinct from the skull fragment identified as Arctic fox. The skull fragment of Arctic fox exhibits striations on the parietal produced by some sharp tool. Thus it likely represents an animal that was butchered. No such striations occur on the fragment identified as red fox.

The Arctic fox often has been cited as an excellent indicator of tundra environments, but there is little supporting faunal evidence to suggest that tundra conditions ever prevailed in the vicinity of Marmes Rockshelter.

Ursus sp. (NISP = 2): Both the black bear (*Ursus americanus*) and the grizzly bear (*Ursus horribilis*) are known to occur in Washington. In the eastern half of the state, the black bear now is restricted to the mountainous regions. Dalquest (1948) gives the probably past distribution of the grizzly bear as the northern Cascades, the Okanogan Highlands, and the Blue Mountains. The grizzly now is extinct over most, if not all, of Washington.

A left third metacarpal from 45GA17 and an ungual phalanx from Marmes Rockshelter definitely belong to the grizzly bear. Two isolated canines from Marmes Rockshelter could represent either of the two species. One of the teeth has a groove carved around the base, indicating that it was some sort of ornament. The wolf metatarsal from 45GA127 has a linear series of transverse cuts along one edge, suggesting that it too had some ceremonial significance. The bear metacarpal from the same site could also be an artifact, although it has not been worked. All of the bones identified as bear could have been carried to the sites from forested areas in the nearby mountains. No fragmentary skeletal parts of bears have been found in any of the sites under consideration, suggesting that bears may not have been killed near these sites.

Martes americana (NISP = 4): The principal habitat of the pine marten is the heavily forested portions of the Olympic, Cascade, and Blue Mountains of the northeastern part of the state, although they occasionally range into the talus slopes above timberline (Dalquest 1948). Because martens are arboreal and never have been observed in non-forested regions at low altitude, the presence of their remains suggests an environment dominated by coniferous forest.

This species is represented at Marmes Rockshelter by a single mandible from the early floodplain sediments. The presence of this specimen can be interpreted in one of two ways: either coniferous forests occurred near the Marmes site 10,000 years ago, or the bone was brought there by humans, or floated in from a forested region. Walker (1971) states that pine marten were taken occasionally by the Indians for their furs. If the single mandible from the early floodplain sediments at Marmes Rockshelter was brought to the site from a remote forested region, it would suggest some use of this species other than simply for furs, because mandibles normally would be removed during the skinning process unless the entire carcass was brought to the site before skinning.

Mustela cf. *frenata* (NISP = 3): There is some difficulty in distinguishing a large male ermine (*M. erminea*) from a small female long-tailed weasel (*M. frenata*) because of the overlap in size of these two species (Guilday 1969: Kurtén 1968). The three mandibles in the study collection have been referred to the species *frenata* because they are consistently larger than any of our comparative specimens of *M. erminea*.

M. frenata is relatively unspecialized in its choice of habitats and occurs throughout the state in all terrestrial environments. The ermine, on the other hand, is found everywhere except in the Columbia Basin. Should the ermine be positively identified from an archaeological site in the Columbia Basin, it certainly would suggest a more mesic environment at some time in the past.

Taxidea taxus (NISP = 1): The peculiar skull and extreme modifications of the forelimbs for digging make this one of the easiest animals to identify from skeletal remains. Badgers do not occur west of the Cascades but are common throughout eastern Washington. Where present, badgers show no particular climatic preference, for they live in habitats from the high mountain meadows to the driest portions of the Columbia Basin. Therefore, they are of little value for ecological interpretation. The presence of a single element assigned to this taxon precludes accurate assessment of the paleoeconomic importance of badgers. However, it is known that the Nez Perce (Walker 1971:55) and Sanpoil-Nespelem (Ray 1932:85) hunted badgers. Whether badgers were pursued for their meat, fat, or fur is unclear.

Mephitis mephitis (NISP = 2): Neither Dalquest (1948) nor Ingles (1965) include the region north of the Snake and Columbia rivers in central and eastern Washington within the range of the striped skunk. Ingles (1965) reports that this species occurs throughout California, Oregon, and Washington except in the hottest and driest deserts and the highest mountains, especially along stream margins. On the other hand, Dalquest (1948) reports striped skunks from the Yakima Valley, some of which is as hot and dry as much of the Columbia Basin.

Mephitis remains were recovered from early postglacial sediments at the Lind Coulee site (Enbysk 1956). Enbysk reported these remains as representing skunks occurring north of their present range during a time when the climate was cooler and moister than present. Gustafson (1972) identified two mandibles from a single individual of Mephitis from late sediments at Marmes Rockshelter, and a complete skull from a late human burial at the base of a talus slope nearby. Striped skunks are observed regularly north of the Snake River in eastern Washington today, and several specimens collected from there are preserved in the Charles R. Conner Museum at Washington State University. Thus the former interpretation of Enbysk must be questioned.

Lynx rufus (NISP = 4): Like the coyote, the bobcat ranges over all of the state of Washington and is of little value as an environmental indicator. Bones of this species in sediments at Marmes Rockshelter may represent the natural deaths of bobcats using the site as temporary shelter, although the presence of a charred fragment of the distal end of a left humerus suggest that bobcats were eaten at least occasionally. The Sanpoil-Nespelem hunted the lynx, cougar, and "wildcat" (Ray 1932:85) and the Flathead hunted the lynx for its hide and occasional meat (Turney-High 1937:115). The Kutenai pursued the lynx primarily for its pelt, but would eat the meat when hungry (Turney-High 1941:41).

Neither Dalquest (1948) nor Ingles (1965) record the related Canadian Lynx (*Lynx canadensis*) from the Columbia Basin. However, Hudson (1964) reports a single specimen of *L. canadensis* from the vicinity of Pullman, Washington. He since has procured a second specimen for this area. Some of the more fragmentary skeletal remains of *Lynx rufus* could possibly be confused with those of *L. canadensis*. Such confusion would in no way affect environmental interpretation unless very large numbers of the forest-loving *L. canadensis* were encountered.

Lynx canadensis (NISP = 1): The Canadian lynx is represented by a distal humerus from the Marmes horizons (Caulk 1988:40). Today, the lynx searches for its prey in the forested Idaho panhandle and Blue Mountains of northeast Oregon and southwest Washington (Hall and Kelson 1959; Nellis 1973). However, two Canadian lynx were killed near Pullman Washington in 1962 and 1963 in open wheat fields and pastures (Hudson 1964, in Caulk 1988:40).

Order Artiodactyla. Medium/Large Artiodactyl (NISP = 481): Small quadrangular (e.g., < 10 mm) and relatively thick (e.g., > 5 mm) long bone fragments from deer-to-elk-size animals are among the most numerous identified specimens in the Marmes Floodplain sediments. Larger spiral-fractured long bone fragments measuring several centimeters in maximum dimension are only present in the rockshelter. The small size of these specimens precludes a more accurate taxonomic assignment. However, the small size of these specimens likely indicates the intensive processing of Medium/Large Artiodactyl bones for the fat and grease they contain.

Cervus canadensis (NISP = 52): Gustafson (1972: Table 5.1) reported an NISP of 51 elk from the Marmes rockshelter. In addition, a single elk first phalanx was present in the bulk samples of bones Caulk (1988) analyzed from the floodplain. This phalanx probably belongs with the vertebrae, ribs and hind limb bones of a large elk exposed as a feature during the 1968 floodplain excavations. Because the latter remains were not part of the sample selected for analysis on the floodplain, they are discussed separately in the section on butchery in the Summary and Conclusions of this chapter.

Elk bones, along with those of deer (*Odocoileus* sp.) and pronghorn antelope (*Antilocapra americana*), are among the most common faunal remains found in archaeological sites throughout the Columbia Basin. The fragmentary, and often fire-charred, remains of this species testify to the fact that elk were a major source of food for people living in this arid region. Elk bones are common in all cultural sediments throughout the last 10,000 years, even in the most arid parts of the Columbia Basin. The Columbia Basin is not included in the range of modern elk, yet its frequency in archaeological sites argues that this species must have been more than a casual wanderer into this region until recent times. Elk remains from archaeological sites include teeth, skull fragments and bones of the feet. These probably would not have been transported any great distance, so they are best interpreted as having come from the vicinity of the site rather than from the mountainous region surrounding the basin proper. Confirmation of the former presence of elk in the Columbia Basis may come through analysis in the future of sediments of the same age as archaeological sites, but unaffected by human habitation.

By the turn of the century, elk were nearly extinct in eastern Washington. The Rocky Mountain elk (*C. canadensis nelsoni*) was introduced into the Blue Mountains in 1911, 1919, and 1931. (Couch 1953). The introductions were successful, and healthy herds are sustaining themselves throughout the region despite heavy hunting pressures.

Odocoileus sp. (NISP = 211): Both the mule deer (*Odocoileus hemionus*) and the white-tailed deer (*O. virginianus*) probably are present in sediments at Marmes Rockshelter. These two species are impossible to distinguish unless antlers or fragments of the facial region containing the characteristic lacrimal fossa are preserved. A single antler (from 45WT39) and a fragment of a left facial region (from Marmes Rockshelter) are from the white-tailed deer. These are the only specimens that we have encountered which are complete enough for species determination.

Deer bones are encountered even more frequently than those of elk in sediments at most archaeological sites in southeastern Washington. We know of no site in the Columbia Basin that does not contain at least a few deer bones. Not only are remains of the genus common to all sites, but they also represent the most abundant remains throughout most sediments from the earliest to the latest culture-bearing horizons. The consistently fragmented condition of the bones and the common occurrence of charred fragments attest to the importance of this genus as one of the primary food resources for people in the Columbia Basin.

In eastern Washington, both the mule deer and the white-tailed deer are found from the higher forested regions to the arid sagebrush flats at lower elevations. Thus this genus is of little value as an environmental indicator.

Antilocapra americana (NISP = 191): According to Einarsen, "Washington has not produced pronghorns in the history of [modern] man although it is obvious that the arid portion of the state offer [sic] suitable habitat" (Einarsen 1948:201), but no documented historic occurrences are known. Because of the nature of vegetation and topography of the Columbia Basin, Dalquest (1948) includes this portion of the state as probable antelope range. Taylor and Shaw (1929) included the pronghorn as one of Washington's mammals now extirpated within the state. Hall and Kelson (1959), Ingles (1965), and others have used the hearsay records to include the previous range of pronghorn antelope within the Columbia Basin of Washington.

Despite the lack of historic evidence for their presence, pronghorn remains are second only to those of deer in archaeological sites throughout the central Columbia Basin. Osborne (1953) reports bones of this species from several sites along the Snake and Columbia rivers. Pronghorn remains have been identified from lower Snake River archaeological sites ranging in age from 10,000 years ago to historic. Grabert (1968) lists pronghorn among the bones identified from the historic Fort Okanogan site, and Gustafson (1972) has identified this species from historic sediments at Fort Walla Walla. In addition, pronghorn remains were recovered from sediments dated tentatively between 1600 and 1800 A.D. at the Wexpusnime site (45GA61) across the river from Wawawai and at 45WT39, one mile north of Wawawai, Washington. Osborne (1953) discusses other archaeological occurrences of pronghorn in Washington, but in most cases the exact provenience is not clear. Thus *Antilocapra americana* must have been present in the Columbia Basin from 10,000 years ago into early historic time. To date, these localities define the limits of the known recent range of the pronghorn antelope in the state of Washington.

Pronghorns prefer open, rolling country covered with sagebrush or similar brushy vegetation, although in some parts of their former range they have been forced into lower timbered areas (Einarsen 1948). Apparently, they never spend much time in heavily forested regions except where there are large expanses of short brush and grassland interspersed with the forest. Their presence in a site thus indicates the absence of closed forest or dense brush.

Ovis canadensis (NISP = 0): Although a single metapodial fragment attributable to *Ovis canadensis* was found in a deep test pit on the Marmes floodplain, it was not included in the samples analyzed from the site. Though generally considered to be representative of the timberless regions in the mountains of Washington, bighorn sheep once were common along the steep cliff faces of the Columbia River (Dalquest 1948). They probably also occurred along the Snake River where suitable habitat prevailed. Remains of this species usually are not abundant in archaeological sites, but they are found sporadically in many sites along the Snake and Columbia rivers. Possibly the sparsity of bighorn sheep bones reflects the difficulty in hunting this cliff-dwelling species.

Because bighorn sheep are known to have occurred from the highest mountains to the lowest reaches of the Columbia Basin, their remains shed little light on past environmental conditions.

Bison bison (NISP = 0): Inclusion of the Columbia Basin within the historic range of the bison is based on somewhat tenuous evidence. According to Dalquest (1948:404), "Gibbs [1972] was told by an Indian hunter in 1853 that a lost bull had been killed in the Grand Coulee (state of Washington) 25 years before but that 'this was an extraordinary occurrence, perhaps before unknown.'" Schroedl (1973) reports only four additional references to the historic occurrence of bison in the Columbia Plateau.

Because the bones of *Bos and Bison* are so difficult to distinguish, most such bones found in historic sites are referred to the genus *Bos* or *Bos/Bison*, and only those bones found in prehistoric context are referred definitely to the genus *Bison*. Until large samples are available, the question of the possible historic occurrence of bison in the state of Washington will not be resolved.

A large form of Bison was present in eastern Washington prior to 7,500 years B.P. (Daugherty 1956b; Irwin and Moody 1977, 1978). By about 3,000 years B.P. modern *Bison bison* definitely was present in the state of Washington. A large number of bison bones dated at ca 2,000 years were recovered from 45GA17. Osborne (1953) reports the presence of bison at a number of sites in the Columbia Basin, but he provides little information as to their chronology. Bison bones from the Tucannon site, 45C01 (Nelson 1966), the Harder site, 45FR40 (Kenaston 1966), 45WT2

(Nance 1966), and Granite Point locality, 45WT41 (Leonhardy 1970), all have been dated at about 2,000 years ago.

Domestic Mammals. *Sus scrofa, Bos taurus* (NISP = 48), *Ovis aries, Equus caballus,* and *Canis familiaris:* Only cattle were identified from the bones known to have come from the Marmes Rockshelter surface, but remains of domestic pigs, cattle, sheep, and horses commonly are strewn over the surface of archaeological sites, particularly rockshelters and caves. If their bones can be assigned to stratigraphic units associated with archaeological materials, they can provide rather accurate dates for recent occupation. Usually, however, they occur in sediments postdating Euro-American settlement. For example, remains of cattle were abundant on or near the surface at Marmes Rockshelter. Many bones of older individuals were associated directly with those of fetuses, suggesting that they may represent cows that sought solitude in the shelter and died.

Although few horse bones were found at Marmes Rockshelter, large numbers were recovered from a burial site (45FR36b) associated with the Palus Indian Village near the mouth of the Palouse River (Sprague 1965). Horse bones have been recovered from numerous sites throughout the Columbia Basin (Osborne 1953). Because modern horses did not reach the Columbia Plateau until after 1760 (Ewers 1955), the presence of this species in sediments of post-Pleistocene age provides direct evidence of very recent occupation. Probably no animal has had such a profound influence on the cultures of American Indians as the horse, although bones from archaeological sites suggest that the basic animal diet was the same before as after the introduction of the horse.

Caulk (1988:79) lists a single first phalanx from the Harrison horizon as belonging to a domestic dog, but this assignment is open to question. The earliest documented records for domestic dog in the Columbia Basin occur in sediments dated at ca 2,000 years ago. A single skull and partial skeleton of that age were found at 45GA17. A second skull and partial skeleton was found at 45WT2, and another at 45GA61. The age of these latter specimens remains questionable. All three specimens from eastern Washington resemble the larger dogs described by Gleeson (1970) from the Ozette Village site on the Olympic Peninsula of Washington. All specimens have a broad skull with steeply sloping frontal region, lack the first premolar in both upper and lower jaws, and possess a massive mandible with deep coronoid fossa and a recurved coronoid process. Because of these features, along with shorter limbs, they are relatively easy to distinguish from the coyote which is of similar size.

Summary of Taxa Identified

Most bones of mammals present in archaeological sites in southeastern Washington are of species that still inhabit the region today. Notable exceptions include the wolf (*Canis lupus*), the Arctic fox (*Alopex lagopus*), and red fox (*Vulpes fulva*), the pine marten (*Martes americana*), the grizzly bear (*Ursus horribilis*), the elk (*Cervus canadensis*), the pronghorn antelope (*Antilocapra americana*), the bighorn sheep (*Ovis canadensis*), and the bison (*Bison bison*). All but the Arctic fox and the pine marten almost certainly were extant until relatively recent times.

Deer, elk, and pronghorn antelope remains are abundant in all sites from which faunal remains have been identified. Cottontail rabbits (*Sylvilagus* sp.) and jackrabbit (*Lepus* sp.) represent the next most abundant remains. With the exception of *Spermophilus* sp. and *Thomomys* sp., which are considered to be intrusive into cultural sediments, the remaining species identified are relatively rare, although remains of *Bison bison* often are abundant locally in sediments averaging about 2,000 years old. The value as indicators of environment of most species identified is low because most are ubiquitous animals, which have broad tolerance to a wide range of habitats.

Remains of large mammals made up the bulk of the fauna in the rockshelter portion of the site (Gustafson 1972). The most obvious feature of the floodplain fauna is the large number of bones from small mammals, particularly rodents, and the paucity of large mammal remains identifiable to taxon. Part of the discrepancy is explained by the fact that the floodplain material was water-screened through one-mm mesh, whereas the rockshelter sample was dry-screened through 1/4 in (.625 mm) mesh. This may partially account for the low percentage of small mammal remains in the rockshelter sample but not for the low proportion of large animal bones on the floodplain. As will be shown, most of the bone fragments from the floodplain were tiny and charred. This contrasts sharply with the faunal remains from the rockshelter where most fragments were large and relatively few were charred. Here, two aspects of the Marmes fauna are presented: 1) the similarities and differences between the Marmes floodplain and rockshelter based on their respective faunas, and 2) the size and condition of the thousands (ca. 60,000) of small bones and indeterminate bone fragments separated from the Marmes Floodplain samples sorted during this study. Where appropriate, data from Gustafson (1972) and Caulk (1988) will be used to supplement data generated during this study.

Comparison Within the Marmes and Harrison Horizons

The most common floodplain faunal remains were snake vertebrae and rodent teeth and jaws. The other large category is "Medium/Large Artiodactyl" which is represented by small fragments of long bones and tooth enamel fragments from deer-to-elk-size animals. Table 10.8 includes identifications made during this study and those made by Caulk (1988) from the Marmes floodplain. Specimens representing economically important large mammals such as deer (n = 6), elk (n = 1), and pronghorn (n = 2) are conspicuously rare in the floodplain sediments.

Small (i.e., < 10 mm) shaft fragments are the most numerous remains from these animals and they were assigned to the "Medium/Large Artiodactyl" category. Cortical bone exceeding five millimeters in thickness distinguished these bone fragments from the remains of smaller animals like rabbits. Together, "Medium/Large Artiodactyl" remains equal 17.5% (n = 192) of the 1,094 specimens assigned to taxon from the floodplain. Nearly all of these specimens (n = 188), or 17.1% of the floodplain NISP percentage total, rested in the Marmes horizons and this suggests that large mammal hunting and processing were important site activities.

Relatively few specimens from medium-sized mammals such as rabbits and marmots are present in the sampled portions of the floodplain. Jackrabbits (*Lepus* sp.) and cottontails (*Sylvilagus* sp.), ethnographically important sources of meat, fat, and furs, are represented by only 53 specimens and equal about five percent of the floodplain NISP total. Jackrabbit remains are nearly seven times as abundant in the Harrison horizons compared to cottontail remains, whereas cottontail and jackrabbit remains are equally represented in the overlying Marmes horizons. Nonetheless, the Harrison horizons contained 27 jackrabbit bone and/or tooth fragments compared to only 10 in the Marmes horizons. Likewise, Marmot (*Marmota* cf. *flaviventris*) remains are three times as numerous in the Harrison horizons (NISP = 15) than in the Marmes horizons (NISP = 5). Though the small number of specimens identifiable to these genera could easily account for these patterns, it is possible that temporal differences existed in the natural abundance or economic importance of these taxa. The frequency of carnivore remains, like those of rabbits and hares, is highest also in the Harrison horizons. Fourteen carnivore taxa have been identified from the floodplain and only five of these—red fox (*Vulpes* cf. *fulva*), badger (*Taxidea* cf. *taxus*), lynx (*Lynx canadensis*), *Canis* sp. and Canid—are found in the Marmes horizons. Severely splintered bones and teeth indentifiable as fox only are the most numerous (NISP = 28) and equal 9.1% of the Harrison horizon NISP percentage total and red fox remains (NISP = 19) are three-times as abundant as Arctic fox (*Alopex lagopus*) remains (NISP = 6) in the Harrison horizons. A single tooth fragment from a coyote, dog, or wolf (e.g., *Canis* sp.) is present in the Marmes and Harrison horizons and canid specimens equal 10.9% of the NISP percentage total for the silts beneath the Harrison horizons. Other carnivore taxa found only in the Harrison horizons include the pine marten (*Martes americana*), long-tailed weasel (*Mustella frenata*), and weasel-sized (*Mustella*-size) carnivores. Specimens that compare favorably with Bobcat (*Lynx rufus*) and badger (*Taxidea* cf. *taxus*) have only been found in the Harrison horizons and the silts beneath the Harrison horizons.

Rodent taxa (n = 24) dominate the floodplain assemblage in all strata and studies. Particularly abundant are the isolated teeth and fragmented postcranial elements of Northern pocket gophers (*Thomomys* cf. *talpoides*), voles (*Microtus* sp.), Great Basin pocket mice (*Perognathus* cf. *parvus*), white-footed mice (*Peromyscus* sp.), and Cricetidae-size (mice, rats, and voles) animals. The vast majority of the "unidentifiable" specimens sorted from the bulk samples represent the severely fragmented postcranial remains of these small animals. Bird remains were absent in the 1998 floodplain sample analyzed; however, Caulk (1988) assigned 12 specimens to this vertebrate class. Thus, avian taxa are rare at the site suggesting they played a minor role in the prehistoric diet of Marmes inhabitants.

Floodplain Subarea Analyses

We judgmentally sampled four floodplain subareas during this study based on their faunal and/or artifact content as depicted on detailed plan view diagrams of the floodplain excavations. Excavation units sampled around the owl's foot first reported by Gustafson (1972:108, Figure 6.1) included N10/W5, N15/E5, N15/W0, N15/W5, N20/E5, and N20/W0. The owl's foot represents the only completely articulated series of bones found on the floodplain and it rested between two retouched chert flakes suggesting that the foot and flakes were bound together—perhaps in a medicine bundle. Excavation units associated with previously identified Arctic fox (*Alopex lagopus*) remains and a diffuse lithic scatter in the northeastern portion of the floodplain included N15/E10, N15/E20, N20/E10-E20, N25/E10-E20, and N30/E10-E20. A discrete concentration of smaller bone fragments rested in the southwestern portion of the floodplain block excavation and we sorted bulk samples from units N10/W55-65, N15/W55-65, N20/W55, and N20/W65 in this area. We also focused on excavation units, N10/W25-W45 and N15/W25-45, surrounding the large semi-articulated elk skeleton exposed near the center of the floodplain excavations.

Identifications made during this study, per-subarea and taxon, are listed in Table 10.9. Similarly, identifications included in Caulk (1988:Appendix B) from excavation units within each subarea analyzed in this study are listed in Table 10.10. Both studies are combined in Table 10.11 and all values presented in this section represent the NISP per-taxon.

In addition to those remains reported in Table 10.9, Table 10.10, and Table 10.11, samples were brought to Gustafson periodically by Marmes staff members during the 1968 floodplain excavations. In most cases, no provenience beyond "Marmes floodplain" was given at the time, and they simply were recorded as examples of the kinds of mammals represented in the early floodplain deposits. The following list is given by Gustafson (1972:100):

Microtus sp.
Martes americana
Marmota cf. *flaviventris*
Taxidea taxus
Citellus cf. *washingtoni*
Vulpes fulva
Thomomys talpoides
Alopex lagopus
Neotoma cinerea
Canis latrans
Perognathus parvus
Cervus canadensis
Lepus sp.
Odocoileus sp.
Sylvilagus sp.
Ovis canadensis
Mustella frenata

The only animal in this list not included in any of the tables is *Ovis canadensis*. This identification is based on a single distal metapodial shown to Gustafson and reported to have come from alluvial deposits beneath the Harrison horizon.

The greatest number of identified specimens (NISP = 160) is found associated with the bone and debitage concentration resting in the southwestern portion of the floodplain. Together, the four floodplain subareas contained 549 (65 %) of the specimens assigned to taxon during this study. Thirty five percent (NISP = 301) of the identified specimens were found in excavation units not in one of the four floodplain subareas. Particularly interesting is the high percentage and number of Medium/Large Artiodactyl bone and tooth fragments present in the subareas compared to the remainder of the sampled floodplain excavation units. Not surprisingly, the greatest percentage (44.2) of Medium/Large Artiodactyl bone and tooth fragments remains were found associated with the large semi-articulated elk skeleton. However, Medium/Large Artiodactyl remains comprise 39.7% of the NISP Total in the area of the Arctic fox remains and 23.8% of the NISP Total in the southwest quadrant. The frequency of these severely fragmented remains representing deer-to-elk-size animals is about three times greater in the Arctic fox and southwest quadrant subareas compared to the other analyzed excavation units not included in one of the four subareas.

In our 1998 study, we discovered that jackrabbit (*Lepus* sp.) remains (NISP = 19) are far more common in the area surrounding the Arctic fox remains compared to the rest of the floodplain excavations, and cottontail (*Sylvilagus* sp.) bones and teeth are concentrated in excavation units surrounding the owl's foot. In addition, yellow-bellied marmot (*Marmota* cf. *flaviventris*) remains also are only found in excavation units surrounding the Arctic fox and semi-articulated elk bones.

Of all sampled areas the owl's foot and southwest quadrant subareas appear most disturbed by burrowing rodents; in that, these

Table 10.9 Number (n) of identified specimens (NISP), percentages and totals, by taxon from each sub-area analyzed during the 1998 study of the Marmes floodplain (45FR50) fauna.

| | \multicolumn{8}{c}{FLOODPLAIN SUB-AREA} | | |
TAXON	Owl's Foot (Area 1)		Arctic Fox Remains (Area 2)		Elk Remains (Area 3)		Southwest Quadrant (Area 4)		Other Analyzed Excavation Units	
	n	%	n	%	N	%	n	%	n	%
AMPHIBIANS										
cf. *Hyla* sp.	--	--	1	(.7)	--	--	--	--	6	(2.0)
cf. *Rana pipiens*	--	--	--	--	1	(.8)	--	--	3	(1.0)
REPTILES[1]										
Unidentified snakes	3	--	2	--	1	--	1	--	6	--
cf. *Pituophis* sp.	1	--	--	--	--	--	2	--	10	--
cf. *Thamnophis* sp.	2	--	2	--	1	--	2	--	9	--
cf. *Crotalus* sp.	1	--	3	--	2	--	2	--	6	--
cf. *Sceloporus* sp.	--	--	--	--	--	--	--	--	1	--
MAMMALS										
Sylvilagus sp.	10	(8.7)	1	(.7)	2	(1.6)	1	(.6)	2	(.7)
Lepus-size	--	--	--	--	--	--	--	--	1	(.3)
Lepus sp.	--	--	19	(13.0)	--	--	1	(.6)	8	(2.7)
Marmota cf. *flaviventris*	--	--	5	(3.4)	3	(2.3)	--	--	3	(1.0)
Spermophilus sp.	2	(1.7)	--	--	2	(1.6)	--	--	4	(1.3)
Spermophilus cf. *columbianus*	--	--	--	--	4	(3.1)	1	(.6)	4	(1.3)
Spermophilus cf. *townsendii*	1	(.9)	--	--	1	(.8)	1	(.6)	1	(.3)
Cricetidae-size	5	(4.3)	3	(2.1)	7	(5.4)	6	(3.8)	52	(17.3)
Thomomys cf. *talpoides*	54	(47.0)	11	(7.5)	11	(8.5)	53	(33.1)	28	(9.3)
cf. *Perognathus* sp.	5	(4.3)	4	(2.7)	7	(5.4)	3	(1.9)	28	(9.3)
Perognathus cf. *parvus*	9	(7.8)	10	(6.8)	19	(14.7)	6	(3.8)	31	(10.3)
Reithrodontomys sp.	--	--	2	(1.4)	--	--	--	--	3	(1.0)
Peromyscus sp.	2	(1.7)	5	(3.4)	3	(2.3)	2	(1.3)	17	(5.6)
Peromyscus cf. *maniculatus*	--	--	1	(.7)	--	--	--	--	4	(1.3)
Onychomys cf. *leucogaster*	1	(.9)	--	--	--	--	--	--	1	(.3)
Neotoma cf. *cinerea*	2	(1.7)	--	--	1	(.8)	13	(8.1)	--	--

283

Table 10.9 Number (n) of identified specimens (NISP), percentages and totals, by taxon from each sub-area analyzed during the 1998 study of the Marmes floodplain (45FR50) fauna.

TAXON	FLOODPLAIN SUB-AREA								Other Analyzed Excavation Units	
	Owl's Foot (Area 1)		Arctic Fox Remains (Area 2)		Elk Remains (Area 3)		Southwest Quadrant (Area 4)			
	n	%	n	%	N	%	n	%	n	%
cf. *Clethrionomys* sp. (?)	--	--	--	--	--	--	1	(.6)	--	--
cf. *Phenacomys* sp. (?)	--	--	1	(.7)	--	--	1	(.6)	1	(.3)
cf. *Microtus* sp.	--	--	--	--	--	--	3	(1.9)	--	--
Microtus sp.	12	(10.4)	7	(4.8)	10	(7.8)	22	(13.8)	70	(23.3)
cf. *Ondatra* sp.	--	--	1	(.7)	--	--	--	--	--	--
Canid-size	--	--	1	(.7)	--	--	7	(4.4)	1	(.3)
Vulpes cf. *fulva*	--	--	14	(9.6)	--	--	--	--	1	(.3)
Mustella-size	--	--	--	--	--	--	--	--	1	(.3)
Taxidea cf. *taxus*	--	--	--	--	--	--	1	(.6)	--	--
Vulpes/Lynx-size	--	--	1	(.7)	--	--	--	--	--	--
cf. *Lynx rufus*	--	--	1	(.7)	--	--	--	--	3	(1.0)
Medium/Large Artiodactyl	11	(9.6)	58	(39.7)	57	(44.2)	38	(23.8)	28	(9.3)
Cervus canadensis	--	--	--	--	1	(.8)	--	--	--	--
TOTALS[2]	114	(99.9)	146	(100.0)	129	(100.1)	160	(100.1)	301	(99.8)

[1] Reptile values represent number of units containing reptile remains and are not included in the NISP total calculations.
[2] Percentage totals do not equal 100.0 due to rounding.

Table 10.10 Number (n) of identified specimens, percentages and totals per-taxon, from Caulk (1988:Appendix B) for each sub-area analyzed during the 1998 study of the Marmes floodplain (45FR50) fauna.

TAXON	FLOODPLAIN SUB-AREA									
	Owl's Foot		Arctic Fox Remains		Elk Remains		Southwest Quadrant		Other Analyzed Excavation Units	
	n	%	n	%	n	%	n	%	n	%
AMPHIBIANS										
cf. *Rana pipiens*	1	(3.6)	--	--	--	--	--	--	--	--
REPTILES[1]										
Unidentified snakes	1	--	--	--	--	--	--	--	7	--
cf. *Pituophis* sp.	10	--	2	--	--	--	--	--	--	--
cf. *Thamnophis* sp.	1	--	--	--	--	--	--	--	1	--
cf. *Crotalus* sp.										
BIRDS										
Bucephala clangula	--	--	--	--	--	--	3	(18.3)	--	--
Ptychoramphus aleuticum	--	--	--	--	--	--	--	--	2	(2.2)
Empidonax sp.	--	--	--	--	1	(2.6)	--	--	--	--
Pica pica	--	--	--	--	--	--	--	--	1	(1.1)
Hylocichla sp.	--	--	--	--	--	--	--	--	2	(2.2)
Sialia sp.	--	--	--	--	--	--	--	--	3	(3.3)
MAMMALS										
Unidentified Lagomorph	--	--	--	--	4	(10.0)	--	--	--	--
Lepus sp.	--	--	5	(7.2)	--	--	--	--	4	(4.4)
Marmota sp.	5	(17.9)	--	--	4	(10.0)	--	--	1	(1.1)
Thomomys cf. *talpoides*	11	(39.3)	8	(11.6)	16	(40.0)	--	--	21	(23.1)
Perognathus cf. *parvus*	3	(10.7)	9	(13.0)	2	(5.0)	4	(25.0)	5	(5.5)
Castor canadensis	--	--	--	--	1	(2.5)	--	--	--	--

Table 10.10 Number (n) of identified specimens, percentages and totals per-taxon, from Caulk (1988:Appendix B) for each sub-area analyzed during the 1998 study of the Marmes floodplain (45FR50) fauna.

	FLOODPLAIN SUB-AREA									
TAXON	Owl's Foot		Arctic Fox Remains		Elk Remains		Southwest Quadrant		Other Analyzed Excavation Units	
	n	%	n	%	n	%	n	%	n	%
Peromyscus sp.	3	(10.7)	7	(10.1)	6	(15.0)	2	(12.5)	15	(16.5)
Neotoma sp.	1	(3.6)	--	--	--	--	2	(12.5)	10	(11.0)
Microtus sp.	1	(3.6)	--	--	1	(2.5)	--	--	2	(2.2)
Microtus montanus	--	--	--	--	--	--	--	--	1	(1.1)
Microtus pennsylvannicus	--	--	--	--	1	(2.5)	--	--	2	(2.2)
Microtus oeconomus	--	--	--	--	--	--	--	--	1	(1.1)
Lagurus curtatus	--	--	1	(1.4)	--	--	--	--	5	(5.5)
Canid	1	(3.6)	1	(1.4)	--	--	4	(25.0)	1	(1.1)
Canis sp.	--	--	--	--	--	--	--	--	3	(3.3)
Canis latrans	--	--	--	--	--	--	--	--	2	(2.2)
Canis familiaris	--	--	--	--	--	--	1	(6.3)	--	--
Unidentified fox	2	(7.1)	24	(34.8)	1	(2.5)	--	--	1	(1.1)
Alopex lagopus	--	--	6	(8.7)	--	--	--	--	--	--
Vulpes cf. *fulva*	--	--	4	(5.8)	--	--	--	--	1	(1.1)
Martes americana	--	--	3	(4.3)	--	--	--	--	--	--
Mustela frenata	--	--	--	--	--	--	--	--	3	(3.3)
Lynx canadensis	--	--	1	(1.4)	--	--	--	--	--	--
Cervus canadensis	--	--	--	--	1	(2.5)	--	--	--	--
Odocoileus sp.	--	--	--	--	2	(5.0)	--	--	3	(3.3)
Antilocapra americana	--	--	--	--	--	--	--	--	2	(2.2)
Totals[2]	28	(100.1)	69	(99.7)	40	(100.0)	16	(100.1)	91	(100.1)

[1] Reptile values represent number of units containing reptile remains and are not included in the NISP total calculations.
[2] Percentage totals do not equal 100.0 due to rounding.

Table 10.11 Number (n) of identified specimens (NISP), percentages and totals per-taxon, from the Marmes floodplain (45FR50) sub-areas analyzed in this study and by Caulk (1988:Appendix B).

TAXON	FLOODPLAIN SUB-AREA										Other Analyzed Excavation Units	
	Owl's Foot		Arctic Fox Remains		Elk Remains		Southwest Quadrant					
	n	%	n	%	n	%	n	%			n	%
AMPHIBIANS												
cf. *Hyla* sp.	--	--	1	(.5)	--	--	--	--			6	(1.5)
cf. *Rana pipiens*	1	(.7)	--	--	1	(.6)	--	--			3	(.8)
REPTILES[1]												
Unidentified snakes	--	--	--	--	4	(2.4)	1	--			7	(1.8)
cf. *Pituophis* sp.	1	(.7)	--	--	--	--	2	--			--	--
cf. *Thamnophis* sp.	10	(6.9)	--	--	--	--	--	--			1	(.3)
cf. *Crotalus* sp.	1	(.7)	--	--	--	--	--	--			--	--
cf. *Sceloporus* sp.	--	--	--	--	--	--	--	--			1	(.3)
BIRDS												
Bucephala clangula	--	--	2	(.9)	--	--	3	(1.7)			--	--
Ptychoramphus aleuticum	--	--	--	--	1	(.6)	--	--			2	(.5)
Empidonax sp.	--	--	--	--	--	--	--	--			--	--
Pica pica	--	--	--	--	--	--	--	--			1	(.3)
Hylocichla sp.	--	--	--	--	--	--	--	--			2	(.5)
Sialia sp.	--	--	--	--	--	--	--	--			3	(.8)
MAMMALS												
Unid. Lagomorph	--	--	--	--	--	--	--	--			1	(.3)
Sylvilagus sp.	10	(6.9)	1	(.5)	2	(1.2)	1	(.6)			6	(1.5)
Lepus sp.	--	--	24	(11.2)	4	(2.4)	1	(.6)			9	(2.3)
Marmota cf. *flaviventris*	5	(3.5)	5	(2.3)	3	(1.8)	--	--			3	(.8)
Spermophilus sp.	2	(1.4)	--	--	2	(1.2)	--	--			4	(1.0)

287

Table 10.11 Number (n) of identified specimens (NISP), percentages and totals per-taxon, from the Marmes floodplain (45FR50) sub-areas analyzed in this study and by Caulk (1988:Appendix B).

TAXON	FLOODPLAIN SUB-AREA								
	Owl's Foot		Arctic Fox Remains		Elk Remains		Southwest Quadrant		Other Analyzed Excavation Units	
	n	%	n	%	n	%	n	%	n	%
Spermophilus cf. *columbianus*	0	--	--	--	4	(2.4)	1	(.6)	4	(1.0)
Spermophilus cf. *townsendii*	1	(.7)	--	--	1	(.6)	1	(.6)	1	(.3)
Cricetidae-size	5	(3.5)	3	(1.4)	23	(13.8)	6	(3.4)	73	(18.6)
Thomomys cf. *talpoides*	65	(45.1)	19	(8.8)	11	(6.6)	53	(30.1)	28	(7.1)
cf. *Perognathus* sp.	5	(3.5)	4	(1.9)	9	(5.4)	7	(4.0)	33	(8.4)
Perognathus cf. *parvus*	12	(8.3)	19	(8.8)	20	(12.0)	6	(3.4)	31	(7.9)
Castor canadensis	--	--	--	--	1	(.6)	--	--	1	(.3)
Reithrodontomys sp.	--	--	2	(.9)	1	(.6)	2	(1.1)	--	--
Peromyscus sp.	5	(3.5)	12	(5.6)	3	(1.8)	2	(1.1)	18	(4.6)
Peromyscus cf. *maniculatus*	--	--	1	(.5)	--	--	--	--	17	(4.3)
Onychomys cf. *leucogaster*	--	--	--	--	--	--	2	(1.1)	4	(1.0)
Neotoma sp.	1	(.7)	--	--	--	--	--	--	11	(2.8)
Neotoma cf. *cinerea*	2	(1.4)	--	--	1	(.6)	13	(7.4)	--	--
cf. *Clethrionomys* sp. (?)	--	--	--	--	--	--	1	(.6)	--	--
cf. *Phenacomys* sp. (?)	--	--	1	(.5)	--	--	1	(.6)	--	--
cf. *Microtus* sp.	--	--	--	--	1	(.6)	3	(1.7)	1	(.3)
Microtus sp.	13	(9.0)	7	(3.3)	10	(6.0)	22	(12.5)	2	(.5)
Microtus montanus	--	--	--	--	1	(.6)	--	--	71	(18.1)
Microtus pennsylvannicus	--	--	--	--	--	--	--	--	2	(.5)
Microtus oeconomus	--	--	--	--	--	--	--	--	1	(.3)
Lagurus curtatus	--	--	1	(.5)	--	--	--	--	5	(1.3)
cf. *Ondatra* sp.	--	--	1	(.5)	--	--	4	(2.3)	--	--
Canid	1	(.7)	2	(.9)	--	--	7	(4.0)	1	(.3)

Table 10.11 Number (n) of identified specimens (NISP), percentages and totals per-taxon, from the Marmes floodplain (45FR50) sub-areas analyzed in this study and by Caulk (1988:Appendix B).

TAXON	FLOODPLAIN SUB-AREA									
	Owl's Foot		Arctic Fox Remains		Elk Remains		Southwest Quadrant		Other Analyzed Excavation Units	
	n	%	n	%	n	%	n	%	n	%
Canis sp.	--	--	--	--	--	--	--	--	4	(1.0)
Canis latrans	--	--	--	--	--	--	1	(.6)	2	(.5)
Canis familiaris	--	--	--	--	1	(.6)	--	--	--	--
Unid. fox.	2	(1.4)	24	(11.2)	--	--	--	--	1	(.3)
Alopex lagopus	--	--	6	(2.8)	--	--	--	--	--	--
Vulpes cf. *fulva*	--	--	18	(8.4)	--	--	--	--	1	(.3)
Mustella-size	--	--	--	--	--	--	--	--	1	(.3)
Martes americana	--	--	3	(1.4)	--	--	--	--	1	(.3)
Mustela frenata	--	--	--	--	--	--	--	--	3	(.8)
Taxidea cf. *taxus*	--	--	--	--	--	--	1	(.6)	--	--
Vulpes/Lynx-size	--	--	1	(.5)	--	--	--	--	--	--
cf. *Lynx rufus*	--	--	1	(.5)	--	--	--	--	3	(.8)
Lynx canadensis	--	--	1	(.5)	--	--	--	--	--	--
Medium/Large Artiodactyl	11	(7.6)	58	(27.0)	59	(35.3)	38	(21.6)	28	(7.1)
Cervus canadensis[2]	--	--	--	--	1	(.6)	--	--	--	--
Odocoileus sp.	2	(1.4)	--	--	--	--	--	--	3	(.8)
Antilocapra americana	--	--	--	--	--	--	--	--	2	(.5)
Totals[3]	144	(101.3)	215	(99.0)	167	(99.7)	176	(100.2)	392	(100.3)

[1] Reptile values represent number of units containing reptile remains and are not included in the NISP total calculations.
[2] Elk (*Cervus canadensis*) NISP is combined, thus a value of "1" is presented here.
[3] Percentage totals do not equal 100.0 due to rounding.

subareas contain the greatest frequency of northern pocket gopher (*Thomomys talpoides*) teeth and bones, but ground squirrel (*Spermophilus* sp.) specimens are concentrated near the large semi-articulated elk skeleton. Voles (*Microtus* sp.) and Cricetidae-size mammals are most numerous in units not included in one of the four floodplain subareas.

Caulk's (1988) analysis of the Marmes floodplain fauna produced similar results (see Table 10.10). In Caulk's sample, jackrabbit remains were present in the Arctic fox subarea only and Marmot remains were confined to the excavation units surrounding the large semi-articulated elk skeleton and owl's foot. Like our study, the vast majority of carnivore remains are found in excavation units surrounding the Arctic Fox remains originally reported by Gustafson (1972:70, Figure 4.1). Bones and teeth from the Arctic Fox (*Alopex lagopus*), red fox (*Vulpes fulva*), pine marten (*Martes americana*), and bobcat (*Lynx rufus*) have been found in this area only. Caulk tentatively identified a single domestic dog (*Canis familiaris*) element from a unit in the Harrison horizon in the southwest quadrant of the floodplain.

Combining the identifications made during this study and by Caulk does not alter the above patterns (see Table 10.11). The majority of specimens from ethnographically important taxa (e.g., Medium/Large Artiodactyl, *Lepus* sp., *Sylvilagus* sp., *Ondatra* sp., and *Spermophilus* sp.) are found in the samples selected from the floodplain subareas. Small fragments from deer-to-elk-size long bones are most numerous in the Arctic fox and southwest quadrant subareas, whereas jackrabbit and cottontail remains are most numerous in the area of the Arctic fox and owl's foot respectively. Wood rat (*Neotoma* sp.) and vole (*Microtus* sp.) bones and teeth are most numerous in the southwest quadrant, whereas northern pocket gopher teeth are most abundant around the owl's foot. Red fox (*Vulpes fulva*) remains are three times as abundant compared to their arctic dwelling counterparts (*Alopex lagopus*), and pine marten (*Martes americana*) and Canadian lynx (*Lynx canadensis*) bones and teeth are found only in association with the Arctic fox remains. Excavation units surrounding the floodplain subareas contain the most, albeit few, deer (*Odocoileus* sp.) and pronghorn (*Antilocapra americana*) remains; Caulk identified two fragmentary deer bones in the owl's foot subarea.

These patterns suggest that animal processing and perhaps cooking took place in, or near, the floodplain subareas and that these were important site activities. Subsequently, the subareas themselves likely represent limited activity loci occupied intermittently for short periods of time. Further, the concentration of carnivore remains in the Arctic Fox subarea is peculiar and implies a cultural origin for these specimens. It is possible that these specimens represent animals brought to the site for their fur and perhaps teeth.

Rockshelter and Floodplain Fauna Comparisons

Although no additional analysis of the faunal specimens from the Marmes rockshelter was conducted during this study, Gustafson's recollections of material being removed during the 1964 and 1968 excavations, and his qualitative assessment of more than 150 additional samples of rockshelter remains viewed during this project, suggest there is little difference between the 1968 rockshelter fauna and that reported in his dissertation for the 1962-1964 excavations (Gustafson 1972).

Table 10.5 summarizes mammal remains from Marmes Rockshelter. Geologic (stratigraphic) units indicated are those of Fryxell and Daugherty (1962). The basal portion of Unit I-II corresponds in time with the floodplain cultural horizons. In Table 10.5, faunal remains from geologic units VI-VIII, IV-V and I-II are combined for reasons given below. Except where natural stratigraphic units could be recognized, vertical excavation levels were .5 ft thick, often encompassing two or more stratigraphic units. Gustafson (1972:Appendix) separated fauna by stratigraphic units wherever possible when presenting the raw data, but they were combined for analysis.

Units VI through VIII are considered together because of the extreme mixing of these deposits. Unit IV is relatively undisturbed volcanic ash from Mt. Mazama and represents an instant in geologic time, and Unit V is the overlying, mixed volcanic ash and loess. Both Units I and II are characterized as coarse basalt rockfall; Unit II is differentiated from Unit I in that the former contained interstitial fill of silt and sand whereas the latter did not. While excavating or during a wind storm or as people moved about the shelter, fine sediment would filter down from Unit II into Unit I making it extremely difficult to detect the transition from one to the other. However, they were relatively easy to distinguish in the sidewalls while exposing stratigraphy, hence the two often are separated in the stratigraphic profiles.

In Table 10.5, taxa are divided roughly into large (upper group), medium (middle) and small (lower) size classes. Two major trends are apparent: 1) taxa remain essentially the same throughout the past 10,000 years, and 2) the largest number of mammal remains occurs in Unit III, which represents the shortest span of time. Gustafson found similar trends at Granite point (45WT41) excavated by Leonhardy (1968, 1970) on the Snake River about 14 km upstream from Lower Granite Dam. However, several additional trends are apparent when the NISP frequencies of economic taxa from the Marmes Rockshelter alone are compared by geologic unit (Table 10.12). Elk remains are common in the oldest geologic units (I, II, and III) only and this suggests elk were more abundant and represented an important resource during the early-Holocene. Deer bones and teeth also dominate the oldest units ranging between 19.6 and 26.4 percent of the NISP total per-geologic unit. Interestingly, the frequency of deer remains reaches its maximum in Unit III, but steadily declines after the catastrophic eruption of Mt. Mazama at ca. 6730 B.P. Pronghorn remains are most numerous in mid-Holocene units IV-VIII, which indicates that pronghorn were likely the most important source of meat, fat and hides throughout the mid-and-late-Holocene at Marmes Rockshelter. Whether the dominance of pronghorn in geologic units IV-VIII represents the advent of communal hunting or effects of Altithermal aridity and/or taphonomic processes is unclear.

Another major trend in Marmes Rockshelter involves leporid remains (Table 10.13). The frequency of cottontail and jackrabbit remains covaries inversely, with cottontail remains increasing through time at the expense of jackrabbits. Further, the total number of leporid bones steadily increases through time from five specimens in units I and II to 16 specimens in units VI and VIII. Though the sample is small, this pattern strongly suggests that medium-sized mammal procurement increased in importance through time, which in turn, perhaps signals an increase in diet-breadth (Bettinger 1991:84; Kelly 1995:78-90). Further, the dominance of cottontail remains raises the possibility that more cottontails were dispatched on an encounter basis compared to jackrabbits. This is an expected result given the behavioral differences of cottontails and jackrabbits — see below in Intersite Analyses for detailed discussion.

We also compared the fauna from geologic units I and II of Marmes Rockshelter to the fauna from the Harrison and Marmes horizons of the Marmes Floodplain (Table 10.13). This is accomplished by combining the identifications made during this study and Caulk's (1988) Master's thesis research and comparing them to Gustafson's dissertation research (1972). Stratigraphic correlation (Huckleberry, Gustafson, and Gibson 1998) and radiometric age determinations (Sheppard et al. 1987) indicate these strata are essentially contemporaneous.

Table 10.12 Number (n = NISP) and percentage of identified economic mammals per-geologic unit and taxon from Marmes rockshelter (45FR50) (adapted from Gustafson 1972:Table 5.1)

	ROCKSHELTER GEOLOGIC UNIT							
	VI-VIII		IV-V		III		I-II	
TAXON[1]	n	%	n	%	n	%	n	%
Odocoileus sp.	26	(33.3)	49	(34.5)	108	(41.8)	22	(45.8)
Antilocapra americana	31	(39.7)	65	(45.8)	79	(30.6)	14	(29.2)
Cervus canadensis	2	(2.6)	8	(5.6)	36	(14.0)	5	(10.4)
Sylvilagus sp.	14	(17.9)	12	(8.5)	10	(3.9)	2	(4.2)
Lepus sp.	2	(2.6)	3	(2.1)	5	(1.9)	3	(6.3)
Canis cf. *latrans*	3	(3.8)	5	(3.5)	20	(7.8)	2	(4.2)
TOTAL[2]	78	(99.9)	142	(100.0)	258	(100.0)	48	(100.1)

[1]Where several tooth fragments were found together, they were counted as one specimen.
[2]Percentage totals do not equal 100.0 due to rounding.

Table 10.13 Number (n) of identified specimens (NISP), percentages and totals, from this study of the Marmes floodplain and Caulk (1988) compared to the rockshelter (Gustafson 1972) by correlated stratigraphic unit and taxon.

Taxon	STRATUM						NISP TOTAL	
	Marmes Horizon		Harrison Horizon		Rockshelter Unit I-II			
	n	%	n	%	n	%	n	%
AMPHIBIANS								
cf. *Hyla* sp.	7	(1.1)	--	--	--	--	7	(.6)
cf. *Rana pipiens*	4	(.6)	1	(.3)	--	--	5	(.4)
REPTILES[1]								
Unidentified snakes	10	--	3	--	--	--	13	--
cf. *Pituophis* sp.	18	--	8	--	--	--	26	--
cf. *Thamnophis* sp.	13	--	14	--	--	--	28	--
cf. *Crotalus* sp.	11	--	3	--	--	--	16	--
cf. *Sceloporus* sp.	1	--	--	--	--	--	1	--
BIRDS								
Bucephala elangula	--	--	3	(1.0)	--	--	3	(.3)
Ptychoramphus aleuticum	--	--	2	(.6)	--	--	2	(.2)
Empidonaz sp.	--	--	1	(.3)	--	--	1	(.1)
Pica pica	--	--	1	(.3)	--	--	1	(.1)
Hylocichla sp.	--	--	2	(.6)	--	--	2	(.2)
Sialia sp.	--	--	3	(1.0)	--	--	3	(.3)
MAMMALS								
Unid. Lagomorph	1	(.1)	4	(1.3)	--	--	8	(.7)
Sylvilagus sp.	10	(1.4)	4	(1.3)	2	(1.8)	16	(1.4)
Lepus sp.	10	(1.4)	27	(8.7)	3	(2.7)	40	(3.6)
Marmota cf. *flaviventris*	5	(.7)	15	(4.9)	3	(2.7)	21	(1.9)
Spermophilus sp.	6	(.9)	2	(.6)	1	(.9)	8	(.7)
Spermophilus cf. *columbianus*	4	(.6)	5	(1.6)	--	--	9	(.8)
Spermophilus cf. *townsendii*	2	(.3)	--	--	--	--	2	(.2)
Cricetidae-size	61	(8.8)	8	(2.6)	10	(1.0)	79	(7.1)
Thomomys cf. *talpoides*	166	(24.0)	38	(12.3)	10	(8.9)	207	(18.6)
cf. *Perognathus* sp.	16	(2.3)	13	(4.2)	3	(2.7)	29	(2.6)
Perognathus cf. *parvus*	54	(7.8)	29	(9.4)	--	--	83	(7.5)
Castor Canadensis	--	--	1	(.3)	--	--	1	(.1)
Reithrodontomys sp.	3	(.4)	1	(.3)	--	--	4	(.4)
Peromyscus sp.	38	(5.5)	23	(7.4)	--	--	61	(5.5)
Peromyscus cf. *maniculatus*	5	(.7)	--	--	--	--	6	(.5)

Table 10.13 Number (n) of identified specimens (NISP), percentages and totals, from this study of the Marmes floodplain and Caulk (1988) compared to the rockshelter (Gustafson 1972) by correlated stratigraphic unit and taxon.

Taxon	STRATUM							
	Marmes Horizon		Harrison Horizon		Rockshelter Unit I-II		NISP TOTAL	
	n	%	n	%	n	%	n	%
Onychomys cf. *leucogaster*	2	(.3)	--	--	1	(.9)	2	(.2)
Neotoma sp.	1	(.1)	12	(3.9)	3	(2.7)	16	(1.4)
Neotoma cf. *cinerea*	2	(.3)	--	--	--	--	2	(.2)
cf. *Clethrionomys* sp. (?)	--	--	1	(.3)	--	--	1	(.1)
cf. *Phenacomys* sp. (?)	3	(.4)	--	--	--	--	3	(.3)
cf. *Microtus* sp.	3	(.4)	--	--	--	--	3	(.3)
Microtus sp.	86	(12.4)	26	(8.4)	--	--	112	(10.1)
Microtus montanus	--	--	1	(.3)	--	--	1	(.1)
Microtus pennsylvannicus	--	--	3	(1.0)	--	--	3	(.3)
Microtus oeconomus	1	(.1)	--	--	--	--	1	(.1)
Lagurus curtatus	1	(.1)	3	(1.0)	--	--	4	(.4)
cf. *Ondatra* sp.	1	(.1)	--	--	2	(1.8)	3	(.3)
Canid	2	(.3)	8	(2.6)	--	--	10	(.9)
Canis sp.	1	(.1)	1	(.3)	--	--	2	(.2)
Canis latrans	--	--	1	(.3)	2	(1.8)	3	(.3)
Canis familiaris	--	--	1	(.3)	--	--	1	(.1)
Unid fox.	--	--	28	(9.1)	--	--	28	(2.5)
Alopex lagopus	--	--	6	(1.9)	--	--	6	(.5)
Vulpes cf. *fulva*	1	(.1)	19	(6.1)	--	--	20	(1.8)
Mustela-size	--	--	1	(.3)	--	--	1	(.1)
Martes americana	--	--	3	(1.0)	--	--	3	(.3)
Mustela frenata	--	--	3	(1.0)	--	--	3	(.3)
Taxidea cf. *taxus*	1	(.1)	--	--	--	--	1	(.1)
cf. *Lynx rufus*	--	--	2	(.6)	--	--	2	(.2)
Lynx canadensis	1	(.1)	--	--	--	--	1	(.1)
Medium/Large Artiodactyl	188	(27.2)	2	(.6)	44	(39.3)	231	(20.8)
Cervus canadensis	1	(.3)	--	--	5	(4.5)	6	(.5)
Odocoileus sp.	2	(.3)	4	(1.3)	22	(19.6)	28	(2.5)
Antilocapra americana	1	(.1)	1	(.3)	14	(12.5)	16	(1.4)
Total	690	(99.6)	309	(99.7)	112	(100.1)	1111	(99.6)

Reptile values represent number of units containing reptile remains and are not included in the NISP total calculations.

Though rockshelter units I-II and the Marmes and Harrison floodplain strata are similar in age, the faunas they contained are markedly different compared by the number and percentage of specimens assigned to specific taxa. Medium/Large Artiodactyl remains comprise a significant portion (27.2%) of the NISP total for the Marmes Horizons, but the rockshelter contained the greatest frequency (39.3%) of specimens assigned to this category. Interestingly, specimens from these deer-to-elk size animals are virtually absent (NISP = 2) in the Harrison Horizons.

Floodplain specimens assigned to the Medium/Large Artiodactyl category are far smaller than their rockshelter counterparts. Few floodplain bone fragments exceeded one centimeter in maximum dimension whereas most of the rockshelter bone fragments assigned to this category measure several centimeters in length. The quarter-inch dry screening of the rockshelter deposits compared to the one-millimeter water screening of the floodplain deposits partially explains this discrepancy. However, screen size does not explain the paucity of larger bone fragments in the bulk samples analyzed during this study. It is possible that most of the larger bones and bone fragments were point-located and collected separately during the 1968 excavations. This would account for the 512 previously sorted mammal bone samples listed in Appendix K that remain unanalyzed.

Taxonomic diversity also distinguishes the floodplain and rockshelter faunas. Thus far, 57 taxa have been identified from the floodplain sediments compared to only 15 from rockshelter units I and II. The one-millimeter water screening of the floodplain sediments is most certainly responsible for these differences. Rodents and small-to-medium-sized mammals dominate the floodplain faunal assemblage and small isolated molariform teeth and tooth fragments (e.g., < 5 mm) are by far the most frequently identified elements (see Appendix J). Northern pocket gophers (*Thomomys talpoides*), voles (*Microtus* sp.), and pocket mice (*Perognathus* sp.) are the most numerous floodplain genera. White-footed mice (*Peromyscus* sp.) and mouse-sized animals (e.g., Cricetidae-size) animals also form a significant part of the floodplain fauna. The relative absence of these animals in the rockshelter is likely a consequence of the stone-laden sediments that characterize rockshelter units I and II. Most of these rodent taxa prefer finer-grained substrates that afford easier construction of the burrows and runways that they rely upon for food and shelter.

Carnivore remains are particularly abundant in the floodplain, especially the Harrison Horizons, compared to the rockshelter. Of the ten carnivore genera identified at the site only one, the coyote (*Canis latrans*), is present in the rockshelter sediments. Deer (*Odocoileus* sp.) and pronghorn (*Antilocapra americana*) bones and teeth are present in floodplain and rockshelter.

However, specimens from these animals are most abundant in the rockshelter where deer comprise 19.6% of the NISP Total and pronghorn comprise 12.5% of the NISP Total for rockshelter units I and II. Elk bones, discounting the large floodplain elk skeleton (Gustafson 1972), are most common in rockshelter units I and II (NISP = 5) compared to the floodplain (NISP = 1).

In summary, the floodplain contains the greatest diversity of taxa, but the rockshelter holds the greatest number of specimens representing economically important genera. The distribution of taxa across the sampled portion of the floodplain suggests that the identified floodplain subareas likely represent loci where people processed and prepared animals and maintained their stone tools. The remains of different carnivores concentrated around the original Arctic fox find (Gustafson 1972), suggests that the inhabitants of the floodplain likely sought these animals for their fur and perhaps teeth. Examination of the rockshelter fauna indicates deer (*Odocoileus* sp.) provided the primary source of animal products prior to the eruption of Mt. Mazama at ca. 6730 B.P. Pronghorn were more important after this time, though deer continued to provide a substantial portion of the prehistoric diet. Elk (*Cervus canadensis*) were most important during the deposition of rockshelter unit III, but likely represented a modest contribution to the diet before and after this.

Bone Breakage and Condition

We did not observe significant differences in the size, shape, or kind and degree of thermal alteration(s) among the bones and teeth we removed from the bulk samples sorted during this analysis. Therefore, we chose a judgmental sample of 11,037 of an estimated 60,000 total bones, teeth and bone fragments to estimate semiquantitatively the size and shape of each specimen and the kind and degree of thermal alteration. Attributes for the specimen lots were recorded after taxonomic identifications had been made and weights had been taken for all lots of bone removed from the processed bulk samples. While analyzing each lot of bones under a dissecting microscope, we observed little variation in overall condition of skeletal elements throughout the entire floodplain sample (ca. 60,000 elements). Our sample consisted of the first 40 bags of bones processed (ca. 18% of total) during the 1998 project.

Most of the fragments appeared to have been burned on the edges as well as on the surface. This led us to believe that most may have been broken into tiny pieces before they were burned. We developed criteria that we thought would reflect cultural practices. Gustafson (1987a, 1987b, 1996) has applied similar criteria elsewhere, but this is the first time that an attempt was made to determine if bones were burned before or after being broken into small pieces.

Three variables were recorded for each analyzed specimen: *size, shape,* and *condition.* We assigned specimens to one of seven ordinal size categories: < 5 mm, 5-15 mm, 25-35 mm, 35-50 mm, 50-100 mm and > 100 mm. The shape of each specimen is characterized by one of four nominal categories. Postcranial specimens that lacked articular ends were coded as *foreshortened* (mid-shaft fragments). Fragments with a long-axis less than twice their width were characterized as *tabular*. Specimens with a long axis that measured more than twice their width are identified as *splintered*. Intact elements were coded as *whole*. Counts were separated into four stratigraphic categories: "Marmes" horizon, "Between Marmes and Harrison" horizons, "Harrison" horizon and "Below Harrison" horizon.

All samples counted and coded included the bones, teeth and fragments that had been identified to taxon as well as the indeterminate or unidentified fraction. Therefore, a fairly large number of long bones and teeth included in the floodplain NISP Total derived during this study, as well as rodent vertebrae and ribs (i.e., "whole" bones) were present; these comprised the majority of remains "unaltered" by burning.

Size. Over 9,200 whole or broken bones (83.9%) were in the 0-5 mm size range, almost 1,800 (15.9%) ranged from 5-15 mm, only 15 (.1%) comprised the 15-25 mm category and one (< .1%) was in the 25-35 mm range. None in the sample analyzed were in the categories 35-50 mm, 50-100 mm or > 100 mm. Thus, more than 99 percent of the remains counted were less than 15 mm in maximum dimension and are practically impossible to assign to specific taxonomic categories. Several taphonomic studies (Binford 1981; Brain 1981; Bunn 1982) suggest that high percentages of "unidentifiable" specimens are a characteristic outcome of human food-processing. Gifford-Gonzalez's (1989:Figure 6) research also indicates that humans produce more "unidentifiable" bone fragments than hyenas, dogs, leopards and wolves. However, natural fires also represent a possible agent for the severe fragmentation observed in the assemblage. Most of the specimens from each stratum are charred on all surfaces. Gilchrist and Mytum (1986) report that burning resulted in the extreme fragmentation of 10-50% of the deer, beaver and muskrat bones used in their experiment. Charred and calcined bone is also more brittle because collagen fibers are destroyed (Johnson 1985). These experiments suggest that burning reduced the probability of finding larger more identifiable vertebrate remains.

Fracture type. Table 10.14 shows the distribution of fragments by shape and stratum. A Chi-square analysis ($n = 11037$; $x^2 = 241.52$; $df = 9$; $p = .001$) comparing strata by specimen shape suggests significant differences exist between strata based on the specimens they contain. Tiny (< 15 mm) splintered and tabular fragments are far more abundant throughout the sequence than are foreshortened and whole bones (Figure 10.3). Midshaft fragments ends are the next most abundant category ($n = 1243$) followed by whole elements ($n = 268$). Splintered specimens are particularly abundant in the Marmes (47.7%) and Harrison (43.6%) horizons, whereas midshaft specimens are nearly twice as abundant in the overbank silts between and below the Marmes and Harrison horizons. These differences account for the significant Chi-square. These results indicate a greater degree of fragmentation exists among the specimens recovered from the Marmes and Harrison horizons.

Several scenarios possibly account for the preponderance of tabular specimens and the significantly greater number of splintered specimens in the Marmes and Harrison horizons. Bone tends to develop longitudinal cracks and break transversely after prolonged subareal weathering. Johnson's (1985:171) research suggests that desiccated bones tend to break perpendicular to rather than along the collagen fiber matrix. Thus, transverse fractures—what

Table 10.14 Number and percentage of bone fragments by shape and stratum.

Shape	Stratum									
	Marmes Horizons		Between Marmes & Harrison Horizons		Harrison Horizon		Below Harrison Horizons		Total	
	n	%	n	%	N	%	n	%	n	%
Midshaft	448	(10.7)	26	(22.6)	675	(10.9)	94	(15.6)	1243	(11.3)
Splintered	1992	(47.7)	33	(28.7)	2680	(43.6)	180	(29.9)	4885	(44.3)
Tabular	1552	(37.1)	53	(46.1)	2723	(44.9)	313	(52.0)	4641	(42.0)
Whole	188	(4.5)	3	(2.6)	62	(1.0)	15	(2.5)	268	(2.4)
Total	4,180	(100.0)	115	(99.9)	6140	(100.4)	602	(100.0)	11,037	(100.0)

Johnson attributes to "horizontal tension failure"—are a characteristic outcome of dry bone breakage. Dry and/or mineralized bone often fractures perpendicular or parallel to the long axis of the diaphysis. Green/fresh bone, on the other hand, should fracture more readily along the collagen fiber matrix in a helical fashion, which in turn results in a curvate fracture margin and shape (i.e., spiral fracture).

The splintered and tabloid specimens in this assemblage are square-to-rectilinear in plan view and the fracture margins are almost always perpendicular to the fragment's long axis. This suggests that the collagen fibers in the bones were compromised prior to breakage. Subsequently, prolonged subareal weathering and/or burning—possibly along with trampling—provide a parsimonious explanation for the breakage patterns observed in the assemblage.

Condition. Table 10.15 shows the number and percentage of fragments by condition for each stratum. Again, Chi-square analysis (n = 11,307; x^2 = 488.85; df = 12; p = .001) comparing strata based on the condition of the bones and bone fragments they contained is significant. However, these test results are tentative because more than 20% of the cells contain values < 5 (Hayes 1973). We assumed that if the bones were burned first, then broken, the broken edges would appear "blackened" or "ashened". On the other hand, if they were broken first, then burned (or burned again), the edges should also appear "blackened" or "ashened".

Bones and bone fragments lacking macroscopic color evidence of exposure to fire are classified as unaltered. Bones blacken between 400° and 500°C and become calcined at temperatures exceeding 600° to 700°C (Buikstra and Swegle 1989:255). Shipman et al. (1984:308-313) placed sheep and goat mandibles and astragali in a kiln for four hours and documented several color stages. Bone heated between 300°C and 500°C is mostly blackened but can appear yellowish-red and red to purple. Intensely heated bones (>600°C) were purplish-blue and blue. When completely incinerated, or calcined, bone becomes bluish-white or gray in color. Here, blackened specimens are classified as charred and ashened specimens are those with gray, blue-gray, white, and occasionally buff colored surfaces indicating almost complete incineration.

In every case, bones "blackened after breakage" (i.e. edges also blackened) and "ashened after breakage": (i.e. edges also ashened) far outnumbered those inferred to have been "blackened before breakage" or "ashened before breakage" (Figure 10.4). The same pattern pertained throughout the strata, though the frequency of these specimens across strata differs. These differences account for the significant Chi-square results. A greater frequency of "blackened after breakage" specimens were found in the Harrison horizons and the silts beneath them. Conversely, unaltered specimens were found in greater frequencies in the Marmes Horizon

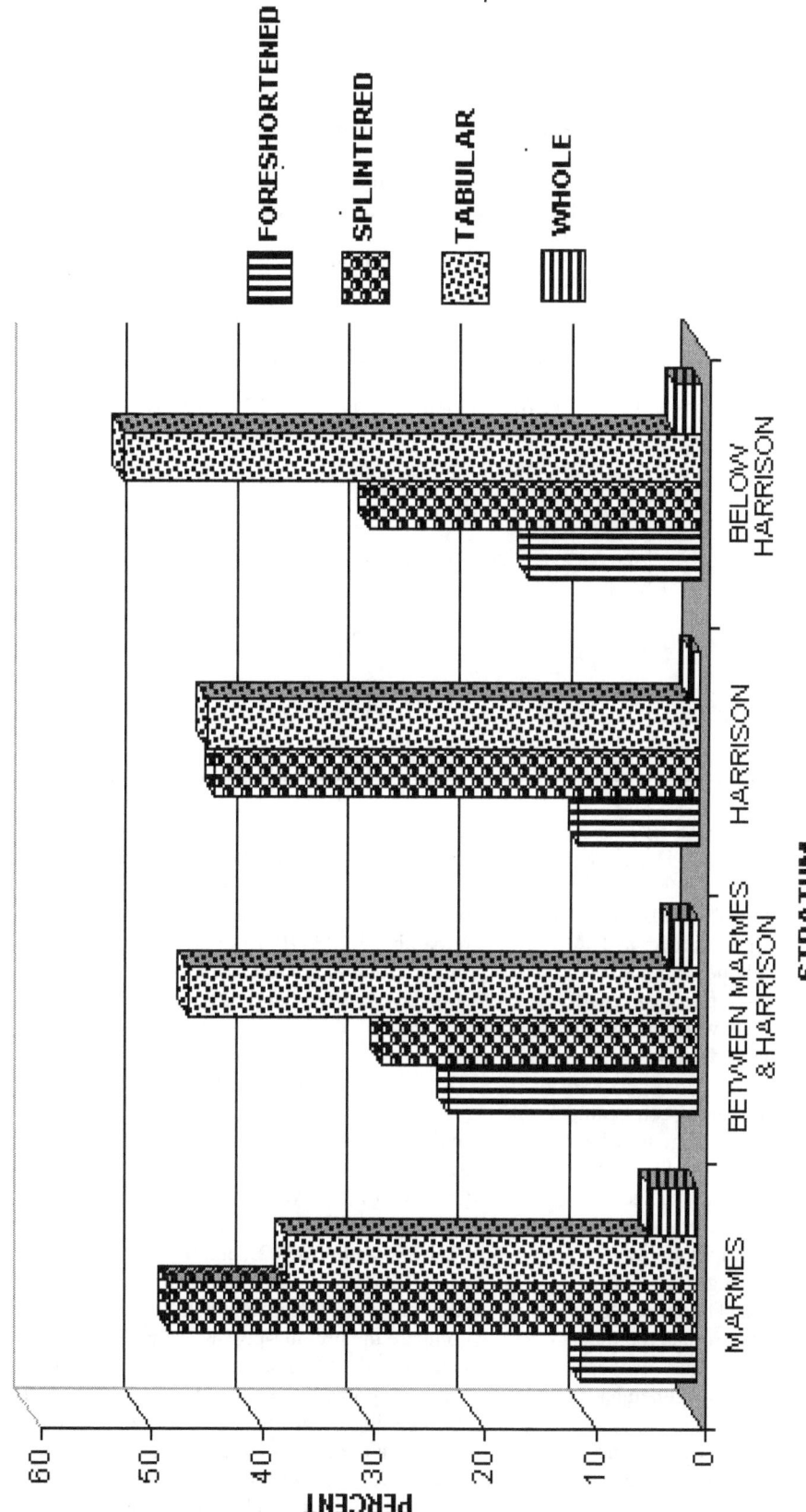

Figure 10.3 Graph showing percentage of ascribed shapes by vertical stratum. Splintered and tabular (center two) are most abundant shapes.

Table 10.15 Number and percentage of bone fragments by condition and stratum.

Condition	Marmes Horizons		Between Marmes & Harrison Horizons		Harrison Horizon		Below Harrison Horizons		Total	
	n	%	n	%	n	%	n	%	n	%
Ashened after breakage	179	(4.3)	3	(2.6)	310	(5.0)	12	(2.0)	504	(4.6)
Ashened before breakage	34	(.8)	1	(.9)	2	(<.1)	--	--	37	(.3)
Blackened after breakage	2,584	(61.8)	76	(66.1)	4,543	(73.8)	476	(79.1)	7,670	(69.5)
Blackened before breakage	28	(.7)	--	--	228	(3.7)	--	--	256	(2.3)
Unaltered	1,335	(32.4)	35	(30.4)	1,066	(17.4)	114	(18.9)	2570	(23.3)
Total	4,180	(100.0)	115	(100.0)	6,140	(99.9)	602	(100.0)	11,037	(100.0)

and the silts between the Marmes and Harrison horizons.

If one assumes that "ashened" means the bone was heated until it became calcined, and "blackened" indicates burned bone, then most—8,194 of 8,497 or 96 percent—of the "blackened" and "ashened" specimens may have been broken into tiny fragments before their edges were burned.

However, our initial assumptions concerning the time of burning relative to breakage may have been too simplistic. We have informally observed juniper and ponderosa pine fueled fires fed overnight blacken and ashen deer metapodials, humeri, and femora all the way through irrespective of weathering stage (*sensu* Behrensmeyer 1978:150, 153). Therefore, the attributes recorded here may not reflect the time of burning relative to breakage as much as the degree and duration of burning a specimen has undergone.

Most "unaltered" remains are whole bones and teeth or midshaft fragments from small rodents, many of which (e.g., most pocket gophers—only 4% burned) probably are from animals intrusive to the site deposits. Many of the burned bones are tabular (3,638 or 32.9%) or splintered (3,699 or 33.5%) fragments. Whole bones or midshaft fragments comprise the remaining 1,133 (10.3%) burned fragments. Thus, a total of 8,470 (76.7%) of the remains counted had been burned.

Intersite Analyses

In this section we compare quantitatively the Marmes fauna to the faunas recovered from Granite Point (45WT41), 45WT2, the Lind Coulee Site (45GR97), the Alpowa Locality (45AS80 and 45AS82), and the Upper Landing site (10IH1017). We found that the only reasonable way to accomplish this task was to compare the pre-and-post-Mazama (6730 B.P., Hallett et al. 1997) faunas of these sites. Though a coarse-grained comparison, we felt that it was the meaningful way to search for possible patterns present in the faunas. We also compare qualitatively the Marmes fauna to the faunas analyzed by Chatters (1986), Deaver and Greene (1978) and Greene (1976) where appropriate. The Sunset Creek Site (45KT28), reported by

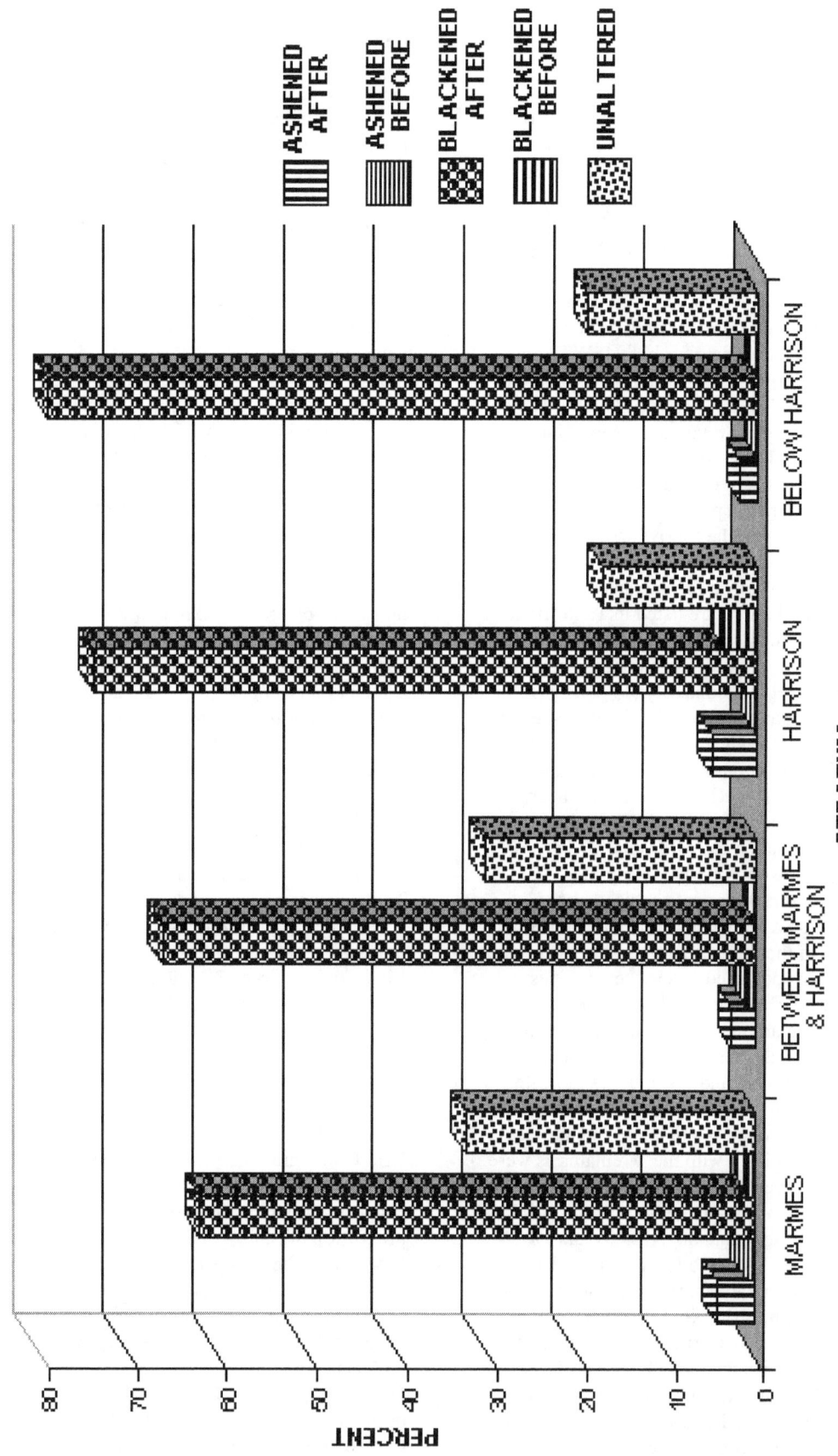

Figure 10.4 Graph showing percentage of each ascribed condition by vertical stratum. Compare ashened after breakage with ashened before, and blackened after with blackened before, and both with unaltered.

Nelson (1969) also was consulted, but the faunal treatment was limited to bone tools and shell remains.

Gustafson performed the initial analysis of both the 45WT2 fauna (Nance 1966) and the Granite Point fauna (Leonhardy 1968). Located at the confluence of the Palouse and Snake rivers, 45WT2 is an open site that extends along a sandy shelf one mile downriver of Marmes Rockshelter. Granite Point is situated on the Snake River below Wawawai Canyon about 14 km upstream from Lower Granite Dam. Gustafson (1972) recompiled the faunal data for both 45WT2 and Granite Point during his dissertation research and the data presented here is derived from his Table 5.2 and Table 5.3. However, and as for Marmes Rockshelter (see Table 10.12), the NISP percentages have been recalculated and summed per-stratum rather than per-taxon in this analysis. This was done so that the following tables can be directly compared to those presented earlier in this report. The NISP and recalculated percentages for 45WT2 are listed in Table 10.16 and Table 10.17 contains the same information for Granite Point, but uses the stratigraphic designations used by Leonhardy (1970).

An NISP listing for the Lind Coulee Site was also derived from Irwin and Moody (1978:227-241). To the best of our knowledge, this is the first time that the Lind Coulee fauna has been synthesized and presented in a NISP and percentages per-taxon format (Table 10.18). We feel that this in itself is a crucial step in understanding this important scabland site.

Situated on a scabland flood channel near the western rim of the Palouse River basin, the Lind Coulee site was occupied at least seven times within a period of between fifty and two hundred years beginning around 9,000 years ago (Irwin and Moody 1978:215-217). High frequencies of fetal and newborn elk and bison bones indicate that the animals were killed during the spring calving season. Windust and lanceolate projectile points were associated with these remains as were bone awls and eyed needles.

Like Marmes Rockshelter (see Table 10.12), large mammals dominate the NISP Totals for 45WT2, Granite Point, and Lind Coulee. Deer (*Odocoileus* sp.) are the most abundant economic taxon at the Marmes site, Granite Point and 45WT2, whereas, elk (*Cervus canadensis*) and bison (*Bison* sp.) dominate the Lind Coulee assemblage. Pronghorn (*Antilocapra americana*) are only abundant in Marmes Rockshelter, but are also present in low-to-moderate numbers at Granite Point and 45WT2, and in several of the Period II housepits at sites 45OK383 and 45OK69 reported by Chatters (1986). Deer and pronghorn were considered together in Deaver and Greene's (1978) analysis of the 45AD2 rockshelter fauna. Located in the Scablands along Cow Creek and first reported by Greene (1976), the frequency of deer and pronghorn remains at 45AD2 increased from 3% of the NISP total in basal level 5 to 24% in level 1. The frequency of rabbit remains parallels the increased use of deer and pronghorn at this site (Deaver and Greene 1978:Table 5). Rabbit (*Sylvilagus* sp.) and hare (*Lepus* sp.) bones exist at all sites discussed herein. However, rabbits are by far the most common leporid at Marmes Rockshelter, Lind Coulee and 45AD2.

Granite Point is the only site discussed in which hare bones outnumbered those of their cottontail cousins. Marmes and Granite Point share several carnivore taxa including the coyote (*Canis latrans*), red fox (*Vulpes fulva*), bobcat (*Lynx rufus*), and badger (*Taxidea taxus*). Lind Coulee lacked coyote bones, but contained numerous badger and skunk (*Mephitis mephitis*) remains throughout the excavated overbank sediments. Rodent remains—particularly ground squirrel (*Spermophilus* sp.) and northern pocket gopher (*Thomomys talpoides*) teeth and jaws—are common at all four sites. Yellow-bellied marmots (*Marmota flaviventris*) are most common in the older sediments of Granite Point, 45WT2, and Lind Coulee. Though, the greatest number of marmot remains at Marmes Rockshelter occurs in geologic units VI-VIII (see Table 10.12).

Table 10.19 compares the pre-Mazama faunas from Marmes rockshelter and floodplain, Lind Coulee, Granite Point, and 45WT2—the only sites in this analysis that contained faunal assemblages of this age. Here, NISP values for the Marmes site represent the NISP Totals presented previously in Table 10.13, which includes the NISP Totals for Marmes and Harrison floodplain horizons and rockshelter units I-II. Combining the faunas from Marmes rockshelter units I-II and the Marmes and Harrison floodplain horizons changes the proportion of rabbits and hares only.

One of the most striking differences between the sites is in the greater number of taxa present at the Marmes and Lind Coulee sites. Marmes and Lind Coulee are the only sites that contain early-Holocene amphibian, reptile, and bird remains. However, the Marmes and Lind Coulee faunas are quite distinct when compared. The large number of bison bones at Lind Coulee distinguish it from the other deer dominated assemblages. Lind

Table 10.16 Number (n) of identified specimens (NISP), and percentages, per-cultural unit and taxon from 45WT2. Adapted from Gustafson (1972:Table 5.2).

¹⁴C yrs. B.P. (Approx.) Cultural Unit TAXON[1]	150 A		1,300 B		6,700 C		7,300 D		? E		NISP TOTAL	
	n	%	n	%	n	%	n	%	n	%	n	%
Odocoileus sp.	2	(7.7)	21	(20.0)	8	(7.5)	7	(15.2)	6	(18.2)	44	(13.9)
Antilocapra americana	--	--	2	(1.9)	1	(.9)	--	--	--	--	3	(.9)
Cervus canadensis	1	(3.8)	3	(2.9)	--	--	2	(4.3)	2	(6.1)	8	(2.5)
Bison sp.	--	--	--	--	1	(.9)	--	--	--	--	1	(.3)
Indeterminate Large Mammal	16	(61.5)	31	(29.5)	29	(27.1)	14	(30.4)	10	(30.3)	100	(31.5)
Subtotal	19	(73.5)	62	(54.3)	38	(35.5)	23	(50.0)	18	(54.5)	156	(49.2)
Lagomorph	--	--	1	(1.0)	4	(3.7)	3	(6.5)	2	(6.1)	10	(3.2)
Canis cf. *latrans*	--	--	1	(1.0)	--	--	1	(2.2)	--	--	2	(.6)
Marmota cf. *flaviventris*	--	--	--	--	--	--	4	(8.7)	--	--	4	(1.3)
Ondatra zibethica	--	--	1	(1.0)	2	(1.9)	--	--	--	--	3	(.9)
Indeterminate Medium Mammal	3	(11.5)	13	(12.4)	15	(14.0)	5	(10.9)	4	(12.1)	40	(12.6)
Subtotal	3	(11.5)	16	(15.2)	21	(19.6)	13	(28.3)	6	(18.2)	59	(18.6)
Spermophilus cf. *washingtoni*	4	(15.4)	28	(26.7)	29	(27.1)	1	(2.2)	3	(9.1)	65	(20.5)
Thomomys talpoides	--	--	4	(3.8)	18	(16.8)	9	(19.6)	6	(18.2)	37	(11.7)
Subtotal	4	(15.4)	32	(30.5)	47	(45.8)	10	(21.7)	9	(27.3)	102	(32.2)
GRAND TOTAL[2]	26	(99.9)	110	(99.9)	106	(99.9)	47	(99.9)	33	(100.1)	322	(100.0)

[1] Where several tooth fragments were found together, they were counted as one specimen.
[2] Percentage totals do not equal 100.0 due to rounding.

Table 10.17 Number (n) of identified specimens (NISP), number and percentages, Per-cultural unit and taxon from Granite Point (45WT41). Adapted from Gustafson (1972:Table 5.3).

^{14}C yrs. B.P. (Approx.)	3,000		5,000		6,850		8,000		NISP TOTAL	
Cultural Unit	A_1, A_2, A_3, B_1		A_5, A_6, B_2, B_3		C_1		B_4, C_2, C_3, C_4			
TAXON	n	%	n	%	n	%	n	%	n	%
Odocoileus sp.	21	(32.3)	82	(50.9)	4	(9.8)	103	(31.7)	210	(35.5)
Antilocapra americana	1	(1.5)	8	(5.0)	--	--	2	(.6)	11	(1.9)
Cervus canadensis	7	(10.8)	8	(5.)	--	--	59	(18.2)	74	(12.5)
Bison sp.	5	(7.7)	--	--	--	--	--	--	5	(.8)
Ovis sp.	1	(1.5)	--	--	--	--	1	(.3)	2	(.3)
Subtotal	35	(53.8)	98	(60.9)	4	(9.8)	165	(50.8)	302	(51.0)
Sylvilagus sp.	1	(1.5)	3	(1.9)	1	(2.4)	10	(3.1)	15	(2.5)
Lepus sp.	--	--	--	--	--	--	26	(8.0)	26	(4.4)
Canis cf. *latrans*	3	(4.6)	7	(4.3)	--	--	15	(4.6)	25	(4.2)
Vulpes cf. *fulva*	--	--	--	--	--	--	1	(.3)	1	(.2)
Lynx rufus	--	--	3	(1.9)	--	--	1	(.3)	4	(.7)
Castor canadensis	--	--	--	--	1	(2.4)	--	--	1	(.2)
Taxidea taxus	--	--	--	--	--	--	1	(.3)	1	(.2)
Marmota flaviventris	--	--	1	(.6)	--	--	4	(1.2)	5	(.8)
Subtotal	4	(6.2)	13	(8.1)	2	(4.9)	53	(16.3)	72	(12.2)
Spermophilus cf. *columbianus*	10	(15.4)	--	--	4	(9.8)	43	(13.2)	57	(9.6)
Thomomys talpoides	14	(21.5)	44	(27.3)	28	(68.3)	51	(15.7)	137	(23.1)
Neotoma cinerea	1	(1.5)	5	(3.1)	--	--	1	(.3)	7	(1.2)
Peromyscus maniculatus	--	--	--	--	3	(7.3)	4	(1.2)	7	(1.2)
Microtus cf. *montanus* .	1	(1.5)	--	--	--	--	2	(.6)	3	(.5)
Lagurus curtatus	--	--	--	--	--	--	1	(.3)	1	(.2)
Subtotal	26	(40.0)	50	(31.1)	35	(85.4)	107	(32.9)	218	(36.8)
GRAND TOTAL[2]	65	(99.8)	161	(100.0)	41	(100.0)	325	(99.9)	592	(100.0)

[1] Where several tooth fragments were found together, they were counted as one specimen.
[2] Percentage totals do not equal 100.0 due to rounding.

Table 10.18 Number (n) of identified specimens (NISP), number and percentages, for the Lind Coulee Site (45GR97) as reported by Moody and Irwin (1978:227-241, Appendix E).

Taxon	Common Name	NISP TOTAL n	%
REPTILES			
Chelonia	turtles	1	(.2)
Squamata	lizards, snakes	1	(.2)
BIRDS			
Eggshell	eggshell	7	(1.4)
Branta-size	geese	6	(1.2)
Branta cf. *bernicla*	Brant	2	(.4)
Anas-sized	mallard-size	4	(.8)
Anas platyrhynchos	mallard	6	(1.2)
Anas carolinensis	Green-winged teal	2	(.4)
Ardea-size	heron-size	3	(.6)
Ardea herodias	Great blue heron	2	(.4)
Turdus-size	robin-size	2	(.4)
MAMMALS			
Lagomorph-size	rabbits, hares, and pikas	1	(.2)
Sylvilagus idahoensis	pygmy rabbit	13	(2.6)
Lepus sp.	jackrabbits	1	(.2)
Indeterminate rodent	mice, voles, rats	5	(1.0)
Marmota cf. *flaviventris*	yellow bellied marmot	24	(4.8)
Spermophilus sp.[1]	ground squirrels	19	(3.8)
Thomomys cf. *talpoides*	northern pocket gopher	12	(2.4)
Perognathus cf. *parvus*	Great Basin pocket mouse	1	(.2)
Castor canadensis	beaver	3	(.6)
Peromyscus cf. *maniculatus*	meadow mouse	2	(.4)
Neotoma cf. *cinerea*	wood rat	2	(.4)
Microtus cf. *montanus*	meadow vole	7	(1.4)
Ondatra zibethica	muskrat	6	(1.2)
Vulpes cf. *fulva*	red fox	1	(.2)
Taxidea cf. *taxus*	badger	17	(3.4)
Mephitis cf. *mephitis*	striped skunk	20	(4.0)
Lutra canadensis	river otter	2	(.4)
Large Artiodactyl	bison, elk	99	(19.6)
cf. *Cervus* sp.	elk	5	(1.0)
Cervus cf. *canadensis*	elk	30	(6.0)
Odocoileus-size	deer	2	(.4)
Odocoileus sp.	deer	7	(1.4)
cf. *Bison* sp.	bison	31	(6.2)
Bison bison	bison	158	(31.3)
TOTALS[2]		504	(100.3)

[1] Specimens are assigned to genus only in Appendix E; however, elements representing *Spermophilus columbianus* and *S. washingtoni* were recovered.
[2] Percentage total does not equal 100.0 due to rounding.

Table 10.19 Total number (n) and percentage of identified specimens (NISP), per-taxon, from pre-Mazama contexts at Marmes Rrockshelter and floodplain (45FR50), Lind Coulee (45GR97), Granite Point (45WT41), and 45WT2.

TAXON	45FR50 NISP	%	45GR97 NISP	%	45WT41 NISP	%	45WT2 NISP	%
AMPHIBIANS								
cf. *Hyla* sp.	7	(.5)	--	--	--	--	--	--
cf. *Rana pipiens*	5	(.3)	--	--	--	--	--	--
REPTILES[1]								
Chelonia	--	--	1	(.2)	--	--	--	--
Squamata	--	--	1	(.2)	--	--	--	--
Unidentified snakes	13	--	--	--	--	--	--	--
cf. *Pituophis* sp.	26	--	--	--	--	--	--	--
cf. *Thamnophis* sp.	28	--	--	--	--	--	--	--
cf. *Crotalus* sp.	16	--	--	--	--	--	--	--
cf. *Sceloporus* sp.	1	--	--	--	--	--	--	--
BIRDS								
Bucephala clangula	3	(.2)	--	--	--	--	--	--
Ptychoramphus aleuticum	2	(.1)	--	--	--	--	--	--
Empidonax sp.	1	(.1)	--	--	--	--	--	--
Pica pica	1	(.1)	--	--	--	--	--	--
Hylocichla sp.	2	(.1)	--	--	--	--	--	--
Sialia sp.	3	(.2)	--	--	--	--	--	--
Branta-size	--	--	6	(1.2)	--	--	--	--
Branta cf. *bernicla*	--	--	2	(.4)	--	--	--	--
Anas-size	--	--	4	(.8)	--	--	--	--
Anas platyrhynchos	--	--	6	(1.2)	--	--	--	--
Anas carolinensis	--	--	2	(.4)	--	--	--	--
Ardea-size	--	--	3	(.6)	--	--	--	--
Ardea herodias	--	--	2	(.4)	--	--	--	--
Turdus-size	--	--	2	(.4)	--	--	--	--
MAMMALS								
Lagomorph size	12	(.8)	--	--	--	--	9	(11.3)
Unidentified Lagomorph	8	(.5)	1	(.2)	--	--	5	(6.3)
Sylvilagus sp.	26	(1.7)	13	(2.6)	10	(3.1)	--	--
Lepus sp.	45	(3.0)	1	(.2)	26	(8.0)	--	--
Marmota cf. *flaviventris*	23	(1.5)	24	(4.8)	4	(1.2)	4	(5.0)
Spermophilus sp.	8	(.5)	19	(3.8)	--	--	--	--
Spermophilus cf. *columbianus*	9	(.6)	--	--	43	(13.2)	--	--
Spermophilus cf. *washingtoni*	2	(.1)	--	--	--	--	4	(5.0)
Spermophilus cf. *townsendii*	2	(.1)	--	--	--	--	--	--
Thomomys cf. *talpoides*	208	(13.7)	12	(2.4)	51	(15.7)	15	(18.8)
Cricetidae-size	84	(5.5)	5	(1.0)	--	--	--	--
cf. *Perognathus* sp.	29	(1.9)	--	--	--	--	--	--
Perognathus cf. *parvus*	83	(5.5)	1	(.2)	--	--	--	--
Castor canadensis	1	(.1)	3	(.6)	--	--	--	--
Reithrodontomys sp.	4	(.3)	--	--	--	--	--	--
Peromyscus sp.	61	(4.0)	--	--	--	--	--	--
Peromyscus cf. *maniculatus*	6	(.4)	2	(.4)	4	(1.2)	--	--

Table 10.19 Total number (n) and percentage of identified specimens (NISP), per-taxon, from pre-Mazama contexts at Marmes Rrockshelter and floodplain (45FR50), Lind Coulee (45GR97), Granite Point (45WT41), and 45WT2.

TAXON	45FR50		45GR97		45WT41		45WT2	
	NISP	%	NISP	%	NISP	%	NISP	%
Onychomys cf. *leucogaster*	2	(.1)	--	--	--	--	--	--
Neotoma sp.	16	(1.1)	--	--	--	--	--	--
Neotoma cf. *cinerea*	4	(.3)	2	(.4)	1	(.3)	--	--
cf. *Clethrionomys* sp. (?)	1	(.1)	--	--	--	--	--	--
cf. *Phenacomys* sp. (?)	3	(.2)	--	--	--	--	--	--
cf. *Microtus* sp.	3	(.2)	--	--	--	--	--	--
Microtus sp.	112	(7.4)	--	--	--	--	--	--
Microtus cf. *montanus*	1	(.1)	7	(1.4)	2	(.6)	--	--
Microtus pennsylvannicus	3	(.2)	--	--	--	--	--	--
Microtus oeconomus	1	(.1)	--	--	--	--	--	--
Lagurus curtatus	4	(.3)	--	--	1	(.3)	--	--
cf. *Ondatra* sp.	7	(.5)	6	(1.2)	--	--	--	--
Canid	10	(.7)	--	--	--	--	--	--
Canis sp.	2	(.1)	--	--	--	--	--	--
Canis latrans	23	(1.5)	--	--	15	(4.6)	1	(1.3)
Canis familiaris	1	(.1)	--	--	--	--	--	--
Unid. fox.	28	(1.8)	--	--	--	--	--	--
Alopex lagopus	6	(.4)	--	--	--	--	--	--
Vulpes cf. *fulva*	20	(1.3)	1	(.2)	1	(.3)	--	--
Mustela-size	1	(.1)	--	--	--	--	--	--
Martes americana	3	(.2)	--	--	--	--	--	--
Mustela frenata	3	(.2)	--	--	--	--	--	--
Taxidea cf. *taxus*	1	(.1)	17	(3.4)	1	(.3)	--	--
Mephitis cf. *mephitis*	--	--	20	(4.0)	--	--	--	--
Lutra canadensis	--	--	2	(.4)	--	--	--	--
cf. *Lynx rufus*	2	(.1)	--	--	--	--	--	--
Lynx canadensis	1	(.1)	--	--	--	--	--	--
Medium/Large Artiodactyl	354	(23.3)	100	(20.0)	--	--	24	(30.0)
cf. *Cervus* sp.	--	--	5	(1.0)	--	--	--	--
Cervus canadensis	43	(2.8)	30	(6.0)	59	(18.2)	4	(5.0)
Odocoileus sp.	128	(8.4)	7	(1.4)	103	(31.7)	13	(16.3)
Antilocapra americana	96	(6.3)	--	--	2	--	--	--
cf. *Bison* sp.	--	--	31	(6.2)	--	--	--	--
Bison bison	--	--	158	(31.3)	--	--	--	--
Ovis sp.	--	--	--	--	1	(.3)	--	--
TOTALS[2]	1521	(99.9)	504	(100.3)	325	(99.9)	80	(99.0)

[1]Reptile values represent number of units containing reptile remains and are not included in the NISP total calculations.

[2]Percentage totals do not equal 100.0 due to rounding.

Coulee was apparently the only site situated in an environment capable of sustaining relatively large numbers of bison around 9,000 years ago. A diversity of carnivore and rodent taxa, coupled with a preponderance of pronghorn bones, distinguish the Marmes fauna from not only Lind Coulee, but also Granite Point and 45WT2. In addition, the higher frequency minimally identifiable specimens assigned to the Medium/Large Artiodactyl also suggest that animal processing was a more important activity at the Marmes and Lind Coulee sites (see Bone Breakage and Condition). Granite Point contained the greatest percentage of elk remains followed by Lind Coulee, whereas, elk formed a minor component of the pre-Mazama faunas at Marmes and 45WT2.

The post-Mazama fauna from Marmes Rockshelter units III, IV-V, and VI-VIII is combined and compared to the faunas from the Alpowa Locality (Lyman 1976:33-56), the Upper Landing site in lower Hells Canyon (Reid et al. 1991:373-432), Granite Point, and 45WT2 in Table 10.20. The Alpowa Locality is located near the mouth of Alpowa Creek below Clarkston and the Upper Landing site is located near the mouth of Klopton Creek in Lower Hells Canyon. Both sites are situated along the Snake River. Faunal identifications from the "Pig Farm" (45AS78) were excluded from the Alpowa data because we were unable to determine if specimens from the pre-Mazama early Cascade component of the site were included in the NISP tabulations. Most of the identified specimens from the Alpowa Locality rested in the Harder Phase houses at Timothy's Village (45AS82) and 45AS80. However, charcoal from House 5 at Timothy's Village was dated to 4060 ± 130 B.P. suggesting that the house was abandoned during the late Cascade subphase. In addition, two semisubterranean Tucannon phase house pits rested above the House 5 floor, but beneath a Harder phase house pit (82-2A) that yielded charcoal dated to 1910 ± 80 B.P. According to Reid et al. (1991:368), occupation of the Upper Landing site (10IH1017) began between 5,500 and 5,000 years ago during the Late Cascade and/or early- Tucannon phases as defined by Leonhardy and Rice (1970). Deer and deer-size animals are the most common faunal remains found in the oldest portions of the site. Rabbit and other small mammal remains are rare in the earliest occupation. The site likely functioned as an intermittent hunting camp until between 1700 and 1600 B.P. when people constructed four semisubterranean Harder phase houses. Bone and antler artifact frequencies increased dramatically after the site began to function as a residential base. House renovation/construction ceased sometime shortly after 1000 B.P. and was likely abandoned around 500 years ago. As with the Alpowa Locality, Lee Lyman (in Reid et al. 1991:373-432) analyzed the Upper Landing site fauna.

The excavated post-Mazama portion of the Marmes fauna is restricted to the rockshelter, thus far fewer taxa are listed in Table 10.20 due to the absence of the earlier floodplain fauna. Pronghorn are the most common artiodactyl at the Marmes site in post-Mazama contexts and they also increased in frequency at Granite Point. Whether the increased proportion of pronghorn remains at these sites suggests differences in climate or hunting techniques (communal hunting?) is unclear. Except for Marmes Rockshelter, deer were likely the most important economic taxon at most other sites. Elk remains are relatively rare at most sites, and bighorn sheep are more abundant at the Alpowa Locality and Upper Landing sites where more suitable habitat presumably existed.

The percentage of minimally identifiable "Medium/Large Artiodactyl" bone fragments increases dramatically compared to the pre-Mazama faunas. This suggests that deer-to-elk-size animal processing was more intensive during post-Mazama times. The "Medium/Large Artiodactyl category represents the greatest percentage of identifications for all sites listed in Table 10.20 except for Granite Point. The Period II (3800 to 4400 B.P.) and Period III (2200 to 3300 B.P.) year-round pithouse camps at Wells Reservoir (Chatters 1986:Table 24) contain considerably higher proportions of "Unidentified" mammal bone fragments compared to earlier occupations and summer or winter camps. Perhaps these fragments indicate that fat/grease rendering and marrow extraction became routine tasks at residential Tucannon and Harder phase sites along the lower Snake River as well as at Period II and III residential sites in and near Wells Reservoir. This could possibly account for the dominance of the many small and minimally identifiable fragments in the "Medium/Large Artiodactyl" categories at such sites.

Coyotes are the only carnivore present at Marmes rockshelter during this time, whereas, wolves (*Canis lupus*), red foxes (*Vulpes fulva*), raccoons (*Procyon lotor*), and smaller cats (*Lynx* sp.) are present at Alpowa and Upper Landing.

Table 10.20 Total number (n) and percentage of identified specimens (NISP) per-taxon from post-Mazama contexts at Marmes Rockshelter (45FR50), the Alpowa Locality (45AS80/45AS82), Granite Point (45WT41), Upper Landing 10IH017) and 45WT2. Adapted from Gustafson (1972:Tables 5.1, 5.2, 5.3), Lyman (1976:33-56) and Reid et al. (1991:373-432).

TAXON	45FR50 NISP	%	45AS80/45AS82 NISP	%	10IH1017 NISP	%	45WT41 NISP	%	45WT2 NISP	%
AMPHIBIANS										
Rana catesbeiana	--	--	--	--	14	(.7)	--	--	--	--
REPTILES[1]										
Indeterminate snake	--	--	--	--	46	--	--	--	--	--
BIRDS										
Pediocetes phasianellus	--	--	15	(.6)	--	--	--	--	--	--
Bonasa umbellus	--	--	--	--	1	(.1)	--	--	--	--
Dendragapus obscurus	--	--	4	(.2)	--	--	--	--	--	--
Anas platyrhynchos	--	--	3	(.1)	--	--	--	--	--	--
Turdus migratorius	--	--	1	(<.1)	--	--	--	--	--	--
Colaptes cafer	--	--	1	(<.1)	--	--	--	--	--	--
Falco sparverius	--	--	1	(<.1)	--	--	--	--	--	--
Bubo virginianus	--	--	21	(.9)	--	--	--	--	--	--
Dryocopus pileatus	--	--	--	--	1	(.1)	--	--	--	--
Indeterminate Bird	--	--	57	(2.4)	18	(.9)	--	--	--	--
MAMMALS										
Lagomorph size	22	(5.0)	--	--	--	--	--	--	5	(2.1)
Unidentified Lagomorph	--	--	19	(.8)	--	--	--	--	--	--
Sylvilagus sp.	26	(5.9)	--	--	145	(7.5)	--	--	--	--
Sylvilagus nuttallii	--	--	62	(2.8)	--	--	--	--	--	--
Lepus sp.	5	(1.1)	16	(.7)	1	(.1)	1	(.4)	--	--
Marmota cf. *flaviventris*	6	(1.4)	--	--	--	--	5	(1.9)	--	--
Spermophilus sp.	--	--	6	(.3)	--	--	14	(5.2)	--	--
Spermophilus cf. *columbianus*	28	(6.3)	--	--	--	--	--	--	61	(25.6)
Spermophilus cf. *washingtoni*	25	(5.6)	--	--	--	--	--	--	--	--
Cricetidae-size	--	--	--	--	1	(.1)	--	--	--	--
Thomomys sp.	7	(1.6)	311	(13.9)	1	(.1)	86	(32.2)	22	(9.2)
Thomomys cf. *talpoides*	--	--	8	(.4)	1	(.1)	1	(.4)	--	--
Castor canadensis	--	--	--	--	--	--	--	--	--	--
Perognathus parvus	--	--	5	(.2)	--	--	3	(1.12)	--	--
Peromyscus cf. *maniculatus*	--	--	3	(.1)	--	--	--	--	--	--
Onychomys cf. *leucogaster*	--	--	--	--	4	(.2)	--	--	--	--
Neotoma sp.	--	--	--	--	8	(.4)	--	--	--	--
Neotoma cf. *cinerea*	8	(1.8)	--	--	--	--	6	(2.2)	--	--

Table 10.20 Total number (n) and percentage of identified specimens (NISP) per-taxon from post-Mazama contexts at Marmes Rockshelter (45FR50), the Alpowa Locality (45AS80/45AS82), Granite Point (45WT41), Upper Landing 10IH017) and 45WT2. Adapted from Gustafson (1972:Tables 5.1, 5.2, 5.3), Lyman (1976:33-56) and Reid et al. (1991:373-432).

TAXON	45FR50 NISP	%	45AS80/45AS82 NISP	%	10IH1017 NISP	%	45WT41 NISP	%	45WT2 NISP	%
Microtinae	--	--	--	--	16	(.8)	--	--	--	--
Microtus sp.	--	--	41	(1.8)	--	--	--	--	--	--
Microtus cf. *montanus*	--	--	--	--	8	(.4)	1	(.4)	--	--
Microtus longicaudus	--	--	--	--	6	(.3)	--	--	--	--
Microtus richardsoni	--	--	--	--	7	(.4)	--	--	--	--
Lagurus curtatus	--	--	98	(4.4)	--	--	--	--	--	--
Ondatra zibethica	3	(.7)	1	(<.1)	--	--	--	--	3	(1.3)
Canis sp.	--	--	29	(1.3)	18	(.9)	--	--	--	--
Canis latrans	8	(1.8)	11	(.5)	--	--	10	(3.7)	1	(.4)
Canis lupus	--	--	1	(<.1)	1	(.1)	--	--	--	--
Canis familiaris	--	--	3	(.1)	1	(.1)	--	--	--	--
Vulpes cf. *fulva*	--	--	4	(.2)	6	(.3)	--	--	--	--
Ursus americanus	--	--	--	--	2	(.1)	--	--	--	--
Procyon lotor	--	--	3	(.1)	1	(.1)	--	--	--	--
Martes pennanti	--	--	--	--	3	(.2)	--	--	--	--
Mephitis mephitis	--	--	7	(.3)	--	--	--	--	--	--
Taxidea cf. *taxus*	--	--	6	(.3)	--	--	--	--	--	--
Lutra canadensis	--	--	5	(.2)	6	(.3)	--	--	--	--
Lynx sp.	--	--	--	--	--	--	--	--	--	--
cf. *Lynx rufus*	--	--	22	(1.0)	--	--	3	(1.1)	--	--
Carnivore	--	--	--	--	--	--	--	--	--	--
Medium/Large Artiodactyl	147	(33.2)	616	(27.4)	1062	(54.9)	--	--	76	(31.9)
Cervidae	--	--	--	--	6	(.3)	--	--	--	--
Cervus canadensis	10	(2.3)	43	(1.9)	6	(.3)	15	(5.6)	4	(1.7)
Odocoileus sp.	75	(16.9)	534	(23.8)	475	(24.6)	107	(40.0)	31	(13.0)
Odocoileus virginianus	--	--	1	(<.1)	--	--	--	--	--	--
Odocoileus hemionus	--	--	2	(.1)	--	--	--	--	--	--
Antilocapra americana	96	(21.6)	29	(1.3)	--	--	9	(3.4)	3	(1.3)
Bison bison	--	--	114	(5.1)	--	--	5	(1.9)	1	(.4)
Ovis sp.	--	--	--	--	--	--	1	(.4)	--	--
Ovis canadensis.	--	--	--	--	--	--	--	--	--	--
TOTALS[2]	444	(99.2)	2245	(99.6)	1932	(100.3)	267	(99.7)	238	(99.9)

[1]Reptile remains are not included in the NISP totals
[2]Percentage totals do not equal 100.0 due to rounding

Carnivore remains seem positively correlated with the house pit sites (e.g., 45AS80/45AS82 and 10IH1017 in Table 10.20) and this suggests that the function of Marmes Rockshelter changed after the eruption of Mt. Mazama at 6730 B.P. (Hallett et al. 1997). The numerous storage pits encountered in the upper portions of Marmes Rockshelter perhaps indicate that the shelter functioned as a food storage locality rather than a residential site during the Tucannon and Harder phases.

Jackrabbits (*Lepus* sp.) were the most numerous leporid in the pre-Mazama components presented in Table 10.19. This trend reverses after the eruption of Mt. Mazama when cottontails (*Sylvilagus* sp.) become the dominant leporid at all sites listed in Table 10.20. Changes in hunting techniques or mobility patterns may explain this shift.

Most of the year, highly mobile Archaic nuclear families likely hunted individual small mammals. Seasonal resource abundance, seasonal anadromous fish runs on the Plateau or fall pinyon nut harvests in the Great Basin for example, allowed several nuclear families to aggregate. At such times, a sufficient number of people made communal net hunting effective and possible as recorded in ethnographic accounts. This scenario is well documented for the Northern Paiute (Couture et al. 1986:Figure 2; Fowler and Liljeblad 1986:82-83; Whiting 1950:19) and Western Shoshone (Steward 1938:122, 176). These Numic groups routinely held communal rabbit drives near their fall camps where abundant seasonal plant resources allowed several families to stay in one place in November when pelts were in their prime. Drive captains—an achieved rather than inherited role—directed the placement of large nets (> 100 m in length) making semicircular or V-shaped enclosures in the brush filled valley bottoms. Men, women and children then systematically drove large numbers of jackrabbits and occasionally cottontails, an important distinction, into the net(s). Szuter (1991) synthesized ethnographic accounts of similar activities for many of the Formative groups occupying the low deserts of Arizona. However, we believe that the dominance of cottontail bones in post-Mazama sediments/contexts in this analysis is the result of prehistoric hunters dispatching more of cottontails on an encounter basis compared to jackrabbits.

It is possible that the increased sedentism and likely larger populations of the Tucannon and Harder phases likely made communal hunting easier to carry out. Scheduling and manpower would be less problematic. Subsequently, the annual number of communal hunts could be increased. This hypothesized shift in the frequency of net hunting should have an archaeological signature.

If residentially oriented Tucannon and Harder Phase people engaged in communal net hunts more frequently than their predecessors, then jackrabbit remains should dominate the small mammal component of faunal assemblages post-dating the Cascade phase. Jackrabbits rely on speed (up to ca. 40 km/hour) and distance to avoid predators; whereas, cottontails rely on cover and rarely flee further than 30 m when flushed. These differences are related to habitat preferences, which are rooted in the ontogeny and evolutionary history of these animals.

Female jackrabbits give birth to precocial offspring in open fur-lined hollows. The young are hopping about in a few days—an adaptation to open, arid environments. Subsequently, juvenile jackrabbits are better able to avoid predators in open terrain compared to the altricial offspring of their cottontail cousins. However, this rapid development and reliance on speed makes them more susceptible to net hunting compared to cottontails. Cottontails prefer rocky hill and canyon country where they rest in rocky crevices and thick brush during the day. Cottontail newborns are very small, somewhat helpless, and require cover for protection and to enable several weeks growth before leaving the "nest." These attributes make cottontails an easier prey to capture for individual hunters. Hence, it is likely that individual hunters would dispatch more juvenile cottontails than jackrabbits throughout the year; the cottontail's proclivity for hiding versus running representing another asset for individual hunters.

Summary and Conclusions

Taxa identified in both the floodplain and rockshelter deposits studied at the Marmes site are quite different, but most occur in the vicinity today. Remains of exotic taxa, including Arctic Fox and pine marten (Figure 10.5), were present in the early floodplain deposits, but have not been identified from the rockshelter. Overall, faunal remains from the rockshelter at the Marmes site differ significantly from those on the floodplain, both in their quantitative taxonomic characteristics and in the condition of the remains.

Figure 10.6 shows a typical array of faunal remains from the rockshelter compared with one

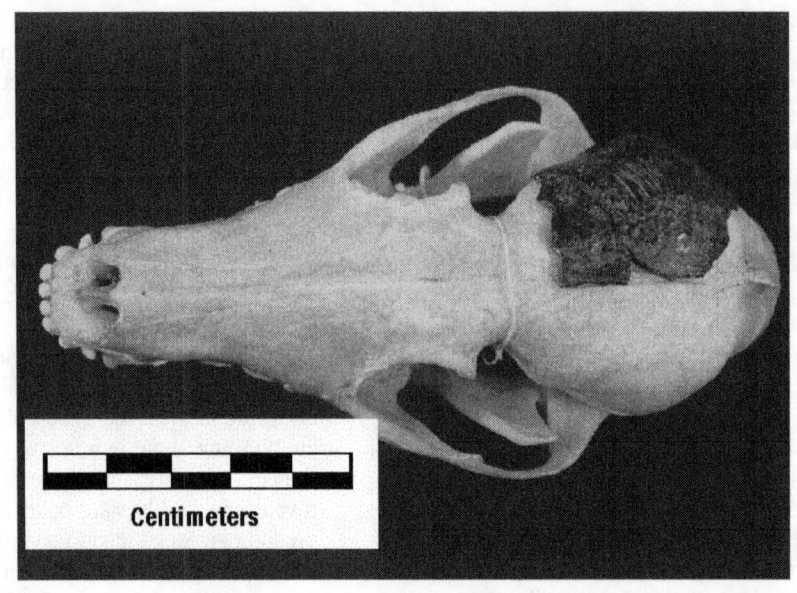

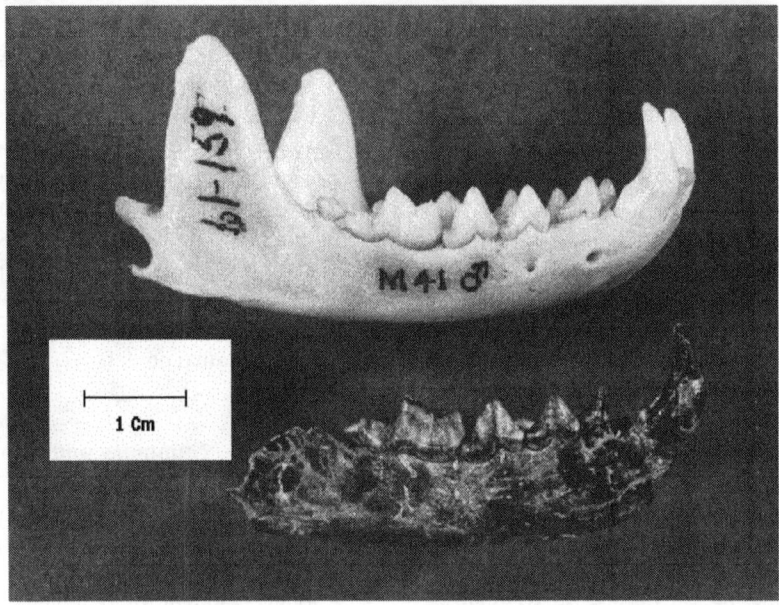

Figure 10.5 Computer scanned images of modern Arctic fox skull with blackened fragment of Marmes floodplain specimen (top). Modern and Marmes jaws of Pine marten (below). Both images are from Gustafson (1972).

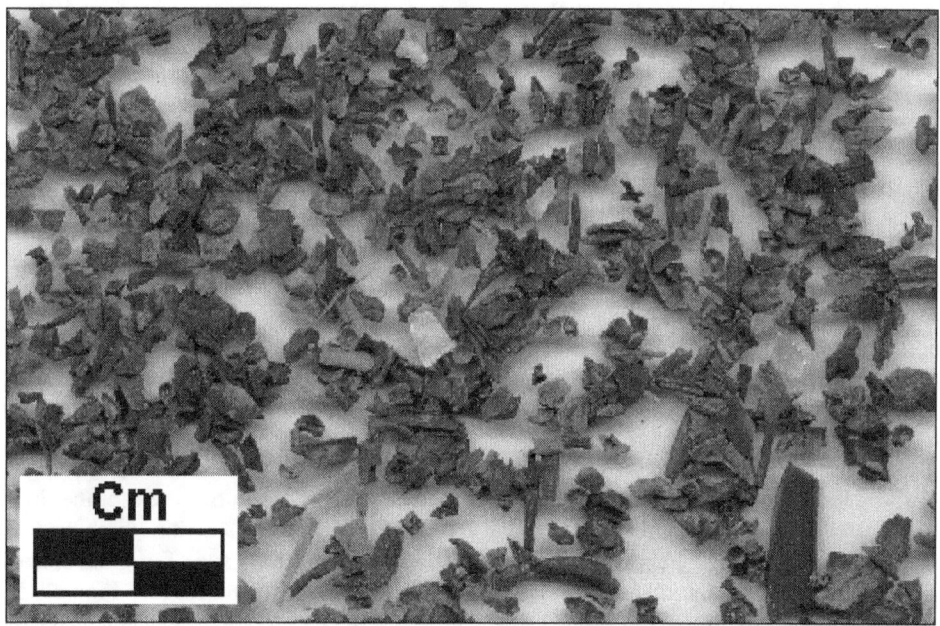

Figure 10.6 Computer scanned images of typical bones from the Marmes Rockshelter (above) compared with a sample of floodplain faunal remains below. Upper photo is modified from Gustafson (1972).

from the early floodplain deposits. Bones, fragments and teeth from the rockshelter are much larger, relatively few are burned, many exhibit spiral fractures and other evidence of butchery. Also most are from deer, pronghorn, elk, rabbits and other animals of known ethnographic and prehistoric importance. Conversely, the vast majority of faunal remains from the floodplain are very small (< 15 mm), most are burned, and few exhibit spiral fractures or evidence of butchery other than being broken into tiny fragments. Deer, elk, and pronghorn are represented in very small numbers on the floodplain (see Table 10.7), but they comprise the largest component of the rockshelter sample (see Table 10.5). Remains of snakes and rodents form the bulk of the identified floodplain fauna. Also, exotic taxa, preferring conditions more mesic than that at Marmes today, have been identified only from the floodplain deposits.

Pocket gopher remains are abundant, 213 of the 1094 specimens identified to taxon in the floodplain deposits (see Table 10.7), but were relatively rare in the rockshelter where they total only 11 of 1013 specimens identified to taxon (see Table 10.5). The number of ground squirrels is about the same in both areas—22 on the floodplain and 30 in the rockshelter. The difference may have something to do with the pocket gopher's preference for roots and tubers growing in open areas and its need for an extensive burrow system. A very high proportion of pocket gophers also occurred at the Granite Point site, 45WT41 (Leonhardy 1968, 1970), which also is an open site (see Table 10.17).

Interpreting the significance of the rockshelter fauna is fairly straightforward. Most rockshelter remains (77.2% of the NISP total, see Table 10.5) are from game animals such as deer, pronghorn and elk, and most long bones were broken above and below the joint in the fashion considered typical of prehistoric butchery in eastern Washington. Although this pattern had been postulated by Gustafson (unpublished) as early as 1966, he was not able to fully document it until 1968 while conducting research at a bison kill site (45GA17) reported by Schroedl (1973).

This typical butchery pattern has been demonstrated repeatedly by Gustafson and students since that time (Gustafson 1987a, 1987b, 1996; Lyman 1976, 1978). It results in intact joint ends where the shafts had been broken off directly above and below the joint. Carpals and tarsals often are complete and show no signs of cuts, scratches or other alterations. In the case of the lower forelimb, breakage occurs above the distal radio-ulna and below the proximal metacarpal. The carpals between are left intact. The lower hind limb is dismembered in the same fashion with breakage above the distal tibia and below the proximal metatarsal leaving the intervening tarsals unaltered.

Although there are no quantitative data available for the 1962-1964 rockshelter bones, Gustafson observed the presence of joint ends resulting from breakage above and below joints; most carpals and tarsals were intact. Shattered shaft fragments, often exhibiting spiral fractures, and split or broken phalanges suggested that further processing, perhaps for marrow extraction, had occurred at the site.

The significance of the floodplain faunal assemblage is more perplexing. The paucity of identifiable game animal bones, numerous carnivore remains, the thousands of tiny charred fragments lacking spiral fractures, and the preponderance of snake and rodent remains stand in strong contrast to the rockshelter assemblage.

A few bones of a very large elk were among the first bones brought to Gustafson by Fryxell shortly after excavation of the 1965 backhoe trench. The elk remains associated with human bone fragments and artifacts from the initial backhoe discovery included a distal left tibia, left calcaneum, left astragalus, left naviculocuboid and a fragmentary left metatarsal with several fresh breaks. Because these were typical of remnants of butchered remains from the rockshelter and other sites on the lower Snake River, Gustafson concluded that the elk had been "clearly butchered" (Fryxell et al. 1968a:177). However, additional remains of this elk exposed in the 1968 floodplain excavations led him later to question the interpretation of butchery (Table 10.21).

Recovered remains of this large elk were concentrated in two areas of the floodplain (see Figure 1.18). Only hind limb bones were found *in situ* near the site of original discovery in the 1965 backhoe trench (see Figure 6.5). Cervical vertebrae, thoracic vertebrae and ribs were found less than 15 ft. away (see Figure 6.6 and Figure 10.7). Lumbar vertebrae, forelimbs and skull were not found.

Table 10.21 Spatial distribution of bone elements from large elk.

UNIT	ELK BONE	1965 Backhoe Trench	Feature 1	Silts Above Marmes Horizons	Marmes A_1 Horizon
15N/35W	8 Ribs or Rib Frags (7 with head)		+		+
15N/35W	3 Rib Frags (1 with head)		+		+
15N/35W	1 Rib Frag		+	?	?
15N/35W	6 Ribs or Rib Frags (4 with head)		+	?	?
15N/30W	1 Rib Frag with Head		+		
30N/20W	2 Rib Frags			+	
?	2 Rib Frags (1 with head)		?		
?	1 Rib Frag		?		
15N/35W	First Thoracic Vertebra		+	?	?
15N/35W	9 Thoracic Vertebrae		+	?	?
15N/35W	Atlas and Axis Vertebrae		+	?	?
15N/35W	3 Cervical Vertebrae		+	?	?
10N/W45	2 Cervical Vertebrae		+	?	?
15N/35W	Right Crest of Tibia[2]		+		+
30N/30W	Anterior Shaft Left Tibia[1]				+
?	Left Crest of Tibia[1]				
30N/25W	Frag Long Bone Shaft			?	?
25N/25W	5 Frags Long Bone Shaft			+	
25N/25W	Distal 1/2 Left Metatarsal			+	
25N/25W	Head of Femur			?	?
25N/25W	Distal Left Femur			+	
25N/25W	Proximal 1/2 Right Tibia (no crest)[2]			+	
25N/25W	Distal 1/2 Right Tibia[2]			+	
?	Distal Left Tibia[3,]	+			
?	Distal Left Fibula[3]	+			
?	Left Astragalus[3]	+			
?	Left Naviculocuboid[3]	+			
?	Left Calcaneum[4]	+			
?	Proximal Left Metatarsal with Broken Posterior Shaft[3,5]	+			
?	2 Metapodial Condyles[5]	+			

+ = location is certain; ? = location is uncertain.
[1]Shaft and crest are parts of the same left tibia.
[2]Proximal and distal halves fit together at old break.
[3]Distal tibia, fibula, astragalus, naviculocuboid and proximal metatarsal articulate. Astragalus was submitted for a ^{14}C date.
[4]Among first bones found *in situ* near base of 1965 backhoe trench.
[5]Left metatarsal and metapodial condyles probably are from the same bone.

Figure 10.7 Photograph of the elk bone feature in the south-central portion of the Marmes floodplain excavation area.

Unlike the majority of bones found on the floodplain, the elk remains were not burned and were found in silts on or very near the surface of the uppermost (Marmes A_1) soil horizon. Most of the ribs were whole, rather than broken into several pieces and the vertebrae were intact. Most bones of the hind limb were broken near mid-shaft and were scattered about. The distal end of the right metatarsal and a first phalanx had been pierced from both sides by carnivore canines. Although the placement and condition of the elk remains do not preclude butchery, they differ from the "typical" pattern, and carnivore activity clearly is indicated. An alternative explanation is that humans had separated the limb bones at the joints. Subsequently, they may have broken the bones at mid-shaft for marrow extraction and later smashed some of them into the tiny fragments found scattered about the floodplain. Later, carnivores may have scavenged the remains.

Environmental Implications

As stated earlier, the vast majority of animals represented in both the floodplain and rockshelter deposits studied at the Marmes site occur in the vicinity today. However, seven taxa (see Table 10.22) identified from the early floodplain deposits at the Marmes site suggest cooler or moister local habitats than previously recognized from eastern Washington archaeological sites.

Fryxell (Fryxell and Keel 1969) noted polygonal fractures that he interpreted as "frost cracks" in the Marmes early floodplain deposits, and his rockfall frequency data (Fryxell 1965; Fryxell et al. 1968a) show increased rockfall during that time suggesting greater freeze-thaw activity further suggesting cooler, moister conditions than those of today. Marshall (1971) found higher terraces containing larger particle sizes than today in the early floodplain deposits along the lower Palouse River. Both Fryxell and Marshall inferred a cooler, moister local climate at the time the Marmes floodplain alluvium was

Table 10.22 Habitat preferences of mammals identified from Marmes floodplain[1]. Includes additional taxa from Gustafson (1972) and Caulk (1988). Shading indicates the modern range of habitats for each taxon.

TAXON	ARID STEPPE	MESIC STEPPE	CONIFEROUS FOREST	RIVERINE	TUNDRA
Spermophilus washingtoni	+	&			
Onychomys leucogaster	+	&			
Reithrodontomys megalotis	+	&			
Antilocapra americana	+	&			
Perognathus parvus	+	+			
Lepus townsendii	+	+			
Sylvilagus nuttallii	+	+	&		
Marmota flaviventris	+	+	&		
Taxidea taxus	+	+	&		
Vulpes fulva	&	+	+		
Mustella frenata	+	+	+		
Neotoma cinerea	+	+	+		
Microtus spp.	+	+	+		
Peormyscus maniculatus	+	+	+		
Thomomys talpoides	+	+	+		
Canis latrans	+	+	+		
Lynx rufus	+	+	+		
Odocoileus spp.	+	+	+		
Cervus canadensis	+	+	+		
Ovis canadensis	+	+	+		
Spermophilus columbianus		+	+		
Martes americana			+		
Phenacomys intermedius			+		
Clethrionomys gapperi			+	&	
Lynx canadensis ?		&	+		
Ondatra zibethica				+	
Castor canadensis				+	
Microtus oeconomus ?					+
Alopex lagopus					+

[1] A "+" indicates preferred habitat, "&" indicates less optimal habitat.

deposited (ca. 10,000 B.P.). The presence on the Marmes floodplain of taxa adapted to cooler or moister conditions than today seems to support these data (Table 10.22). These more mesic adapted, northern taxa are rare in the sample and may have coexisted with a predominant fauna composed of extant taxa.

Gustafson (1972) identified only the Arctic fox (*Alopex lagopus*) and pine marten (*Martes americana*) as being truly exotic to the region. Caulk (1988) found additional remains of these species and added Canadian lynx (*Lynx canadensis*) and the tundra vole (*Microtus oeconomus*) as potential members of the early Holocene fauna at Marmes. In the present sample, we have identified tentatively a larger ground squirrel (*Spermophilus* cf. *columbianus*), the heather vole (*Phenacomys* cf. *intermedius*), and the red-backed mouse (*Clethrionomys* cf. *gapperi*). Members of all these taxa live today in environments cooler and/or moister than present (see Table 10.22).

Spermophilus columbianus inhabits the mesic steppe and lower forest zones of eastern Washington and adjacent Idaho. *Phenacomys intermedius* prefers the boreal forests of the Cascade Range and higher mountains of extreme northeastern and southeastern Washington (Dalquest 1948). *Cleithrionomys gapperi* occupies the same habitats as *Phenacomys intermedius*, but its range extends lower in the mountains and into adjacent chaparral (Dalquest 1948, Ingles 1965). Members of both genera possess rooted molars with a prismatic crown pattern. This trait readily distinguishes them from their ubiquitous microtine relatives that possess high crowned, rootless molars.

Unfortunately, these are rare specimens, and more remains with more reliable features would be necessary to confirm the identifications. Nonetheless, the fact that now there are several taxa implicated gives one more confidence in the interpretations that follow. It is common knowledge among Quaternary vertebrate paleontologists that late Pleistocene/early-Holocene faunal assemblages often contained extinct taxa and both northern (cool summers) and southern (mild winters) faunal elements superimposed on the extant fauna (various authors in Hibbard et al. 1965; Martin and Wright 1967; Martin and Klein 1984). Rensberger et al. (1984) found remains of extinct mammals, northern faunal elements, southern faunal elements and modern fauna in Pleistocene deposits at the Kennewick Roadcut south of Kennewick, Washington.

The northern elements found in the early Marmes floodplain deposits, suggest that summers might have been cooler there around 10,000 B.P. Lyman (1986) has calculated that lowering temperatures in the Marmes vicinity by as little as 1° C could have provided marginal conditions for Arctic Fox migrations into the region. Although negative evidence, there is some suggestion that winters may not have been milder, or no more severe than today; we found no southern faunal elements represented in the floodplain assemblage, and arid steppe fauna makes up the bulk of the remains.

Still, the prevalence of arid steppe and grassland fauna precludes the presence of closed forest in the vicinity. However, judging from what is known of modern forests adjacent to steppe communities [Daubenmire and Daubenmire 1968; Daubenmire 1970], scattered forests may have existed within a steppe vegetation zone on north-facing slopes, in protected riparian areas, or at the base of talus slopes where effective precipitation is greater.

> . . . Just as closed forests could not have covered the Columbia Basin in postglacial time, neither could true tundra, for the only indicator of tundra conditions is the single skull fragment of Arctic Fox. No other tundra dwellers have been found in archaeological sites or natural sediments in southeastern Washington (Gustafson 1972:130-131).

Based on the Marmes fauna, a cold, sagebrush steppe environment with local patches of forest would seem to fit the data best for early Holocene time. There is little in the faunal record of eastern Washington to support the geological and botanical evidence of a period of mid-Holocene drought; however, the fauna may not have been affected as directly or animals may not be as sensitive to change from arid to more arid conditions.

Gustafson (1972:34) has summarized the inferred climatic scheme as interpreted by Fryxell (1965, 1968a) and Marshall (1971) as cool and moist before about 8,000 B. P. Fryxell also inferred a climate warmer and drier than today between about 7,000 and 4,500 B. P. Only extant taxa have been identified during the latter period in eastern Washington; i.e., to date there are no known qualitative differences from extant fauna at any eastern Washington archaeological site. However, there are some quantitative changes that may reflect a warming trend beginning at about the time of deposition of the Mazama volcanic ash (ca. 6,800 B. P.). These changes were alluded to earlier in the discussion of Table 10.12. Gustafson (1972) noted a slight increase in pronghorn remains and corresponding decrease in elk remains at both Marmes Rockshelter and the Granite Point site after about 6,800 B.P. (Table 10.23).

Gustafson (unpublished) believes that the complex mosaic of habitats, topography and soils in eastern Washington have allowed the smaller mammals that are more sensitive to environmental change simply to shift position in the landscape as climate and environments have fluctuated. For example, during a warmer, drier period animals living on a south-facing slope could shift to a north-facing slope where effective precipitation is greater. Under cooler conditions, animals occupying north-slope habitats with deep, loamy soils might remain on that slope in places where gravely soils could support a flora more like that present under the previous drier regime. Thus, mid-postglacial climatic fluctuations from hot/dry to warm/dry may not be reflected qualitatively in archaeological faunal remains.

Table 10.23 Pre-Mazama and post-Mazama ratios of pronghorn to elk at Marmes Rockshelter and Granite Point.

	Pronghorn/Elk	
	Marmes Rockshelter	Granite Point
Post-6,800	9.6/1	0.6/1
Pre-6,800	2.3/1	0.3/1

11
FISH REMAINS

Virginia L. Butler

Several regional syntheses (e.g., Ames et al. 1998; Schalk and Cleveland 1983) have noted the presence of fish remains at Marmes Rockshelter. However, until the current project, fish bones from the site had not been studied in any detail. Given that many questions about regional human subsistence change turn on establishing the relative importance of fish vs. terrestrial mammals, the changing role of salmon, and whether Holocene environmental change affected salmon abundance and distribution, studying the fish remains from this early site is extremely worthwhile. The recovery of fish remains from the Marmes Rockshelter and adjacent floodplain provides an opportunity to explore some of these questions. The remains described represent some of the earliest, well documented archaeological fish remains from the Columbia Plateau and as such, provide a very important record of fishes in regional streams as well as human resource use patterns.

Methods and Materials

The fish remains collected from the rockshelter and floodplain areas of the site were sampled in very different ways. Remains from the rockshelter were recovered during all four of the 1960s excavation field seasons. Excavated matrix was sieved dry, through ¼-inch mesh screens. However, during 1968, sediments in the upper strata were quickly removed and not screened and faunal recovery was limited. Given these recovery concerns as well as time and budgetary constraints, only a sample of the fish remains from the rockshelter were selected for study (see Chapter Two, *Sampling* for description of the sample).

At the floodplain locale, detailed excavation focused only on the Late Pleistocene/early Holocene sediments. Larger artifacts and faunal items recovered by the excavators were bagged separately, but most of the excavated sediment was water screened through 1-mm mesh. The screen residue retained from each unit level was dried and bagged as a bulk sample. For the current study, many of the bulk samples were sorted (Gustafson and Wegener 1998) and found to include fish remains along with other cultural material.

Stratigraphic unit information was used in this analysis where available. In many cases, however, field assignments were not made. Using depth, spatial coordinates and redrawn stratigraphic profiles for reference, Hicks assigned these materials to stratigraphic units and suggests these represent only provisional records of vertical position because they are not based on field observations (see Chapter Two).

Remains were identified to the finest taxonomic level using Butler's comparative collections from the Columbia Basin and adjacent regions in western North America. Because of their very small size, most of the floodplain faunal remains were identified using low power magnification (10-20x). Fish remains were quantified using Number of Identified Specimens (NISP) (Grayson 1984). For Cypriniformes (minnow and sucker) vertebrae, the first two and last vertebrae on the column were distinguished; definition of abdominal and caudal vertebrae, which represent most vertebrae on the column, follows Wheeler and Jones (1989). Salmonid vertebrae were assigned to four categories based on morphological difference associated with position on column (Butler 1993).

To estimate variation in body size of fishes represented, the width (measure labeled as diameter, shown in Casteel (1976:84) was measured using digital calipers or a micrometer within the microscope.

All of the specimens were examined for evidence of burning. For several of the specimens from the floodplain, it was difficult to determine whether a dark color reflected burning or staining that might result from absorption of minerals in the surrounding matrix. The decision to call a specimen burned was based on conservative criteria: only those specimens which were uniformly black or calcined (white/blueish cast) were called burned.

Descriptive Summary of Fish Remains

The following information describes the criteria used in assigning the specimens to taxonomic category as well as information on ecology and habitats. Descriptions are provided for each order of fishes identified. Appendix N presents the analysis data.

Class Osteichthyes - Bony Fishes.

Order Acipenseriformes
Family Acipenseridae - sturgeons
Acipenser sp. – sturgeon
Materials: *Rockshelter*: 1 scute.

Remarks: Two species of sturgeon are known for western North America, *A. medirostris* (green sturgeon) and *A. transmontanus* (white sturgeon). Green sturgeon are known today only from brackish waters of the lower 40 miles of the Columbia and fully marine environments, while white sturgeon are documented throughout the river system, including the upper Columbia and Snake Rivers. Based on habitat preference and modern distribution, the scute (or bony body scale) is most probably from the white sturgeon.

Order Salmoniformes
Family Salmonidae—Salmons, Trouts, and Whitefish
Materials: *Floodplain*: 1 indeterminate vertebra type, 2 vertebra type-3.

Oncorhynchus sp. – Salmon and Trout
Materials: *Rockshelter*: 1 basipterygium, 2 vertebra type-1 or atlas, 13 vertebra type-2, 10 vertebra type-3, 2 vertebra type-4, 1 indeterminate vertebra type, 65 vertebra fragments: 94 specimens.
Floodplain: 2 gill rakers, 10 vertebra fragments: 12 specimens

Prosopium williamsoni—Mountain Whitefish
Materials: *Floodplain*: 1 vertebra type-3

Remarks: There are ten species of salmon, trout and whitefish with records for the Columbia basin upriver of the estuary. The genus *Oncorhynchus* is represented by six species of anadromous and resident forms. The *Oncorhynchus* specimens from the site were assigned based on their large size, or, for the vertebrae, distinctive shape and fenestration pattern. A single vertebra was assigned to *Prosopium williamsoni*; the centrum of this species bears a series of horizontal struts, rather than the fenestrations characteristic of *Oncorhynchus*. A second species of *Prosopium* known for the Columbia basin, *P. coulteri* (Pygmy whitefish), reaches much smaller adult size than *P. williamsoni*. The archaeological specimen was assigned to *P. williamsoni* based on its large size.

Two vertebrae (inventory numbers 18987 and 18953) of very similar shape and size could not be identified below the family level. They are from a small salmonid (vertebra diameter is about 1.5 mm). As an indication of body size, a modern fish of *Salvelinus malma* (Dolly Varden) measuring 216 mm in total length has vertebra type-3 with widths that range between 2.5 and 3.0 mm, suggesting that the archaeological samples are from fish smaller than 200 mm. Both vertebrae have very large openings for the notochord (diameter of opening is about .5 mm), which is not characteristic of *Oncorhynchus* or *Prosopium*. The specimens most closely match *Salvelinus*, which includes the Bull trout, *S. confluentus*, and Dolly Varden, *S. malma*. However, the notochord opening for the archaeological specimens is larger than that found on available comparative specimens, and thus the materials were assigned to the family level.

Historically, the Snake River and several tributary rivers and streams provided extensive spawning habitat for migratory species of *Oncorhynchus*, particularly *O. tshawytscha* (chinook), *O. nerka* (sockeye), and *O. mykiss* (steelhead or rainbow trout) (Parkhurst 1950). Vast numbers of spring and summer runs of chinook used the Snake River upriver from the Palouse River as a passage way to spawning grounds in tributary rivers and streams; the fall run chinook used extensive spawning habitat in the main stem Snake River between the Palouse River and Hells Canyon (Fulton 1968). According to fishery documents, the Palouse River itself, however, was, at least historically, not used by spawning salmon (Fulton 1968). In his fisheries survey and historic overview of Snake River Basin anadromous fishes, Parkhurst (1950) did not survey the Palouse River. He notes that a high falls (Palouse Falls) located about 10 km above the mouth "renders the stream inaccessible to migratory fish" (1950:5).

Ethnographic records show that salmon fishing was extremely productive at the confluence of the Palouse and Snake Rivers (Ray 1975). Lewis and Clark and later explorers describe a very large village at the mouth of the

Palouse River. Ross Cox, who spent time in the village around 1812, noted that in early August, people there were engaged in catching and drying salmon in large numbers (Ray 1975). Historic documents also show fishing camps and villages along the Palouse River itself (Ray 1975). Ray refers to two such locales in particular: *A'patap*, which was located at the foot of Palouse Falls, and *Claxo'pa*, about four miles above the mouth. Unfortunately, the documents do not indicate whether the Native American fishery along the Palouse River targeted resident freshwater fishes or anadromous salmon and trout.

The falls have been a barrier to migratory salmonids at least since the late Pleistocene and Holocene periods, thus it is clear that the Palouse River itself was not a passageway for fish migrating to headwater areas to spawn. It is possible of course that the Palouse River channel below the falls was used by salmon for spawning and thus a fishery might have developed to target such fishes. One would need to evaluate the potential stream conditions (bed morphology, temperature, flow patterns) to establish whether the lower Palouse once provided suitable spawning habitat. For now, the pre-dam records for fish distribution lead to the conclusion that the Palouse River never supported large salmonid populations.

Order Cypriniformes
Family Cyprinidae—Minnows
Materials: *Rockshelter*: 1 basioccipital, 1 basiphenoid, 1 exoccipital, 1 interopercle, 4 endopterygoids, 3 metapterygoids, 3 opercles, 3 parasphenoids, 3 preopercles, 1 prootic, 1 pterosphenoid, 6 cleithra, 1 coracoid, 3 basiptergia, 11 pharyngeals, 5 1st vertebrae (atlas), 4 2nd vertebrae, 4 abdominal vertebrae: 56 specimens.
Floodplain: 3 articulars, 2 basioccipitals, 2 ceratohyals, 2 epihyals, 1 hyomandibula, 4 opercles, 1 parasphenoid, 1 pterotic, 2 quadrates, 1 urohyals, 6 cleithra, 97 pharyngeals and pharyngeal teeth, 17 1st vertebrae (atlas), 9 2nd vertebrae, 11 1st or 2nd vertebrae, 2 abdominal vertebra: 161 specimens.

Ptychocheilus oregonensis—Northern Pike Minnow
Materials: *Rockshelter*: 2 basioccitals, 4 ceratohyals, 1 dentary, 8 hyomandibulae, 2 maxillae, 1 palatine, 1 urohyal, 1 vomer, 9 pharyngeals: 29 specimens.
Floodplain: 5 dentaries, 3 pharyngeals: 8 specimens.

Acrocheilus alutaceus---Chiselmouth
Materials: *Rockshelter*: 1 dentary, 2 hyomandibulae, 1 prootic.
Floodplain: 1 dentary, 1 pharyngeal.

Mylocheilus caurinus—Peamouth
Materials: *Floodplain*: 5 pharyngeals.

Richardsonius balteatus—Redside shiner
Materials: *Floodplain*: 1 pharyngeal.

Rhinichthys sp.—Dace
Materials: *Floodplain*: 2 pharyngeals.

Rhinichthys/Richardsonius
Materials: *Floodplain*: 2 pharyngeals.

Family Catostomidae
Catostomus sp.—suckers
Materials: *Rockshelter*: 1 basioccipital, 1 exoccipital, 1 hyomandibula, 3 endopterygoids, 3 metapterygoids, 1 palatine, 1 parasphenoid, 1 parietal, 1 preopercle, 2 prootics, 1 sphenotic, 1 urohyal, 1 vomer, 9 cleithra, 2 coracoids, 5 pharyngeals, 3 1st vertebrae, 6 2nd vertebrae: 42 specimens.
Floodplain: 3 basioccipitals, 3 ceratohyals, 2 dentaries, 1 epihyal, 1 hyomandibula, 1 metapterygoid, 2 opercles, 1 palatine, 1 quadrate, 1 urohyal, 1 vomer, 8 cleithra, 15 pharyngeals and pharyngeal teeth, 15 1st vertebrae, 5 2nd vertebrae: 60 specimens.

C. macrocheilus—Largescale sucker
Materials: *Rockshelter*: 2 dentaries, 1 hyomandibula, 2 maxillae, 3 palatines, 1 quadrate: 9 specimens.
Floodplain: 1 maxilla.

C. columbianus—Bridgelip sucker
Materials: *Floodplain*: 1 maxilla.

Cyprinidae/Catostomidae
Materials: *Rockshelter*: 105 abdominal vertebrae, 66 caudal vertebrae, 3 indeterminate vertebra type, 11 vertebra fragments: 185 specimens.
Floodplain: 2 hyomandibulae, 2 scapulae, 5 pharyngeals and pharyngeal teeth, 5 1st or 2nd vertebrae, 418 abdominal vertebrae, 374 caudal vertebrae, 14 ultimate vertebrae, 24 indeterminate vertebrae type, 659 vertebra fragments: 1,503 specimens.

Remarks: Nine species of Cyprinidae are known historically for the Columbia Basin (Lee et al. 1980). Several elements of the jaw, pharyngeal arch (toothed bone at the rear of the mouth) and lateral face are extremely distinctive and species or generic assignments of sufficiently complete specimens was possible. Six species of *Catostomus* are known for the Columbia-Snake River Basin (*C. macrocheilus, C. columbianus, C. platyrhynchus, C. catostomus, C. discobolus, C. ardens*). The specimens assigned to species very closely matched the reference material. However, reference material was lacking for *C. discobolus* (known in the Snake River only above Great Falls) and C. ardens (found in the Snake River above Shoshone Falls) so the assignment of remains to *C. macrocheilus* and *C. columbianus* is somewhat provisional.

Except for the first and second vertebra on the column, which can be distinguished as Cyprinidae or *Catostomus*, vertebrae from these taxonomic groups cannot be distinguished, so the joint category, Cyprinidae/Catostomidae, was used. In a few cases, abdominal vertebrae were fused to the second vertebra and these abdominal vertebrae could be identified to a finer taxonomic level. Finally, several extremely eroded cranial and postcranial specimens could not be identified more precisely than Cyprinidae/Catostomidae.

Fishes in the Cypriniformes order occupy a wide range of habitats but in general terms tend to occupy more slowly moving, warmer waters than salmonids. Of the cyprinids identified at Marmes, *Ptychocheilus oregonensis* (northern pike minnow) favors slow to moderate currents in streams and prefers the warmest temperatures in the waters it occupies (Wydoski and Whitney 1979). *Acrocheilus alutaceus* (chiselmouth) also prefers warmer areas of streams in moderately fast to fast moving waters (Wydoski and Whitney 1979). Most studies of *Mylocheilus caurinus* (peamouth) have focused on lake populations. In Lake Washington (Seattle, WA), peamouth tend to occupy the warmest water, favoring deep water during the winter and moving inshore during spring and summer (Wydoski and Whitney 1979).

Catostomus species are bottom fishes, feeding on algae or bottom dwelling invertebrates. Both *C. macrocheilus* and *C. columbianus* occupy quiet areas in the backwaters or edges of the main current of streams (Wydoski and Whitney 1979). During spawning season of largescale sucker, which occurs usually in April or May, large schools are found along river edges. Spawning usually occurs in shallow water along the downstream ends of pools.

Results

Before reviewing results of analysis, it is important to consider several aspects of archaeological recovery and curation that have biased the fish assemblage available for this study. Early reports on excavation procedures note that ¼-inch mesh screens were used during each of the four field seasons at the site (Gustafson 1972). The effects of screen size on fish faunal recovery in the rockshelter cannot be directly documented given that bulk samples are not available for analysis. However, the presence of *very small* fish remains from the floodplain bulk samples (see below) and abundant research elsewhere (Butler 1987a; Casteel 1972; Gordon 1993), suggests that use of large mesh screens in the rockshelter has biased the sample in favor of relatively large-bodied fishes. Besides screen size bias, however, which can be estimated to some extent, other recovery practices suggest the collected sample may bear a very poor relationship to the target population of fish remains in the rockshelter deposit. An early document describing field procedures notes that "not all bone fragments were saved during the early seasons of excavation" (Gustafson 1972:54). Unfortunately, field records are not available that indicate the extent of this practice or which excavated areas are particularly affected, thus it is difficult to control for this problem.

Also, there are discrepancies between this author's records of fish taxa and relative abundance and those of previous researchers (Gustafson 1972), which suggest that the sample available for this study may be different from that originally excavated. Gustafson notes "Salmonid vertebrae and other fish remains sometimes are abundant (particularly in the storage pit areas—Units VI and VII)" (1972: 106). As noted below, salmonid remains in the recently analyzed sample are most common in Stratum V, where they represent over 60% of the fish fauna; in Stratum VI, *a single salmonid* specimen was identified and in Stratum VII, about 15% of the fish remains are from salmonids (emphasis mine). In addition, there is a discrepancy in the species reported in Gustafson (1972) and identified in the current study sample. Gustafson sent a sample (from unknown provenience) of fish remains to

William Taylor (U.S. National Museum) who identified two species positively: *Ptychocheilus oregonense* (= *oregonensis*) and *Mylocheilus caurinus*. This author identified *Ptychocheilus* in the Rockshelter sample and other cyprinids, but not *Mylocheilus*. *Mylocheilus* has extremely distinctive jaw and pharyngeal morphology; it is unlikely that the discrepancy reflects analyst identification error. Given all of the other documented losses—both of specimens and provenience—associated with the rockshelter assemblage (Hicks and Moura 1998), it appears likely that the differences in reporting result from the current study's sample not including all of the fish remains that Taylor and Gustafson examined. It appears that parts of the recovered fish assemblage have become misplaced or lost over the 40+ years since the inception of the project.

In sum, given the documented and indirectly suggested biases associated with the fish bone sample for this study, it would be unwise to assume the materials reported on here are representative of the fish bones in the rockshelter deposits. In terms of quantification, it is most appropriate to treat the fish assemblage at the nominal scale only rather than ordinal or ranked scale. That is, it is best to view the fish record as a list of taxa present rather than use NISP values to examine relative taxonomic importance. Similarly, given that there is no control over field recovery loss or curatorial loss across the site deposits, interpretations about changing frequencies of fish taxa within the site will be tentative, since the changes could result from sampling problems.

Rockshelter Fish Remains

A total of 688 fish remains were identified from the rockshelter (Table 11.1). About 60% of these (420 specimens) could be identified to a taxonomic level below fish and eight taxa are represented. Freshwater minnows and suckers (Cyprinidae, Catostomidae) dominate the assemblage. *Ptychocheilus oregonensis* (northern pike minnow) is the dominant freshwater species represented; *Catostomus macrocheilus* (largescale sucker) is the sole species of sucker present. Large-bodied salmonids (*Oncorhynchus*), almost certainly from migratory runs, represent 13.7% of the collection and a single specimen from sturgeon is present. As noted above, it is problematic to assume the relative abundance of fish taxa in the rockshelter is representative of the target population of fish remains in the site deposits. For now and until better control of the biases is achieved, it is best to treat these data at the nominal scale.

Table 11.1 Frequency of fish remains by taxon, rockshelter.

Taxon	Rockshelter	
	NISP	%
Acipenser sp.	1	0.1
Oncorhynchus sp.	94	13.7
Cyprinidae	56	8.1
Ptychocheilus oregonensis	29	4.2
Acrocheilus alutaceus	4	0.6
Catostomus sp.	42	6.1
Catostomus macrocheilus	9	1.3
Cyprinidae/Catostomidae	185	26.9
Unidentifiable	268	39.0
Total	688	100

Taphonomy and Origin of Fish Remains. In evaluating the significance and meaning of a faunal assemblage in cultural terms, it is first necessary to establish that the remains in fact result from human activity (Grayson 1991; Lyman 1994). Geological study of the site matrix suggests that the source of the site deposits is primarily eolian, with endogenous roof fall and human occupational debris contributing matrix as well (see Chapter Five). There is no evidence for fluvial deposition in Stratum I and above, so the possibility that fish remains represent flood rafted carcasses can be eliminated. Potentially nonhuman terrestrial or avian scavengers or predators of fish could have brought whole fish or fish parts to the rockshelter. It is unlikely that the salmon (*Oncorhynchus* sp.) remains entered the rockshelter this way, however. As noted previously, the Palouse River probably never supported anadromous salmon runs; the nearest source of these fish would be over a mile away in the Snake River. It is unlikely that nonhuman scavengers would have transported the salmonid carcasses or parts this distance. Perhaps significantly, Gustafson (1972) recorded specimens from *Canis* cf. *latrans* (probably coyote) in each of the rockshelter strata (1972:Table 5.1), which suggests that at least one carnivore that is known to eat fish (Butler and Schroeder 1998) lived in the vicinity of the rockshelter. However, Gustafson does not note any evidence for carnivore damage on the large mammal bones (Gustafson 1972; Gustafson and Wegener 1998), although, the absence of this observation may indicate this surface attribute was not examined, rather than a real absence of carnivore damage. Time and budgetary constraints prevented carrying out a detailed study of surface modification of fish bone for the current study. However, a small grab sample of fish remains (about 15 specimens) was examined under 10-20 power magnification and did not reveal any patterns indicative of digestive process (Butler and Schroeder 1998—dark staining, rounding, erosion, vertebra compression). One specimen showed sign of rodent gnawing but none exhibited sign of carnivore processing.

Sometimes burning is used to link a faunal assemblage to human activity (Balme 1980; Butler 1990). Burned bone can result from natural fires, but if widespread burning is not indicated (multiple artifacts and sediments do not show evidence of burning), then it is reasonable to argue that cultural agents burned the bone in question. It is of course possible that even if cultural practice is responsible for the burning, the burning *postdates* the natural deposition of faunal remains. These concerns are moot for the Rockshelter fish remains; only one specimen from the rockshelter (Inventory No. 9199, in Stratum I) was burned, so this in itself does not provide particular support either way.

Another potential way to examine fish bone origin and site formation processes would examine spatial distribution of fish bones relative to other classes of animal remains and clearly modified objects or feature context. If fish bone abundance varied in concert with abundance or distribution of modified objects or feature context, then it would be possible to argue for a common source, humans. Unfortunately, given the variation in collection procedures across excavation units and strata, it will be difficult to use fish bone abundance and distribution to evaluate taphonomic questions.

In sum, while a definitive conclusion about the origin of the fish bones in the rockshelter is not possible, several factors and conditions support the human role in forming the fish deposit, including the lack of carnivore damage on the sample examined, the lack of discussion of carnivore damage in previous faunal reports (Gustafson 1972; Gustafson and Wegener 1998), and the presence of fish taxa that were likely not available in the nearby river.

Intra-site variation. Table 11.2 shows the frequency of taxa across the 14 strata (and aggregate strata as assigned in the field and laboratory) in the rockshelter. Table 11.3 shows the frequency of taxa for the three stratum units that contained most of the fish remains—V, VI, and VII. For temporal reference, Mazama tephra underlies Stratum V, which is a mixed deposit containing some Mazama tephra along with other windblown sediments and provided a single radiocarbon date (4250 ± 300 B. P.). A number of radiocarbon dates were obtained from Strata VI and VII, providing radiocarbon ages ranging between 1940 and 660 B. P. (Sheppard et al. 1987). In general terms, the earliest Stratum V likely represents a time span from about 6,800 years ago to about 4,000 years ago. Stratum VI likely spans the time period from 1300 to 1940 B. P. and Stratum VII represents an age of between 660 and 1600 B. P. (Sheppard et al. 1987; Gustafson and Wegener 1998).

As shown in Table 11.3, there is some striking variation in taxonomic representation across the three stratigraphic units, particularly in the relative frequency of *Oncorhynchus* sp. Unit V is dominated by *Oncorhynchus*, which represents over 60% of the fish remains in the unit. In Unit VI, *Oncorhynchus* is represented by a single specimen and resident freshwater fishes comprise most of the collection. In Stratum VII, freshwater fishes still dominate (with about 85%), but the abundance of *Oncorhynchus* is greater than in Stratum VI. A Chi Square contingency test shows that these differences are significant ($X^2 = 129.70$, $p < .0001$); all of the resident freshwater taxa in each stratum were aggregated to compensate for small sample sizes. These differences might reflect site based or region-wide changes in the organization of subsistence and settlement activities. It is widely accepted that sometime after 5,000 years ago, people made increasing use of salmon as a stored resource. Perhaps the higher frequency of salmon in Unit V signals the stored use of this resource. The presence of storage features in Units V and above suggests that the rockshelter was serving as a place where resources were cached. If it is accepted that the Palouse River never supported much of an anadromous salmon population, then the prominence of salmon remains may reflect the capture of salmon in the main stem Snake River and transport of dried fish to the Rockshelter. This reasoning follows Chatters (1987) use of "geographic displacement" as an indicator of food storage. When remains of organisms are recovered some distance from their known habitat or distribution, their presence indicates resource transport. Given the costs associated with food transport, efforts to reduce weight, through bulk processing and drying, would be promoted. Interpreting the changing taxonomic frequency in this way, however, could be in error, given the sampling problems discussed earlier, and the suggestion remains tentative.

Body size. Vertebrae were measured to roughly estimate the sizes of fishes present and whether there were any changes in the body size of fish over time. Casteel (1976) showed that vertebral size and linear dimensions like length were highly correlated. This study did not rely on regression analysis, but rather simple comparisons of vertebra measures of fishes of known length to identify change in body size represented.

For *Oncorhynchus*, only eight vertebrae were complete enough to measure. Values range between 6.2 mm to 10.7 mm (Table 11.4). The small sample size prevents examining any temporal changes in body sizes represented. Based on comparison of vertebrae from fish of known length (Table 11.5), the rockshelter samples come from a range of body sizes, ranging between about 350 and 800 mm in standard length (the distance from the tip of the snout to the base of the tail or end of hypural).

Over 100 vertebrae from minnows and suckers were complete enough to measure. As seen in Table 11.6, mean diameter varies little (between 5.4 and 7.1 mm) from the lower strata to the upper units, suggesting there is little variation in the sizes of freshwater fish that were deposited throughout the rockshelter. Taking into account the standard deviation, most of the vertebrae range between 4 and 8 mm in size. Based on comparison with modern fish body size and vertebra size (Table 11.7), vertebra measures this size are from fish that range in size between about 250 and 500 mm standard length.

Summary. The rockshelter fish fauna shows that a range of fish taxa were used by people occupying the rockshelter. Fish remains were identified in each of the stratigraphic units spanning the Holocene record of occupation, indicating that fish played some part in subsistence activities for the 10,000 years the site was used. Both resident freshwater and anadromous fishes are present, suggesting that past people were generalized in their fishing practices. The resident freshwater fish would have been available in the river adjacent to the site. The Palouse River, however, probably did not support a migratory salmon run throughout the Holocene. The presence of large-bodied salmon, likely representing migratory fish in the site, suggests that people traveled to the main stem Snake River for this resource.

There has been little study of the life history and habitat requirements of the resident freshwater fish species in Plateau rivers and streams that are prominent in the rockshelter.

Table 11.2 Frequency (NISP) of fish taxa by strata/unit[1], rockshelter.

Taxon	<I	I	II	I-II	III	IV-V	V	V-VI	VI	VI-VII	VII	VII-VIII	VIII	Intru VII	Unass
Acipenser sp.								1							
Oncorhynchus sp.		3			6	8	62		1		13	1			
Cyprinidae			1	1			5	3	26		17		3		
Ptychocheilus oregonensis						1		4	16	1	4		3		
Acrocheilus alutaceus							1	2	1						
Catostomus sp.		2		1	5		1	3	17		10		2	1	
Catostomus macrocheilus								3	5				1		
Cyprinidae/Catostomidae	1	16	1	3	3	1	27		66		44		7	11	5
Unidentifiable	4		5		15		64	6	61		84		29		
Total	5	21	7	4	30	9	161	22	193	1	172	1	45	12	5

Note: some stratum assignments were made after fieldwork was completed and should be considered provisional

Table 11.3 Frequency (NISP) of fish taxa across three strata/units, rockshelter.

Taxon	V		VI		VII	
	NISP	%	NISP	%	NISP	%
Oncorhynchus sp.	62	63.9	1	0.7	13	14.8
Cyprinidae	5	5.2	26	19.7	17	19.3
Ptychocheilus oregonensis	1	1.0	16	12.1	4	4.5
Acrocheilus alutaceus	1	1.0	1	0.7		
Catostomus sp.	1	1.0	17	12.9	10	11.4
Catostomus macrocheilus			5	3.8		
Cyprinidae/Catostomidae	27	27.8	66	50.0	44	50.0
Total	97	100.0	132	100.0	88	100.0

Note: some stratum assignments were made after fieldwork was completed and should be considered provisional

Table 11.4 Mean width of *Oncorhynchus* vertebrae across strata/unit, rockshelter.

Strata	Mean Vertebra Width (mm)	N	Std. Deviation
VII-VIII	7.8	1	--
VII	10.7	1	--
V	9.6	3	1.4
III	6.2	1	--
I	7.8	2	4.2

Note: some stratum assignments were made after fieldwork was completed and should be considered provisional

Table 11.5 Standard length[1] and vertebra widths of selected species of *Oncorhynchus*.

Taxon	Vertebra Width (mm)[2]	Body Size Standard Length (mm)
Oncorhynchus tshawytscha (VLB 92-6-8)	11.5	825
O. tshawytscha (VLB 86-20-4)	10.1	630
O. kisutch (VLB86-4-1)	8.7	575
O. mykiss (VLB 86-6-12)	8.2	670[3]
O. clarki (VLB 91-10-1)	6.1	365

[1] length from tip of snout to end of hypural bone
[2] measure is an average width of six vertebrae arbitrarily selected from each skeleton
[3] measure is fork length

Thus it is difficult to glean detailed insight on the paleoenvironmental significance of the archaeological remains. The species identified occupy a range of river, stream and lake habitats, including slow and fast moving water, deep pools and shallows; their preference is relatively warm water. Perhaps the most that can be said is that the presence of several species of freshwater fishes throughout the Holocene sequence of human occupation at the site suggests the adjacent river provided adequate habitat for these fish throughout the Holocene. Given that the record for region-wide and local environmental change is clear, it may be significant to note that the river was capable of supporting fish populations for the 10,000-year period.

Floodplain Fish Remains

As noted previously, the fish remains come from a large number of bulk samples that were retained in 1-mm mesh in the field and sorted under a controlled setting in the laboratory. Use of fine mesh and controlled sorting insures minimal loss of small bone specimens, and in turn, minimal loss of remains of small-bodied fishes if they are present in the site deposits. Significantly, during the excavation and field screening process, some items, including faunal remains, were removed from the screen and bagged separately. Unfortunately, if this included fish remains, none of this material was located during this study, nor were any records

Table 11.6 Mean width of cyprinidae/catostomidae vertebrae across strata/unit, rockshelter.

Strata	Mean Vertebra Width (mm)	N	Std. Deviation
VIII	5.5	6	1.5
VII	5.7	33	1.6
VI	6.4	36	1.6
V	5.9	14	1.1
III	7.1	1	--
I-II	5.9	1	--
I	5.4	2	0.1

Note: some stratum assignments were made after fieldwork was completed and should be considered provisional

Table 11.7 Standard length and vertebra widths of selected *Cypriniformes*.

Taxon	Vertebra Width (mm)[1]	Body Size Standard Length (mm)
Catostomus macrocheilus (VLB 92-5-6)	7.4	425
Catostomus columbianus (VLB92-7-10)	5.1	315
Catostomus fumeiventris (UMMZ181667)	4.2	260
Catostomus macrocheilus (VLB 92-5-9)	3.8	239
Mylocheilus caurinus (VLB87-10-3)	2.6	192
Gila bicolor (VLB89-1-25)	2.7	160
Catostomus platyrhynchus (VLB92-10-7)	2.5	147
Rhinichthys cataracte (VLB92-7-5)	0.9	65

[1] measure is an average width of three abdominal and three caudal vertebrae arbitrarily selected from each skeleton

found which indicate the extent of this practice. Presumably, if *fish* remains were selectively bagged during field screening and ultimately lost, they would have been relatively large specimens. In short, there exists the real possibility that the floodplain fish sample available for this study is biased *against large-bodied fishes* and in *favor of small-bodied fishes*. Obviously, these problems introduce difficulties in interpreting the fish faunal record in the deposit. As with the rockshelter fish sample, sampling and curation problems mean that the fish remains provide a nominal record of fish taxa and body sizes present. As relevant, the following discussion reviews the extent to which these biases affect this study's interpretations of the spatial and vertical patterning in fish remains.

Results. As shown in Table 11.8, 2,481 fish remains were recovered from the floodplain bulk samples. About 70% of these (1,762 specimens) could be identified below class fish. Fourteen taxa were identified; resident freshwater fishes absolutely dominate the assemblage, with seven different species represented. Remains of cyprinids (identified to family and species) are more abundant than catostomids (suckers). The salmonids represent a very small fraction (less than 1%) of the assemblage.

Taphonomy and Fish Bone Origin. How did the fish bones come to be in the floodplain deposit? Do they represent the remains of past human subsistence or could they reflect natural deposition of fish by floodwaters or nonhuman scavengers or carnivores? As noted previously, sorting out the agents responsible for a faunal deposit is essential before one can identify cultural activity patterns. Given the very old age of the Marmes floodplain deposits as well as the potential and realized significance of the site to our understanding of human occupation of the region (e.g., Ames et al. 1998), it is necessary to carefully review the agents responsible for the fish remains found there.

Geoarchaeological analysis of the early Holocene sediments that were the focus of study shows that they are horizontally stratified fluvial deposits left by low energy flooding (see Chapter Five). Radiocarbon dates from the deposits suggest they were deposited about the same time as Unit I and II in the Rockshelter (between 10,000 and 8500 radiocarbon years ago). Unequivocal human use of the riverside location is indicated by stone tools, several features, and human remains found in the stratified deposits.

Table 11.8 Frequency (NISP) of fish taxa, floodplain.

Taxon	Floodplain	
	NISP	%
Salmonidae	3	0.1
Oncorhynchus sp.	12	0.5
Prosopium williamsoni	1	0.0
Cyprinidae	161	6.5
Ptychocheilus oregonensis	8	0.3
Acrocheilus alutaceus	2	0.1
Mylocheilus caurinus	5	0.2
Richardsonius balteatus	1	0.0
Rhinichthys sp.	2	0.1
Richardsonius/Rhinichthys	2	0.1
Catostomus sp.	60	2.4
Catostomus macrocheilus	1	0.0
Catostomus columbianus	1	0.0
Cyprinidae/Catostomidae	1,503	60.6
Unidentifiable	719	29.0
Total	2,481	100.0

Overbank deposition was episodic, as the sedimentary record shows a series of weakly developed A horizons (see Chapter Five). A very general site formation model suggests that humans engaged in a variety of activities in the area below the rockshelter and next to the river, organic horizons formed, and then low-energy flood water periodically over topped the bank allowing fine sediments to settle out of suspension and bury remains on the surface. Given such a model, it is conceivable that fish were swept by floodwaters onto the bank where they became trapped and ultimately buried. Several factors and lines of evidence suggest this is not the case, however.

One challenge to the fluvial origin considers the likely season of the floods, the body size of the fish in the deposit and the life history and demography of the source fish populations that would generate such a deposit. First, river flooding along streams and rivers in the Plateau is associated with the melting snow pack, which occurs in the spring and early summer. The source population for a natural fluvial fish deposit would be fish in the river during spring and early summer. Importantly, across western North America, the numerous species of cyprinids and catostomids spawn during this time of year (Moyle 1976; Sigler and Sigler 1987; Wydoski and Whitney 1979). While specific mating behavior and habitat preference varies from species to species, for all taxa, spawning entails reproductively mature adults aggregating for days or weeks at a time, in pools or shallows in the river. I suggest that the source population for a natural flood-rafted deposit of freshwater fish should include adults – representing a range in age or size classes. The body size and age of reproductively viable individuals as well as the maximum size and age reached is highly variable across species; several small-bodied, short-lived cyprinids *Rhinichthys, Richardsonius*, become mature when they are less than 100 mm and then rarely reach over 100 mm in size. Several cyprinids and catostomids (*Ptychocheilus oregonenis, Mylocheilus caurinus, Catostomus macrocheilus*) become sexually mature when they are over 100 mm in length and often attain lengths of over 300 mm and ages of over 15 years. The Marmes Rockshelter vertebra measurements provide independent evidence that relatively large cyprinids and catostomids were present in the early Holocene in the site vicinity. As noted previously, vertebra from cyprinid/catostomid fishes over 300 mm in standard length are most commonly represented.

In short, if the floodplain fish remains are the remnants of flood-rafted carcasses, a range in body size should be present in the fish assemblage, reflecting the range in body size of a spawning population. As shown in Table 11.9, about 400 vertebrae from the floodplain were measured, with samples from each horizon documented. The mean size is very small (1.9 mm) and remarkably consistent for each horizon or aggregate horizon, suggesting first that the body size represented is quite small and second overall uniformity in size class represented. Comparing these vertebra sizes to modern fishes (Table 11.7) suggests that the mean size of fish is considerably smaller than 200 mm in standard length. It might be suggested that the small vertebra size results from the deposition primarily of the small-bodied genera such as *Richardsonius* and *Rhinichthys*. While these genera are noted in the deposit, their remains are much less frequent than the larger-bodied cyprinids (Table 11.8). Overall, the scarcity of large fishes in the floodplain deposit does not match the expectation for a fluvial deposit of spawning adults, which should include a range in body sizes, including relatively large individuals. The best explanation for this narrow and relatively small body size is some kind of selective mortality, notably human selection of small fishes.

Of course this explanation presumes that the scarcity of "large" vertebrae in the deposits is real and not a sampling problem (e.g., selective field bagging and subsequent loss of such vertebrae). Given that it is not possible to rule out this bias, statements about body size distribution and frequency must remain tentative.

It would be useful to have supporting evidence for a human role in the form of cut marks or patterned burning. Cut marks were not seen on any of the specimens. Importantly, cut marks have only rarely been noted on archaeological fish remains (cf. Barrett 1997), and none to this analyst's knowledge have been reported on fish remains in Plateau or coastal archaeological sites. Thus, the absence of evidence for cut marks should not be used to undermine the argument for a human source for the fish bones.

As noted previously, presence or absence of burning can be useful in sorting out taphonomic agents responsible for a bone deposit. In the case of the floodplain faunal remains, however, burning is not helpful largely because the incidence of burning across the site and vertebrate classes is so high. Reporting on the nonfish assemblage, Gustafson and Wegener (see Chapter Ten) note that most of the bone shows sign of burning. This includes an estimated 60,000 fragments from unidentified taxa as well as bones from animals that are likely noncultural in origin (i.e. surface dwelling snakes, burrowing rodents). Some of this burning may result from human fires (cooking or trash fires), but the ubiquity of the burning across most of the vertebrate classes (including taxa likely of noncultural origin) makes it difficult to use burning per se as good evidence for a human source.

Regarding the fish remains, over 30% of the remains altogether show clear sign of burning (Table 11.10). This proportion is much lower than that noted for other vertebrate classes by Gustafson and Wegener, which quite likely reflects different criteria used in identifying burning. Table 11.10 also shows the frequency of burning across the horizons in the floodplain. For the three horizons with sizeable counts, Harrison, Marmes, and joint Marmes/Harrison, the frequency of burning is very similar, suggesting that whatever agents are responsible for the burning, they acted consistently over the course of site formation. Fish body part representation also might be useful in sorting out taphonomic origins (Butler 1990, 1996). In particular, if the fish represent natural, flood-rafted carcasses, the entire skeleton should be represented. Further, if burial was swift, specimen fragmentation should be minimal and the specimens should be in good condition. The first expectation is met, but not the second. A review of skeletal element representation (Table 11.11) shows that numerous elements of the head, fins and vertebral column are present. Given that many elements of the skeleton could not be identified below the combined family taxon, Cyprinidae-Catostomidae, this review considers all of the elements from these freshwater fishes together. [The quantity listed in Table 11.11 is the minimum number of elements (MNE; Bunn 1982) which selects the best represented section of each element and counts the number of times it occurs. Using this quantity rather than NISP controls for the problem introduced by specimen fragmentation. With MNE, a single skeletal element will be counted one time, no matter how many fragments it may have been broken into; with NISP, a single element could be broken into several fragments and if all of the fragments were recognizable, all of them would be counted]. One element from the head

Table 11.9 Mean width of *Cyprinidae/Catostomidae* vertebrae across horizon, floodplain.

Horizon	Mean Vertebra Width (mm)	N	Std. Deviation
Marmes	1.9	28	.6
Mixed Marmes-Harrison	2.0	51	.4
Harrison	1.9	264	.5
Beneath Harrison	4.6	1	--
Unassigned	1.9	40	.4
Total	1.91	384	.5

Note: some stratum assignments were made after fieldwork was completed and should be considered provisional

Table 11.10 Frequency of burned specimens across the floodplain horizons.[1]

Horizon	burned	% burned[2]
Marmes	47	33.8
Mixed Marmes-Harrison	47	31.8
Harrison	616	33.2
Beneath Harrison	0	0
Unassigned	73	22.7
Total	785	31.6

[1] some stratum assignments were made after fieldwork was completed and should be considered provisional
[2] represents the percent of bones in each horizon or horizon aggregate that are burned

(pharyngeal) is represented by 39 specimens; given that two pharyngeals are in an individual fish, this indicates a minimum of 18 individuals are represented in the deposit. The minimum number of fish represented by the abdominal vertebrae is 22 (based on an average of 19 abdominal vertebrae per individual) and the minimum number of fish represented by the caudal vertebrae is 21 (based on an average of 18 caudal vertebrae per individual). These very similar values of both head and trunk elements (controlling for the number of times the elements occur in the skeleton) strongly indicate that the entire skeleton was initially deposited on the floodplain. While whole body deposition is indicated, there is evidence from these data as well for significant specimen breakage and fragmentation, especially of the head elements. Except for the pharyngeal, all of the head elements are represented by *very few* specimens. This pattern most likely is explained by bone destruction that has rendered the remains unidentifiable or so small that they passed through the 1-mm mesh. While degree of specimen fragmentation was not recorded during analysis, the author's impression of the assemblage was that except for vertebral centra, specimens tended to be fragmentary. The very low frequency of most head elements (Table 11.11) certainly supports this notion.

Table 11.11 Frequency of skeletal elements with landmarks (MNE) from *Cyprinidae* and *Catostomidae*, floodplain.

Element	MNE	%
Cranial		
articular	3	0.3
basioccipital	4	0.4
ceratohyal	4	0.4
dentary	8	0.8
epihyal	3	0.3
hyomandibula	3	0.3
maxilla	2	0.2
metapterygoid	1	0.1
opercle	6	0.6
palatine	1	0.1
parasphenoid	1	0.1
pharyngeal	39	4.0
pterotic	1	0.1
quadrate	3	0.3
urohyal	2	0.2
vomer	1	0.1
Paired Fins		
cleithrum	13	1.3
scapula	2	0.2
Vertebral Column		
1st vertebra	30	3.1
2nd vertebra	11	1.1
1st or 2nd vertebra	12	1.2
abdominal vertebra	420	42.8
caudal vertebra	374	38.1
ultimate vertebra	14	1.4
Total	764	100.0

The fragmentary condition of the fish remains is comparable to that described for much of the nonfish fauna (see Chapter Ten). Moreover, patterning in fragmentation across vertebrate groups may in fact offer clues as to the source of the destruction. Gustafson and Wegener point out that most of the approximately 1,300 nonfish specimens that they were able to identify to some taxonomic level were rodents and reptiles, taxa which they also note, are least likely to reflect cultural use and deposition. Further, it was these specimens that showed the highest degree of integrity. In other words, fragmentation is not evenly distributed across vertebrate classes, which implies that the source of fragmentation is not a ubiquitous force (like sediment chemistry or post-depositional trampling), but a more discriminating force that was targeting particular animal groups. Humans would seem to be a likely candidate for this, in the form of cooking and butchering activities.

It is also important to consider whether nonhuman carnivore or scavenging activities have deposited or modified the fish assemblage from the floodplain. As noted for the rockshelter fauna, Gustafson and Wegener (Chapter Ten) do not discuss any sign of carnivore gnawing or other markings on the nonfish assemblage on the floodplain. Budgetary and time constraints prevented undertaking a detailed study of surface morphology that might be used to identify digestive process. The general impression is that characteristic patterns associated with digestive process (rounding, pitting, and vertebra deformation) were not present. Most of the specimens were either darkly stained or burned. Based on available information, the faunal assemblage does not appear to have been ravaged by carnivores.

In sum, several factors suggest that humans are most likely responsible for the fish deposit. If it is assumed that there was minimal removal of large vertebrae during field screening, the narrow range and small body size present suggests a form of selective mortality was operating and humans are the most likely candidate for this. Further, the fish remains are fragmentary which would be expected if the fish were subjected to various cooking and processing activities. Certainly nonhuman agents can cause bone destruction, but the fact that fragmentation is unevenly distributed across the vertebrate assemblage suggests indirectly that the agent of modification was selective and again, humans are a likely agent for this selection. Additional support for the human role in generating the fish deposit would examine whether the intra-site frequency of fish remains is correlated with other clearly cultural items (e.g., lithics); as project reports for these data are generated it is recommended that these studies be conducted. As well, it would be useful to carry out some test excavations "off-site" to evaluate the likelihood for such natural accumulations of freshwater fish remains to occur.

Intra-site variation. Table 11.12 shows the distribution of identified taxa across horizons in the floodplain deposits. As shown, the bulk of the assemblage (over 75%) is from the Harrison Horizon. Given that freshwater fishes are the overwhelming dominant taxa in the locale, they of course dominate each of the horizons.

Examination of fish distribution across bulk samples shows that the distribution of fish bone is extremely uneven. Of the 240 bulk samples processed only 84 (35%) had fish bone. Furthermore, 73% of the fish bone recovered was from just six samples (Table 11.13). These data suggest that fish bone was deposited in discrete areas rather than widely dispersed across the site deposits. The pattern of clumping can be seen in the distribution of fish bone across horizons as well. As shown in Table 11.13, roughly similar numbers of bulk samples contained fish bones in the two main horizons (Marmes and Harrison), but the overall frequency of fish bone in the two horizons is very different. The 34 samples with fish bones in the Harrison unit provided over 1,800 fish remains, while the 38 samples in Marmes provided only 139 fish specimens. Differences in volumes of excavated or sorted matrix from the two horizons do not explain this striking pattern. Rather, these data suggest that during the accumulation of the Harrison Horizon, there was greater use and deposition of fish than in the Marmes Horizon.

While these data suggest that fish use per se may have varied over time, there is remarkable consistency in the body size of fish present (Table 11.9). As noted in the taphonomy section, vertebra size is uniformly small across horizons, suggesting a consistency in fish capture and fishing activities (capture methods, and use patterns) over time.

Summary. The floodplain fish assemblage provides a number of striking patterns that should be reviewed. First, the record shows a dominance of freshwater fishes and extreme scarcity of salmon remains. All of the faunal remains show a high degree of fragmentation, and it is reasonable to be concerned that the scarcity of salmon bone is in part due to the high degree of bone destruction (Butler 1987b). Salmon bone is not as "dense" in terms of mineral content per volume as minnow and sucker bone (Butler 1996; Butler and Schroeder 1998). If bony tissues of both salmon and minnows/suckers were subjected to similar destructive process, then salmon bone would degrade more readily and less likely be part of the faunal record in the site. In the case of the floodplain assemblage, this scenario is not supported. All of the matrix was processed through 1-mm mesh, which is an extremely small mesh size. Fragmentary remains of salmon vertebrae or other durable cranial elements (e.g., gill rakers, teeth) should have been recovered in this fine screen matrix in greater numbers if, in fact, they had been present in higher numbers in the deposits. The virtual absence of salmon bones and teeth in the deposit suggests in fact that salmon were not used much by people that occupied the floodplain.

Table 11.12 Frequency (NISP) of fish taxa by horizon[1], Marmes floodplain.

Taxon	Beneath Harrison NISP	%	Harrison NISP	%	Between Harrison & Marmes NISP	%	Mixed Harrison & Marmes NISP	%	Marmes NISP	%	Unassigned NISP	%	Total NISP	%
Salmonidae									3	3.4			3	0.2
Oncorhynchus sp.			8	0.6					4	4.6			12	0.7
Prosopium williamsoni			1	0.1									1	0.1
Cyprinidae			122	9.1	1	50.0	3	2.1	8	9.2	27	14.8	161	9.1
Ptychocheilus oregonensis			5	0.4							3	1.6	8	0.5
Acrocheilus alutaceus			1	0.1							1	0.5	2	0.1
Mylocheilus caurinus			5	0.4									5	0.3
Richardsonius balteatus							1	0.7					1	0.1
Rhinichthys sp.			2	0.1									2	0.1
Richardsonius/Rhinichthys			2	0.1									2	0.1
Catostomus sp.			49	3.7			3	2.1			8	4.4	60	3.4
Catostomus macrocheilus			1	0.1									1	0.1
Catostomus columbianus			1	0.1									1	0.1
Cyprinidae/Catostomidae	5	100.0	1144	85.3	1	50.0	137	95.1	72	82.8	144	78.7	1503	85.3
Total	5	100.0	1341	100.0	2	100.0	144	100.0	87	100.0	183	100.0	1762	100.0

[1] some of these assignments were made after field work was completed and should be considered provisional

Table 11.13 Frequency of fish remains by horizon, floodplain.

Horizon	# Bulk Samples with Fish	Fish count	%
Unassigned	2	322	13.0
Above Marmes	0		
Marmes	38	139	5.6
Mixed Harrison & Marmes	2	148	6.0
Between Harrison & Marmes	2	3	0.1
Harrison	34	1,858	74.9
Beneath Harrison	6	11	0.4
Total	84	2,481	100.0

Note: some stratum assignments were made after fieldwork was completed and should be considered provisional

There are two obvious reasons why salmon may not have been used. First, the site is located over a mile and a half from the closest source of migratory fish, in the main stem Snake River. Perhaps scheduling of other economic activities meant that local residents were only able to take advantage of very local fishery resources that were in the adjacent river. It is also possible that the lack of salmonid remains reflects the season of site occupation. Migratory salmonids would have been available chiefly in the late spring, summer and fall. If the occupation of the floodplain did not coincide with the timing of migratory runs—perhaps people occupied the site in the winter and spring—then the absence of salmon would be explained.

A third factor could be environmental, suggesting that salmonids were simply not very abundant in regional streams, because of poor spawning habitat or perhaps oceanic conditions. Chatters and others (Chatters et al. 1995) have suggested that conditions later in the Holocene would have been poor for salmon, due to warmer water temperature in spawning habitat and overall poor stream conditions. To properly evaluate this explanation for the early Holocene record on the Snake River, we need more fine-scale environmental data as well as studies of contemporary fish assemblages from other sites in the region (especially on the Snake River). Notably, given the Marmes site's location on a non-salmon producing river, it will be difficult to use the site's fish record per se to identify the role of environmental change in affecting salmonid abundance.

The second major pattern regarding the fish remains is the small body size present. Fish body size estimation has not been carried out for other Plateau archaeological sites for any time period, so we lack a comparative basis for examining the pattern. Additionally, there is little detailed information on the life history, seasonal movements, and schooling behavior for the many freshwater species on the Plateau that could be used to model human fishing strategies. Prehistoric fishers may have used spears, hook and line, or mass harvesting such as nets, traps, or poison. Given the narrow size range, a form of selective mortality is indicated, particularly the use of mass harvesting that was targeting a particular size class of fish (Butler 1996; Greenspan 1998). There are no bone points or net weights in the site's deposits that might be used as independent evidence for fishing methods used. To develop a more comprehensive understanding of Plateau fishing strategies, especially for freshwater fishes for which we have little knowledge, the kinds of strategies used to catch certain species and body sizes, and the more general factors that affect decisions about fishing or technology selection, it would be very useful to carry out fishing experiments in local environments (e.g., Kirch and Dye 1979; Raymond and Sobel 1990).

Comparison of Rockshelter and Floodplain Fish Fauna

Table 11.14 shows the frequency of fish remains identified in the floodplain and the lowest units of the rockshelter, which are contemporaneous. The records show a number of differences. First, the floodplain has a much richer fish fauna with 14 taxa, while the rockshelter only has four. This difference at least in part must be linked to the major difference in archaeological recovery between locales. Use of 1-mm mesh screen in the floodplain led to the recovery of several small-bodied taxa, remains of which would not be caught in ¼" (6.4 mm) mesh. The use of finer mesh also generated an overall larger sample size, which in itself would tend to produce higher richness values (Grayson 1984; Gordon 1993). Both show the presence of anadromous and resident freshwater fishes. Comparison of taxonomic identifications does not reveal any striking difference that could not be explained by recovery practices.

One can also compare the two locales based on vertebra size. Only three vertebrae from the lower units of the rockshelter were complete enough to measure (Table 11.6); these suggest a mean width of 5.6 mm. The almost 400 vertebrae from the floodplain provide a mean width of 1.9 mm (sd=0.5). Granting the very small sample size from the rockshelter, the vertebrae sizes represented in the two locales are extremely different. Recovery practices, particularly the use of ¼-inch mesh screens can be used to explain the lack of small vertebrae in the rockshelter. Possible selective removal (and subsequent loss) of "large", vertebrae during screening of floodplain sediments might help explain the scarcity of such vertebrae in the faunal assemblage. On the other hand, if the dominance of small vertebrae (and scarcity of large vertebrae) in the floodplain deposits is real, that large vertebrae are truly scarce in the floodplain deposits, then it is reasonable to examine why the two locales have such different representation of fish body sizes.

One explanation would suggest that the difference reflects differences in cultural processing between the two areas. Perhaps the same sized resident freshwater fish were captured and deposited in the two areas and that differences in butchering and cooking patterns between the areas led to higher rates of bone destruction in the floodplain. Gustafson and Wegener (see Chapter Ten) point out that the character of the mammalian fauna is very

Table 11.14 Frequency (NISP) of fish taxa in the floodplain and contemporary units of the rockshelter.

Taxon	Rockshelter				Floodplain	
	Below I	I	II	I-II	NISP	%
Salmonidae					3	0.1
Oncorhynchus sp.		3			12	0.5
Prosopium williamsoni					1	0.0
Cyprinidae				1	161	6.5
Ptychocheilus oregonensis					8	0.3
Acrocheilus alutaceus					2	0.1
Mylocheilus caurinus					5	0.2
Richardsonius balteatus					1	0.0
Rhinichthys sp.					2	0.1
Richardsonius/Rhinichthys					2	0.1
Catostomus sp.		2	1		60	2.4
Catostomus macrocheilus					1	0.0
Catostomus columbianus					1	0.0
Cyprinidae/Catostomidae	1	16	1	3	1,503	60.6
Unidentifiable	4		5		719	29.0
Total	5	21	7	4	2,481	100.0

Note: some stratum assignments were made after fieldwork was completed and should be considered provisional

different in the two areas. The rockshelter mammal bones tend to be much more complete (and unburned) than the floodplain mammal remains which are extremely fragmented and burned. They suggest that cultural processing differences may help explain this.

This explanation does not account for the fish pattern, however. The first and second vertebrae on the vertebral column of the minnows and suckers are among the densest in the skeleton (Butler 1996). Even if the larger-bodied fishes had been more aggressively processed (which would lead to a higher incidence of bone destruction), the first and second vertebrae should be present in some numbers, if such fish were caught and deposited on the floodplain. These vertebrae from the floodplain fish fauna were included with other vertebra measures and the data show overall that mainly small individuals are represented. In short, failure to record more large vertebrae in the floodplain is not easily explained by increased fragmentation of bones of larger-bodied fishes.

The small sample size of the rockshelter fish fauna in the earliest stratigraphic units (I, II) will make it difficult to isolate the particular cultural or natural mechanisms that account for the differences between locales. Also, because the faunal records from the two areas were sampled in such different ways, it may be difficult to sort how much of the "patterning" is simply a product of our sampling. Additional comparative study of the feature and artifact record between the two locales will perhaps help shed additional light on the cultural and natural factors that account for the striking differences in faunal records.

12
INVERTEBRATE FAUNA (SHELLFISH)

Pamela J. Ford

Materials and Methods

The Marmes site was excavated in the 1960's under different paradigms than those guiding Plateau archaeology today (see Chapter Two). There is little information available about recovery strategies specific to shellfish; there is no indication that shells were recovered either systematically or with an eye to preserving information about the various uses to which shell has been put in the prehistoric past. The problem this poses to the current analysis is simple: it's hard to apply meaning to the numbers.

That said, this analysis is based upon numbers: counts and weights of shells from various archaeological deposits, and comparisons of relative abundances (percents) based on those shell weights. In this analysis, the sample was determined prior to analysis (see Chapter Two). Concerns about adequate and representative sample size plague all of archaeological research. All of the shells received for analysis were examined and none of the bags were sampled. Descriptions of relative abundances and described changes over time at Marmes must be treated with caution. These descriptions merit comparison with other sites in the region from which representative samples of the shells have been collected in a systematic way.

Artificial Breakage

The majority of the valves of identifiable freshwater mussels (*Margaritifera* sp., and *Gonidea*) at Marmes are broken. This breakage appears consistently on the posterior margin of the valve whether it is a right or a left valve. This breakage is angular either as a straight line break or as an indented break. Both types of breaks (and all the variation in this breakage pattern) appear to be consistent with the use of a pry tool for popping open the closed valve. Many valves are broken on other margins and in other ways, but the majority of the broken valves show some form of breakage on the posterior margin.

Identifiability of Shells

This analyst has noticed in previous analyses that the condition of shell determines how "well" it can be identified. It always seems like a "better" identification if the identification is to the genus- or species-level. Those shells that are in better condition are more readily identified to these taxonomic levels. If a set of shells in better condition are compared with a set of shells in poorer condition, it cannot be concluded that the differences in the taxonomic identifications between the two groups are culturally meaningful. The differences may be the result of the shells' condition or "identifiability."

In this analysis, identifiability was measured by comparing the relative abundance of materials identified to finer taxonomic levels with the relative abundance of materials identified to grosser taxonomic levels. The finest—or most exclusive—level is the species; the grossest—or most inclusive—level is the phylum. This measurement has been used in other analyses to measure the variable condition of shell, at least in part due to erosion of surface features on the shell (Nelson et al. 1986). If the sets of shell coming from various archaeological deposits show the same identifiability, then other differences observed between them are less likely to be due to differential post-depositional factors.

At Marmes, identifiability for shell is distinctively different for the rockshelter deposits when compared with the floodplain deposits. On the floodplain, 100 percent of the shell collected from the Marmes Horizon deposits and 100 percent of the shell collected from the Harrison Horizon deposits is shell that has been identified to the more inclusive taxonomic groupings: superfamily, class, and phylum. That means that the shells in these two deposits exhibit few characteristic attributes that would enable a finer level identification. All of the stratigraphic units in the rockshelter have many shells identified to the more exclusive taxonomic groupings: family, genus and species. That means that the shells themselves are in such condition that they exhibit enough characteristic attributes to enable the finest levels of identification. In other words, shells from the floodplain deposits are more fragmented, sometimes dirtier, possibly more frequently burnt, and perhaps more corroded than the shells from the rockshelter. (A grain-size analysis of the shell would probably show that the floodplain shells are smaller in size.)

Shell Size at Death

There are many reasons to assess shell size at death. This measurement can inform on shellfish population changes over time (perhaps even stress). The measurement may provide clues about resource utilization including overexploitation. It may also suggest explanations about human preferences. A reasonable measurement for predicting the valve size at death is a measure of the hinge width on archaeological specimens. Using the hinge instead of the entire valve allows the use of the frequently fragmented valves from a variety of depositional environments. To establish the reliability and validity of this measure for a particular taxon, a preliminary comparison of valve width with hinge width is carried out for modern specimens within the taxon (Ford 1985).

An assessment of shell size at death for the deposits at Marmes would provide some very interesting information to address the questions Lyman proposed about freshwater mussels for this region (1984). However, this measurement was not taken for several reasons. The taxon with the most securely robust and least friable valve at this site is *Margaritifera*. Yet the majority of the archaeological specimens of *Margaritifera* were broken at the hinge, so that few of the hinges were complete. The archaeological specimens of *Gonidea* were frequently very friable with layers of nacre coming off of the specimens even as they were being handled, thus altering the ability to reliably measure the hinge.

It is recommended that archaeologists in the region continue to collect modern shellfish specimens in order to establish the reliability and validity of particular measurements as indicators of valve size at death, and that this kind of analysis be carried out on archaeological shells whenever possible.

Results

Description of Shellfish Taxa

The following presents the criteria used for specimen identification, tabulation, and includes comments on modern distribution of taxa present in the archaeological deposits at 45FR50. Clarke (1981) is followed for classification of the Phylum Mollusca. The total weight of shells analyzed for this study is 46,686.5g.

Phylum Mollusca—Molluscs
Material: Valve fragments – 145.8g.
Remarks: The molluscs are soft-bodied invertebrates, many with hard valves covering all or some portion of the body. It is the valves that are recovered in archaeological sites. Valves of calcium carbonate break in indeterminate ways, leaving variously shaped pieces. Those included under this taxon are, for the most part, very small in size. Class (Gastropoda or Pelecypoda) is not determined for these fragments. These make up 0.3 percent of the shells in the site.

Class Gastropoda—Snails and Slugs
Material: Valve fragments and columellae – 5.3g, (43 MNI).
Remarks: The gastropods with calcareous valves include the freshwater snails. These molluscs have a single external valve. Archaeological specimens that are identified to class level are in such condition that surface features upon which identifications are usually based are missing. These make up less than 0.1 percent of the shells in the site.

Subclass Pulmonata –Lung-breathing Snails
Order Basommatophora
Superfamily Planorbacea
Family Planorbidae –Ramshorn Snails
Family Planorbidae
Material: Columellae – 0.2g (13 MNI).
Remarks: The shells of this taxon are small and flatly coiled. Many species of this family inhabit regions of inland Western North America where water is present in permanent streams and lakes, and some inhabit temporary water sources. These are a very tiny fraction of the shells in the site: less than 0.01 percent.

Class Pelecypoda—Pelecypods, Bivalves
Material: Indistinct valve fragments – 4,548.2g.
Remarks: The bivalves are molluscs with two hinged valves covering all or most of the soft body. All unidentifiable bivalve fragments are included here. Many of these are moderately small because they do not retain enough identifying characteristics to indicate that they belong to the superfamily Unionacea but they are

not small enough to be confused with fragments of Gastropoda. These make up 9.7 percent of the shells in the site.

Order Eulamellibranchia
Superfamily Unionacea—Freshwater Mussels

Material: Valve fragments, some with partial hinge – 14,127.1g.

Remarks: This superfamily contains the family Margaritiferidae (Pearly river-mussels) and the family Unionidae (Pearly mussels). These are relatively large bivalves with pearly nacre. They occupy freshwater environments exclusively. The valve fragments identified to this taxonomic level exhibit the pearly nacre characteristic of these mussels and are large enough in size to be distinguished from nacreous gastropods. These make up 30.3 percent of the shells in the site.

Family Margaritiferidae

Margartifera falcata –Western Pearly Mussels

Material: Valve fragments with hinges and/or particular mussel attachment scars – 2,315.5g; 611 left valves, 256 right valves, 19 whole valves.

Remarks: Clarke (1981) distinguishes *M. falcata* (the western-river pearl mussel) from *M. margaritifera* (the eastern-river pearl mussel) quite clearly but archaeological reports have not been so clear. What is distinct about the genus is the very visible muscle scar in the beak cavity and the solid structure of the nacreaous valve (less friable than *Gonidea* and more robust than *Anodonta*).

The western pearly mussel is found in Pacific drainages from California and New Mexico north to the southern interior of British Columbia and even further to the north. These bivalves occupy streambeds where the stream is wider than four meters and will live in both hard and soft water. Taylor describes this species' habitat simply as "trout streams" (1981). There are modern riverine locations in which the mussels are so closely packed that the bottom of the riverbed is obscured from view (Clarke 1981). It should be noted that host fishes for this bivalve include chinook salmon, rainbow trout, brown trout and brook trout (Clarke 1981). (Note: *Margaritifera margaritifera* does not appear in the western drainages of North America.) These make up five percent of the shells from the entire site.

Family Unionidae –Pearly Mussels
Subfamily Ambleminae –Button Shells and Relatives

Gonidea angulata –Rocky Mountain Ridged Mussel

Materials: Valve fragments with hinges, complete valves, some fragments with distinctive attributes – 25,505.9g; 2,338 Left valves, 2,382 Right valves, 62 whole valves.

Remarks: The valves of *Gonidea angulata* are approximately the same size as those of *Margaritifera falcata*. What readily distinguishes *Gonidea* is its very friable nacre which chips away almost like the mineral mica. Some of the archaeological specimens were very thickly formed with many layers of nacre forming a thick (but not dense) valve. Some were very small (presumably very young) but most were within the size range of mature *Gonidea* and mature *Margaritifera*. Another valve characteristic that allows identification of this taxon is its distinctively geometric posterior ridge.

This river mussel is distributed in the Columbia River system from southern British Columbia and south in the Pacific drainage to southern California. It prefers creeks and rivers to lakes. In lakes, *Gonidea* specimens occur in muddy sand but have also been observed in a variety of substrates. These make up 54.7 percent of the shells from the site.

Subfamily Anodontinae –Floater Mussels

Anodonta spp.

Materials: Valve fragments that include the distinctive portions of the hinge – 16.5g; 5 left valves, 5 right valves, no whole valves

Remarks: Two species of *Anodonta* are distributed across the region that includes the Marmes site. The two species have been known to coexist in the same habitat (Clarke 1981). Although it is easy to distinguish the species when examining modern specimens, the fragmented nature of the archaeological specimens precludes species-level identification.

Both *A. nuttalliana* (Winged floater) and *A. kennerlyi* (Western floater) occur in rivers and lakes on muddy and sandy substrates. The host fish for these floaters are unknown, however, others of this genus rely on some anadramous species as host fish (Clarke 1973, 1981).

The valves of *Anodonta* are very thin and none were recovered whole. These mussels make up only 0.04 percent of the shells from the site.

Superfamily Sphaeriacea
Family Sphaeriidae –Fingernail Clams and Pea Clams

Materials: whole valves – 0.1g; 1 valve
Remarks: These clams are small bivalves. The members of this family are distributed worldwide in fresh water. Several species of *Sphaerium* and of *Pisidium* are distributed within the site area. Some of them will not occur in swamps or stagnant water but others prefer that. Most prefer a muddy substrate (Clarke 1981). These tiny clams make up less than 0.01 percent of the shells in the site.

Discussion of Results by Stratum

Results of shell identification are presented in Table 12.1 and in Appendix O.

Floodplain Shell

The samples of shell from these floodplain deposits tended to be small in size: the total weight of shell from Marmes Horizon deposits is 14.08 g and the total weight of shell from Harrison Horizon deposits is 4.9g.

The shells themselves were very fragmented. There were no shells from either deposit that could be identified to a level more exclusive than the superfamily. In fact, 77.4% of the shell from the Marmes Horizon and 89.8% of the shell from the Harrison Horizon were identified as Pelecypoda which is simply the taxonomic level of Class (the equivalent of identifying a piece of bone as Mammal).

Marmes Horizon. The shells collected and identified from this stratum are broken and sometimes mixed with dirt. The shellfish could be identified as Mollusca (20.3%), Pelecypoda (77.4%), and Unionacea (1.4%). Since the superfamily Unionacea includes all three of the freshwater mussels that have been recognized at the Marmes site, that identification indicates that freshwater mussel were present, but the condition of the shell precludes a family-level or genus-level identification.

Harrison Horizon. The particular shells, their condition and taxonomic identifications are similar to those of the Marmes Horizon. These shellfish could be identified as Pelecypoda (89.9%) and Mollusca (10.2%).

Rockshelter Shell

Stratum I. Stratum I is the lowest cultural stratum. *Gonidea* makes up 61.5% of the shell in this deposit. *Margaritifera* is present but makes up only 0.5% of the deposit. In this unit we see no *Anodonta* however, this is the provenience of the only specimen of Sphaeriidae. Judging by identifiability, the shells in this deposit are in good condition.

Stratum II. Stratum II resembles Stratum I. *Gonidea* is present as 67.1% of the shell in the unit. *Margaritifera* makes up only 2.9% of the shell. *Anodonta* and Sphaeriidae are not present. Since valves of *Gonidea* make up a substantial portion of the shells present in both Stratum I and Stratum II, it appears that the species is established as a healthy population in the vicinity of the Marmes site.

Stratum III. Stratum III along with Stratum I exhibits the most taxonomic variation present at any of the Marmes site deposits. All three of the large freshwater mussels are present: *Gonidea* (42.6%), *Margaritifera* (1.1%), and *Anodonta* (0.01%). The rest of the shells are made up of the more inclusive classifications: Unionacea (45.9%), Pelecypoda (9.8%), Mollusca (0.4%), and Gastropoda (0.01%).

Stratum IV. Stratum IV is described as relatively undisturbed pumicite (ash from the eruption of Mount Mazama). However, some shell is labeled with this provenience and may be from intrusive pits into the ash layer. The shell content of these intrusions is as follows: Unionacea (50.9%), Gonidea (28%), Pelecypoda (18.6%), Margaritifera (2.1%) and Gastropoda (0.01%).

Stratum V. This stratum lies just above the ashfall and would represent resource utilization patterns in an environment once affected by volcanic ash. Stratum V contains specimens of all three of the freshwater mussels identified at Marmes: *Margaritifera* (30.4%), *Gonidea* (6%), and *Anodonta* (0.2%). An additional 51.6% of the shells represent the superfamily Unionacea. Also present are those identified as Pelecypoda (8.4%) and Mollusca (3.3%).

It is within the shells from Stratum V that the single marine specimen was recovered. One valve of *Olivella biplicata* was identified. *Olivella* shells from the coast were traded throughout western North America in prehistoric and protohistoric times. Many more have been identified from the burials at Marmes (Erickson 1990).

Table 12.1 Summary of invertebrate remains from 45FR50.

Unit	Mollusca	Gastropoda	Pelecypoda	Representation - % Unionacea	Margaritifera	Gonidea	Anodonta	Percentage Identifiable to Genus/Family Level	Percentage Less Identifiable
Rockshelter Stratum									
VIII	0	0	0	28	72	0	0	60	40
VII	0.01	0	8.8	37.4	53.8	5.5	0	53.9	46.1
VI	0.6	0	0.8	53.9	29.7	15	0	40.1	58.8
V	3.3	0	8.4	51.6	30.4	6	0.2	27.3	72.7
IV	0	0.01	18.6	50.9	2.1	28	0	20.4	79.6
III	0.4	0.01	9.8	45.9	1.1	42.6	0.001	41.94	58.11
II	0.3	0.05	9.1	20.5	2.9	67.1	0	70.7	29.3
I	0	0.01	8	30	0.5	61.5	0	69.2	30.72
Floodplain Horizon									
Marmes	20.3	0	77.4	1.4	0	0	0	0	100
Harrison	10.2	0	89.8	0	0	0	0	0	100

Stratum VI. This stratum contains an impressive amount of *Margaritifera* (29.7% of the total shell) and quite a bit of *Gonidea* (15%). In comparison with the earlier deposits, it is apparent by these percentages that the relative abundances of the two mussel species have changed.

Stratum VII. *Margaritifera* shells make up 53.8% of the shells in the stratum. The other shells present are in such condition that they are identifiable only to the superfamily Unionacea (37.4%), the class Pelecypoda (8.8%) or the phylum Mollusca (0.01%).

Stratum VIII. This is the youngest stratum in the rockshelter. Using measures of identifiability, the shells in this deposit are in good condition and represent only *Margaritifera* (72%) and the superfamily Unionacea (28%) that includes all three of the freshwater mussel taxa present at this site.

Discussion

The shellfish recovered from the deposits at the Marmes site present no surprises. The three most common shellfish are the river mussels that have been recovered from many other Columbia Basin archaeological sites (Chatters 1998; Lyman 1980a, 1980b, 1984, 1985, 1988, 1989, 1992, 1995, 1997). *Margaritifera*, *Gonidea*, and *Anodonta* are all river mussels with nacreaous shells that occupy the sediments of freshwater habitats throughout the Columbia Basin (Clarke 1973, 1981; Pennak n.d.; Taylor 1981). The shells of *Gonidea* and *Anodonta* are thin and relatively fragile while, in comparison, the shells of *Margaritifera* are more robust and less friable.

There is a distinction between the floodplain deposits and the rockshelter deposits at the Marmes site. The floodplain deposits contain shell that is less easily identified to the finest-level taxonomic classifications. The shells in the floodplain deposits are more frequently identified to the more inclusive phylum, class and superfamily levels than are the shells in the rockshelter deposits. There may be many reasons for this. The two kinds of deposits may represent different kinds of activity areas. The shells left in the floodplain deposits may have been incorporated into those deposits after a very different sequence of processes than the sequence of processes affecting the rockshelter shells. If that is the case, then the sequence affecting the rockshelter deposits apparently involves processes less likely to break, damage, or corrode the shells in the deposits. Additionally, post-depositional processes on the floodplain are likely to differ significantly from those in the rockshelter. This difference appears to be reflected in the different condition of the shells identified from the two proveniences and the measure used to indicate condition is "identifiability".

A second major distinction in the deposits at the Marmes site is restricted to the Rockshelter deposits. In the earliest strata, the river mussel *Gonidea*, appears in greater numbers than the pearly river mussel *Margaritifera* (Table 12.1). Beginning with Stratum II, the relative abundance (percentage) of *Gonidea* begins to decline and the decline is dramatic, particularly in Stratum IV, which marks the eruption of Mt. Mazama. This raises questions about the ability of *Gonidea* to adjust to the condition for waterways when the ash covered the area. It is possible that the amount of ash incorporated into freshwater environments was enough to begin to affect the health of river mussels available to local human inhabitants of the region. The adverse effects would have been felt in a number of ways. The ashfall may have affected bottom sediments occupied by the shellfish. The ashfall may have been harmful to the anadromous host fish (Clarke 1981). The ashfall may have affected the oxygen levels in the waterways. Adverse conditions could have reduced the availability of *Gonidea* for exploitation by the human inhabitants of the region. If reduction in populations of *Gonidea* did occur, the relative increase in *Margaritifera* use can be explained. But perhaps the decrease in *Gonidea* is the result of culture change that would have occurred with or without the volcanic eruption.

Caution should be exercised in drawing conclusions from the co-occurrence of this change in cultural shell abundance with the ashfall. In particular, there are two problems inherent in this analysis which have bearing on the issue. First, the archaeological recovery strategies for shell at Marmes are so poorly known that we cannot assess the representativeness of the shells in comparison with what was present in the site. Second, the comparisons being made are based upon percentages which form a closed array. Every time that the relative abundance (percent) of one taxon changes, then by necessity, the relative abundances (percents) of all other taxa also change, whether or not the absolute abundances of these other taxa have actually changed. Because of these two problems, it cannot be suggested that the numbers of *Gonidea* and *Margaritifera* identified here represent a real change.

However, this evidence prompts research into other Plateau sites containing shellfish from both before and after the Mazama eruption in order to investigate the possibility that the ashfall adversely affected the populations of *Gonidea* in the region. Lyman has proposed a sequence of shellfish utilization for Columbia Basin sites (1980a). The shellfish taxa included in this sequence are *Margaritifera*, *Gonidea*, and *Anodonta*. Completed shellfish analyses from additional sites are added to the summary in Table 12.2, for comparison with the Marmes analysis. None of the evidence is conclusive, but the reduction in numbers of *Gonidea* appears to be consistent with the occurrence of the Mt. Mazama eruption. Further investigation of this apparent pattern is needed. Future systematic analyses of shellfish from regional archaeological sites and a better understanding of the biological requirements of certain species would allow us to compare the situation from one site to another.

Table 12.2 Columbia Basin bivalve sequence.*

Years B.P.	Pelecypoda Classes Present	Marmes Strata	Marmes Percentages
1,000	*Margaritifera*, *Gonidea*	VI-VIII	*Margaritifera*, (56.3%) *Gonidea* (4.9%)
2,000	*Margaritifera*, *Gonidea*		
3,000	*Margaritifera* *Gonidea*	IV-V	*Margaritifera* (13.4%) *Gonidea* (19.3%) *Anodonta* (<0.001%)
4,000	*Margaritifera*		
5,000	*Margaritifera*		
6,000	*Margaritifera* (high) *Gonidea* (low)		
6,730	----------Eruption of Mount Mazama------------		
7,000	*Margaritifera* (low) *Gonidea* (high)	III	*Margaritifera* (1.1%) *Gonidea* (42.6%) *Anodonta* (0.001%)
8,000	*Margaritifera* (low) *Gonidea* (high)		
9,000		II	*Margaritifera* (2.9%) *Gonidea* (67.1%)
10,000		I	*Margaritifera* (0.5%) *Gonidea* (61.5%)

* Modified from Lyman 1980a:126, Table 4. Modifications include identifications in Lyman 1985, 1988, 1990, 1992, 1994, 1995, and 1997.

13
BOTANICAL MATERIALS

Joy Mastrogiuseppe

The principal focus of the Marmes site investigations was on older deposits (see Chapter One). In 1968 in particular, excavation activity was intense and rushed, and documentation was poor except for levels directly associated with human burials. There was little interest in plant materials. Because the upper levels contained most of the botanically-rich features, and those levels were shoveled out and discarded in the rush to reach older cultural deposits, the plant materials in the resulting collection are a relatively small percentage of the total botanical materials that occurred in the site. As such, the materials selected for this analysis may not provide an accurate representation of the relative importance of various plant materials used by the native people at the Marmes site.

The botanical materials that were collected during the Marmes site excavations consist mostly of wood, with some seeds and a few remnants of textiles and pit linings. The provenience of botanical samples is listed in Table 13.1. Most of the plant materials were collected from sediment layers within the Marmes rockshelter; only 13 samples were from the flood plain area (Table 13.2). In this report the term "sediment samples" refers to those soil samples collected during the Marmes site excavations from which plant materials were recovered for this project by sieving. The term "botanical samples" refers to all the Marmes plant materials analyzed for this project, including those plant materials recovered from sediment samples.

Methods of Analysis

Recognition/identification of archaeobotanical materials requires a thorough background knowledge of the plant species currently occurring in the area and of changes which have occurred in vegetation since the time when the materials were deposited, as well as familiarity with characteristics of these plants which would make them more likely or less likely to be used and/or stored by humans or animals. Also important is information on materials obtained through trade with other groups. The Marmes site botanical samples were studied through comparison with known materials including the author's ethnobotanical reference collection and specimens in the herbaria at Washington State University and the University of Idaho.

Fruits and Seeds

Each sample containing fruits or seeds was placed in a glass Petri dish and sorted. The fruits and seeds were identified by comparison with known material and with standard references (Martin and Barkley 1973; Schopmeyer 1974). Fine features were examined under a dissecting microscope (10X to 70X).

Cordage and Matting

These materials were studied under a dissecting microscope (10X to 70X) and identified by comparison with known samples.

Wood and Bark

Wood samples were examined under a dissecting microscope and the anatomy compared with samples in the author's personal reference collection and wood photomicrographs as well as descriptions in standard references (Friedman 1978; Greguss 1955; Panshin et al. 1964). Where necessary, a small area of the wood was moistened and a razor blade was used to create an even surface or remove a thin wood section. Wood sections were examined under the compound microscope at 100-200×.

Twigs were identified with standard references (Gilkey and Packard 1962; Hayes and Garrison 1960) and by comparison with known material. Bark was identified by comparison with known samples.

Fungi

The three fungi from Marmes Rockshelter were brought for analysis to Jack Rogers, Washington State University, Pullman.

Table 13.1 Provenience of botanical samples Marmes archaeological site, 45FR50.
Quotation marks indicate comments from the sample bags.

Inventory #	Excavation Unit	Depth	Depositional Unit	Comments
675	N95-100/W25-30	90.75-89.96	Unit III	"disturbed"
816	N5/E15	61.09	Marmes	
866	N100-105/W35-40	97.06-96.5	unknown	"from storage pit"
872	N100-105/W35-40	96.0-95.5	probably[1] Unit VII	"bag 3 of 3, from level bag"
903	N95-100, W45-52	93.45-93.27	Unit III	"floor"
936	N95-100, W45-52	94.88	Unit VI	
942	N95-100, W45-52	94.88-94.35	Unit VI	
946	N95-100, W45-52	94.37-93.88	Unit VI	
947	N95-100, W45-52	94.37-93.88	Unit VI	
972	N95-100, W45-52	94.93-94.37	Unit VII	
1059	N10/W40	59.21		"Level XII," "unsorted"
1260	N10/W40	56.39-55.98	Marmes A10-11 (?)	sediment sample, "Level XVI"
1272	N10/W55	none	Marmes A8	sediment sample
1327	N10/W60	58.39	Marmes A7,8	
2080	N10/W35	none	Marmes A	
3474	N25/E15	none	Harrison #2	
4019	N15/W65	none	"likely Harrison 3,4,5"	sediment sample
4031	N75/W50	84.50-84.00	probably[1] Unit 1	sediment sample
4034	N15/W0	60.3	A 3 & 4	sediment sample
4140	N90-95/W10-15	91.31-90.50	Unit III	
4160	N90-95/W35-40	89.83-89.38	Unit II	"disturbed"
4187	N90-95/W30-35	90.36-90.00	Unit II	
4189	N90-95/W30-35	90.36-90.00	Unit II	
4551	N78-80/W10-15	98.58-97.22	Unit VII	from level bag
4698	N95-102/W43-47	96.00-94.93	Unit VII	"associated with earth"
4699	N95-102/W43-47	96.00-94.93	Unit VII	"associated with earth"
4705	none	none	unknown	
5134	N75-80/W30-35	92.18-91.27	Unit III	
5231	N85-90/W20-25	98.0-97.5	probably[1] Unit VII	from level bag
5323	N85-90/W35-40	97.42-97.00	Unit VII or VIII	from level bag
5586	N80-85/W10-15	98.81	Unit VIII	from level bag
5590	N80-85/W10-15	98.81	Unit VIII	from level bag
5692	N80-85/W10-15	98.81	Unit VIII	from level bag
5862	N101.5/W37.5	93.50-93.00	probably[1] Unit III	
6457	N100-105/W30-35	97.00-96.50	probably Unit VIII	
6953	N90-95/W15-20	89.65	probably Unit II	sediment sample from level bag from hearth
7289	N80-85/W10-15	98.81	probably[1] Unit VIII	
7355	N95-100/W25-30	93.5-93.0	probably[1] Unit III	

[1] Depositional Units noted as "probably" were extrapolated by comparing excavation unit and depth with the excavation unit and depth of other (usually non-botanical) samples for which the Depositional Unit had been recorded. If no samples from the same excavation unit and depth had the Depositional Unit recorded, the extrapolation was based on samples of similar depth from adjacent excavation units, or identified in the stratigraphic profiles from the site.

Table 13.1 Provenience of botanical samples Marmes archaeological site, 45FR50. (cont.)
Quotation marks indicate comments from the sample bags

Inventory #	Excavation Unit	Depth	Depositional Unit	Comments
7402	N10-15/W35	none	Marmes A	
8433	N90-95/W35-36	92.5-91.0	probably[1] Unit III	"trench excavated east side for casts"
8730	N90-95/W10-15	94.5-94.0	probably[1] Unit V	
8822	N85-90/W45-52	97.29-97.00	Unit VII	
9239	N100-105/W35-40	93.65-93.50	probably[1] Unit III	sediment sample
9284	N95-100/W30-35	89.40-84.88	Unit II	
9526	N100-105/W35-40	95.0	probably[1] Unit VII	from level bag, "pit"
9551	N10-15/W30	none	between A5 & A6	
9690	N95-100/W25-30	90.75-89.96	Unit III	"unit in disturbed area"
9958	N80-85/W20-25	101.15-100.45	Unit VII	
10522	N100-105/W30-35	none	? probably Unit VII	"strat of storage pit" from level bag
10712	N85-90/W30-35	87.26-86.62	Unit I	"disturbed"
10783	N20/W30	none	between A1-2 & A3	"silt layer"
10823	N90/W45	82.93	probably[1] Unit I	
10827	N90-95/W10-15	90.0-89.5	probably[1] Unit II	
11165	N85-90/W30-35	88.59-88.07	Unit I	
11169	N85-90/W30-35	88.59-88.07	Unit I	
11316	N30/E5	62.53	Harrison 3 & 4	
11322	N10/W35	none	Marmes A	
11343	N15/W35	none	Harrison 3	"and silt above 4th A"
11394	N15/W30	none	A	
12097	N85-90/W30-35	90.55-90.09	probably[1] Unit II	
12232	N95-100/W10-15	97.91-96.92	Unit VIII	
12303	N95-100/W45-52	95.80-94.18	Unit VII	"Feature 3"
12337	N90-95/W10-15	96.66-96.53	probably[1] Unit VII	"from mat area"
12475	N95-100/W15-20	96.61-95.77	Unit VII	from level bag
12495	N90-95/W10-15	95.0-94.5	probably[1] Unit V	
12498	N100-105/W35-40	96.5-96.0	probably[1] Unit VII	from level bag
12499	N100-105/W35-40	96.5-96.0	probably[1] Unit VII	from level bag
12721	N85-90/W40-45	97.25-96.65	Unit VII	
12762	N85-90/W40-45	94.56-93.78	Unit V	
12797	N85-90/W40-45	97.82-97.18	Unit VIII	
12836	N95-100/W15-20	97.34-97.05	Unit VIII	from level bag
12901	N95-100/W35-40	94.5-94.0	probably[1] Unit V	
12932	N95-100/W25-30	93.5-93.0	probably[1] Unit III	from level bag
12958	N95-100/W25-30	96.21	probably[1] Unit VII	"on surface of 96.21"
12974	N100-105/W35-40	97.06-96.5	probably[1] Unit VIII	from level bag
12975	N100-105/W35-40	97.06-96.5	probably[1] Unit VIII	from level bag
12976	N100-105/W35-40	97.06-96.5	probably[1] Unit VIII	from level bag
12979	N100-105/W35-40	97.06-96.5	probably[1] Unit VIII	
12998	N90-95/W15-25	96.0-95.0	probably[1] Unit VI-VII	"surface to bottom of pit layer"

Table 13.1 Provenience of botanical samples Marmes archaeological site, 45FR50. (cont.)
Quotation marks indicate comments from the sample bags

Inventory #	Excavation Unit	Depth	Depositional Unit	Comments
13215	N95-100/W35-40	94.0-93.5	probably[1] Unit V	from level bag
13333	N90-95/W45-50	97.27-94.82	Unit VIII	
13450	N90-95/W10-15	97.5-97.0		
13462	N90-95/W10-15	98.0-97.5	probably[1] Unit VIII	from level bag
13464	N90-95/W10-15	98.0-97.5	probably[1] Unit VIII	
13466	N90-95/W10-15	96.0-95.5	probably[1] Unit VI	
13467	N90-95/W10-15	96.0-95.5	probably[1] Unit VII	
13492	N90-95/W10-15	96.5-96.0	probably[1] Unit VII	
13806	N90-95/W10-15	97.0-96.5	probably[1] Unit VII	
13890	N100-105/W35-40	93.5-93.0	probably[1] Unit III	
13901	N100-105/W35-40	96.5-96.0	probably[1] Unit VII	
13905	N100-105/W35-40	96.5-96.0	probably[1] Unit VII	from level bag
13908	N100-105/W35-40	96.5-96.0	probably[1] Unit VII	from level bag
13973	N90-95/W10-15	95.5-95.0	probably[1] Unit VI	
13987	N100-105/W35-40	95.0	probably[1] Unit VII	"95' to level of pit bottom"
14240	N90-95/W30-35	96.0-95.5	probably[1] Unit VII	from level bag
14298	N90-95/W30-35	84.56-84.28	Unit IB	
14477	N90-95/W30-35	97.05-96.50	probably[1] Unit VII	from level bag
14541	N95-100/W25-30	96.69	probably[1] Unit VII	
14602	N95-100/W45-52	95.80-94.18	Unit VII	"Feature 3"
14609	N95-100/W45-52	95.80-94.18	Unit VII	"Feature 3"
14610	N95-100/W45-52	95.80-94.18	Unit VII	
14655	N95-100/W25-30	96.5-96.0	probably[1] Unit VII	from level bag
14656	N95-100/W25-30	96.5-96.0	probably[1] Unit VII	
14930	N85-90/W30-35	95.0-94.5	probably[1] Unit VI	
14932	N85-90/W30-35	95.0-94.5	probably[1] Unit VI	
15037	N100-105/W20-25	96.10-93.14	Unit VII	"Feature 1"
15053	N85-90/W30-35	93.0-92.5	probably[1] Unit III	
15068	N85-90/W30-35	97.5-97.0	probably[1] Unit VII	from level bag
15157	N95-100/W45-51	96.06	Unit VIII	from level bag
15399	N85-90/W40-45	90.93-90.35	Unit II	
15611	N85-90/W30-35	93.96-93.00	probably[1] Unit V	"trench around burial"
15721	N85-90/W30-35	94.0-93.5	probably[1] Unit V	
15775	N85-90/W30-35	97.5	probably[1] Unit VII	from pit, from level bag
15809	N90-95/W40-45	94.5-94.0	probably[1] Unit V or VI	from level bag
16365	N95-100/W25-30	96.69	probably[1] Unit VII	from level bag
16843	N90-95/W10-15	97.0-96.5	probably[1] Unit VII	from level bag
16859	N90-95/W10-15	97.5-97.0	probably[1] Unit VIII	
16930	N90-95/W10-15	96.66-96.53	probably[1] Unit VII	from level bag
17246	N95-100/W45-52	94.37-93.88	Unit VI	
17309	N95-100/W30-35	91.17-89.40	Unit III	"disturbed"
17351	N90-95/W20-25	87.55-85.67	Unit I	"dist. above Unit 1"
17456	N90-95/W40-45	95.0-94.5	probably[1] Unit VII	

Table 13.2 Botanical samples from the Palouse River floodplain Marmes site.

Inv. No.	Unit	Depth	Level	Plant Materials Identified	Comments
816	N5/E15	61.09	"Marmes"	commercial wood matchstick	from recent times
1059	N10/W40	59.21	"Level XII"	wood knot cf. *Rhus glabra*	locally available
1260	N10/W40	56.39-55.98	Marmes A10-11 (?)	conifer wood, *Rhus glabra* wood (both charred)	sediment sample
1272	N10/W55	none	Marmes A8	grass fragments (charred and uncharred)	sediment sample
1327	N10/W60	58.39	Marmes A7,8	grass stem fragments cf. *Elymus cinereus*	
2080	N10/W35	none	Marmes A	*Pseudotsuga menziesii* wood, *Celtis reticulata* pit fragments	
4019	N15/W65	none	"likely Harrison 3,4,5"	*Artemisia tridentata* and conifer wood (both charred), grass leaf epidermis, *Equisetum* epidermis, rootlets	sediment sample
4034	N15/W0	60.3	A 3 & 4	Angiosperm wood, bark, carbonized sugar (all charred), grass vein	sediment sample
10783	N20/W30	none	between A1-2 & A3	*Celtis reticulata* pit fragments	
11316	N30/E5	62.53	Harrison 3 & 4	*Thuja plicata* wood, grass leaves, cf. *Pseudotsuga* leaf fragments	charred
11322	N10/W35	none	Marmes A	cf. *Polygonum* fruit fragment	
11343	N15/W35	none	Harrison 3 & silt below	*Polygonum* fruit, *Chenopodium* fruits, wood and leaf fragments	all charred
11394	N15/W30	none	Marmes A	*Hordeum* fragments	

Sediment Samples

Sediment samples from the Marmes flood plain were mixed with water and gently flushed through nested Tyler sieves with mesh openings of 2.36 mm, 1.4 mm, 0.71 mm, and 0.5 mm. Each recovered fraction was air-dried on paper and collected into a vial. Samples from the rockshelter were sieved dry. For the two coarser fractions recovered from each sample (caught by the 2.36 and 1.4 mm mesh sieves) plant materials were separated from all other materials (stone, bone, feather, shell, insect fragments, etc.). The plant materials were intensively examined under a dissecting microscope (10-70×), and charred or partially charred plants were separated from uncharred plants. The two finer fractions were studied less intensively; they were quickly scanned for seeds and a general list was prepared of other plant materials present. Wood from sediment samples was identified to the extent possible considering degree of preservation and size of fragments.

Results and Discussion

Common names for plants found in Marmes site samples are used in text (see Table 13.3 for correlation of common and Latin names). Plant materials identified from the botanical samples are listed in Tables 13.4, 13.5 and 13.6, by sample inventory number and the Latin name of the plant. The amount of materials recovered from sediment samples is detailed in Table 13.7, and identifications for these samples are in Table 13.8.

Table 13.3 Common names of plants.

Formal Name	Common Name
Angiosperm	A flowering plant
Arctium minus	Burdock
Artemisia tidentata	Big sagebrush
Boraginaceae	The borage family
Carex pellita	Wooly sedge
Carix vesicaria	Bladder sedge
Celtis reticulata	Netleaf hackberry
Chenopodium	Lambsquarters
Elymus cinereus	Great basin wildrye
Holodiscus discolor	Oceanspray
Hordeum	Barley (wild and cultivated)
Philadelphia lewisii	Syringa or mock orange
Phragmites australis	Broomgrass or reedgrass
Pinus ponderosa	Ponderosa pine
Polygonum	Knotweed
Populus	Black cottonwood and quaking aspen
Prunus	Wild cherries (chokecherry, bitter cherry)
Pseudotsuga menziesii	Douglas-fir
Quercus alba	Eastern white oak
Rhus glabra	Smooth sumac
Salix	Willow
Salix amygdaloides	Peach willow
Salix lasiandra	Whiplash willow
Scirpus acutus	Tules
Sisymbrium altissimum	Tumble mustard
Sitanion	Bottlebrush squirreltail
Thuja plicata	Western redcedar
Zea mays	Corn, maize

Table 13.4 Botanical samples containing seeds from the Marmes site.

Inv. #	Identification	Comments	Analysis
942	5 *Salix/Populus* wood fragments	Very decayed, locally available	
	4 *Rhus glabra* twig fragments	Sumac was sometimes used medicinally. Locally available	
	2 *Thuja plicata* wood fragments	One piece is very decayed.	From commercial molding
	2 *Holodiscus discolor* wood fragments	Oceanspray was an important industrial wood. Locally available	
946	1 *Celtis reticulata* pit	Hackberry fruits were an important food, pits rich in calcium. Locally available, ripen in late summer/fall	
	1 *Celtis reticulata* pit + 5 fragments	Hackberry fruits were an important food, pits rich in calcium. Locally available, ripen in late summer/fall	
2080	1 *Pseudotsuga menziesii* wood fragment		
	Celtis reticulata pit fragments	Chip	
4187	1 *Celtis reticulata* pit + 1 fragment	Hackberry fruits were an important food, pits rich in calcium. Locally available, ripen in late summer/fall	
7355	1 *Celtis reticulata* pit fragment	Hackberry fruits were an important food, pits rich in calcium. Locally available, ripen in late summer/fall	
8730	2 *Celtis reticulata* pits + 1 fragment	Hackberry fruits were an important food, pits rich in calcium. Locally available, ripen in late summer/fall	
10783	2 *Celtis reticulata* pit fragments	Hackberry fruits were an important food, pits rich in calcium. Locally available, ripen in late summer/fall	
10823	2 very weathered *Celtis reticulata* pit halves	Hackberry fruits were an important food, pits rich in calcium. Locally available, ripen in late summer/fall	
10827	1 *Celtis reticulata* pit	Hackberry fruits were an important food, pits rich in calcium. Locally available, ripen in late summer/fall	
11322	1 cf. *Polygonum* fruit fragment	Locally available	
11343	1 *Polygonum* fruit (cf. *P. aviculare* or *P. douglasii*)	Locally available	
	2 *Chenopodium* fruits	Lambsquarters fruits were sometimes eaten. Locally available Ripe in late summer/fall	
	2 tiny wood fragments	Too small for identification	
	2 Boraginaceae leaf fragments	Cf. *Amsinckia*, *Cryptantha*	
	1 rootlet		
12495	3 *Celtis reticulata* pits + 4 fragments	Hackberry fruits were an important food, pits rich in calcium. Locally available, ripen in late summer/fall	

353

Table 13.4 Botanical samples containing seeds from the Marmes site (*cont*).

Inv. #	Identification	Comments	Analysis
12762	1 *Celtis reticulata* pit + 2 fragments	Hackberry fruits were an important food, pits rich in calcium. Locally available, ripen in late summer/fall	
12932	1 *Celtis reticulata* pit half	Hackberry fruits were an important food, pits rich in calcium. Locally available, ripen in late summer/fall	
12979	Rodent nest fragments: macerated grass leaves tiny charcoal fragments		Probably initially from a pit lining
	1 *Chenopodium* fruit	Locally available, ripen in mid/late summer	
13492	2 *Prunus* pit fragments	Wild cherries were one of the favored fruits. Locally available Ripe in late summer/fall	
	1 fecal pellet + 2 fragments		
13987	1 *Celtis reticulata* pit fragment	Hackberry fruits were an important food, pits rich in calcium. Locally available, ripen in late summer/fall	
14930	2 *Celtis reticulata* pits + 14 fragments	Hackberry fruits were an important food, pits rich in calcium. Locally available, ripen in late summer/fall	
15611	6 *Celtis reticulata* pits + 6 fragments	Hackberry fruits were an important food, pits rich in calcium. Locally available, ripen in late summer/fall	
15721	4 *Celtis reticulata* pits + 4 fragments	Hackberry fruits were an important food, pits rich in calcium. Locally available, ripen in late summer/fall	
15809	1 *Celtis reticulata* pit half	Hackberry fruits were an important food, pits rich in calcium. Locally available, ripen in late summer/fall	
	2 Mushrooms	Immature	Brought into rockshelter, not growing there

Table 13.5 Botanical samples containing cordage and matting from the Marmes site.

Inv. #	Identification	Comments	Analysis
947	4 tiny *Scirpus acutus* fragments	Tules locally available, collected in late summer/autumn	From matting
	5 *Salix/Populus* wood fragments	Including four twigs, locally available	
	2 pieces *Pseudotsuga menziesii* wood		
	2 pieces *Rhus glabra* wood	Sumac was sometimes used medicinally.	
		Locally available	
4705	*Carex pellita/C. vesicaria* cordage twist	Sedge cordage was abundant in other Palouse River caves (McGregor, Porcupine). Diameter ca. 1 cm, two-strand, Z-	
		Sedges locally available, collected in mid to late summer	
12337	*Scirpus acutus*	Tules locally available, collected in late summer/autumn	Twined selvage from mat
	Tules were collected in late summer/fall.		
14609	Mouse nest, including:	(Most of the nest is fibers & feathers.)	Matting fragment
	Scirpus acutus	Tule loop, tules locally available, collected in late summer/autumn	
	tiny *Scirpus acutus* fragments	Tules, not numerous. Locally available, collected in late summer/autumn	
	Scirpus acutus fibers	From decayed tules. Tules locally available, collected in late summer/autumn	
	Hordeum fragments and fibers	Locally available	Probably brought in by rodents
	Sitanion fragments	Locally available	Probably brought in by rodents
	Grass leaf fragments	Some are crimped.	Possibly from grass mats
	Grass fibers		Possibly from grass mats
	Charcoal fragments		
	Bark	Curled, very thin	
	Paper fragments		
	Feathers	Abundant	
	Shell fragments		
	Rootlets		
	Hair		

Table 13.6 Botanical samples containing wood and other materials from the Marmes site.

Inv. #	Identification	Comments	Analysis
675	*Sisymbrium altissimum* pod & stem fragments	Non-native plant	Probably blew into cave
816	(Labeled bag 2 of 2) Broken matchstick	Commercially made	Possibly from pit "floor"
866	*Salix amygdaloides/S. lasiandra* wood	One end freshly broken. Locally available	
872	*Salix* bark strips	Willow bark had industrial and medicinal uses.	
903	3 *Thuja plicata* wood fragments	Largest piece	Sharp point beveled on one side, awl or knife? Charred, one side smoothed
936	Puffball fragments, possibly *Scleroderma*	Medium size piece Probably not edible	Brought into rockshelter (not growing there)
942	5 *Salix/Populus* wood fragments	Very decayed, locally available	
	4 *Rhus glabra* twig fragments	Sumac was sometimes used medicinally. Locally available	
	2 *Thuja plicata* wood fragments	One piece is very decayed.	
	2 *Holodiscus discolor* wood fragments	Oceanspray was an important industrial wood. Locally available	From commercial molding
	1 *Celtis reticulata* pit	Hackberry fruits were an important food, pits rich in calcium. Locally available, ripen in late summer/fall	
947	4 tiny *Scirpus acutus* fragments	Tules locally available, collected in late summer/autumn	From matting
	5 *Salix/Populus* wood fragments	Including four twigs, locally available	
	2 pieces *Pseudotsuga menziesii* wood		
	2 pieces *Rhus glabra* wood	Sumac was sometimes used medicinally. Locally available	
972	Grass stem fragment, cf. *Elymus cinereus*	Locally available	
	Thuja plicata wood fragment		
1059	Wood knot, cf. *Rhus glabra*	Sumac was sometimes used medicinally. Locally available	
1260	Conifer wood (charred)	Sediment sample	
	Rhus glabra wood (charred)	Locally available	
1272	Grass fragments	Charred and uncharred. Sediment sample	
1327	Grass stem fragment, cf. *Elymus cinereus*	Charred, locally available	
2080	1 *Pseudotsuga menziesii* wood fragment	Chip	
	Celtis reticulata pit fragments	Hackberry fruits were an important food, pits rich in calcium. Locally available, ripen in late summer/fall	
3474	Twisted fibers of sinew		Thread
4019	*Artemisia tridentata* wood	Charred, locally available. Sediment sample	From firewood?
	Conifer wood	Charred	
	Grass leaf fragments		
	Equisetum epidermis	Locally available	
	Rootlets		Normal soil component
4031	*Rhus glabra* wood (charred)	Locally available. Sediment sample	
	Unidentified fibrous stem tissue		
	Fish bone (2 small vertebrae)		
	Rootlets		Normal soil component

Table 13.6 Botanical samples containing wood and other materials from the Marmes site (*cont*).

Inv. #	Identification	Comments	Analysis
4034	Angiosperm wood (charred)	Sediment sample	
	Bark (charred)		
	Carbonized sugar,	Probably from charred fruit flesh	
	Grass leaf vein		
4140	Grass fragments, cf. *Sitanion hystrix*	Leaves, inflorescence fragment. Locally available	Probably recent
4160	3 pieces *Thuja plicata* wood		
	1 match stick	Commercially produced	Contaminant
4189	*Thuja plicata* wood	Sliver, decayed, one end charred	
4551	*Thuja plicata* wood	Carved zigzag ca. 1.5 cm long Lightning symbol?	Smoothed/polished on flat sides, both sides of point are beveled.
4698	*Philadelphus lewisii*	Three curved pieces (fit together). Locally available	Cradleboard hoop fragments
4699	*Salix amygdaloides/lasiandra*	Two pieces. Locally available	Cradleboard back fragments with lacing holes
5134	*Pseudotsuga menziesii* wood	Beveled end was inserted in something.	Handle, perhaps for a basket or box lid
5231	*Pseudotsuga menziesii* wood	Milled appearance	Stained with blue and red (synthetic colors?)
5323	2 pieces *Pseudotsuga menziesii* wood	One has milled appearance.	
	2 pieces *Thuja plicata* wood	Milled appearance	One is sapwood.
	3 pieces *Salix/Populus* wood	Decayed, locally available	
5586	5 pieces *Rhus glabra* twigs	Sumac stems were sometimes used medicinally. Locally available	Two pieces with cut ends (1 cut in three steps).
5590	1 piece *Thuja plicata* wood		Shaped
5692	2 branch bases, cf. *Arctium minus*	Not native	Probably recent
	Fragments of mushroom caps	"Glued" together by rodent urine	Brought in to rockshelter, not growing there
5862	Grass leaves and other embedded materials	Locally available	Probably from pit lining or separating layer
	Elymus cinereus and possibly *Phragmites australis*		
6457	*Quercus alba* wood	Exotic; white oak is important commercially.	Fragment of commercially milled molding
6953	Wood fragments	Too small to identify. Sediment sample	
	Scirpus acutus stem fragments	Tiny. Tules	Brought in to rockshelter
	Grass leaf fragments		
7289	*Salix/Populus* twig	Ends broken off, locally available	
7402	Grass fragments		
8433	*Pseudotsuga menziesii* wood	Beveled sides, arched shape	From animal dung
8822	*Thuja plicata* wood	Decayed, stained pinkish at one end Locally available. Sediment sample	Shaped, perhaps a part of something bigger? perhaps for piercing?
9239	*Rhus glabra* wood (charred)	Tiny fragments. Locally available	Shaped, perhaps a pestle?
	Cf. *Salix/Populus* wood (charred)	Abundant, perhaps from pit lining?	
	Adhering grass stem and leaf fragments	Locally available	
	Equisetum epidermis		
	Rootlets, mycorrhizal sclerotium	Normal soil components	

Table 13.6 Botanical samples containing wood and other materials from the Marmes site (*cont*).

Inv. #	Identification	Comments	Analysis
9284	*Thuja plicata* wood		Shaped
9526	Non-plant		Fragments of cocoon?
9551	1 wood fragment cf. *Thuja plicata*	Tiny	
	2 cf. *Salix/Populus* wood fragments	Tiny. Locally available	
	1 grass leaf fragment		
9690	1 piece *Pseudotsuga menziesii* wood	Splinter	
	1 matchstick	Charred at end, commercially manufactured	Contaminant
9958	Dense fiber bundle	Includes bark fragments	Insect cocoon/nest
10522	*Thuja plicata* wood	Very soft (decaying)	
10712	*Thuja plicata* wood	Split, one end is flat, curving grain	
11165	*Thuja plicata* wood	Branch splinter	
11169	Part of mouse nest	No recognizable plants	
11316	Cf. *Thuja plicata* charcoal	Tiny fragment	
	Grass leaf fragments and lemma		
	Wood fragments	Tiny	
	Plant tissue fragment	Unrecognizable	
	Cf. *Pseudotsuga menziesii* leaf tip		Solidified with urine
11343	1 Polygonum fruit (cf. *P. aviculare* or *P. douglasii*)	Locally available	
	2 *Chenopodium* fruits	Lambsquarters fruits were sometimes eaten. Locally available Ripe in late summer/fall	
	2 tiny wood fragments	Too small for identification	
	2 Boraginaceae leaf fragments	Cf. *Amsinckia*, *Cryptantha*	
	1 rootlet		
11394	*Hordeum*	Three inflorescence fragments, 1 glume, lemmas	
	2 cf. *Pseudotsuga menziesii* leaf fragments		
	Unrecognizable plant material		
	Coiled animal droppings		
12097	Grass leaf/stem fragments		
12232	*Salix* wood	Grooved stick, one end has been rounded; the other end is broken. Locally available	Appears to be recent Perhaps a game piece (e.g. counting stick).
12303	*Salix/Populus* twig fragment	End is cut. Locally available	From cordage
	Phragmites australis leaves	Twisted. Locally available	Smoothed, shaped, with fresh breaks
12475	*Thuja plicata* wood		Possible pit "floor" pieces
12498	Mixed wood and other materials	Several pieces are stained pink, from berries?	
	4 pieces *Salix amygdaloides/S. lasiandra* wood	3 pieces have been worked. Locally available	The beaver-worked piece was subsequently cut off by humans.
	1 piece *Celtis reticulata* wood	Both ends are cut. Locally available	One end broken after cutting
	1 piece *Thuja plicata* wood		Split and shaped, one side smooth
	Fragments from these pieces	Including *Salix* bark	
	3 small pieces non-plant material	Bone, shell, rock	

Table 13.6 Botanical samples containing wood and other materials from the Marmes site (*cont*).

Inv. #	Identification	Comments	Analysis
12499	1 *Celtis reticulata* branch piece	Upright branch, curving base, lengthwise crack. Locally available	Cut off at base (knot) while branch was alive.
	2 *Salix amygdaloides*/*S. lasiandra* branch pieces	One piece with long tapering cut at end. Locally available	Lever, or perhaps shaped to slide into something else.
	5 small wood pieces—*Salix sp.*	Locally available	
12721	*Thuja plicata* wood	Chip	Splinter
12797	*Thuja plicata* wood	Milled appearance	Shaped
	Salix/*Populus* wood	Locally available	
12836	1 piece *Pseudotsuga menziesii*	Milled appearance, charred	Shaped to a blunt point
12901	3 pieces *Thuja plicata* wood	Two pieces are from a single broken piece.	Perhaps split to shape
		One piece with curving grain around small knot	Small chip
12958	*Thuja plicata* wood fragment		
12974	2 pieces *Salix amygdaloides*/*S. lasiandra* wood	One is worked. Locally available	Sides shaped to a flat surface; possibly a peg
		Other piece has end fibers bent over.	Possibly tapped on rock or other hard surface
	4 pieces *Salix sp.* wood including one twig	1 very small piece is worked.	End of that piece is shaped to a dull point.
12975	*Zea mays*, 2 stem pieces	One stem 25.5 cm long, including two nodes, bent at ± right angle, had also been bent just below first bend but straight now.	Possibly bent while used to poke at something
		Second stem 6.5 cm long, charred at end.	Poking fire, or tinder?
	Zea mays, 3 leaf pieces	Narrow strip is looped, other 2 are twisted.	
12976	1 piece *Pseudotsuga menziesii* wood	Milled appearance, charred	Shaped to a point
	1 piece *Thuja plicata* wood	Milled appearance	Sapwood, with insect hole
12979	Rodent nest fragments: macerated grass leaves tiny charcoal fragments		Probably initially from a pit lining
	1 *Chenopodium* fruit	Locally available, ripen in mid/late summer	
12998	2 pieces wood cf. *Thuja plicata*	Very small	
	Bone fragment		
13215	*Thuja plicata* wood		
13333	3 pieces *Thuja plicata* wood	Largest piece is charred.	All worked (largest piece worked on only 1 side) Shallow groove crossways, a measuring device or game piece?
	1 piece *Pseudotsuga menziesii* wood	Smallest piece	Splinter

Table 13.6 Botanical samples containing wood and other materials from the Marmes site (*cont*).

Inv. #	Identification	Comments	Analysis
13450	Insect pupae		
13462	2 fragments *Pseudotsuga menziesii* wood		
	Grass stem fragment		
13464	Fecal pellet		
13466	Fecal pellet + fragments		
13467	Flattened grass stem fragments		
13806	4 decayed pieces *Philadelphus lewisii* wood	Syringa was an important industrial wood. Locally available	
	4 pieces *Holodiscus discolor* wood	Oceanspray was an important industrial wood. Locally available	
	4 pieces angiosperm wood, probably *Salix*	Small, decayed	
	2 cf. *Salix* bark fragments	Locally available	
	1 grass stem fragment		
13890	Conifer wood cf. *Thuja plicata*	Thin decayed fragments	
	4 pieces of bone		
13901	Macerated grass stem/leaf fragments		From dung or rodent nest
	1 *Salix* wood fragment	Locally available	
	A few other small pieces of wood		
13905	1 piece *Philadelphus lewisii* wood	Syringa was an important industrial wood. Locally available	
	3 pieces *Celtis reticulata* wood	Two pieces have a very smoothly cut end. Locally available	These two from a very short piece that was cut off at one end and then recently cut in two
	1 *Thuja plicata* wood fragment	Locally available	
	2 *Phragmites australis* stem fragments	Non-native plant	
	1 fragment *Sisymbrium altissimum* inflorescence	Smooth surfaces	Probably blew into cave
13908	4 pieces *Thuja plicata* wood	Tiny flakes	Smaller two probably fragments of the larger 2.
13973	Decayed angiosperm wood fragments		
14240	2 pieces *Thuja plicata* wood		
14298	Matted coarse paper		Corrugated cardboard, a contaminant
14477	*Pinus ponderosa* wood		Shaped
14541	*Thuja plicata* wood		
14602	33 pieces *Salix/Populus* wood	At least some is *Salix*. 1 piece charred	Some pieces are stained purplish-red (from berries?)
	17 pieces *Thuja plicata* wood	Locally available	Some pieces are stained purplish-red (from berries?)
	13 pieces *Philadelphus lewisii* wood	Syringa was an important industrial wood. Locally available	One piece was partly cut and then snapped off.
	7 *Scirpus acutus* stem fragments	Tules. Locally available	From matting
	2 pieces of bone		

360

Table 13.6 Botanical samples containing wood and other materials from the Marmes site (*cont*).

Inv. #	Identification	Comments	Analysis
14609	Mouse nest, including:	(Most of the nest is fibers & feathers.)	
	Scirpus acutus	Tule loop, tules locally available, collected in late summer/autumn	Matting fragment
	tiny *Scirpus acutus* fragments	Tules, not numerous. Locally available, collected in late summer/autumn	
	Scirpus acutus fibers	From decayed tules. Tules locally available, collected in late summer/autumn	
	Hordeum fragments and fibers	Locally available	Probably brought in by rodents
	Sitanion fragments	Locally available	Probably brought in by rodents
	Grass leaf fragments	Some are crimped.	Possibly from grass mats
	Grass fibers		Possibly from grass mats
	Charcoal fragments		
	Bark	Curled, very thin	
	Paper fragments		
	Feathers	Abundant	
	Shell fragments		
	Rootlets		
	Hair		
14610	Large animal dung, mostly grass stems/leaves		
14655	*Pseudotsuga menziesii*	Milled appearance, end charred	Firewood?
	Pinus ponderosa	Milled appearance	Shaped
14656	Grass fragments		In fecal fragments
	Wood fragment cf. *Thuja plicata*		
14932	Grass fragments		
15037	*Pseudotsuga menziesii* wood		In fecal material
			Splinter recently split off smooth wood, stained with same synthetic blue as Sample 5231
			Narrow compression marks as if piece was bound to something else
15053	Non-plant	Hemispherical "concretion"	
15068	2 *Pseudotsuga menziesii* wood fragments	1 piece with milled appearance	
	Salix/Populus wood fragment	Very small. Locally available	
	6 *Sambucus* twig fragments	Locally available	
	Several smaller wood fragments		Four pieces were recently split
	Grass leaf fragment		

Table 13.6 Botanical samples containing wood and other materials from the Marmes site (*cont*).

Inv. #	Identification	Comments	Analysis
15157	4 pieces *Pseudotsuga menziesii* wood	1 carved in shape of stemmed point Milled appearance Milled appearance	For training, toy, or hunting practice? Largest piece curves around knothole Rounded on one side
	1 piece conifer bark cf. *Pinus*	Compressed, with melted pitch, with shallow grooves across outside	Part of a boat or other waterproof equipment? apparently was bound to something else
	13 pieces *Thuja plicata* wood	Two pieces have milled appearance, some charred.	Cut or cut and then snapped off
	1 *Phragmites australis* stem piece	Locally available.	End cut
	1 *Zea mays* stem piece	½ short piece of corn stem, including node	Ends cut
	1 small twig segment	First-year twig	
	2 long-bone fragments		
15399	4 pieces *Thuja plicata* wood		Smoothed
15775	2 pieces *Philadelphus lewisii* wood	Root. Locally available	
15809	1 *Celtis reticulata* pit half	Hackberry fruits were an important food, pits rich in calcium. Locally available, ripen in late summer/fall	
16365	2 Mushrooms	Immature	Brought into rockshelter, not growing there
16843	*Thuja plicata* wood	Decayed	Bored out by insects?
	Philadelphus lewisii wood fragments	Largest one hollow, insect remains inside Locally available	
16859	1 *Salix/Populus* twig fragment	Quite decayed. Locally available	
	1 *Holodiscus discolor* twig fragment	Oceanspray was an important industrial wood. Locally available	
16930	*Rhus glabra* wood fragments	Sumac was sometimes used medicinally. Locally available	
17246	*Philadelphus lewisii* twig fragment	Locally available	
17309	1 piece *Pseudotsuga menziesii* wood		End sharpened, other end has been hammered. Two sides very smooth
	3 pieces *Pinus* wood, cf. *P. ponderosa*		1 piece has hack marks.
	3 pieces *Thuja plicata* wood	Narrow strips, smooth sides	1 piece shaped to a point, another cut
17351	2 pieces *Thuja plicata* wood	Smooth surfaces	Larger piece appears deliberately split off. Smaller piece is bent, fibers broken. As if end stepped on while piece held
17456	Manure and bone (no wood)		

Table 13.7 Plant materials recovered from sediment samples (larger fractions) from the Marmes site.

Sample #	Dry Volume (in milliliters)	Dry Volume of plant materials recovered with 2.36 and 1.4 mm mesh sieves (in milliliters)		
		Charred	Uncharred	Total
1260	5	0.1	0	0.1
1272	25	0	0	0
4019	19	0.1	0	0.1
4031	65	5.0	trace	5.0
4034	3.3	trace	0	trace
6953	30	0	0	0
9239	11	trace	0.3	0.3

Marmes Floodplain Samples

Plant materials do not persist as long in soil as rock or bone, unless they are protected from decay by relatively dry conditions (as in caves) or continuously wet anaerobic conditions (as in bogs). Hackberry pits may persist for a long time in the soil due to their chemical structure. All the hackberry pits in the floodplain samples are fragments, and some of them are from deposits more than 1,000 years old. They may be from hackberry trees that were growing on the Palouse River floodplain.

The seeds/fruits of lambsquarters and knotweed are probably from plants growing *in situ*. Great Basin wildrye and sumac may also have been growing on the Palouse River floodplain before it was altered by agricultural activities (certainly they would have been growing nearby). These plants were all culturally important, but their presence in the floodplain samples may or may not have any direct cultural context.

It is more likely that the conifer wood in the samples does represent cultural activity. Both Douglas-fir and western red cedar (*Thuja plicata*) were important industrial woods and were used to make a variety of tools and structures. They probably did not grow on the Palouse River floodplain within at least the last 7,000$^+$ years. These woods were likely brought to the site by water, by humans, or both. Drift logs would have been common on the Snake River (ca. one mile to the south) during times of high water. They would be less common on the lower Palouse River because it drains much less heavily forested areas and because drift logs on the Palouse would be likely to break up tumbling 198 feet over Palouse Falls. The people who were at Marmes probably towed Snake River drift logs up the Palouse to the Marmes area. The presence of possible Douglas-fir leaf fragments in the Marmes floodplain samples would support the idea that drift logs were the source of the Douglas-fir wood in the samples. The charcoal (big sagebrush, conifer wood including western red cedar, and angiosperm wood) and the carbonized sugar from Marmes floodplain samples were likely from campfires or cooking fires (Table 13.2). The matchstick in one sample obviously dates from a time after Euroamerican contact in the area and at the time of excavation had probably not been in the soil very many years (the soft wood it is made from would decay rapidly).

Marmes Rockshelter Samples

Fruits and Seeds. The only fruits/seeds collected from Marmes Rockshelter are hackberry pits (*Celtis reticulata*) and two fragments of wild cherry pit (*Prunus emarginata* or *P. virginiana*) (Table 13.4). Hackberry fruits were eaten by early people in the area, usually ground up whole. The pits are rich in calcium (Yanovsky et al. 1932) and are believed to be primarily responsible for the strong teeth of the Nez Perce people. Hackberry is frequent along the Snake River and probably was also frequent along the lower Palouse, so an abundant local supply would have been available. The fruits ripen in late summer and persist on the trees into the winter. This author believes that many or most of the hackberry pits in Marmes Rockshelter were brought there by humans, based on their number and broad areal distribution. The nearby McGregor Cave (45FR201) rockshelter also contained more hackberry pits than any other kind of fruit or seed (Mastrogiuseppe 1994).

Table 13.8 Plant materials in sediment samples from the Marmes site.

Sample	Charred materials	Comments	Uncharred materials	Comments
1260	Conifer wood *Rhus glabra* wood	Locally available		
1272	Grass fragments		Grass fragments	
4019	*Artemisia tridentata* wood Conifer wood	Locally available, from firewood?	Grass leaf fragments *Equisetum* epidermis Rootlets	Locally available
4031	*Rhus glabra* wood	Locally available	Unidentified fibrous stem tissue Fish bone (2 small vertebrae) Rootlets	Normal soil component
4034	Angiosperm wood Bark Carbonized sugar	Probably from charred fruit flesh	Grass leaf vein	
6953	Wood fragments	(Too small to identify)	Tiny *Scirpus acutus* stem fragments Grass leaf fragments	Tules
9239	*Rhus glabra* wood	Locally available	Adhering grass stem and leaf fragments	Abundant, perhaps from pit lining?
	Cf. *Salix/Populus* wood	Tiny fragments. Locally available	*Equisetum* epidermis Rootlets Mycorrhizal sclerotium	

The presence of two wild cherry pit fragments probably has no cultural significance, although cherries were eaten by early people in the area and cherry pits were present in large numbers in other Palouse River rockshelters (Mastrogiuseppe 1994, 1995). Wherever they grow, wild cherries have been gathered for food. They were usually ground whole and made into a pudding or sauce (the grinding process oxidizes the cyanide in the pits [Timbrook 1982]).

Textiles. Only one fragment of cordage was collected from Marmes Rockshelter (Sample 4705; Figure 13.1 h, Table 13.5). Unfortunately it was not catalogued during excavation so its provenience is unknown (it may have been on the surface and out of context). This piece of cordage is 13 cm long and 1.4 cm in diameter. It is very similar to much of the cordage from other nearby rockshelters: Squirt Cave (45WW25), McGregor Cave (45FR201), Porcupine Cave (45FR202) (Endacott 1992; Mallory 1966; Mastrogiuseppe 1994, 1995). The cordage is made from twisted stems/attached leaves of sedges (*Carex pellita* or *C. vesicaria*), two strands twisted together in the typical fashion for coarse cordage, a Z-twist (Mallory 1966). The abundance of sedge cordage found in certain Columbia Plateau sites may be unique (Mallory 1966), but sedges were the most common material used to create the medium width and coarse cordage found at these sites (Endacott 1992; Mallory 1966; Mastrogiuseppe 1994, 1995). It is likely that many fragments of cordage were discarded during later Marmes excavations.

No basketry fragments are present in the materials collected from Marmes, and only one fragment of actual matting is present (although other samples do contain tule fragments that appear to have come from matting). The matting fragment is a twined selvage from a tule mat (Figure 13.1 g, Table 13.5). A comment with the specimen ("from mat area") implies that there was more matting present in the rockshelter. Tule mats were the all-purpose textiles along the lower Palouse and Snake Rivers, being used for everything from roofing and floor mats to plates. They were also used to separate layers in food storage pits (probably worn mats were used for this purpose [Hicks and Morgenstein 1994]). Tules were locally available and their air-filled stems make them excellent for insulation and cushioning.

Wood. Marmes rockshelter contained a considerable quantity of wood. The wood was well distributed through the excavated areas and through different excavation levels.

Western red cedar (*Thuja plicata*). The most frequent wood from Marmes is western red cedar, found in 35 samples (Table 13.6). This is nearly twice as many samples as any other wood. Fourteen of these samples contained western red cedar that showed evidence of working or shaping. One particularly interesting piece is a small, carved zigzag resembling a lightning symbol (Sample 4551; Figure 13.1 a, Table 13.6). Its function is unknown, perhaps an ornament or spiritual symbol. One piece of red cedar from Sample 13333 is 4.6 x 0.5 x 0.3 cm and has been modified by smoothing and shallow grooves are apparent across one surface. This could possibly be a measuring device or game piece. Western red cedar has been called the plant with the greatest total number of indigenous uses (Moerman 1998), and it was a very important industrial material on the Plateau. One of the reasons the wood was favored is that the chemistry of its heartwood makes it resistant to decay. At the same time, the wood is soft and easy to shape. Although no western red cedar has grown anywhere near Marmes Rockshelter within the past few thousand years, uprooted trees commonly drifted down the Clearwater and Snake Rivers. Because these trees float high in the water (the wood is lightweight) they were frequently pulled out of the rivers by indigenous people. It is probable that the Marmes people towed cedar logs up the Palouse River and then ashore where they could make things with them—ranging from dugout canoes to small awls and ornaments.

Willow (*Salix*) and cottonwood/quaking aspen (*Populus*). Willow, cottonwood, and aspen woods are quite soft, structurally weaker and faster-decaying than locally-used coniferous woods, and fast-burning. These woods were used in many ways. The wood is easily shaped and the supply was abundant. Some willows have special characteristics that make them particularly useful. For example, the stems of sandbar willow (*Salix exigua*) are unusually flexible and they were used for lashing, binding, and making strong coarse water-resistant cordage. It is not possible to distinguish the various species of willow on the basis of wood anatomy, but several species would have been locally available at Marmes, including

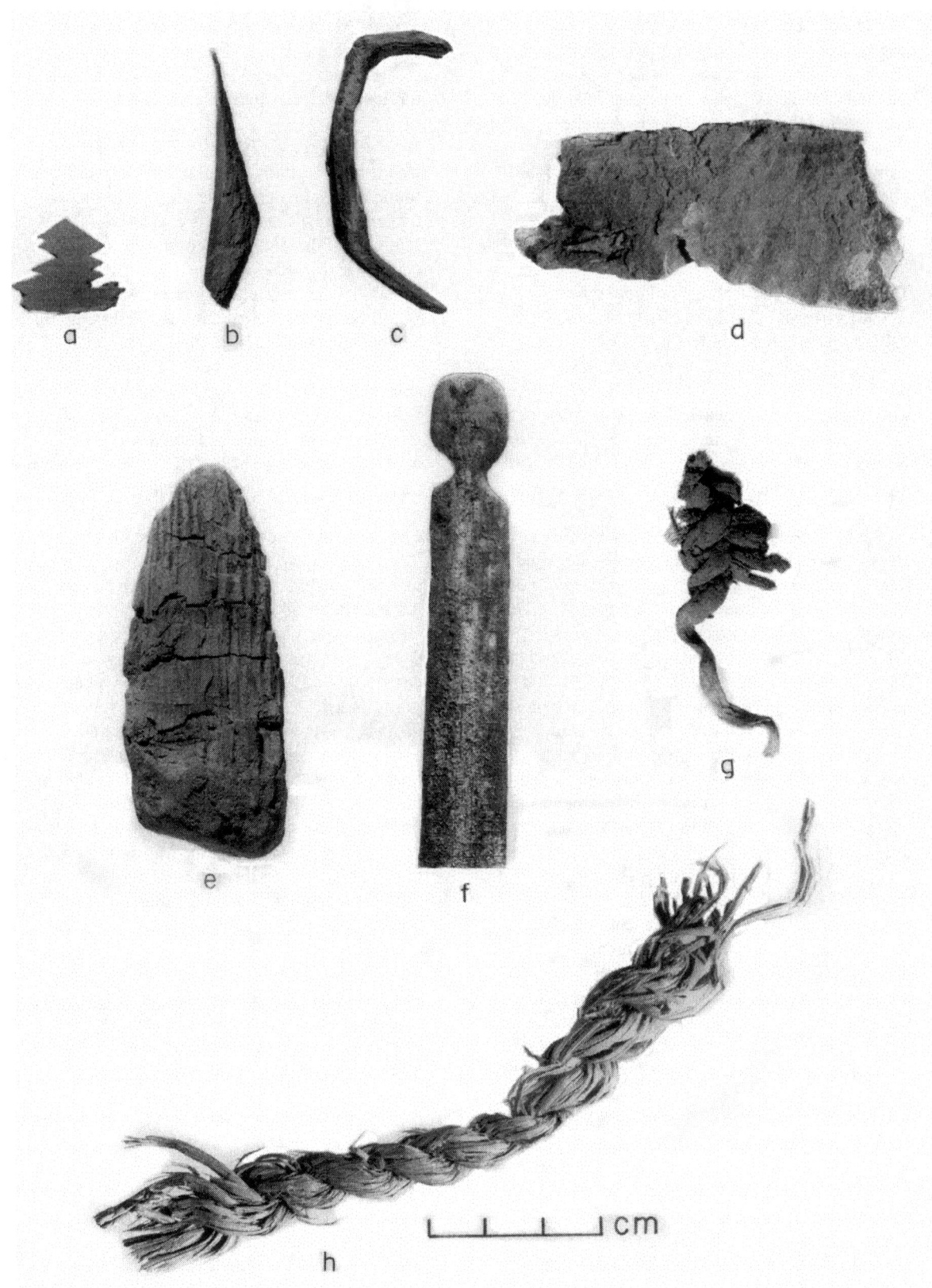

Figure 13.1 Botanical artifacts from the Marmes site (a=Inv. No. 4551, b=#8433, c=#5134, d=#15157, e=#8822, f=#4698, g=#12337, h=#4705).

the large trees peachleaf willow (*S. amygdaloides*) and/or whiplash willow (*S. lasiandra*). It is also not possible to distinguish cottonwood and aspen by wood anatomy, and often not possible to distinguish cottonwood/aspen from willow (especially if preservation is not good or the sample piece is small).

Eighteen of the Marmes samples contain wood of willow and/or cottonwood/aspen (Table 13.6). Six of these samples are relatively large pieces that would have come from peachleaf or whiplash willow. Most of the Marmes willow is unmodified or minimally modified. Interesting pieces include one from Sample 12499, which may be a lever or pry bar, and a grooved stick 23.5 cm long by 1.2 cm in diameter (Sample 12232) which might be a game piece. Perhaps the most significant willow artifact from Marmes is the cradleboard back (Sample 4699, Table 13.6). Two pieces of the cradleboard cover back are present, one 28 x 8.2 cm and the other 12.5 x 5.5 cm. Both pieces show the holes used for lacing on a cradleboard cover.

Douglas-fir (*Pseudotsuga menziesii*). Douglas-fir wood was present in 16 Marmes samples (Table 13.6). Some pieces have a milled appearance. Several of the hand-worked Douglas-fir pieces are quite interesting. One may be a piercing device; it is 4.5 cm long and long-tapered to a point (Sample 8433; Figure 13.1 b, Table 13.6). Sample 15157 includes a piece of Douglas fir carved to the shape of a stemmed point 4.4 x 1.3 cm (stem 2 cm long). This might have been used for training or as a toy. Sample 5134 is a small handle carved from Douglas-fir, 4.7 cm long, 1.4 cm wide, and with the ends bent over for about 2 cm (Figure 13.1 c, Table 13.6). The ends are beveled and were apparently inserted in something else. This object may be the handle of a lid from a box or basket. Sample 8822 is a thick piece of red cedar shaped into an oval 6.5 x 2.5 cm. The narrower end is stained pinkish, and the implement may be a pestle (Figure 13.1 e, Table 13.6).

Ponderosa pine (*Pinus ponderosa*). Three samples contain pine wood (Table 13.6). Sample 15157 includes a piece of conifer bark that may be pine bark. This bark piece (6.5 x 3 cm) appears to be somewhat compressed and contains a lot of melted pitch between the bark layers. The top surface is impressed in lines where apparently binding material passed over it (Figure 13.1 d, Table 13.6). Thus it appears that the bark was bound to something else. It is possible that this piece was part of a boat or other equipment used in or with water.

Native shrubs. Wood of at least five native shrubs is present in the Marmes samples (Table 13.6). Syringa and smooth sumac are in seven samples each. None of the sumac is worked, and sumac would not be a good candidate for working because the wood is extremely soft with a large pith. Sumac thickets are abundant in the vicinity of the Marmes site, and the rhizomes and twigs have medicinal uses (Broncheau-McFarland 1992; Harbinger 1964; Turner et al. 1990). It is possible that people brought sumac into the rockshelter for medicinal use, or it may have been brought in by rodents. Syringa, on the other hand, is a strong hard wood that does not splinter or warp, and it was used for making items requiring structural strength: snowshoes, bows, digging sticks, and other implements (Turner 1979). Most of the syringa in the samples has not been worked, with the notable exception of the fragments of a cradleboard hoop (Sample 4698; Figure 13.1 f, Table 13.6). The Marmes cradleboard hoop is made from syringa wood bent and twisted into the arching hoop shape. The sample consists of three articulating pieces, 26, 25, and 21 cm long, with a maximum width of 1.7 cm. The use of syringa for cradleboard hoops has been previously documented for other Columbia Plateau groups including the Okanogan and the Thompson (Turner 1979; Turner et al. 1990).

The other native shrub woods present are hackberry and oceanspray (two samples each), and blue elderberry (one sample). The hackberry wood pieces have cut ends. Two of these pieces (in Sample 13905) are from a short piece (originally 4.6 x 1.3 cm) that was very smoothly and recently cut in two; the cut ends fit closely together. The original piece had been cut at at least one end, but not recently. It is curious that such a short piece would be cut in two.

The oceanspray wood in the Marmes samples was apparently not worked. Some of the elderberry twigs were split lengthwise, possibly through natural processes. Elderberry stems have a very large pith, which breaks down, leaving a hollow center. This is why elderberry is such a good material for flutes, whistles, and pipe stems.

Grasses. Great Basin wildrye stems and leaves were in three samples (Table 13.6). In two of these samples the stems/leaves were crisscrossed at various angles. In one of these two samples, small "bundles" of grass are cemented into a paperlike layer, apparently by rodent urine; the other is a sediment sample. It is believed both of these represent pit lining material. A few remnants of broomgrass (*Phragmites australis*) stems and leaves are also present in the Marmes samples. Some of the broomgrass leaves are twisted, suggesting that they are remnants of cordage.

Pieces of corn stems (*Zea mays*) are in two samples, both taken from about the same depth. This suggests that corn may have been grown at or near Marmes, presumably during post-contact times. At another Palouse River site (45WT2, at the river's confluence with the Snake River), there were charred corncobs in a rock oven that was in the upper site sediments (Nance 1966:39,105). Charred wood from the same oven yielded a radiocarbon date of 150 ± 70 B. P., which is probably after Euroamericans arrived in the area (Nance 1966). Nance notes that corn was grown on the lower Snake River by 1839 and was probably first planted at the Spalding mission (on the lower Clearwater River near its confluence with the Snake River) within two years after the mission's establishment in 1836 (Nance 1966:39).

Other grass fragments are scarce in the collection and may not be culturally significant.

Fungi. Three fungus samples are included in the Marmes botanical materials (Table 13.6). One includes two small immature white mushrooms at the "button" stage (Sample 15809). A second sample (5692) is fragments of a different, larger, mature mushroom cap. The third sample (936) contains puffball fragments which might possibly be a *Scleroderma* species. These fungi can only be accurately identified by a specialist in the specific group of fungi, and that could take days of analysis (J. Rogers, personal communication 1999). It is quite unlikely that any of these fungi actually grew in the rockshelter, as they are types that usually grow in shrub steppe habitat, not in caves. It is also unlikely that they were brought in for food (especially not the puffball if it is a *Scleroderma*—most *Scleroderma* species are poisonous [J. Rogers, personal communication 1999]). These three fungi may have been used for medicine or for spiritual/ceremonial purposes.

Other plant materials. Sample 872 contains strips of willow bark (Table 13.6). These strips were probably deliberately removed from the tree, as there are cut ends on three pieces. Cordage made from willow bark has been found in McGregor Cave (Mastrogiuseppe 1994), and the bark was also used medicinally. Other plant materials at Marmes include corrugated cardboard, commercial matchsticks, and remnants of the weedy introduced plants tumblemustard (*Sisymbrium altissimum*) and burdock (*Arctium minus*) (Table 13.6).

Temporal Considerations

Archaeologists have defined eight major depositional units at Marmes Rockshelter (Fryxell and Daugherty 1962) with the oldest Stratum I radiocarbon dated at 9280–10,810 B.P. uncorrected radiocarbon years (Sheppard et al. 1984, 1987). Some plant materials have apparently been used by Marmes people throughout the time periods represented by the Marmes excavations. Wood of western red cedar was recovered from each of the depositional strata at the Marmes site (Table 13.9). Western red cedar does not now occur anywhere near Marmes Rockshelter and is unlikely to have occurred near there during the time periods represented by cultural materials at Marmes. However, far upstream on the Palouse and especially on the Clearwater/Snake River drainages there are stands of western red cedar. Since this tree prefers moist habitats, many of these stands occur on the river floodplains, and drift logs would have been relatively common. All of the western red cedar at Marmes would have been carried in from distant areas either by the rivers or by humans. Marmes people may have obtained some western red cedar through trade, and after Euroamericans came to the inland Northwest the Native people had access to milled lumber.

Douglas-fir wood was collected from Stratum III and Strata VI-VII (Table 13.9). Ponderosa pine wood was collected from Stratum VI and VII. There may have been some ponderosa pine (and possibly even some Douglas-fir) growing on cooler slopes/floors of canyons near Marmes during less xeric climatic phases, although the scarcity of these woods in Marmes deposits suggests otherwise. It is more likely that these woods, like western red cedar, were carried to the area by water or by people. Douglas-fir wood was used on the Plateau for many purposes (Aoki 1994; Blankinship 1905; Broncheau-McFarland 1992; Coville 1904; Harbinger 1964; Hunn 1990; Lepofsky et al. 1996; Moerman 1998; Spinden 1908). It is strong and relatively durable. Ponderosa pine wood is not as strong as Douglas-fir but is suitable for many uses, including firewood (Aoki 1994; Blankinship 1905; Broncheau-McFarland 1992; Churchill 1983;

Table 13.9 Plant materials by age/depositional unit, Marmes Rockshelter.

Unit	Characteristics (Leonhardy and Rice 1970; Sheppard et al. 1984, 1987)	Age (Radiocarbon Years) from Sheppard et al. 1984	Cultural Phase (Leonhardy & Rice 1970)	Plant materials collected
Unit VIII	Surface/highly disturbed Horses arrived ca. 1730 (Haines 1938) Direct Euroamerican contact 1805	< ca. 600 B.P.	Numipu	*Thuja plicata* wood (some shaped including stemmed point) *Thuja plicata* wood, with some very smooth surfaces shaped *Salix* wood (including possible game piece) *Salix amygdaloides/S. lasiandra* wood (some worked) *Salix/Populus* wood (some shaped) *Rhus glabra* twigs (cut ends) *Pseudotsuga menziesii* wood *Pseudotsuga menziesii* wood with milled appearance worked conifer bark *Celtis reticulata* wood (ends cut) *Holodiscus discolor* wood *Philadelphus lewisii* wood *Sambucus* twigs (some split) *Quercus alba* wood (commercially milled molding) Mushrooms *Chenopodium* fruit *Phragmites australis* stems/leaves Other native grasses *Zea mays* stems/leaves Non-native weeds (*Arctium, Sisymbrium*)
Unit VII	Fire hearths, storage pits Moderate rockfall, windblown sediments People lived in winter pithouse villages. Increased use of root foods Food storage pits were used.	660 B.P. – 1600 B.P.	Harder, Piquinin (sample bags say "poss. disturbed")	*Thuja plicata* wood, some shaped (including pestle) *Pinus ponderosa* wood with milled appearance *Pseudotsuga menziesii* wood with milled appearance *Philadelphus lewisii* wood (cradleboard hoop) *Salix amygdaloides/S. lasiandra* cradleboard back *Salix amygdaloides/S. lasiandra* wood (some worked) *Salix/Populus* wood, *Salix* bark *Philadelphus lewisii* wood (some cut) *Holodiscus discolor* wood *Rhus glabra* wood *Scirpus acutus* stems from and in matting *Phragmites australis* leaves from cordage *Elymus cinereus* stem Other native grass fragments *Celtis reticulata* pits *Prunus* pits

Table 13.9 Plant materials by age/depositional unit, Marmes Rockshelter (cont).

Unit	Characteristics (Leonhardy and Rice 1970; Sheppard et al. 1984, 1987)	Age (Radiocarbon years) from Sheppard et al. 1984	Cultural Phase (Leonhardy & Rice 1970)	Plant materials collected
Unit VI	Fire hearths, filled excavations Moderate rockfall, windblown sediments People began to stay in villages for winter. Root foods became more important. Collector strategies rather than foraging Food storage pits were used.	1300 B.P. – 1940 B.P.	Tucannon but sample bags say "disturbed" (presumably by later burials/ storage pit digging)	*Scirpus acutus* stem fragments from matting *Thuja plicata* wood *Pseudotsuga menziesii* wood *Rhus glabra* twigs/wood *Salix/Populus* wood *Holodiscus discolor* wood *Phiadelphus lewisii* twig *Celtis reticulata* pits Puffball
Unit V	Windblown sand, sparse rockfall, much mixing Climate somewhat cooler & more mesic	4250 B.P.	Late Cascade	*Thuja plicata* wood *Celtis reticulata* pits
Unit IV	Mazama ash, undisturbed	6730 B.P.		
Unit III	Occupation level. Windblown sediment, Moderate rockfall, shell/bone midden lenses People made greater use of plant foods. Warming climate	6200 B.P. – 7870 B.P.	Early Cascade but sample bags say "disturbed" (presumably by later burials/ storage pit digging)	Non-native weedy plant (*Sisymbrium altissimum*) *Pseudotsuga menziesii* wood (including handle, possible piercing object). *Thuja plicata* wood *Pinus ponderosa* wood with some very smooth surfaces *Celtis reticulata* pits *Elymus cinereus* stems/leaves Other native grass fragments commercially made matchstick (contaminant)
Unit II	Coarse rockfall, windblown sediment Somewhat warming climate	8525 B.P. – 9540 B.P.	Windust but sample bags say "disturbed" (presumably by later burials/ storage pit digging)	*Thuja plicata* wood (some shaped *Celtis reticulata* pits commercially produced matchstick (contaminant)
Unit I	Coarse heavy rockfall, little fill People were broad-spectrum foragers.	9820 B.P. - 10,810 B.P.	Windust	*Thuja plicata* wood *Celtis reticulata* pits *Rhus glabra* wood Matted corrugated cardboard (contaminant)

Coville 1904; Cutright 1969; Endacott 1992; Harbinger 1964; Hart 1976; Hunn 1990; James 1996; Lepofsky et al. 1996; Malouf 1969; Moerman 1998; Newberry 1887; Spinden 1908; Striker 1995; Turner 1979, 1988; Turner et al. 1990).

One sediment sample probably from Stratum I contained sumac wood (Table 13.9). Other than that sample, wood of native shrubs is not represented in the Marmes rockshelter botanical samples until Stratum VI, where there was wood of at least four shrubs—willow, syringa, oceanspray, and sumac (Table 13.9). All these woods were also present in the more recent Stratum VII and VIII (Table 13.9). It is not presently known if this apparent diversification in woods used is also indicated at sites in other areas of the Columbia Plateau (not enough wood has been analyzed). However, it is possible that the diversity of wood from more recent Marmes deposits reflects an increasing focus on technological use of plant materials beginning with the Tucannon cultural phase (Leonhardy and Rice 1970). Although much of the Marmes wood is apparently not worked by humans (except for some cut ends) its presence in the rockshelter at least suggests that people were bringing it in. These shrub species (except willow) may have grown within rodent foraging range of the rockshelter, but based on their association with cultural materials and their consistent presence in these Units, it is more likely that at least some of them were brought into the rockshelter by humans. Willow would have been the most abundant locally available wood, growing on the river floodplains and in other moist areas. Willow wood is soft and easily worked. It was used for many purposes including construction, implements, and firewood. Syringa and oceanspray woods are harder and more durable than willow and were used where strength is more important. The sumac may have been used as kindling. Sumac was also used for medicinal purposes by various Columbia Plateau groups including the Nez Perce (Broncheau-McFarland 1992; Harbinger 1964), the Wanapum and associated groups (Hunn 1990), the Thompson (Turner et al. 1990), and the Blackfoot (Hart 1976, 1979).

Among the few textile materials collected from Marmes, tules occurred in Stratum VI and VII; undoubtedly they were also originally present in Stratum VIII. With a local supply of tules readily available, it is unlikely that Marmes people stopped using tules before Euroamerican contact. After contact, indigenous Columbia Plateau peoples quickly adopted new materials like canvas and other machine-woven fabrics. The lack of tule remnants below Marmes Unit VI may be due to lack of preservation in these older deposits or to an actual difference in the cultural importance of tules, or both. Materials recovered from McGregor Cave (45FR201), Porcupine Cave (45FR202), and Squirt Cave (45WW25) did not extend back to times older than those represented in Marmes Stratum VI.

Little is known about the storage pits constructed during the late prehistoric period use times in Marmes Rockshelter except that they were present and that during Marmes excavations they were largely destroyed and their materials discarded (C. Gustafson personal communication 1997). However, two samples that are probably from Stratum III (5862, 9239) contained matted Great Basin wildrye stems and leaves that are in crisscrossing bundles. This is not a depositional pattern characteristic of rodents, and this author believes these grasses represent pit lining/layering material. The pits they represent are probably intrusive into Stratum III, i.e. the pits would have been constructed during a more recent period and dug down into deposits representing earlier times. Sample 12498 (Table 13.6) (probably from Marmes Stratum VII) contains pieces of willow/poplar stems that may have formed the "floor" of a storage pit (Hicks and Morgenstein 1994).Although hackberry pits collected from Marmes are not numerous (almost certainly due to excavation methods such as the use of ¼-inch screens that would not catch most seeds), their consistent presence in Marmes Rockshelter sediments suggests that hackberry fruits were probably used for food throughout the history of the site. Hackberry pits were plentiful in the nearby McGregor Cave and Porcupine Cave, in numbers too large to be accounted for solely by rodent activity (Mastrogiuseppe 1994, 1995).

Botanical materials collected from the Marmes floodplain are too few to draw any meaningful temporal conclusions (Table 13.10). Most could have been from plants growing on the floodplain, except western red cedar and Douglas-fir, the presence of which are discussed above.

Contaminants

When interpreting materials from Marmes excavations, it must be kept in mind that contamination of Depositional Units occurred

Table 13.10 Plant materials by depositional unit from the Marmes floodplain.

Stratigraphic Unit	Plant materials collected
Marmes	broken commercial matchstick
Marmes A1 & A2	grass, including *Hordeum*, *Pseudotsuga menziesii* wood, cf. *Pseudotsuga menziesii* leaves, *Celtis reticulata* pits, cf. *Polygonum* fruit
between Marmes & Harrison	*Celtis reticulata* pits
Harrison A3, A4, A5	*Thuja plicata* wood, *Artemisia tridentata* wood, Cf. *Pseudotsuga menziesii* leaves, bark, grass leaves, *Polygonum* and *Chenopodium* fruits, carbonized sugar
Between A5 & A6	*Thuja plicata* wood, *Salix/Populus* wood, bark, grass leaves
Marmes A6	no documented samples
Marmes A7 & A8	*Elymus cinereus*
Marmes A8	Grass fragments
Marmes A9	*Rhus glabra* wood, coniferous wood
Marmes A 10-11 (?)	*Rhus glabra* wood, coniferous wood

both in ancient times through reworking of strata during construction of Marmes food storage and burial pits and in contemporary times through possible vandalism/pothunting and Marmes excavation procedures. Marmes Rockshelter was excavated in 1962, 1963, and 1968. Materials that blew in to units under excavation, were accidentally kicked in, or were deliberately placed in these units during early excavations were sometimes collected during later excavations. Clear examples of contamination by plant materials include the presence of commercial matchsticks in Units II and III, of corrugated cardboard in Unit I (cardboard was placed on the exposed surface of excavation units as a sitting/kneeling surface for excavators), and of commercially milled wood at any level below Unit VIII. Sometimes hand-worked wood resembles milled wood, especially when surfaces have been smoothed and the splits made formed right angles. However, the appearance of some of the wood from Marmes Rockshelter strongly suggests that it may have originated from a commercial mill.

14
SUMMARY OF RESULTS

Brent A. Hicks

Results of the different analyses conducted as part of this study of the Marmes Rockshelter excavations are summarized to the extent necessary to place them in an interpretive format that addresses the study's research goals as presented in Chapter Two. Each of the following summary discussions is derived from the previous chapters and the analysts' reports that were their basis.

Background Research

Chapter Two recounted the research efforts into the site records and materials and how the information collected was used in refining the study's goals and expectations. Research conducted throughout the period of this study addressed the collected materials from the site artifacts and other materials gathered during the excavations. Project records were examined for information related to items that either are missing from the collection (n=695) or for which provenience information is no longer present (n=1,497). The records also were examined to determine the methods used in fieldwork and to establish the extent of post-field analysis conducted and the results of those analyses; this information also contributed to development of this study's sampling plan. Field notes and logs were examined for detail on features and other areas of the site's deposits of particular interest to the current study. Unfortunately, the field catalogs are incomplete with the kind of detail needed for matching bags to their collected locations generally lacking. The detail in the excavators' field notes varies considerably and rarely provides enough information to correlate with the catalogs or the few forms that were prepared. Level forms were not used in the early years' excavations. In 1968, the level forms consist largely of text statements that resemble field excavator's notes with the attendant variability in information types and detail usually found in such notes; as such, they lack uniform information presentation from one unit to the next. Feature forms were only used for burial features. Stratigraphic profiles, however, were consistently prepared. In summary, much that happened archaeologically is not reported and this hampered attempts to assign provenience to materials in the collection. The lack of provenience information for many formed tools and other material collected in level bags had a profound effect on site interpretation for this study.

The single other greatest effect was the choice of collection methods by the original investigators, in particular the vertical emphasis of the excavations in pursuit of information that would contribute to the culture-historical focus of the archaeology field in the Plateau in the 1960s. Related to this is the effect of variation in the methods of collection of different material types in the early years of excavation, including a lack of a consistent collection policy for botanical materials and perhaps shell as well, although the effect on interpretation of the former is much greater. Another collection method problem for this study was the loss of small-size materials that would drop through the ¼-inch (6-mm) mesh screens (especially fish bone) used in all years of the rockshelter excavations. These last two factors impact interpretation of subsistence resource-related research questions the most, further bolstering the decision to prioritize interpretation of culture-historical themes.

Site Setting

Chapter Three includes a recounting of the changing climatic regimes in the Plateau during the time of Marmes Rockshelter occupation and suggests vegetation changes that would have accompanied them. In addition, Chapter Five presents the hypothesized changes to the site setting that occurred since the glacial floodwaters drained from the site area as exhibited in the sedimentological record. Each of these affects how the inhabitants used the site and surrounding area.

The Marmes site contains three landscape elements: rockshelter, colluvial slope, and floodplain. Each has its own distinct assemblage of deposits that have formed since the last of the Missoula floods approximately 13,000 years ago.

The rockshelter contains basaltic rooffall rock blocks (éboulis) mixed with windblown silts. These sediments have been modified since deposition by freeze-thaw, pedogenic, and cultural processes. The main period of soil formation (as evidenced by salt translocation and field descriptions of structural development) appears to have occurred within the rockshelter during the early Holocene prior to the eruption of Mt. Mazama approximately 6,700 years ago. This corresponds to a time of reduced sedimentation in the rockshelter based on an evaluation of the radiocarbon chronology (Sheppard et al. 1987). The rate of sedimentation in the rockshelter was maximum prior to 10,000 B.P. and then decreased substantially until approximately 7000 B.P. when it began to increase again but at a lesser rate than in the Pleistocene. This is somewhat different from Fryxell's (Fryxell et al. 1968b) estimate of relatively high rockfall frequency extending from approximately 12,000 B.P. to as recently as 8000 B.P. Fryxell's (1964) interpretation that increased rockfall during the latest Pleistocene and earliest Holocene reflected colder conditions more amenable to mechanical weathering is reasonable and matches the climatic trends presented in Chapter Three through ca. 10,000 B.P. But Fryxell's ending date of 8000 B.P. for deposition as a result of cold climate is probably too late as conditions had reached the warmest and driest of the Holocene by that time. Other factors than cold condition weathering may have affected the rate of deposition, including changes in the exposure of well-jointed bedrock in the ceiling and the rate of eolian sedimentation which when increased (ca. after 9000 B.P. [Chatters 1998:44]) can reduce the density of rock spalls in the sedimentary matrix without decreasing the rate of rockfall.

During times of increased weathering, the effect within the rockshelter would have been expansion of its cavern back into the bedrock and at its margins, quicker at locations of weaker rock and at surfaces more exposed to changing temperature. In particular, the overhead dripline would have receded more quickly than the interior contributing both bedrock and colluvium to the berm that would have grown in height and breadth. The berm also may have benefited from larger-scale failures of bedrock at the dripline early in the depositional sequence of the rockshelter (see stratigraphic drawings in Appendix B). But its height relative to the rockshelter floor increases gradually after the advent of the period of greater eolian sedimentation and particularly with the deposition of Mazama ash. The period of reduced sedimentation and soil development in the rockshelter corresponds to soils developed into early Holocene alluvium along the lower Snake and Palouse Rivers that are estimated to have formed 8000 to 6700 B.P. (Hammatt 1977; Marshall 1971). The calcic soils record a period of landscape stability during this period that reflects the relatively dry conditions that correspond to the warmest and driest of the Holocene until after ca. 6500 B.P. (see Chapter Three).

The colluvial slope below the Marmes rockshelter contains an assemblage of basaltic rock debris and silty loess that traces down and interfingers with floodplain deposits of the Palouse River. The colluvium contains a buried paleosol capped by Mazama tephra that is traceable into the rockshelter and down into the floodplain and is correlatable to the other early Holocene soils in the region. Of climatic significance within the colluvial deposits are saline and sodic soils. Given the well-drained position of the colluvial slope, it is believed that saline soils formed during relatively dry periods of the Holocene whereas the alkaline soils formed during subsequent shifts to more mesic conditions and greater effective leaching. The frequency and chronology of these pedogenic events is uncertain, as is the magnitude of climate shifts necessary to bring about such changes. The fact that saline soils are presently forming at the surface suggests modern climate is adequate for the accumulation of salts. The fact that sodic soils occur at depth but within Holocene sediments provides supporting evidence that conditions previously were more moist than today, or possibly cooler allowing for more effective moisture and greater leaching. It is likely that the effective moisture threshold between the two conditions was crossed several times in the Holocene and need not require large magnitude changes in climate. This data can contribute to further refinement of the Columbia Plateau's Holocene climatic sequence.

Below the colluvial slope at the Marmes site is an early Holocene alluvial terrace previously mapped by Marshall (1971) as Terrace 1. The terrace contains both low-energy, silty and sandy overbank and high-energy, gravelly channel deposits. After the Glacier Peak ashfall (ca. 11,250 B.P.), the area experienced relatively low-energy, overbank sedimentation during which weakly developed, organic soils formed in the floodplain. These include the Marmes and Harrison Horizons, the uppermost incipient A-horizons defined by Fryxell as paleosurfaces with good potential for preservation of in situ archaeological and faunal materials, as well as eleven deeper weakly-developed soils. Sometime between 9,500 and 6,700 years ago, the main channel of the Palouse River encroached northward either through avulsion or migration towards the rockshelter and eroded some of the earlier, overbank deposits, but did not cut deeply into the floodplain. The floodplain continued to aggrade until sometime following the deposition of Mazama tephra approximately 6,700 years ago when the Palouse River downcut forming the terrace. Subsequent deposition on the terrace has been limited to eolian sediments and alluvial fan and colluvial sediments towards the margins of the floodplain. This terrace correlates in time and lithology to other terraces along the lower Snake and middle Columbia Rivers and indicates a regional lowering of base level, ostensibly due to climate change, about the time of the eruption of Mount Mazama. This may be the cooler, moister regime that brings an end to the mid-Holocene dry period after ca. 6500 B.P. After this

All cultural materials from the Marmes site have been modified to some degree by post-depositional processes. However the amount of modification varies from relatively low in the rockshelter and in the overbank alluvial silts and sands of the floodplain to high on the colluvial slope and channel gravels of the floodplain. Whereas in the floodplain there is likely to be geological processes biasing the visibility and preservation of different aged cultural horizons, the rockshelter has a continuous depositional sequence from the latest Pleistocene through the Holocene that allows for a more realistic measure of past human activity. [Huckleberry, Gustafson, and Gibson 1998:97]

Feature Review Results

There is a striking contrast in feature types between the rockshelter and the floodplain portions of the Marmes site during the correlated time periods investigated (Stratum I). The floodplain contains no fire-related features (i.e. hearths) but otherwise appears to represent an occupational use area where lithic reduction and butchering occurred. While floodplain bone has been shown to be blackened and even charred (See Chapter Ten), there is no mention in the field records of fire-cracked rock in any density that would indicate a fire-related feature. Instead, all cooking and other domestic activities that used fire occurred inside the rockshelter or off-site. This separation of activities between the two areas of the site appears to continue for the rest of the site's history as the floodplain portion of the site sees relatively little use after the Marmes horizon. While this conclusion reflects the lack of excavation of post-Marmes horizon floodplain strata, observations of the project staff clearly indicate that any cultural materials in the floodplain did not approach the rockshelter materials in quantity or elaboration.

Examination of the four subareas of the floodplain defined by this study has indicated that none of them rise to the level of typical definition of archaeological features. Clearly, some associated archaeological materials within the subareas are features (e.g., the two sections of elk carcass, the owl foot artifact, the human remains concentrations). But the subareas themselves include a diversity of materials with varying density between the Marmes and Harrison horizons for the units that make up each subarea. Still, these subareas were defined to aid in describing the activities that may have occurred in the floodplain in the absence of the utility of the feature definition that was done by the excavators. There appears to have been some utility to our definition of the subareas because collectively they contain the greatest amount of faunal bone and lithic materials recovered from the floodplain excavations despite representing less than half of the area that was excavated. These subareas also contained the largest number of large-size faunal bones (especially the two central concentrations that included the elk carcass). This indicates some concentration of activities in the four subareas. However, the relative concentration of lithic debitage in the east portion of the excavated floodplain does not appear related to a concentration of other resources types. There is some higher representation of faunal bone fragments here, but it appears more likely that the people conducting the activities that produced both the lithics and bone were attracted to the large boulders that occur in this area. These boulders could have provided shade, windbreaks, or just backrests for the site's inhabitants while working on the myriad of domestic activities of early Holocene foraging lifeway. Similarly, the relative concentration of small faunal bone fragments in the southwest portion of the floodplain excavation does not appear to correlate with numbers of other resource types, although there is a slight increase in lithics in this subarea compared with the floodplain as a whole when the eastern lithic concentration is excluded.

More significantly perhaps, is the differential use of the floodplain and the rockshelter during this period. Chapter Eight noted that while debitage is more numerous in the floodplain, only 5% of the formed stone tools from this period occur there with 95% found in the rockshelter. In addition, all but one of the bone pins from this period were recovered from the rockshelter. Conversely, all of the eyed bone needles occur in the floodplain during this same period. This describes a clear division of tasks associated with these implements between the floodplain and the rockshelter. Root and Gustafson (Chapter Nine) indicate that eyed needles would be used for sewing clothing and that these needles are so small that their presence where found must be primary deposition. This would appear to indicate a strong pattern of tailoring activities being conducted outside the rockshelter, perhaps due to the need for better lighting, which also might explain the greater proportion of debitage in the floodplain. Root and Gustafson also make the point that the holes in the sewed material would have to be punched with an awl, that the needles are too delicate for this. But the awls with known provenience are found throughout the site, including in later periods, perhaps indicating the multiple functions of this tool type.

Within the rockshelter, the overall number of features increases greatly from the early to the late stratigraphic units. There is a gradual horizontal movement of cultural features from the front of the rockshelter in the earliest strata towards the interior of the rockshelter through time that likely reflects the expansion of the cavern through erosion and the growth of the berm at the front of the rockshelter. Prior to the period associated with Stratum VII, most of the features found in the rockshelter (other than burials) are fire-related features and likely represent activities related to occupation of the rockshelter (cooking and warming), perhaps in tandem with resource processing (food preservation by drying or smoking, heat-treating stone). Stratum VII features also are dominated by fire-related features but also include numerous storage pits that may have been burned in the process of preparation for re-use. The (minimal) descriptions of the fire-related features in Strata V, VI, and VII do not indicate a transition in the use of the site from a habitation site to a predominantly storage facility as might be expected in the forager-to-collector settlement and subsistence pattern transition hypothesized for the Plateau during this time period. It may be that the field records just don't offer the information. Contemporary excavations of such features would target data types that could produce such information. The small number of Stratum VIII features identified by Rice include only a single "recent" fire pit (1969:73); the rest of the features in this stratum are a result of historic disturbance, generally related to pothunting activities.

While a few storage pit features are present in pre-Stratum VII periods, the great increase in this feature type from Stratum VI to VII is significant in that it probably reflects an increase in collector-like activity, in this case intensive collection and processing of food resources for later consumption. Unfortunately, the numbers of storage features in Marmes Rockshelter cannot be used as a proxy for describing the timing of this transition in economic behavior because the excavation strategy leaves it uncertain how many such features were in the deposits. Chapter Six describes the high content of botanical materials associated with rockshelter storage features observed in other rockshelters in Palouse Canyon and the lack of such materials in the Marmes site collection. In this case, the lack of botanical materials was not due to their absence in the site, as they are noted in the field records, but the fact that they were not collected consistently. Many more storage features likely were present in Marmes Rockshelter than the records indicate, especially given that the upper strata remaining in the rockshelter were shoveled out at the beginning of the 1968 field season.

That Marmes Rockshelter was used for storage is no surprise as all of the large rockshelters in the Palouse Canyon area exhibit such features. Recent work (Hicks and Morgenstein 1994; Hicks 1995) has indicated that rockshelter pits maintain much cooler temperatures than the outside air and temperature decreases with depth. In addition, rockshelter sediments exhibit relatively high humidity compared to the open air. In fact, it appears that rockshelters are good locations for the storage of organics not because they are dry, but because they are cool and have relatively low temperature and humidity flux through the annual cycle (Hicks 1995:83-84). It is believed that rockshelters with prominent berms along the dripline were identified as especially desirable because these features reduce the air exchange between the rockshelter cavity and the exterior.

Additional features may have existed in Marmes Rockshelter that were not detected in the course of excavation. In particular, the startling percentage (95%) of formed stone tools recovered from the earliest rockshelter strata compared with the floodplain suggests that some of these tools may have been cached in the cave by early Holocene inhabitants. These caches may have been dispersed by rodent activity or not recognized by the excavators if they occurred at the edges of multiple units.

Human Remains

The best estimate of the minimum number of individuals represented by the human remains features observed at the Marmes site is 36, although this discounts the many small skeletal elements and bone fragments found scattered throughout the rockshelter deposits not associated with documented human remains features. In addition, it cannot be certain that the removal of rockshelter sediments above the Mazama ash by backhoe in 1968 did not displace burial features for which no other evidence had been observed. The same may also be true of any post-Marmes horizon burials that may have existed in the floodplain sediments that were removed by bulldozer and backhoe in 1968. The total of 36 human remains features is arrived at by accounting for the 26 features in the rockshelter, accepting Krantz's estimate of a minimum of six individuals in the cremation hearth, and the four individuals from the floodplain (see Chapter Seven).

However, a potential relationship of the floodplain remains with the cremation hearth bears consideration and could affect the estimated total of human remains from the site. It was noted in Chapter One that Fryxell's investigation of the site stratigraphy found the surface of the Marmes floodplain cultural deposit to be approximately consequent with Stratum I-II in the rockshelter (Fryxell and Keel 1969:28, 55-56). It was Daugherty and Fryxell's belief that the burned cranium fragments on the floodplain probably derived from the area of the cremation hearth and animals or human action was the cause of the displacement (R. Daugherty, personal communication, 1997; C. Gustafson, personal communication, 1998). Krantz also suggested that "some of the floodplain material [may have been] carried, thrown, or rolled out of the shelter at the time of the "cremation" and came to rest on the edge of the floodplain at the bottom of the slope leading out of the rockshelter" (1979:174). In examining that possibility, Krantz compared the condition of the floodplain remains with that of the cremation hearth remains he analyzed.

> "Marmes I received the same kind of burning and breakage as in the "cremation hearth" but with no evidence of second burning….The child skull, Marmes II, was also burned and may originally have come from the shelter too.
>
> Marmes III was not burned and may have come to rest on the floodplain by a number of possible means, but in any case was not involved in the cremation. Marmes IV is too poorly represented for speculation." [Krantz 1979:174)

This observation should be viewed cautiously since much of the faunal bone from the floodplain examined for the current study was burned and even charred and one suggestion for this was wildfires sweeping over the floodplain (see Chapter Ten). This may have resulted in burning of the crania prior to their burial in the floodplain by natural sedimentation. This does not explain the variation in burning of the floodplain crania that Krantz notes however.

There is little direct evidence of actions that could have displaced the floodplain remains from the rockshelter, but a correlation of the provenience of these remains and the hearth feature can be examined. The stratigraphy correlated by Fryxell indicates similar time periods for these two surfaces. Radiocarbon dating of samples from both stratigraphic units suggested to Fryxell that the cremation hearth appeared "almost equally early" (Fryxell and Keel 1969:67) as the Marmes horizon. Examination of the reported conventional radiocarbon ages of dated samples from the Marmes horizon and the corresponding rockshelter strata (Stratum I and II) does not resolve the question since dates for Stratum I and II cover a period from ca. 10,800 to ca. 8500 B.P. (see Table 2.1).

The current study sought to examine the possible correlation between the floodplain remains and the cremation hearth deposits. This required more refined dating of the cremation feature and the Marmes horizon in which all four of the floodplain remains were found. Two samples of wood charcoal picked from feature sediments from two different areas of the cremation hearth and two bone samples from the Marmes A1 horizon, including a piece of the elk found adjacent to one of the floodplain human remains features, were processed using the AMS method of dating. The samples from the cremation hearth returned conventional dates of 9,430 ± 40 B.P. (Beta-120803; wood charcoal; $\delta^{13}C/^{12}C$ = -26.0%) and 9,360 ± 60 B.P. (Beta-156696; wood charcoal; $\delta^{13}C/^{12}C$ = -24.4%) which correlate with previous dates associated with Stratum I/II (Sheppard et al. 1987:122). When calibrated the two new dates overlap within 2 sigma. In comparison, the two Marmes A1 samples returned conventional dates of 9870 ± 50 B.P. (Beta-120802; faunal bone; $\delta^{13}C/^{12}C$ = -21.7%) and 9710 ± 40 B.P. (Beta-156699; faunal bone; $\delta^{13}C/^{12}C$ = -16.8%) which also correlate with the two Marmes horizon dates submitted by Fryxell in the 1960s (see Table 2.1). When calibrated these two new Marmes horizon dates also overlap within 2 sigma.

But neither of the new cremation hearth AMS dates overlap with either of the new Marmes A1 horizon AMS dates at 2 sigma. Based on this (very) limited information (additional dating of the hearth sediments could very well produce an older date), the Marmes horizon floodplain sediments in which the human remains occur developed prior to the dated use of the cremation hearth. This does not mean that the floodplain human remains did not derive from the rockshelter, it just means that they appear to be unrelated to the cremation hearth. More dates of cremation hearth charcoal would be required to solidify this conclusion.

Certainly the fact that most of the floodplain human remains are not burned and broken to the extent of the human remains in the cremation hearth suggests a difference in treatment. In addition, Marmes I was "at least a large part of a human body, not just the head..." (1979:163) suggesting it burned to its observed extent in place on the floodplain. In response to Krantz's (1979) suggestion that the human bones in the cremation hearth are a result of cannibalism, Sprague (2000:3) states that this "shows a profound lack of understanding of the typical cremation process where several times the bones are gathered after the fire has died down and then burned again or broken into small pieces for further disposal." In fact, Krantz's description of the human bones occurring in the hearth as only very small fragments would seem to support Sprague's contention. Other than Marmes II, the size of the other three floodplain crania after reconstruction would appear to differentiate them in size and condition from the cremation hearth remains. For the floodplain remains to have derived from the cremation hearth would seem to suggest that they somehow avoided treatment similar to the rest of the human remains observed there. Finally, while there is little evidence of cannibalism in the archaeological record of the Plateau, there are numerous examples of cremation as part of burial rituals, including perhaps Burial 19 from Stratum V at Marmes Rockshelter (cf. Sprague 2000).

While the origin of the floodplain human remains is unknown, the two additional Marmes A1 horizon dates run by this study confirms their significance as some of the oldest ever found in the Plateau. Only Buhl Woman from southern Idaho (and probably attributable more to the northern Great Basin cultural area), dated at 10,675 ± 95 B.P. (Green et al. 1998), predates the Marmes human remains in the Northwest. Kennewick Man is the next oldest in the Northwest after the initial Marmes human remains at between 8130 ± 40 B.P. and 8410 ± 40 B.P. (U.S. Department of the Interior 2000). Nationwide there are fewer than 25 other sets of human remains confirmed as older than 9,000 years old (Roberts 1999). But perhaps more significant is that the Marmes site exhibits human remains features throughout its deposits. In addition to the floodplain remains and the cremation hearth, five human remains features occurred in Stratum III (ca. 6200-7870 B.P.; Sheppard et al. 1987:Table 2), 18 are attributable to Stratum V (ca. 4250 B.P.; ibid), and three more are attributed to Stratum VII (ca. 660-1600 B.P.; ibid) (see Chapter Seven). Two of the human remains features in Stratum III are definite interments, while ten of the eighteen features in Stratum V are definite burials, nine primary and one secondary; two of the three human remains features from Stratum VII are considered burials. Other human remains features may have been interments but the records are not clear. Collectively, these burials inform on burial patterns of the lower Snake River through time.

Sprague (2000) prepared an overview of burial patterns for the Plateau Culture area based on his 45 years of work in the area that included a considerable number of burial removals; this is essentially an update of the conclusions of his doctoral dissertation (1967). Sprague reviews the archaeological data on hundreds of human remains features from the lower Snake River and the mid-Columbia River to below the confluence of the Columbia and Snake Rivers. While some are described as late prehistoric (and others that had no associated dateable materials also may be from the prehistoric period), by far the greatest number are attributed to the protohistoric and historic periods (Sprague 2000). Only Marmes Rockshelter (and Kennewick Man) has human remains that predate the Mazama ashfall (Sprague 2000). Kennewick Man's date places him between the dates on the cremation hearth (Stratum I-II) and the five human remains features attributed to Stratum III, three of which were not burials; two appear to have perished or were left on the surface of the ground. The ten burials associated with Stratum V at Marmes Rockshelter, which was assigned a single date of 4250 B.P. and probably dates to before 3,000 years ago (Sheppard et al. 1987), are matched in Sprague's record by only a single burial from Rabbit Island below the mouth of the Snake River on the Columbia River that may be up to 3,000 years old (Sprague 2000:5).

Because of the paucity of burials dating to prior to the late-prehistoric period (ca. the last 2,000 years), it should be no surprise that attempts at defining a chronology of burial patterns for the Columbia Plateau area are restricted to the late period. Sprague (1959) first proposed a burial sequence for the lower Snake River that was modified slightly by Combes (1968) and refined with more detailed criteria by Rodeffer (1973). Including modification, this sequence consists of:

1. a late prehistoric pattern of burials with orientation varying from north to south generally through the westerly half of the compass, some degree of flexure, placed on the side, and accompanied by grave goods consisting largely of dentalia;
2. a brief period of rock cairns with associated fire and an occasional cedar cist, increase in grave goods, placement in a flexed position on the back, variable orientation but generally west to south, and a heavy painting of the body with red ochre; and
3. an early historical period of extended burials on the back, oriented east and placed in cedar burial boxes, and inclusion of large numbers of trade items in the grave. [modified from Sprague 2000:8]

Sprague notes that "The basic pattern would appear to be generally applicable to the whole of the ethnographic Plateau region, becoming stronger and more unifying later in time. The universal pattern in the late historic period was one of extended burials on the back in rough boxes" (2000:8). Burials recovered north of the Columbia Plateau along the Columbia River in recent years included three burials with dentalia included as grave goods; all of the burials (n=6) recovered by the project were oriented to the south or downriver (Roulette 1997). It is suggested that at least in prehistory orientation may follow less a compass direction than the direction of the flow of the nearby major river.

The Marmes burials appear to extend the late prehistoric pattern described above further back in time, at least in terms of inhumation, orientation, and presence of grave goods. Inhumation in pits is established in Marmes Rockshelter beginning in Stratum III (early to mid-Cascade Phase) and is the pattern for definite interments thereafter. For those burial features that retained the information, orientation fits the description of varying "from north to south generally through the westerly half of the compass" (see modification of Sprague 2000 above) for all but one of the primary burials in Stratum V; orientation information was not available for the rest of the burials. Given the westerly flow of the Snake River here, a downriver orientation, as suggested above, would appear to be supported for these prehistoric burials. Interestingly, for the one secondary burial in this stratum, Breschini suggested a primary burial may have been disturbed by a later pit excavation (burial or storage) and then given a ceremonial reburial, a similar scenario as that interpreted for a burial salvaged near Kettle Falls that dated to 890 ± 50 B.P. (Roulette 1997). Placement of grave goods with burials at Marmes Rockshelter extends back to the Stratum III burials, but instead of large numbers of dentalia shell, olivella shell is common; no dentalia were identified in any of the burials from the Marmes site. Lithic tools (especially projectile points) also were common inclusions in Marmes Rockshelter burials beginning with the earliest confirmed burials in pits (Stratum III). Red ochre, sometimes in considerable quantities, becomes a common burial inclusion by Stratum V along with large numbers of olivella shell beads, various lithic, bone and antler formed tools, and hackberry seeds. Lithic debitage and ornaments such as bear teeth occur in some Stratum V burials. The two Stratum VII burials, which date to the late prehistoric period, also reflect Sprague's (2000) burial pattern with the continued presence of formed tools, olivella shell beads, and other cultural items.

Material Analysis Results

Summaries of the results of the different cultural material analyses (lithic, fauna bone tools, botanical, fish and shellfish) are presented.

Lithic Analysis

A total of 10,793 artifacts, including 9,230 pieces of the debitage and 1,563 tools were analyzed for information on reduction technology and function. In addition, twenty of the obsidian artifacts were selected from the assemblage to determine volcanic glass trace element composition and geological sources. Also, twenty-nine artifacts and four soil samples were analyzed for possible blood residues.

The Marmes site's lithic assemblage, from the Windust through the Harder Phases, reflects a habitation site where people performed a wide range of domestic tasks. The technological profile for the diagnostic debitage from the Numipu Phase deposits is quite similar to that of the Harder Phase debitage but data on site use during this period is more limited than for the previous phases. This may reflect the Marmes site's occupants move to an open village site, the regional pattern for habitation sites. The lithic artifacts represent tool manufacturing and maintenance (of stone tools and tools of hard organic materials such as bone, antler, and wood), food procurement (projectile points, bola stones, possible net weights), and processing of food and other materials, probably including hides and fiber. These combinations of activities characterize a residential camp and given its location in the canyon the site may represent winter occupations.

This general pattern appears to be repeated through each period of occupation, including that associated with Stratum V which has been interpreted by some as a time period when the site functioned solely as a cemetery and that artifacts associated with the deposit are grave goods (Fryxell and Daugherty 1962:28). The abundance and variety of utilitarian artifacts identified in this stratum would seem to dispel this suggestion. Ceremonial use of the shelter may be represented during some portion of the period associated with Unit V, but does not appear to characterize its entire duration.

Although the occupations at the Marmes site may have been continuous as suggested by Fryxell and Keel (1969:viii), the level of activity as measured by the number of lithic artifacts recovered from the rockshelter, reflect significant fluctuations in the intensity of the occupations. The artifact distribution data suggest very intensive occupations during the Cascade Phase from about 7900 to about 6200 radiocarbon years B.P. and again during the Harder Phase from about 1600 to about 700 radiocarbon years B.P. Projectile points are most abundant in the Cascade (Stratum II and III), and Harder (Stratum VII) subassemblages. Flake tools, choppers, hammerstones, and scrapers are also most abundant in these stratum. The high frequencies of the butchery or animal processing tools and faunal remains found in these strata suggest intensive hunting activities during these periods.

The proportions of lithic tools and debitage vary through time in some aspects that may reflect change in site function. For example, in the rockshelter the ratio of debitage to tools is relatively low for the Windust and Early Cascade period strata by comparison with the later strata. This change may reflect more intensive tool use, tool exhaustion, and tool disposal during the early periods and more intensive tool manufacturing during the later periods. Likewise, the percentage of tools that are byproducts of manufacturing (cores, blanks, and preforms), pieces broken and rejected during the tool manufacturing process, also increases for the later periods. It may be that Windust and Early Cascade people generally arrived at the site with stone tools sufficient for planned activities while later people more often manufactured stone tools while occupying the site.

Reanalysis of the lithic artifact collections from the Marmes site confirms the general outline of Leonhardy and Rice's (1970) model of a culture-historical sequence for the lower Snake River region. A frequency distribution of the typologically distinctive projectile points identified within the subassemblages shows each type modality peaks in the rockshelter stratum predicted by the Leonhardy and Rice (1970) chronology. The distributions around these modes, however, suggest that some types were long-lived, most notably the Cascade type. Cascade points occur in every stratum with typeable points. Other diagnostic artifact types cross-cut stratigraphic and temporal boundaries for the assemblages and periods that they characterize. The reasons for this probably include both mixing of the archaeological deposits and survival of technological traditions through long periods of time.

The evolution of technologies at the Marmes site appears to be deliberate and conservative. That is, while there are differences between successive occupations, they seem gradual and incremental. What distinguishes one period from another is simply the relative frequency of certain artifact types and traits. Some more subtle aspects of the lithic technologies appear to change within the periods representing the putative cultural phases or style zones. For example, the earliest Windust phase assemblage at the site from Stratum I in the rockshelter shows more emphasis on the later stages of percussion bifacial thinning, suggesting a biface-based technology akin to Clovis and other early Paleoindian technologies. The slightly later Windust materials from the floodplain and Stratum II in the rockshelter contain proportionately fewer late-stage percussion bifacial thinning flakes indicating a shift to thin, linear flake blanks for the production of projectile points. These blanks are produced by different core reduction techniques and do not require systematic percussion thinning. Patterns of Windust and Cascade projectile point reworking following breakage observed in the Marmes assemblage also are similar to those described elsewhere for Paleoindian projectile points.

There is a notable contrast in lithic material density in the floodplain that may overlap in time with the transition in technology described above and/or site use. Twice as much debitage is noted in the Harrison horizon as in the Marmes horizon. This correlates with the ratio of stone and bone tool counts cited by Fryxell and Keel (1969:34-35) of 45 tools in the Harrison horizon and 19 in the Marmes horizon. While a number of the tools recovered from bulldozer backfill piles could have derived from the Marmes horizon (see Chapter One) they would not significantly alter the ratio of tools between the horizons. In addition, the backfill tools could be from strata above the Marmes horizon. A brief exploration of a shell concentration below Mazama ash located in a bulldozer trench west of the main floodplain excavation represents the only excavation of post-Marmes horizon sediments in the floodplain. Shell from this area was dated to 7980 ± 300 B.P. (Sheppard et al. 1987:Table 1). There is virtually no reporting of this excavation in the site records, but correlating the "field Specimen Numbers" listed in a table of 17 "Artifacts from the "Cascade" Component" (Fryxell and Keel 1969:32) with the "List of Artifacts – Floodplain" (Marmes site records, WSU Museum of Anthropology Archives) made after all the bags had been processed in 1969 indicates that these artifacts came from various units of the floodplain excavation area and were not restricted to the western trench units. It is expected that these Cascade component tools were encountered in places where crew members shoveled off the riverine overburden sediments that elsewhere were removed by bulldozer. The seventeen tools noted in Fryxell and Keel's table (1969:32) probably do not represent the entirety of Cascade Phase cultural materials that were in the floodplain deposits. The 17 tools noted include four projectile points (three crypto-crystalline silicate and one basalt with serrated edges), two cobble choppers, two scrapers, six retouched flakes, a chert core fragment, a basalt hammerstone, and a piece of worked long bone interpreted as a "flaker". This is a relatively diverse assemblage of tool types given the restrictive collection method for this stratum and suggests continued use of the floodplain in tandem with the rockshelter at least into the Cascade Phase associated with Strata III-V.

Another underlying trend in the lithic technologies involves proportional use of different lithic material compositions. Basalt (or andesite) and CCS materials predominate and obsidian is present in small amounts throughout the site occupations. CCS materials are overwhelmingly predominant in the earliest occupations, but the use of basalt increases steadily through the late Cascade occupation reaching equal importance, and then declining again thereafter. The heaviest use of basalt materials appears to be linked with production of basalt projectile points. In this respect, the Windust and Cascade assemblages both contribute to a trend that suggests continuity.

The predominant sources of lithic raw materials used during the early Cascade Phase occupation can be evaluated based on cortex remaining on flakes and tools. This indicates that alluvial gravels, available from nearby Palouse and Snake River gravel deposits, were used for tool-making at the site. However, the relatively small amount of bifacial reduction debitage compared to the large number of bifacial projectile points indicate that most of the bifacial tools were made elsewhere although discarded at the Marmes site. This relationship between bifacial tools and bifacial debitage may indicate that the raw materials preferred for manufacturing bifacial tools were not available in the local area.

At least three obsidian sources were used during the early Cascade Phase occupation, two previously known sources and a source of unknown location. The two known sources are possibly the closest available obsidian of good quality but both are more than 220 km south of the Marmes site in high elevation areas. During the early Cascade Phase, obsidian entered the Marmes site primarily in the form of bifacial tools that were reworked, resharpened, or discarded there.

One lanceolate projectile point from the rockshelter resembles Knife River Flint, the quarries of which are located approximately 1,300 km (806 miles) east of the Marmes site. If indeed it is Knife River Flint, the artifact represents a long-distance connection with groups of the northwestern high plains ca. 10,000 years ago.

There were eight positive reactions to three antisera on seven artifact samples by the blood residue analysis. However, the control soil samples for six of the positive reactions also reacted to the same antisera; the corresponding control soil samples for the remaining two positive reactions were negative suggesting that these tools came into contact with the blood of the identified species. A basalt hammerstone from Stratum III tested positive to the chicken antiserum, which may indicate any number of birds that occurred in the site vicinity (e.g., quail, pheasant, ducks, geese, swan). A basalt projectile point from Stratum III reacted to trout antiserum, which can indicate salmon and steelhead as well as trout. That so many control soil samples tested positive to deer and trout antiserum suggests that residues from the corresponding species may be widespread in these deposits, especially given the varied tool types that were tested. However, the potential for extensive previous handling of the tested artifacts, and lack of direct correlation of proveniences of the control soil samples used weakens the use of these results for site interpretation.

Bone Tools

Sixty of the 78 pieces of modified bone are of known provenience and were used in comparative analyses. Just over one-half (32 of 60) of the bone artifacts of known provenience come from the Windust Phase deposits.

The modified bone collection shows a clear selection for dense long bones and metapodials from deer-sized or larger animals. Most tools were made by careful grinding, and perhaps carving as well. Manufacturing debris and needle blanks broken in manufacture indicate that people made bone tools at the site. Unpatterned tools, such as splinter awls, that require little effort in manufacture, may also have been made on the site and discarded after use. Abraders, bifacial knives, utilized flakes, burins, and drills are stone tool types likely used in the manufacture and maintenance of the bone tools.

The presence of sewing and hide working tools, such as needles and awls, suggests that family groups occupied the floodplain and rockshelter. The presence of additional artifacts such as bone tubes and pendants also suggests Marmes inhabitants wore tailored leather clothing, and likely decorated their dress as well. Bone pins may have been worn in clothing, and perhaps even in hair.

Eyed needles and needle blanks only occur in the floodplain Windust phase deposits. At least one needle was complete when it entered the archaeological record. It is unlikely that this was discarded, and it was more likely lost. Small items are rarely picked up and discarded in secondary trash dumps (Schiffer 1987), suggesting the needles and needle blanks were lost or discarded near the places that they were used or made. The floodplain may have been an area where people sewed clothing, an activity that would leave little archaeological signature. The presence of needle blanks also suggests that tool manufacture took place on the floodplain. The Windust deposits contain a single broken atlatl spur, suggesting weapon repair. This also indicates that at least some Windust phase points were atlatl darts. A splitting wedge from the rockshelter Windust phase deposits suggests woodworking was linked to those occupations. Bone pins are concentrated in the Windust phase deposits, where 8 of 12 were recovered.

Antler/bone rods are strongly associated with human remains at the Marmes site. The four antler rod fragments from the floodplain may be part of the same artifact, and all are in the vicinity of one of the human remains concentrations there (although these remains do not appear to have been interred). Six of the rods are associated with Stratum V burials. Single fragments were recovered from the Windust and early Cascade rockshelter deposits. Thus, ten of 12 rods have associations with human remains and, at least at the time of interments in Stratum V (ca. the Tucannon Phase), are related to intentional burial. Discussions of the functions of bone rods should take such associations into account, and it is possible that these are not utilitarian artifacts, at least by the Tucannon Phase. However, bone rods have been observed in non-burial circumstances throughout North America and various hypotheses for functions of this class of artifact have been advanced in the archaeological literature, including as levered hafting wedges for tightening sinew binding on cutting tools (Lyman, O'Brien, and Hayes 1998). Under this scenario, tool maintenance activities may be posited also to have occurred in the floodplain portion of the site.

Spatulate tools occur in the earliest and the latest deposits. No bone pins were recovered from the Stratum VI-VIII deposits. Pointed bone tools, at least some of which may be projectile points, occur in the floodplain and in early Cascade (Stratum III) and late deposits (Stratum VI-VIII) in the rockshelter. Ornaments and polished bone also occur throughout the rockshelter deposits. Bone awls are common and occur in all geologic units, suggesting that basketry or hide working were important activities throughout the occupational history of the site.

A striking contrast between the rockshelter and floodplain areas of the site can be seen in that needles and needle blanks only occur in the floodplain. However, such items could have fallen through the ¼-inch mesh screens used in the rockshelter. The needles are small and were probably lost where recovered archaeologically, implying that people sewed and made clothing on the floodplain. The presence of awls in both areas suggests that hide or basket working was carried out in both parts of the site. The tools that are certainly utilitarian from the rockshelter are complete and still usable. These include two awls, an antler wedge, and a spatulate tool. These implements may have been stored in the rockshelter for subsequent use and never retrieved.

Faunal Analysis

The faunal remains from the rockshelter at the Marmes site differ significantly from those on the floodplain, both in their quantitative taxonomic characteristics and in the condition of the remains, but most taxa identified occur in the vicinity today. Exotic taxa (i.e. Arctic Fox, pine marten) were present in the floodplain deposits but not in the rockshelter.

Animal bones, bone fragments and teeth from the rockshelter are much larger than those observed from the floodplain, relatively few are burned, and many exhibit spiral fractures and other evidence of butchery. In addition, most are from deer, pronghorn, elk, rabbits and other animals documented archaeologically throughout the Plateau as prey species. Conversely, the vast majority of faunal remains from the floodplain are very small, most are burned, and few exhibit spiral fractures or evidence of butchery. Deer, elk, and pronghorn are represented in very small numbers on the floodplain. Remains of snakes and rodents form the bulk of the identified floodplain fauna. For example, pocket gopher remains are abundant in the floodplain deposits but were relatively rare in the rockshelter. The difference may have something to do with the pocket gopher's preference for roots and tubers growing in open areas and its need for an extensive burrow system.

Most rockshelter faunal remains are from game animals such as deer, pronghorn and elk, and most long bones were broken above and below the joint in the fashion considered typical of prehistoric butchery in eastern Washington (Gustafson unpublished). Although there are no quantitative data available for the 1962-1964 rockshelter bones ("not all fragments of bone were saved during the early seasons of excavation" [Gustafson 1972:54]), Gustafson observed the presence of joint ends resulting from breakage above and below joints, and the current study observed this butchery pattern in the faunal assemblage from the rockshelter's earliest strata. Large bird bones with both joint ends missing were recovered from atop the glacial lake sediments deep in the rockshelter. Another large bird bone, identified as a swan femur (see Appendix P), was recovered from the Marmes Horizon in the floodplain. This bone not only lacked both joint ends but also exhibited cut and chop marks indicating considerable effort had been expended to separate the tendons that connected to this bone. Shattered shaft fragments from game animals, often exhibiting spiral fractures, and split or broken phalanges suggest that further processing, perhaps for marrow extraction, had occurred at the site. The significance of the floodplain faunal assemblage is more perplexing, including the large elk remains found on the surface of the Marmes horizon in association with human remains.

While the vast majority of animals represented by the faunal assemblage from the Marmes site occur in the vicinity today, a number of taxa from the early floodplain deposits suggest cooler or moister local habitats than previously recognized from eastern Washington archaeological sites. Wigand's climatic reconstruction (see Chapter Three) provides a firm context for Gustafson's (1972), Caulk's (1988), and this study's (see Chapter Ten) observations that more mesic adapted, northern taxa may have coexisted with a predominant fauna composed of extant taxa in the early years of the site's occupation. Identification of a femur of a swan, a species that today nests well to the north, from the Marmes Horizon also supports their observations. However, there is little in the faunal record of eastern Washington to support the geological and botanical evidence of a period of mid-Holocene drought, but the fauna may not have been affected as directly or animals may not be as sensitive to change from arid to more arid conditions.

Fish

Perhaps more than any other analysis conducted for this study, the fish remains analysis sample was greatly impacted by the collection methods employed during the fieldwork and apparent subsequent losses of fish fauna from the collection. Despite considerable evidence throughout the Plateau of intensive late prehistoric use of fish as a principal subsistence resource, fish remains are not abundant in the Marmes Rockshelter collection and, in particular, salmonid remains are very scarce. This scarcity is attributable to collection methods and loss rather than a real lack of use by the prehistoric occupants of the site. This is documented by Gustafson's statement that a considerable number of salmonid bones were observed in the site deposits during excavation (personal communication, 1998). In addition, Gustafson (1972) notes that "numerous salmon vertebrae were recovered—primarily from sediments distrubed [sic] through aboriginal excavation of storage pits" (1972:102), and "Salmonid vertebrae and other fish remains sometimes are abundant (particularly in the storage pit areas—Units VI and VII at Marmes Rockshelter). These undoubtedly served as food" (1972:106). The reference to disturbed sediments is in keeping with Fryxell and Daugherty's (1962) judgement that the digging of pits for storage disturbed the upper stratigraphic layers in the rockshelter, rendering the remaining materials in those layers of little value for interpretation. While fish remains apparently were abundant in the site, they are not abundant in the site collection housed at WSU and either were not uniformly collected in the field or have been misplaced or lost. As such, the sample developed for this analysis does not reflect the full picture of fish use by the Marmes Rockshelter inhabitants and these results should be viewed as the taxa present and not in terms of relative taxonomic importance. Similarly, interpretations about changing frequencies of fish taxa must be considered tentative.

The Marmes fish faunal sample drawn for this study from the rockshelter and floodplain locales shows that a range of freshwater species and anadromous salmonids were used by people throughout the Holocene in the Snake River Basin. Both resident freshwater and anadromous fishes are present in the rockshelter deposits, suggesting that past people were generalized in their fishing practices, probably throughout the last 10,000 years. The resident freshwater fish would have been available in the Palouse River adjacent to the site. The Palouse River, however, probably did not support a migratory salmon run throughout the Holocene. The presence of large-bodied salmon, likely representing migratory fish in the site, suggests that people traveled to the main stem Snake River for this resource.

Freshwater fishes dominate the early Holocene record from the floodplain (for which we have greatest controls over sampling and recovery). To the extent taphonomic and sampling concerns can be controlled, this suggests that salmon were not used much by people that occupied the floodplain. Reasons for this may include: scheduling of other economic activities meant that the site's residents were only able to take advantage of fish available in the adjacent Palouse River; or, occupation of the floodplain did not coincide with the seasons of the migratory salmonids availability (late spring through fall); or, perhaps salmonids were not abundant in regional streams during this time period.

Shellfish

With one exception, the invertebrate remains examined from the Marmes site (45FR50) represent freshwater species of molluscs. The two most common shellfish (*Margaritifera* and *Gonidea*) are the river mussels that have been recovered from many other Columbia Plateau archaeological sites. This indicates that the people living in the region were utilizing freshwater resources that could have been found in many of the rivers, streams and lakes or ponds of the region. Modern descriptions indicate that in certain places, freshwater mussels may exhibit very large populations. The identified shellfish would have served well as part of a Plateau subsistence system because of their size and regional availability. The use of shellfish is consistent with a subsistence system that utilizes riverine resources.

The relative abundance of the shellfish observed in this analysis suggest a decline in abundance of *Gonidea* following the eruption of Mount Mazama, perhaps related to the presence of volcanic ash in waterways. It is possible that the influx of the ash into freshwater environments affected the health of *Gonidea* river mussels resulting in a decrease in their use by the human inhabitants of the site. This decrease is consistent with observations made by Lyman for many Plateau assemblages. Differing environmental conditions also may have contributed to fluctuations in relative abundance of *Margaritifera* and *Gonidea* in the Marmes site.

There is a distinction between the floodplain deposits and the rockshelter deposits at the Marmes site that is marked by the differing condition of the shell. This may be a result of different cultural activities, weathering differences, or both.

Botanical Remains

The botanical materials from the rockshelter are not as diverse as might be expected. Wood is the most abundant material, but there are fewer clearly modified/manufactured wood items than were recovered from nearby Squirt Cave (45WW25) on the Snake River (Endacott 1992). Squirt Cave was used for caching food and tools, and while Marmes Rockshelter was also a storage facility during later parts of its history, in earlier times it probably was a habitation site. Squirt Cave and other nearby rockshelters such as McGregor Cave (45FR201) and Porcupine Cave (45FR202) also contained a rich assemblage of textiles, especially cordage and matting (Mallory 1966; Mastrogiuseppe 1994, 1995).

The Marmes site excavations resulted in the collection of very few prepared textiles. Only one cordage fragment, a few leaves probably from cordage, one matting fragment, and a few tule fragments probably from matting are present in the samples analyzed for this project. Seeds and fruits from the Marmes site also are much less diverse than those from McGregor Cave and Porcupine Cave. Only two kinds of fruits (hackberry and wild cherry) were collected from Marmes Rockshelter, and one of these is represented by only two fragments.

Botanical materials present in samples collected from the floodplain portion of the Marmes site are too few to draw any meaningful temporal conclusions. Most of them could have been from plants growing on the flood plain where they were deposited.

The differences between plant materials collected from the Marmes site and those recovered from other sites nearby probably represent differences in excavation methods rather than actual qualitative differences in materials present. In the rockshelter, ¼-inch screens were used to sift sediments, and these would not catch most seeds. Furthermore, there was not great interest in plant materials at the time of the Marmes site excavations and they often were simply not collected; much of the textile material (and probably wood) was simply shoveled out of excavation units to quickly get down to the older cultural levels. It is also possible that unknown factors have caused poorer preservation of botanical materials at Marmes than in nearby rockshelters (e.g., a higher water table). The nature of site use may also contribute to the relative lack of Marmes site plant remains. However, the most likely explanation is methods of excavation.

Despite the limited sample of materials, the Marmes botanical assemblage provides important information on the cultural importance of plants, particularly woody plants. Some of the artifacts are unusual and/or particularly interesting. Marmes site materials also suggest that after Euroamerican contact pieces of commercially-milled wood were gathered and brought to the rockshelter, perhaps primarily for firewood. Some of these pieces, however, were apparently reshaped by Marmes people. Milled wood also may have entered the site record by historic visitors (e.g., the burned cow carcass observed the first year on the surface) and by the excavators in the 1960s. Photographs in the collection indicate that milled boards of varying sizes were used to hoist the plastered burial casts out of the rockshelter, for construction of the different screens used, for the backing of the soil columns removed by Fryxell, and for general purposes (e.g., walkways, supports).

The consistent presence of western red cedar wood and hackberry pits in Marmes sediments suggests that these materials were probably used throughout the history of the site.

Conclusions

The limitations of the records available from the site have been noted previously. The overall effect of those limitations on the ultimate accuracy of the material analysis results in reflecting human use of the site would make for an interesting statistical analysis exercise. Instead, I offer my perceptions derived from having reviewed draft and final reports of the different analyses, from having worked with the analysts' reports in preparing the preliminary reports (Hicks 1998, 2000) as well as in compiling that reporting into this final report, and in particular, having worked with the data in assessing its interpretive fit.

The quality and quantity of the lithic analysis data is the best of all of the material remains data types for offering representation through time. Despite the many missing formed tools, the lithic data represents the most complete and broadest picture across all the stratigraphic units. It also appears that the conclusions of the lithic analysis have not been weakened by the lack of inclusion of the missing items. The lithic analysis concluded that the lithic assemblage suggests the site was used for habitation, the most complex site type with the highest diversity of tool types, throughout its cultural use history. If lithic artifacts are missing from the collection because they were deliberately taken, then we may assume they were supreme types or styles such as large and/or well-made projectile points (and we know such items are among the missing) rather than less spectacular items such as utilized flakes. If items are missing due to misplacement over the years since the collection was made, we should expect they include a range of fine to poor tool types. In all likelihood these items have gone missing both due to pilfering and misplacement. But in either scenario, the functions of the missing items will be less-represented (or even unrepresented) in the tool kit for a given period. And yet the lack of these functions has not reduced the diversity of the lithic assemblage to the extent that an interpretation of a less complex use of the site was made. Lithic items that have lost their provenience information, despite representing some 40% of the lithic tools collected, similarly have not impacted this interpretation of a complex use of the rockshelter and floodplain.

The faunal data is applicable to interpreting site use, but limited by the different collection methods used between the rockshelter (¼-inch mesh screen) and floodplain (1-mm mesh) portions of the site. In addition, the lack of quantitative data for the fauna recovered from the rockshelter portion of the site is a concern. For this study, Gustafson revisited the data from his dissertation (1972) without reexamining the bones or creating new sample lots. Gustafson's dissertation was zoological (rather than archaeological) in focus and concerned more with what animals were present and their implications for past environments rather than explicating patterns of cultural use. Gustafson's (1972) dissertation remains the only analysis of the Marmes Rockshelter fauna.

The lack of quantitative data for the faunal remains collected in 1962-1964 restricts comparison of species representation through time and weakens the contrasts cited between rockshelter and floodplain fauna. In comparison to these data problems, faunal items that have gone missing from the collection or that no longer retain provenience information have had a minimal effect on interpretation. Similarly, the shell analysis results are applicable in terms of relative densities, but due to uncertainties about collection emphasis in 1962-1964 we cannot be sure that all shellfish that occurred in the site are represented or that spatial patterning of shellfish is accurately represented. Still, the shell types identified and the trends found match the few other shellfish studies in the Plateau.

The same also can be said of the bone tool analysis results where there must be concern for what is not represented in the analysis sample as some of these items are missing or have no provenience, or were excluded from the sample for this study.

The fish and botanical analyses results have more limited applicability – they indicate some of the species that were used, but not likely all of them (especially for the botanical materials), and no confidence can be assigned to the proportions or apparent changes through time. The field sampling was too limited in terms of intentional non-collection, materials that have gone missing (especially larger fish bones), and/or what may never have been collected due to screen size.

Features were given more attention in the records than many of the other material remains (other than the human remains). The feature data offers good information on the extent of use of the rockshelter through time when correlated with the other material remains representation. However, the feature documentation generally lacks detail on contents that could assist interpretation of subsistence and site function at different times of the site's use.

388

15
INTERPRETATION

Brent A. Hicks

Interpretation of the results of this study are presented in the context of the study's research themes and questions. The categories of research questions presented in Chapter Two include cultural chronology, past environment and climate change, site function, trade, human remains/burials, and the regional trends to which the Marmes site data may contribute.

Cultural Chronology

Temporal questions have been the major focus of previous investigations at Marmes Rockshelter. Data types collected by this and previous studies that can address temporal questions include radiocarbon dates, stratigraphy, and volcanic ash identifications. In addition, such things as projectile point types/styles, stone material type proportion trends, and fauna species utilization trends through time can inform on cultural chronology aspects of time-related questions.

Numerous radiocarbon dates and two volcanic tephra deposits are correlated with the geological strata defined for the site by Fryxell (Fryxell and Daugherty 1962; see Chapter Five). These data, together with additional radiocarbon dates obtained by Sheppard et al. (1987), establish the chronological outline of the cultural use of the site. Eight new charcoal and bone samples from specific contexts have been processed for this study to address several chronology questions. Seven of the resulting dates are added to those presented in Sheppard et al. (1987:Table 1; and adapted as Table 2.1 herein) (Table 15.1); the eighth sample contained only degraded collagen and did not return a sound date. These data provide firm radiocarbon ages for all of the geological strata with the exception of Strata V and VI. This serves as the background for discussion of this research theme. The new dates (all processed AMS) focused on materials representative of the earlier uses of both the floodplain and rockshelter portions of the site. A number of observations are offered about the results of the dating,

The dates obtained for the Marmes horizon by this study overlap with the Harrison horizon dates processed previously, confirming Fryxell's belief that the Marmes and Harrison horizons were nearly contemporaneous. As noted in Chapter Seven (*Human Remains*), the new dates from the Marmes horizon obtained by this study and the dates of the cremation hearth in rockshelter Stratum I/II do not overlap at 2 sigma when calibrated. (All calibrated ages based on Calib 4.3, Stuiver and Reimer 2000.) This supports the notion that the cultural use of the floodplain strata was contemporaneous with the use of the rockshelter during Stratum I, but not Stratum II. Indeed, this study's oldest conventional date for the floodplain is solidly within the range of conventional dates from Stratum I in the rockshelter. Sample I-638, collected from beneath the Mazama ash layer (dated to 6730 B.P.; Hallett et al. 1997), appears somewhat out of position with a conventional date of 6200 ± 475 B.P., but its calibrated range at 2 sigma indicates that its conventional date assignment is at the early end of that range. Digging of the burial pit for Human Remains Feature 15 may account for the stratigraphically anomalous date of 7550 ± 300 B.P. for sample WSU-120; this date does not overlap with any of the Stratum I dates at 2 sigma when calibrated. The location of this sample is one foot north and at the same elevation as the location recorded for this burial, which is described as being in a definite burial pit dug from Unit III down into Unit I.

The dating of Stratum V is considered unresolved. Only one sample from this stratum was processed by Fryxell (no appropriate Stratum V samples could be found to submit for dating by this study) and it may be that Stratum V was not actually dated. Sheppard et al. (1987:Table 1) note that the shell sample cited for Stratum V was originally assigned to Stratum VII, but based on the resulting age and depth where collected, they designated it Stratum V and note the potential for disturbance from aboriginal storage (and burial) pit excavation. It is noted here that the many burial pits initiated during the time period associated with Stratum V nearly always intruded into Stratum IV and in some cases to Stratum III (see Chapter Seven). In addition, burials initiated in Stratum VII intruded to Stratum V. Each intrusion introduces

Table 15.1 All radiocarbon dates from Marmes Rockshelter[a].

Sample #	Sample Material[b]	Age Years B.P.	Depth (feet)	North	West	Stratum
Rockshelter Dates						
WSU-362	charcoal	Modern	Surface			
WSU-3034	charcoal	660 +/- 75	96.5	97.5	12.5	VII[c]
WSU-206	charcoal	1,100 +/- 300	97.7	82.5	44	VII
WSU-212	charcoal	1,300 +/- 300	97.7	82.5	44	VII
WSU-3033	shell	1,600 +/- 100	96.5	97.5	12.5	VII[c]
WSU-205	charcoal	1,300 +/- 300	95.9	92.5	25.8	VI
WSU-3035	bone	1,940 +/- 70	94.3	102.5	42.5	VI[c]
WSU-207	shell	4,250 +/- 300	94	82.5	44	V[d]
------------------------------Mazama Tephra 6730 B.P. (Stratum IV)----------------------------						
I-638	shell	6,200 +/- 475	92.7	97.5	27.5	III[e]
WSU-3036	shell	7,070 +/- 110	93.9	83.5	25	III[c,e]
WSU-209	shell	7,400 +/- 300	92.4	85	45.5	III
WSU-3037	shell	7,805 +/- 130	92.1	84	25	III[c]
WSU-210	shell	7,870 +/- 300	89.7	87.5	40	III
WSU-3038	shell	8,525 +/- 100	89.8	82.5	27.5	II[c]
W-2208	charcoal	8,700 +/- 300	85.6	75	50	II/I
W-2207	shell	9,010 +/- 300	85.6	75	50	II/I
Y-2482	charcoal	9,200 +/- 110	?	78	20	II/I
Beta-156696	charcoal	9,360 +/- 60	82	70	45	II/I[i]
Beta-120803	charcoal	9,430 +/- 40	84	85	40	II/I[i]
W-2210	shell	9,540 +/- 300	?	78	20	II/I
WSU-120	shell	7,550 +/- 300	89	90	25	I[f]
Beta-168491	shell	9,610 +/- 40	79	85	30	I[i]
WSU-366	shell	10,475 +/- 300	89	74.5	20	I
WSU-211	shell	10,750 +/- 300	88.3	87.5	24	I
WSU-363	shell	10,810 +/- 300	89	74.5	20	I
Beta-156698	bone	11,230 +/- 50	79	85	30	I[i]
----------------------------------Unidentified Tephra Layer----------------------------------						
Floodplain Dates						
W-2213	shell	7,980 +/- 300	69	0	71	III[e]
Beta 156699	bone	9,710 +/- 40	62.7	25	55	I[g,i]
W-2209	shell	9,820 +/- 300	?	-20	24.7	I[g]
Beta-120802	bone	9,870 +/- 50	62.7	25	25	I[g,i]
Y-2481	shell	9,970 +/- 110	?	-20	24.7	I[g]
W-2212	charcoal	9,840 +/- 300	?	-10	-30	I[h]
W-2218	charcoal	10,130 +/- 300	?	-10	-30	I[h]
Beta-156697	charcoal	10,570 +/- 70	61.96-62.18	25	65	I[i,j]

[a] C^{13} isotopic ratios were not determined for the radiocarbon dates obtained in the 1960s or 1980s.
[b] All shell dates are on western-river pearl mussel (*Margaritifera falcata*) from which the outer 20% has been removed with hydrochloric acid before generation of the counting gas. Bone collagen, obtained by a modified Longin procedure (1971), was dated (Sheppard et al. 1987).
[c] Sheppard et al. (1987) dates.
[d] Stratigraphically anomalous, originally assigned to Stratum VII, Sheppard et al 1987 placed it in Stratum V based on its age and depth, noting that Strata V through VII deposits were disturbed by numerous storage pits.
[e] Immediately beneath Mazama tephra (intrusive?).
[f] Stratigraphically anomalous, probably from Stratum III.
[g] Marmes horizon.
[h] Harrison horizon.
[i] New dates; C^{13} ratios were measured for these dates (see Appendix Q).
[j] a composite charcoal sample from soil A horizons A10-A16.

an element of uncertainty regarding the derivation of the cultural materials recovered. Citing the same source of potential uncertainty in the upper deposits as Sheppard et al. (1987) and their willingness to reassign samples based on the result of dating, it is suggested that the date range assigned to Stratum VI is too limited. If the relatively imprecise date of sample WSU-205 is reassigned to Stratum VII (where it fits firmly within the other Stratum VII dates), then the more precise date of sample WSU-3035 can be seen as terminus of the Stratum VI period. This doesn't afix sample WSU-207 any better, but it could be asserted that 4250 B.P. could represent the early part of the Stratum VI period or the late part of the Stratum V period.

The uncertainty in the dating of most of the Marmes Rockshelter strata arises out of the inability to provide bracketing dates. It would take many more dates from each strata to provide higher resolution to the dating of each stratum. In some cases, however, many more dates would not improve the resolution of the data since some of the strata are not very thick, but represent several thousand years of deposition without any intervening temporal markers. The lack of high resolution dating within the strata affects the degree to which conclusions can be offered for many of the research questions addressed by this study.

- What is the archaeologically demonstrable first use of the site?

Previous dating of the rockshelter deposits had established cultural use in Stratum I by 10,810 ± 300 B.P. (see Table 15.1). The next earliest date, 10,750 ± 300 B.P., was obtained from a location some 15 feet closer to the mouth of the rockshelter. Two similar dates from such dispersed locations in the rockshelter is a strong argument for cultural use at this time. These two dates and the other previously submitted Stratum I date (WSU-336) all overlap at 1 sigma (sample WSU-120 is probably intrusive from Stratum III). However, cultural materials were recovered from much deeper in the deposits than the samples associated with those three dates, including a few items recovered from on top of the glacial lake sediments that form the basal sediments in the rockshelter. These items were never dated, and while there is no indication in the progress reports as to why, it may have been because the expense of radiocarbon dating limited that kind of analysis to broader questions, such as establishing the periods of the strata rather than its earliest extent.

One of the samples submitted for this study sought to establish the earliest use of the rockshelter by dating the deepest demonstrably cultural material. A piece of bird bone with indications of butchering similar to the swan bone dated from the Marmes horizon (see Appendix Q) was selected. This bone was described in the field notes as lying directly on the lake sediments at the base of excavations in the rockshelter. Description of the find and the sediments it was found in strongly indicate that this item did not end up on the lake sediments as a result of the action of the excavators as was suspected higher in the profile where excavation consisted largely of removing rooffall rock. It is possible that some small cultural items trickled down from above through time, sifting down through open spaces between the coarse rooffall rock deposits. But this bone was found at the bottom of a level described as containing sandy sediment, meaning the bone would also have to have found its way down through sand in addition to the overlying rooffall rock. Even if that were the case, the date still represents the earliest cultural use of the rockshelter, even if it doesn't represent a use that took place directly on top of the lake sediments where the bone was found. This bone returned a conventional date of 11,230 ± 50 B.P. (Beta-156698; faunal bone; $\delta^{13}C/^{12}C$ = -17.9‰), nearly half a millenium older than the previous oldest conventional date obtained from the site. When this 11,230 B.P. date (11502-11053 BC) and the 10,810 B.P. date (11505-9934 BC) are calibrated they do overlap at 2 sigma. For our purposes, this indicates that the new early date represents only an earlier extension of the assemblage described for Stratum I. As noted in Chapter Two, attempts to obtain a confirming radiocarbon date of material from the same location failed. But recovery of the dated bone from on top of glacial lake sediments with volcanic ash believed to be from the Glacier Peak eruption that dates to virtually the same date (11,250 B.P.; Johnson et al. 1994) lends credence to this earliest radiocarbon date.

This date offers two additional items for temporal interpretation. Both Leonhardy (1970) and Hammatt believed the lake that resulted from the last glacial-era flood drained out of the Snake River between 11,000 and 10,000 years ago. Hammatt (citing Cole 1968) sets the availability of riverine landforms at Wildcat Canyon on the Columbia River at *no later than* 10,600 ± 200 B.P. (1977:173). Similar landforms in the lower Palouse Canyon would have become available (drained) perhaps a hundred years earlier (the flow rates in the Columbia River probably would have dictated how quickly the lowest elevations in the lower Snake River could drain). Using the Wildcat Canyon as the lower limiting date, this would make the possible earliest use of similar landforms in the vicinity of the Marmes site at ca. 10,700+ B.P. (ignoring error range for the moment).

The earliest date from Marmes Rockshelter (11,230 ± 50 B.P.) would push back further the time period when landforms in the Palouse Canyon (and the lower Snake River drainage) became available to human use. The limiting age of 11,200 ^{14}C years B.P. for the base of Unit I is based on an assumption that the unidentified tephra (see Table 15.1) found in the upper part of the water-deposited sediments by Fryxell is Glacier Peak (Foit et al. 1992). Although Sheppard et al. (1987) and Huckleberry, Gustafson, and Gibson (1998) consider this unconfirmed, this new radiocarbon date assures that the ash is not a more recent ashfall. The 12,800 B.P. date obtained by Lenz et al. (2002) for the last Missoula Flood, which established the last time glacial lake sediments could be deposited in Palouse Canyon, occurs after the next most likely tephra that could have been deposited, that of the Mt. St. Helens S series which occurred ca. 13,000 B.P. (Mullineaux 1986). Further, this new AMS radiocarbon date, from a bone collected in direct association with the glacial lake sediments in the rockshelter, is virtually the conventional age assigned to Glacier Peak tephra (11,250 B.P.; Johnson et al. 1994). The calibrated 2 sigma range of error for the bone date is fairly narrow at less than five hundred years, which if considered associated with the ash deposit would mean that the ash could not be St. Helens S series which occurred ca. 1,800 years before the Glacier Peak eruptions. For the ash in the lake sediments to be St. Helens S, no deposition of sediment could have occurred over the ca. 1,600 years between the last glacial flood and the deposition of the swan bone.

The swan bone was found atop the glacial lake sediments and there is not a perfect association between the bone and the volcanic ash in the lake sediments; we can only state that the bone was deposited after the ash deposition, which appears to have occurred with the final draining of the glacial lake from these elevations in Palouse Canyon. Given that both Leonhardy's (1970) and Hammatt's (1977) estimates of the time when the lake drained away (and therefore the last of the lacustrine sediments in the rockshelter could be deposited) can now be seen as conservative (see Chapter Three), it strengthens the conclusion that the tephra in the lacustrine sediments is Glacier Peak and the date of that eruption is a reasonable limiting date for the base of the rockshelter deposits. Therefore, given the date of the bird bone, a conclusion that Marmes Rockshelter saw some level of cultural activity soon after the lake drained away from these elevations in the Palouse Canyon is reasonable.

It is unfortunate that the deepest deposits did not get greater treatment to establish them as more than an ephemeral, poorly defined initial use of the site. Without a larger horizontal exposure of the materials found at this depth, together with more extensive dating of the substrata directly above the sediments lying atop the glacial lake silts that Fryxell discerned and labeled IA through IE, the inevitable critique of these materials as occurring as a result of mixing of younger deposits with older, deeper strata by animal burrowing or some other natural agency cannot be positively refuted. But Dillehay (2000) notes that archaeologists may disregard such materials too readily. "Light or sporadic human activity is not seriously considered. Yet is not this light scatter of material exactly what we would expect from an initial exploratory phase of human entry? …too many archaeologists have used taphonomy as a handy excuse to reject ephemeral records rather than test them more thoroughly…to determine their geological integrity and archaeological validity. ….(I)n the absence of modern site formation studies, we can only ponder the status of these records" (2000:217). Surely this lament applies to Marmes Rockshelter, but it is a lament aimed at an effort and techniques already over three decades old.

- Does the initial assemblage correlate with other early site assemblages in the region?

As noted above, the date of 11,230 B.P. does not necessarily establish an older date of occupation of Marmes Rockshelter, but it does alter slightly temporal comparison with other early site assemblages in the Columbia Basin. In particular, the dated sample occurred after deposition of Glacier Peak ash (ca. 11,250 B.P.; Johnson et al. 1994), which has been used to date the Richey-Roberts Clovis Cache materials found lying atop sediments impregnated with this ash (Mehringer and Foit 1990). Based solely on these materials' provenience in relation to Glacier Peak ash, initial cultural use of Marmes Rockshelter and the cache event at the Richey-Roberts Clovis site occurred at roughly the same time. A single date in Marmes Rockshelter is not adequate evidence to indicate that these two sites were contemporaneous, especially since an effort to obtain a confirming date was not successful. But it raises interesting questions about the relatedness of the earliest Marmes component, identified as Windust Phase and included as early Period IB in Ames et al. (1998) Plateau cultural chronology, and the Clovis-era peoples represented as Period IA.

This raises more questions than it resolves since there is a dearth of Plateau Clovis assemblages to compare with Windust (or Windust consequent) assemblages, and not many Windust Phase occupation sites either. This discussion will focus on three sites with assemblages that are contemporaneous with Unit I in Marmes Rockshelter: Richey-Roberts in East Wenatchee, Washington; Cooper's Ferry (10IH73) along the Lower Salmon River in west-central Idaho; and, Paulina Lake (35DS34) in central Oregon. The Cooper's Ferry and Paulina Lake sites occur near the margins of the Plateau/Great Basin physiographic regions/culture areas and can be expected to exhibit influences from both areas through time. However, the early Holocene occupants in both regions pursued a mobile foraging settlement and subsistence pattern, and the point styles are considered to be local cognates of the Windust Phase (Leonhardy and Rice 1970; Rice 1972), which fall within the technological framework of the Western Stemmed Tradition (Connolly 1999; Davis 2001; Willig and Aikens 1988).

The Richey-Roberts materials are believed to represent an intentional cache and therefore probably do not reflect the whole assemblage utilized at a Clovis habitation site. Materials recovered from the Richey-Roberts site include "Formed bone objects (spear shaft spacers and foreshafts), large bifaces and bifacial blades, fluted points, unifacial implements, and debitage…" (Ames et al. 1998:103), some of which are likely to be the "chopper, a graver…four sidescrapers, five sharp-edged prismatic blades and flake-blades, and three flaked stone axes or adzes" (Gramly n.d.:4). The Windust Phase assemblage at Marmes Rockshelter, Stratum I in particular, is dominated by lithic flake and bifacial tools, bone tools (pins, eyed needles, rods, awls and other pointed implements, ornaments), stone projectile points (both stemmed Windust and lanceolate Cascade), hammerstones, choppers, and much lithic manufacturing debris. General similarity likely would exist between any Windust and Clovis assemblages given that they are both believed to reflect early mobile foraging populations (although the bone rods continue to raise speculation as to their functions [e.g., Lyman et al. 1998]). In this case, the differences likely reflect considerable assemblage function differences: Marmes Rockshelter is a habitation site and the Clovis materials may have been a cache for a fall-winter hunt (Gramly n.d.:4). In particular, the Marmes bone technology appears focused on domestic activities such as tailoring, while Gramly describes the Clovis assemblage as a tool kit for procuring and butchering animals.

Clearly there is a difference between the "classic" Clovis fluted bifaces and Windust phase bifaces/knives, with different manufacturing processes required for hafting the two styles. But Willig and Aikens (1988:10) state that the gradation from Clovis to their Western Stemmed points that include the Windust style is striking. This may be supported by the observation that the earliest Windust phase assemblage from Stratum I in Marmes Rockshelter shows more emphasis on the later stages of percussion bifacial thinning, suggesting a biface-based technology akin to Clovis and other early Paleoindian technologies, implying in situ transition from paleo technologies to Windust and later in the Plateau (e.g., projectile point shown on cover of this report - this point is among the missing). "Patterns of Windust and Cascade projectile point breakage and reworking observed in the Marmes assemblage also are similar to those described elsewhere for Paleoindian projectile points" (Ozbun et al. 2000:83; Chapter Eight). The breakage pattern in Windust and Cascade points is inferred to be a result of high velocity impacts (ibid), which are presumed to indicate the use of atlatls or other throwing boards in the delivery system. If these breakage patterns are also similar to those observed in Paleoindian points, then use of atlatls is indicated for Paleoindian peoples as well.

The Cooper's Ferry and Paulina Lake assemblages provide a better comparison with the earliest uses of Marmes Rockshelter than does Richey-Roberts. Like Marmes, they have projectile points attributable to the Western Stemmed Tradition (Willig and Aikens 1988) in strata with similar dates. Component I at Paulina Lake produced a single conventional radiocarbon age of 9920 ± 470 B.P. (Connolly 1999:98); despite the large error range this date, when calibrated at 2 sigma (10990-8240 BC), does not overlap with the earliest Marmes Rockshelter date. However, Connolly's sample does overlap with the calibrated date ranges for several other

Marmes Rockshelter Stratum I dates. Two samples from Component A1 at the Cooper's Ferry site produced conventional radiocarbon ages of 11,370 ± 40 B.P. and 11,410 ± 130 B.P. (Davis 2001:291); both were processed using the AMS method. When calibrated to 2 sigma, these two dates give ranges of 11830-11100 BC and 11860-11080 BC, respectively; these overlap solidly with the calibrated date range of the earliest date from Marmes Rockshelter. (Note: there is some concern regarding the Cooper's Ferry dates due to the presence of considerably later dates [8710 B.P. and 7300 B.P.] in association with the early dates. Davis (2001:293-295) provides reasonable explanations for how the later samples may be intrusive.)

Davis (2001:281, 287) recovered only four projectile points from Component A1 at Cooper's Ferry; all are stemmed Lind Coulee style points. Three are whole and one is broken at the base in the manner described in Chapter Eight as a result of high-velocity impacts. Connolly (1999:112-116) recovered five points from Component 1 at Paulina Lake that retained enough features to identify by style; two are stemmed (Windust) and three are foliate (Cascade) points. Most are broken medially, which deviates from the breakage pattern described in Chapter Eight for Marmes Rockshelter points. However, Connolly collapses his Components 1 and 2 (lower pre-Mazama) when discussing the site's assemblage, and numerous Component 2 stemmed points are broken at the base. The Paulina Lake projectile point assemblage also is similar to that of Marmes Rockshelter in that the early strata contain both stemmed (Windust) and foliate (Cascade) points. Component 1 and Stratum I, respectively, contain both point styles in low numbers. By the time of Component 2/Stratum II, both assemblages contain more foliate points than stemmed. The trend continues in Component 3/Stratum III, both of which represent the time period just before the Mazama ashfall, where foliate points outnumber stemmed 4:1. Gradation in the relative representation of these two point styles through the pre-Mazama Holocene supports Ames' et al. (1998) combination of the Windust and Cascade Phases into a single period in their cultural chronology (see Chapter Four). The gradation in these two point styles representation through time also is evident at Cooper's Ferry where "small leaf-shaped projectile points were recovered [from Component A3] and are similar to types grouped under the late Windust and Early Cascade Phases" (Davis 2001:290). Davis notes that the projectile point assemblage at the Cooper's Ferry site describes "a continuous evolutionary progression of non-fluted technology beginning at a period contemporaneous (at minimum) with the appearance of Clovis fluted point technology" (2001:304). Given the limited size of the assemblage from the earliest component at Cooper's Ferry, and the suggestion of problems in the dating of that assemblage, correlating the much larger earliest component assemblages from Marmes Rockshelter and Paulina Lake provide a stronger basis for Davis' statement.

- What is the temporal relationship between the initial cultural materials in the rockshelter and on the floodplain? What is the continuing relationship between the two areas of the site through time?

As can be seen in Table 15.1, the three new AMS floodplain dates stretch the conventional dates obtained from the floodplain both earlier and later. A new AMS date of 10,570 ± 70 B.P. (Beta-156697; wood charcoal; $\delta^{13}C/^{12}C$ = -24.9‰) was obtained from a charcoal rich sediment sample collected from the soil A10-A16 horizons deep in the floodplain excavation. This is somewhat older than the previously older date of 10,130 ± 300 B.P. obtained from the Harrison horizon, but when both dates are calibrated they overlap at 2 sigma (as do the new most recent date and the previous most recent date). Still, the more precise AMS date, and the deeper elevation of the new sample, would suggest that there was a somewhat older use of the floodplain deposits than previously documented, although this earlier floodplain date does not exceed Fryxell's (Fryxell and Keel 1969) assumption about the time period when the terrace would have become available for human use. The small number of cultural items associated with these deeper soil A horizons are consistent in terms of type and material with the cultural materials recovered from the better documented Marmes and Harrison horizons. The new older date for the Marmes horizon (A1-A2) is between the two previous dates while the new more recent date (9710 ± 40 B.P. [Beta-156699; $\delta^{13}C/^{12}C$ = -16.8‰]) obtained from a swan bone with cultural butcher marks lengthens use of the Marmes horizon during this time period slightly.

The new early floodplain date fits firmly within the range of dates obtained by Fryxell for Stratum I in the rockshelter (see Table 15.1). And while the new early date for the rockshelter suggests earlier initial use of the rockshelter than the floodplain (the two dates do not overlap at 2 sigma when calibrated), it isn't certain that older materials may not exist in the floodplain. In particular, Caulk notes that "18 of these incipient "A" horizons containing flaked stone materials were identified" (1988:13). For this study, these earliest cultural items are not distinguished from the assemblages that date soon after and that are better represented. As such, Stratum I is considered to represent both the earliest geological strata in the rockshelter and the floodplain materials up to and including the Marmes horizon.

Several new dates were processed by this study to more closely examine the correlation between the human remains in the floodplain and the cremation hearth in the rockshelter as part of a larger attempt to solidify the assumption that the floodplain deposits are of the same age as Stratum I in the rockshelter (Fryxell and Daugherty 1962). As noted there, the most recent floodplain dates do not overlap with dates from the cremation hearth. The cremation hearth (and other deposits at this elevation in the rockshelter) was consistently assigned a designation of Stratum I/II because while a distinction between these two geological strata could be readily seen in profile, the excavators had difficulty distinguishing between them when working vertically down through the transition between these strata. However, by designating all floodplain deposits as Stratum I, and citing a discrepancy in dates between the most recent Marmes horizon date and the cremation hearth, perhaps we have assigned the cremation hearth to Stratum II. However, the calibrated date range of other Stratum I/II samples (see Appendix Q) do overlap with the most recent new Marmes horizon date. This may be due to the imprecision of their initial dating (i.e. sigma 1 ranges of 300 years), but a systemic assignation of Stratum I/II materials to Stratum II is not merited.

It is apparent that the floodplain did see some continuing use after the time period associated with the Marmes horizon. Cascade Phase materials were observed in floodplain sediments above the Marmes horizon. But other than those collected from around a shell concentration underlying Mazama ash west of the main floodplain excavation area (dated at 7980 ± 300 B.P.; see Table 15.1), none of these were investigated through excavation. As noted in Chapter Fourteen, the 17 tools collected from the "Cascade Component" in the floodplain (Fryxell and Keel 1969:32) represent a diverse assemblage of tool types but probably do not represent the entirety of Cascade Phase cultural materials that were in the floodplain deposits. Given this diversity of tool types despite the lack of consistent collection of cultural materials from this stratum, some continued use of the floodplain in tandem with the rockshelter is indicated at least into the Cascade Phase associated with Strata III-V. Use of the floodplain after this time is uncertain as associated sediments were removed with a bulldozer.

- Was the site utilized continuously throughout the whole of its dated record? If so, was it utilized by the same cultural group? If not, what explanations for a break in site use are suggested by the site data?

Continuous use is a difficult argument to demonstrate using archaeological materials. Material densities at the Marmes site are not high enough to assert continuous, year-round occupation during the time period associated with any of the strata or cultural phases; the assemblages represent palimpsests of multiple periods of use during each period. In addition, continuous use by tethered foragers during the early- to mid-Holocene, as was suggested in Chapter Two, may mean a three-week stay once each year. Conversely, continuous use as a residential base by people using a collector strategy could be year-round and for many years in a row and would produce larger and more diverse assemblages; as noted above, this is not demonstrated for any period of the Marmes site's use. However, use as a short-term camp in a collector strategy may produce an assemblage similar to that of forager base camps.

It usually is easier to demonstrate discontinuity, given that a 50-cm thick strata can represent 2,000 years of time in the Marmes site. But while there may be good representation of materials assigned to a specific phase, it doesn't mean the site wasn't unoccupied for a hundred years during the 1,500-year period of that phase. Periodic abandonment of this kind in a site with a long history is generally not identifiable in the archaeological record outside fluvial environments. It is even harder to demonstrate periods of abandonment by one cultural group and reoccupation by another cultural group where the assemblage contents don't vary or exhibit gradation from one phase to the next.

Cultural material is found from the earliest dated deposits to the historic era at Marmes Rockshelter. In general, and citing in particular the lithic assemblage, all of the assemblage constituents associated with each of the phases defined for the lower Snake River by Leonhardy and Rice (1970) are found in the site. All of the point styles defined by Leonhardy and Rice and found in other sites in the lower Snake River region are found in the deposits at the site (see Figures 8.1 and 8.26). However, the period immediately after the Mazama ashfall was considered a period of abandonment of Marmes Rockshelter with the exception of using the thick ash deposit for burials (Fryxell and Keel 1969). While Stratum IV certainly lacks occupational surfaces, this is a geologically-defined stratum that was deposited over a short period, perhaps a few days to a few weeks. For some time immediately after the eruption, ground surfaces throughout the Palouse Canyon also would have been covered with this ash and wind-blown deposition into the cave would have consisted mostly of ash as well. While the latter would be a secondary deposit, it wouldn't be discernible in profile. Some period of time passed before the next identifiable sediment type began to be deposited in the rockshelter. This next sediment type (Stratum V) consisted of Mazama ash mixed with aeolian silts (which must have contained a considerable amount of the fine tephra as well) indicating that while Mazama ash was still being blown around in the immediate environment, non-tephra sediment was again available. It is difficult to say how much time passed between the primary ashfall deposit and re-initiation of silt in the wind-blown environment, but this is a consideration for the refining of the dating of Strata V and VI discussed above. Refinement of climate chronologies may have the most to offer to this question as the amount of silt available to aeolian depositional regimes is somewhat dependent on available moisture and extent of vegetation cover; this is considered below.

The perception that Stratum V was used only for ritualistic (burial) purposes would appear to be refuted by consistent counts of lithic tools and debitage in the strata that postdate Stratum IV. In particular, if the site was largely abandoned during Stratum V except for ritualistic uses, debitage counts would be expected to fall off dramatically. Instead, the debitage count is within 10% of that in Stratum III, which contains the best evidence of habitation of any period of the site's use. In addition, a change from habitation to ritualistic use should be reflected in dramatically changed functions of formed tools. But while the number of formed stone tools falls by half from Stratum III to Stratum V, roughly the same proportions of tool types occur in both strata suggesting the same activities occurred in both periods. This would also appear to forestall an argument that the Stratum V tools are intrusive as a result of storage pit excavation from upper strata. Also, the combined Stratum IV/V bone tool counts are the highest in the rockshelter after Stratum II, although some of these are certainly grave goods. It would seem a reasonable assumption to make that a thick tephra deposit would severely impact how a site with restricted space such as Marmes Rockshelter could be used. Data supports such an assumption only for the period associated with the ash deposit itself. By the time the floor sediments became Mazama ash mixed with aeolian silts (Stratum V), archaeological materials indicate that use of the cave for habitation resumed.

The Stratum VI assemblage is similar to Stratum V with a small increase in formed stone tools and roughly the same number of bone tools (once items associated with burials are discounted). Stratum VII is marked by a period of digging of pits, but its assemblage also reflects an increase in largely the same activities as those in the previous two periods. There are adjustments in the proportions of tool types (addressed below under Site Function), but the same types of tools, and therefore probably the same kinds of activities are represented and in considerably higher numbers. Stratum VIII is mostly associated with historic era disturbances, and sediments assigned to this strata contain late prehistoric materials brought to the surface by aboriginal and historic digging in the site all the way through to items contemporary with the excavations in the 1960s. Because of the extent of disturbance, little can be said of continuity of site use into the period associated with this stratum. It is considered representative of the ethnographic Numipu phase and may reflect the site's relations to other similarly dated sites nearby. Euroamerican artifacts were recovered from the site (Collins and Andrefsky 1995:55), indicating some use of the site during the period associated with Stratum VIII. Most of these items appear to be grave goods (some have ochre staining), but no burial features were recorded in

Stratum VIII by the excavators, while Stratum VII predates the protohistoric period on the Plateau (Campbell 1985). This may indicate that protohistoric burials in Stratum VIII were disturbed by pothunters prior to the archaeological investigations in the 1960s. If Euroamerican items were only included in burials it could be seen as an indication that the site was not used for habitation during the time period associated with the most recent strata. The list of Euroamerican items does not include functional artifacts (e.g., pots, knives) and instead appear to be decorative (e.g., beads, textiles) or ceremonial (i.e. bell) items, furthering the impression that the site was not used for occupation in the last 200 years.

As the above discussion indicates, the Marmes site exhibits evidence of use in every period of the site's post-glacial lake history as defined by geological strata with the exception of the short period associated with deposition of Mazama ash. However, as noted at the beginning of the discussion on this research question, this does not demonstrate continual use by the same cultural group. The continuous presence of tool types with recognizably distinct styles tied to dated contexts can be used for this purpose. In the Plateau, projectile point styles are the only tool type with a long history that have been used in this way.

Leonhardy and Rice (1970) established the initial cultural chronology of the lower Snake River region, defining phases largely based on tool assemblages (including especially that from Marmes Rockshelter) including projectile points. While their chronology has been tweaked some by subsequent researchers, the one thing that has not changed much is the recognition of the applicability of the point styles through time. To this day, most researchers can readily identify Windust and Cascade points and will describe small dart point styles as "Tucannon-like"; beginning with the advent of bow-and-arrow technology, styles are less sub-regionally distinct. As noted above, all of the point types defined in the Leonhardy and Rice (1970) chronology are present in dateable contexts in Marmes Rockshelter. Ozbun et al. (in Chapter Eight) note that "many of the diagnostic artifact types cross-cut stratigraphic and temporal boundaries for the assemblages and periods they characterize...[due to] both mixing of the archaeological deposits and survival of technological traditions through long periods of time." Figure 8.22 shows that the different point types' peak of representation fits the Leonhardy and Rice chronology well, but that each point type overlaps with those identifiable as earlier and later types. There is no break in any of the styles relative to each other (which could suggest cultural group replacement) and the Cascade point style is very long-lived. These trends are not flukes introduced as a result of deposit mixing, since other tool types that complete Leonhardy and Rice's phase assemblage definitions are found in Marmes Rockshelter in tandem with these point styles (see Chapter Eight).

Comparison of the proportional use of different lithic materials through time in Marmes Rockshelter and the region may be used to identify potential breaks in use by the same cultural group. Basalt (or andesite) and CCS materials predominate throughout the site occupations. CCS is overwhelmingly predominant in the earliest occupations, but the use of basalt increases steadily throughout the Cascade subphases reaching equal importance, and then declining again thereafter. The increase in basalt use in the Cascade Phase appears to be a regional trend (Ames 2000:50-51).

Table 15.2 represents a modification of Sheppard's et al. (1987) Table 2 to include the addition of a column that shows Leonhardy and Rice's (1970) cultural chronology phases assigned using the dates originally proposed by Leonhardy and Rice (see Chapter Four). In this modified table, the age range of each of the Marmes site's geological strata are shown, reflecting all radiocarbon dates including the six new dates introduced by this study. Other modifications to this table include adjustments of some radiocarbon dates' stratum assignments as discussed above (i.e. moving the 1300 B.P. date for sample WSU-205 to Stratum VII, moving the 6200 B.P. date originally assigned to Stratum III [thought to be intrusive] to post-Mazama ash Stratum V) and the addition of the floodplain area dates within Stratum I.

Matching up Leonhardy and Rice's (1970) cultural phases with the Marmes Rockshelter geological strata indicate little direct correlation. One could infer from this that the environmental changes that contributed to development of geologically distinct strata in the rockshelter did not effect changes in the cultural materials at the site. But these dates are not direct dating of the

Table 15.2 Correlation of radiocarbon dates of geological strata in Marmes Rockshelter with Leonhardy and Rice's (1970) cultural chronology phases.

Stratum	radiocarbon age range	Phase[1]
VIII	modern	Numipu/historic
VII	660 – 1600 B.P.	Harder/Piqunin
VI	1940 - ? B.P. (one date)	Tucannon/Harder
V	4250 (?) – 6200 B.P	late Cascade/Tucannon[2]
IV	6730 B.P. (ash, no dates)	break between early and late Cascade
III	7070 - 7870 B.P.	Cascade
II	8525 – 9540 B.P.	Windust/early Cascade
I (and FP)[3]	9610 – 11,230 B.P.	Windust

[1] Leonhardy and Rice (1970), Leonhardy (1975): Windust 10,000-9000, Cascade 8000-5000, Tucannon 5000-2500, Harder 2500-250, Numipu 250-present B.P.

[2] base of this phase marked by primary Mazama (ca. 6730 B.P.), assemblage includes Cascade and Cold Springs point types; break between SV and SVI uncertain.

[3] all floodplain dates are older than cremation hearth (SI/II) and the other SII date; new date from floodplain A10-16 horizon fits within the range of rockshelter SI dates.

geological stratum (i.e. dates of the earliest and latest cultural manifestation of each strata), but spot dating of the cultural materials within them. As such, the beginning and ending dates of each phase's date range are extendable. But even given that flexibility, only one of the proposed breaks between phases falls near a break between geological strata (Tucannon/Harder ≅ Strata V/Strata VI), and perhaps then only because there are few dates for either of these strata. The 4250 B.P. date for Stratum V may not be close to the termination of this stratum's true date range, but no other dates are available. Likewise, the single date for Stratum VI (1940 B.P.) is a reasonable termination date for this stratum but the beginning of the date range for this stratum is uncertain. Since the Tucannon Phase is hypothesized to transition gradually into the Harder Phase (Leonhardy and Rice 1970) changes in the assemblages between the two phase's representation in the Marmes site would not be expected to provide a discernible marker to excavators.

The later date for Stratum V of 4250 B.P. does not correlate well with the late Cascade point styles found in Stratum V, if the date of the beginning of the Tucannon Phase (ca. 5,000 B.P.) as defined by Leonhardy and Rice is accepted. Recall that the sample that returned the conventional radiocarbon date of 4250 B.P. was originally assigned to Stratum VII but Sheppard et al. (1987) reassigned it to Stratum V based on its depth and age. But Stratum V contains none of the corner-notch and stemmed Tucannon dart points (which are poorly represented in this site). Instead, Stratum V is marked by lanceolate Cascade and Cold-Springs side-notched point types. The coexistence of these two point types typify the late Cascade Phase, defined by Leonhardy and Rice as a post-Mazama ashfall subphase of the preceding Cascade Phase. But Bense (1972) has noted that both of these forms were used by Tucannon peoples before largely being replaced by other forms. Thus the difficulty of dating Stratum V based on its assemblage matches with the region's observed record. Leonhardy (1975) revised the terminating date for the Tucannon Phase to 4500 B.P. instead of 5000 B.P. (Leonhardy and Rice 1970), but the earlier date appears justified based on housepits at Hatwai and Hatiupuh that date to near 5000 B.P. (Ames et al 1998:109); those terminating dates are within the calibrated 2 sigma range for the conventional 4250 B.P. date. (The situation may be different in the northern Columbia Plateau where Cascade points are associated with housepits dated to ca. 4000 B.P. at Cassimer Bar [Chatters 1986].) Assigning the 4250 B.P. date to Stratum VI, the assemblage from which better reflects the Tucannon Phase, instead of Stratum V would resolve the only discrepancy in the site's point style series.

However, reassigning a sample date does not clear up all that is questionable about the assemblages associated with Strata V and VI and their reference to Leonhardy and Rice's (1970) cultural phases. While the geological strata are well marked in profiles, the cultural material assigned to Stratum VI is not so distinguished that a clear demarcation between cultural phases can be made within these two strata or with the stratum that follows. Leonhardy and Rice were hampered in their definition of the cultural phases associated with Marmes Rockshelter's Strata V (late Cascade subphase) and VI (Tucannon Phase) in that they identified a hiatus in their data occurring ca. 5000 B.P. that they interpreted as indicating historical unrelatedness of these two phases. Identification of a hiatus appears to be due to the small number of assemblages on which they based these phase definitions (three each for late Cascade and Tucannon). Subsequent research (e.g., Harder 1998; Kennedy 1976; Lucas 1994) has compared many more assemblages along the lower Snake River and found no hiatus in cultural use of the area, but these researchers have found a clear change in assemblages. This is not borne out in the lithic assemblages at Marmes Rockshelter. Other than an increase in the proportion of biface blanks and preforms to completed projectile points, the proportion of formed tool types is similar. While there is an increased emphasis on CCS over basalt in Stratum VI compared with Stratum V, the proportions of stage reduction of the debitage do not vary significantly enough to define historical unrelatedness. As noted above, roughly the same number of bone tools (once items associated with burials are discounted) occur in both strata. The lack of distinction between the Stratum VI and Stratum VII assemblages can be attributed to the technological relatedness of their associated phases (Tucannon and Harder Phases) as described by Leonhardy and Rice (1970) and supported by subsequent research.

The change in technology between the late Cascade subphase and the Tucannon Phase has been well demonstrated in other sites from the lower Snake River region (cf. Harder 1998) and appears to coincide with the inception of a changed settlement and subsistence system from mobile foraging to a semi-sedentary collecting system where fish and plants became increasingly important resources (see below in Site Function). In this regard, the relatively crude lithic technology described for the Tucannon Phase (Leonhardy and Rice 1970:11) may represent the declining investment in maintenance of tool types that were losing their importance or were being refocused for different prey species. The change in point styles from the late Cascade small, leaf-shaped lanceolate points and the large, side-notched (Cold Springs) points to the smaller and less well-made side-notched and stemmed points of the Tucannon Phase is the one demonstrable change in Marmes Rockshelter assemblages associated with the different technologies asserted elsewhere for these phases (see Chapter Eight – the latter are referred to collectively as "Dart Points"). The diminished size of the Tucannon Phase points is not likely due to a change in the technology of the delivery system since both the small Cascade and Cold Springs side-notched points are acknowledged as dart points and atlatls appear to have come into use sometime in the Windust Phase (Ames et al. 1998), as is demonstrated at Marmes Rockshelter by the presence of a broken atlatl spur in the Windust deposits. At that time, the change to atlatl thrown darts implies the need to strike prey from further away, perhaps due to a changed focus of hunting faster or more aware animals than previously were targeted. But there is much overlap between Windust and early (large foliate) Cascade points and the two delivery systems these very different point styles imply co-exist in the early Holocene. The change in delivery systems appears to run its course by the time of the Mazama ashfall, with Cascade points decreasing in size, marking late Cascade subphase assemblages; no Windust points are found with late Cascade foliate or large side-notched (Cold Springs) points at Marmes Rockshelter.

The change in point styles from the late Cascade subphase to the Tucannon Phase is not abrupt as Tucannon Phase people continued to use both of the late Cascade subphase point styles in decreasing numbers (Yent 1976). The accompanying change to a more sedentary settlement pattern and collector subsistence strategy also does not appear to have been abrupt. Reid and Gallison (1995:2.26-2.30) note that plant processing tools associated with the use of roots rather than seeds is noted first in the Cascade Phase, as is evidence of the exploitation of salmon and steelhead, all thought to be particularly demonstrative of Tucannon Phase adaptation (Leonhardy and Rice 1970; Harder 1998). In fact, Bense (1972) drew a correlation between the subsistence base of Cascade Phase peoples and the ethnographic Nez Perce. Chatters (1989, 1995) has hypothesized that initial attempts at semi-sedentary lifeway in the Plateau failed despite having included a considerable element of the previous foraging lifestyle; after several hundred years, semi-sedentary living appears again, focused more on a collector lifestyle. All of these trends could account for the lack of distinguishing characteristics in the Marmes Rockshelter Strata V and VI assemblages.

Change to the smaller, more roughly fashioned Tucannon Phase dart points is demonstrated in the region's record, and the trend of continued decrease in size of dart points that began in the Cascade Phase may be indicative of further refocusing of prey species and/or technological responses to environmental changes that may have altered the location on the landscape where preferred species were found. Smaller points (and shafts) may indicate the hunting of animals in more constricted topography (perhaps in the canyon bottoms or upland forests instead of the basin plains) where larger points on longer shafts are not an advantage. The projectile points found in the Marmes Rockshelter assemblages associated with the Harder and Numipu Phases (see Table 15.2) exhibit the points well established in the region's record as typical of these time periods. These represent finer dart points and the inception of bow-and-arrow as a delivery system.

In concluding the discussion on the last cultural chronological research question, it appears that the only good argument for a break in the use of Marmes Rockshelter is following the Mazama ashfall, but there is good representation of cultural material in all the geological strata that post-date that stratum. The regional hiatus between the Cascade and Tucannon Phases suggested by Leonhardy and Rice (1970) is not borne out in this site. Instead, "The evolution of technologies at the Marmes site appears to be deliberate and conservative. That is, while there are differences between successive occupations, they seem gradual and incremental. What distinguishes one period from another is simply the relative frequency of certain artifact types and traits" (Ozbun et al. 2000:80, see Chapter Eight). Subsequent research in the region shows that there was no hiatus in cultural use of the area. While Leonhardy and Rice's perception of a hiatus appears based on the limited data available to them at the time, it did point other researchers toward identifying a transition in technologies in response to a change in lifeway (Ames 2000:51).

Trade

Several research questions anticipated that certain cultural items or the material type of cultural items would derive from outside the local catchment area of the Marmes Rockshelter inhabitants.

- What materials/items were obtained outside the Lower Snake River vicinity?
- Where were these items/materials obtained from and what does this indicate about social contacts/networks through time?
- Can the site's data inform on whether the trade was motivated by technological or subsistence needs, or are other cultural factors indicated?

Out-of-area items or materials were found in the Marmes site collection and efforts were made to collect information about them to address these questions. In general, the nature of the information obtained makes it difficult to separate the above questions. Lithic material, marine shells, and, beginning in the protohistoric period, Euroamerican goods, were commonly traded items in the Plateau; all are represented in the Marmes deposits. In addition, western red cedar and Douglas fir wood items occur, particularly in the later deposits (see Chapter Thirteen), but probably were developed from drift logs pulled from the Snake River. All of these materials have technological rather than subsistence applications; the use of shell as decorative items speaks more to social motivations than either technological or subsistence uses.

Several kinds of toolstone occur in the lithic materials that are not known to occur in the Lower Snake River area, in particular, obsidian and petrified wood. Obsidian is prominent in trade networks in the Plateau because there are few known sources close to the Columbia Basin/central Plateau. [However, to maintain perspective, Galm (1994:283) notes that obsidian rarely accounts for even one percent of toolstone in Interior Plateau assemblages.] Most of the sources of obsidian found in archaeological sites in the Columbia Basin are located in south-central Oregon, southern Idaho, and in British Columbia. These sources are widely used not only because of their high quality stone, but also because they are large sources with high quantities of accessible stone. Additional sources of obsidian were included in prehistoric trade networks but in lesser quantities, probably as a result of distance (e.g., Yellowstone, Wyoming source; Roulette 1997:26), quality (e.g., central Washington Cascade "Blue"; Fennelle Miller, personal communication, 2000), or periphery of source location to major trade centers (see below).

Obsidian sourcing of select materials found in the Marmes Rockshelter collection, both by this study (see Appendix I) and Hess (1997), found that five distinct sources are indicated as having been accessed for use at the site. Artifacts traced to the Indian Creek obsidian source (on a tributary of the Snake River in eastern Oregon) are found in all of the geological stratum at Marmes Rockshelter and in multiple examples in each stratum (thus mixing of deposits is probably not responsible for its presence throughout). Items from the Whitewater Ridge source (in eastern Oregon between small tributary streams of the Malheur River and the John Day River) are found in Strata III, VI and VII; a single example in Stratum VII could be a result of mixing of Stratum VI deposits where this stone source is well-represented. Two samples from Stratum I have been sourced to the Gregory Creek source (a short distance up the Malheur River from the Snake River in eastern Oregon; Unknown 7 in Hess 1997). One item each from Strata IV/V and VII were sourced to the Timber Butte source (a short distance up the Payette River from the Snake River in western Idaho). The fifth source remains unidentified (Hess 1997).

All four identified sources are located in highland areas south of the Marmes site and three of these are accessible short distances up primary tributary rivers that join the Snake River within 50 river kilometers of each other. This is probably not indicative of how the materials reached Marmes Rockshelter (i.e. river-based trade), however, as Hell's Canyon would have proved a significant barrier to loaded canoes. This appears to be a striking concentration of sources given the long history represented in Marmes Rockshelter's deposits. However, Ozbun et al. (see Chapter Eight) note that these four identified sources may be the closest obsidian deposits of good quality to Marmes Rockshelter (between 220 and 320 km). A long history (at least 9,700 years; see Table 15.2) of travel to and/or trade for these materials is indicated. This may indicate one of the directions that the Marmes site occupants traveled in the part of the annual round not spent at the site.

In addition to obsidian, certain non-local CCS materials occur in Plateau sites, likely obtained through a combination of exchange and direct acquisition through time. At the Marmes site, both petrified wood and a material tentatively identified as Knife River Flint were found (see Figure 8.25 a). The nearest known quantity of petrified wood to Marmes Rockshelter is the Ginko materials along the mid-Columbia River in Washington. Petrified wood in Plateau sites is widespread, but not usually in great numbers except near to the source area. It may be that petrified wood was not obtained/exchanged in any greater amounts than other fine CCS materials that occur around the Columbia Basin. But the ease of recognition of petrified wood compared with the multi-colored and multi-hued (especially with heat-treating) cherts that commonly occur alongside the Columbia basalt flows insures that trade will be assumed when it is found in a site. If the identification of Knife River Flint is correct, this indicates some trading relationship between the Plateau and the Plains. Unfortunately, this item is one that has not retained its provenience information and the stratum from which it was collected is unknown. The point style resembles Agate Basin points from Wyoming and has attributes that could place it in both Windust and Cascade Phase assemblages in the Plateau indicating that it probably came from one of the deeper strata in Marmes Rockshelter.

Two kinds of marine shell have been identified in Marmes site deposits, Olivella (*O. biplicata*) (see Chapter Twelve) and a single item made from abalone (*Haliotis* sp.) (Erickson 1990). Olivella shells (Figure 15.1) are found in all of the strata and are strongly associated with human remains features, although not necessarily only with burials. Both of the human remains features in Stratum III that were not interred in pits had Olivella shells in direct association. In Feature 2 they occur near a hand and may have been worn as a bracelet. Many other Olivella are found not in direct association with human remains features but likely represent items displaced from human remains features by rodents or other sources of disturbance (as has been suggested for individual human bones found unassociated with such features). But since Olivella shell appear to have been worn as a decorative item, given the holes drilled for stringing (and the consistent removal of their spires in the earliest strata), some of these probably entered the deposits as lost items given the extensive use of the site for habitation. A single abalone item was found, an extensively ground, square outlined pendant, in Stratum III (Erickson 1990:116).

Olivella shell is found in a number of other sites with older components along the lower

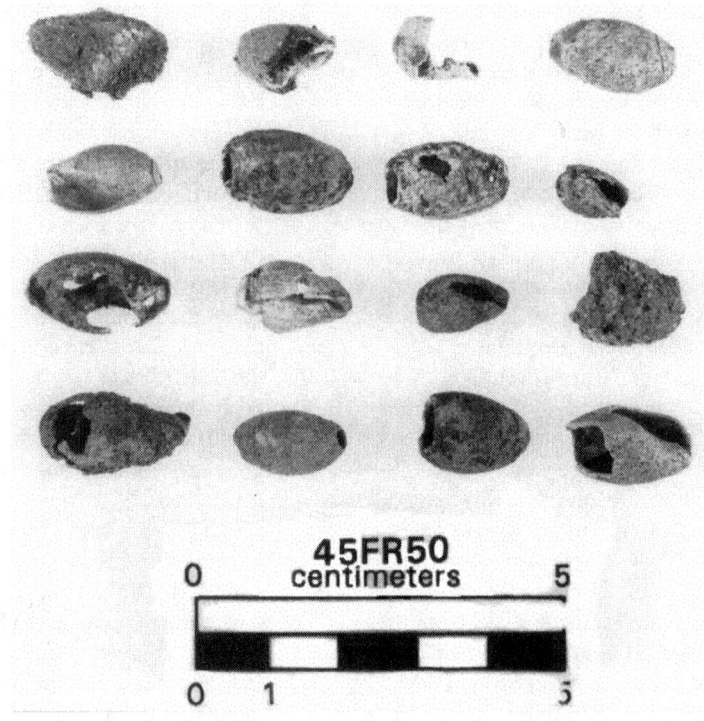

Figure 15.1 Olivella shell beads (print found in Marmes site records, WSU Museum of Anthropology archives).

Snake River. In particular, Olivella shells with the spires ground off were recovered at Alpowa/45-AS-78 (Early Cascade subphase, ca. 8,000 – 7,000 BP)(Brauner 1976), Granite Point/45-WT-41 (Area C, ca. 9,000 – 6,700 BP)(Leonhardy 1968, 1970), Tucannon/45-CO-1 (Assemblage 2, ca. 6,500-4,000 BP)(Nelson 1966), and in the Orondo Rockshelter/45-DO-59 (ca. 6,500 – 2,000 BP)(Gunkel 1961). At the Tucannon Site, the pattern of grinding off the spire continues through the late prehistoric period (Erickson 1990). Thus, this particular method of preparing this specific artifact type is found in multiple sites and over a long period of time.

Ames et al. (1998:119) note that the trade of obsidian and marine shell expands greatly in their Period II (late Cascade and Tucannon Phases of the lower Snake River chronology) as indicated by greater representation at sites throughout the region. But they don't increase proportionally within assemblages. This is borne out in the Marmes site deposits with only a small percentage increase of obsidian in Strata V and VI; obsidian is consistently a small part of the Marmes Rockshelter lithic assemblage. Olivella shells, because of their strong association with human remains features, were not analyzed for this study except where they occur in the bulk shell bags (non-feature related deposits). But based on review of the field notes, once they are identified in the site's deposits in Unit I/II, they are relatively consistently mentioned through to the uppermost deposits. Interestingly, dentalium shell, a common item in Period II sites throughout the Plateau, do not occur at all at Marmes Rockshelter.

The presence of marine shell and out-of-area toolstones indicate direct acquisition at some distance from Marmes Rockshelter, acquisition through trade, or both. Regarding the acquisition of obsidian, Hess (1997) examined the differences in changes of artifact form as distance from material source location increases between Late Pre-Mazama Period and Late Prehistoric Period items. He hypothesized that direct acquisition of source material in the earlier period would allow investigation of mobility patterns based on the changes in tool form (size reduction as a result of retooling through time) over distance. In the later period, when indirect acquisition was expected, this correlation of form and distance was not expected to hold, also as a result of changed scheduling of tasks (semi-sedentary collector vs. mobile forager). The hypothesis holds well for the earlier period with direct acquisition more fully demonstrated and distance from source roughly equaling a proxy for time passed.

In the late prehistoric period, it is clear that inter-regional trade routes were well established, particularly between the Plateau and the Pacific Coast (cf. Anastosio 1972). With increasing sedentism and an increased focus on certain food resources (and therefore certain resource locations), long trips (prior to obtaining the horse in ca. 1730) to obtain out-of-area non-subsistence resources at the source may have become less practical than obtaining those items at trading centers (e.g., the Dalles, Kettle Falls) in the course of trading subsistence items. Still, while Olivella shell probably was obtained by the Marmes site's occupants through the large trading centers along the lower Columbia River, obsidian from mountain sources to the south and southeast of Marmes Rockshelter likely was not; not over the 9,700 years it is found in the site's cultural strata. Walker's (1967) hypothesis that the ecological variation in the Plateau fostered interareal movement of localized resources appears to be a better fit and can better accommodate some "chaos" in the regional data. Since Indian Creek obsidian and Olivella shells are found from the earliest to the latest deposits in Marmes Rockshelter, the patterning of Walker's "interareal movement of localized resources" may have been long-standing on the lower Snake River. This could suggest that there has been no interruption in the cultural groups that participated in the processes of that movement of resources. This consistency would seem to apply both to the south, from where the Marmes occupants obtained obsidian, and the west from where the marine shell must have come.

Intuitively, it seems that the transition from direct procurement of obsidian to obtaining obsidian through trade (a reflection of the transition from mobile foragers to a more sedentary residence pattern [Hess 1997]) would have markedly changed the presence of obsidian in archaeological sites. As obsidian was traded more and more through the major trading centers, one would expect that the sources closest to those trading centers would become the predominant material types traded. Over time this should have shown up in the regional archaeological record as a decrease in the number of sources used or at least an increase in the percentage of certain obsidian types in more sites through time.

The Gregory Creek obsidian source is 300 kilometers south of Marmes Rockshelter and the nearest source of Olivella shell is ca. 500 kilometers west. In addition, northern abalone (*Haliotis* sp.), while found along the Oregon coast, becomes abundant only in the warmer waters off California (Erickson 1990). All three of these materials are found in Marmes deposits that date to the Late Pre-Mazama period during which direct procurement of obsidian is hypothesized by Hess. It cannot be asserted that all of these materials came to the Marmes site by direct procurement. On the other hand, the lack of any obsidian from the large Oregon Cascade Mountain sources could be used as an assertion of direct procurement (too distant). But while material from that source and others in central Oregon probably was available at The Dalles (Five-Mile Rapids) throughout later prehistory, it did not make its way to the Marmes site through exchange as should have occurred under Hess's hypothesis. This may indicate that upstream trade of toolstone did not occur, but clearly upstream trade of shell did. Data from other sites along the lower Snake River can clarify this observation. Such a discrepancy underscores the complexity of behavior that is indicated by long-distance trade, and the presence of items at Marmes Rockshelter indicate that complex human interactions were occurring in the earliest periods represented by the Plateau archaeological record.

Past Environment and Climate Change

Reid, in discussing the possible effect on the archaeological record of a drought that occurred in the late prehistoric period, suggests that there may be "patterned relationships between severe regional droughts, settlement pattern shifts, and perhaps even changes in the organization of subsistence and material culture" (1991:31). This appears likely, but may not be explicable in the region's archaeological record given the lack of precision of this kind of archaeological data to track short-term events, even several centuries-long events.

The excavation strategy used at Marmes Rockshelter does not allow investigation of short-term climatic and cultural material correlations that could address uncertainties in the region's cultural chronology. Instead, the research questions related to past environments focus on longer term trends that must be correlated with similarly long term trends in the cultural data as a means for interpreting cultural use of Marmes Rockshelter.

- What is the sequence of climatic change in the Columbia Basin over the course of aboriginal use of Marmes Rockshelter?

Chapter Three presents a summary of the Holocene environmental picture for the Plateau based on summary information presented in the recent Smithsonian Handbook for the Plateau (Chatters 1998) and Wigand's extrapolation of more detailed Great Basin vegetation-related data trends. While there are some discrepancies between the climatic and vegetation trends described by Wigand and Chatters, for the most part they are a matter of timing of transitions rather than disagreements about the trends themselves. (An attempt to analyze pollen from Marmes Rockshelter for this study was not successful – see Appendix R.) Figure 15.2 presents a tabular summary of several different kinds of information that reflect the environmental conditions on the Columbia Basin through the Holocene to which human groups would have been adapted. Figure 15.2 builds on Gustafson and Wegener's (1998) modification of Gustafson's (1972:34,120) pioneering work that inferred the distribution of big game mammals in the region based on Marmes Rockshelter faunal data and other inferences of past environments (Fryxell 1965; Marshall 1971) as one of the earlier presentations on the effects of past climate on human use of the Plateau.

One of the sources of climatic inference Gustafson used was by Fryxell (1965) who presented trends in rooffall rock frequency that he observed at Marmes Rockshelter as a proxy for climatic change. In general, more recent information follows the broad climatic trends inferred by Fryxell, but with changes in the timing of the inception and waning of the trends. While Fryxell had depicted the early Holocene cool, moist period terminating about 8,000 B.P., both Wigand (Chapter Three) and Chatters (1998) infer the subsequent mid-Holocene warming and drying trend beginning as early as ca. 9,500 B.P. (Figure 15.2). In addition, both Wigand and Chatters see the warm and dry period terminating earlier than does Fryxell and with continued warm and increasing moisture ca. 6000-4000 B.P. After 4000 B.P. all inferred trends correlate again although new information suggests the trends come earlier than Fryxell suggested: a cooler, wet period from ca. 4,000 to 3,000 years ago, followed by a gradual drying to largely present day conditions by ca. 2,000 years ago.

The shellfish analysis for this study may contribute to describing climatic trends in the region. It is doubtful that the site occupants carried unshelled clams a great distance, so the shells in Marmes Rockshelter are inferred to have originated in the Palouse River. However, *Margaritifera* and *Gonidea* have different preferred environments: *Margaritifera* prefers cooler water and rocky river bottoms, while *Gonidea* prefers warmer and slower moving water. For both to occur in the Marmes deposits in varying percentages probably indicates that water conditions changed in the nearby harvest areas through time. In this scenario, higher air temperatures should lead to reduced vegetation along streams providing less shade and warmer water temperatures. A reduction in vegetation also would lead to increased runoff resulting in silty water and muddy river bottoms, the preferred *Gonidea* habitat). The opposite scenario would favor *Margaritifera*: lower air temperatures and increased precipitation leads to increased vegetation providing more shade along streams cooling the water; increased vegetation would reduce runoff resulting in clear, cooler water and rocky river bottoms. It is unlikely that the conditions in the Snake and Palouse Rivers were substantially different (i.e. that one had a better *Gonidea* habitat while the other had a better *Margaritifera* habitat), so variation in percentages of the two shellfish may be a reflection of changes in climate and environment. Results presented in Chapter Twelve show *Gonidea* dominating the pre-Mazama strata, the two shellfish fairly even for the 4,000 years following the Mazama ashfall, then *Margaritifera* dominating in the later strata. This means that *Gonidea* predominates through the early Holocene warm-dry period as would be expected, and *Margaritifera* makes gains in the cooler period that follows. Sampling biases may intrude where *Margaritifera* predominates despite a warming climate in the last 2,000 years. These trends are considered further below under *Site Function*.

Review of the region's environmental trends shown in Figure 15.2 indicates that the date ranges of the Marmes Rockshelter strata (based on Table 15.2) roughly correlate with some of

Figure 15.2. Regional environmental trends.

B.P.	Climate — Wigand	Climate — Chatters	River Environment — Chatters, Butler, & Schalk	Vegetation Trends — Wigand, Mehringer	Vegetation Trends — Chatters	Rock Fall Freq.
1,000	Warmer and drier to present conditions	Drought @ ~600 B.P. Drought to 1,600 B.P. Warmer to present conditions	Modern conditions Rivers agrade during drought, salmon decrease	Increasing fire frequency: increases complex of canyon/ upland habitats.	Human use of fire contributes to maintenance of modern vegetation Shrub-steppe declines in favor of bunchgrass	Low Med. Low Med
2,000	Gradual drying					
3,000	Cooler, Wet	Both summer & winter are cooler, wettest	Later freshet, optimal for salmon, but short, intense runs	Increasing vegetation productivity	Dense shrub steppe ground cover	High
4,000	Warm, but increasing moisture	Abrupt cooling, still wet winters.	Gravel bottoms Decrease in river temperature and stays low after		Extensive meadows	Low
5,000		Further increasing precipitation	Water remains warm	Sagebrush steppe retreats to SE Increase in vegetation density	Increase in productivity of vegetation	
6,000	Driest	Increasing effective moisture, cooling.	Increased flooding, a later freshet		Increase in vegetation density	Lowest
Mazama		Driest			Further increase in S+agebrush and open ground flora	
7,000		Warm, dry summer	Longer periods of warm water	Decrease in vegetation cover		Med
8,000		Warmer, wetter winters. (maritime) decrease in rock fall.	Fish runs severely decrease Short, early spring freshet	Sagebrush steppe	Grasses replaced by sagebrush /shrubs Dune formation	High High
9,000	Increasingly warmer and drier	Increasingly more arid	Decreasing flows			
10,000		Decreasing effective precipitation	Sandy, shifting beds	Increase in grasses, sagebrush	Grasses, steppe plants	Highest
	Cool, increasing moisture		At least some salmon in the Columbia	Forests expand	Forests recede	
11,000		Warm, moist summers				Med
12,000	Cool, dry	Cold, dry winters (continental)		Sagebrush & grass dominate		High

the climatic trends identified for the region. This would be expected since environmental changes should be the principal instrument in change to the sediments.

Stratum I ends about the time of increasing aridity. the climatic trends identified for the region. This would be expected since environmental changes should be the principal instrument in change to the sediments. Stratum I ends about the time of increasing aridity. Stratum II, the end of which is not well determined by dates and could extend to ca. 8,000 B.P., ends roughly at the beginning of the driest period. The time periods of Strata III and IV are tied to the Mazama ashfall rather than climatic conditions. Stratum V ends about the time of a period of abrupt cooling and increased precipitation. The end of Stratum VI could be as early as ca. 1900 B.P. or as late as ca. 1600 B.P. The early date would be during the transition to modern conditions; the later date would be at the end of a drought. As such, the latter date would be expected to more reflect a change in sediment deposition. Stratum VII then occurs during the time period associated with largely modern conditions but that includes relatively short periods of drought.

In interpreting the importance of these trends for investigating the archaeological record of Marmes Rockshelter and the region, a summary statement by Chatters is applicable:

> Resource productivity of the Plateau environment varied according to two dimensions of climate: effective moisture and temperature. Typically warmth and drought and cold and dampness coincided, but at least one warm wet period and several cooler, drier intervals occurred, demonstrating the relative independence of these variables. [Chatters 1998:46]

The length of seasons in the annual cycle also have a great effect on the volume of many economic resources and the length of time that they are available to people. In particular, cool water temperature in tandem with large spring freshets that strip sands from stream bottoms provide the most ideal river conditions for salmon production. But large spring freshets are usually a result of a relatively short period of mountain snowmelt. Thus, under cool conditions, salmon runs are large in volume (even immense as noted in early historic records when river conditions would have been only above average), but short in extent. Warmer water temperatures and a protracted freshet means salmon will be in the river for a longer period of time, but in smaller numbers and if these conditions persist, as during the mid-Holocene warm and dry period, salmon can become susceptible to infection and eggs could be affected by fungal outbreaks lowering the overall numbers of fish. The overall trend of cooling focusing the resource and warming dispersing it is applicable to root and berry plants (short harvest during cooler conditions) and large game animals (longer lowland residence during cooler periods) as well (Butler and Schalk 1986; Chatters 1998; Chatters et al. 1995). These trends certainly directed the focus of human adaptation over time. Unfortunately, the lack of tools in the Marmes collection commonly associated with plant processing (e.g., pestles, hopper mortars) hinders discussion of how these climatic trends may have changed the relative contribution of plants to prehistoric subsistence.

A topic of particular interest for future research is the initial human use of the Snake River canyon bottomlands following the draining of the last glacial floodwaters. The earliest date obtained by this study (11,230 B.P.) sets a new limiting date for the latest that landforms at this elevation became available for human use. All other similar elevations in the lower Snake River drainage would have become available at this time as well, and if people found and began to use Marmes Rockshelter at this time then they likely were exploring/foraging over much of the Snake River bottomlands. The newly exposed landforms, draped with fine-grained, lake-deposited sediments, would have been a relatively productive environment for plant proliferation. The newly exposed soils would have been seeded by the adjacent plant communities, perhaps with a few exotic plants introduced from the upper reaches of the river during flood events. The climate at this time was much cooler than at present and the relatively protected river bottom may have quickly become one of the more biologically diverse areas in the Plateau region. Animals and humans alike probably made use of the landforms and plant communities that developed as quickly as they became established (Gustafson 1972; see Chapter Three). Research into this period of initial use of the bottomlands, and the period when conditions in the Snake River became amenable to anadromous fish, may be important to understanding early Plateau forager settlement patterns.

- What effects did the Mazama ashfall have on vegetation and soil development?
- What is the geological and sedimentary history of the Marmes site area and how might that have dictated/impacted its cultural use?

These two questions are related and will be discussed together; the second question largely was addressed in Chapter Five. Mazama ash is the most identifiable geological strata in the site and is important as a time marker. Intuitively, it is easy to believe that it affected the cultural use of the cave. Two sets of human remains were found on the surface of Stratum III, which coincides with the deposition of Stratum IV in geological terms. We cannot demonstrate that they died at the same time as the primary ashfall, but given that interment in burial pits is established prior to this time, it is possible that the eruption and ashfall somehow suffocated the two people. Given the thickness of the deposit it would have formed the floor of the rockshelter for some time; there is no indication that the occupants sought to remove it, which indicates something about their adaptability in terms of selection of occupation sites. The site was thought to have been abandoned as an occupation site for the period associated with Stratum IV and the cultural materials recovered from within this stratum are believed to be intrusive from strata above and below (Fryxell and Daugherty 1962). No burials were placed directly into the ash deposit, but by the time the site was re-occupied (during the period associated with Stratum V) numerous pits were dug through Stratum V into the ash deposit for interments. Other than direct impacts of the ashfall on the use of the Marmes site area, there are a number of possible indirect effects of the ashfall that could have altered the cultural use of the site.

While Mazama ash is found in virtually all unaltered soil profiles in the region and clearly represented a primary deposition of some thickness in many areas, it appears from the Mazama stratigraphy that after the initial deposition the ash joins other sediments available in the region (loess) for aeolian re-deposition. Stratum V is a combination of the ash and loess, the latter of which is associated with Plateau aeolian sediments that occur to the present. The extent of mixing is thorough to the point that Stratum V appeared as a slightly browner version of the Stratum IV primary ash deposit when viewed in profile (but often could not be discerned in sediment samples in the lab – see Chapter Five). This thorough mixing cannot be attributed only to mixing of the rockshelter deposits after deposition. As noted earlier, it is unknown how much time passed after the ashfall before the area's aeolian sediment base included both ash that was being redeposited from wind-susceptible landforms and loess. If that period of time could be discerned, potential effects on vegetation (and economic mammals) would be easier to infer.

Data from this study can narrow down the time period in which Mazama ash continued to be a major constituent of aeolian deposition within Marmes Rockshelter and probably the Snake River canyon. Stratum IV, the primary Mazama ash deposit in Marmes Rockshelter, was deposited ca. 6730 B.P. (Hallett et al. 1997). Unfortunately, Stratum V is the most poorly dated stratum in Marmes Rockshelter, but if the inferred range of dates for this stratum (see Table 15.2) are accurate, then conditions which made Mazama ash available for wind-transport persisted for at least 2,600 years and probably longer. The sediments associated with Stratum VI, which appears to lack much pumicite, mark the end of these conditions, but the inception of Stratum VI also is uncertain. An estimate of ca. 3,000 years in which Mazama ash contributed to the aeolian base of the area seems reasonable.

Pollen cores from the region do not indicate a change in vegetation associated with the ashfall; this may be beyond the precision of pollen core analysis. Pollen evidence shows that vegetation cover is at a minimum at the time of the Mazama ashfall due to the climate being the warmest and driest of the Holocene. Low vegetation cover would have provided little barrier to the ash being moved by the wind. Considerable re-deposition probably occurred with open landforms largely being stripped of the ash while constricted landforms and other barriers to wind (e.g., at the base of north-facing rock walls, in east-west tending coulees) would have accumulated greater amounts. Sometime around 6,000 B.P. conditions begin to ameliorate and vegetation cover increases, which would have reduced the amount of ash available for wind-transport (see Figure 15.2). However, it appears to have taken over a 1,000 years to significantly increase the density of ground cover plants. If the beginning of Stratum VI is correlated with the period of dense vegetation cover associated with an abrupt period of cooling (ca. after 4,500 B.P.), then our assigning of the 4250 B.P. date to Stratum V rather than Stratum

VI (see Cultural Chronology above) would appear inaccurate. However, the timing of this cooling period and therefore when vegetation cover reached its highest density varies between researchers as post-4500 B.P. (Chatters 1998) and post-4000 B.P. (see Chapter Three). Based on this, the 4250 B.P. date may be a good relative date for the break between Strata V and VI.

The lack of change in regional vegetation as a result of the Mazama ashfall may not indicate much for our purposes of interpretation of the use of Marmes Rockshelter. A great increase of pumice in the environment probably introduced health risks to people, not the least of which would have been lung health, but this may have been very restricted in time. Some adaptation probably was required as a result of the ashfall's effects on the food resources they used. Economic mammal numbers may have declined as pumice impacted their health, both through breathing and ingestion of vegetation. Humans may have been able to adapt to the problem of ingesting the ash alongside many of the resources they used through the development of new methods of processing. It is expected that animals and perhaps vegetal resources would have been more successful in parts of the environment that received the least direct ashfall, creating short-term relative "clumping" of certain prey populations that may have geographically focused mobile foragers' seasonal rounds differently than prior to the ashfall. Such short-term (e.g., decades-long?) adaptive change is not visible in the Marmes Rockshelter record given the nature of the deposits and the lack of precision of the excavations. If people changed only a part of the seasonal round, perhaps extending their stay in certain locales or environments but still moving to most catchment areas throughout the annual round, this too would not be visible in the Marmes Rockshelter record. Only abandonment of the site for a considerable length of time would be detectable as a means for interpreting the canyon's occupants' adaptation to the ashfall. And while there is no abandonment suggested by the site's record, other changes to the site's inhabitants annual round also cannot be described.

Effects of the Mazama ashfall on soil development at the Marmes site are variable and probably fairly limited in time. As noted in Chapter Five, soil formation within the rockshelter is uncertain and probably spatially variable as a factor of available moisture. Since the amount of moisture available to the interior of the cavern was very limited, vegetation growth would have been negligible and the ash would have had little effect. There would have been moisture available at the mouth of the rockshelter, as indicated by the development of the berm, which occurred in large part due to overland flow of sediment carried by groundwater from above. Ash would have been a constituent of the berm deposits as a result of direct deposition and wind- and overland re-deposition. Primary deposits of ash at the mouth of the cave would have been susceptible to removal by wind; this appears to have been the case (see Appendix B, 30W profile). But re-deposited ash in the form of Stratum V occurs as a thick stratum in the berm and natural soil formation is likely to have occurred (Hicks and Morgenstein 1994:80-81; Hicks 1995:82). The naturally-formed soils of the Palouse Canyon and by extension the Snake River Canyon areas supported vegetation through time, even during the warm, dry period the region was in following the Mazama ashfall (cf. Thompson 1985).

Site Function

- What is the nature of the site's cultural materials, how does it change through time, and what does it indicate about how the site was used?

Chapter Two described the settlement and subsistence strategies inferred for the Plateau through time and the different hypotheses for the timing of, and the factors leading to, the hypothesized change from mobile foraging to logistical collecting strategies. It was noted that Marmes Rockshelter contributes applicable information to the debate because of its lengthy archaeological record, and perhaps could help describe, in archaeological terms, the transition from mobile foraging to logistical collecting.

A model was proposed with the Marmes site representing a residential base of "tethered" mobile foragers during the early period of its use, and a residential base or field camp (after Binford 1980) of collectors during the late prehistoric period. Contrasting assemblages would be expected if these two strategies had been implemented in their distinct forms as defined by Binford (1980) during different time periods. In an idealized scenario, the mobile forager occupation would represent utilization of a limited set of resources and the resulting assemblage would be relatively tightly focused functionally. The later collector occupation would be represented by a more generalized assemblage exhibiting a wider range of functions and resources. However, these represent two ends of Binford's (1980) continuum and the likelihood that functionally similar activities occur in both strategies can make interpretation difficult. In particular for the Marmes site, tethered foragers may have visited the site several times in a year as different resources reached their prime; this would make the resulting assemblage less tightly focused and could resemble that which would result from a field camp within a collecting strategy. Given the long sequence of cultural use of Marmes Rockshelter, the vicinity of the site may have offered enough different kinds of resources to have made it an attractive location despite normal variation in resource availability from year to year, as well as through greater cycles of resource variation such as in response to climate change. Adaptability to resource variation would be expected of foragers living a generalized hunter-gatherer existence. Differentiation of subsistence strategies would then rest on whether resources were brought to the site for processing and consumption (as expected of forager residential bases), or were processed elsewhere and only the latter products of such activities are present in the site (as expected of collector residential bases). This would require refined data sets for the food resources present in the site's strata; this aspect of the proposed model for the site will be examined in Subsistence below.

Despite the oft-stated problems with collection methods and sampling that have limited the Marmes site materials available to this study, a large collection of diverse materials has been examined and collectively indicate a considerable and continuing use of the site largely for habitation. The lithic materials are the best represented part of the site's collection, and while interpretation of this site can include conclusions about the other material types, the lithic tools and debitage patterns through time offer the greatest confidence for interpretation of the site's use. The lithic analysis has demonstrated an abundance and variety of stone formed tools representing cultural activities associated with tool manufacturing and maintenance, food procurement, and processing of food and other materials. These combinations of activities, viewed against the background of all the other kinds of materials that occur in the site, characterize the site type as a residential camp for all but the periods associated with Stratum IV and Stratum VIII. While the mixing of deposits, particularly in Strata V and VII, reduces the efficacy of inferences of the cultural use of the site during the time periods associated with these strata, the consistent presence of tools, manufacturing by-products, and remnants of the resources these cultural activities addressed in all of the site's strata supports the interpretation of habitation. Given its location in the relatively warm and protected canyon, high density of artifacts and fire-related features in the natural shelter (but not on the open floodplain part of the site), and the heavy wear on the tools, the site may represent winter occupations.

Manufacturing and maintenance activities addressed stone tools as well as tools of hard organic materials such as bone, antler, and wood. Use-wear observed on stone and bone tools indicates that processing activities probably addressed such non-food materials as hides and fibers, the latter of which certainly included botanical materials although they are generally lacking in the collection. Some of these processed items or material types are present in the site collection and were included in analyses for this study, such as the antler and bone tools. Being able to infer what the materials that were being worked were can compensate somewhat for their being missing from or only poorly represented in the site (i.e. non-provenienced items in the collections, and items and materials noted in the field notes that apparently were not collected).

Other analyses have demonstrated the focus of food procurement activities at the site. While projectile points and the faunal data point to hunting, the presence of shellfish and fish remains in every strata of the rockshelter and in the floodplain cultural strata as well indicate the importance of these resources to the occupants' subsistence. While the variation in fish and shellfish remains' representation in the site from stratum to stratum is to some extent a product of collection and perhaps sampling methods, it is clear that cultural activities that addressed these resources occurred in all time periods. The presence of high densities of shell, especially in Stratum III, indicates that shellfish were processed in the rockshelter and implies a nearby source, probably the Palouse River. The extent of shellfish use by the site's occupants may be underrepresented by the shells observed in the site if shellfish collected at a greater distance were not hauled to the site for extraction of the edible portion. The relative representation of small fishes over anadromous fish may be further indication of the winter season use of the site. Alternatively, small fish may have been processed and consumed at the site, while large fish may generally have been processed elsewhere.

Botanical food resources are poorly represented at Marmes Rockshelter and conclusions as to how much was used and in what time periods cannot be offered. But generalized hunter-gatherers worldwide made use of plant foods (and optimal foraging theory aside probably as an essential portion of the diet) so it is likely that evidence of plant use by the earlier foragers is underrepresented. The collection of many kinds of plants would not require durable artifacts. Berries and seeds are picked by hand, and roots can be obtained with a digging stick that when damaged probably was discarded at the collection location rather than in the rockshelter. In Plateau assemblages, the few durable artifact types associated with plant foods are related to processing activities (e.g., mortars and pestles), rather than collection of the resources. But these too are probably underrepresented in comparison to the proportion of the diet that plant resources are believed to have represented. In part this is because roots and berries, rather than seeds or hard-shelled nuts, were the principal plant foods available in high volume. Despite this, hackberry seeds are present in every stratum in Marmes Rockshelter suggesting they were consumed or stored there. The presence of wood artifacts, and field notes that talk about grasses and cordage and matting fragments, particularly associated with storage pits, suggests that similar representation of botanical materials as has been observed in other rockshelter sites in the Palouse Canyon may have occurred in Marmes Rockshelter (e.g., Hicks and Morgenstein 1994; Hicks 1995, 1996; Mallory 1966). There is no good reason to think that the extensive use of botanical materials in the late prehistoric period (and probably earlier) documented for Plateau peoples was excluded from the Marmes site.

For the earliest use of the site, that associated with the floodplain and Strata I and II in the rockshelter, the overall impression is that some generalized activities occurred in both areas. Comparison of the density of the floodplain and rockshelter materials indicate that the rockshelter was the focus of most activities that left durable remnants. Lack of hearths, FCR, or other indications of cooking features attributable to cultural activities suggests the floodplain activities were limited but included activities related to specialized tools (e.g., needles, bone pins); activities involving fire hearths only occurred in the rockshelter. Overall, conclusions about spatial separation of activities involving non-lithic items are weak due to low numbers. Representation of activities involving lithics is better with higher numbers of items and the fact that 95% of the stone tools were found in the rockshelter during this period. This may indicate selection of the rockshelter for activities that utilized formed tools (and fire-related features), or that the rockshelter served as a cache location for these items. Strong patterning of the food resource data cannot be asserted, given the collection issues and the lack of quantitative faunal data from the early rockshelter excavations. The large elk carcass seems to indicate some resource processing was done on the floodplain, but Gustafson is not sure that butchering is represented. In all likelihood, these minimal data patterns between the floodplain and rockshelter reflect such things as group size, the length and season of occupation, as well as the restricted set of activities and the resources they addressed.

The cremation hearth indicates some level of socio-religious structure that at least extended to treatment of the dead; cannibalism as suggested by Krantz (1975) has not been demonstrated. But the many dispersed hearth areas that make up the cremation hearth feature complex were used for fire-related cultural activities in addition to the burning of human bone. The dispersed nature of the collective hearths suggests that these activities took place over some time and reflect the day-to-day occupation of the rockshelter during its initial use. The presence of non-human mammal bone broken and burned and lithic tools and debitage in association with a hearth feature would be expected in any occupation site. That this complex of hearths occurs over a large horizontal area and its use continued through the break between Strata I and II, indicates long-term generalized use in addition to the activities associated with the burned human remains (i.e. ritualistic cremation). Clearly, cultural use of the rockshelter during the period associated with Unit I was for occupation, and cremation was not the principal focus of cultural activities at the site.

As noted in Chapter Two, there are few early Plateau habitation sites with which to compare the Marmes site. The Coopers Ferry and Paulina Lake sites are the only sites with firm dates as early as the Marmes site. But the small size of the assemblage from the earliest components at Cooper's Ferry would offer a poor comparison with the Marmes site assemblage; the Paulina Lake assemblage offers a stronger comparison of apparent site function.

Connolly (1999) collapses his Components 1 and 2 when presenting his interpretation of the Paulina Lake site. This Early Pre-Mazama period is dated between ca. 11,000 and ca. 8,500 calibrated years ago (1999:126), which roughly correlates with Strata I and II at the Marmes site. Paulina Lake site features of this period include several elements associated with a domestic structure, possibly a windbreak, wickiup, or tepee-like shelter, that included a central fire hearth, and an unassociated fire-hearth. The tool assemblage is dominated by projectile points and other bifaces, scrapers and utilized flakes, and various cobble tools (e.g., hammerstones, choppers, abraders, and edge-ground/faceted cobbles) reflecting hunting, processing, and manufacturing tasks. Connolly sees the Early Pre-Mazama period materials as representing "a base camp from which a variety of subsistence, processing, maintenance, and procurement tasks were performed" (1999:127).

The tool types in Strata I and II at the Marmes site are similar to those in the Early Pre-Mazama period assemblage at Paulina Lake with the addition of various bone tools that reflect domestic tasks. Comparison of the numbers of tools favors the Paulina Lake site (although many tools are missing from the Marmes collection or lack provenience). The larger number of tools may be due to the Paulina Lake site's proximity to obsidian sources. There are no known lithic sources near to the Marmes site (although chert does occur as veins and nodules in the basalt flows in the area). Instead, the density of shell and animal bone in Strata I and II may indicate the resources that attracted use. The rockshelter, and perhaps the large boulders in the floodplain that appear to have concentrated some of the floodplain activities, represent the domestic structures at Marmes; there probably was no need for a constructed shelter such as that at the Paulina Lake site. The Marmes site has a much higher density of features than the Paulina Lake site, and the features appear to indicate more diverse activities. Many more fire-related features occur at Marmes and the length and intensity of use appears greatly exaggerated compared to Paulina Lake. In addition, the concentrations of shell and animal bone, both in the rockshelter and on the floodplain suggest more intensive use of Marmes than the Paulina Lake site. Some of these differences may be attributable to Paulina Lake being an open site where preservation of these materials is poorer.

The overall impression is that both are forager base camps at this time, but the Marmes site was used more intensively than the Paulina Lake site. Given the low resolution of the data within these strata at the Marmes site, as was noted early in this chapter, the impression that Marmes was used more intensively can be a result of many factors. Both assemblages are almost certainly palimpsests representing multiple, uncountable occupations at these sites. The greater apparent intensity of use at the Marmes site may be a result of more regular use or use for longer periods of time (or both) than the Paulina Lake site. Variation in site use through time is not discernible for the period being compared. Projectile point styles would not be demonstrative as several periods of intense utilization, perhaps several decades long, separated by centuries of disuse, could be responsible for the relatively low numbers of Windust and Cascade points at both sites. Likewise, activities representing very different functions could have been the focus of site utilization at different times, but the material remains for all the activities for the whole period are represented (palimpsests). As such, the impression that the Marmes site was more intensively used than the Paulina Lake site must be viewed with caution.

After the earliest period of use, cultural activities appear to largely have been restricted to the rockshelter, although collection issues intrude again as little excavation of the upper floodplain deposits occurred. Inside the rockshelter, although the use of the Marmes site was at least recurrent if not continuous (with the exception of Stratum IV), the level of activity as measured by the number of lithic artifacts recovered from the rockshelter reflect significant fluctuations in the intensity of the occupations (however, note Ames' [2002] caution about assemblage size/richness effect on measures of representation). These fluctuations do not appear to be a reflection of the missing and non-provenienced tools since the trends in tool numbers between strata are matched by the numbers of fire-related features. Variation in fire-related features may reflect changing use of the site through the hypothesized transition from mobile foraging to semi-sedentary collecting. In the generalized foraging scenario, fire-related features indicate occupation-related activities (e.g., warming and cooking) and resource processing (including tool manufacture). In a logistical scenario, where the site should be used for activities that support an occupation site nearby, fire-related features largely may be resource processing with some secondary warming and cooking functions depending on the distance to the occupation site and the nature of the resources being addressed by the task group at Marmes Rockshelter. The differences between fires used for the different strategies are not discernible under the excavation strategies used in the rockshelter and might not be demonstrable, beyond citing variation.

Stratum III and Stratum VII appear to reflect the most intensive occupations based on artifact distribution, faunal, shellfish, and feature data suggesting very intensive occupations during the early Cascade Phase and the later Harder Phase. The high frequencies of the butchery or animal processing lithic tools and faunal remains found in these strata suggest intensive hunting activities during these periods, but the lack of other kinds of data probably do not fully represent the suite of resource activities that occurred at the site.

Variation in the proportions of lithic tools and debitage between strata also may reflect change in site function (see Chapter Eight). These changes suggest more intensive tool use, tool exhaustion, and tool disposal during the early periods, and more intensive tool manufacturing during the later periods. Likewise, the percentage of tools that are byproducts of manufacturing (cores, blanks, and preforms), pieces broken and rejected during the tool manufacturing process, also increases for the later periods. It may be that Windust and Early Cascade people generally arrived at the site with stone tools sufficient for planned activities while later people more often manufactured stone tools while occupying the site. This may reflect the transition from mobile foraging to semi-sedentary collector settlement and subsistence strategy.

Subsistence

Several research questions addressed subsistence patterns in the Marmes site.

- What is the relative dietary importance of different terrestrial faunal species through time?
- What is the contribution of fish to the diet and how does species utilization change through time, in particular, the changing role of salmon and the cause?
- What is the dietary contribution of plant foods, shellfish, etc. at different time periods?
- What indications of seasonality of use are present in the data?

Many of these questions ask about the relative importance of one food resource type in comparison to others, and are referable to examination of the proposed model of site use through time described in *Site Function* above. Under that model, the Marmes site would represent a residential base for mobile foragers and food resources would be gathered nearby and processed and consumed on site. Given that this site is expected to be just one site in an extensive seasonal round, and the timing of site utilization is based on the timing of local resources, then the breadth of resources found in the site should be relatively narrow. After the change to a logistical collector strategy that is hypothesized for the region, the site is expected to represent a field camp of collectors, a base while away from the residential base. It certainly was used as a cache site as marked by the many storage pits observed in the upper strata. But the many fire-hearths in these strata also indicate that habitation and/or resource processing using fire or smoke occurred. In addition, as noted under *Site Function*, a diversity of functional tool types occurs in all strata. This would not be expected if the site were used solely as a cache by collectors, or if the stored resources were processed on site. A wide range of food resources could be expected in a collector field camp, especially the target resources that led to the site visit.

Setting aside the field collection and sampling issues for the moment and looking only at the abundance of the different food resources at the Marmes site, provides the following generalized picture. During the early period of site use (Strata I, II and III), when mobile foraging is the hypothesized strategy in the Plateau, emphasis is on large and medium mammals, fish and shellfish. The blood residue analysis results would appear to support this as deer and anadromous fish (trout) residue were found in all samples from Stratum I, including the soil control samples. During the later period (Strata V, VI, and VII), during which the logistical collecting strategy is hypothesized to develop into the dominant pattern, large and medium mammals decrease and small mammals increase, fish remains high and increases, but shellfish decreases considerably.

Atwell (1989) conducted a similar examination of subsistence resources for the Columbia Plateau. In an attempt to overcome the potential sources of bias in individual faunal assemblages from sites in the region Atwell created aggregate assemblages based on number of individual specimens (NISP) for tracking changes in subsistence fauna through time. Atwell's results (1989:160-162) are almost in direct contrast to the patterns described above for Marmes. The early period shows an increase in large mammals (deer) that reaches its peak at a time equivalent to Marmes Strata V when few other subsistence resources appear in abundance. Small mammals are important in the early period as well, but decline by the beginning of the later period. Fish are not found in significant numbers until the time of Marmes Stratum VI. Atwell did not examine shellfish abundance. Each of these statements are almost opposite of the trends in Marmes described above, which may indicate the lack of dependability of the Marmes assemblages, or the inapplicability of an analysis based on NISP for intersite comparisons.

That warning stated, general patterns of the food resources at the Marmes site follow what could be expected of a mobile forager residential base during the pre-Mazama period. The higher counts of large and medium mammal remains (increasing from Stratum I to Stratum III), high fish counts on the floodplain (where higher resolution data are available compared with the rockshelter), and density of shellfish remains collectively describe processing and consumption activities occurring on site. The shellfish are particularly assertive for foraging, indicating that gathering occurred nearby and the shellfish were consumed on site. Large mammal remains in the rockshelter are dominated by deer, elk, and pronghorn bones (Gustafson 1972), which exhibit evidence of butchering and are highly fragmented. The floodplain strata also exhibited many small fragments of bone, with much higher incidence of burning than the rockshelter bone fragments. Whether these were fragmented and burned to this extent in hearths located in unexcavated areas of the floodplain, or represent cleanout from the rockshelter hearths at this time, the level of processing appears related to consumption and reflects a forager use of the site. The high counts of fish bone found throughout the strata are of small fish species, which may be a sampling bias, but probably reflects consumption of fish taken from the adjacent Palouse River during the early period, and processing and/or storage during the later period.

The representation of food resources during the post-Mazama strata, including the time of the hypothesized change to a logistical collector strategy when the site is expected to represent a field camp, is poorer than that for the forager residential base in the early period. A field camp in a collector strategy is the focus of processing of resources that largely are consumed elsewhere (at a residential base). As such, the inedible parts of the resources should be discarded here, features used during resources processing should be present, and evidence of the maintenance (and some manufacturing) of tools used for processing the resources should be in the deposits. At the Marmes site, the presence of a diverse assemblage of tools in each of the strata associated with the post-Mazama use of the site indicates habitation activities, such as might be expected at a field camp that was used often or for long periods of time. This diverse assemblage may represent a palimpsest of many regular, but very short-term occupations of the site while processing and storing resources.

In addition to a field camp, the presence of numerous storage pits in Marmes Rockshelter indicates that the site served as a cache for a collector group occupying a nearby residential base. Initially, this may have been 45FR36c, a site with eight housepit depressions located about five kilometers southwest of the Marmes site, and dated to the early Harder Phase (Schalk 1983). Residence appears to have shifted from 45FR36c to the Palus Village site (45FR36a) which was occupied during the last 2,000 years (Rice 1968). Both of these residential sites are considered to mark the transition to semi-sedentary collecting lifeway in this part of the lower Snake River drainage. The presence of increasing counts of small mammal remains throughout the post-Mazama period may not be a result of increasing storage of these resources at the site, since much of what is stored in the site would be expected to be removed to the residential base for consumption. This is probably an overall subsistence trend due to resource intensification and/or over-exploitation of larger mammals. Small mammals may have been consumed during the short-term occupations of Marmes Rockshelter when preparing resources for storage. The relative lack of shell, compared to the forager base camp, is expected since collectors would not have stored the shells. If shellfish meat was cached in the site, perhaps even dried over the many fire hearths present in the rockshelter strata at this time, the shells likely were removed near to the river to save hauling the extra weight to the rockshelter.

The minimal presence of salmon bone in the later period is likely a result of sampling bias as noted previously. The confluence of the Snake and Palouse Rivers is described in the historic record as a good fishing place, presumably for salmon. Since Marmes is the nearest rockshelter to the two habitation sites noted above, it is difficult to believe that salmon was not stored there. Small numbers of salmon bones have been found at several other rockshelter storage sites located further from Palus Village in the Palouse River canyon (Blukis Onat, Hicks, and Stump 1996). It is possible that preparation of salmon for storage was conducted elsewhere and such processing activities removed most of the bones. Early historic-era accounts of visits to Palouse Canyon and other Plateau fishing sites describe fish being wind-dried on racks (Trafzer and Scheurman 1986; Hunn et al. 1998). If consumption occurred at the residential base, little evidence of what was stored in the site would be represented. Some small representation of the stored resource would still be expected in the site's deposits, perhaps from spoilage (Hicks and Morgenstein 1994). Blood residue analysis results suggest that anadromous fish species may have been present throughout the early and middle strata; if the results are accurate, this may imply processing of salmon, steelhead, and/or trout occurred in the site at this time. Immunological analysis conducted on materials from other sites in the vicinity of Marmes Rockshelter also found indications of salmon, as well as rabbit, camas, sheep, deer, dog, chicken (Hicks 19995, 1996; Hicks and Morgenstein 1994).

Resource processing is indicated in most of the strata at the Marmes site and appears to have been the focus of at least some activity at the site by both foragers and collectors. The extent of deposits that resulted from resources processing in Strata I – III supports a description of use by tethered foragers during this time. The extent of these deposits, especially consumption-related debris, also implies that a relatively high volume of these resources (i.e. large mammals, shellfish, fish) were available nearby, since foragers are hypothesized to move residence to the location of resources. A ready supply of resources may have been a major reason the site saw continued utilization through the long period of the site's use, and may have encouraged longer stays during the later period associated with the collector strategy, resulting in assemblages indicating continued habitation.

Marmes Rockshelter in a Regional Perspective

The *Subsistence* discussion above would appear to support the model of Marmes Rockshelter's changed use as predicted by the hypothesized change from mobile foraging to logistical collecting strategies, but representation of the food resources from the site is in question due to collection and sampling issues discussed previously. It cannot be demonstrated which food resource had a higher value than another at the Marmes site. In this regard, it is best to view the food resources record as a list of the resources present at the time associated with each strata, rather than trying to determine their relative importance. As a means to use this information, Figure 15.3 shows the variation in each food resource type at Marmes Rockshelter against inferred regional trends for the same resources. Due to the paucity of botanical remains and tools commonly associated with plant processing, the extent of use of plant foods is not demonstrable for Marmes Rockshelter and are not included. This leaves a large hole in the subsistence picture for the site as plant foods are expected to have represented a substantial part of the diet throughout human history. Shellfish trends in Marmes Rockshelter also cannot be asserted as representative because of potential collection problems, but they largely fit those previously presented by Lyman (1980a).

In comparing Figure 15.3 with Figure 15.2 in Past Environments above, the adjustments to the timing of the climatic trends asserted by Gustafson (1972) require reassessment of his suggestions for faunal patterning. Gustafson and Wegener noted no known qualitative differences from extant fauna at any eastern Washington archaeological site after the end of the early Holocene cool period, but that quantitative differences between certain taxa may reflect both long-term and brief climatic fluctuations (see Chapter Ten). As an example, they suggest that the increase in pronghorn remains and corresponding decrease in elk remains observed at both Marmes Rockshelter and the Granite

Figure 15.3 Some Snake River region economic resource trends.

B.P.	Distribution of Big Game Mammals		Fish		Shellfish		Medium size mammals
	(Gustafson, Chatters)	Marmes	Region (Butler, Schalk, Chatters	Marmes	Region (Lyman, Chatters)	Marmes	Marmes
1,000	Slight increase	Lower	Droughts at 600 and 1600 BP, decrease in salmon	Freshwater fish 4:1 over salmon	Gonidea increase during warm conditions & sandy bed	Margaritifera very dominant	Moderate
2,000	Decrease		Gradual decrease				Moderate
3,000	Bison common						
	Increased proliferation						
4,000	Gradual decline		Increase in salmon productivity				
5,000		Moderate		Most freshwater fish in VI-VII	Margaritefera increases considerably	Nearly equal Marg. & Gonidea to ~2000 B.P.	Moderate
6,000	Maximum			Salmon 2:1 over freshwater fish			
Mazama							
7,000	Abrupt increase	Highest		Salmon increase, freshwater decrease			Highest
8,000	Increase in pronghorn relative to elk		Anadromous fish runs severely reduced		Continued Gonidea dominance		
9,000							
10,000	Elk, bison, deer, mtn. sheep, pronghorn	Lowest		Small, freshwater fish	Gonidea dominant	Gonidea dominant by far to Mazama	Lowest
	Minimum						
11,000	Increase elk, decrease pronghorn		Some salmon in Columbia				
12,000							

Point site after about 6,800 B.P. (see Tables 10.19 and 10.20) may reflect the mid-Holocene warming trend. However, the newer information would appear to suggest this change comes with an increase in available moisture after the height of the dry period, rather than in response to the dry period itself.

These environmental trends certainly directed the focus of human resources adaptation over time and probably on an annual basis. Mobile foragers would have been very knowledgeable about the available resource base. They would have known where the best resources were within their annual common use area (home territory) and probably were able to infer how the conditions observed within their home territory were playing out in adjacent areas that were accessed less often. Based on the health and timing of certain plants, surface water/springs, the extent and pace of melt of lowland and mountain snows, river conditions, etc. they could make predictions about the status of important resources in the coming months and make adjustments to the seasonal round early in the year and on an ongoing basis throughout the year. Living close to and depending on the signs of the current annual cycle would have made such signs a topic of much concern and discussion. These concerns may have served as an incentive to seek out other foraging groups to compare notes, encouraging social discourse, a pattern observed through to the early historic period.

Given this, the hypothesis that resource pressures stemming from climatic and environmental change was a principal factor for a change from foraging to collecting makes intuitive sense. On the Columbia Plateau, intensive collection/harvest and storage are often paired with the appearance of pithouse villages as indicators of the advent of the semi-sedentary logistically-organized collecting settlement and subsistence pattern. The advent of storage and its implications for intensive collection of resources is seen as a key feature of cultural complexity. The occurrence of storage pits in increasing numbers after ca. 4,000 B.P. is associated with semi-sedentism and the need to store large volumes of foodstuffs for use in seasons when fewer food resources were available or the availability of resources was less certain. This pattern also has been cited as an indication of increased population.

In Marmes Rockshelter, fish representation is in disagreement with the regional trend (cf Chatters 1995, 1998; Chatters et al. 1995) (see Figure 15.3). The regional trend is evident in environmental data rather than archaeological representation of fish because much more attention has been given to collecting the former rather than the latter. An attempt was made by Johnston (1987) to provide an overview of fishing evidence for the southern Plateau but it is severely hampered by the poor data available from excavated sites; in particular, fish remains generally have not been distinguished by species in archaeological investigations. Chatters (1995:382) notes that from 3900 to 2400 B.P. there is evidence for an increase in available moisture and a decline in temperature that persisted until at least 2200 B.P. (1995:387). This would have led to changes in river temperatures and stream flows that Butler and Schalk (1986; Schalk 1977) inferred would have had significant restricting effects on the seasonal concentration of fish. It is during this period (especially 3300 to 2200 B.P.) that the percentage of salmon remains in archaeological sites on the mid-Columbia climbs to greater than 90% of fish remains, compared with only 30% between 4400 and 3800 B.P. (Chatters 1995:383,387). But salmonids are the dominant fish at Marmes Rockshelter (Stratum V) well before they peak in regional assemblages. Salmon are poorly represented at Marmes Rockshelter by the time they become a dominant stored resource in regional sites. Also at odds with the regional pattern is the presence of salmon bone in the earliest strata at Marmes, during the Windust and early Cascade Phases prior to the Mazama ashfall. The cooling period Butler and Schalk (1986) and Chatters (1995) refer to does have the highest numbers of fish for any period of the use of Marmes Rockshelter (Stratum VI – Tucannon/ Harder Phases), but only a single example is salmonid (see Chapter Eleven). Caution is warranted in considering these numbers because of field collection methods issues (especially the lack of 1962-1964 quantitative data), but ¼"-mesh screens should have selected in favor of salmon bones versus the smaller freshwater fishes, so credence can be given to the identified trend at Marmes Rockshelter, especially since it is fairly dramatic.

Researchers (e.g., Butler and Schalk 1986, Chatters 1995) have cited these cooling climatic conditions as the impetus for more intensive procurement and increased storage of salmon and, by extension, favoring a logistically-focused, semi-sedentary collector settlement and subsistence system over a mobile foraging one. This makes sense in reference to a more restricted, but plentiful salmon run. When the water temperatures were warmer, the salmonids were available for a longer time in the rivers, albeit in lower numbers than during cooler periods, but still probably in numbers that could have contributed to sustenance of the lower group sizes predicted for this period of prehistory. Having salmon predictably available for much of the year would be of great benefit to mobile foragers although Schalk (1977:241) suggests that dependence on fish would be limited prior to the adoption of storage. Ames (1988:356) says early foragers may have aggregated at prime fishing locations a few times a year, but suggests that the social gathering was more important than gathering fish. With cooling temperatures the salmon runs became more restricted in time, but also larger and more predictable in terms of their timing over the annual cycle. If forager diet had become somewhat dependent on fish, adaptations that could include high volume harvest and storage to maintain the relative amount of fish in the annual diet can be hypothesized as part of a transition to a collector strategy.

Similar to the effects on salmon, Chatters (1998) also suggests that terrestrial resources (interpreted here as large-game mammals since smaller mammals maintain a relatively small home territory) would have become more seasonally restricted with increased moisture and decreased temperature. For large mammals, seasonal restriction meant staying at lower elevations longer as higher elevation plants would have become available later in the year due to an increased snow pack. If resource pressures were a principal factor for the change from a forager to collector strategy, this would have kept large economic game animals closer to residential bases along the larger rivers and streams. This may have allowed for a longer period of residence at these sites, delaying the need to move to a new catchment area (following the large animals to higher ground and to locations of early spring plant foods).

Ames argues that there was no transition from foraging to collecting strategies; rather that there was a period of mixed strategies (2002:29-30). In either case, while changes in subsistence may be detectable in the regional archaeological record, the timing of the transition from mobile foraging to semi-sedentary living may vary across the Plateau. Ames et al. (1998) note the appearance of an initial semi-sedentary adaptation after ca. 5000 B.P. in the southeast part of the basin along the Snake and Clearwater Rivers; this pattern persists a few hundred years then ends (Ames 1991:129). Ames (1991:128-129) describes two additional episodes of sedentism on the Plateau, each with different residential mobility and subsistence patterns. Chatters (1995) describes another episode of sendentary residence, this in the northwest part of the basin, where housepits appear ca. 4500 B.P. along the Columbia River and appear to reflect a foraging subsistence strategy rather than a collector strategy. This adaptation did not persist, fading away by ca. 4000 B.P. But semi-sedentary residence returns there after several hundred years, this time with a more logistically-oriented subsistence strategy.

Environmental change has been posited as a principal factor in the change in settlement and subsistence pattern for the Plateau. But the environmental reconstruction presented in this study does not appear to correlate with cultural chronological periodization. As an example, the early Holocene warm-dry period took place almost entirely within the time to which the Cascade Phase is attributed. This would appear to suggest that the most extreme climatic event of the Holocene had a negligible effect on the region's inhabitants' adaptation. Even Ames' (1991) initial episode of semi-sedentary settlement (ca. 5000 B.P.) occurs nearer the end of this climatic pattern rather than at its inception. Conversely, the second semi-sedentary settlement attempt after ca. 4000 B.P. may be strongly correlated with the cooler, wetter period that followed the early Holocene dry period. But after 2000 B.P. at least two periods of severe drought occur that seem to have little impact on the archaeological record in sites along the lower Snake River. As suggested in Chapter Four, the Plateau peoples' subsistence strategy may have varied considerably within the data parameters that we recognize archaeologically as semi-sedentary settlement, revealing it as a truly adaptive strategy that compensated for localized environmental and resource fluctuations on an ongoing basis.

This invites a return to Reid's statement that there may be "patterned relationships between severe regional droughts, settlement pattern shifts, and perhaps even changes in the organization of subsistence and material culture" (1991:31). Replacing "severe regional droughts" with almost any climatic and environmental variation would not undermine the applicability of Reid's statement, but it should be noted that he was addressing the archaeological record of upland sites. Much of the region's archaeological record is described based on excavated sites along the major river drainages and much of that record generally lacks the precision to track short-term events, even several centuries-long events. The effect on interpretation of cultural horizons can be seen in the disagreements between the region's different (but generally overlapping) cultural chronologies (for example, those in Ames et al. 1998, Campbell 1985, Chatters and Pokotylo 1998, Leonhardy and Rice 1970, Nelson 1969, and Sappington 1994), all of which should be a reflection of people's adaptation to environmental change to some extent. Investigation of numerous sites of similar function that overlap in time (and perhaps even in similar locations/placement within a given catchment area) would be needed to overcome the variation any one site may exhibit that prevent discernment of the patterned relationships Reid seeks. The potential for patterned relationships to account for variation in the Plateau archaeological record would seem to be a worthy focus of future investigations using contemporary methods of investigation.

Marmes Rockshelter, like every site, had a role in the hypothesized changes to settlement and subsistence patterns for the Plateau. Unlike most sites, it reflects these patterns, to some degree, since ca. 11,000 B.P. But a single site cannot characterize a settlement and subsistence pattern and understanding the utilization of any site must include reference to its role in relation to its surroundings including other nearby archaeological sites. As noted above, the earliest date obtained by this study (11,230 B.P.) sets a new limiting date for the latest that landforms at this elevation became available for human use following the creation of the scablands by glacial floods. A relatively small area of the Marmes Rockshelter was excavated to as deep as the basal lake sediments. It is unclear how many of the cultural items found below Fryxell's original Stratum I (i.e. in Strata IA through IE) may have derived from upper strata by trickling down during or before excavation. But all three of the earliest dates from Stratum I, including the 11,230 B.P. date overlap at 2 sigma when calibrated. It is clear that the Stratum I assemblage represents occupation of the rockshelter and the site was used for habitation from that time on.

Over 90 prehistoric cultural sites had been recorded in the area of the Palouse River canyon and the confluence of the Palouse and Snake Rivers. Thirteen of these have had cultural materials radiocarbon dated, but none exhibit cultural use as far back as the early cultural strata in Marmes Rockshelter. Only three sites (45CO1, 45WT2, and 45FR202) have demonstrated use prior to the Harder Phase (after ca. 2500 B.P.); this limits our ability to interpret the Marmes site's role in the early period of the local (Palouse River canyon vicinity) settlement complex.

Site 45WT2 is an open site near the mouth of the Palouse River that was first used prior to 7300 B.P. and has extensive Cascade Phase deposits, both below and above a Mazama ash deposit (Nance 1966:41-42). It may be that this became the local habitation site of the Marmes Rockshelter residents after the Mazama ashfall, although both sites saw considerable use prior to the eruption. After the Cascade Phase, this site was used intermittently to the historic period. Porcupine Cave (45FR202), located several hundred meters north of Marmes Rockshelter, is a fairly large rockshelter that faces southwest and has virtually no raised berm, making it susceptible to wind and the heat of the afternoon sun. A date of 5305 ± 50 B.P. was obtained from wood charcoal recovered ca. 3 m deep at the front of the cave (Hicks 1995:83); the great depth of deposits prevented the excavations from reaching older sediments and the initial use of this site is undetermined. Based on limited excavations, the site appears to only have been used for storage within the last thousand years and the presence of two hearth features that predate the storage pits suggest the cave was utilized for other activities prior to at least the middle Harder Phase (Hicks 1995). 45CO1 was investigated only with two test pits that found cultural material estimated to date between 4000 and 6000 B.P. (Nelson 1966:17), although a deeper deposit at 4 m below the surface contained only non-diagnostic materials (1966:18).

These three sites demonstrate the obvious: that landforms in the vicinity of the habitation site that Marmes Rockshelter represents also were being used. Given the foraging subsistence pattern assumed for much of the history of the Marmes site's use, many other landforms in the vicinity are likely to have been used as well. Many of the 90 sites that have been recorded in the Palouse River canyon area are derivative of cultural activities that support a residential site (e.g., hunting blinds, resource processing locations, storage sites). And while most of these are believed to be more recent in time than the periods represented by the three sites presented above, some of the activities that these sites represent also may have been used by tethered foragers based at Marmes Rockshelter, 45CO1, or 45WT2. Hunting techniques change through time, as demonstrated by the advent of the use of atlatls and diminishing point size through time for example, but the use of such things as blinds for surprising prey is based on the hunter's awareness of the prey's habits and preferences. That awareness is developed through familiarity with a person's environment and also was transmitted socially. It is expected that the resident foragers were familiar with the local geography and how it conditioned their prey species behavior, and used that awareness to increase their ability to obtain prey. Good hiding places, particularly at narrow passage points, would have been good hiding places regardless of the hunter's subsistence and settlement organization (although collectors may have emphasized game drives and multiple simultaneous kills).

Resource processing site locations are largely dependent on where the resource is obtained. While in the logistical collector strategy these may need to be more 'ideal' to obtain higher volumes and to accommodate more complex processing activities, good resource locations were probably used by people under both subsistence strategies. A better understanding of the archaeological signatures of "locations" (Binford 1980), and awareness of the potential variability in their archaeological content based on the extent of resource processing that occurs there is needed. Higher productivity locations may have been occupied by foragers as residential bases, and by collectors as field camps, the latter only long enough to prepare the resource for transportation and perhaps for storage (Binford 1980). Such uses may account for the many non-specific "camp" sites recorded in the site database for the Lower Snake River region (Hicks 1999), sometimes defined on little more than a collection of eroding fire-cracked rock. Here, the environmental changes identified as a principal factor in the shift in adaptive strategy from foraging to collecting may contribute to archaeological definitions for these site types, where changes in predation focus brought about by the shift to cooler and wetter climatic conditions should be more definitive than residency preference in discerning between them (Reid and Gallison 1995). This may help account for the variation in assemblages and subsistence data in sites along the Lower Snake River if their role in the catchment area addressed by the inhabitants' predation focus is considered over data thought to fix a site's settlement type.

Storage sites are associated strongly with a logistical/collector strategy (and the dates of Palouse Canyon rockshelter sites affirm this), but foragers also stored things in pits (including in Marmes Rockshelter). Food storage is a central question in reconstructing settlement and subsistence patterns of any region. But in the Plateau, despite the importance of storage and surplus to the principal change in adaptation acknowledged for the region, our knowledge and understanding of the phenomenon is limited. Much of this is due to the difficulty of recognizing cultural remains associated with storage strategies, which has inhibited investigation into storage behavior. Early storage is assumed to have involved food resources but archaeological data is limited and we cannot rule out that these pits were used for caching tools and other non-food items. Early storage events certainly represent much smaller volumes than were stored by later logistically-organized inhabitants. Storage pits are found in caves and rockshelters both early and late in prehistory, but one of the best sources of evidence of the increasing use of storage in late prehistory is finding storage pits in many more places (e.g., in the floors of structures, in talus slopes, in pits built against bedrock walls). In the Plateau region, the lithic focus of most past archaeology makes excavating storage sites unattractive because they don't produce a lot of lithics. In addition, small storage sites are not considered "significant" in comparison to lithic or structural sites and thus are not addressed by most compliance-focused work. Also, most of the storage sites that have been looked at produced relatively recent dates that fit the much discussed late prehistoric subsistence pattern so comfortably that it didn't stir much debate. Maybe because of this, not enough of them have been examined in detail.

Considering social explanations for the continued appearance of habitation related assemblages into the late prehistoric period, one could posit that group members may have been excluded from the larger group at times and resided in the rockshelter. Other rockshelters in the Palouse River canyon do not have such a collection of tools and debitage as does Marmes Rockshelter, but Marmes is the closest rockshelter to the two late prehistoric residential sites. Use of the site for burials did not prevent its use for other activities. Avoidance of burial sites is a common element of traditional spirituality of contemporary western Native Americans, but the rockshelter saw considerable use for storage and activities associated with the many fire hearths, so short-term residence also may have occurred. It also may be that the site was not strongly associated with burials, since only two burials were placed in the rockshelter during the period associated with Stratum VII. By the time of the Harder Phase, Plateau burials are placed in open sites and in talus slopes (the use of the terrace cemetery 45FR36b comes much later). The association of Marmes Rockshelter with mortuary activities may largely be an archaeological focus. Given the number of people that must have lived in the Palouse River canyon over the 11,000 years of the site's use, 38 sets of human remains (not all of which were interments) may not have made Marmes Rockshelter a 'burial ground' in the eyes of the area's inhabitants. If there was an exception to this, it probably would have been Stratum V, but this dates to several thousand years before the inception of the Stratum VII period. Native Americans might dispute this suggestion. As Fallon Shoshone-Paiute representatives noted in a presentation on the Spirit Cave site in Nevada (NAGPRA Review Committee, 2002), their people knew the cave had burials in it even though it had not received an interment in some 9,000 years.

This lack of correlation of Marmes Rockshelter's apparent function in the late prehistoric period and the expectations posited from the regional subsistence and settlement pattern indicates that the cultural material in the later strata do not fit a classic logistical collector model. Potential social and economic explanations have been forwarded; some combination of these explanations also may apply. But a 'classic' fit to a model is not necessarily expected, and given that the early strata at the site exhibit a tethered forager adaptation so well, one couldn't hope for a classic fit for the entire 11,000 years of the Marmes site's history.

Recommendations

The Marmes site is important to Archaeology and the northwest, not only because it is a National Historic Landmark site with a long, continuous archaeological record, but because the excitement generated by archaeologists during the investigations in the 1960s translated to the public. The public's curiosity about the antiquity of human use of Marmes Rockshelter and the region and how that curiosity rose beyond an interest in Native American history to questions about all of human history is an example of why cultural resources laws exist and are widely supported. The United States, through both the National Park Service and the Army Corps of Engineers, has invested a considerable sum of money in this site in recognition of the widespread support historic topics have with the public they serve. One of those investments was the protection of the site. Although the coffer-dam failed to keep water from the site it has protected the area from the effects of wave action, which can be extreme along the north shore of the embayment formed by the reservoir at the bottom of the Palouse River canyon. In addition, the extensive efforts to cover the site in plastic sheeting and tons of clean sand before the water rose to the level of the site has protected the site for future research, an appropriate management action for a National Historic Landmark.

Deposits remain at this site that are important for addressing a number of research questions. Some of these questions have been asked previously but have gone wanting due to a lack of data or due to a lack of resolution in the available data that can provide the conclusion desired. Human remains may be present in the deposits that were not excavated. Large areas of the floodplain portion of the site were not excavated and may retain artifacts and features associated with the earliest uses of the Snake River bottom lands by people. Only a few units were excavated to the deepest habitable levels in the rockshelter. The few items found there provide tantalizing clues about how both the rockshelter and the floodplain area contributed to these earliest occupations. This study has made it clear how rare these early site are and how hard they are to find.

Should a substantial drawdown of the reservoir occur that uncovers the area of the Marmes site, the site must be protected from those with a personal interest in what lies within the deposits. If the site is to become exposed for an adequate length of time for the sediments to drain, further excavation at this site is recommended. However, further excavation should not be contemplated without adequate time to conduct careful, detailed work. Such work should be implemented under an Archaeological Resources Protection Act permit with a peer-reviewed, interdisciplinary research design, which includes consultation with and involvement of affected Native American tribes. In the course of the fieldwork, the public must be allowed to participate in an appropriate manner.

Continued study of the Marmes site and its relationship with other sites in the Palouse River canyon should engage Native American tribes with interests in the history of their people at this location. Relating the traditional knowledge of cultural use of the area with the different site types represented will provide corresponding data of use to archaeology and to the tribes.

The Marmes site collection contains human remains and grave goods, which must be repatriated under the Native American Graves Protection and Repatriation Act if a claim is made by a Native American tribe. Because the Marmes human remains were recovered from strata that span some 10,000 years, the results of the Kennewick Man court case will need to be weighed in the Corps of Engineers' decision-making and repatriation process on the Marmes site collection.

The Marmes collection would benefit from additional curatorial work. Efforts to locate missing items that may be incorrectly boxed with other site collections should be continued, as should seeking records that can provide provenience information for those materials that lack it. The Corps should continue to gather outstanding records related to Marmes Rockshelter and previous research into the collection; such records should be appropriately curated with the collection. This study sorted the cultural material from numerous screened, bulk samples that were created during the 1968 excavations in the floodplain area of the site. The samples were not sorted in the field due to the limited time for excavation and the need to focus the available manpower on excavating. Many more such samples remain in the collection unsorted. Most of the contents of the unsorted bulk samples are non-cultural gravels and root casts that would not pass through the screen mesh. These non-cultural sediments represent a considerable volume of material that is being maintained as if it is of archaeological value. The cultural material should be sorted out of the bulk sediments and accessioned into the collection and the non-cultural remainder discarded, as it would have been if it had been sorted in 1968.

Until such time that further archaeological work may be possible at the Marmes site, the current collection and the data from this study represent the site. One of the principal goals of this study was to make the data from the site available for use by Plateau researchers, and as expected there is a lot of it. The author's intent since the inception of this study has been to make the data resulting from this study available digitally for ease of use. This has been made possible by the Corps of Engineers and Washington State University, where the University will host a website that contains the data in a downloadable format. It is recommended that the Corps of Engineers and WSU enter into an agreement that will ensure that the website is maintained well into the future. Many research questions not addressed by this study, and further inquiry into those that were addressed here, can be examined using this data. Correlation of material types between sites in the region, to examine all manner of hypotheses, has the potential to make significant contributions to our understanding of human use of the Plateau. In the end, this is likely to be the principal contribution of this study.

REFERENCES

Ackerman, L. A.
  1998 Kinship, Family, and Gender Roles. In *Handbook of North American Indians, Plateau,* vol. 12, edited by D. E. Walker, pp. 515-524. Smithsonian Institution Press, Washington, D.C.

Adams, W. H.
  1972 *Component I at Wawawai (45WT39): The Ethnographic Period Occupation.* Report to the National Park Service, Seattle. Laboratory of Anthropology, Washington State University, Pullman.

Ahler, S. A.
  1986 *The River Flint Quarries: Excavations at Site 32DU508.* State Historical Society of North Dakota, Bismark.

Aikens, C. M.
  1986 *Archaeology of Oregon*, 2nd edition. U.S. Department of the Interior, Bureau of Land Management, Oregon State Office, Portland.
  1993 *Archaeology of Oregon,* 3rd edition. U.S. Department of the Interior, Bureau of Land Management, Oregon State Office, Portland.

Ames, K. M.
  1985 Hierarchies, Stress and Logistical Strategies among Hunter-gatherers in Northwestern North America. In *Prehistoric Hunter-gatherers: The Emergence of Cultural Complexity*, edited by T. D. Price and J. D. Brown, pp. 55-80. Academic Press, New York.
  1988 Early Holocene Forager Mobility Strategies on the Southern Columbia Plateau. In *Early Human Occupation in Far Western North America: The Clovis-Archaic Interface*, edited by Judith A. Willig, C. Melvin Aikens, and John L. Fagan, pp. 325-360. Nevada State Museum Anthropological Papers Number 21, Carson City, Nevada.
  1991 Sedentism: A Temporal Sift or a Transitional Change in Hunter-Gatherer Mobility Patterns? In *Between Bands and States*, edited by S. A. Gregg, pp. 108-134. Center for Archaeological Investigations, Occasional Paper No. 9. Southern Illinois University.
  2000 *Cultural Affiliation Study of the Kennewick Human Remains: Review of the Archaeological Data.* Report prepared for the Department of the Interior, National Park Service, Washington, D.C.
  2002 Comments on the Draft Marmes Report. Report prepared for the Confederated Tribes of the Colville Reservation, History/Archaeology Program, Nespelem, Washington.

Ames, K. M., and A. G. Marshall
  1980 Villages, Demography and Subsistence Intensification on the Southern Columbia Plateau. *North American Archaeologist* 2(1):25-52. Farmingdale, New York.

Ames, K. M., J. P. Green, and M. Pfoertner
  1981 *Hatwai (10NP143): Interim Report.* Boise State University Archaeological Reports No. 9, Boise, Idaho.

Ames, K. M., D. E. Dumond, J. R. Galm, and R. Minor
  1998 Prehistory of the Southern Plateau. In *Plateau*, edited by Deward E. Walker, Jr., pp. 103-119. Handbook of North American Indians, vol. 12, W. C. Sturtevant, general editor. Smithsonian Institution, Washington, D.C.

Anastasio, A.
  1972 The Southern Plateau: An Ecological Analysis of Intergroup Relations. *Northwest Anthropological Research Notes* 6(2).

Andrefsky, W., Jr.
  1995 Cascade Phase Lithic Technology: An Example from the Lower Snake River. *North American Archaeologist* 16(2):95-115.

Antevs, E.
  1948 The Great Basin, with Emphasis on Glacial and Post-Glacial Times: Climatic Changes and Pre-White Man. *Bulletin of the University of Utah* 38(20):168-191.

Aoki, H.
  1994 *Nez Perce Dictionary.* University of California Press, Berkeley.

Asherin, D. A., and J. J. Claar
  1976 *Inventory of Riparian Habitats and Associated Wildlife Along the Columbia and Snake Rivers.* College of Forestry, Wildlife and Range Sciences, University of Idaho, Moscow.

Atwell, R. G.
  1989 Subsistence Variability on the Columbia Plateau. Unpublished Master's thesis, Department of Anthropology, Portland State University, Portland.

Baker, V. R.
  1978a Quaternary Geology of the Channeled Scabland and Adjacent Areas. In *The Channeled Scablands: A Guide to the Geomorphology of the Columbia Basin, Washington*, edited by V. R. Baker and D. Nummedal, pp. 17-35. National Aeronautics and Space Administration, Washington, D.C.
  1978b Paleohydraulics and Hydrodynamics of the Scabland Floods. In *The Channeled Scablands: A Guide to the Geomorphology*

of the Columbia Basin, Washington, edited by V. R. Baker and D. Nummedal, pp. 59-79. U.S. National Aeronautics and Space Administration, Washington, D.C.

Baker, V. R., and D. Nummedal
1978 *The Channeled Scablands: A Guide to the Geomorphology of the Columbia Basin, Washington*. Planetary Geology Program, Office of Space Sciences, U.S. National Aeronautics and Space Administration, Washington, D.C.

Baker, V. R., B. N. Bjornstad, A. Busacca, K. R. Fecht, E. P. Kiver, U. Moody, J. G. Rigby, D. F. Stradling, and A. M. Tallman
1991 Quaternary Geology of the Columbia Plateau. In *Quaternary Nonglacial Geology; Conterminous U.S.*, edited by R. B. Morrison, pp. 215-246. Geological Society of America, Boulder.

Bakewell, E. F.
1993 Shades of Gray: Lithic Variation in Pseudobasaltic Debitage. *Archaeology in Washington* V:23-32.

Balme, J.
1980 An Analysis of Charred Bone from Devil's Lair, Western Australia. *Archaeology and Physical Anthropology in Oceania* 15:81-85.

Banfield, A. W. F.
1974 *The Mammals of Canada*. University of Toronto Press, Toronto, Canada.

Barrett, J. H.
1997 Fish Trade in Norse Orkney and Caithness: A Zooarchaeological Approach. *Antiquity* 71:616-638.

Bartholomew, M. J.
1982 Pollen and Sediment Analysis of Clear Lake, Whitman County, Washington: The Last 600 Years. Unpublished Master's thesis, Washington State University, Pullman.

Baumhoff, M. A., and R. F. Heizer
1965 Postglacial Climate and Archaeology in the Desert West. In *The Quaternary of the United States: A review volume for the VII Congress of the International Association for Quaternary Research*, edited by H. E. Wright and D. G. Frey, pp. 697-707. Princeton University Press, Princeton.

Beall, T.
1917 Pioneer Reminiscences. *The Washington Historical Quarterly* Volume VIII, No. 2:83-90.

Behrensmeyer, A. K.
1978 Taphonomic and Ecologic Information from Bone Weathering. *Paleobiology* 4:150-162.

Bense, J.
1972 The Cascade Phase: A Study in the Effect of the Altithermal on a Cultural System. Ph.D. dissertation, Washington State University, Pullman. University Microfilms, Ann Arbor.

Bettinger, R. L.
1991 *Hunter-Gatherers: Archaeological and Evolutionary Theory*. Plenum Press, New York.

Binford, L. R.
1968 Archaeological Perspectives. In *New Perspectives in Archaeology*, edited by S. R. Binford and L. Binford, pp. 5-32. Aldine, Chicago.
1972 Archaeology and Anthropology. *American Antiquity* 28:217-225.
1978 *Nunamuit Ethnoarchaeology*. Academic Press, New York.
1980 Willow Smoke and Dog's Tails: Hunter-Gatherer Settlement Systems and Archaeological Site Formation. *American Antiquity* 45:4-20.
1981 *Bones: Ancient Men and Modern Myths*. Academic Press, New York.

Birkeland, P. W.
1984 *Soils and Geomorphology*. Academic Press, New York.

Blair, W. F., A. P. Blair, P. Brodkorb, F. R. Cagle, and G. A. Moore
1968 *Vertebrates of the United States*. McGraw-Hill Co., New York.

Blankinship, J. W.
1905 *The Native Economic Plants of Montana*. Montana Agricultural College Experiment Station Bulletin 56.

Blinman, E., P. J. Mehringer, Jr., and J. C. Sheppard
1979 Chapter 13. Pollen Influx and the Deposition of Mazama and Glacier Peak Tephra. In *Volcanic Activity and Human Ecology*, edited by P. D. Sheets and D. K. Grayson, pp. 393-425. Academic Press, San Francisco.

Blomgren, P. S.
1996 Late Prehistoric Bone Awl Morphology and Use in the Rio Grande Valley, New Mexico. Unpublished Master's thesis, Department of Anthropology, Washington State University, Pullman.

Blukis Onat, A. R., B. A. Hicks, and S. A. Stump
1996 *A Cultural Resources Management Document for the Palouse Canyon Archaeological District*. BOAS Research Report No. 9212-4.2, BOAS Inc., Seattle.

Bohn, H., B. McNeal, and G. O'Connor
1985 *Soil Chemistry*, 2nd ed. John Wiley and Sons, New York.

Boggs, S., Jr.
1987 *Principles of Sedimentology and Stratigraphy*. Merrill Publishing Company, Columbus, Ohio.

Boldurian, A. T., and S. M. Hubinsky
1994 Preforms in Folsom Lithic Technology: A

View from Blackwater Draw, New Mexico. *Plains Anthropologist* 39(150):445-464.

Bonnichsen, R., and R. Will
1980 Cultural Modification of Bone: the Experimental Approach in Faunal Analysis. In *Mammalian Osteology*, by B. M. Gilbert, pp. 7-30. B. Miles Gilbert Publisher, Laramie.

Borden, C. E.
1952 Results of Archaeological Investigations in Central British Columbia. *Anthropology in British Columbia* 3:31-43. Victoria.
1953 Some Aspects of Prehistoric Coastal-Interior Relations in the Pacific Northwest. *Anthropology in British Columbia* 4:26-32. Victoria.

Born, S. M.
1972 *Late Quaternary History, Deltaic Sedimentation, and Mudlump Formation at Pyramid Lake, Nevada*. Center for Water Resources Research, Desert Research Institute, Reno, Nevada.

Borziyak, I. A.
1993 Subsistence Practices of Late Paleolithic Groups along the Dnestr River and its Tributaries. In *From Kostenki to Clovis: Upper Paleolithic -- Paleo-Indian Adaptations*, edited by O. Soffer and N. D. Praslov, pp. 67-84. Plenum Press, New York.

Bouyoucos, G.
1962 Hydrometer Method Improved for Making Particle Size Analysis of Soils. *Agronomy Journal* 54:464-465.

Boyd, R. T.
1985 The Introduction of Infectious Diseases among the Indians of the Pacific Northwest, 1774-1874. Unpublished Ph.D. dissertation, Department of Anthropology, University of Washington, Seattle.

Bradley, B. A.
1982 Flaked Stone Technology and Typology. In *The Agate Basin Site: A Record of the Paleoindian Occupation of the Northwestern High Plains*, edited by George C. Frison and Dennis J. Stanford, pp. 181-208. Studies in Archaeology, edited by Stuart Struever, Northwestern University. Academic Press, New York.

Brain, C. K.
1981 *The Hunters or the Hunted? An Introduction to South African Cave Taphonomy*. University of Chicago Press, Chicago.

Brauner, D.
1976 Alpowa: The Culture History of the Alpowa Locality. Unpublished Ph.D. dissertation, Washington State University, Pullman.

Bremner, J. M., and C. S. Mulvaney
1980 Nitrogen-Total. In *Methods of Soil Analysis, Part 2, Chemical and Microbiological Properties*, edited by A. L. Page, pp. 595-624. Soil Science Society of America, Madison.

Breschini, G.
1975 The Marmes Burial Casts. Unpublished Master's thesis, Department of Anthropology, Washington State University, Pullman.
1979 The Marmes Burial Casts. *Northwest Anthropological Research Notes* 13(2):111-158.

Bretz, J. H.
1923 Channeled Scablands of the Columbia Plateau. *Journal of Geology* 31:617-549.
1959 Washington's Channeled Scablands. *Washington Division of Mines and Geology* No. 45.
1969 The Lake Missoula Floods and the Channeled Scabland. *Journal of Geology* 77 (5):505-543.

Broncheau-McFarland, S.
1992 Tsoop-Nit-Pa-Lu and a Corridor of Change: Evolution of an Ancient Travel Route Nee-Me-Poo Trail. Master's thesis, University of Idaho, Moscow.

Bryan, K.
1927 The Palouse Soil Problem. *U.S.G.S. Bulletin* 790:21-46.

Buikstra, J. E., and M. Swegle
1989 Bone Modification Due to Burning: Experimental Evidence. In: *Bone Modification* edited by R. Bonnichsen and M. H. Sorg, pp. 247-248. Orono: Center for the Study of the First Americans, University of Maine.

Bull, W. B.
1991 *Geomorphic Responses to Climatic Change*. Oxford University Press, New York.

Bullock, M. E.
1992 Morphological and Functional Analysis of a Pueblo III Anasazi Bone Tool Assemblage from Southwestern Colorado. Unpublished Master's thesis, Department of Anthropology, Washington State University, Pullman.

Bunn, H.
1982 Animal Bones and Archaeological Inference. *Science* 215:494-495.

Burtchard, G. C., K. A. Simmons, E. Adams-Rasmussen, and B. D. Cochran
1981 *Test Excavations at Box Canyon and Three Other Side Canyon Sites in the McNary Reservoir*. Laboratory of Archaeology and History, Project Report Number 10. Washington State University, Pullman.

Busacca, A. J.
1989 Long Quaternary Record in Eastern

Washington, U.S.A., Interpreted from Multiple Buried Paleosols in Loess. *Geoderma* 45:105-122.

1991 Loess Deposits and Soils of the Palouse and Vicinity. In Chapter 8: Quaternary Geology of the Columbia Plateau, *The Geology of North America. Vol. K-2, Quaternary Nonglacial Geology: Coterminous U.S.*, edited by R. B. Morrison, pp. 216-228. The Geological Society of America.

Butler, B. R.
1958 *Archaeological Investigations on the Washington Shore of The Dalles Reservoir, 1955-1957*. Two volumes. Prepared for the National Park Service by the University of Washington, Seattle.
1961 *The Old Cordilleran Culture in the Pacific Northwest*. Occasional Papers of the Idaho State University Museum 5, Pocatello.

Butler, V. L.
1987a Fish Remains. In *The Duwamish No. 1 Site: 1986 Data Recovery*, pp. 10-1 through 10-37. Report Submitted to METRO (Municipality of Metropolitan Seattle). URS Corporation, Seattle, BOAS, Inc., Seattle.
1987b Review of Explanations for Salmonid Bone Scarcity in the Pacific Northwest. Paper presented at the 52nd Annual Meeting of the Society for American Archaeology, Toronto.
1990 *Distinguishing Natural from Cultural Salmonid Deposits in Pacific North America*. Ph.D. Dissertation, University of Washington, Seattle, University Microfilms, Ann Arbor.
1993 Natural Versus Cultural Salmonid Remains: Origin of The Dalles Roadcut Bones, Columbia River, Oregon, U.S.A. *Journal of Archaeological Science* 20:1-24.
1996 Tui Chub Taphonomy and the Importance of Marsh Resources in the Western Great Basin of North America. *American Antiquity* 61:699-717.

Butler, V. L., and R. A. Schalk
1986 Holocene Salmonid Resources of the Upper Columbia. In *The Wells Reservoir Archaeological Project*, pp. 232-252, edited by J. E. Chatters. Central Washington University Archaeological Survey Report 86-6. Ellensburg, Washington.

Butler, V. L., and R. A. Schroeder
1998 Do Digestive Processes Leave Diagnostic Traces on Fish Bones? *Journal of Archaeological Science* 25:957-971.

Butzer, K.
1971 *Environment and Archaeology: An Ecological Approach to Prehistory*. 2nd ed. Aldine, Chicago.

Campbell, S.
1985 *Summary of Results, Chief Joseph Dam Cultural Resources Project, Washington*. University of Washington, Office of Public Archaeology. Prepared for the U.S. Army Corps of Engineers, Seattle District, Seattle.
1989 *Postcolumbian Culture History in the Northern Columbia Plateau: A.D. 1500-1900*. Ph.D. dissertation. Department of Anthropology, University of Washington, Seattle. University Microfilms, Ann Arbor.

Carlson, R. L.
1996 Early Namu. In *Early Human Occupation in British Columbia*, edited by R. L. Carlson and L. D. Bona, pp. 83-102. University of British Columbia Press, Vancouver.

Casteel, R. W.
1972 Some Biases in the Recovery of Archaeological Faunal Remains. *Proceedings of the Prehistoric Society* 36:832-388.
1976 *Fish Remains in Archaeology*. Academic Press, New York.

Caulk, G.
1988 Examination of Some Faunal Remains from the Marmes Rockshelter Floodplain. Unpublished Master's thesis, Washington State University, Pullman.

Chadwick, O. A., D. M. Hendricks, and W. D. Nettleton
1985 Silica in Duric Soils: A Depositional Model. *Soil Science Society of America Journal* 51:975-982.

Chang, K.
1967 *Rethinking Archaeology*. Random House, New York.

Chatters, J. E.
1986 *The Wells Reservoir Archaeological Project, Volume 1, Summary of Findings*. Central Washington Archaeological Survey, Archaeological Report 86-6, Central Washington University, Ellensburg.
1987 Hunter-Gatherer Adaptations and Assemblage Structure. *Journal of Anthropological Archaeology* 6:336-375.
1989 Resource Intensification and Sedentism on the Southern Plateau. *Archaeology in Washington*, volume I:1-19.
1995 Population Growth, Climatic Cooling, and the Development of Collector Strategies on the Southern Plateau, Western North America. *Journal of World Prehistory* 9(3):341-400.
1998 Environment. In *Plateau*, edited by Deward E. Walker, Jr., pp. 29-48. Handbook of North American Indians, vol. 12, W. C. Sturtevant, general editor. Smithsonian Institution, Washington, D.C.

Chatters, J. E., and K. Hoover

1992 The Response of the Columbia River Fluvial System to Holocene Climatic Change. *Quaternary Research* 37:42-59.

Chatters, J. E., and D. L. Pokotylo
1998 Prehistory: Introduction. In *Plateau*, edited by Deward E. Walker, Jr., pp. 73-80. Handbook of North American Indians, vol. 12, W. C. Sturtevant, general editor. Smithsonian Institution, Washington, D.C.

Chatters, J. E., V. L. Butler, M. J. Scott, D. M. Anderson, and D. A. Neitzel
1995 A Paleoscience Approach to Estimating the Effects of Climatic Warming on Salmonid Fisheries of the Columbia River Basin. In Climate Change and Northern Fish Populations, edited by R. J. Beamish, pp. 489-496. *Canadian Special Publication of Fisheries and Aquatic Sciences* 121. Ottawa.

Churchill, T. E.
1983 Inner Bark Utilization: A Nez Perce Example. Master's thesis, Oregon State University, Corvallis.

Clark, J. G. D.
1971 *Excavations at Starr Carr*. Cambridge University Press, Cambridge.

Clarke, A. H.
1973 The Freshwater Molluscs of the Canadian Interior Basin. *Malacologia* 13(1-2).
1981 *The Freshwater Molluscs of Canada*. National Museums of Canada, Ottawa.

Clarke, D.
1968 *Analytical Archaeology*. Metheun and Company, London.

Cleveland, G.
1977 Experimental Replication of Butchered Artiodactyla Bone with Special Reference to Archaeological Features at 45FR5. In *Preliminary Archaeological Investigations at the Miller Site, Strawberry Island, 1976: A Late Prehistoric Village near Burbank, Franklin County, Washington*, by G. C. Cleveland, J. J. Flenniken, D. R. Huelsbeck, R. Mierendorf, S. Samuels, and F. Hassan, pp. 30-48. Washington Archaeological Research Center Project Report Number 46, Pullman.

Cohen, A. S., M. Palacios, R. M. Negrini, P. E. Wigand, and D. B. Erbes.
2000 A Paleoclimate Record for the Past 250,000 Years from Summer Lake, Oregon, U.S.A.: II. Sedimentology, Paleontology, and Geochemistry. *Journal of Paleolimnology* 24(2): 151-182.

COHMAP Members
1988 Climatic changes of the last 18,000 years: observations and model simulations. *Science* 241:1043-1052.

Cole, D. L.
1968 *Archaeological Investigations in Area 6 of Site 35GM9, The Wildcat Canyon Site*. University of Oregon, Museum of Natural History.

Collins, M. B.
1993 Comprehensive Lithic Studies: Context, Technology, Style, Attrition, Breakage, Use-wear and Organic Residues. *Lithic Technology* 18(1&2):87-94.

Collins, M. B., and W. Andrefsky, Jr.
1995 *Archaeological Collections Inventory and Assessment of Marmes Rockshelter (45FR50) and Palus Sites (45FR36A, B, C)*. Center for Northwest Anthropology Project Report Number 28, Department of Anthropology, Washington State University, Pullman.

Combes, J. D.
1963 *The Archaeology of Ford Island: Section 1; A Report on the Excavation of a Late Indian Burial Site in the Ice Harbor Region, Washington*. Washington State University, Laboratory of Anthropology, Report of Investigations, Number 18. Pullman.
1968 Burial Practices as an Indicator of Cultural Change in the Lower Snake River Region. Master's thesis, Department of Anthropology, Washington State University, Pullman.

Connolly, T.J.
1999 Newberry Crater: A Ten-Thousand Year Record of Human Occupation and Environmental Change in Basin Plateau Borderlands. University of Utah Press Anthropological Papers 121.

Couch, L. K.
1927 Migration of the Washington Black-tailed Jackrabbit. *Journal of Mammalogy* 8:313-314.
1953 Chronological Data on Elk Introductions in Oregon and Washington. *Murrelet* 16:3-6.

Couture, M. D, M. F. Hicks, and L. Housley
1986 Foraging Behavior of a Contemporary Northern Great Basin Population. *Journal of California and Great Basin Anthropology* 8(2):150-160.

Coville, F.
1904 Plants Used in Basketry. *In Indian Basketry; Studies in a Textile Art Without Machinery* by O. T. Mason, pp. 19-43. Doubleday, Page, and Company, New York.

Cox, R.
1957 *The Columbia River*. University of Oklahoma Press, Norman.

Crabtree, D. E.
1974 The Obtuse Angle as a Functional Edge. *Tebiwa* 16(1):46-53.

1982 *An Introduction to Flintworking.* 2nd Ed. Occasional Papers of the Idaho State University Museum, No. 28, Pocatello.

Crabtree, D. E., and E. H. Swanson, Jr.
1968 Edge-Ground Cobbles and Blade-Making in the Northwest. *Tebiwa* 11(2):50-58.

Crandell, D.R., D. R. Mullineaux, M. Rubin, E. Spiker, and M. L. Kelley
1979 Radiocarbon Dates from Volcanic Deposits at Mount St. Helens, Washington: U.S. *Geological Survey Open-File Report* 81-844.

Cressman, L. S.
1960 Cultural Sequences at The Dalles, Oregon. *Transactions of the American Philosophical Society* 50(10).

Culin, S.
1907 Games of the North American Indians. *24th Annual Report of the Bureau of American Ethnology* XXXIX:1-846.

Culliford, B. J.
1971 *The Examination and Typing of Bloodstains in the Crime Laboratory.* National Institution of Law Enforcement and Criminal Justice, U.S. Printing Office, Washington, D.C.

Cutright, P. R.
1969 *Lewis and Clark: Pioneering Naturalists.* University of Illinois Press, Urbana.

Dalquest, W. W.
1941 Distribution of Cottontail Rabbits in Washington State. *Journal of Wildlife Management* 5:408-411.
1948 *The Mammals of Washington, Volume 2.* Museum of Natural History, University of Kansas Publications.

Daubenmire, R. F.
1942 An Ecological Study of the Vegetation of Southeast Washington and Adjacent Idaho. *Ecological Monographs* 12:53-79.
1970 *Steppe Vegetation of Washington.* Washington Agricultural Experiment Station, Technical Bulletin 62, Pullman.

Daubenmire, R. F., and J. B. Daubenmire
1968 *Forest Vegetation of Eastern Washington and Northern Idaho.* Washington Agricultural Experiment Station, Technical Bulletin 60.

Daugherty, R. D.
1956a Early Man in the Columbia Intermontane Province. *University of Utah Anthropological Papers* No. 24. Salt Lake City.
1956b Archaeology of the Lind Coulee Site, Washington. *Proceedings of the American Philosophical Society* 100:223-278.
1959 *Early Man in Washington.* Division of Mines and Geology Information Circular, State of Washington Department of Conservation.
1960 *Archaeological Excavations in the Ice Harbor Reservoir, 1959.* Washington State University, Laboratory of Anthropology, Report of Investigations Number 3. Pullman.
1962 The Intermontane Western Tradition. *American Antiquity* 28(2):144-150.

Davies, K. G., A. M. Johnson, and D. O. Johansen (eds.)
1961 *Peter Skene Ogden's Snake Country Journal 1826-27.* The Hudson's Bay Record Society, London.

Davis, J. O.
1982 Bits and Pieces: The Last 35,000 Years in the Lahontan Area. In Man and Environment in the Great Basin, edited by D. B. Madsen and J. F. O'Connell, pp. 53-75. *SAA Papers* No. 2. Society for American Archaeology, Washington, D.C.
1983 Geology of Gatecliff Shelter: Sedimentary Facies and Holocene Climate. In The Archaeology of Monitor Valley, 2: Gatecliff Shelter, by D. H. Thomas, pp. 64-87. *Anthropological Papers* Volume 59, Part 1. American Museum of Natural History, New York.

Davis, O. K., D. A. Kolva, and P. J. Mehringer, Jr.
1978 Pollen Analysis of Wildcat Lake, Whitman County, Washington: The Last 1000 Years. *Northwest Science* 51(1):13-30.

Davis, W. B.
1939 *The Recent Mammals of Idaho.* The Caxton Printers, Ltd., Caldwell, Idaho.

Deaver, K., and G. S. Greene
1978 Faunal Utilization at 45AD2: A Prehistoric Archaeological Site in the Channeled Scablands. *Tebiwa* 14:1-21.

de Smet, P. J.
1905 *Life, Letters, and Travels of Father Pierre-Jean de Smet, S. J., 1801-1873*, edited by H. M. Chittenden and A. T. Richardson, vols. 1-4. Francis P. Harper, New York.

Dillehay, T. D.
2000 *The Settlement of the Americas.* Basic Books, New York, NY.

Dixon, E. J.
1999 *Boats, Bones and Bison: Archaeology and the First Colonization of Western North America.* University of New Mexico Press, Albuquerque.

Dodd, W.
1979 The Wear and Use of Battered Tools at Amigo Rockshelter. In *Lithic Use-Wear Analysis*, edited by B. Hayden, pp. 231-242. Academic Press, New York.

Draper, J. A., and M. Morgenstein
1993 *Archaeological Testing in the Palouse Canyon Archaeological District: 1992 Field Season.* Report prepared for the Walla Walla District, U.S. Army Corps of Engineers, by BOAS, Inc., Seattle.

Drury, C. M.
1958 *The Diaries and Letters of Henry H. Spaulding and Asa Bowen Smith Relating to the Nez Perce Mission, 1838-1842*. Arthur H. Clark, Glendale.

Dugas, D. P.
1998 Late Quaternary Variations in the Level of Paleo-Lake Malheur, Eastern Oregon. *Quaternary Research* 50(3):276-282.

Dunnell, R. C.
1979 Archaeological Potential of Anthropological and Scientific Models of Function. In *Essays in Honor of Irving B. Rouse*, edited by R. Dunnell and E. Hall, pp. 41-73. The Hague, Mouton.

Eidt, R. C.
1977 Detection and Examination of Anthrosols by Phosphate Analysis. *Science* 197:1327-33.

Einarsen, A. S.
1948 *The Pronghorn Antelope and its Management*. Wildlife Management Institute, Washington, D.C.

Elliott, T. C. (editor)
1909 Journal of John Work. *Oregon Historical Quarterly* 10(3):296-313.
1914 Journal of John Work, June-October 1925. *Washington Historical Quarterly* 5(2):83-115.

Enbysk, B. J.
1956 Vertebrates and Mollusca from Lind Coulee, Washington. In Archaeology of the Lind Coulee Site by Richard D. Daugherty. *Proceedings of the American Philosophical Society* 100:223-278.

Endacott, N.
1992 The Archaeology of Squirt Cave: Seasonality, Storage, and Semisedentism. Unpublished Master's thesis, Washington State University, Pullman.

Erickson, K.
1990 Marine Shell Utilization in the Plateau Cultural Area. *Northwest Anthropological Research Notes* 24(1):91-144.

Ewers, J. C.
1955 The Horse in Blackfoot Indian Culture. *Smithsonian Institution, Bulletin of American Ethnology, Bulletin* 159.

Farrand, W. R.
1985 Rockshelter and Cave Sediments. In *Archaeological Sediments in Context*, edited by J. K. Stein and W. R. Farrand, pp. 21-40. Center for the Study of Early Man, Orono, Maine.

Ferring, C. R.
1992 Alluvial Pedology and Geoarchaeological Research. *In Soils in Archaeology: Landscape Evolution and Human Occupation*, edited by V. T. Holliday, pp. 1-40. Smithsonian Institution Press, Washington, D.C.

Flenniken, J. J.
1978 The Experimental Replication of Paleo-Indian Eyed Needles from Washington. *Northwest Anthropological Research Notes* 12 (1):61-71.
1981 *Replicative Systems Analysis: A Model Applied to the Vein Quartz Artifacts from the Hoko River Site*. Laboratory of Anthropology Reports of Investigation No. 59. Washington State University, Pullman.
1991 The Diamond Lil Site: Projectile Point Fragments as Indicators of Site Function. *Journal of California and Great Basin Anthropology* 13(2):180-193.

Flenniken, J. J., and J. Haggerty
1979 Trampling as an Agency in the Formation of Edge Damage: An Experiment in Lithic Technology. *Northwest Anthropological Research Notes* 13:208-214.

Flenniken, J. J., and P. J. Wilke
1989 Typology, Chronology, and Technology of Great Basin Dart Points: An Anthropological Perspective. *American Anthropologist* 19:149-158.

Fletcher, W. F.
1982 *Starbuck: 1882-1982*. Ye Galleon Press, Fairfield, Washington.

Foit, N., P. J. Mehringer, Jr., and J. C. Sheppard
1991 Age, Distribution, and Stratigraphy of Glacier Peak Tephra in Eastern Washington and Western Montana, United States. *Canadian Journal of Earth Science* 30:535-552.

Ford, P. J.
1985 Shellfish Harvesting and the Available Food Supply. Paper presented at the 50th Annual Meeting of the Society for American Archaeology, Denver.

Fowler, C. S., and S. Liljeblad
1986 Northern Paiute. In *Handbook of North American Indians*, vol. 11, edited by William C. Sturtevant, pp. 435-466. Smithsonian Institution, Washington, D.C.

Franklin, J., and C. T. Dyrness
1972 *Natural Vegetation of Oregon and Washington*. Oregon State University Press, Corvallis.
1973 *Natural Vegetation of Oregon and Washington*. U.S. Department of Agriculture, General Technical Report PNW-8.

Freeman, O., J. D. Forrester, and R. L. Lupher
1945 Physiographic Divisions of the Columbia Intermontane Province. *Annals of the Association of American Geographers* 35(2):53-75.

Friedman, J.
 1978 Wood Identification by Microscopic Examination. A Guide for the Archaeologist on the Northwest Coast of North America. British Columbia Provincial Museum Heritage Record 5.

Frison, G. C.
 1982 The Sheaman Site: A Clovis Component. In *The Agate Basin Site: A Record of the Paleoindian Occupation of the Northwestern High Plains*, edited by G. C. Frison and D. J. Stanford, pp. 143-157. Academic Press, New York.
 1991 *Prehistoric Hunters of the High Plains*. Academic Press, San Diego.

Frison, G. C., and B. A. Bradley
 1980 *Folsom Tools and Technology at the Hanson Site, Wyoming*. University of New Mexico Press, Albuquerque.

Frison, G. C., and C. Craig
 1982 Bone, Antler and Ivory Artifacts and Manufacture Technology. In *The Agate Basin Site: A Record of the Paleoindian Occupation of the Northwestern High Plains*, edited by G. C. Frison and D. J. Stanford, pp. 157-173. Academic Press, New York.

Frison, G. C., and G. Zeimens
 1980 Bone Projectile Points: An Addition to the Folsom Cultural Complex. *American Antiquity* 45:231-237.

Fryxell, R.
 1963a *Late Glacial and Post-Glacial Geological and Archaeological Chronology of the Columbia Plateau, Washington*. Washington State University Report of Investigations Number 23. Pullman.
 1963b Through a Mirror, Darkly. *The Record*, pp. 1-18. Washington State University Library, Pullman.
 1964 Regional Patterns of Sedimentation Recorded by Cave and Rock-shelter Stratigraphy in the Columbia Plateau, Washington. In *Geological Society of America Special Paper* 76:272. Geological Society of America, Boulder.
 1965 Figure 41. In *Guidebook for Field Conference E, Northern and Middle Rocky Mountains*. 7th Congress of the International Association for Quaternary Research, Lincoln.
 1971 The Contribution of Interdisciplinary Research to Geologic Investigation of Prehistory, Eastern Washington. Unpublished Ph.D. thesis, University of Idaho, Moscow.

Fryxell, R., and E. F. Cook
 1964 *A Field Guide to the Loess Deposits and Channeled Scablands of the Palouse Area, Eastern Washington*. Washington State University Laboratory of Anthropology Report of Investigations Number 27. Pullman.

Fryxell, R., and R. D. Daugherty
 1962 *Interim Report: Archeological Salvage in the Lower Monumental Reservoir, Washington*. Reports of Investigations Number 21. Washington State University, Laboratory of Archeology and Geochronology, Pullman.

Fryxell, R., and B. Keel
 1969 *Emergency Salvage Excavations for the Recovery of Early Human Remains and Related Scientific Material from the Marmes Rockshelter Archaeological Site, Southeastern Washington, May 3-December 15, 1968*. Report to the U.S. Army Corps of Engineers, Walla Walla District, by the Laboratory of Anthropology, Washington State University, Pullman.

Fryxell, R., T. Bielicki, R. Daugherty, C. Gustafson, H. Irwin, B. Keel, and G. Krantz.
 1968a Human Skeletal Material and Artifacts from Sediments of Pinedale (Wisconsin) Glacial Age in Southeastern Washington, United States. In *Proceedings VIIIth International Congress of Anthropological and Ethnological Sciences, Volume 3: Ethnology and Archaeology*, pp. 176-181.

Fryxell, R., T. Bielicki, R. Daugherty, C. Gustafson, H. Irwin, and B. Keel
 1968b A Human Skeleton from Sediments of Mid-Pinedale Age in Southeastern Washington. *American Antiquity* 33(4):511-514.

Fulton, L. A.
 1968 *Spawning Areas and Abundance of Chinook Salmon (Oncorhynchus tshawytscha) in the Columbia River Basin--Past and Present*. U.S. Fish and Wildlife Service Special Scientific Report: Fisheries No. 571.

Galm, J. R.
 1994 Prehistoric Trade and Exchange in the Interior Plateau of Northwestern North American. In *Prehistoric Exchange Systems in North America*, pp. 275-305, edited by T. G. Baugh and J. E. Ericson. Plenum Press, New York and London.

Gee, G. W., and J. W. Bauder
 1986 Particle-Size Analysis. *In Methods of Soil Analysis, Part 1: Physical and Mineralogical Methods,* 2nd edition, edited by A. Kunze, pp. 383-412. Soil Science Society of America, Madison.

Gehr, K. D.
 1980 Late Pleistocene and Recent Archaeology and Geomorphology of the South Shore of Harney Lake, Oregon. Unpublished Masters Thesis, Portland State University.

Gerasimov, I. P., and M. A. Glazovskaya
 1965 Fundamentals of Soil Science and Soil

Geography (Translated from Russian). Israel Program for Scientific Translations, Jerusalem.

Getz, L. L.
1985 Habitats. In *Biology of New World Microtus*, edited by R. H. Tamarin. Special Publication No. 8, The American Society of Mammalogists.

Gibbs, G.
1972 *Indian Tribes of Washington Territory*. Reprinted by Ye Galleon Press, Fairfield, Washington.

Gifford-Gonzalez, D. P.
1989 Ethnographic Analogues for Interpreting Modified Bones: Some Cases from East Africa. In *Bone Modification*, edited by R. Bonnichsen and M. H. Sorg, pp. 179-246. Orono: University of Maine Center for the Study of the First Americans.

Gilchrist, R., and H. C. Mytum
1986 Experimental Archaeology and Burnt Animal Bones from Archaeological Excavations. *Journal of Mammalogy* 30:163-169.

Gilkey, H. M., and P. L. Packard
1962 *Winter Twigs. Northwestern Oregon and Western Washington*. Oregon State University Press, Corvallis.

Gleeson, P. F., Jr.
1970 Dog Remains from the Ozette Village Archaeological Site. Unpublished Master's thesis, Department of Anthropology, Washington State University.

Gordon, E. A.
1993 Screen Size and Differential Faunal Recovery: A Hawaiian Example. *Journal of Field Archaeology* 20:453-460.

Gould, R. A.
1977 Ethno-archaeology; or, Where Do Models Come From? In *Stone Tools as Cultural Markers; Change, Evolution, and Complexity*, edited by R. V. S. Wright, pp. 162-168. AIAS, Canberra.

Grabert, G. F.
1968 *The Astor Fort Okanogan*. University of Washington Department of Anthropology, Reports in Archaeology, No. 2. Seattle.

Gramly, R.
n.d. Blood Residues Upon Tools from the East Wenatchee Clovis Site, Douglas County, Washington. Great Lakes Artifact Repository, Buffalo, New York.

Graybill, D. A., M. R. Rose, and F. L. Nials.
1994 Tree-rings and Climate: Implications for Great Basin Paleoenvironmental Studies. In Proceedings of the International High-level Radioactive Waste Management Conference and Exposition, May 1994, pp. 2569-2573. Las Vegas, Nevada.

Grayson, D. K.
1984 *Quantitative Zooarchaeology*. Academic Press, New York.
1991 The Small Mammals of Gatecliff Shelter: Did People Make a Difference? In *Beamers, Bobwhites, and Blue-Points: Tributes to the Career of Paul W. Parmalee*, edited by J. R. Purdue, W. E. Klippel, and B. W. Styles, pp. 99-109.
1993 *The Desert's Past: A Natural Prehistory of the Great Basin*. Smithsonian Institution Press, Washington, D.C.

Green, T., B. Cochran, T. Fenton, J. Woods, G. Titmus, L. Tieszen, M. A. Davis, and S. Miller
1998 The Buhl Burial: A Paleoindian Woman from Southern Idaho. *American Antiquity* 63(3):437-456.

Greene, G. S.
1976 Prehistoric Utilization in the Channeled Scablands of Eastern Washington. Unpublished Ph.D. dissertation, Department of Anthropology, Washington State University, Pullman.

Greenspan, R. L.
1998 Gear Selectivity Models, Mortality Profiles, and the Interpretation of Archaeological Fish Remains: A Case Study from the Harney Basin, Oregon. *Journal of Archaeological Science* 25:973-984.

Greguss, P.
1955 *Identification of Living Gymnosperms on the Basis of Xylotomy*. Akademiai Kiado, Budapest.

Grolier, M.
1965 *Geology of Part of the Big Bend Area of the Columbia, in the Columbia Plateau, Washington*. Ph.D. dissertation, John Hopkins University. University Microfilms, Ann Arbor.

Grosso, G. H.
1967 Cave Life on the Palouse. *Natural History* February:38-43.

Guilday, J. E.
1969 Small Mammal Remains from the Wasden Site (Owl Cave), Bonneville County, Idaho. *Tebiwa* 12:47-57.

Gunkel, A.
1961 A Comparative Cultural Analysis of Four Archaeological Sites in the Rocky Reach Reservoir Region. Theses in Anthropology No. 1, Washington State University, Pullman.

Gustafson, C. E.
1972 Faunal Remains from the Marmes Rockshelter and Related Archaeological Sites in the Columbia Basin. Unpublished Ph.D. dissertation, Department of Zoology, Washington State University, Pullman.
1987a An Analysis of Mammal Remains from

Layser Cave. In *A Data Recovery Study of Layser Cave (45-LE-223) in Lewis County, Washington*, by R. D. Daugherty, J. J. Flenniken, and J. M. Welch. USDA Forest Service, Gifford Pinchot National Forest.

1987b An Analysis of Mammal Remains from Judd Rock Rockshelters. In *A Data Recovery Study of Judd Peak Rockshelters in Lewis County, Washington*, by R. D. Daugherty, J. J. Flenniken, and J. M. Welch. USDA Forest Service, Gifford Pinchot National Forest.

1990 *Faunal Remains from Pittsburg Landing (10 IH 1639)*. Report Submitted to the Hells Canyon National Recreation Association, December 1990.

1996 Faunal Remains from Palouse Falls. In *Archaeological Studies in the Palouse Canyon Archaeological District: 1995 Field Season*, by B. A. Hicks, Appendix D. BOAS Research Report No. 9212.4, BOAS, Inc., Seattle.

Gustafson, C. E., and S. Gibson
1998 Marmes Stratigraphy, Sedimentology, and Inferred Formational Processes. Draft Manuscript on file at History/Archaeology Department, Confederated Tribes of the Colville Reservation, Nespelem.

Gustafson, C. E., and R. Wegener
1998 Faunal Remains Analysis. In *Marmes Rockshelter (45FR50) Preliminary Report: 1998 Results*, edited by B. A. Hicks, pp. 99-152. Confederated Tribes of the Colville Reservation, History/Archaeology Department. Prepared for U.S. Army Corps of Engineers.

Haines, F.
1938 The Northward Spread of Horses Among the Plains Indians. *American Anthropologist* 40(3):429-437.

1955 *The Nez Perces: Tribesmen of the Columbia*. The University of Oklahoma Press, Norman.

Hall, E. R., and K. R. Kelson
1959 *The Mammals of North America*, Vol. I and II. Ronald Press, New York.

Hallett, D. J., L. V. Hills, and J. J. Clague
1997 New Accelerator Mass Spectrometry Radiocarbon Ages for the Mazama Tephra Layer from Kootenay National Park, British Columbia, Canada. *Canadian Journal of Earth Sciences* 34:1202-1209.

Hammatt, H.
1977 *Late Quaternary Stratigraphy and Archaeological Chronology in the Lower Granite Reservoir Area, Lower Snake River, Washington*. Ph.D. dissertation, Washington State University. University Microfilms, Ann Arbor.

Hansen, H. P.
1947 Postglacial Vegetation of the Northern Great Basin. *American Journal of Botany* 34(3):164-171.

Harbinger, L.
1964 The Importance of Foods in the Maintenance of the Nez Perce Cultural Identity. Unpublished Master's thesis, Washington State University, Pullman.

Harder, D. A.
1998 A Synthetic Overview of the Tucannon Phase in the Lower Snake River Region of Washington and Idaho. Unpublished Master's thesis, Washington State University, Pullman.

Hart, J. A.
1976 *Montana--Native Plants and Early Peoples*. Montana Historical Society and Montana Bicentennial Administration.

1979 The Ethnobotany of the Flathead Indians of Western Montana. *Botanical Museum Leaflets, Harvard University* 27(10):263-307.

Hayes, D. W., and G. A. Garrison
1960 *Key to Important Woody Plants of Eastern Oregon and Washington*. USDA Agriculture Handbook 148.

Hays, W. L.
1973 *Statistics for the Social Sciences*. Holt, Rinehart and Winston, Inc., New York.

Hess, S.
1997 Rocks, Range, and Renfrew: Using Distance-Decay Effects to Study Late Pre-Mazama Period Obsidian Acquisition and Mobility in Oregon and Washington. Unpublished Ph.D. dissertation, Washington State University, Pullman.

Hewes, G. W.
1998 Fishing. In *Handbook of North American Indians, Plateau*, vol. 12, edited by D. E. Walker, pp. 620-640. Smithsonian Institution Press, Washington, D.C.

Hibbard, C. W., C. E. Ray, D. E. Savage, D. W. Taylor, and J. E. Guilday
1965 Quaternary Mammals of North America. In *The Quaternary of the United States*, ed. by H. E. Wright, Jr., and D. G. Frey, pp. 509-525. Princeton University Press, Princeton.

Hicks, B. A.
1995 *Archaeological Studies in the Palouse Canyon Archaeological District: 1994 Field Season*. BOAS Research Report No. 9212.3, prepared for the Walla Walla District, U.S. Army Corps of Engineers, by BOAS, Inc., Seattle.

1996 *Archaeological Studies in the Palouse Canyon Archaeological District: 1995 Field Season*. BOAS Research Report No. 9212.4-1, prepared for the Walla Walla District, U.S.

Army Corps of Engineers, by BOAS, Inc., Seattle.
1998 *Marmes Rockshelter (45FR50) Preliminary Report: 1998 Results.* Confederated Tribes of the Colville Reservation, History/Archaeology Department. Prepared for the U.S. Army Corps of Engineers, Walla Walla District, Washington.
1999 Lower Snake River Reach Cultural Resources Management Plan. Confederated Tribes of the Colville Reservation, History/Archaeology Department. Prepared for the U.S. Army Corps of Engineers, Walla Walla District, Washington.
2000 *Marmes Rockshelter (45FR50) Preliminary Report: 1999 Results.* Confederated Tribes of the Colville Reservation, History/Archaeology Department. Prepared for the U.S. Army Corps of Engineers, Walla Walla District, Washington.

Hicks, B. A., and M. E. Morgenstein
1994 *Archaeological Studies in the Palouse Canyon Archaeological District: 1993 Field Season.* BOAS Research Report No. 9212.2, prepared for the Walla Walla District, U.S. Army Corps of Engineers, by BOAS, Inc., Seattle.

Hicks, B. A., and G. Moura
1998 Project Background and Collections Research (Chapter Three). In *Marmes Rockshelter (45FR50) Preliminary Report: 1998 Results,* edited by B. A. Hicks, pp. 38-43. Confederated Tribes of the Colville Reservation report to U.S. Army Corps of Engineers.

Hitchcock, L. C., and A. Cronquist
1981 *Flora of the Pacific Northwest: An Illustrated Manual.* University of Washington Press, Seattle.

Hoffman, R. S., and J. W. Koeppl
1985 Zoogeography. In *Biology of New World Microtus,* edited by R. H. Tamarin. Special Publication No. 8, The American Society of Mammalogists.

Holmes, W. H.
1919 *Handbook of Aboriginal American Antiquities.* Part 1. Bureau of American Ethnology Bulletin 60.

Huckleberry, G., C. E. Gustafson, and S. Gibson
1998 Stratigraphy and Site Formation Processes. In *Marmes Rockshelter (45FR50) Preliminary Report: 1998 Results,* edited by B. A. Hicks, pp. 44-98. Confederated Tribes of the Colville Reservation, History/Archaeology Department. Prepared for the U.S. Army Corps of Engineers, Walla Walla District, Washington.

Huckleberry, G., T. W. Stafford, Jr., and J. C. Chatters
1998 *Preliminary Geoarchaeological Studies at Columbia Park, Kennewick, Washington, U.S.A.* Manuscript on File, U.S. Army Corps of Engineers, Walla Walla District, Washington.

Hudson, G. E.
1964 Canada Lynx (*Lynx canadensis*) in the Palouse Wheat Country. *The Murrelet* 45(2):29.

Hughes, R. E.
1993 *Analysis of Seven Obsidian Artifacts from the Hells Canyon Area, Oregon and Idaho.* Geochemical Research Laboratory Letter Report. Submitted to Rain Shadow Research, Inc., Pullman, Washington.

Hunn, E. S.
1990 *Nch'i-Wana, The Big River: Mid-Columbia Indians and Their Land.* University of Washington Press, Seattle.

Hunn, E. S., N. J. Turner, and D. H. French
1998 Ethnobiology and Subsistence. In *Plateau,* edited by Deward E. Walker, Jr., pp. 525-545. Handbook of North American Indians, vol. 12, W. C. Sturtevant, general editor. Smithsonian Institution, Washington, D.C.

Ingles, L. G.
1965 *Mammals of the Pacific States.* Stanford University Press, Stanford.

Irwin, A. M., and U. Moody
1977 *The Lind Coulee Site (45GR97) 1974 Field Season.* Washington Archaeological Research Center, Project Report Number 53. Washington State University, Pullman.
1978 *The Lind Coulee Site (45GR97).* Washington Archaeological Research Center, Project Report Number 56, Washington State University, Pullman.

Irwin-Williams, C., H. Irwin, G. Agogino, and C. V. Haynes
1973 Hell Gap: A Paleo-Indian Occupation on the High Plains. *Plains Anthropologist* 18:40-53.

James, C.
1996 *Nez Perce Women in Transition.* University of Idaho Press, Moscow.

Jewett, S. G., W. P. Taylor, W. T. Shaw, and J. W. Aldrich
1953 *Birds of Washington State.* University of Washington Press, Seattle.

Johnson, C. G., Jr., R. R. Clausnitzer, P. J. Mehringer, Jr., and C. D. Chadwick
1994 *Biotic and Abiotic Processses of Eastside Ecosystems: The Effects of Management on Plant and Community Ecology, and on Stand and Landscape Vegetation Dynamics.* USDA Forest Service, PNW-GTR 322.

Johnson, D. L.
1979 Archaic Biface Manufacture: Production Failures, A Chronicle of the Misbegotten. *Lithic Technology* 8(2):25-35.

Johnson, E.
1985 Current Developments in Bone Technology. In *Advances in Archaeological Method and Theory,* edited by M. B. Schiffer, vol. 8, pp. 157-235. Academic Press, New York.

Johnson, M. L., and S. Johnson
1982 Voles (*Microtus* species). In *Wild Mammals of North America: Biology, Management and Economics*, edited by J. A. Chapman and G. A. Fieldhammer, pp. 326-354. The Johns Hopkins University Press, Baltimore and London.

Johnston, R. T.
1987 Archaeological Evidence of Fishing in the Southern Plateau, A Cultural Area of the Columbia Plateau. Unpublished Master's thesis, Department of Anthropology, University of Idaho, Moscow.

Jorgensen, J. G.
1980 *Western Indians: Comparative Environments, Languages, and Cultures of 172 Western American Indian Tribes*. W.H. Freeman, San Francisco.

Kane, P.
1925 *Wanderings of an Artist Among the Indians of North America*. Radisson Society, Toronto.

Kappler, C. J.
1904 *Indian Affairs, Laws, and Treaties*, 2nd edition, vol. 2. 58th Congress, 2nd Session, Senate Documents, vol. 39, No. 319 (Serial Nos. 4623, 4624). Washington, D.C.

Keel, B. C., and R. Fryxell
1969 *Recovery of Early Human Remains from the Marmes Rockshelter Archaeological Site, Southeastern Washington, May 3-Dec. 15, 1968: A Progress Report to U.S. Army Corps of Engineers District, Walla Walla, Corps of Engineers Concerning Contract No. DACW68-68-C-0107.* Laboratory of Anthropology, Washington State University, Pullman.

Keeley, L. H.
1980 *Experimental Determination of Stone Tool Uses: A Microwear Analysis*. University of Chicago Press, Chicago and London.

Kelly, I.
1978 Coast Miwok. In *California,* edited by Robert F. Heizer, pp. 414-425. Handbook of North American Indians, vol. 8, W. C. Sturtevant, general editor. Smithsonian Institution, Washington, D.C.

Kelly, R. L.
1995 *The Foraging Spectrum: Diversity in Hunter Gatherer Lifeways*. Plenum Press, New York.

Kelly, R., and L. Todd
1988 Coming into the Country: Early Paleoindian Hunting and Mobility. *American Antiquity* 53:231-249.

Kelso, G.
1970 Hogup Cave, Utah: Comparative Pollen Analysis of Human Coprolites and Cave Fill. In *Hogup Cave,* edited by C. M. Aikens, pp. 251-262. University of Utah Anthropological Papers 93, Salt Lake City.

Kenaston, M. R.
1966 *The Archaeology of the Harder Site, Franklin County, Washington.* Laboratory of Anthropology, Report of Investigations Number 35. Washington State University, Pullman.

Kennedy, H.
1976 An Examination of the Tucannon Phase as a Valid Concept: Step One. Unpublished Master's thesis, University of Idaho, Moscow.

Kietzman, D.
1985 Paleomagnetic Survey of the Touchet Beds in Burlingame Canyon of Southwestern Washington. Unpublished Masters thesis, Eastern Washington University, Cheney.

Kirch, P. V., and T. S. Dye
1979 Ethnoarchaeology and the Development of Polynesian Fishing Strategies. *Journal of the Polynesian Society* 88:53-76.

Kirk, R.
1970 *The Oldest Man In America: An Adventure in Archaeology*. Harcourt Brace Jovanovich, New York.

Kirk, R., and R. D. Daugherty
1978 *Exploring Washington Archaeology*. University of Washington Press, Seattle.

Kiver, E. P., U. L. Moody, J. G. Rigby, and D. F. Stradling
1991 Late Quaternary Stratigraphy of the Channeled Scabland and Adjacent Areas. In Chapter 8: Quaternary Geology of the Columbia Plateau, *The Geology of North America. Vol. K-2, Quaternary Nonglacial Geology: Coterminous U.S.,* edited by R. B. Morrison, pp. 238-245. The Geological Society of America.

Klein, J., J. C. Lerman, P. E. Damon, and F. K. Ralph
1982 Calibration of radiocarbon dates. *Radiocarbon* 24(2):103-150.

Krantz, G.
1979 Oldest Human Remains from the Marmes Site. *Northwest Anthropological Research Notes* 13(2):159-174.

Kurten, B.
- 1968 *Pleistocene Mammals of Europe*. Aldine Publishing Company, Chicago.

Kurten, B., and E. Anderson
- 1980 *Pleistocene Mammals of North America*. Columbia University Press, New York.

Lahren, L. A., and R. Bonnichsen
- 1974 Bone Foreshafts from A Clovis Burial in Southwestern Montana. *Science* 186:147-150.

LaMarche, V. C., Jr.
- 1974 Paleoclimatic Inferences from Long Tree-Ring Records. *Science* 183(4129):1043-1048.

Lau, B., B. A. Blackwell, H. P. Schwarcz, I. Turk, and J. I. Blickstein
- 1997 Dating a Flautist? Using ESR (Electron Spin Resonance) in the Mousterian Cave Deposits at Divje Babe I, Slovenia. *Geoarchaeology: An International Journal* 12: 507-536.

LeBlanc, R.
- 1992 Wedges, Pièces Esquillées, Bipolar Cores, and other Things: An Alternative to Shott's View of Bipolar Industries. *North American Archaeologist* 13:1-14.

Lee, D. S., C. R. Gilbert, C. H. Hocutt, R. E. Jenkins, D. E. McAllister, and J. R. Stauffer, Jr.
- 1980 *Atlas of North American Freshwater Fishes*. Publication No. 1980-12, North Carolina Biological Survey.

Leechman, D.
- 1951 Bone Grease. *American Antiquity* 4:335-356.

Lenz, B., H. Gentry, and D. Clingman
- 2002 Numeric Ages of Final Cross-Scabland Pleistocene Flooding, Columbia Plateau. Paper presented at the 2002 Annual Meeting of the Geological Society of America, Denver.

Leonhardy, F. C.
- 1968 *Artifact Assemblages from Granite Point Locality I (45WT41): The First Analysis*. Report to the National Park Service. Laboratory of Anthropology, Washington State University, Pullman.
- 1970 Artifact Assemblages and Archaeological Units at Granite Point Locality (45WT41), Southeastern Washington. Unpublished Ph.D. dissertation, Washington State University, Pullman.
- 1975 The Lower Snake River Culture Typlogy-1975. In *Northwest Anthropological Research Notes*, Vol. 10, edited by R. Sprague and D. E. Walker, p. 61. Laboratory of Anthropology, University of Idaho, Moscow.

Leonhardy, F. C., and D. Rice
- 1970 A Proposed Cultural Typology for the Lower Snake River Region, Southeastern Washington. *Northwest Anthropological Research Notes* 4(1):1-29.

Leonhardy, F. C., G. Schroedl, J. Bense, and S. Beckerman
- 1971 *Wexpusnime (45GA61): Preliminary Report*. Laboratory of Anthropology Reports of Investigations No. 49, Washington State University, Pullman. Submitted to the National Park Service.

Lepofsky, D., K. D. Kusmer, D. Hayden, and K. P. Lertzman
- 1996 Reconstructing Prehistoric Socioeconomies from Paleoethnobotanical and Zooarchaeological Data: An example from the British Columbia Plateau. *Journal of Ethnobiology* 16(1):31-62.

Levy, R.
- 1978 Eastern Miwok. In *California*, edited by Robert F. Heizer, pp. 398-413. Handbook of North American Indians, vol. 8, W. C. Sturtevant, general editor. Smithsonian Institution, Washington, D.C.

Lewis, A. B.
- 1906 *Tribes of the Columbia Valley and the Coast of Washington and Oregon*. Memoirs of the American Anthropological Association, vol. 1, Lancaster.

Lohse, E. S., and D. Sammons-Lohse
- 1986 Sedentism on the Columbia Plateau: A Matter of Degree Related to the Easy and Efficient Procurement of Resources. *Northwest Anthropological Research Notes* 20(2):115-136.

Lohse, E. S., and R. Sprague
- 1998 History of Research. In *Plateau*, edited by Deward E. Walker, Jr., pp. 8-28. Handbook of North American Indians, vol. 12, W. C. Sturtevant, general editor. Smithsonian Institution, Washington, D.C.

Lothson, G. A.
- 1989 A Model for Prehistoric Bighorn Sheep Procurement, Processing, and Utilization Along the Middle Reach of the Columbia River, Washington. Unpublished Ph.D. dissertation, Department of Anthropology, Washington State University, Pullman.

Lovett, C. K.
- 1984 Paleomagnetism of the Touchet Beds of the Walla Walla River Valley, Southeastern Washington. Unpublished Masters thesis, Eastern Washington University, Cheney.

Lowie, R. H.
- 1954 *Indians of the Plains*. University of Nebraska Press, Lincoln.

Lucas, S. W.
- 1994 Origin of the Tucannon Phase in Lower Snake River Prehistory. Unpublished Master's Thesis, Department of Anthrop-

ology and the Department Geography, Oregon State University, Corvallis.

Lyman, R. L.
1976 A Cultural Analysis of Faunal Remains from the Alpowa Locality. Unpublished Master's thesis, Washington State University, Pullman.
1978 Prehistoric Butchery Techniques in the Lower Granite Reservoir, Southeastern Washington. *Tebiwa* 13:1-25.
1980a Freshwater Bivalve Molluscs and Southern Plateau Prehistory: A Discussion and Description of Three Genera. *Northwest Science* 54(2):121-136.
1980b The Hatwai Molluscan Fauna. Appendix in *Hatwai (10NP143): Interim Report*, edited by K. M. Ames, J. P. Green and M. Pfoertner. Boise State University Archaeological Reports No. 9: 158-163.
1984 A Model of Large Freshwater Clam Exploitation in the Prehistoric Southern Columbia Plateau Culture Area. *Northwest Anthropological Research Notes* 18(1):97-107.
1985 The Paleozoology of the Avey's Orchard Site. In *Avey's Orchard: Archaeological Investigations of a Late Prehistoric Columbia River Community*, edited by J. R. Galm and R. A. Masten, pp. 243-319. Eastern Washington University reports in Archaeology and History 100-42, Cheney.
1986 On the Analysis and Interpretation of Species List Data in Zooarchaeology. *Journal of Ethnobiology* 6(1):67-81.
1988 Zooarchaeology of 45DO189. In *Archaeological Investigations at River Mile 590: The Excavations at 45DO189*, by J. R. Galm and R. L. Lyman, pp. 97-141. Eastern Washington University Reports in Archaeology and History 100-61, Cheney.
1989 Zooarchaeology. In *Archaeological Data Recovery at Hatiuhpuh, 45WT134, Whitman County, Washington*, by D. Brauner et. al., pp. 98-138. Report of the Department of Anthropology, Oregon State University, Corvallis, to the U. S. Army Corps of Engineers, Walla Walla District.
1992 Rocky Reach Archaeofauna. In *Cultural Resource Investigations Along the Rocky Reach Reservoir: The 1990 Test Excavations*, K. Boreson, pp. 94-96. Eastern Washington University Reports in Archaeology and History 100-75, Cheney.
1994 *Vertebrate Taphonomy*. Cambridge University Press, Cambridge.
1995 Zooarchaeology of the Moses Coulee Cave (45DO 331) Spoils Pile. *Northwest Anthropological Research Notes* 29:141-176.
1997 Zooarchaeology of 45CH302. In *Archaeological Investigations at the Stemilt Creek Village Site (45CH302), Chelan County, Washington*, ed. by K. Boreson and J. R. Galm. Eastern Washington University Reports in Archaeology and History 100-82, Cheney.

Lyman, R. L., M. J. O'Brien, and V. Hayes
1998 A Mechanical and Functional Study of Bone Rods from the Richey-Roberts Clovis Cache, Washington, U.S.A. *Journal of Archaeological Science* 25:887-906.

Mack, R. N., N. W. Rutter, V. M. Bryant, Jr., and S. Valastro
1978a Late Quaternary Pollen Record from Big Meadow, Pend Oreille County, Washington. *Ecology* 59(5):956-965.

Mack, R. N., N. W. Rutter and S. Valastro
1978b Late Quaternary Pollen Record from the Sanpoil River Valley, Washington. *Canadian Journal of Botany* 56(14):1642-1650.

Mack, R. N., N. W. Rutter, S. Valastro, and V. M. Bryant, Jr.
1978c Late Quaternary vegetation History at Waits Lake, Colville River Valley, Washington. *Botanical Gazette* 139(4):499-506.

Mack, R. N., N. W. Rutter, V. M. Bryant, Jr., and S. Valastro
1978d Reexamination of Postglacial Vegetation History in Northern Idaho: Hager Pond, Bonner County. *Quaternary Research* 10(2):241-255.

Mack, R. N., N. W. Rutter, and S. Valastro
1979 Holocene Vegetation History of the Okanogan Valley, Washington. *Quaternary Research* 12(2):212-225.

Mack, R. N., N. W. Rutter, and S. Valastro
1983 Holocene Vegetational History of the Kootenai River Valley, Montana. *Quaternary Research* 20(2):177-193.

Mackin, J. H.
1961 *A Stratigraphic Section in the Yakima Basalt and the Ellensburg Formation in South-central Washington*. Reports of Investigations 19, Washington Division of Mines and Geology.

Mallory, O.
1966 A Comparative Cultural Analysis of Textiles from McGregor Cave, Washington. Unpublished Master's thesis, Department of Anthropology, Washington State University, Pullman.

Malouf, C.
1969 The coniferous forests and their Uses in the Northern Rocky Mountains through 9,000 Years of Prehistory. In *Coniferous Forests of the Northern Rocky Mountains: Proceedings of the 1968 Symposium*, edited by R. A.

Taber, pp. 271-290. Center for Natural Resources, University of Montana Foundation, Missoula.

Manring, B.
1975 *Conquest of the Coeur D'Alenes, Spokanes, and Palouses.* Ye Galleon Press, Fairfield, Washington.

Marshall, A. G.
1971 An Alluvial Chronology of the Lower Plouse River Canyon and Its Relation to Local Archaeological Sites. Unpublished Master's thesis, Department of Anthropology, Washington State University, Pullman.

Martin, A. C., and W. D. Barkley.
1973 *Seed Identification Manual* (Second Printing). University of California Press, Berkeley.

Martin, P. S., and R. G. Klein (ed.)
1984 *Quaternary Extinctions: A Prehistoric Revolution.* University of Arizona Press, Tucson.

Martin, P. S., and H. E. Wright, Jr. (ed.)
1967 *Pleistocene Extinctions: The Search for a Cause.* Proceedings of the VII Congress of the International Association for Quaternary Research. Yale University Press, New Haven.

Marwitt, J. P.
1973 Median Village and Fremont Culture Regional Variation. *University of Utah Anthropological Papers* 95. University of Utah Press, Salt Lake City.

Mastrogiuseppe, J. D.
1994 Report on Archaeological Plant Materials, McGregor Cave, Franklin County, Washington. Submitted to BOAS, Inc., Seattle.
1995 Report on Archaeological Plant Materials, Porcupine Cave, Franklin County, Washington. Submitted to BOAS, Inc., Seattle.

McGregor, J.
1982 *Counting Sheep: From Open Range to Agribusiness on the Columbia Plateau.* University of Washington Press, Seattle and London.

McKee, B.
1972 *Cascadia: The Geologic Evolution of the Pacific Northwest.* McGraw-Hill Book Company, New York.

Mehringer, P. J., Jr.
1985 Late Quaternary Pollen Records from the Interior Pacific Northwest and Northern Great Basin. In *Pollen Records of Late-Quaternary North American Sediments,* edited by V. M. Bryant and R. G. Holloway, pp. 167-190. American Association of Stratigraphic Palynologists Foundation, Dallas.
1986 Prehistoric Environments. In *Volume 11: Great Basin, Handbook of North American Indians,* edited by W. L. D'Azevedo, pp. 31-50. Smithsonian Institution, Washington, D.C.
1988 Clovis Cache Found: Weapons of Ancient Americans. *National Geographic* 174:500-503.
1989a *Age of the Clovis Cache at East Wenatchee, Washington.* Submitted to the Washington State Historic Preservation Office. Department of Anthropology, Washington State University, Pullman.
1989b Of Apples and Archaeology. *Universe* 1(2):2-8. Washington State University, Pullman.
1996 *Columbia River Basin Ecosystems: Late Quaternary Environments.* Manuscript on file, U.S. Forest Service and Bureau of Land Management, Walla Walla.

Mehringer, P. J., Jr., and W. J. Cannon
1994 Volcaniclastic Dunes of the Fort Rock Valley, Oregon: Stratigraphy, Chronology, and Archaeology. In *Archaeological Research in the Great Basin: Fort Rock Archaeology Since Cressman* edited by C. M. Aikens and D. L. Jenkins, pp. 283-327. University of Oregon Anthropological Papers, Eugene.

Mehringer, P. J., Jr., and F. Foit, Jr.
1990 Volcanic Ash Dating of the Clovis Cache at East Wenatchee, Washington. *National Geographic Research* 6(4):495-503.

Mehringer, P. J., Jr., and P. E. Wigand
1987 Western Juniper in the Holocene. In *Proceedings of the Pinyon-Juniper Conference, Reno, Nevada, Jan. 13-16, 1986,* pp. 109-119. General Technical Report INT-215.
1990 Comparison of Late Holocene Environments from Woodrat Middens and Pollen In *Packrat Middens: The Last 40,000 years of Biotic Change,* edited by J. L. Betancourt, T. R. Van Devender, and P. S. Martin, pp. 294-325. University of Arizona Press, Tucson.

Miller, R. F., and P. E. Wigand
1991 Holocene Changes in Semiarid Piñon-Juniper Woodlands: Response to Climate, Fire, and Human Activities in the U.S. Great Basin. *BioScience* 44(7):465-474.

Miss, C. J., and B. D. Cochran
1980 *Archaeological Evaluations of the Riparia (45WT1) and Ash Cave (45WW61) Sites on the Lower Snake River.* Laboratory of Archaeology and History, Project Report Number 14, Washington State University, Pullman.

Moerman, D. E.
   1998   *Native American Ethnobotany*. Timber Press, Portland, Oregon.

Moody, U. L.
   1978   *Microstratigraphy, Paleoecology, and Tephrochronology of the Lind Coulee Site, Central Washington*. Ph.D. Dissertation, Washington State University. University Microfilms, Ann Arbor.

Moody, U. L., and W. Dort
   1990   Microstratigraphic Analysis of Sediments and Soils; Wasden Archaeological Site, Eastern Snake River Plain, Idaho. In *Archaeological Geology of North America*, edited by N. P. Lasca and J. Donahue, pp. 361-382. Geological Society of America Centennial Special Volume 4, Boulder, CO.

Moore, S. M.
   1985   Modified Bone and Antler from On-A-Slant Village (32MO26). *Journal of the North Dakota Archaeological Association* 2:37-66.

Moyle, P. B.
   1976   *Inland Fishes of California*. University of California Press, Berkeley.

Mullineaux, D. R.
   1986   Summary of Pre-1980 Tephra-fall Deposits Erupted from Mount St. Helens, Washington State, U.S.A. *Bulletin of Volcanology* 48:17-26.

Mullineaux, D. R., J. Hyde, and M. Rubin
   1975   Widespread Late Glacial and Postglacial Tephra Deposits from Mt. St. Helens Volcano, Washington. *Journal of Research of the U.S. Geological Survey* 3(3):329-335.

Mullineaux, D. R., R. E. Wilcox, W. F. Ebaugh, R. Fryxell, and M. Rubin
   1978   Age of the Last Major Scabland Flood of the Columbia Plateau in Eastern Washington. *Quaternary Research* 10:171-178.

Muto, G.
   1976   *The Cascade Technique: An Examination of a Levallois-like Reduction System in Early Snake River Prehistory*. Ph.D. dissertation, Washington State University. University Microfilms, Ann Arbor.

Nakonechny, L. D.
   1998   Archaeological Analysis of Area A, Wexpusnime Site (45GA61). Unpublished Master's thesis, Department of Anthropology, Washington State University, Pullman.

Nance, C.
   1966   45WT2: An Archaeological Site on the Lower Snake River. Unpublished Master's thesis, Department of Anthropology, Washington State University, Pullman.

Negrini, R. M., and J. O. Davis
   1992   Dating late Pleistocene Pluvial Events and Tephras by Correlating Paleomagnetic Secular Variation Records from the Western Great Basin. *Quaternary Research* 38:46-59.

Negrini, R. M., D. B. Erbes, K. Faber, A. M. Herrera, A. P. Roberts, A. S. Cohen, P. E. Wigand, and F. F. Foit, Jr.
   2000   A Paleoclimate Record for the Past 250,000 years from Summer Lake, Oregon, U.S.A.: I. Chronology and Magnetic Proxies for Lake Level. *Journal of Paleolimnology* 24(2):125-149.

Nellis, C. H.
   1973   The Lynx in the Northwest. In *Proceedings of a Symposium on the Native Cats of North America: Their Status and Management*, edited by S. E. Jorgensen and L. David Mech, pp. 23-28. Published by the U.S. Department of the Interior, Fish and Wildlife Service, Twin Cities, Minnesota.

Nelson, C.
   1966   *A Preliminary Report on 45CO1, a Stratified Open Site in the Southern Columbia Plateau*. Washington State University, Laboratory of Anthropology, Report of Investigations, Number 39, Pullman.
   1969   *The Sunset Creek Site (45KT28) and Its Place in Plateau Prehistory*. Washington State University, Laboratory of Anthropology, Report of Investigation Number 47, Pullman.

Nelson, D. W., and L. E. Sommers
   1982   Total Carbon, Organic Carbon, and Organic Matter. *In Methods of Soil Analysis Part 2: Chemical and Microbiological Properties*, edited by A. L. Page, pp. 539-580. Soil Science Society of America, Madison.

Nelson, M. A., P. J. Ford, and J. K. Stein
   1986   Turning a Midden into Mush: Evidence of Acidic Conditions in a Shell Midden. Paper presented at the 51st Annual Meeting of the Society for American Archaeology, New Orleans.

Netboy, A.
   1980   *The Columbia River Salmon and Steelhead Trout, Their Fight for Survival*. University of Washington Press, Seattle.

Newberry, J. S.
   1887   Food and fiber plants of the North American Indians. *Popular Science Monthly* 32:31-46.

Newman, M. E.
   1990   The Hidden Evidence from Hidden Cave, Nevada: An Application of Immunological Techniques to the Analysis of Archaeological Materials. Unpublished Ph.D. dissertation, Department of Anthropology, University of Toronto, Ontario.

Noe-Nygaard, N.
1977 Butchering and Marrow Fracturing as a Taphonomic Factor in Archaeological Deposits. *Paleobiology* 3:218-237.

Nowak, C. L., R. S. Nowak, R. J. Tausch, and P. E. Wigand
1994a A 30,000-Year Record of Vegetation Dynamics at a Semi-arid Locale in the Great Basin. *Journal of Vegetation Science* 5:579-590.
1994b Tree and Shrub Dynamics in Northwestern Great Basin Woodland and Shrub Steppe During the Late-Pleistocene and Holocene. *American Journal of Botany* 81(3):265-277.

Nussbaum, R. A., E. D. Brodie, Jr., and R. M. Storm
1983 *Amphibians and Reptiles of the Pacific Northwest*. University Press of Idaho, Moscow, Idaho.

Odell, G. H., and F. Odell-Vereecken
1980 Verifying the Reliability of Lithic Use-wear Assessments by 'Blind Tests": The Low-Power Approach. *Journal of Field Archaeology* 7:87-120.

Olsen, S. R., and L. E. Sommers
1982 Phosphorus. In *Methods of Soil Analysis Part 2: Chemical and Microbiological Properties*, edited by A. L. Page, pp. 403-430. Soil Science Society of America, Madison.

Osborne, D. D.
1953 Archaeological Occurrence of Pronghorn Antelope, Bison and Horse in the Columbia Plateau. *Scientific Monthly* 77:260-269.
1957 Excavations in the McNary Reservoir Basin Near Umatilla, Oregon. *Bureau of American Ethnology Bulletin* 166; River Basin Surveys Paper 8, Washington, D.C.

Owen, J.
1927 *Journals and Letters of Major John Owen*, edited by S. Dunbar and P. C. Phillips, Volumes 1 and 2. E. Eberstadt, New York.

Ozbun, T. L., and J. L. Fagan
1998 *Technological and Functional Analysis of the Lithic Assemblages from Four Sites in the Wind River Nursery Fields*. Archaeological Investigations Northwest, Inc. Report No. 137. Submitted to USDA Forest Service, Mt. Adams Ranger District, Trout Lake, Washington.

Ozbun, T. L., J. L. Fagan, and D. Stueber
1998 A Rejuvenation Function for Clovis Point Fluting. Paper presented at the Twenty-sixth Great Basin Anthropological Conference, Bend, Oregon.

Ozbun, T. L., E. E. Forgeng, J. S. Chapman, M. Zehendner, J. J. Wilt, and J. L. Fagan
1997 *Archaeological Data Recovery at the Private Trust Site (35GR1680) and the East Big Boulder Creek Site (35GR1730), and Treatment of Two Segments of the Sumpter Valley Railway Middle Fork John Day River Highway Project, Grant County, Oregon*. Archaeological Investigations Northwest, Inc. Report No. 130, Portland. Prepared for David Evans and Associates, Inc., Portland, and The Federal Highway Administration, Western Federal Lands Highway Division, Vancouver, Washington.

Ozbun, T. L., D. O. Stueber, M. Zehendner, and J. L. Fagan
2000 *Technological Analysis of Lithic Debitage and Tools from the Marmes Site, 45FR50*. Archaeological Investigations Northwest, Inc. Report No. 184, Portland. Prepared for Confederated Tribes of the Colville Reservation, History/Archaeology Department, Nespelem.

Panshin, A. J., C. DeZeeuw, and H. P. Brown
1964 *Textbook of Wood Technology Volume 1*, 2nd Edition. McGraw-Hill, New York.

Parkhurst, Z. E.
1950 *Survey of the Columbia River and its Tributaries, Part 6, Area V - Snake River System from the Mouth through the Grand Rhonde River*. Special Scientific Report -- Fisheries No. 39, Washington, D.C.

Pennak, R. W.
n.d. *Fresh-water Invertebrates of the United States*. John Wiley and Sons, New York.

Peterson, F. F.
1961 Solodized Solonetz Soils Occurring on the Uplands of the Palouse Loess. Unpublished Ph.D. dissertation, Washington State University, Pullman.

Peterson, R. T.
1961 *A Field Guide to Western Birds*. Houghton Mifflin Co., Boston.

Porter, S. P.
1971 Fluctuations in Late Pinedale Alpine Glaciers in Western North America. In *Late Coenozoic Glacial Ages,* edited by K. K. Turekian, pp. 307-329. Yale University Press, New Haven.

Post, R. H.
1938 The Subsistence Quest. In the Sinkaietcj or Southern Okanogan of Washington (L. Spier, ed.). *General Series in Anthropology* 6, *Contributions from the Laboratory of Anthropology* 2:9-34. George Banta Publishing Company, Menasha.

Ray, V.
1933 *The Sanpoil and Nespelem: Salishan Peoples of Northeastern Washington*. University of Washington Publications in Anthropology, v. 5, University of Washington Press, Seattle.
1936 Native Villages and Groupings of the Columbia Basin. *Pacific Northwest*

Quarterly 27:99-152.
1939 *Cultural Relations in the Plateau of Northwestern America*. Publications of the F. W. Hodge Anniversary Publication Fund 3. Southwest Museum, Los Angeles.
1975 *Visitors Facilities Cultural Report: Chief Joseph Dam*. Manuscript on file at the Colville Confederated Tribes, History/Archaeology Department, Nespelem.

Raymond, A. W., and E. Sobel
1990 The Use of Tui Chub as Food by Indians of the Western Great Basin. *Journal of California and Great Basin Anthropology* 12:2-18.

Reid, K. C.
1991(ed.) *Prehistory and Paleoenvironments at Pittsburg Landing: Data Recovery and Test Excavations at Six Sites in Hells Canyon National Recreation Area, Western Idaho*. Center for Northwest Anthropology Project Report Number 15, Washington State University, Pullman.
1992 Housepits and Highland Hearths: Comparative Chronologies for the Snake River and Blue Mountains. Paper presented at the 45th Northwest Anthropological Conference, Burnaby.
1995(ed.) *An Overview of Cultural Resources in the Snake River Basin: Prehistory and Paleoenvironments (1st Update)*. Rainshadow Research Inc. Project Report No. 31, Pullman.
1997a Lithic Raw Materials. In *Kirkwood Bar: Passports in Time Excavations at 10IH699 in the Hells Canyon National Recreation Area, Wallowa-Whitman NationalForest*, by K. C. Reid and J. C. Chatters, Appendix E. Rainshadow Research Project Report No. 28 and Applied Paleoscience Project Report No. F-6. Submitted to USDA Forest Service, Wallowa-Whitman National Forest.
1997b Gravels and Travels: A Comment on Andrefsky's "Cascade Phase Lithic Technology." *North American Archaeologist* 18(1):67-81.

Reid, K. C., and J. Gallison
1995 The Lower Snake Basin: Hells Canyon to the Columbia. In *An Overview of Cultural Resources in the Snake River Basin: Prehistory and Paleoenvironments (1st Update)*, pp. 2.1-2.135. Rainshadow Research Inc. Project Report No. 31, Pullman.
1996 Lower Snake River Basin. *Northwest Anthropological Research Notes* 30:15-104.

Reineck, H. E., and I. B. Singh
1980 *Depositional Sedimentary Environments*, 2nd edition, Springer-Verlag, Berlin.

Rensberger, J. M., A. D. Barnowsky, and P. Spencer
1984 *Geology and Paleontology of a Pleistocene-to-Holocene Loess Succession, Benton County, Washington*. Archaeological and Historical Services Report No. 100-39, Eastern Washington University, Cheney.

Rhoades, J. D.
1982a Soluble Salts. In *Methods of Soil Analysis, Part 2, Chemical and Microbiological Properties*, edited by A. L. Page, pp. 167-180. Soil Science Society of America, Madison.
1982b Cation Exchange Capacity. In *Methods of Soil Analysis, Part 2, Chemical and Microbiological Properties*, edited by A. L. Page, pp. 149-158. Soil Science Society of America, Madison.

Rice, D.
1968 Archaeological Activities of the Mid-Columbia Archaeological Society – 1968. In *Mid-Columbia Archaeological Society Annual Report*. Richland.
1969 *Preliminary Report, Marmes Rockshelter Archaeological Site, Southern Columbia Plateau*. Report to the National Park Service, San Francisco. Laboratory of Anthropology, Washington State University, Pullman.
1972 The Windust Phase in Lower Snake River Region Prehistory. Unpublished Ph.D. dissertation, Washington State University, Pullman.

Rice, H. S.
1965 The Cultural Sequence at Windust Caves. Unpublished Master's thesis, Washington State University, Pullman.
1985 Native American Dwellings and Attendant Structures of the Southern Plateau. Unpublished Ph.D. dissertation, Washington State University, Pullman.

Richards, L. A., editor
1954 *Diagnosis and Improvement of Saline and Alkali Soils*. U.S. Department of Agriculture Handbook 60. U.S. Government Printing Office, Washington, D.C.

Ringe, D.
1970 Sub-loess Basalt Topography in the Palouse Hills, Southeastern Washington. *Geological Society of America Bulletin* 81:3049-3060.

Roberts, J.
1999 Review of Native American Human Remains Dated to Earlier Than 7,000 Years BP in North America (United States). In Department of the Interior administrative record for Bonnichsen et al. vs. The United States: DOI 05534-05548.

Rodeffer, M.
1973 A Classification of Burials in the Lower Snake River Region. *Northwest Anthropological Research Notes* 7(1):101-131.

Root, M. J.
1991 The Knife River Flint Quarries: The Organization of Stone Tool Production. Unpublished Ph.D. dissertation, Department of Anthropology, Washington State University, Pullman.

Root, M. J., J. D. William, M. Kay, and L. K. Shifrin
1999 Folsom Ultrathin Biface and Radial Break Tools in the Knife River Flint Quarry Area, North Dakota. In *Current Research on Folsom Technology*, edited by D. S. Amick, pp. 144-168. International Monographs in Prehistory, Archaeological Series, Ann Arbor, Michigan.

Roulette, B. R., ed.
1997 *Archaeological Recovery of a Portion of 45FE38*. Report submitted to the Colville Confederated Tribes, Applied Archaeological Research, Portland.

Sappington, R. L.
1996 The Clearwater River Region. In *An Overview of Cultural Resources in the Snake River Basin: Prehistory and Paleoenvironments*. Northwest Anthropological Research Notes 30(1&2):117-165.

Schalk, R. S.
1977 The Structure of an Anadromous Fish Resource. In *For Theory Building in Archaeology*, edited by L. R. Binford, pp. 207-249. Academic Press, Orlando.

1980 (ed.) *Cultural Resource Investigation for the Second Powerhouse Project at McNary Dam, Near Umatilla, Oregon*. Laboratory of Archaeology and History, Project Report Number 1, Washington State University, Pullman.

1983a (ed.) *Cultural Resource Investigations for the Lyons Ferry Fish Hatchery Project, Near Lyons Ferry, Washington*. Laboratory of Archaeology and History, Project Report Number 8, Washington State University, Pullman.

1983b (ed.) *The 1978 and 1979 Excavations at Strawberry Island in the McNary Reservoir*. Laboratory of Archaeology and History, Project Report Number 19, Washington State University, Pullman.

Schalk, R. S., and G. Cleveland
1983 A Chronological Perspective on Hunter-Gatherer Land Use Strategies in the Columbia Plateau. In *Cultural Resource Investigations for the Lyons Ferry Fish Hatchery Project, Near Lyons Ferry, Washington*, edited by R. Schalk, pp. 11-56. Laboratory of Archaeology and History, Project Report Number 8, Washington State University, Pullman.

Scheuerman, R. D., and C. E. Trafzer
1980 The First People of the Palouse Country. *The Bunchgrass Historian* 8:20-23.

Schiffer, M. B.
1987 *Formation Processes of the Archaeological Record*. University of New Mexico Press, Albuquerque.

Schneider, J. S.
1996 Quarrying and Production of Milling Implements at Antelope Hill, Arizona. *Journal of Field Archaeology* 23:299-311.

Schopmeyer, C. S. (Technical Coordinator)
1974 *Seeds of Woody Plants in the United States*. USDA Agriculture Handbook 450.

Schroedl, G.
1973 *The Archaeological Occurrence of Bison in the Southern Plateau*. Report of Investigations No. 51, Laboratory of Anthropology, Washington State University, Pullman.

Schwede, M.
1966 An Ecological Study of Nez Perce Settlement Patterns. Unpublished Master's thesis, Washington State University, Pullman.

1970 The Relationship of Aboriginal Nez Perce Settlement Patterns to Physical Environment and to Generalized Distribution of Food Resources. *Northwest Anthropological Research Notes* 4(2):129-135.

Scott, S. L., ed.
1983 *Field Guide to the Birds of North America*. National Geographic Society, Washington, D.C.

Sellet, F.
1993 Chaine Operatoire: The Concept and Its Applications. *Lithic Technology* 18(1&2):106-112.

Semenov, S. A.
1964 *Prehistoric Technology: An Experimental Study of the Oldest Tools and Artefacts from Traces of Manufacture and Wear*. Translation by M. Thompson. Adams and Dart, Bath, England.

Sheppard, J., P. E. Wigand, and M. Rubin
1984 The Marmes Site Revisited: Dating and Stratigraphy. *Tebiwa* 21:46-49.

Sheppard, J., P. E. Wigand, C. Gustafson, and M. Rubin
1987 A Reevaluation of the Marmes Rockshelter Radiocarbon Chronology. *American Antiquity* 52(1):118-125.

Shipman, P., G. Foster, and M. Schoeniger
1984 Burnt Bones and Teeth: An Experimental Study of Color, Morphology, Crystal Structure and Shrinkage. *Journal of Archaeological Science* 11:170-172.

Sigler, W. F., and J. W. Sigler
  1987 *Fishes of the Great Basin: A Natural History.* University of Nevada Press, Reno.

Smith, A. H.
  1986 Kalispel Ethnography and Ethnohistory. In, *Calispel Valley Archaeological Interim Report for the 1984 and 1985 Field Seasons*, edited by A. V. Thoms and G. C. Burtchard, pp. 139-286. Center for Northwest Anthropology Contributions in Cultural Resource Management No. 10. Washington State University, Pullman.
  1988 *Ethnography of the North Cascades.* Prepared for North Cascades National Park Service, contract number CX-9000-4-E076. Center for Northwest Anthropology, Washington State University, Project Report 7, Pullman.

Smith, C. S.
  1983 A 4300-Year History of Vegetation, Climate, and Fire from Blue Lake, Nez Perce County, Idaho. Unpublished Master's thesis, Washington State University, Pullman.

Smith, G. I., and J. L. Bischoff (eds.)
  1997 An 800,000-Year Geologic and climatic Record from Owens Lake, California: Core OL-92. *Geological Society of America Special Paper* 317.

Smith, H. W., and C. D. Moodie
  1948 Collection and Preservation of Soil Profiles. *Soil Science* 64:61-69.

Smith, P. W.
  1965 Recent Adjustments in Animal Range. In *The Quaternary of the United States*, edited by H. E. Wright, Jr. and David G. Frey, pp. 633-642. Princeton University Press, Princeton.

Spier, L.
  1936 Tribal Distributions in Washington. *General Series in Anthropology, No. 3.* George Banta Publishing Company, Menasha.

Spinden, H.
  1908 The Nez Perce Indians. *Memoirs of the American Anthropological Association II* 8:167-275.

Sprague, R.
  1959 A Comparative Cultural Analysis of an Indian Burial Site in Southeast Washington. Master's Thesis, Washington State University, Pullman.
  1963 *The Descriptive Archaeology of the Palus Burial Site, Lyons Ferry, Washington.* Laboratory of Anthropology, Washington State University, Report of Investigations 32, Pullman.
  1967 Aboriginal Burial Practices in the Plateau Region of North America. Unpublished Ph.D. dissertation, University of Arizona, Tucson.
  1983 The Historic Background of the Joso Bridge Construction Camp. In *Cultural Resource Investigations for the Lyons Ferry Fish Hatchery Project, Near Lyons Ferry. Washington*, edited by R. F. Schalk, pp. 69-78. Laboratory of Archaeology and History Project Report No. 8, Washington State University, Pullman.
  1998 Palouse. In *Plateau*, edited by D. E. Walker, Jr., pp. 352-359. Handbook of North American Indians, vol. 12, W.C. Sturtevant, general editor. Smithsonian Institution, Washington, D.C.
  2000 A Review of Stability in Southern Plateau Burial Practices. In Department of the Interior's administrative record for Bonnichsen et al. vs. the United States: DOI 07639-07656.

Stanford, D.
  1991 Clovis Origins and Adaptations: An Introductory Perspective. In *Clovis Origins and Adaptations*, edited by R. Bonnichsen and K. L. Turnmire, pp. 1-13. Center for the Study of the First Americans, Oregon State University, Corvallis.

Stebbins, R. C.
  1954 *Amphibians and Reptiles of Western North America.* McGraw-Hill Co., New York.
  1966 *A Field Guide to the Western Reptiles and Amphibians.* Houghton Mifflin Co., Boston.

Steffen, A., E. J. Skinner, and P. W. Ainsworth
  1998 A View to the Core: Technological Units and Debitage Analysis. In *Unit Issues in Archaeology,* by A. F. Ramenofsky and A. Steffen, pp. 131-146. Foundations of Archaeological Inquiry, University of Utah Press, Salt Lake City.

Stein, J. K.
  1985 Interpreting Sediments in Cultural Settings. In *Archaeological Sediments in Context*, edited by J. K. Stein & W. R. Farrand, pp. 5-20. Center for the Study of Early Man, Orono, Maine.
  1992 Organic Matter in Archaeological Contexts. In *Soils in Archaeology*, edited by V. T. Holliday, pp. 193-216. Smithsonian Institution Press, Washington, D.C.

Steinberger, J.
  1897 Letter from Justus Steinberger to Headquarters, District of Oregon, Fort Vancouver, Washington Territory, 23 June 1862. War of the Rebellion, Series 1, vol. 50, Part 1, pp. 1154-1155, 55th Congress, 1st Session, House Documents, vol. 13, No. 59, Part 1 (Serial No. 3583), Washington, D.C.

Steward, J. H.
  1938 Basin-Plateau Sociopolitical Groups. *Bureau of American Ethnology Bulletin* 120.

Washington, D.C. (Reprinted: University of Utah Press, Salt Lake City, 1970).

Stine, S.
1990 Late Holocene Fluctuations of Mono Lake, Eastern California. *Palaeogeography, Palaeoclimatology,Palaeoecology* 78:333-381.

Stratton, D., and G. Lindeman
1976 *A Cultural Resource Survey for the United States Army Corps of Engineers, Walla Walla District, Historic Resources Study*. Washington Archaeological Research Center, Project Report No. 238, Pullman.
1978 *A Study of Historical Resources of the Hells Canyon National Recreation Area, Vols. 1 and 2*. National Heritage, Inc., Pullman.

Striker, M.
1995 Sedentism and Resource Availability of the Coeur d'Alene at or Before European Contact. Unpublished Master's thesis, University of Idaho, Moscow.

Stuiver, B., and P. J. Reimer
2002 Radiocarbon Calibration Program Calib 4.3. To be used in conjunction with *Radiocarbon* 35:215-230.

Stuiver, M., B. Kromer, B. Becker, and C. W. Ferguson
1986 Radiocarbon Age Calibration Back to 13,300 years B.P. and the ^{14}C Age Matching of the German Oak and US Bristlecone Pine Chronologies. *Radiocarbon* 28(2B):969-979.

Swanson, D. A., and T. L. Wright
1976 Bedrock Geology of the Northern Columbia Plateau and Adjacent Areas. In *The Channeled Scablands: A Guide to the Geomorphology of the Columbia Basin, Washington*, edited by V. R. Baker, and D. Nummedal, pp. 37-57. U.S. National Aeronautics and Space Administration, Washington, D.C.

Swanton, J. R.
1952 Indian Tribes of North America. *Bulletin of the Bureau of American Ethnology* 145, Smithsonian Institution, Washington, D.C.

Szuter, C. R.
1991 *Hunting by Prehistoric Agriculturalists of the North American Southwest*. Garland Publishing, New York.

Tausch, R. J., N. E. West, and A. A. Nabi
1981 Tree Age and Dominance Patterns in Great Basin Piñon-Juniper Woodlands. *Journal of Range Management* 34:259-264.

Taylor, D. W.
1981 Freshwater Mollusks of California: A Distributional Checklist. *California Fish and Game* 67(3):140-163.

Taylor, W. P., and W. T. Shaw
1929 Provisional List of the Land Mammals of the State of Washington. *Occasional Papers of the Charles R. Conner Museum* 2:1-32.

Thompson, R. S.
1985 *Paleoenvironmental Investigations at Seed Cave, Franklin County, Washington*. Eastern Washington University, Reports in Archaeology and History 100-41, Cheney.

Thompson, R. S., C. Whitlock, P. J. Bartlein, S. P. Harrison, and W. G. Spaulding
1993 Climatic Changes in the Western United States Since 18,000 yr B.P. In *Global Climates Since the Last Glacial Maximum*, edited by H. E. Wright Jr., J. E. Kutzbach, T. Webb III, W. F. Ruddiman, F. A. Street-Perrott, and P. J. Bartlein, pp. 468-513. University of Minnesota Press, Minneapolis.

Thoms, A. V.
1983 *Archaeological Investigations in Upper McNary Dam Project: 1981-1982*. Laboratory of Archaeology and History, Project Report Number 15, Washington State University, Pullman.

Thwaites, R. G. (editor)
1959 *Original Journals of the Lewis and Clark Expedition 1804-1806*, vol. III. Antiquarian Press Ltd., New York.

Timbrook, J.
1982 Use of Wild Cherry Pits as Food by the California Indians. *Journal of Ethnobiology* 2(2):162-176.

Titmus, G. L.
1985 Some Aspects of Stone Tool Notching. In *Stone Tool Analysis: Essays in Honor of Don E. Crabtree*, edited by M. G. Plew, J. C. Woods, and M. G. Pavesic, pp. 243-263. University of New Mexico Press, Albuquerque.

Tomenchuk, J., and P. L. Storck
1997 Two Newly Recognized Paleoindian Tool Types: Single- and Double-Scribe Compass Gravers and Coring Gravers. *American Antiquity* 62(3):508-522.

Trafzer, C., and R. Scheuerman
1986 *Renegade Tribe: Palouse Indians and the Invasion of the Inland Pacific Northwest*. Washington State University Press, Pullman, Washington.

Treasher, R. C.
1925 Origin of the Loess of the Palouse Region, Washington. *Science* 61:469.

Turner, N. J.
1979 *Plants in British Columbia Indian Technology*. British Columbia Provincial Museum Handbook 38. Victoria.
1988 Ethnobotany of coniferous trees in Thompson and Lilliooet Interior Salish of British

Columbia. *Economic Botany* 42(2):177-194.

Turner, N. J., L. C. Thompson, M. T. Thompson, and A. Z. York
1990 *Thompson Ethnobotany: Knowledge and Usage of Plants by the Thompson Indians of British Columbia*. Royal British Columbia Museum Memoir No. 3.

Turney-High, H. H.
1937 The Flathead Indians of Montana. *Memoirs of the American Anthropological Association* 48. Menasha.
1941 Ethnography of the Kutenai. *Memoirs of the American Anthropological Association* 48. Menasha.

United States Department of Agriculture
1951 *Soil Survey Manual*. U.S. Department of Agriculture Handbook 18. U.S. Government Printing Office, Washington, D.C.
1975 *Soil Taxonomy*. U.S. Department of Agriculture Handbook 436. U.S. Government Printing Office, Washington, D.C.

United States Department of the Interior/Geological Survey
1982 *The Channeled Scablands of Eastern Washington: The Geologic Story of the Spokane Floods*. U.S. Government Printing Office, Washington, D.C.

United States Department of the Interior
2000 Memorandum: Results of Radiocarbon Dating the Kennewick Human Skeletal Remains. Obtained from National Park Service website: www.cr.nps.gov/aad/kennewick/c14memo.htm.

Van Devender, T. R., R. S. Thompson, and J. L. Betancourt
1987 Chapter 15: Vegetation History of the Deserts of Southwestern North America: The Nature and Timing of the Late Wisconsin-Holocene Transition. In *North America and Adjacent Oceans During the Last Deglaciation, The Geology of North America K-3*, edited by W. F. Ruddiman and H. E. Wright, Jr., pp. 323-352. Geological Society of America, Boulder.

Vaughan, P. C.
1985 *Use-wear Analysis of Flaked Stone Tools*. University of Arizona Press, Tucson.

Vehik, S.
1977 Bone Fragments and Bone Grease Manufacture: A Review of Their Archaeological Use and Potential. *Plains Anthropologist* 22:169-182.

Waitt, R.
1985 Case for Periodic, Colossal Jokulhlaups from Pleistocene Glacial Lake Missoula. *Geological Society of America Bulletin* 96:1271-1286.

Walker, D. E., Jr.
1967 *Mutual Cross-Utilization of Economic Resources in the Plateau: An Example from Aboriginal Nez Perce Fishing Practices*. Washington State University, Laboratory of Anthropology, Report of Investigations 41, Pullman.
1971 American Indians of Idaho. *Anthropological Monographs of the University of Idaho* 2. Moscow.
1973 American Indians of Idaho. *Anthropological Monographs of the University of Idaho* 2, Moscow.
1978 *Indians of Idaho*. University of Idaho Press, Moscow.

Walker, D. E., Jr., editor
1998 *Plateau*. Handbook of North American Indians, Volume 12. Smithsonian Institution, Washington, D.C.

Walkley, A., and I. A. Black
1934 An Examination of the Degtjareff Method for Determining Soil Organic Matter and a Proposed Modification of the Chromic Acid Titration Method. *Soil Science* 37:29-38.

Walter, H.
1954 Le facteur eau dans les regions arides et sa signification pour l'organisation de la vegetation dans les contrees sub-tropicales. In *Les Divisions Ecologiques du Monde*, pp. 27-39. Centre Nationale de la Recherche Scientifique, Paris.

Waters, M. R.
1992 *Geoarchaeology: A North American Perspective*. University of Arizona Press, Tucson.

Webster, G. D., M. J. P. Kuhns, and G. L. Waggoner
1982 Late Cenozoic Gravels in Hells Canyon and the Lewiston Basin, Washington and Idaho. In Cenozoic Geology of Idaho, edited by B. Bonnichsen, and R. M. Breckenridge, pp. 669-683. *Idaho Bureau of Mines and Geology Bulletin* 26:669-683.

Westgate, J. A.
1977 Identification and Significance of Late-Holocene Tephra from Otter Creek, Southern British Columbia, and Localities in West-Central Alberta. *Canadian Journal of Earth Sciences* 14:2593-2600.

Wharton, R. A., P. E. Wigand, M. R. Rose, R. L. Reinhardt, D. A. Mouat, H. E. Klieforth, N. L. Ingraham, J. O. Davis, C. A. Fox, and J. T. Ball
1990 Chapter 9: The North American Great Basin: A Sensitive Indicator of Climatic Change. In *Plant Biology of the Basin and Range*, edited by C. B. Osmond. Ecological Studies Series 80. Springer-Verlag.

Wheeler, A., and A. K. G. Jones
1989 *Fishes*. Cambridge University Press, Cambridge.

Whiting, B. B.
1950 *Paiute Sorcery*. Viking Fund Publications in Anthropology 15. New York.

Whitlock, C., and P. J. Bartlein
1997 Vegetation and Climate Change in Northwest America During the Past 125 kyr. *Nature* 388:57-61.

Wigand, P. E.
1987 Diamond Pond, Harney County, Oregon: Vegetation History and Water Table in the Eastern Oregon Desert. *Great Basin Naturalist* 47:427-458.
1989 Vegetation, Vegetation History, and Stratigraphy. In *Prehistory and Paleoenvironments of the Silvies Plateau, Harney Basin, Southeastern Oregon*, edited by K. C. Reid, J. A. Draper and P. E. Wigand, pp. 37-85. Project Report Number 8, Center for Northwest Anthropology, Washington State University, Pullman.
1997 A Late-Holocene Pollen Record from Lower Pahranagat Lake, Southern Nevada, USA: High Resolution Paleoclimatic Records and Analysis of Environmental Responses to Climate Change. In *Proceedings of the Thirteenth Annual Pacific Climate (PACLIM) Workshop, April 15-18, 1996. Interagency Ecological Program, Technical Report 53*, edited by C. M. Isaacs and V. L. Tharp, pp. 63-77. California Department of Water Resources, Sacramento.

Wigand, P. E., and M. R. Rose
1990 *Calibration of High Frequency Pollen Sequences and Tree-Ring Records*. Proceedings of the International Highlevel Radioactive Waste Management Conference and Exposition, April 1990, Las Vegas.

Wigand, P. E., and C. L. Nowak
1992 Dynamics of Northwest Nevada Plant Communities During the Last 30,000 Years. In *The History of Water: Eastern Sierra Nevada, Owens Valley, White-Inyo Mountains*, edited by C. A. Hall, V. Doyle-Jones, and B. Widawski. White Mountain Research Station Symposium Volume 4:40-62.

Wigand, P. E., M. L. Hemphill, S. E. Sharpe, and S. Patra
1995 Great Basin Woodland Dynamics During the Holocene. In *Proceedings of the Workshop-Climate Change in the Four Corners and Adjacent Regions: Implications for Environmental Restoration and Land-Use Planning*, edited by W. J. Waugh. CONF-9409325, U.S. Department of Energy, Grand Junction, Colorado.

Wigand, P. E., and D. Rhode
2002 Great Basin Vegetation History and Aquatic Systems: The Last 150,000 years. In *Great Basin Aquatic Systems History,* edited by R. Hershler, D. B. Madsen, and D. R. Currey, pp. 309-367. Smithsonian Contributions to Earth Sciences 33. Smithsonian Institution Press, Washington, D.C.

Wilfong, C.
1990 *Following the Nez Perce Trail: A Guide to the Nee-Mee-Poo National Historical Trail with Eyewitness Accounts*. Oregon State University Press, Corvallis.

Wilke, P. J., J. J. Flenniken, and T. L. Ozbun
1991 Clovis Technology at the Anzick Site, Montana. *Journal of California and Great Basin Anthropology* 13:242-272.

Wilkes, C.
1856 *Narrative of the United States Exploring Expedition*, vols. 1-6. G. P. Putnam, New York.

Willey, G., and P. Phillips
1958 *Method and Theory in American Archaeology*. University of Chicago Press, Chicago.

Williams, S. B.
1990 Immunology and Archaeology: Blood Residue Analysis of Three Sites. Unpublished Master's thesis, Department of Anthropology, Portland State University.

Willig, J. A., and C. M. Aikens, eds.
1988 *Early Human Occupation in Far Western North America: The Clovis-Archaic Interface*. Nevada State Museum Anthropological Papers Number 21, Carson City, Nevada.

Wilmsen, E. N., and F. H. H. Roberts, Jr.
1984 *Lindenmeier, 1934-1974: Concluding Report on Investigations*. Originally published 1978, Smithsonian Contributions to Anthropology, Number 24, reprint edition, Washington, D.C.

Wood, R. W., and D. L. Johnson
1978 A Survey of Disturbance Processes in Archaeological Site Formation. In *Advances in Archaeological Method and Theory*, Volume 1, edited by M. B. Schiffer, pp. 315-370. Academic Press, New York.

Woods, J. C.
1987 Manufacturing and Use Damage on Pressure-Flaked Stone Tools. Unpublished Master's thesis. Department of Sociology, Anthropology, and Social Work, Idaho State University, Pocatello.

Wormington, H. M.
- 1964 *Ancient Man in North America*. Denver Museum of Natural History Popular Series No. 4, Denver, Colorado.

Wydoski, R. S,. and R. R. Whitney
- 1979 *Inland Fishes of Washington*. University of Washington Press, Seattle.

Yamaguchi, D. K.
- 1983 New Tree-Ring Dates for Recent Eruptions of Mount St. Helens. *Quaternary Research* 20:246-250.

Yanovsky, E. K., E. K. Nelson, and R. M. Kingsbury
- 1932 Berries Rich in Calcium. *Science* 75 (1952): 565-566.

Yent, M.
- 1976 The Cultural Sequence at Wawawai (45WT39) Lower Snake River Region, Southeastern Washington. Unpublished Master's thesis, Washington State University, Pullman.

Young, S. P., and E. A. Goldman
- 1944 *The Wolves of North America*. Dover Publications, Inc., New York.

Ziegler, A.
- 1965 The Role of Faunal Remains in Archaeological Investigations. *Symposium on Central California Archaeology: Problems, Programs, and Interdisciplinary Approaches*, pp. 47-75. Central California Archaeological Foundation and the Sacramento Anthropological Society of Sacramento State College.

Zweifel, M.
- 1994 A Guide to the Identification of the Molariform Teeth of Rodents and Lagomorphs of the Columbia Basin. Unpublished Master's thesis, Washington State University, Pullman.

www.ingramcontent.com/pod-product-compliance
Lightning Source LLC
Chambersburg PA
CBHW060417300426
44111CB00018B/2876